5th International Symposium on Quantum Theory and Symmetries 2007

Journal of Physics: Conference Series Volume 128

Valladolid, Spain
22-28 July 2007

ISBN: 978-1-60560-974-4

Printed from e-media with permission by:

Curran Associates, Inc.
57 Morehouse Lane
Red Hook, NY 12571
www.proceedings.com

Some format issues inherent in the e-media version may also appear in this print version.

Printed by Curran Associates, Inc. (2009)

Institute of Physics Publishing
Dirac House, Temple Back
Bristol BS1 6BE UK

Phone: +44 (0)117 929 7481
Fax: +44 (0)117 929 4318

custserv@iop.org

TABLE OF CONTENTS

Journal of Physics: Conference Series **128** (2008) 011001 doi:10.1088/1742-6596/128/1/011001

Preface

This volume of Journal of Physics: Conference Series contains the Proceedings of the 5th International Symposium in Quantum Theory and Symmetries (QTS5), held in Valladolid, Spain, July 22-28, 2007. This is the fith of a series of conferences previously held in Goslar (Germany) 1999, QTS1; Cracow (Poland) 2001, QTS2; Cincinnati (USA) 2003, QTS3, and Varna (Bulgaria) 2005, QTS4.

The QTS5 symposium gathered 181 participants from 39 countries working in different fields on Theoretical Physics. The spirit of the QTS conference series is to join researchers in a wide variety of topics in Theoretical Physics, as a way to make accessible recent results and the new lines of different fields. The QTS5 conference offered the following list of topics:

1. Symmetries in String Theory, Quantum Gravity and related topics
2. Symmetries in Quantum Field Theories, Conformal and Related Field Theories, Lattice and Noncommutative Theories, Gauge Theories
3. Quantum Computing, Information and Control
4. Foundations of Quantum Theory
5. Quantum Optics, Coherent States, Wigner Functions
6. Dynamical and Integrable Systems
7. Symmetries in Condensed Matter and Statistical Physics
8. Symmetries in Particle Physics, Nuclear, Atomic and Molecular Physics
9. Nonlinear Quantum Mechanics
10. Time Asymmetric Quantum Mechanics
11. SUSY Quantum Mechanics, PT symmetries and pseudo-Hamiltonians
12. Mathematical Methods for Symmetries and Quantum Theories
13. Symmetries in Chemistry Biology and other Sciences

Papers accepted for publication in the present issue are based on the contributions from the participants in the QTS5 conference after a peer reviewing process. In addition, a special issue of Journal Physics A contains contributions from plenary speakers, some participants as well as contributions from other authors whose works fit into the topics of the conference.

The organization of the conference had the following pattern. In the morning there were five plenary or general sessions for all the participants, which aimed to give a state of the art of different topics. The afternoon was devoted to 6 parallel sessions, each one including the most afine topics. Also there was a poster session that could be visited in the intermediates and during the coffe breaks. A total of 23 plenary conferences, 130 parallel communications and 25 posters were presented in QTS5. All the sessions were given in the Palace of Conferences Conde Ansúrez which among other facilities also had a computer room. As social activities we had a guided walk around the historical part of the city, a visit to the near city of Segovia, as well as a wellcome party and farewell dinner. These activities were intended to favor the contact and informal discussions among the participants in the meeting.

We are much indebted to several institutions that without their support the organization of the QTS5 symposium would not have been possible. In this respect we greatly acknowledge to Ministerio de Educación of Spain and Junta de Castilla y León for general financial support, to Fundación Universidades de Castilla y León for a number of grants to young researchers who otherwise would not have attended the conference, to the European Physical Society that provided a number of grants for eastern countries and to the University of Valladolid where the event took place.

We want to thank all the members of the Conference Board and the International Advisory Committee of QTS5 for honoring us with their help and support.

Here we must mention our gratitude to the other members of the Local Organizing Committe of QTS5, Manuel Gadella, José Manuel Izquierdo and Şengül Kuru, that are not Editors of this JPCS issue but shared with us the complex organization of this event.

We acknowledge also to the IOP Publishing for the facilities in the publication of these Proceedings.

Finally, on behalf of the Local Organizing Committee, we would like to thank all the participants in the QTS5 conference for their presence, contributions, and for the good atmosphere achieved during their stay. We hope that the experience of spending these days in Valladolid has been most fruitful for all of them.

O Arratia, J A Calzada, F Gómez-Cubillo, J Negro and M A del Olmo
Universidad de Valladolid, Spain
Editors of the QTS5 Proceedings

Conference Board

- S.T. Ali (Montreal)
- L.L. Boyle (Canterbury)
- M.A. del Olmo (Valladolid)
- V.K. Dobrev (Sofia)
- H.D. Doebner (Clausthal), chair
- E. Kapuscik (Cracow)
- V.I. Man'ko (Moscow)
- G. Marmo (Naples)
- G.S. Pogosyan (Yerevan and Dubna)
- T.H. Seligman (Cuernavaca)
- A.I. Solomon (Paris and Open University)
- P. Suranyi (Cincinnati)
- L.C.R. Wijewardhana (Cincinnati)

International Advisory Committee

- L. Accardi, (Roma)

Journal of Physics: Conference Series **128** (2008) 011001 doi:10.1088/1742-6596/128/1/011001

- M. Asorey, (Zaragoza)
- M. T. Batchelor, (Canberra)
- C.M. Bender, (Washington)
- A. Bohm, (Texas)
- E. Celeghini, (Firenze)
- I. Cirac, (Garching)
- S. Ferrara, (CERN)
- J. P. Gazeau, (Paris)
- G. Goldin , (Rutgers)
- F. Iachello, (Yale)
- T. Janssen, (Nijmegen)
- J. Klauder, (Gainesville)
- P. Kulish, (St. Petersburg)
- B. Mielnik, (Mexico DF)
- W. Miller, (Minneapolis)
- M. Plyushchay, (Santiago de Chile)
- O. Ragnisco, (Roma)
- S. Randjbar-Daemi, (ICTP)
- M. Santander, (Valladolid)
- G. Sierra, (Madrid)
- P. Townsend, (Cambridge)
- S. Twarock, (York)
- F. Wilczek, (Boston)
- P. Winternitz, (Montreal)
- K.B. Wolf, (Cuernavaca)

Local Organizing Committee (University of Valladolid)

- Oscar Arratia
- Juan A. Calzada
- Manuel Gadella
- Fernando Gómez-Cubillo
- Jos Manuel Izquierdo
- Şengül Kuru
- Javier Negro
- Mariano A. del Olmo (Chairman)

Official Foto

FACULTAD DE CIENCIAS

Journal of Physics: Conference Series **128** (2008) 011002 doi:10.1088/1742-6596/128/1/011002

Opening Talk

Ladies and Gentlemen
Dear Friends and Colleagues

I welcome you at the 5th International Symposium "Quantum Theory and Symmetries, QTS5" in Valladolid as Chairman of the Conference Board of this biannual series.

The aim of the series is to arrange an international meeting place for scientists working in theoretical and mathematical physics, in mathematics, in mathematical biology and chemistry and in other sciences for the presentation and discussion of recent developments in connection with quantum physics and chemistry, material science and related further fields, like life sciences and engineering, which are based on mathematical methods which can be applied to model and to understand *microphysical and other systems* through inherent *symmetries in their widest sense.* This systems include, e.g., foundations and extensions of quantum theory; quantum probability; quantum optics and quantum information; the description of nonrelativistic, finite dimensional and chaotic systems; quantum field theory, particle physics, string theory and quantum gravity.

Symmetries in their widest sense describe properties of a system which could be modelled, e.g., through geometry, group theory, topology, algebras, differential geometry, noncommutative geometry, functional analysis and approximation methods; numerical evaluation techniques are necessary to connect such symmetries with experimental results. If you ask for a more detailed characterisation of this notion a hand waving indirect answer is: Collect titles and contents of the contributions of the proceedings of QTS4 and get a characterisation through semantic closure.

Quantum theory and its Symmetries was and is a diversified and rapidly growing field. The number of and the types of systems with an internal symmetry and the corresponding mathematical models develop fast. This is reflected in the content of the five former international symposia of this series: The first symposium, QTS1–1999, was organized in Goslar (Germany) with 170 participants and 89 contributions in the proceedings; it was centred on the foundations and extensions of quantum theory, on quantisation methods and on q-algebras. In QTS2–2001 in Cracow (Poland) with 175 participants and 81 contributions; the main topics were applications of quantum mechanics, representations of algebras and group theoretical techniques in physics. In the symposium QTS3–2003 in Cincinnati (USA) with 145 participants and 92 contributions, quantum field theory, loop quantum gravity, string and brane theory was discussed. The focus in QTS4–2005 in Varna (Bulgaria) with 228 participant and 105 contributions, was on conformal field theory, quantum gravity, noncommutative geometry and quantum groups. Three proceedings volumes were published with World Scientific and one volume with Heron Press. The promising and interesting programme for QTS5–2007 in Valladolid (Spain) attracted more than 200 participants; the contributions will be published in a Special Issue of Journal of Physics A and a volume of Journal Physics of Conference Series. This shows the wide scope of symmetry in connection with quantum physics and related sciences.

In the background of the symposia series is the Conference Board with presently 13 members. The Board encourage scientists and Institutions to present detailed proposals for a QTS symposium; it agrees to one proposal and is prepared to assist in matters of organisation; the local organisers are responsible for the scientific programme and for the organisation, including the budget. The Board decided that the next symposium QTS6 will be held 2009 at the University of Kentucky in Lexington (USA); Alan Shapere is the chairman of the Local Organizing committee.

In the name of all of you I express my appreciation and my thanks to the members of the Local Organizing Committee of QTS5, especially to Mariano del Olmo. The programme is outstanding; it covers recent and new developments in our field. The organization is very effective and complete. We have all the necessary condition for a successful and smooth meeting. Thank you again Mariano.

H.-D. Doebner
Chairman of the Conference Board of QTS5

Journal of Physics: Conference Series **128** (2008) 012001 doi:10.1088/1742-6596/128/1/012001

Supersymmetric 3D gravity with torsion: asymptotic symmetries and black hole stability

Branislav Cvetković and Milutin Blagojević
Institute of physics, P. O. Box 57, 11001 Belgrade, Serbia
E-mail: cbranislav@phy.bg.ac.yu, mb@phy.bg.ac.yu

Abstract. We show that $N = 1+1$ supersymmetric extension of 3D gravity with torsion, with suitable boundary conditions, has the asymptotic superconformal symmetry. The existence of exact supersymmetries implies the stability conditions for the black hole solutions.

1. Introduction

Since the basic understanding of quantum gravity is still missing, the researchers are naturally oriented toward models that are technically simpler but share the same conceptual features with general relativity (GR). In this sense, the study of three-dimensional (3D) gravity within the realm of *Riemannian* geometry of spacetime was particulary fruitful [1, 2, 3, 4, 5]. However, 3D gravity can also be used to explore a more general, gauge-theoretic conception of gravity, based on *Riemann-Cartan* geometry of spacetime—the geometry that is characterized by both the *curvature* and the *torsion* [6]. Nowadays, fifteen years after a general topological model for 3D gravity with torsion was proposed by Mielke and Baekler [7], it is clear that this approach has led to a number of respectable results [8, 9, 10, 11, 12, 13, 14].

It was shown recently that $N = 1 + 1$ supersymmetric extension of 3D gravity with torsion, with a suitable generalization of the anti-de Sitter (AdS) asymptotic conditions, possesses superconformal symmetry at spatial infitity [13]. In this paper, we use this result to explore exact supersymmetries and the stability properties of the black hole with torsion. Our results represent a natural generalization of those obtained for the BTZ black hole in [15].

The paper is organized in the following way. In section 2, we review some basic aspects of the topological 3D gravity with torsion. Section 3 is devoted to exploring the supersymmetric extension of the Mielke-Baekler model with $N = 1 + 1$ gravitini and its asymptotic structure. In section 4, we explore exact supersymmetries of the AdS and black hole configurations and combine them with the asymptotic superconformal algebra, with the conclusion that the zero-energy black hole can be interpreted as the ground state of the Ramond sector, while the AdS solution is the ground state of the Neveau-Schwartz sector, with periodic and antiperiodic boundary conditions, respectively. Finally, section 5 is devoted to concluding remarks.

We use the same conventions as in [13]: the Latin indices $(i, j, k, \dots)$ refer to the local orthonormal frame, the Greek indices $(\mu, \nu, \rho, \dots)$ refer to the coordinate frame, and both run over $0, 1, 2$; the metric components in the local Lorentz frame are $\eta_{ij} = (+, -, -)$; totally antisymmetric object ε^{ijk} is normalized by $\varepsilon^{012} = +1$; gamma matrices are pure imaginary and Majorana spinors are real.

2. 3D gravity with torsion

Theory of gravity with torsion can be naturally described as Poincaré gauge theory (PGT), with an underlying spacetime structure corresponding to Riemann-Cartan geometry [6]. Basic gravitational variables in PGT are the triad field b^i and the Lorentz connection $A^{ij} = -A^{ji}$ (1-forms), and the corresponding fields strengths are the torsion T^i and the curvature R^{ij} (2-forms). In 3D, we use the simplified notation $A^{ij} =: -\varepsilon^{ij}{}_k\omega^k$, $R^{ij} := -\varepsilon^{ij}{}_k R^k$, so that

$$T^i = db^i + \varepsilon^i{}_{jk}\omega^j \wedge b^k\,, \qquad R^i = d\omega^i + \frac{1}{2}\varepsilon^i{}_{jk}\omega^j \wedge \omega^k\,. \tag{2.1}$$

Gauge symmetries of the theory are local translations and local Lorentz rotations. The covariant derivative $\nabla \equiv \nabla(\omega)$ acts on a general tangent-frame spinor/tensor in accordance with its spinorial/tensorial structure; when X is a form, $\nabla X := \nabla \wedge X$.

The metric structure of PGT is defined by $g = \eta_{ij} b^i \otimes b^j$. Metric and connection are related to each other by the *metricity condition*, $\nabla g = 0$, which defines *Riemann-Cartan geometry* of spacetime. In PGT, we have a useful identity

$$\omega^i \equiv \tilde{\omega}^i + K^i\,, \tag{2.2}$$

where $\tilde{\omega}^i$ is the Levi-Civita connection, and K^i is the contortion defined by $T^i = \varepsilon^i{}_{mn} K^m \wedge b^n$.

Generalized dynamics. The *topological* Mielke-Baekler model for 3D gravity [7], represents a natural generalization of general relativity (GR) with a cosmological constant:

$$I_0 = 2a\int b^i R_i - \frac{\Lambda}{3}\int \varepsilon_{ijk} b^i b^j b^k + \alpha_3 \int \left(\omega^i d\omega_i + \frac{1}{3}\varepsilon_{ijk}\omega^i\omega^j\omega^k\right) + \alpha_4 \int b^i T_i\,, \tag{2.3}$$

where the wedge product sign $\wedge$ is omitted for simplicity. The first term with $a = 1/16\pi G$ is the usual Einstein-Cartan action, the second term is a cosmological term, the third term is the Chern-Simons action for the Lorentz connection, and the last term is a torsion counterpart of the first one.

The vacuum field equations, in the sector $\alpha_3\alpha_4 - a^2 \neq 0$, take the simple form

$$2T^i = p\varepsilon^i{}_{jk}\, b^j b^k\,, \qquad 2R^i = q\varepsilon^i{}_{jk}\, b^j b^k\,, \tag{2.4}$$

where

$$p := \frac{\alpha_3\Lambda + \alpha_4 a}{\alpha_3\alpha_4 - a^2}\,, \qquad q := -\frac{(\alpha_4)^2 + a\Lambda}{\alpha_3\alpha_4 - a^2}\,.$$

Thus, the vacuum solutions are characterized by constant torsion and constant curvature.

In Riemann-Cartan spacetime, one can use the identity (2.2) to express the curvature in terms of its *Riemannian* piece $\tilde{R}^i = R^i(\tilde{\omega})$ and the contortion. The resulting identity, combined with the equations of motion (2.4), leads to the on-shell relation

$$\tilde{R}^{ij} = -\Lambda_{\text{eff}}\, b^i \wedge b^j \qquad \Lambda_{\text{eff}} := q - \frac{1}{4}p^2\,, \tag{2.5}$$

where Λ_{eff} is the effective cosmological constant. The form of $\tilde{R}^{ij}$ implies that our spacetime has maximally symmetric metric.

The black hole with torsion. For negative Λ_{eff} (AdS sector), the Mielke-Baekler model has an exact vacuum solution, the black hole with torsion [8, 9], which is a natural generalization

Journal of Physics: Conference Series **128** (2008) 012001 doi:10.1088/1742-6596/128/1/012001

of the well-known BTZ black hole [4]. In static coordinates $x^\mu = (t, r, \varphi)$ (with $0 \leq \varphi < 2\pi$), the BTZ metric is given as

$$\begin{aligned} ds^2 &= N^2 dt^2 - N^{-2} dr^2 - r^2 (d\varphi + N_\varphi dt)^2 \,, \\ N^2 &= \left(-8Gm + \frac{r^2}{\ell^2} + \frac{16G^2 J^2}{r^2} \right) , \qquad N_\varphi = \frac{4GJ}{r^2} \,. \end{aligned} \tag{2.6}$$

For the black hole with torsion, the triad field is taken as

$$b^0 = N dt \,, \qquad b^1 = N^{-1} dr \,, \qquad b^2 = r \left(d\varphi + N_\varphi dt \right) \,, \tag{2.7a}$$

while the connection, in accordance with (2.2), has the form

$$\omega^i = \tilde{\omega}^i + \frac{p}{2} b^i \,, \tag{2.7b}$$

where the Riemannian connection $\tilde{\omega}^i$ reads:

$$\tilde{\omega}^0 = -N d\varphi \,, \qquad \tilde{\omega}^1 = N^{-1} N_\varphi dr \,, \qquad \tilde{\omega}^2 = -\frac{r}{\ell} \frac{dt}{\ell} - r N_\varphi d\varphi \,. \tag{2.7c}$$

Energy and angular momentum of the black hole are given by the expressions

$$E = m + \frac{\alpha_3}{a} \left(\frac{pm}{2} - \frac{J}{\ell^2} \right) , \qquad M = J + \frac{\alpha_3}{a} \left(\frac{pJ}{2} - m \right) . \tag{2.8}$$

Locally isometric but globally distinct AdS solution (AdS_3) can be formally obtained from the black hole by taking $J = 0, 8Gm = -1$.

3. Supersymmetric extension

There exists a simple locally supersymmetric extension of 3D gravity with torsion, based on the action (2.3), which includes two gravitini fields and is usually referred to as $N = 1 + 1$ AdS supergravity [2, 12, 13]. Consider the action:

$$I = I_0 - g \int \left(\bar{\psi} \nabla \psi - i\mu \bar{\psi} b^i \gamma_i \psi \right) - g' \int \left(\bar{\psi}' \nabla \psi' - i\mu' \bar{\psi}' b^i \gamma_i \psi' \right) , \tag{3.1}$$

where $\psi^\Pi = (\psi, \psi')$ are the gravitini fields (1-forms) and $\nabla \psi^\Pi = (d - \frac{i}{2} \omega^m \gamma_m) \psi^\Pi$ are their covariant derivatives. The action is invariant under the following locally sypersymmetry transformations with spinorial parameters $\varepsilon^\Pi = (\varepsilon, \varepsilon')$:

$$\begin{aligned} \delta_S b^i &= i \bar{\varepsilon} \gamma^i \psi + i \bar{\varepsilon}' \gamma^i \psi' \,, \\ \delta_S \omega^i &= -2i\mu' \bar{\varepsilon} \gamma^i \psi - 2i\mu \bar{\varepsilon}' \gamma^i \psi' \,, \\ \delta_S \psi^\Pi &= 4a \left(\nabla \varepsilon^\Pi - i\mu^\Pi b^k \gamma_k \varepsilon^\Pi \right) , \end{aligned} \tag{3.2}$$

provided the coupling constants g, g' and $\mu^\Pi = (\mu, \mu')$ satisfy the relations

$$\begin{aligned} 2ag &= a - 2\mu' \alpha_3 \,, \qquad & 2ag' &= a - 2\mu \alpha_3 \,, \\ 2\mu + \frac{p}{2} &= \frac{1}{\ell} \,, \qquad & 2\mu' + \frac{p}{2} &= -\frac{1}{\ell} \,. \end{aligned}$$

Here, ℓ is the AdS radius, and for μ and μ' real and different from each other, the effective cosmological constant is negative: $\Lambda_{\text{eff}} = -(\mu - \mu')^2 = -1/\ell^2 < 0$.

The variation of the action with respect to $b^i, \omega^i, \bar\psi$ and $\bar\psi'$ yields the supersymmetric field equations. As one can easily verify, the black hole and AdS3 are *exact solutions of the supersymmetric field equations with zero gravitini*, $\psi = \psi' = 0$.

Asymptotic symmetries. For $\Lambda_{\text{eff}} < 0$, the asymptotic structure of 3D gravity with torsion is well understood [10]. It can be generalized to the supersymmetric theory (3.1) by a suitable completion of the asymptotic conditions in the fermionic sector. The asymptotic symmetry is characterized by four chiral parameters: two of them are bosonic, $T^\mp(x^\mp)$, while the other two are fermionic, $\epsilon^\mp(x^\mp)$. The asymptotic Poisson bracket algebra of the canonical generators $\tilde G(T^\mp, \epsilon^\mp)$ is given (in the quantum-mechanical notation) by two independent copies of the super-Virasoro algebra [13]:

$$\begin{aligned}
[L_n^\mp, L_m^\mp] &= (n-m)L_{n+m}^\mp + \frac{c^\mp}{12} n^3 \delta_{m+n,0}\,, \\
[L_n^\mp, Q_m^\mp] &= \left(\frac{1}{2}n - m\right) Q_{m+n}^\mp\,, \\
\{Q_n^\mp, Q_m^\mp\} &= 2L_{n+m}^\mp + \frac{c^\mp}{3} n^2 \delta_{m+n,0}\,,
\end{aligned} \tag{3.3}$$

where $L_n^\mp$ and $Q_n^\mp$ are Fourier modes of $\tilde G(T^\mp, \epsilon^\mp)$, and $c^\mp$ are classical central charges: $c^- = 12 \cdot 2\pi\ell a g\,,\ c^+ = 12\cdot 2\pi\ell a g'\,.$

In the sectors with periodic (Ramond) or anti-periodic (Neveau-Schwartz) boundary conditions for fermions, index of $Q_m^\mp$ takes on integer or half-integer values, respectively. The reality properties of the modes are: $(L^\mp)_n^\dagger = L_{-n}^\mp$, $(Q^\mp)_n^\dagger = Q_{-n}^\mp$.

The eigenvalues of the bosonic operators $L_0^\mp$ can be expressed in terms of the energy E and angular momentum M of the system. For the black hole/AdS3 solution, we have:

$$L_0^\mp = \frac{\ell E \pm M}{2} = (\ell m \pm J)\frac{c^\mp}{48\pi a\ell}\,. \tag{3.4}$$

4. The black hole stability

Killing spinors. Since the black hole and AdS3 are maximally symmetric, their symmetries are locally identical, but the difference in their global structures implies completely different symmetries in the large. In the supersymmetric theory (3.1), the supersymmetry transformations that leave the black hole or AdS3 configuration with $\psi = \psi' = 0$ invariant, are called exact supersymmetries. Since $\delta_S b^i = \delta_S \omega^i = 0$ for $\psi = \psi' = 0$, the spinor parameters $(\varepsilon, \varepsilon')$ of exact supersymmetries are defined solely by the requirements $\delta_S\psi = \delta_S\psi' = 0$. Expressing ω^i in terms of the Levi-Civita connection $\tilde\omega^i$ as in (2.7b), these requirements take the form

$$\tilde\nabla\varepsilon = \frac{i}{2\ell} b^k \gamma_k \varepsilon\,, \qquad \tilde\nabla\varepsilon' = -\frac{i}{2\ell} b^k\gamma_k\varepsilon'\,, \tag{4.1}$$

where $\tilde\nabla$ is Riemannian covariant derivative. Equations (4.1) are called the Killing spinor equations; they define the spinor parameters $(\varepsilon, \varepsilon')$ of the exact supersymmetries. Among their solutions, only those that are in agreement with the global properties of the black hole/AdS3 are acceptable.

One can show that AdS3 possesses four Killing spinors, two for ε and two for ε':

$$\begin{aligned}
\varepsilon &= \left(\sqrt{\frac{N_*+1}{2}}\sigma^3 - \sqrt{\frac{N_*-1}{2}}\right)\left(\cos\frac{x^-}{2} + i\sigma^2\sin\frac{x^-}{2}\right)\zeta\,, \\
\varepsilon' &= \left(\sqrt{\frac{N_*+1}{2}}\sigma^3 + \sqrt{\frac{N_*-1}{2}}\right)\left(\cos\frac{x^+}{2} + i\sigma^2\sin\frac{x^+}{2}\right)\zeta'\,.
\end{aligned} \tag{4.2}$$

Journal of Physics: Conference Series **128** (2008) 012001 doi:10.1088/1742-6596/128/1/012001

Here, $N_* = \sqrt{1 + r^2/\ell^2}$ is the value of N at AdS_3, $x^\pm = t/\ell \pm \varphi$ and ζ, ζ' are constant spinors.

The black hole possesses locally the same number of Killing spinors as AdS_3. However, since the black hole is obtained from AdS_3 by a process of identifications, spinors on the black hole manifold, compatible with these identifications, can be either periodic or anti-periodic [4, 15]. The form of this restriction depends on the values of the parameters m and J.

1. For a *generic black hole* with $|J| < m\ell$, there are no periodic/anti-periodic solutions of (4.1), and no Killing spinors.

2. Consider now the *extreme black hole* with $|J| = m\ell$ and $m \neq 0$.

(a) For $J = m\ell$, the periodic/anti-periodic solution for ε' does not exist, while the solution for ε, compatible with the periodicity in φ, has the form ($u = r/\ell - 4Gm\ell/r$):

$$\varepsilon = \left[\left(|u|^{1/2} + |u|^{-1/2}\right) - \text{sgn}u\left(|u|^{1/2} - |u|^{-1/2}\right)\sigma^3\right]\begin{pmatrix} 0 \\ 1 \end{pmatrix}. \tag{4.3a}$$

(b) Similarly, in the case $J = -m\ell$, there are no periodic/anti-periodic solutions for ε, while globally acceptable solution for ε' takes the form

$$\varepsilon' = \left[\left(|u|^{1/2} + |u|^{-1/2}\right) + \text{sgn}u\left(|u|^{1/2} - |u|^{-1/2}\right)\sigma^3\right]\begin{pmatrix} 1 \\ 0 \end{pmatrix}. \tag{4.3b}$$

3. The case $m = J = 0$ corresponds to the *zero-energy black hole*, for which both E and M vanish. The related Killing spinors can be obtained from the extreme black hole solutions (4.3) in the limit $m \to 0$. The periodicity requirement implies the existence of *two* Killing spinors:

$$\varepsilon \sim \sqrt{\frac{r}{\ell}}\begin{pmatrix} 0 \\ 1 \end{pmatrix}, \qquad \varepsilon' \sim \sqrt{\frac{r}{\ell}}\begin{pmatrix} 1 \\ 0 \end{pmatrix}. \tag{4.4}$$

Thus, *a generic black hole has no Killing spinors, the extreme black hole has only one, while the zero-energy black hole has two.*

The existence of Killing spinors, combined with the asymptotic supersymmetry algebra, has serious implications on the stability of the black hole/AdS_3.

Ramond sector. In the Ramond sector (with periodic boundary conditions), the subalgebra of (3.1) with vanishing central charge, generated by $(L_0^\mp, Q_0^\mp)$, yields:

$$L_0^\mp = Q_0^\mp Q_0^\mp = Q_0^\mp (Q_0^\mp)^\dagger \geq 0\,. \tag{4.5}$$

These relations are equivalent to $|M| \leq \ell E$, which implies $E \geq 0$.

The zero-energy black hole has two Killing spinors, which are the parameters of two exact supersymmetries, generated by the action of $Q_0^\mp$. Moreover, this configuration *saturates* the bounds (4.5), which means that it has the lowest values of the conserved charges $L_0^\mp$, compared to any other black hole configuration. Consequently, *the zero-energy black hole can be naturally interpreted as the black hole ground state* (the ground state of the Ramond sector).

One should note that for an extremal black hole, only one of the bounds in (4.5) is saturated, while for a generic black hole, both L_0^- and L_0^+ are strictly positive.

Neveu-Schwarz sector. The AdS solution belongs to the Neveu-Schwarz sector (anti-periodic boundary conditions), with (the value of $L_0^\mp$) $= -c^\mp/24$. If one makes the shift $L_0^\pm \to L_0^\pm - c^\pm/24$, the value of the new $L_0^\mp$ on AdS_3 vanish. The $osp(1|2) \oplus osp(1|2)$ subalgebra with vanishing charge is generated by $(L_{\mp 1}, L_0, Q_{\mp 1/2})^\mp$, and it implies

$$2L_0^\mp = Q_{1/2}^\mp Q_{-1/2}^\mp + Q_{-1/2}^\mp Q_{1/2}^\mp = Q_{1/2}^\mp (Q_{1/2}^\mp)^\dagger + Q_{-1/2}^\mp (Q_{-1/2}^\mp)^\dagger \geq 0\,. \tag{4.6}$$

The AdS solution has four Killing spinors, which are the parameters of four exact supersymmetries, generated by the action of $(Q_{1/2}^\mp, Q_{-1/2}^\mp)$, and it saturates the bounds (4.6). Thus, *we are led to interpret AdS_3 as the ground state of the Neveu-Schwarz sector.*

Journal of Physics: Conference Series **128** (2008) 012001 doi:10.1088/1742-6596/128/1/012001

5. Concluding remarks

In conclusion, our results can be summarized as follows.

- The asymptotic symmetry of $N = 1+1$ sypersymmetric extension of 3D gravity with torsion is described by two independent super-Virasoro algebras with different central charges.
- The zero-energy black hole and AdS_3 can be naturally interpreted as the ground states of the sectors with periodic and anti-periodic boundary conditions, respectively.

These results are a natural generalization of those found earlier in Riemannian GR [15].

Acknowledgments

This work was supported by the Serbian Science Foundation.

References

[1] Brown J D and Henneaux M 1986 Central charges in the canonical realization of asymptotic symmetries: an example from three-dimensional gravity *Commun. Math. Phys.* **104** 207

[2] Achucarro A and Townsend P A 1986 Chern-Simons action for three-dimensional anti-de Sitter supergravity theories *Phys. Lett. B* **180** 89

[3] Witten E 1988 2+1 dimensional gravity as an exactly soluble system *Nucl. Phys. B* **311** 46

[4] Bañados M, Teitelboim C and Zanelli J 1993 The black hole in three-dimensional spacetime *Phys. Rev. Lett.* **16** 1849

Bañados M, Teitelboim C and Zanelli J 1993 Geometry of 2+1 black hole *Phys. Rev. D* **48** 1506

[5] For a review, see: Carlip S 2005 Conformal field theory, (2+1)-dimensional gravity, and the BTZ black hole *Class. Quant. Grav.* **22** R85-R124

[6] Blagojević M 2002 *Gravitation and gauge symmetries* (Bristol: Institute of Physics Publishing)

Blagojević M 2003 Three lectures on Poincaré gauge theory *SFIN* A **1** 147

[7] Mielke E W, Baekler P 1991 Topological gauge model of gravity with torsion *Phys. Lett. A* **156** 399

Baekler P, Mielke E W, Hehl F W 1992 Dynamical symmetries in topological 3D gravity with torsion *Nuovo Cim. B* **107** 91

[8] García A, Hehl F W, Heinecke C and Macías A 2003 Exact vacuum solutions of (1+2)-dimensional Poincaré gauge theory: BTZ solution with torsion *Phys. Rev. D* **67** 124016

Mielke E W and Maggiolo A R 2003 Rotating black hole solution in a generalized topological 3D gravity with torsion *Phys. Rev. D* **68** 104026

[9] Blagojević M and Vasilić M 2003 3D gravity with torsion as a Chern-Simons gauge theory *Phys. Rev. D* **68** 104023

Cacciatori S, Caldarelli, Giacomini A, Klemm D and Mans D 2006 Chern-Simons formulation of three-dimensional gravity with torsion and nonmetricity *J. Geom. Phys.* **56** 2623

[10] Blagojević M and Cvetković B 2006 Canonical structure of 3D gravity with torsion in: *Progress in General Relativity and Quantum Cosmology* vol. 2 ed. Ch. Benton (New York: Nova Science Publishers) 103

Blagojević M and Cvetković B 2006 Asymptotic charges in 3d gravity with torsion *Journal of Physics: Conf. Series* **33** 248

[11] Blagojević M and Cvetković B 2006 Black hole entropy in 3D gravity with torsion *Class. Quantum Grav.* **23** 4781

Blagojević M and Cvetković B 2006 Black hole entropy from the boundary conformal structure in 3D gravity with torsion *J. High Energy Phys. JHEP***10** 005

Blagojević M and Cvetković B 2007 Covariant description of the black hole entropy in 3D gravity *Class. Quant. Grav.* **24** 129

[12] Giacomini A, Troncoso R and Willison S 2007 Three-dimensional supergravity reloaded *Class. Quantum Grav.* **24** 2845

[13] Cvetković B and Blagojević M 2007 Supersymmetric 3D gravity with torsion: asymptotic symmetries *Class. Quantum Grav.* **24** 3933

[14] Klemm D, Tagliabue G 2007 The CFT dual of AdS gravity with torsion *Preprint* arXiv:0705.3320 [gr-qc].

[15] Coussaert O and Henneaux M 1994 Supersymmetry of 2+1 black hole *Phys. Rev. Lett.* **72** 183

Bañados M, Bautier K, Coussaert O, M Henneaux and Ortiz M 1998 Anti-de Sitter/CFT Correspondence in Three-Dimensional Supergravity *Phys. Rev. D* **58** 085020

Journal of Physics: Conference Series **128** (2008) 012002 doi:10.1088/1742-6596/128/1/012002

Fermion-fermion and boson-boson amplitudes: surprising similarities

Valeri V. Dvoeglazov
Universidad de Zacatecas, Apartado Postal 636, Suc. UAZ, Zacatecas 98062, Zac., México
E-mail: valeri@planck.reduaz.mx

Abstract. Amplitudes for fermion-fermion, boson-boson and fermion-boson interactions are calculated in the second order of perturbation theory in the Lobachevsky space. An essential ingredient of the model is the Weinberg's $2(2j+1)-$ component formalism for describing a particle of spin j. The boson-boson amplitude is then compared with the two-fermion amplitude obtained long ago by Skachkov on the basis of the Hamiltonian formulation of quantum field theory on the mass hyperboloid, $p_0^2 - \mathbf{p}^2 = M^2$, proposed by Kadyshevsky. The parametrization of the amplitudes by means of the momentum transfer in the Lobachevsky space leads to same spin structures in the expressions of $T-$ matrices for the fermion case and the boson case. However, certain differences are found. Possible physical applications are discussed.

1. Introduction

The problem of correct description of quarkonium (the bound state of quark and antiquark – the mesons) and of barions were hot topics since long ago. However, we are now faced at the correct description of a bound state of spin-1 particles (and higher spins). Instead of developing other methods we suggest to think about the general methods of describing higher-spin particles (and their bound states) on an equal footing with fermions. The pioneer ideas have been proposed in refs. [8, 10]. The amplitudes in the Lobachevsky space are calculated in the next Section.

2. Amplitudes

The scattering amplitude for the two-fermion interaction had been obtained in the 3-momentum Lobachevsky space [1] in the second order of perturbation theory long ago [2a,Eq.(31)]:

$$\begin{aligned} T_V^{(2)}(\mathbf{k}(-)\mathbf{p},\mathbf{p}) &= -g_v^2\frac{4m^2}{\mu^2+4\text{æ}^2} - 4g_v^2\frac{(\sigma_1\text{æ})(\sigma_2\text{æ})-(\sigma_1\sigma_2)\text{æ}^2}{\mu^2+4\text{æ}^2} - \\ &- \frac{8g_v^2 p_0\text{æ}_0}{m^2}\,\frac{i\sigma_1[\mathbf{p}\times\text{æ}]+\mathbf{i}\sigma_2[\mathbf{p}\times\text{æ}]}{\mu^2+4\text{æ}^2} - \frac{8g_v^2}{m^2}\,\frac{p_0^2\text{æ}_0^2+2p_0\text{æ}_0(\mathbf{p}\cdot\text{æ})-m^4}{\mu^2+4\text{æ}^2} - \\ &- \frac{8g_v^2}{m^2}\,\frac{(\sigma_1\mathbf{p})(\sigma_1\text{æ})(\sigma_2\mathbf{p})(\sigma_2\text{æ})}{\mu^2+4\text{æ}^2}\quad, \end{aligned} \tag{1}$$

g_v is the coupling constant. The additional term (the last one) has usually *not* been taken into account in the earlier Breit-like calculations of two-fermion interactions. This consideration is based on use of the formalism of separation of the Wigner rotations and parametrization of currents by means of the Pauli-Lubanski vector, developed long ago [3]. The quantities

$$
æ_0 = \sqrt{\frac{m(\Delta_0 + m)}{2}} \quad , \quad \mathbf{æ} = \mathbf{n}_\Delta \sqrt{\frac{m(\Delta_0 - m)}{2}}
$$

are the components of the 4-vector of a momentum half-transfer. This concept is closely connected with a notion of the half-velocity of a particle [4]. The 4-vector Δ_μ:

$$
\mathbf{\Delta} = \mathbf{\Lambda}_{\mathbf{p}}^{-1}\mathbf{k} = \mathbf{k}(-)\mathbf{p} = \mathbf{k} - \frac{\mathbf{p}}{\mathbf{m}}(\mathbf{k_0} - \frac{\mathbf{k}\cdot\mathbf{p}}{\mathbf{p_0} + \mathbf{m}}), \tag{2}
$$

$$
\Delta_0 = (\Lambda_p^{-1}k)_0 = (k_0 p_0 - \mathbf{k}\cdot\mathbf{p})/m = \sqrt{m^2 + \mathbf{\Delta^2}} \tag{3}
$$

can be regarded as the momentum transfer vector in the Lobachevsky space instead of the vector $\mathbf{q} = \mathbf{k} - \mathbf{p}$ in the Euclidean space.[1] This amplitude had been used for physical applications in the framework of the Kadyshevsky's version of the quasipotential approach [1, 2].

On the other hand, in ref. [8] an attractive $2(2j + 1)$ component formalism for describing particles of higher spins has been proposed. As opposed to the Proca 4-vector potentials which transform according to the $(\frac{1}{2}, \frac{1}{2})$ representation of the Lorentz group, the $2(2j + 1)$ component functions are constructed via the representation $(j, 0) \oplus (0, j)$ in the Weinberg formalism. This description of higher spin particles is on an equal footing to the description of the Dirac spinor particle, whose field function transforms according to the $(\frac{1}{2}, 0) \oplus (0, \frac{1}{2})$ representation. The $2(2j + 1)$- component analogues of the Dirac functions in the momentum space are

$$
\mathcal{U}(\mathbf{p}) = \sqrt{\frac{M}{2}} \begin{pmatrix} D^J(\alpha(\mathbf{p}))\, \xi_\sigma \\ D^J\left(\alpha^{-1\,\dagger}(\mathbf{p})\right) \xi_\sigma \end{pmatrix} \quad , \tag{4}
$$

for the positive-energy states; and[2]

$$
\mathcal{V}(\mathbf{p}) = \sqrt{\frac{M}{2}} \begin{pmatrix} D^J\left(\alpha(\mathbf{p})\Theta_{[1/2]}\right) \xi^*_\sigma \\ D^J\left(\alpha^{-1\,\dagger}(\mathbf{p})\Theta_{[1/2]}\right) (-1)^{2J} \xi^*_\sigma \end{pmatrix} \quad , \tag{5}
$$

for the negative-energy states, ref. [5, p.107], with the following notations being used:

$$
\alpha(\mathbf{p}) = \frac{p_0 + M + (\sigma\cdot\mathbf{p})}{\sqrt{2M(p_0 + M)}}, \quad \Theta_{[1/2]} = -i\sigma_2 \quad . \tag{6}
$$

[1] I keep a notation and a terminology of ref. [2]. In such an approach all particles (even in the intermediate states) are on the mass shell (but, spurious particles present). The technique of construction of the Wigner matrices $D^J(A)$ can be found in ref. [5, p.51,70,English edition]. In general, for each particle in interaction one should understand under 4-momenta p_i^μ and k_i^μ $(i = 1, 2)$ their covariant generalizations, $\breve{p}_i^\mu$, $\breve{k}_i^\mu$, *e.g.*, refs. [3, 6, 7]:

$$
\breve{\mathbf{k}} = (\mathbf{\Lambda}_{\mathcal{P}}^{-1}\mathbf{k}) = \mathbf{k} - \frac{\mathcal{P}}{\sqrt{\mathcal{P}^2}}\left(\mathbf{k_0} - \frac{\mathcal{P}\cdot\mathbf{k}}{\mathcal{P}_0 + \sqrt{\mathcal{P}^2}}\right) \quad ,
$$

$$
\breve{k}_0 = (\Lambda_{\mathcal{P}}^{-1}k)_0 = \sqrt{m^2 + \breve{\mathbf{k}}^{\,2}},
$$

with $\mathcal{P} = p_1 + p_2$, $\Lambda_{\mathcal{P}}^{-1}\mathcal{P} = (\mathcal{M}, \mathbf{0})$. However, we omit the circles above the momenta in the following, because in the case under consideration we do not miss physical information if we use the corresponding quantities in c.m.s., $\mathbf{p}_1 = -\mathbf{p}_2 = \mathbf{p}$ and $\mathbf{k}_1 = -\mathbf{k}_2 = \mathbf{k}$.

[2] When setting $\mathcal{V}(\mathbf{p}) = S^c_{[1]}\mathcal{U}(\mathbf{p}) \equiv \mathcal{C}_{[1]}\mathcal{K}\mathcal{U}(\mathbf{p}) \sim \gamma_5\mathcal{U}(\mathbf{p})$, like the Dirac $j = 1/2$ case we have other type of theories [9, 10, 11]. $S^c_{[1]}$ is the charge conjugation operator for $j = 1$. $\mathcal{K}$ is the operation of complex conjugation.

Journal of Physics: Conference Series **128** (2008) 012002 doi:10.1088/1742-6596/128/1/012002

These functions obey the orthonormalization equations, $\mathcal{U}^\dagger(\mathbf{p})\gamma_{00}\mathcal{U}(\mathbf{p}) = M$, M is the mass of the $2(2j+1)-$ particle. The similar normalization condition exists for $\mathcal{V}(\mathbf{p})$, the functions of "negative-energy states".

For instance, in the case of spin $j=1$, one has

$$D^1(\alpha(\mathbf{p})) = 1 + \frac{(\mathbf{J}\cdot\mathbf{p})}{M} + \frac{(\mathbf{J}\cdot\mathbf{p})^2}{M(p_0+M)} \quad , \tag{7}$$

$$D^1\left(\alpha^{-1\dagger}(\mathbf{p})\right) = 1 - \frac{(\mathbf{J}\cdot\mathbf{p})}{M} + \frac{(\mathbf{J}\cdot\mathbf{p})^2}{M(p_0+M)} \quad , \tag{8}$$

$$D^1\left(\alpha(\mathbf{p})\Theta_{[1/2]}\right) = \left[1 + \frac{(\mathbf{J}\cdot\mathbf{p})}{M} + \frac{(\mathbf{J}\cdot\mathbf{p})^2}{M(p_0+M)}\right]\Theta_{[1]} \quad , \tag{9}$$

$$D^1\left(\alpha^{-1\dagger}(\mathbf{p})\Theta_{[1/2]}\right) = \left[1 - \frac{(\mathbf{J}\cdot\mathbf{p})}{M} + \frac{(\mathbf{J}\cdot\mathbf{p})^2}{M(p_0+M)}\right]\Theta_{[1]} \quad , \tag{10}$$

($\Theta_{[1/2]}, \Theta_{[1]}$ are the Wigner operators for spin 1/2 and 1, respectively). Recently, much attention has been paid to this formalism [12].

In refs. [5, 8, 13, 14, 15] the Feynman diagram technique was discussed in the above-mentioned six-component formalism for particles of spin $j=1$. The Lagrangian is the following one:[3]

$$\begin{aligned}\mathcal{L} = \nabla_\mu\overline{\Psi}(x)\Gamma_{\mu\nu}\nabla_\nu\Psi(x) - M^2\overline{\Psi}(x)\Psi(x) - \frac{1}{4}F_{\mu\nu}F_{\mu\nu} + \\ + \frac{e\lambda}{12}F_{\mu\nu}\overline{\Psi}(x)\gamma_{5,\mu\nu}\Psi(x) + \frac{e\kappa}{12M^2}\partial_\alpha F_{\mu\nu}\overline{\Psi}(x)\gamma_{6,\mu\nu,\alpha\beta}\nabla_\beta\Psi(x)\,.\end{aligned} \tag{11}$$

In the above formula we have $\nabla_\mu = -i\partial_\mu \mp eA_\mu$; $F_{\mu\nu} = \partial_\mu A_\nu - \partial_\nu A_\mu$ is the electromagnetic field tensor; A_μ is the 4-vector of electromagnetic field; $\overline{\Psi}, \Psi$ are the six-component field functions of the massive $j=1$ Weinberg particle. The following expression has been obtained for the interaction vertex of the particle with the vector potential, ref. [13, 14]:

$$-e\Gamma_{\alpha\beta}(p+k)_\beta - \frac{ie\lambda}{6}\gamma_{5,\alpha\beta}q_\beta + \frac{e\kappa}{6M^2}\gamma_{6,\alpha\beta,\mu\nu}q_\beta q_\mu(p+k)_\nu \quad , \tag{12}$$

where $\Gamma_{\alpha\beta} = \gamma_{\alpha\beta} + \delta_{\alpha\beta}$; $\gamma_{\alpha\beta}$; $\gamma_{5,\alpha\beta}$; $\gamma_{6,\alpha\beta,\mu\nu}$ are the $6\otimes 6$-matrices which have been described in ref. [16, 8]:

$$\gamma_{ij} = \begin{pmatrix} 0 & \delta_{ij}\mathbb{1} - J_iJ_j - J_jJ_i \\ \delta_{ij}\mathbb{1} - J_iJ_j - J_jJ_i & 0 \end{pmatrix} \quad , \tag{13}$$

$$\gamma_{i4} = \gamma_{4i} = \begin{pmatrix} 0 & iJ_i \\ -iJ_i & 0 \end{pmatrix} \quad , \quad \gamma_{44} = \begin{pmatrix} 0 & \mathbb{1} \\ \mathbb{1} & 0 \end{pmatrix} \quad , \tag{14}$$

and

$$\gamma_{5,\alpha\beta} = i[\gamma_{\alpha\mu}, \gamma_{\beta\mu}]_- \quad , \tag{15}$$

$$\gamma_{6,\alpha\beta,\mu\nu} = [\gamma_{\alpha\mu}, \gamma_{\beta\nu}]_+ + 2\delta_{\alpha\mu}\delta_{\beta\nu} - [\gamma_{\beta\mu}, \gamma_{\alpha\nu}]_+ - 2\delta_{\beta\mu}\delta_{\alpha\nu} \quad . \tag{16}$$

J_i are the spin matrices for a $j=1$ particle, e is the electron charge, λ and κ correspond to the magnetic dipole moment and the electric quadrupole moment, respectively.

[3] In the following I prefer to use the Euclidean metric because this metric got application in a lot of papers on the $2(2j+1)$ formalism.

Journal of Physics: Conference Series **128** (2008) 012002 doi:10.1088/1742-6596/128/1/012002

In order to obtain the 4-vector current for the interaction of a boson with the external field one can use the known formulas of refs. [2, 3], which are valid for any spin:

$$\mathcal{U}^\sigma(\mathbf{p}) = \mathbf{S_p}\mathcal{U}^\sigma(\mathbf{0}) \quad , \quad \mathbf{S_p^{-1}S_k} = \mathbf{S_{k(-)p}} \cdot \mathbf{I} \otimes \mathbf{D^1}\left\{\mathbf{V^{-1}(\Lambda_p, k)}\right\} \quad , \tag{17}$$

$$W_\mu(\mathbf{p}) \cdot D\left\{V^{-1}(\Lambda_p, k)\right\} = D\left\{V^{-1}(\Lambda_p, k)\right\} \cdot \left[W_\mu(\mathbf{k}) + \frac{p_\mu + k_\mu}{M(\Delta_0 + M)} p_\nu W_\nu(\mathbf{k})\right], \tag{18}$$

$$k_\mu W_\mu(\mathbf{p}) \cdot D\left\{V^{-1}(\Lambda_p, k)\right\} = -D\left\{V^{-1}(\Lambda_p, k)\right\} \cdot p_\mu W_\mu(\mathbf{k}) \quad . \tag{19}$$

W_μ is the Pauli-Lubanski 4-vector of relativistic spin.[4] The matrix $D^{(j=1)}\left\{V^{-1}(\Lambda_p, k)\right\}$ is for spin 1:

$$\begin{aligned} D^{(j=1)}\left\{V^{-1}(\Lambda_p, k)\right\} &= \frac{1}{2M(p_0 + M)(k_0 + M)(\Delta_0 + M)} \left\{[\mathbf{p} \times \mathbf{k}]^2 + \right.\\ &+ \left[(p_0 + M)(k_0 + M) - \mathbf{k} \cdot \mathbf{p}\right]^2 + 2i\left[(p_0 + M)(k_0 + M) - \mathbf{k} \cdot \mathbf{p}\right]\{\mathbf{J} \cdot [\mathbf{p} \times \mathbf{k}]\} - \\ &- \left. 2\{\mathbf{J} \cdot [\mathbf{p} \times \mathbf{k}]\}^2\right\} \quad . \end{aligned} \tag{23}$$

The formulas have been obtained in ref. [15]:

$$\mathbf{S_p^{-1}}\gamma_{\mu\nu}\mathbf{S_p} = \gamma_{44}\left\{\delta_{\mu\nu} - \frac{1}{M^2}\chi_{[\mu\nu]}(\mathbf{p}) \otimes \gamma_5 - \frac{2}{M^2}\Sigma_{[\mu\nu]}(\mathbf{p})\right\} \quad , \tag{24}$$

$$\mathbf{S_p^{-1}}\gamma_{5,\mu\nu}\mathbf{S_p} = 6i\left\{-\frac{1}{M^2}\chi_{(\mu\nu)}(\mathbf{p}) \otimes \gamma_5 + \frac{2}{M^2}\Sigma_{(\mu\nu)}(\mathbf{p})\right\} \quad , \tag{25}$$

where

$$\chi_{[\mu\nu]}(\mathbf{p}) = p_\mu W_\nu(\mathbf{p}) + p_\nu W_\mu(\mathbf{p}) \quad , \tag{26}$$

$$\chi_{(\mu\nu)}(\mathbf{p}) = p_\mu W_\nu(\mathbf{p}) - p_\nu W_\mu(\mathbf{p}) \quad , \tag{27}$$

$$\Sigma_{[\mu\nu]}(\mathbf{p}) = \frac{1}{2}\left\{W_\mu(\mathbf{p})W_\nu(\mathbf{p}) + W_\nu(\mathbf{p})W_\mu(\mathbf{p})\right\} \quad , \tag{28}$$

$$\Sigma_{(\mu\nu)}(\mathbf{p}) = \frac{1}{2}\left\{W_\mu(\mathbf{p})W_\nu(\mathbf{p}) - W_\nu(\mathbf{p})W_\mu(\mathbf{p})\right\} \quad , \tag{29}$$

lead to the 4- current of a $j = 1$ Weinberg particle more directly:[5]

$$j_\mu^{\sigma_p\nu_p}(\mathbf{p}, \mathbf{k}) = j_{\mu\,(S)}^{\sigma_p\nu_p}(\mathbf{p}, \mathbf{k}) + j_{\mu\,(V)}^{\sigma_p\nu_p}(\mathbf{p}, \mathbf{k}) + j_{\mu\,(T)}^{\sigma_p\nu_p}(\mathbf{p}, \mathbf{k}) \quad , \tag{35}$$

[4] It is usually introduced because the usual commutation relation for spin is not covariant in the relativistic domain. The Pauli-Lubanski 4-vector is defined as

$$W_\mu(\mathbf{p}) = (\Lambda_\mathbf{p})_\mu^\nu W_\nu(\mathbf{0}), \tag{20}$$

where $W_0(\mathbf{0}) = 0$, $\mathbf{W}(\mathbf{0}) = M\sigma/\mathbf{2}$. The properties are:

$$p^\mu W_\mu(\mathbf{p}) = 0, \quad W^\mu(\mathbf{p})W_\mu(\mathbf{p}) = -M^2 j(j+1). \tag{21}$$

The explicit form is

$$W_0(\mathbf{p}) = (\mathbf{S} \cdot \mathbf{p}), \quad \mathbf{W}(\mathbf{p}) = M\mathbf{S} + \frac{\mathbf{p}(\mathbf{S} \cdot \mathbf{p})}{p_0 + M}. \tag{22}$$

[5] *Cf.* with a $j = 1/2$ case:

$$\mathbf{S_p^{-1}}\gamma_\mu\mathbf{S_k} = \mathbf{S_p^{-1}}\gamma_\mu\mathbf{S_p}\mathbf{S_{k(-)p}}\mathbf{I} \otimes \mathbf{D^{1/2}}\{\mathbf{V^{-1}(\Lambda_p, k)}\}, \tag{30}$$

$$j^{\sigma_p\nu_p}_{\mu\,(S)}(\mathbf{p},\mathbf{k}) = -g_S\xi^\dagger_{\sigma_p}\left\{(p+k)_\mu\left(1+\frac{(\mathbf{J}\cdot\mathbf{\Delta})^2}{M(\Delta_0+M)}\right)\right\}\xi_{\nu_p} \quad , \tag{36}$$

$$j^{\sigma_p\nu_p}_{\mu\,(V)}(\mathbf{p},\mathbf{k}) = -g_V\xi^\dagger_{\sigma_p}\left\{(p+k)_\mu+\frac{1}{M}W_\mu(\mathbf{p})(\mathbf{J}\cdot\mathbf{\Delta})-\frac{\mathbf{1}}{\mathbf{M}}(\mathbf{J}\cdot\mathbf{\Delta})\mathbf{W}_\mu(\mathbf{p})\right\}\xi_{\nu_p} \quad , \tag{37}$$

$$\begin{aligned} j^{\sigma_p\nu_p}_{\mu\,(T)}(\mathbf{p},\mathbf{k}) &= -g_T\xi^\dagger_{\sigma_p}\left\{-(p+k)_\mu\frac{(\mathbf{J}\cdot\mathbf{\Delta})^2}{M(\Delta_0+M)}+\right. \\ &\left.+\frac{1}{M}W_\mu(\mathbf{p})(\mathbf{J}\cdot\mathbf{\Delta})-\frac{\mathbf{1}}{\mathbf{M}}(\mathbf{J}\cdot\mathbf{\Delta})\mathbf{W}_\mu(\mathbf{p})\right\}\xi_{\nu_p} \quad . \end{aligned} \tag{38}$$

Next, let me now present the Feynman matrix element corresponding to the diagram of two-boson interaction, mediated by the particle described by the vector potential, in the form [2, 14] (read the remark in the footnote # 1):

$$\begin{aligned} &< p_1,p_2;\,\sigma_1,\sigma_2|\hat{T}^{(2)}|k_1,k_2;\,\nu_1,\nu_2 >= \\ &= \sum_{\sigma_{ip},\nu_{ip},\nu_{ik}=-1}^{1} D^{\dagger\,(j=1)}_{\sigma_1\sigma_{1p}}\left\{V^{-1}(\Lambda_{\mathcal{P}},p_1)\right\}D^{\dagger\,(j=1)}_{\sigma_2\sigma_{2p}}\left\{V^{-1}(\Lambda_{\mathcal{P}},p_2)\right\}\times \\ &\times\ T^{\nu_{1p}\nu_{2p}}_{\sigma_{1p}\sigma_{2p}}(\mathbf{k}(-)\mathbf{p},\mathbf{p})D^{(j=1)}_{\nu_{1p}\nu_{1k}}\left\{V^{-1}(\Lambda_{p_1},k_1)\right\}D^{(j=1)}_{\nu_{1k}\nu_1}\left\{V^{-1}(\Lambda_{\mathcal{P}},k_1)\right\}\times \\ &\times\ D^{(j=1)}_{\nu_{2p}\nu_{2k}}\left\{V^{-1}(\Lambda_{p_2},k_2)\right\}D^{(j=1)}_{\nu_{2k}\nu_2}\left\{V^{-1}(\Lambda_{\mathcal{P}},k_2)\right\} \quad , \end{aligned} \tag{39}$$

where

$$T^{\nu_{1p}\nu_{2p}}_{\sigma_{1p}\sigma_{2p}}(\mathbf{k}(-)\mathbf{p},\mathbf{p}) = \xi^\dagger_{\sigma_{1p}}\xi^\dagger_{\sigma_{2p}}\,T^{(2)}(\mathbf{k}(-)\mathbf{p},\,\mathbf{p})\,\xi_{\nu_{1p}}\xi_{\nu_{2p}} \quad , \tag{40}$$

$\xi^\dagger$, ξ are the 3-analogues of 2-spinors. The calculation of the amplitude (40) yields ($p_0 = -ip_4$, $\Delta_0 = -i\Delta_4$):

$$\begin{aligned} \hat{T}^{(2)}(\mathbf{k}(-)\mathbf{p},\mathbf{p}) &= g^2\left\{\frac{[p_0(\Delta_0+M)+(\mathbf{p}\cdot\mathbf{\Delta})]^2-M^3(\Delta_0+M)}{M^3(\Delta_0-M)}+\right. \\ &+\ \frac{i(\mathbf{J}_1+\mathbf{J}_2)\cdot[\mathbf{p}\times\mathbf{\Delta}]}{\Delta_0-M}\left[\frac{p_0(\Delta_0+M)+\mathbf{p}\cdot\mathbf{\Delta}}{M^3}\right]+\frac{(\mathbf{J_1}\cdot\mathbf{\Delta})(\mathbf{J_2}\cdot\mathbf{\Delta})-(\mathbf{J_1}\cdot\mathbf{J_2})\mathbf{\Delta^2}}{2M(\Delta_0-M)}- \\ &-\ \left.\frac{1}{M^3}\frac{\mathbf{J}_1\cdot[\mathbf{p}\times\mathbf{\Delta}]\ \mathbf{J}_2\cdot[\mathbf{p}\times\mathbf{\Delta}]}{\Delta_0-M}\right\} \quad . \end{aligned} \tag{41}$$

$$\mathbf{S_p^{-1}}\gamma_\mu\mathbf{S_p} = \frac{\mathbf{1}}{\mathbf{m}}\gamma_0\left\{\mathbb{1}\otimes\mathbf{p}_\mu+\mathbf{2}\gamma_\mathbf{5}\otimes\mathbf{W}_\mu(\mathbf{p})\right\}, \tag{31}$$

$$\mathbf{S_p^{-1}}\sigma_{\mu\nu}\mathbf{S_p} = -\frac{\mathbf{4}}{\mathbf{m^2}}\mathbb{1}\otimes\mathbf{\Sigma}_{(\mu\nu)}(\mathbf{p})+\frac{\mathbf{2}}{\mathbf{m^2}}\gamma_\mathbf{5}\otimes\chi_{(\mu\nu)}(\mathbf{p})\,. \tag{32}$$

Of course, the product of two Lorentz boosts is *not* a pure Lorentz transformation. It contains the rotation, which describes the Thomas spin precession (the Wigner rotation $V(\Lambda_\mathbf{p},\mathbf{k})\in SU(2)$)). And, then,

$$j^{\sigma_p\nu_p}_\mu(\mathbf{k}(-)\mathbf{p},\mathbf{p}) = \frac{1}{m}\xi^\dagger_{\sigma_p}\left\{2g_v æ_0 p_\mu + f_v æ_0 q_\mu + 4g_\mathcal{M}W_\mu(\mathbf{p})(\sigma\cdot\mathbf{æ})\right\}\xi_{\nu_p} \quad , \quad (g_\mathcal{M}=g_v+f_v)\,. \tag{33}$$

The indices $\mathbf{p}$ indicate that the Wigner rotations have been separated out and, thus, all spin indices have been "resetted" on the momentum $\mathbf{p}$. One can re-write [2b] the electromagnetic current (33):

$$j^{\sigma_p\nu_p}_\mu(\mathbf{k},\mathbf{p}) = -\frac{e\,m}{æ_0}\xi^\dagger_{\sigma_p}\left\{g_\mathcal{E}(q^2)\,(p+k)^\mu+g_\mathcal{M}(q^2)\left[\frac{1}{m}W_\mu(\mathbf{p})(\sigma\cdot\mathbf{\Delta})-\frac{\mathbf{1}}{\mathbf{m}}(\sigma\cdot\mathbf{\Delta})\mathbf{W}_\mu(\mathbf{p})\right]\right\}\xi_{\nu_p}\,. \tag{34}$$

$g_\mathcal{E}$ and $g_\mathcal{M}$ are the analogues of the Sachs electric and magnetic form factors. Thus, if we regard $g_{S,T,V}$ as effective coupling constants depending on the momentum transfer one can ensure ourselves that the forms of the currents for a spinor particle and those for a $j=1$ boson are the same (with the Wigner rotations separated out).

We have assumed $g_S = g_V = g_T$ above. The expression (41) reveals the advantages of the $2(2j+1)$- formalism, since it looks like the amplitude for the interaction of two spinor particles with the substitutions

$$\frac{1}{2M(\Delta_0 - M)} \Rightarrow \frac{1}{\mathbf{\Delta^2}} \quad \text{and} \quad \mathbf{J} \Rightarrow \sigma \quad .$$

The calculations hint that many analytical results produced for a Dirac fermion could be applicable to describing a $2(2j+1)$ particle. Nevertheless, an adequate explanation is required for the obtained difference. You may see that

$$\frac{1}{\mathbf{\Delta}^2} = \frac{1}{2M(\Delta_0 - M)} - \frac{1}{2M(\Delta_0 + M)} \tag{42}$$

and

$$(p+k)_\mu (p+k)^\mu = 2M(\Delta_0 + M)\,. \tag{43}$$

Hence, if we add an additional diagramm of another channel ($\mathbf{k} \rightarrow -\mathbf{k}$), we can obtain the *full* coincidence in the T-matrices of the fermion-fermion interaction and the boson-boson interaction. But, of course, one should take into account that there is no the Pauli principle for bosons, and additional sign " $-$ " would be related to the indefinite metric.

3. Conclusions

The conclusions are: The main result of this paper is the boson-boson amplitude calculated in the framework of the $2(2j+1)-$ component theory. The separation of the Wigner rotations permits us to reveal certain similarities with the $j = 1/2$ case. Thus, this result provides a ground for the conclusion: if we would accept the description of higher spin particles on using the Weinberg $2(2j+1)-$ scheme many calculations produced earlier for fermion-fermion interactions mediated by the vector potential can be applicable to processes involving higher-spin particles. Moreover, the main result of the paper gives a certain hope at a possibility of the unified description of fermions and bosons. One should realize that all the above-mentioned is not surprising. The principal features of describing a particle on the basis of relativistic quantum field theory are *not* in some special representation of the group representation, $(1/2,0)\oplus(0,1/2)$, or $(1,0)\oplus(0,1)$, or $(1/2,1/2)$, but in the Lorentz group itself. However, certain differences between denominators of the amplitudes are still not explained in full.

Several works dealing with phenomenological description of hadrons in the $(j,0) \oplus (0,j)$ framework have been published [17, 18, 19].

Acknowledgments

This paper is based on the talks given at the 5th International Symposium on "*Quantum Theory and Symmetries*", July 22-28, 2007, Valladolid, Spain and the 10th Workshop "*What comes beyond the Standard Model?*", July 17-27, 2007, Bled, Slovenia. I am grateful to participants of recent conferences for discussions.

References

[1] Kadyshevsky V G 1964 *ZhETF* **46** 654, 872 [1964 *JETP* **19** 443, 597]
1968 *Nuovo Cim.* **55**A 233
1968 *Nucl. Phys.* B**6** 125
Kadyshevsky V G, Mir-Kasimov R M and Skachkov N B 1972 *Fiz. Elem. Chast. At. Yadra* **2** 635 [1972 *Sov. J. Part. Nucl.* **2** 69]
[2] Skachkov N B 1975 *TMF* **22** 213 [1975 *Theor. Math. Phys.* **22** 149]
1975 ibid. **25** 313 [1975 *Theor. Math. Phys.* **25** 1154]
Skachkov N B and Solovtsov I L 1978 *Fiz. Elem. Chast. At. Yadr.* **9** 5 [1978 *Sov. J. Part. and Nucl.* **9** 1]
[3] Shirokov Yu M 1951 *ZhETF* **21** 748

1954 *DAN SSSR* **99** 737
1957 *ZhETF* **33** 1196, 1208 [1958 *Sov. Phys. JETP* **6** 919, 929]
1958 *ZhETF* **35** 1005 [1959 *Sov. Phys. JETP* **8** 703]
Chou Kuang Chao and Shirokov M I 1958 *ZhETF* **34** 1230 [1958 *Sov. Phys. JETP* **7** 851]
Cheshkov A A and Shirokov Yu M 1962 *ZhETF* **42** 144 [1962 *Sov. Phys. JETP* **15** 103]
1963 *ZhETF* **44** 1982 [1963 *Sov. Phys. JETP* **17** 1333]
Cheshkov A A 1966 *ZhETF* **50** 144 [1966 *Sov. Phys. JETP* **23** 97]
Kozhevnikov V P, Troitskiĭ V E, Trubnikov S V and Shirokov Yu M 1972 *Teor. Mat. Fiz.* **10** 47
[4] Chernikov N A 1957 *ZhETF* **33** 541 [1958 *Sov. Phys. JETP* **6** 422]
1973 *Fiz. Elem. Chast. At. Yadra* **4** 773
[5] Novozhilov Yu V 1971 *Vvedenie v teoriyu elementarnykh chastitz.* (Moscow, Nauka, 1971); [English translation: *Introduction to Elementary Particle Theory* (Pergamon Press, Oxford, 1975)]
[6] Faustov R N 1973 *Ann. Phys. (USA)* **78** 176
[7] Dvoeglazov V V *et al.* 1991 *Yadern. Fiz.* **54** 658 [1991 *Sov. J. Nucl. Phys.* **54** 398]
[8] Weinberg S 1964 *Phys. Rev.* B**133** 1318
1964 ibid. **134** 882
1969 ibid. **181** 1893
[9] Wigner E P 1962 In *Group Theoretical Concepts and Methods in Elementary Particle Physics.* ed. F. Gürsey. (Gordon and Breach, 1962)
[10] Sankaranarayanan A and Good Jr R H 1965 Nuovo Cim. **36** 1303
1965 *Phys. Rev.* **140**B 509
Sankaranarayanan A 1965 *Nuovo Cim.* **38** 889
[11] Ahluwalia D V, Johnson M B and Goldman T 1993 Phys. Lett. B**316** 102
Ahluwalia D V and Goldman T 1993 Mod. Phys. Lett. A**8** 2623
[12] Dvoeglazov V V 1998 *Int. J. Theor. Phys.* **37** 1915
2003 *Turkish J. Phys.* **27** 35
2006 *Int. J. Modern Phys.* B**20** 1317
[13] Hammer C L, McDonald S C and Pursey D L 1968 *Phys. Rev.* **171** 1349
Tucker R H and Hammer C L 1971 *Phys. Rev.* D**3** 2448
Shay D and Good Jr R H 1969 *Phys. Rev.* **179** 1410
[14] Dvoeglazov V V and Skachkov N B, 1984 *JINR Communications* P2-84-199, in Russian
1987 ibid. P2-87-882, in Russian.
[15] Dvoeglazov V V and Skachkov N B 1988 *Yadern. Fiz.* **48** 1770 [1988 *Sov. J. of Nucl. Phys.* **48** 1065]
[16] Barut A O, Muzinich I and Williams D 1963 *Phys. Rev.* **130** 442
[17] Dvoeglazov V V and Khudyakov S V 1994 *Izvestiya VUZov:fiz.* No. 9 110 [1994 *Russian Phys. J.* **37** 898].
[18] Dvoeglazov V V 1994 *Rev. Mex. Fis. Suppl. (Proc. XVII Oaxtepec Symp. on Nucl. Phys., Jan. 4-7, 1994, México)* **40** 352
[19] Dvoeglazov V V, Khudyakov S V and Solganik S B 1998 *Int. J. Theor. Phys.* **37** 1895
Dvoeglazov V V and Khudyakov S V 1998 *Hadronic J.* **21** 507

Journal of Physics: Conference Series **128** (2008) 012003 doi:10.1088/1742-6596/128/1/012003

On some aspects of nocommutative pure Yang-Mills theory

Stefan Giller[†], **Cezary Gonera**[‡], **Piotr Kosiński**[‡] **and Paweł Maślanka**[‡]

[†]Jan Dlugosz Academy in Czestochowa, Institute of Physics, ul. Armii Krajowej 13/15, 42-200 Czestochowa, Poland

[‡]Theoretical Physics Department II, University of Łódź, Pomorska 149/153, 90-236 Łódź, Poland

E-mail: sgiller@uni.lodz.pl, cgonera@uni.lodz.pl, pkosinsk@uni.lodz.pl, pmaslan@uni.lodz.pl

Abstract. Two types of gauge transformations of noncommutative pure gauge theory are discussed. It is shown that Yang-Mills theory with the so called twisted gauge symmetry is consistent provided it also enjoys the standard noncommutative $*$-gauge symmetry.

1. Introduction

The standard approach to gauge symmetry on noncommutative space-time [1,2,3,] basically consists in replacing the product by the star one in both the definitions of gauge transformations as well as their action on gauge potentials. This appears to be quite restrictive in a sense that practically only star counterpart of $U(N)$ group transformations close to form a group [3,4,5]. Such restrictions on gauge groups are absent in the so called twisted gauge transformations approach [6-13] Here, contrary to the previous case, there is no star product involved in the definitions of guge transformations and their action on fields. Noncommutativity of transformations is reflected in modified Leibnitz rule. So, by construction, twisted gauge transformations unlike standard $*$-ones impose rather mild restrictions on gauge groups. However, it appears that twisted gauge theory is consistent provided it also posses the standard noncommutative gauge symmetry. It is just an aim of this short note to show that, in general, the current conservation in the theory with twisted gauge symmetry implies some constraints which are not automatically fulfilled by virtue of field equations (more specifically, these constraints follow by taking the divergence of field equations). If one demands no such constraints exist, the range of gauge potentials must be enlarged so that the action exhibits also standard noncommutative gauge symmetry [14].

The note starts with a flash definitions of noncommutative Yang-Mills theory and both types of gauge transformations; standard noncommutative ones and twisted ones. Then the consistency of theory with twisted gauge symmetry is discussed.

2. Noncommutative pure Yang-Mills theory

Noncommutative pure Yang-Mills theory is defined by the following action

$$S = -\frac{1}{4}\int d^4x Tr(F_{\mu\nu} * F_{\mu\nu}) \tag{1}$$

where

$$F_{\mu\nu} \equiv \partial_\mu A_\nu - \partial_\nu A_\mu - i[A_\mu, A_\nu]_*, [A_\mu, A_\nu]_* \equiv A_\mu * A_\nu - A_\nu * A_\mu \tag{2}$$

$$\begin{aligned} e^{\frac{i}{2}\theta^{\mu\nu}\partial_\mu^x\partial_\nu^y} A_\mu(x)A_\nu(y)\mid_{x=y} = m(F^{-1}A_\mu \otimes A_\nu) \equiv m_*(A_\mu \otimes A_\nu) \\ m(A_\mu \otimes A_\nu) = A_\mu A_\nu \\ F = e^{\frac{-i}{2}\theta^{\mu\nu}\partial_\mu \otimes \partial_\nu} \end{aligned} \tag{3}$$

and $A_\mu(x)$ is $N \times N$ matrix valued function.

This action can be considered as being obtained from usual commutative pure Yang-Mills one by replacing ordinary product of fields by star product defined by eq.(3). Consequently, ordinary commutators are replaced by corresponding $*$-ones given by eq. (2).
Varying the action with respect to gauge potentials gives the following field equations

$$\partial_\mu F^{\mu\nu} - i[A_\mu, F^{\mu\nu}]_* = 0 \tag{4}$$

These equations look as their commutative counterparts however one has to remember that commutator is now $*$-one. It appears that it is possible to define two types of gauge transformations which leave the action invariant. There are standard noncommutative gauge transformations for gauge potentials belonging to $u(N)$ Lie algebra and twisted gauge ones for potentials belonging a priori to any Lie algebra.

3. Standard $*$-$U_*(N)$ gauge transformations

Standard noncommutative gauge transformations are given by following equations

$$\begin{aligned} A_\mu \longmapsto A_\mu^{'} = U_* * A_\mu * U_*^{-1} - i\partial_\mu U_* * U_*^{-1} \\ U_*(x) = e_*^{i\alpha^a(x)T^a} \equiv I + i\alpha^a T_a + \frac{i^2}{2}\alpha^a * \alpha^b T_a T_b + ... \end{aligned} \tag{5}$$

where $T_a \in u(N)$. Two points are really important here. The first is the use of $*$-product in definition of gauge transformations as well as their action on fields. The second point related to the first one is the observation that gauge transformations close to form a group provided matrices T_a span $u(N)$ Lie algebra. So practically, only unitary group $U(N)$ can be $*$ gauged.

Now, invariance of the action S under infinitesimal $*$-gauge transformations

$$\begin{aligned} \delta A_\mu \equiv A_\mu^{'} - A_\mu = i[\alpha, A_\mu]_* + \partial_\mu \alpha \\ \delta\partial_\nu A_\mu = \partial_\nu \delta A_\mu \end{aligned} \tag{6}$$

implies (via noncommutative counterpart of second Noether theorem) that

$$[A_\mu, \frac{\delta L}{\delta A_\mu}]_* + \partial_\nu [A_\mu, F^{\nu\mu}]_* = 0 \tag{7}$$

$$\partial_\mu \frac{\delta L}{\delta A_\mu} = i[A_\mu, \frac{\delta L}{\delta A_\mu}]_* \tag{8}$$

where $\frac{\delta L}{\delta A_\mu} \equiv \partial_\nu F^{\nu\mu} - i[A_\nu, F^{\nu\mu}]_*$ is the Euler - Lagrange expression.

Eq.(7) provides a conserved current $j^\nu \equiv [A_\mu, F^{\nu\mu}]_*$, $\partial_\mu j^\mu = 0$. The conservation of this current (which can also be checked directly using field equations) is crucial for consistency of the theory. In fact, eq.(8) confirms that no constraints on A_μ follow by taking divergence of field equations.

Journal of Physics: Conference Series **128** (2008) 012003 doi:10.1088/1742-6596/128/1/012003

4. Twisted gauge transformations

Twisted gauge transformations and their action on fields are defined exactly as in ordinary commutative situation i.e.

$$A_\mu \longmapsto A'_\mu = UA_\mu U^{-1} - i(\partial_\mu U)U^{-1}$$
$$U(x) = e^{i\alpha^a(x)T^a}. \tag{9}$$

Contrary to the first kind of transformations there is no $*$-product involved in these definitions. However, an important assumption is added that derivatives entering $*$-products do not act on gauge parameters. This assumption results in twisted Leibnitz rule (see eq(10) below) which can be expressed in terms of twisted coproduct Δ_F of α which explains the name of symmetry.

$$\delta(\psi * \phi) = i\alpha^a(x)(T_a\psi * \phi + \psi * T_a\phi) = m_*(\Delta_F(i\alpha)\psi \otimes \phi)$$
$$\Delta_F = F\Delta F^{-1} \quad , \quad \Delta(i\alpha) = i\alpha \otimes I + I \otimes i\alpha \tag{10}$$

The above twisted Leibnitz rule is given for functions transforming as matter fields. In the case of gauge fields there would be an extra terms coming from derivative of gauge parameter entering gauge potentials transformation rule.

5. Pure Yang - Mills theory with twisted gauge symmetry

By the very definition twisted gauge transformations impose rather mild restrictions on gauge groups. So, a natural question arises whether it is possible to construct noncommutative gauge theory with some compact Lie group G different from unitary one, playing the role of gauge group and with gauge potentials belonging to a Lie algebra representation of G. To answer this question it is convenient to select some N-dimensional unitary irreducible representation of G (and the corresponding irreducible representation of g, denoted also by g) and to assume that the gauge potential takes its values in g. The Lie algebra g is a subalgebra of $u(N)$. The latter viewed as linear space is equipped with the scalar product $(A, B) \equiv Tr(AB)$ invariant under the adjoint action of $U(N)$. If B_i, $i = 1, ..., a$ span an orthonormal bases in the orthogonal complement $g_\perp$ of g in $u(N)$($g_\perp = \{B \in u(N); Tr(AB) = 0, A \in g\}$) then the condition that the potential A_μ belongs to the representation g reads $A \in g \Longleftrightarrow Tr(AB_i) = 0$, $i = 1, ..., a$.

To define Yang-Mills theory with twisted gauge group G, Lagrange multiplier method can be used. The relevant action reads

$$S = -\frac{1}{4}\int d^4x Tr(F_{\mu\nu} * F_{\mu\nu}) + \int d^4x \rho_i^\mu Tr(A_\mu B_i) \tag{11}$$

where $\rho_i^\mu(x)$ are the Lagrange multipliers and where it is assumed that the gauge potential A_μ *a priori* takes its values in $u(N)$.

Eq. (11) implies the following equations of motion

$$Tr(A_\mu B_i) = 0 \tag{12}$$

$$\partial_\mu F^{\mu\nu} - i[A_\mu, F^{\mu\nu}]_* + \rho_i^\nu B_i = 0, \tag{13}$$

hence

$$\rho_i^\nu = -Tr(B_i(\partial_\mu F^{\mu\nu} - i[A_\mu, F^{\mu\nu}]_*)) \tag{14}$$

Eq.(13) means that in general, $f^\nu \equiv \partial_\mu F^{\mu\nu} - i[A_\mu, F^{\mu\nu}]_* \in g_\perp$ (in the case of ordinary commutative gauge theory $f^\mu \in g$ by construction and one deals with ordinary field equations $\partial_\mu F^{\mu\nu} - i[A_\mu, F^{\mu\nu}]_* = 0$, similarly, in standard (i.e. not twisted) approach to gauge theory on noncommutative space-time g must basically be $u(N)$ and $g_\perp = 0$; again one arrives at ordinary form of field equations).

Due to antisymmetry of $F^{\mu\nu}$ tensor the field eqs.(13) imply the followig consistency condition

$$\partial_\nu(i[A_\mu, F^{\mu\nu}]_* - \rho_i^\nu B_i) = 0 \tag{15}$$

ie.

$$\frac{i}{2}[\partial_\nu A_\mu - \partial_\mu A_\nu, F^{\mu\nu}]_* + i[A_\mu, \partial_\nu F^{\mu\nu}]_* - \partial_\nu \rho_i^\nu B_i = 0. \tag{16}$$

Using field equations, Jacobi identities and taking into account that $\partial_\mu A_\nu - \partial_\nu A_\mu = F_{\mu\nu} + i[A_\mu, A_\nu]_*$ eq.(16) can be rewritten in the form

$$[A_\nu, \rho_i^\nu B_i]_* + \partial_\nu \rho_i^\nu B_i = 0 \tag{17}$$

which finally implies

$$Tr(\Gamma[A_\nu, \rho_i^\nu B_i]_*) = 0 \tag{18}$$

for any $\Gamma \in g$.

Now, $*$-commutator consists of two pieces, one proportional to matrix commutator and the second one involving matrix anticommutator;

$$[A_\nu, \rho_i^\nu B_i]_* = \frac{1}{2}(\{A_\nu^a, \rho_i^\nu\}_*[T_a, B_i] + [A_\nu^a, \rho_i^\nu]_*\{T_a, B_i\}). \tag{19}$$

However, commutators of $B's$ and $T's$ belong to orthogonal complement of Lie algebra g

$$[T_a, B_i] \in g_\perp \tag{20}$$

(Eq.(20) results because if $U \in G$ then $U^+\Gamma'U \in g$ and $Tr(\Gamma'UB_iU^+) = Tr(U^+\Gamma'UB_i) = 0$ so that $UB_iU^+ \in g_\perp$). Consequently, consistency condition (see eq.(18)) reads

$$Tr(\Gamma[A_\nu, \rho_i^\nu B_i]_*) = Tr(\Gamma[A_\nu, \rho_i^\nu]_*\{T_a, B_i\}) = 0 \tag{21}$$

It is clear that eq.(21) does not produce any further constraints on A_μ provided $\{T_a, B_i\} \in g_\perp$ but then eq.(20) would imply

$$T_a B_i \in g_\perp \qquad and \qquad B_i T_a \in g_\perp \tag{22}$$

so that

$$\Sigma B_i \in g_\perp \qquad and \qquad B_i \Sigma \in g_\perp \tag{23}$$

for each Σ belonging to the enveloping algebra $U(g)$ of g which due to its assumed irreducibility is an algebra M_N of all $N \times N$ matrices (this is Burnside theorem). So, equations (23) imply that orthogonal complement $g_\perp$ of g is a two-sided ideal in M_N. However, it is known (by Wedderburn theorem) that M_N has only two ideals: $I = \phi$ or $I = M_N$. $g_\perp = M_N$ would imply no gauge symmetry, so one is left with $g_\perp = \phi$ ie. g must be $u(n)$ Lie algebra.

In this way one concludes that gauge theory with twisted gauge group G imposes no extra constraints on gauge fields A_μ provided G is unitary group $U(N)$. But then theory also enjoys standard $*$-$U_*(N)$ symmetry which controls consistency of the theory and provides conserved currents $j^\nu \equiv [A_\mu, F^{\nu\mu}]_*$.

References

[1] Douglas M R, Nekrasov N A 2001 *Rev. Mod. Phys.* **73** 977
[2] Szabo R J 2003 *Phys. Rep.* **378** 207
[3] Chaichian M, Presnajder P, Sheikh-Jabbari M M, Tureanu A 2002 *Phys. Lett.* B **526** 132
[4] Bonora L, Schnabl M, Sheikh-Jabbari M M, Tomasiello A 2000 *Nucl. Phys.* B **589** 461
[5] Bars J, Sheikh-Jabbari M M, Vasiliev M A 2001 *Phys. Rev.* D **64** 086004
[6] Vassilevich D V 2006 *Mod. Phys. Lett.* A **21** 1279
[7] Aschieri P, Dmitrijevich M, Meyer F, Schraml S, Wess J 2006 *Lett. Math. Phys.* **78** 61
[8] Wess J, *Deformed Gauge Theories* (hep-th/0608135)
[9] Garcia-Bondia J M , Lizzi F , Ruiz Ruiz F , Vitale P 2006 *Phys. Rev.* D **74** 025015
[10] Chaichian M, Tureanu A 2006 *Phys. Lett.* **B637** 199
[11] Alvarez-Gaume L, Meyer F, Vazguez-Moro M A 2006 *Nucl. Phys.* B **753** 92
[12] Banerjee R, Samanta S, *Gauge Generators, Transformations and Identities on a Noncommutative Space* (hep-th/0608214)
[13] Banerjee R, Samanta S, *Gauge Symmetries on* Θ *-Deformed Spaces* (hep-th/0611249)
[14] Giller S, Gonera C, Kosinski P, Maslanka P 2007 *Phys. Lett.* B **655** 80 (hep-th/0701014)

Journal of Physics: Conference Series **128** (2008) 012004 doi:10.1088/1742-6596/128/1/012004

Classical integrable $N = 1$ and $N = 2$ super sinh-Gordon models with jump defects

J F Gomes, L H Ymai and A H Zimerman
Instituto de Física Teórica, Universidade Estadual Paulista
Rua Pamplona, 145 01405-900 - São Paulo, SP, Brazil
E-mail: jfg@ift.unesp.br, leandroy@ift.unesp.br, zimerman@ift.unesp.br

Abstract. The structure of integrable field theories in the presence of jump defects is discussed in terms of boundary functions under the Lagrangian formalism. Explicit examples of bosonic and fermionic theories are considered. In particular, the boundary functions for the $N = 1$ and $N = 2$ super sinh-Gordon models are constructed and shown to generate the Backlund transformations for its soliton solutions. As a new and interesting example, a solution with an incoming boson and an outgoing fermion for the $N = 1$ case is presented. The resulting integrable models are shown to be invariant under supersymmetric transformation.

1. Introduction

The classical Lagrangian formulation of a class of relativistic integrable field theories admiting certain discontinuities (defects) has been studied recently [1] - [3]. In particular, in ref. [1] the authors have considered a field theory in which different soliton solutions of the sine-Gordon model are linked in such a way that the integrability is preserved. The integrability of the total system imposes severe constraints specifying the possible types of defects. These are characterized by Backlund transformations which are known to connect two different soliton solutions.

The presence of the defect indicates the breakdown of space isotropy and henceforth of momentum conservation. The key ingredient to classify integrable defects is to impose certain first order differential relations between the different solutions (Backlund transformation). This introduces certain boundary functions (BF) which are specific of each integrable model and leads to the conservation of the total momentum.

Here, we analize the structure of the possible boundary terms for various cases. We first consider, for pedagogical purposes, the pure bosonic case studied in refs. [1] - [3] and derive, according to them, the border functions by imposing conservation of the total momentum. Next, we consider a pure fermionic theory and propose Backlund transformation in terms of an auxiliary fermionic non local field. Such structure is then generalized to include both bosonic and fermionic fields. In particular we construct boundary functions for the $N = 1$ supersymmetric sinh-Gordon model and show that it leads to the Backlund transformation proposed in [4].

We extend the formalism to the $N = 2$ supersymmetric sinh-Gordon model [5]. This system was originally proposed in [6] and [7] (see also [8]). Later in [9] and [10] a systematic algebraic aproach was developed.

2. General formalism: bosonic case

In this section we introduce the Lagrangian approach proposed in [1]. Consider a system described by

$$\mathcal{L} = \theta(-x)\mathcal{L}_{p=1} + \theta(x)\mathcal{L}_{p=2} + \delta(x)\mathcal{L}_D, \tag{1}$$

where $\mathcal{L}_p(\phi_p, \partial_\mu \phi_p) = \frac{1}{2}(\partial_x \phi_p)^2 - \frac{1}{2}(\partial_t \phi_p)^2 - V(\phi_p)$ describes a set of fields denoted by ϕ_1 for $x < 0$ and ϕ_2 for $x > 0$. A defect placed at $x = 0$, is described by

$$\mathcal{L}_D = \frac{1}{2}(\phi_2 \partial_t \phi_1 - \phi_1 \partial_t \phi_2) + B_0 \tag{2}$$

where B_0 is the border function. The equations of motion are therefore given by

$$\begin{aligned} \partial_x^2 \phi_1 - \partial_t^2 \phi_1 &= \partial_{\phi_1} V(\phi_1), \qquad x < 0 \\ \partial_x^2 \phi_2 - \partial_t^2 \phi_2 &= \partial_{\phi_2} V(\phi_2), \qquad x > 0 \end{aligned} \tag{3}$$

and $x = 0$,

$$\begin{aligned} \partial_x \phi_1 - \partial_t \phi_2 &= -\partial_{\phi_1} B_0, \\ \partial_x \phi_2 - \partial_t \phi_1 &= \partial_{\phi_2} B_0, \end{aligned} \tag{4}$$

The momentum is

$$P = \int_{-\infty}^{0} \partial_x \phi_1 \partial_t \phi_1 + \int_0^\infty \partial_x \phi_2 \partial_t \phi_2. \tag{5}$$

Acting with time derivative and inserting eqns. of motion (3) we find

$$\begin{aligned} \frac{dP}{dt} &= \int_{-\infty}^{0} (\frac{1}{2}\partial_x(\partial_t \phi_1)^2 + \frac{1}{2}\partial_x(\partial_x \phi_1)^2 - \partial_x \phi_1 \frac{\delta V_1}{\delta \phi_1})dx \\ &+ \int_0^\infty (\frac{1}{2}\partial_x(\partial_t \phi_2)^2 + \frac{1}{2}\partial_x(\partial_x \phi_2)^2 - \partial_x \phi_2 \frac{\delta V_2}{\delta \phi_2})dx \end{aligned} \tag{6}$$

Using eqns. (4) after integration, we find

$$\frac{dP}{dt} = [-\frac{\partial B_0}{\partial \phi_+}\dot{\phi_+} + \frac{\partial B_0}{\partial \phi_-}\dot{\phi_-} + \frac{1}{2}(\frac{\partial B_0}{\partial \phi_1})^2 - \frac{1}{2}(\frac{\partial B_0}{\partial \phi_2})^2 - V_1 + V_2]|_{x=0} \tag{7}$$

where $\phi_\pm = \phi_1 \pm \phi_2$ and $\dot{\phi_\pm} = \partial_t \phi_\pm$. If the border function factorizes into $B_0 = B_0^+(\phi_+) + B_0^-(\phi_-)$, the modified momentum $\mathcal{P} = P + \left(B_0^+ - B_0^-\right)|_{x=0}$ is conserved provided B_0 satisfies its defining condition, i.e.,

$$[\frac{1}{2}(\frac{\partial B_0}{\partial \phi_1})^2 - \frac{1}{2}(\frac{\partial B_0}{\partial \phi_2})^2 - V_1 + V_2]|_{x=0} = 0 \tag{8}$$

Let us illustrate the above structure by first considering the free massive bosonic theory for which $V_p = \frac{1}{2}m^2\phi_p^2$.

$$\frac{1}{2}(\frac{\partial B_0}{\partial \phi_1})^2 - \frac{1}{2}(\frac{\partial B_0}{\partial \phi_2})^2 = \frac{m^2}{2}(\phi_1^2 - \phi_2^2) = \frac{m^2}{2}\phi_+\phi_- \tag{9}$$

Journal of Physics: Conference Series **128** (2008) 012004 doi:10.1088/1742-6596/128/1/012004

The solution is easely found to be

$$B_0 = -\frac{m\beta^2}{4}\phi_-^2 - \frac{m}{4\beta^2}\phi_+^2 \tag{10}$$

and β^2 denotes a free (spectral) parameter.

As second example, consider the sinh-Gordon model for which $V_p = 4m^2\cosh(2\phi_p)$. The defining eqn. (8) indicates the natural decomposition

$$\begin{aligned} \frac{1}{2}(\frac{\partial B_0}{\partial \phi_1})^2 - \frac{1}{2}(\frac{\partial B_0}{\partial \phi_2})^2 &= 4m^2(\cosh(2\phi_1) - \cosh(2\phi_2)) \\ &= 8m^2\sinh(\phi_+)\sinh(\phi_-) \end{aligned} \tag{11}$$

yielding

$$B_0 = -m\beta^2\cosh(\phi_-) - \frac{4m}{\beta^2}\cosh(\phi_+) \tag{12}$$

and hence we rederive from (4) the Backlund transformation for the sinh-Gordon model

$$\begin{aligned} \partial_x\phi_1 - \partial_t\phi_2 &= m\beta^2\sinh(\phi_-) + \frac{4m}{\beta^2}\sinh(\phi_+), \\ \partial_x\phi_2 - \partial_t\phi_1 &= m\beta^2\sinh(\phi_-) - \frac{4m}{\beta^2}\sinh(\phi_+), \end{aligned} \tag{13}$$

3. Fermions and the $N = 1$ super sinh-Gordon model

Before discussing the Super sinh-Gordon Model let us consider the pure fermionic prototype described by the Lagrangian density

$$\mathcal{L}_p = \bar\psi_p\partial_t\bar\psi_p - \bar\psi_p\partial_x\bar\psi_p + \psi_p\partial_t\psi_p + \psi_p\partial_x\psi_p + W(\psi_p, \bar\psi_p) \tag{14}$$

where, for the free fermionic theory, $W(\psi_p, \bar\psi_p) = 2m\bar\psi_p\psi_p$. For the half line, $x < 0$ or $x > 0$ the equations of motion are given by

$$\partial_x\psi_p + \partial_t\psi_p = -\frac{1}{2}\partial_{\psi_p}W_p, \qquad \partial_x\bar\psi_p - \partial_t\bar\psi_p = \frac{1}{2}\partial_{\bar\psi_p}W_p \tag{15}$$

according to $p = 1$ or 2 respectively.

Let us propose the following Backlund transformation

$$\psi_1 + \psi_2 = -i\beta\sqrt{m}f_1 = \partial_{\psi_1}B_1, \qquad \bar\psi_1 - \bar\psi_2 = -\frac{2i\sqrt{m}}{\beta}f_1 = -\partial_{\bar\psi_1}B_1, \tag{16}$$

where f_1 satisfies

$$\begin{aligned} \dot f_1 &= -\frac{i\beta\sqrt{m}}{2}(\psi_1 - \psi_2) + \frac{i}{2\beta}\sqrt{m}(\bar\psi_1 + \bar\psi_2) = -\frac{1}{4}\partial_{f_1}B_1, \\ \partial_x f_1 &= \frac{i\beta\sqrt{m}}{2}(\psi_1 - \psi_2) + \frac{i}{2\beta}\sqrt{m}(\bar\psi_1 + \bar\psi_2) \end{aligned} \tag{17}$$

written in terms of a border function $B_1 = B_1(\bar\psi_1, \bar\psi_2, \psi_1, \psi_2, f_1)$, which now, due to the Grassmanian character of the fermions, depends upon the non local fermionic field f_1. By considering the Lagrangian system (1) with $\mathcal{L}_p$ given by (14) and

$$\mathcal{L}_D = -\psi_1\psi_2 - \bar\psi_1\bar\psi_2 + 2f_1\partial_t f_1 + B_1(\bar\psi_1, \bar\psi_2, \psi_1, \psi_2, f_1) \tag{18}$$

construct the momentum to be

$$P = \int_{-\infty}^{0}(-\bar{\psi}_1\partial_x\bar{\psi}_1 - \psi_1\partial_x\psi_1)dx + \int_0^{\infty}(-\bar{\psi}_2\partial_x\bar{\psi}_2 - \psi_2\partial_x\psi_2)dx \tag{19}$$

In analising its conservation, we find

$$\frac{dP}{dt} = [-W_1 + W_2 - \bar{\psi}_1\partial_t\bar{\psi}_1 + \bar{\psi}_2\partial_t\bar{\psi}_2 - \psi_1\partial_t\psi_1 + \psi_2\partial_t\psi_2]|_{x=0} \tag{20}$$

after using equations of motion (15) to eliminate time derivatives and integrating over x. Using the Backlund transformation (16) and (17), eqn. (20) becomes

$$\begin{aligned}\frac{dP}{dt} &= -[W_1 - W_2 - \partial_{\bar{\psi}_1}B_1\dot{\bar{\psi}}_1 + \partial_{\psi_1}B_1\dot{\psi}_1 \\ &- \partial_{\bar{\psi}_2}B_1\dot{\bar{\psi}}_2 + \partial_{\psi_2}B_1\dot{\psi}_2 + \partial_t(\bar{\psi}_2\bar{\psi}_1) - \partial_t(\psi_2\psi_1)]|_{x=0}\end{aligned} \tag{21}$$

If we assume the border function to decompose as $B_1 = B_1^+(\bar{\psi}_+, f_1) + B_1^-(\psi_-, f_1)$, where $\bar{\psi}_\pm = \bar{\psi}_1 \pm \bar{\psi}_2$, $\psi_\pm = \psi_1 \pm \psi_2$, the modified momentum

$$\mathcal{P} = P + [\bar{\psi}_2\bar{\psi}_1 - \psi_2\psi_1 + B_1^+ - B_1^-]_{x=0} \tag{22}$$

is conserved provided

$$[W_2 - W_1 - \partial_{f_1}B_1^-\dot{f}_1 + \partial_{f_1}B_1^+\dot{f}_1]|_{x=0} = 0 \tag{23}$$

For the free fermi fields system (14)-(15) eqn. (23) becomes

$$\frac{1}{2}(\partial_{f_1}B_1^+)(\partial_{f_1}B_1^-) = 2m(\bar{\psi}_1\psi_1 - \bar{\psi}_2\psi_2) \tag{24}$$

The solution is

$$B_1 = -\frac{2i}{\beta}\sqrt{m}f_1\bar{\psi}_+ + i\beta\sqrt{m}f_1\psi_- \tag{25}$$

Let us now consider the super sinh-Gordon model described by

$$\begin{aligned}\mathcal{L}_p &= \frac{1}{2}(\partial_x\phi_p)^2 - \frac{1}{2}(\partial_t\phi_p)^2 + \bar{\psi}_p\partial_t\bar{\psi}_p - \bar{\psi}_p\partial_x\bar{\psi}_p + \psi_p\partial_t\psi_p \\ &+ \psi_p\partial_x\psi_p + V(\phi_p) + W(\phi_p, \psi_p, \bar{\psi}_p)\end{aligned} \tag{26}$$

and

$$\begin{aligned}\mathcal{L}_D &= \frac{1}{2}(\phi_2\partial_t\phi_1 - \phi_1\partial_t\phi_2) - \psi_1\psi_2 - \bar{\psi}_1\bar{\psi}_2 + 2f_1\partial_t f_1 \\ &+ B_0(\phi_1, \phi_2) + B_1(\phi_1, \phi_2, \bar{\psi}_1, \bar{\psi}_2, \psi_1, \psi_2, f_1)\end{aligned} \tag{27}$$

where $V_p = 4m^2\cosh(2\phi_p)$ and $W_p = 8m\bar{\psi}_p\psi_p\cosh(\phi_p)$.
Propose the following Backlund transformation [11],

$$\begin{aligned}\partial_x\phi_1 - \partial_t\phi_2 = -\partial_{\phi_1}(B_0 + B_1), \qquad \partial_x\phi_2 - \partial_t\phi_1 = \partial_{\phi_2}(B_0 + B_1) \\ \psi_1 + \psi_2 = \partial_{\psi_1}B_1, \quad \bar{\psi}_1 - \bar{\psi}_2 = -\partial_{\bar{\psi}_2}B_1, \quad \dot{f}_1 = -\frac{1}{4}\partial_{f_1}B_1\end{aligned} \tag{28}$$

Assuming the decomposition

$$B_0 = B_0^+(\phi_+) + B_0^-(\phi_-), \qquad B_1 = B_1^+(\phi_+, \bar{\psi}_+, f_1) + B_1^-(\phi_-, \psi_-, f_1) \tag{29}$$

we find that the modified momentum

$$\mathcal{P} = P + [B_0^+(\phi_+) - B_0^-(\phi_-) + B_1^+(\phi_+, \bar{\psi}_+, f_1) - B_1^-(\phi_-, \psi_-, f_1) - \bar{\psi}_1\bar{\psi}_2 + \psi_1\psi_2]_{x=0} \tag{30}$$

is conserved provided the border functions B_0 and B_1 satisfy

$$\begin{aligned} W_1 - W_2 &= \frac{1}{2}(\partial_{f_1} B_1^+)(\partial_{f_1} B_1^-) + 2(\partial_{\phi_+} B_0^+)(\partial_{\phi_-} B_1^-) + 2(\partial_{\phi_-} B_0^-)(\partial_{\phi_+} B_1^+) \\ V_1 - V_2 &= \frac{1}{2}(\frac{\partial B_0}{\partial \phi_1})^2 - \frac{1}{2}(\frac{\partial B_0}{\partial \phi_2})^2 \end{aligned} \tag{31}$$

The solution for (31) is found to be

$$\begin{aligned} B_0 &= -m\beta^2 \cosh(\phi_-) - \frac{4m}{\beta^2}\cosh(\phi_+), \\ B_1 &= -\frac{4i}{\beta}\sqrt{m}\cosh(\frac{\phi_+}{2}) f_1 \bar{\psi}_+ + 2i\beta\sqrt{m}\cosh(\frac{\phi_-}{2}) f_1 \psi_- \end{aligned} \tag{32}$$

The boundary functions B_0 and B_1 in (32) generate the Backlund transformation which agrees with the one proposed in [4] for the $N = 1$ super sinh-Gordon model. The equations of motion obtained from (26) and (27) are verified to be invariant under the supersymmetry transformation

$$\delta\bar{\psi}_p = \epsilon\partial_z\phi_p, \quad \delta\phi_p = \epsilon\bar{\psi}_p, \quad \delta\psi_p = 2\epsilon m \sinh\phi_p \tag{33}$$

where $\partial_z = 1/2(\partial_x + \partial_t)$.

In ref. [12] the general soliton solutions for the $N = 1$ super sinh-Gordon model were constructed using vertex functions techniques and in [11] different solutions were analysed in the context of the Backlund transformation. Besides those examples discussed in [11], a new and interesting case containing *an incoming boson and an outcoming fermion* is given when $\phi_1 \neq 0, \psi_1 = \bar{\psi}_1 = 0$ and $\phi_2 = 0, \psi_2, \bar{\psi}_2 \neq 0$. Under such conditions the Backlund transformation (28) reads,

$$\begin{aligned} \partial_x\phi_1 &= -2(\sigma + \frac{1}{\sigma})\sinh\phi_1, \qquad \partial_t\phi_1 = -2(\sigma - \frac{1}{\sigma})\sinh\phi_1, \\ \psi_2 &= 2\sqrt{\frac{2}{\sigma}}\cosh\left(\frac{\phi_1}{2}\right) f_1, \qquad \bar{\psi}_2 = 2\sqrt{2\sigma}\cosh\left(\frac{\phi_1}{2}\right) f_1, \\ \partial_t f_1 &= 2(\sigma - \frac{1}{\sigma})\cosh^2\left(\frac{\phi_1}{2}\right) f_1, \qquad \partial_x f_1 = 2(\sigma + \frac{1}{\sigma})\cosh^2\left(\frac{\phi_1}{2}\right) f_1 \end{aligned} \tag{34}$$

where we took $m = 1$ and $\sigma = -\frac{2}{\beta^2}$. As solution of eqns. (34) we find

$$\phi_1 = ln\left(\frac{1 + \frac{b_1}{2}\rho}{1 - \frac{b_1}{2}\rho}\right), \qquad f_1 = \epsilon\rho^{-1}\sqrt{1 - \frac{b_1^2}{4}\rho^2} \tag{35}$$

where $\rho = \exp\left(-2\sigma(x+t) - \frac{2}{\sigma}(x-t)\right)$, ϵ is a grassmaniann constant and ψ_2 and $\bar{\psi}_2$ are obtained from the second eqn. (34) after inserting the above ϕ_1 and f_1.

The integrability of the model is verified by construction of the Lax pair representation of the equations of motion. This is achieved by splitting the space into two overlapping regions, namely, $x \leq b$ and $x \geq a$ with $a < b$ and defining a corresponding Lax pair within each of them. The integrability is ensured by the existence of a gauge transformation relating the two sets of Lax pairs within the overlapping region. In ref. [11] we have explicitly constructed such gauge transformation for the $N = 1$ super sinh-Gordon model in terms of the $SL(2,1)$ affine Lie algebra.

4. $N = 2$ super sinh-Gordon model

The Lagrangian density for the $N = 2$ Super sinh-Gordon Model is given by (see for instance [8])

$$\begin{aligned}
\mathcal{L}_p &= \frac{1}{2}(\partial_x\phi_p)^2 - \frac{1}{2}(\partial_t\phi_p)^2 + 2\bar{\psi}_p\partial_t\bar{\psi}_p + 2\bar{\psi}_p\partial_x\bar{\psi}_p + 2\psi_p\partial_t\psi_p - 2\psi_p\partial_x\psi_p \\
&- \frac{1}{2}(\partial_x\varphi_p)^2 + \frac{1}{2}(\partial_t\varphi_p)^2 - 2\bar{\chi}_p\partial_t\bar{\chi}_p - 2\bar{\chi}_p\partial_x\bar{\chi}_p - 2\chi_p\partial_t\chi_p + 2\chi_p\partial_x\chi_p \\
&- 16(\psi_p\bar{\psi}_p + \chi_p\bar{\chi}_p)\cosh\varphi_p\cosh\phi_p - 4\cosh(2\varphi_p) \\
&+ 16(\psi_p\bar{\chi}_p + \chi_p\bar{\psi}_p)\sinh\varphi_p\sinh\phi_p + 4\cosh(2\phi_p)
\end{aligned} \tag{36}$$

whose equations of motion are invariant under the supersymmetry transformations

$$\delta(\phi_p \pm \varphi_p) = 2(\psi_p \mp \chi_p)\epsilon_\pm, \quad \delta(\psi_p \pm \chi_p) = -\partial_z(\phi_p \mp \varphi_p)\epsilon_\pm \tag{37}$$

and

$$\delta(\bar{\psi}_p \pm \bar{\chi}_p) = 2\sinh(\phi_p \pm \varphi_p)\epsilon_\mp \tag{38}$$

Inspired from the $N = 1$ case (see eqn. (27)) we propose the following Lagrangian description for the defect

$$\begin{aligned}
\mathcal{L}_D &= \frac{1}{2}(\phi_2\partial_t\phi_1 - \phi_1\partial_t\phi_2) - 2\psi_1\psi_2 - 2\bar{\psi}_1\bar{\psi}_2 + f_1\partial_t f_2 \\
&- \frac{1}{2}(\varphi_2\partial_t\varphi_1 - \varphi_1\partial_t\varphi_2) + 2\chi_1\chi_2 + 2\bar{\chi}_1\bar{\chi}_2 + f_2\partial_t f_1 \\
&+ B_0(\phi_p, \varphi_p) + B_1(\phi_p, \varphi_p, \psi_p, \chi_p, \bar{\psi}_p, \bar{\chi}_p, f_1, f_2)
\end{aligned} \tag{39}$$

leading to the Backlund transformation

$$\begin{aligned}
\partial_x\phi_1 - \partial_t\phi_2 &= -\partial_{\phi_1}B, \\
\partial_x\phi_2 - \partial_t\phi_1 &= \partial_{\phi_2}B, \\
\partial_x\varphi_1 - \partial_t\varphi_2 &= \partial_{\varphi_1}B, \\
\partial_x\varphi_2 - \partial_t\varphi_1 &= -\partial_{\varphi_2}B, \\
\psi_1 - \psi_2 &= -\frac{1}{2}\partial_{\psi_1}B = -\frac{1}{2}\partial_{\psi_2}B \\
\chi_1 - \chi_2 &= \frac{1}{2}\partial_{\chi_1}B = \frac{1}{2}\partial_{\chi_2}B \\
\bar{\psi}_1 + \bar{\psi}_2 &= \frac{1}{2}\partial_{\bar{\psi}_1}B = -\frac{1}{2}\partial_{\bar{\psi}_2}B, \\
\bar{\chi}_1 + \bar{\chi}_2 &= -\frac{1}{2}\partial_{\bar{\chi}_1}B = \frac{1}{2}\partial_{\bar{\chi}_2}B, \\
\dot{f}_1 &= -\frac{1}{2}\partial_{f_2}B, \\
\dot{f}_2 &= -\frac{1}{2}\partial_{f_1}B
\end{aligned} \tag{40}$$

Journal of Physics: Conference Series **128** (2008) 012004 doi:10.1088/1742-6596/128/1/012004

where $B = B_0 + B_1$. The canonical momentum

$$\begin{aligned} P &= \int_{-\infty}^{0} (\partial_x\phi_1\partial_t\phi_1 - 2\bar{\psi}_1\partial_x\bar{\psi}_1 - 2\psi_1\partial_x\psi_1 \\ &- \partial_x\varphi_1\partial_t\varphi_1 + 2\bar{\chi}_1\partial_x\bar{\chi}_1 + 2\chi_1\partial_x\chi_1)\,dx \\ &+ \int_{0}^{\infty} (\partial_x\phi_2\partial_t\phi_2 - 2\bar{\psi}_2\partial_x\bar{\psi}_2 - 2\psi_2\partial_x\psi_2 \\ &- \partial_x\varphi_2\partial_t\varphi_2 + 2\bar{\chi}_2\partial_x\bar{\chi}_2 + 2\chi_2\partial_x\chi_2)\,dx \end{aligned} \tag{41}$$

is no longer conserved. Proposing the modified momentum to be

$$\mathcal{P} = P + [B_0^{(+)} - B_0^{(-)} + B_1^{(+)} - B_1^{(-)} + 2\bar{\psi}_1\bar{\psi}_2 - 2\psi_1\psi_2 - 2\bar{\chi}_1\bar{\chi}_2 + 2\chi_1\chi_2]_{x=0}$$

which is conserved provided the border function satisfies

$$\partial_{\phi_+}B_0^{(+)}\partial_{\phi_-}B_0^{(-)} - \partial_{\varphi_+}B_0^{(+)}\partial_{\varphi_-}B_0^{(-)} = 4\sinh\phi_+\sinh\phi_- - 4\sinh\varphi_+\sinh\varphi_-, \tag{42}$$

and

$$\begin{aligned} &\partial_{\phi_+}B_0^{(+)}\partial_{\phi_-}B_1^{(-)} + \partial_{\phi_-}B_0^{(-)}\partial_{\phi_+}B_1^{(+)} - \partial_{\varphi_+}B_0^{(+)}\partial_{\varphi_-}B_1^{(-)} - \partial_{\varphi_-}B_0^{(-)}\partial_{\varphi_+}B_1^{(+)} \\ +\;& \partial_{\phi_+}B_1^{(+)}\partial_{\phi_-}B_1^{(-)} - \partial_{\varphi_+}B_1^{(+)}\partial_{\varphi_-}B_1^{(-)} - \frac{1}{2}(\partial_{f_1}B_1^{(-)}\partial_{f_2}B_1^{(+)} + \partial_{f_2}B_1^{(-)}\partial_{f_1}B_1^{(+)}) \\ =\;& -2(\psi_+\bar{\psi}_+ + \psi_-\bar{\psi}_- + \chi_+\bar{\chi}_+ + \chi_-\bar{\chi}_-)\Lambda_- - 2(\psi_+\bar{\psi}_- + \psi_-\bar{\psi}_+ + \chi_+\bar{\chi}_- + \chi_-\bar{\chi}_+)\Lambda_+ \\ +\;& 2(\psi_+\bar{\chi}_+ + \psi_-\bar{\chi}_- + \chi_+\bar{\psi}_+ + \chi_-\bar{\psi}_-)\Delta_- + 2(\psi_+\bar{\chi}_- + \psi_-\bar{\chi}_+ + \chi_+\bar{\psi}_- + \chi_-\bar{\psi}_+)\Delta_+ \end{aligned} \tag{43}$$

where

$$\chi_\pm = \chi_1 \pm \chi_2, \qquad \bar{\chi}_\pm = \bar{\chi}_1 \pm \bar{\chi}_2, \qquad \varphi_\pm = \varphi_1 \pm \varphi_2, \cdots \text{etc}$$

and

$$\begin{aligned} \Lambda_\pm &= \cosh(\frac{\phi_+ + \phi_-}{2})\cosh(\frac{\varphi_+ + \varphi_-}{2}) \pm \cosh(\frac{\phi_+ - \phi_-}{2})\cosh(\frac{\varphi_+ - \varphi_-}{2}), \\ \Delta_\pm &= \sinh(\frac{\phi_+ + \phi_-}{2})\sinh(\frac{\varphi_+ + \varphi_-}{2}) \pm \sinh(\frac{\phi_+ - \phi_-}{2})\sinh(\frac{\varphi_+ - \varphi_-}{2}), \end{aligned}$$

The solution of (42) and (43) is

$$B_0^{(+)} = B_0^{(+)}(\phi_+,\varphi_+) = \frac{2\alpha_3}{\alpha_2}(\cosh\phi_+ - \cosh\varphi_+), \tag{44}$$

$$B_0^{(-)} = B_0^{(-)}(\phi_-,\varphi_-) = \frac{2\alpha_2}{\alpha_3}(\cosh\phi_- - \cosh\varphi_-), \tag{45}$$

and

$$\begin{aligned} B_1^{(+)} &= B_1^{(+)}(\phi_+,\varphi_+,\psi_+,\chi_+,f_1,f_2) \\ &= \frac{i}{\sqrt{2}}f_1\left(-\alpha_3(\psi_+ - \chi_+)\cosh\frac{1}{2}(\phi_+ + \varphi_+)\right) \\ &+ \frac{i}{\sqrt{2}}f_2\left(\frac{8}{\alpha_2}(\psi_+ + \chi_+)\cosh\frac{1}{2}(\phi_+ - \varphi_+)\right), \end{aligned} \tag{46}$$

$$
\begin{aligned}
B_1^{(-)} &= B_1^{(-)}(\phi_-, \varphi_-, \bar{\psi}_-, \bar{\chi}_-, f_1, f_2) \\
&= \frac{i}{\sqrt{2}} f_1 \left(\alpha_2 (\bar{\psi}_- + \bar{\chi}_-) \cosh \frac{1}{2}(\phi_- - \varphi_-) \right) \\
&+ \frac{i}{\sqrt{2}} f_2 \left(-\frac{8}{\alpha_3} (\bar{\psi}_- - \bar{\chi}_-) \cosh \frac{1}{2}(\phi_- + \varphi_-) \right)
\end{aligned} \tag{47}
$$

where α_2 and α_3 are arbitrary constants.

These results are consistent with the Backlund transformation for the $N = 2$ super sinh-Gordon obtained from a superfield formalism [5]. We should mention that the Backlund transformation (40) with B_0 and B_1 given by (44)-(47) are invariant under the supersymmetry transformation (37)-(38). In [5] some examples of Backlund solutions are discussed also.

Acknowledgments
We are grateful to CNPq and FAPESP for financial support.

References

[1] Bowcock P, Corrigan E and Zambon C 2004 *Int. J. Mod. Phys.* **A19** (82) , hep-th/0305022;
[2] Bowcock P, Corrigan E and Zambon C 2004 *J. High Energy Phys.* JHEP **0401** 056, hep-th/0401020;
[3] Corrigan E and Zambon C 2004 *J. Physics* **A37** (L471) , hep-th/0407199
[4] Chaichian M and Kulish P 1978 *Phys. Lett.* **78B** (413)
[5] Gomes J F, Ymai L H and Zimerman A H 2008 *J. High Energy Phys.* JHEP **0803** 001, e-Print: arXiv:0710.1391 [hep-th]
[6] Inami T and Kanno 1990 Nucl. Phys. B359 201
[7] Kobayashi K and Uematsu T 1991 Phys. Lett. B264 107
[8] Nepomechie R 2001 *Phys. Lett.* **509B** (183)
[9] Aratyn H, Gomes J F and Zimerman A H 2004 Nucl. Phys. B676 537
[10] Gomes J F, Ymai L H and Zimerman A H *$N = 2$ and $N = 4$ Supersymmetric mKdV and sinh-Gordon Hierarchies*, hep-th/0409171
[11] Gomes J F, Ymai L H and Zimerman A H 2006 *J. Physics* **A39** (7471) , hep-th/0601014
[12] Gomes J F, Ymai L H and Zimerman A H 2006 *Phys. Lett.* **359A** (630) , hep-th/0607107

Journal of Physics: Conference Series **128** (2008) 012005 doi:10.1088/1742-6596/128/1/012005

Higgsless electroweak model due to the spherical geometry

Nikolay A Gromov

Department of Mathematics, Komi Science Center UrD RAS, Kommunisticheskaya st., 24, Syktyvkar, 167982, Russia

E-mail: gromov@dm.komisc.ru

Abstract. A new formulation of the Electroweak Model with 3-dimensional spherical geometry in the target space is suggested. The *free* Lagrangian in the spherical field space along with the standard gauge field Lagrangian form the full Higgsless Lagrangian of the model, whose second order terms reproduce the same fields with the same masses as the Standard Electroweak Model. The vector bosons masses are automatically generated, so there is no need in special mechanism of spontaneous symmetry breaking.

1. Introduction

The Standard Electroweak Model (SEWM) based on gauge group $SU(2) \times U(1)$ gives a good description of electroweak processes. One of the unsolved problems is the origin of electroweak symmetry breaking. In the standard formulation the scalar field (Higgs boson) performs this task via Higgs mechanism, which generates a mass terms for vector bosons. However, it is not yet experimentally verified whether electroweak symmetry is broken by such a Higgs mechanism, or by something else.

The emergence of large number Higgsless models [1]–[10] was stimulated by difficulties with Higgs boson. These models are mainly based on extra dimensions of different types or larger gauge groups. The construction given in [11] is based on an observation: the underlying group of SEWM can be represented as a semidirect product of $U(1)$ and $SU(2)$.

In the present paper a new formulation of the Higgsless Electroweak Model is suggested. Firstly we observe that the quadratic form $\phi^\dagger\phi = \phi_1^*\phi_1 + \phi_2^*\phi_2 = R^2$ of the complex matter field $\phi \in \mathbf{C}_2$ is invariant with respect to gauge group transformations $SU(2) \times U(1)$ and we can restrict fields on the quadratic form without the loss of gauge invariance of the model. This quadratic form define three dimensional sphere in four dimensional Euclidean space of the real components of ϕ, where the noneuclidean spherical geometry is realized.

Secondly we introduce the *free* matter field Lagrangian in this spherical field space, which along with the standard gauge field Lagrangian form the full Higgsless Lagrangian of the model. Its second order terms reproduce the same fields as the Standard Electroweak Model but without the remaining real dynamical Higgs field. The vector bosons masses are automatically generated and are given by the same formulas as in the Standard Electroweak Model, so there is no need in special mechanism of spontaneous symmetry breaking.

The fermion Lagrangian of SEWM are modified by replacing of the fields ϕ with the restricted on the quadratic form fields in such a way that its second order terms provide the electron mass

Journal of Physics: Conference Series **128** (2008) 012005 doi:10.1088/1742-6596/128/1/012005

and neutrino remain massless. The preliminary versions are in [12],[13].

2. Standard electroweak model

The bosonic sector of SEWM is $SU(2) \times U(1)$ gauge theory in the space $\Phi_2(\mathbf{C})$ of fundamental representation of $SU(2)$. The bosonic Lagrangian is given by the sum

$$L_B = L_A + L_\phi, \tag{1}$$

where

$$L_A = \frac{1}{8g^2}\mathrm{Tr}(F_{\mu\nu})^2 - \frac{1}{4}(B_{\mu\nu})^2 = -\frac{1}{4}[(F^1_{\mu\nu})^2 + (F^2_{\mu\nu})^2 + (F^3_{\mu\nu})^2] - \frac{1}{4}(B_{\mu\nu})^2 \tag{2}$$

is the gauge field Lagrangian for $SU(2) \times U(1)$ group and

$$L_\phi = \frac{1}{2}(D_\mu\phi)^\dagger D_\mu\phi - V(\phi) \tag{3}$$

is the matter field Lagrangian (summation on the repeating Greek indexes is always understood). Here $\phi = \begin{pmatrix} \phi_1 \\ \phi_2 \end{pmatrix} \in \Phi_2(\mathbf{C})$, D_μ are the covariant derivatives

$$D_\mu\phi = \partial_\mu\phi - ig\left(\sum_{k=1}^{3} T_k A^k_\mu\right)\phi - ig'YB_\mu\phi, \tag{4}$$

where

$$T_1 = \frac{1}{2}\sigma_1 = \frac{1}{2}\begin{pmatrix} 0 & 1 \\ 1 & 0 \end{pmatrix}, \quad T_2 = \frac{1}{2}\sigma_2 = \frac{1}{2}\begin{pmatrix} 0 & -i \\ i & 0 \end{pmatrix}, \quad T_3 = \frac{1}{2}\sigma_3 = \frac{1}{2}\begin{pmatrix} 1 & 0 \\ 0 & -1 \end{pmatrix},$$

with σ_k being Pauli matrices, are linear representation of generators in the complex space $\Phi_2(\mathbf{C})$ of the fundamental representation of $SU(2)$ and $Y = \frac{1}{2}\mathbf{1}$ is generator of $U(1)$ in the same space. The gauge fields

$$A_\mu(x) = -ig\sum_{k=1}^{3} T_k A^k_\mu(x), \quad B_\mu(x) = -ig'B_\mu(x)$$

take their values in Lie algebras $su(2)$, $u(1)$ respectively, and the stress tensors are

$$F_{\mu\nu}(x) = \mathcal{F}_{\mu\nu}(x) + [A_\mu(x), A_\nu(x)], \quad B_{\mu\nu} = \partial_\mu B_\nu - \partial_\nu B_\mu.$$

The potential $V(\phi)$ in (3) is introduced by hand in a special form (sombrero)

$$V(\phi) = \frac{\lambda}{4}\left(\phi^\dagger\phi - v^2\right)^2, \tag{5}$$

where λ, v are constants.

The Lagrangian L_B (1) describe massless fields. To generate mass terms for the vector bosons without breaking the gauge invariance one uses the Higgs mechanism. One of L_B ground states

$$\phi^{vac} = \begin{pmatrix} 0 \\ v \end{pmatrix}, \quad A^k_\mu = B_\mu = 0$$

is taken as a vacuum state of the model, and small field excitations

$$\phi_1(x), \quad \phi_2(x) = v + \chi(x), \quad A^a_\mu(x), \quad B_\mu(x)$$

with respect to the vacuum are regarded. The matrix $Q = Y + T^3 = \begin{pmatrix} 1 & 0 \\ 0 & 0 \end{pmatrix}$, which annihilates the ground state $Q\phi^{vac} = 0$, is the generator of the electromagnetic subgroup $U(1)_{em}$. The new fields

$$W_\mu^\pm = \frac{1}{\sqrt{2}}\left(A_\mu^1 \mp iA_\mu^2\right), \quad Z_\mu = \frac{1}{\sqrt{g^2+g'^2}}\left(gA_\mu^3 - g'B_\mu\right), \quad A_\mu = \frac{1}{\sqrt{g^2+g'^2}}\left(g'A_\mu^3 + gB_\mu\right)$$

are introduced, where $W_\mu^\pm$ are complex $(W_\mu^-)^* = W_\mu^+$ and Z_μ, A_μ are real.

For small fields the Lagrangian L_B (1) can be rewritten in the form

$$L_B = L_0 + L_{int},$$

where $L_0 = L_B^{(2)}$ is the usual second order Lagrangian for free vector and scalar bosons, and higher order terms L_{int} are regarded as field interactions. The second order terms of the Lagrangian (1) are as follows

$$L_B^{(2)} = -\frac{1}{2}\mathcal{W}_{\mu\nu}^+\mathcal{W}_{\mu\nu}^- + m_W^2 W_\mu^+ W_\mu^- - \frac{1}{4}\mathcal{F}_{\mu\nu}\mathcal{F}_{\mu\nu} - \frac{1}{4}\mathcal{Z}_{\mu\nu}\mathcal{Z}_{\mu\nu} + \frac{1}{2}m_Z^2 Z_\mu Z_\mu + \frac{1}{2}\left(\partial_\mu\chi\right)^2 - \frac{1}{2}m_\chi^2\chi^2, \tag{6}$$

where $\mathcal{Z}_{\mu\nu} = \partial_\mu Z_\nu - \partial_\nu Z_\mu$, $\mathcal{F}_{\mu\nu} = \partial_\mu A_\nu - \partial_\nu A_\mu$, $\mathcal{W}^\pm{}_{\mu\nu} = \partial_\mu W_\nu^\pm - \partial_\nu W_\mu^\pm$ and describe massive vector fields $W_\mu^\pm$ with identical mass $m_W = \frac{1}{2}gv$ (W-bosons), massless vector field A_μ, $m_A = 0$ (photon), massive vector field Z_μ, with the mass $m_Z = \frac{v}{2}\sqrt{g^2+g'^2}$ (Z-boson) and massive scalar field χ, $m_\chi = \sqrt{2\lambda}v$ (Higgs boson).

W- and Z-bosons have been observed and have the masses $m_W = 80GeV$, $m_Z = 91GeV$. Higgs boson has not been experimentally verified up to now.

3. Higgsless electroweak model with 3D spherical matter space

The complex 2D space Φ_2 can be regarded as 4D real Euclidean space $\mathbf{R}_4$. Let us introduce the real fields r, $\bar\psi = (\psi_1, \psi_2, \psi_3)$ by

$$\phi_1 = r(\psi_2 + i\psi_1), \quad \phi_2 = r(1 + i\psi_3). \tag{7}$$

It is easy to see that the quadratic form

$$\phi^\dagger\phi = \phi_1^*\phi_1 + \phi_2^*\phi_2 = R^2 \tag{8}$$

is invariant with respect to group transformations $SU(2)\times U(1)$ and we can restrict fields on this quadratic form without the loss of gauge invariance of the model. Similar restriction was appeared in the unified conformal model for fundamental interactions [14] in the case when the length scale is defined in such a way that elementary particle masses are the same for all times and in all places.

For the real fields the form (8) is written as $r^2(1+\bar\psi^2) = R^2$, where $\bar\psi^2 = \psi_1^2 + \psi_2^2 + \psi_3^2$, therefore

$$r = \frac{R}{\sqrt{1+\bar\psi^2}}. \tag{9}$$

Hence there are three independent real fields $\bar\psi$. These fields belong to the space Ψ_3 with noneuclidean spherical geometry which is realized on the 3D sphere in 4D Euclidean space $\mathbf{R}_4$. The fields $\bar\psi$ are intrinsic Beltrami coordinates on Ψ_3.

The potential (5) is the constant $V(\phi) = \lambda\left(R^2 - v^2\right)^2/4$ on the sphere and what is more $V(\phi) = 0$ for $R = v$. Therefore, let us define the *free* gauge invariant matter field Lagrangian L_ψ with the help of the metric tensor [12] of the spherical space Ψ_3

$$g_{kk}(\bar{\psi}) = \frac{1 + \bar{\psi}^2 - \psi_k^2}{(1 + \bar{\psi}^2)^2}, \quad g_{kl}(\bar{\psi}) = \frac{-\psi_k \psi_l}{(1 + \bar{\psi}^2)^2}$$

in the form

$$L_\psi = \frac{R^2}{2} \sum_{k,l=1}^{3} g_{kl} D_\mu \psi_k D_\mu \psi_l = \frac{R^2\left[(1 + \bar{\psi}^2)(D_\mu \bar{\psi})^2 - (\bar{\psi}, D_\mu \bar{\psi})^2\right]}{2(1 + \bar{\psi}^2)^2}. \tag{10}$$

Let us note that the Lagrangian on 3D sphere in 4D Euclidean space was considered in [15], where the invariant perturbative theory was developed which does not depend on some particular parametrization.

The covariant derivatives (4) are obtained using the representations of generators for the algebras $su(2)$, $u(1)$ in the space Ψ_3 [12]

$$T_1 \bar{\psi} = \frac{i}{2}\begin{pmatrix} -(1 + \psi_1^2) \\ \psi_3 - \psi_1 \psi_2 \\ -(\psi_2 + \psi_1 \psi_3) \end{pmatrix}, \quad T_2 \bar{\psi} = \frac{i}{2}\begin{pmatrix} -(\psi_3 + \psi_1 \psi_2) \\ -(1 + \psi_2^2) \\ \psi_1 - \psi_2 \psi_3 \end{pmatrix},$$

$$T_3 \bar{\psi} = \frac{i}{2}\begin{pmatrix} -\psi_2 + \psi_1 \psi_3 \\ \psi_1 + \psi_2 \psi_3 \\ 1 + \psi_3^2 \end{pmatrix}, \quad Y \bar{\psi} = \frac{i}{2}\begin{pmatrix} -(\psi_2 + \psi_1 \psi_3) \\ \psi_1 - \psi_2 \psi_3 \\ -(1 + \psi_3^2) \end{pmatrix}$$

and are as follows:

$$D_\mu \psi_1 = \partial_\mu \psi_1 + \frac{g}{2}\left[-(1 + \psi_1^2)A_\mu^1 - (\psi_3 + \psi_1 \psi_2)A_\mu^2 - (\psi_2 - \psi_1 \psi_3)A_\mu^3\right] - \frac{g'}{2}(\psi_2 + \psi_1 \psi_3)B_\mu,$$

$$D_\mu \psi_2 = \partial_\mu \psi_2 + \frac{g}{2}\left[(\psi_3 - \psi_1 \psi_2)A_\mu^1 - (1 + \psi_2^2)A_\mu^2 + (\psi_1 + \psi_2 \psi_3)A_\mu^3\right] + \frac{g'}{2}(\psi_1 - \psi_2 \psi_3)B_\mu,$$

$$D_\mu \psi_3 = \partial_\mu \psi_3 + \frac{g}{2}\left[-(\psi_2 + \psi_1 \psi_3)A_\mu^1 + (\psi_1 - \psi_2 \psi_3)A_\mu^2 + (1 + \psi_3^2)A_\mu^3\right] - \frac{g'}{2}(1 + \psi_3^2)B_\mu. \tag{11}$$

The gauge fields Lagrangian (2) does not depend on the fields ϕ and therefore remains unchanged.

For small fields, the second order Lagrangian (10) is written as

$$L_\psi^{(2)} = \frac{R^2}{2}\left[(D_\mu \bar{\psi})^{(1)}\right]^2 = \frac{R^2}{2}\sum_{k=1}^{3}\left[(D_\mu \psi_k)^{(1)}\right]^2, \tag{12}$$

where linear terms in covariant derivates (11) have the form

$$(D_\mu \psi_1)^{(1)} = \partial_\mu \psi_1 - \frac{g}{2}A_\mu^1 = -\frac{g}{2}\left(A_\mu^1 - \frac{2}{g}\partial_\mu \psi_1\right) = -\frac{g}{2}\hat{A}_\mu^1,$$

$$(D_\mu \psi_2)^{(1)} = \partial_\mu \psi_2 - \frac{g}{2}A_\mu^2 = -\frac{g}{2}\left(A_\mu^2 - \frac{2}{g}\partial_\mu \psi_2\right) = -\frac{g}{2}\hat{A}_\mu^2,$$

$$(D_\mu \psi_3)^{(1)} = \partial_\mu \psi_3 + \frac{g}{2}A_\mu^3 - \frac{g'}{2}B_\mu = \frac{1}{2}\sqrt{g^2 + g'^2}Z_\mu.$$

For the new fields

$$W_\mu^\pm = \frac{1}{\sqrt{2}}\left(\hat{A}_\mu^1 \mp i\hat{A}_\mu^2\right), \quad Z_\mu = \frac{gA_\mu^3 - g'B_\mu + 2\partial_\mu\psi_3}{\sqrt{g^2+g'^2}}, \quad A_\mu = \frac{g'A_\mu^3 + gB_\mu}{\sqrt{g^2+g'^2}}$$

Lagrangian (12) is rewritten as follows

$$L_\psi^{(2)} = +\frac{R^2g^2}{4}W_\mu^+W_\mu^- + \frac{R^2(g^2+g'^2)}{8}(Z_\mu)^2.$$

The quadratic part of the full Lagrangian

$$L_0 = L_A^{(2)} + L_\psi^{(2)} = -\frac{1}{2}\mathcal{W}_{\mu\nu}^+\mathcal{W}_{\mu\nu}^- + m_W^2 W_\mu^+W_\mu^- - \frac{1}{4}(\mathcal{F}_{\mu\nu})^2 - \frac{1}{4}(\mathcal{Z}_{\mu\nu})^2 + \frac{m_Z^2}{2}(Z_\mu)^2,$$

where

$$m_W = \frac{Rg}{2}, \quad m_Z = \frac{R}{2}\sqrt{g^2+g'^2}, \tag{13}$$

and $\mathcal{Z}_{\mu\nu} = \partial_\mu Z_\nu - \partial_\nu Z_\mu$, $\mathcal{F}_{\mu\nu} = \partial_\mu A_\nu - \partial_\nu A_\mu$, $\mathcal{W}^\pm{}_{\mu\nu} = \partial_\mu W_\nu^\pm - \partial_\nu W_\mu^\pm$, describes all the experimentally verified parts of SEWM but does not include the scalar Higgs field. For $R = v$ the masses (13) are identical to those of the SEWM (6).

The interaction part of the full Lagrangian in the first degree of approximation is given by

$$L_{int}^{(1)} = L_A^{(3)} + L_\psi^{(3)},$$

where the third order terms of the gauge field Lagrangian (2) are

$$L_A^{(3)} = -\frac{g}{\sqrt{g^2+g'^2}}\left\{i\left(\mathcal{W}_{\mu\nu}^-W_\mu^+ - \mathcal{W}_{\mu\nu}^+W_\mu^-\right)(g'A_\nu + gZ_\nu) - \frac{\sqrt{2}}{g}(g'A_\mu + gZ_\mu)\times\right.$$
$$\times\left[\mathcal{W}_{\mu\nu}^+(\partial_\nu\psi_2 - i\partial_\nu\psi_1) + \mathcal{W}_{\mu\nu}^-(\partial_\nu\psi_2 + i\partial_\nu\psi_1)\right] - \frac{2ig}{\sqrt{g^2+g'^2}}\left(\mathcal{W}_{\mu\nu}^-W_\mu^+ - \mathcal{W}_{\mu\nu}^+W_\mu^-\right)\partial_\nu\psi_3 -$$
$$-\frac{2\sqrt{2}}{\sqrt{g^2+g'^2}}\left[\mathcal{W}_{\mu\nu}^+(\partial_\nu\psi_2 - i\partial_\nu\psi_1) + \mathcal{W}_{\mu\nu}^-(\partial_\nu\psi_2 + i\partial_\nu\psi_1)\right]\partial_\nu\psi_3 + (g'\mathcal{F}_{\mu\nu} + g\mathcal{Z}_{\mu\nu})\times$$
$$\left.\times\left\{\frac{i}{4}\left[\left(W_\mu^+\right)^2 - \left(W_\mu^-\right)^2\right] + \frac{4}{g^2}\partial_\mu\psi_1\partial_\nu\psi_2 + \frac{\sqrt{2}}{g}\left[W_\mu^+(\partial_\nu\psi_2 - i\partial_\nu\psi_1) + W_\mu^-(\partial_\nu\psi_2 + i\partial_\nu\psi_1)\right]\right\}\right\}$$

and those of the matter field Lagrangian (10) are

$$L_\psi^{(3)} = \frac{R^2g}{2\sqrt{2}}\left\{W_\mu^+\left[\psi_3(\partial_\mu\psi_2 - i\partial_\mu\psi_1) - \frac{g^2-g'^2}{g^2+g'^2}(\psi_2 - i\psi_1)\partial_\mu\psi_3 + \frac{g'(gA_\mu - g'Z_\mu)}{\sqrt{g^2+g'^2}}(\psi_2 - i\psi_1)\right] +\right.$$
$$+W_\mu^-\left[\psi_3(\partial_\mu\psi_2 + i\partial_\mu\psi_1) - \frac{g^2-g'^2}{g^2+g'^2}(\psi_2 + i\psi_1)\partial_\mu\psi_3 + \frac{g'(gA_\mu - g'Z_\mu)}{\sqrt{g^2+g'^2}}(\psi_2 + i\psi_1)\right] +$$
$$\left.+\frac{1}{g}\sqrt{g^2+g'^2}Z_\mu(\psi_1\partial_\mu\psi_2 - \psi_2\partial_\mu\psi_1)\right\}.$$

The fermion Lagrangian of SEWM is taken in the form [16]

$$L_F = L_l^\dagger i\tilde{\sigma}_\mu D_\mu L_l + e_r^\dagger i\sigma_\mu D_\mu e_r - h_e[e_r^\dagger(\phi^\dagger L_l) + (L_l^\dagger\phi)e_r], \tag{14}$$

where $L_l = \begin{pmatrix} \nu_{e,l} \\ e_l^- \end{pmatrix}$ is the $SU(2)$-doublet, e_l^- the $SU(2)$-singlet, h_e is constant and e_r, e_l, ν_e are two component Lorentzian spinors. Here σ_μ are Pauli matricies, $\sigma_0 = \tilde{\sigma_0} = \mathbf{1}$, $\tilde{\sigma_k} = -\sigma_k$. The covariant derivatives $D_\mu L_l$ are given by (4) with L_l instead of ϕ and $D_\mu e_r = (\partial_\mu + ig' B_\mu)e_r$. The convolution on the inner indices of $SU(2)$-doublet is denoted by $(\phi^\dagger L_l)$.

The matter fields ϕ appear in Lagrangian (14) only in mass terms. With the use of (7) and (9), these mass terms are rewritten in the form

$$h_e[e_r^\dagger(\phi^\dagger L_l) + (L_l^\dagger \phi)e_r] = \frac{h_e R}{\sqrt{1+\bar{\psi}^2}} \left[e_r^\dagger e_l^- + e_l^{-\dagger} e_r + \right.$$

$$\left. + i\psi_3 \left(e_l^{-\dagger} e_r - e_r^\dagger e_l^-\right) + i\psi_1 \left(\nu_{e,l}^\dagger e_r - e_r^\dagger \nu_{e,l}\right) + i\psi_2 \left(\nu_{e,l}^\dagger e_r + e_r^\dagger \nu_{e,l}\right)\right].$$

Its second order terms $h_e R\left(e_r^\dagger e_l^- + e_l^{-\dagger} e_r\right)$ provide the electron mass $m_e = h_e R$, and neutrino remain massless.

4. Conclusion

The suggested formulation of the Electroweak Model with the gauge group $SU(2) \times U(1)$ based on the 3-dimensional spherical geometry in the target space describes all experimentally observed fields, and does not include the (up to now unobserved) scalar Higgs field. The *free* Lagrangian in the spherical matter field space is used instead of Lagrangian (3) with the potential (5) of the special form (sombrero). The gauge field Lagrangian is the standard one. There is no need in Higgs mechanism since the vector field masses are generated automatically. The renormalizability of the model is the open question at present and will be a subject of further investigations.

The motion group of the spherical space Ψ_3 is isomorphic to $SO(4)$. It is known [17], that the bosonic sector of SEWM has the custodial $SU(2)$-symmetry, and potential (5) is globally invariant with respect of $SO(4)$ which is locally isomorphic to the group $SU(2)_L \times SU(2)_R$, where $SU(2)_L$ is the gauge group of SEWM. The appearance of the extra $SU(2)_R$-symmetry is interpreted as transformation of $SU(2)$-doublet by pseudoreal representation, that is the complex conjugate doublet is equivalent to the initial one. The introduction of the real fields $\bar{\psi}$ is connected in a certain sense with $SO(4)$-symmetry.

Acknowledgments

The author is grateful to A.V. Zhubr, V.V. Kuratov and I.B. Khriplovich for helpful discussions. I am indebted to V.N. Pervushin and Yu.M. Zinoviev for pointing out the papers [14], [15], correspondingly.

References

[1] Simmons E H, Chivukula R S, He H-J, Kurachi M and Tanabashi M 2006 Higgsless models: lessons from deconstruction *Preprint* hep-ph/0606019 v1
[2] Casalbuoni R 2005 Deconstructed Higgsless models *Preprint* hep-ph/0509165 v1
[3] Chivukala R S, Simmons E H, He H-J, Kurachi M and Tanabashi M 2004 The structure of corrections to electroweak interactions in Higgsless models *Preprint* hep-ph/0406073 v3
[4] Slavnov A A 2006 Renormalizable electroweak model without fundamental scalar meson *Preprint* hep-th/0601125 v1
[5] Slavnov A A 2006 Higgs mechanism as a collective effect due to extra dimension *Preprint* hep-th/0604052 v1
[6] Cacciapaglia G, Csáki C, Marandella G and Terning J 2006 A new custodian for a realistic Higgsless model *Preprint* hep-ph/0607156 v1
[7] Csáki C, Hubisz J and Meade P 2005 TASI lectures on electroweak symmetry breaking from extra dimensions *Preprint* hep-ph/0510275 v1

[8] Csáki C 2004 Higgsless electroweak symmetry breaking *Preprint* hep-ph/0412339 v1
[9] Csáki C, Grojean C, Pilo L and Terning J 2003 Towards a realistic model of Higgsless electroweak symmetry breaking *Preprint* hep-ph/0308038 v2
[10] Csáki C, Grojean C, Murayama H, Pilo L and Terning J 2003 Gauge Theories on an Interval: Unitarity without a Higgs *Preprint* hep-ph/0305237 v1
[11] Sevostyanov A 2007 Towards standard model without the Higgs boson *Preprint* hep-th/0702149 v1
[12] Gromov N A 2007 *Spherical Space permits to go without Higgs Boson* (Syktyvkar, Komi Science Center RAS) ("'Scientific Reports"', N. 493) (in Russian)
[13] Gromov N A 2007 Higgsless electroweak model following from the spherical geometry *Preprint* arXiv:0705.4575v1 [hep-th]
[14] Pawlowski M and Raczka R 1994 A unified conformal model for fundamental interactions without dynamical Higgs field *Found. of Phys.* **24** 1305 (*Preprint* hep-th/9407137)
[15] Slavnov A A and Faddeev L D 1971 *Teor. Mat. Fiz.* **8** 297 (in Russian)
[16] Rubakov V A 1999 *Classical Gauge Fields* (Moscow, Editorial URSS) (in Russian)
[17] Grojean C 2007 *Usp. Fiz. Nauk* **177** 3 (in Russian).

V International Symposium on Quantum Theory and Symmetries IOP Publishing
Journal of Physics: Conference Series **128** (2008) 012006 doi:10.1088/1742-6596/128/1/012006

Nonlinear completion of massive gravity of the Fierz-Pauli type

Shinji Hamamoto
Department of Physics, University of Toyama, Toyama 930-8555, Japan
E-mail: `hamamoto@sci.u-toyama.ac.jp`

Abstract. A possible nonlinear completion of massive gravity of the Fierz-Pauli type is proposed. The theory describes a system consisting of a massive tensor field of the Fierz-Pauli type and an additional massive vector field. Massless limit as well as flat-spacetime limit can be taken smoothly. Constructing a nonlinear version of the physical-state condition which drives an extra scalar ghost from physical states is still unsettled.

1. Introduction
In a series of papers [1, 2, 3, 4, 5], attempts to construct the theory of massive gravity with smooth massless limit were made.

We studied infrared regularization of linearized massive tensor fields in [1, 2, 3, 4]. Two model theories were considered: one is of the pure-tensor (PT) type, which describes an ordinary massive tensor field of five degrees of freedom; the other is of the additional-scalar-ghost (ASG) type, which contains a scalar ghost in addition to the pure tensor. The ASG model shows second-order massless singularities in two-point functions, whereas the PT model contains fourth-order singularities. It turns out that two procedures, the BRS one and the Nakanishi one, are effective in regularizing such singularities. The BRS procedure produces transparent structures to the resulting theories, as compared with the Nakanishi one. So we studied the former in detail. In order to drive away the second-order infrared singularities in the ASG model, we introduce an auxiliary vector-like field, and promote the original theory to the one that is invariant under the vector BRS transformation. On the other hand, to carry out infrared regularization of the fourth-order singularities in the PT model, we need to introduce an auxiliary scalar field in addition to the vector-like one, and make the resulting theory invariant under the scalar BRS transformation as well as the vector one.

When we try to perform nonlinear completion, the ASG model is easier to deal with than the PT model. This is because only the vector BRS transformation is involved there. The nonlinear form of this transformation is simply the quantum version of the general coordinate transformation. The scalar BRS, on the other hand, has no classical counterpart. Constructing its proper nonlinear generalization is not an easy task. A possible nonlinear completion of the BRS model of the ASG-type massive tensor was proposed in [5] (See also [6].). We also pointed out there that ghost condensation mechanism may work well for making innocuous the additional scalar ghost.

The purpose of the present paper is to put into practice nonlinear completion of the infrared-regularized PT model. In order to avoid introducing scalar BRS, we ask for the help of the

Nakanishi procedure. Then it is found straightforward to construct nonlinear Lagrangian for that BRS+Nakanishi model of the PT-type massive tensor.

However, this is not the end of the story. In the present formulation, there occurs a new trouble concerning physical-state condition of the Nakanishi-type. Finding a nonlinear version of such condition requires further studies.

In section 2, we review the case of Abelian vector field. This is to see how the BRS and the Nakanishi procedures work for regularizing massless singularities contained in the original massive theory. Stress is put on the fact that choosing massive gauge in the BRS procedure gives simple pole structure to two-point functions and makes it easy to investigate particle contents of physical states. In section 3, we treat linear theories of massive tensor field. Second-order massless singularities in the ASG model are regularized by the BRS procedure, whereas fourth-order singularities in the PT model are regularized by the use of both the BRS *and* the Nakanishi procedures. Emphasis is laid also here on the usefulness of adopting massive gauge. Section 4 treats nonlinear completion of the BRS+Nakanishi model of the PT-type massive tensor. After introducing nonlinear BRS transformation and basic BRS invariants, we propose possible nonlinear forms for the Lagrangian. Difficulties of finding a nonlinear version of the Nakanishi-type physical-state condition are also pointed out. Summary and discussion are given in section 5.

2. Massive vector

2.1. Massless vector

Let us begin with massless vector. The Lagrangian is given by[1]

$$L_0 = -\frac{1}{4}F_{\mu\nu}F^{\mu\nu} + L^{\alpha}_{\mathrm{GF+FP}}, \tag{1}$$

where $L^{\alpha}_{\mathrm{GF+FP}}$ is the gauge-fixing and Faddeev-Popov (GF+FP) Lagrangian

$$\begin{aligned} L^{\alpha}_{\mathrm{GF+FP}} &= b\left(\partial_\mu A^\mu + \frac{\alpha}{2}b\right) + \mathrm{i}\bar{c}\,\Box\, c \\ &= -\mathrm{i}\delta\left[\bar{c}\left(\partial_\mu A^\mu + \frac{\alpha}{2}b\right)\right] \end{aligned} \tag{2}$$

with the gauge parameter α. The theory is invariant under the BRS transformation

$$\delta A_\mu = \partial_\mu c, \qquad \delta\bar{c} = \mathrm{i}b. \tag{3}$$

For $\alpha = 1$, two-point functions take simple forms:[2]

$$\langle A^\mu A^\nu\rangle = \frac{\eta^{\mu\nu}}{\Box}\delta, \qquad \langle A^\mu b\rangle = \frac{\partial^\mu}{\Box}\delta, \qquad \langle bb\rangle = 0, \qquad \langle c\bar{c}\rangle = -\mathrm{i}\frac{1}{\Box}\delta. \tag{4}$$

Physical states are defined by the use of the conserved BRS charge Q_{B}:

$$Q_{\mathrm{B}}|\mathrm{phys}\rangle = 0. \tag{5}$$

In order to clarify the particle contents of the physical states, we expand any field $\Phi_A(x)$ as

$$\Phi_A(x) = \frac{1}{(2\pi)^{3/2}}\int \mathrm{d}^4p\,\theta(p^0)\left[\Phi_A(p)\mathrm{e}^{\mathrm{i}px} + \Phi^\dagger_A(p)\mathrm{e}^{-\mathrm{i}px}\right]. \tag{6}$$

[1] The flat spacetime metric used in the present paper is $\eta_{\mu\nu} = (-1, +1, +1, +1)$.

[2] Here and hereafter the spacetime coordinates are omitted in the field variables as well as in the δ-functions.

Journal of Physics: Conference Series **128** (2008) 012006 doi:10.1088/1742-6596/128/1/012006

In the axial coordinate $(p^1 = p^2 = 0,\ p^3 > 0)$, let us define as

$$\varphi^1(p) \stackrel{\mathrm{d}}{\equiv} A^1(p), \qquad \varphi^2(p) \stackrel{\mathrm{d}}{\equiv} A^2(p), \qquad \chi(p) \stackrel{\mathrm{d}}{\equiv} \frac{1}{p^3} A^3(p). \tag{7}$$

Then we find that $\{\varphi^i(p)\,(i = 1, 2)\}$ are BRS singlets and physical, whereas $\{\chi(p), b(p), c(p), \bar{c}(p)\}$ constitute a BRS quartet.

2.2. Massive vector: naive model

Mass is introduced through the Proca Lagrangian

$$L_m\,[A^\mu] = -\frac{1}{4}F_{\mu\nu}F^{\mu\nu} - \frac{m^2}{2}A_\mu A^\mu. \tag{8}$$

This naive model of massive vector field gives two-point functions with the second-order massless singularities like

$$\langle A^\mu A^\nu \rangle = \frac{1}{\Box - m^2}\left(\eta^{\mu\nu} - \frac{\partial^\mu \partial^\nu}{m^2}\right)\delta. \tag{9}$$

The field equations

$$\begin{cases} (\Box - m^2)A^\mu = 0, \\ \partial_\mu A^\mu = 0 \end{cases} \tag{10}$$

ensure that the physical degrees of freedom count three, $\{\varphi^i(p)\ (i = 1, 2),\ \chi(p)\}$, in this case.

2.3. Massive vector: Nakanishi model

In order to remove the massless singularities involved in the naive model, Nakanishi [7] proposed the following type of Lagrangian:

$$L_{\mathrm{N}} = L_m\,[A^\mu] + L^\alpha_{\mathrm{GF'}}, \qquad L^\alpha_{\mathrm{GF'}} = b\left(\partial_\mu A^\mu + \frac{\alpha}{2}b\right), \tag{11}$$

where $L_m\,[A^\mu]$ is the Proca Lagrangian and $L^\alpha_{\mathrm{GF'}}$ is the gauge-fixing-like one.[3] For $\alpha = 1$, two-point functions are then

$$\langle A^\mu A^\nu \rangle = \frac{\eta^{\mu\nu}}{\Box - m^2}\delta, \qquad \langle A^\mu b \rangle = \frac{\partial^\mu}{\Box - m^2}\delta, \qquad \langle bb \rangle = -\frac{m^2}{\Box - m^2}\delta, \tag{12}$$

which show the massless singularities have disappeared. Note that the Nakanishi-Lautrup field $b(x)$ is a ghost for $m \neq 0$. Physical states are picked out by the condition

$$b^{(+)}(x)|\mathrm{phys}\rangle = 0, \tag{13}$$

where $b^{(+)}(x)$ denotes the positive frequency part of $b(x)$. In the massive case, we can introduce the field

$$\tilde{\chi}(p) \stackrel{\mathrm{d}}{\equiv} \chi(p) + \mathrm{i}\frac{1}{m^2}b(p), \tag{14}$$

and find that the set of the fields $\{\varphi^i(p)\ (i = 1, 2),\ \tilde{\chi}(p)\}$ is physical and the ghost $b(p)$ is unphysical. In the massless case, on the other hand, the set of the fields $\{\varphi^i(p)\ (i = 1, 2),\ b(p)\}$ becomes physical and the field $\chi(p)$ becomes unphysical. Note that the field b is zero-normed in this case. That means b is not observable, there remaining only two observable degrees of freedom.

[3] The Proca Lagrangian does not show any gauge invariance. Therefore, adding the term $L^\alpha_{\mathrm{GF'}}$ has nothing to do with gauge-fixing procedure.

2.4. Massive vector: BRS model

The BRS model Lagrangian is constructed as follows: introduce an auxiliary scalar field θ, promote the Proca Lagrangian (8) to a gauge invariant one by replacing A^μ with $A^\mu - \frac{1}{m}\partial^\mu\theta$, and append a certain GF+FP Lagrangian. The total Lagrangian is given as

$$\begin{aligned} L_{\mathrm{BRS}} &= L_m\left[A^\mu - \frac{1}{m}\partial^\mu\theta\right] + L^{m\alpha}_{\mathrm{GF+FP}} \\ &= L_m\left[A^\mu\right] - m\theta\partial_\mu A^\mu - \frac{1}{2}\partial_\mu\theta\partial^\mu\theta + L^{m\alpha}_{\mathrm{GF+FP}}. \end{aligned} \tag{15}$$

For the GF+FP Lagrangian, we adopt here the following massive type instead of the massless one (2):

$$\begin{aligned} L^{m\alpha}_{\mathrm{GF+FP}} &= b\left(\partial_\mu A^\mu - m\theta + \frac{\alpha}{2}b\right) + \mathrm{i}\bar{c}\left(\Box - m^2\right)c \\ &= -\mathrm{i}\delta\left[\bar{c}\left(\partial_\mu A^\mu - m\theta + \frac{\alpha}{2}b\right)\right]. \end{aligned} \tag{16}$$

The choice of (16) gives simple pole structure to two-point functions and makes it easy to investigate particle contents of physical states. The BRS transformation which keeps the theory invariant is

$$\delta A_\mu = \partial_\mu c, \qquad \delta\theta = mc, \qquad \delta\bar{c} = \mathrm{i}b. \tag{17}$$

For $\alpha = 1$, two-point functions are calculated to give

$$\begin{gathered} \langle A^\mu A^\nu\rangle = \frac{\eta^{\mu\nu}}{\Box - m^2}\delta, \quad \langle A^\mu b\rangle = \frac{\partial^\mu}{\Box - m^2}\delta, \quad \langle bb\rangle = 0, \quad \langle c\bar{c}\rangle = -\mathrm{i}\frac{1}{\Box - m^2}\delta, \\ \langle A^\mu\theta\rangle = 0, \quad \langle b\theta\rangle = \frac{m}{\Box - m^2}\delta, \quad \langle\theta\theta\rangle = \frac{1}{\Box - m^2}\delta. \end{gathered} \tag{18}$$

Physical states are defined as

$$Q_{\mathrm{B}}|\mathrm{phys}\rangle = 0. \tag{19}$$

If we introduce the field

$$\tilde{\theta}(p) \stackrel{\mathrm{d}}{\equiv} \theta(p) + \mathrm{i}m\chi(p) + \frac{m}{(p^3)^2}b(p), \tag{20}$$

we can find the following particle contents: $\{\varphi^i(p)\ (i = 1,2),\ \tilde{\theta}(p)\}$ are BRS singlets and physical; and $\{\chi(p), b(p), c(p), \bar{c}(p)\}$ make up a BRS quartet.

3. Massive tensor: linear theories

3.1. Massless tensor

A massless tensor field is described by the Lagrangian

$$L_0 = \frac{1}{2}h^{\mu\nu}\Lambda_{\mu\nu,\rho\sigma}h^{\rho\sigma} + L^{\alpha}_{\mathrm{GF+FP}} \tag{21}$$

with

$$\Lambda_{\mu\nu,\rho\sigma} \stackrel{\mathrm{d}}{\equiv} (\eta_{\mu\rho}\eta_{\nu\sigma} - \eta_{\mu\nu}\eta_{\rho\sigma})\Box - (\eta_{\mu\rho}\partial_\nu\partial_\sigma + \eta_{\nu\sigma}\partial_\mu\partial_\rho) + (\eta_{\rho\sigma}\partial_\mu\partial_\nu + \eta_{\mu\nu}\partial_\rho\partial_\sigma). \tag{22}$$

For the GF+FP Lagrangian, we choose

$$\begin{aligned} L^{\alpha}_{\mathrm{GF+FP}} &= b_\mu\left(\partial_\nu h^{\mu\nu} - \frac{1}{2}\partial^\mu h + \frac{\alpha}{2}b^\mu\right) + \mathrm{i}\bar{c}_\mu\Box c^\mu \\ &= -\mathrm{i}\delta\left[\bar{c}_\mu\left(\partial_\nu h^{\mu\nu} - \frac{1}{2}\partial^\mu h + \frac{\alpha}{2}b^\mu\right)\right], \end{aligned} \tag{23}$$

where α is the gauge parameter and h is defined by $h \stackrel{\rm d}{\equiv} h^\mu_\mu$. The theory is invariant under the BRS transformation

$$\delta h^{\mu\nu} = \partial^\mu c^\nu + \partial^\nu c^\mu, \qquad \delta \bar{c}_\mu = \mathrm{i} b_\mu. \tag{24}$$

For $\alpha = \frac{1}{2}$, we have simple forms of two-point functions:

$$\begin{aligned} \langle h^{\mu\nu} h^{\rho\sigma} \rangle &= \frac{1}{\Box} \frac{1}{2} \left(\eta^{\mu\rho}\eta^{\nu\sigma} + \eta^{\mu\sigma}\eta^{\nu\rho} - \eta^{\mu\nu}\eta^{\rho\sigma} \right) \delta, \\ \langle h^{\mu\nu} b^{\rho} \rangle = \frac{1}{\Box} \left(\eta^{\mu\rho}\partial^\nu + \eta^{\nu\rho}\partial^\mu \right) \delta, &\qquad \langle b^\mu b^\nu \rangle = 0, \qquad \langle c^\mu \bar{c}_\nu \rangle = -\mathrm{i}\frac{1}{\Box}\, \delta^\mu_\nu \delta. \end{aligned} \tag{25}$$

Physical states are defined by the condition

$$Q_{\rm B} |{\rm phys}\rangle = 0. \tag{26}$$

In the axial coordinate ($p^1 = p^2 = 0,\ p^3 > 0$) we define

$$\begin{aligned} \phi^1(p) \stackrel{\rm d}{\equiv} \frac{1}{2} \left[h^{11}(p) - h^{22}(p) \right], &\qquad \phi^2(p) \stackrel{\rm d}{\equiv} h^{12}(p), \\ \chi^0(p) \stackrel{\rm d}{\equiv} \frac{1}{2p^0} h^{00}(p), \quad \chi^1(p) \stackrel{\rm d}{\equiv} \frac{1}{p^0} h^{01}(p), &\quad \chi^2(p) \stackrel{\rm d}{\equiv} \frac{1}{p^0} h^{02}(p), \quad \chi^3(p) \stackrel{\rm d}{\equiv} \frac{1}{2p^3} h^{33}(p). \end{aligned} \tag{27}$$

Particle contents are then as follows: $\{\phi^i(p)\ (i = 1, 2)\}$ are BRS singlets and physical; $\{\chi^\mu(p), b_\mu(p), c^\mu(p), \bar{c}_\mu(p)\}$ form BRS quartets.

3.2. Massive tensor: naive model

Naive introduction of mass is carried out through the Lagrangian

$$L^a_m \left[h^{\mu\nu} \right] = \frac{1}{2} h^{\mu\nu} \Lambda_{\mu\nu,\rho\sigma} h^{\rho\sigma} - \frac{m^2}{2} \left(h^{\mu\nu} h_{\mu\nu} - a h^2 \right). \tag{28}$$

The parameter a has two choices of interest: $a = 1$ and $a = \frac{1}{2}$, corresponding to the PT model and the ASG one respectively. In the case of $a = 1$, the Lagrangian has the Fierz-Pauli type mass term, and gives the field equations

$$\begin{cases} \left(\Box - m^2 \right) h^{\mu\nu} = 0, \\ \partial_\nu h^{\mu\nu} = 0, \\ h = 0. \end{cases} \tag{29}$$

Therefore, this model does describe an ordinary massive tensor field of five degrees of freedom. The two-point functions

$$\begin{aligned} \langle h^{\mu\nu} h^{\rho\sigma} \rangle = \frac{1}{\Box - m^2} \Bigg\{ & \frac{1}{2} \left(\eta^{\mu\rho}\eta^{\nu\sigma} + \eta^{\mu\sigma}\eta^{\nu\rho} - \eta^{\mu\nu}\eta^{\rho\sigma} \right) \\ & - \frac{1}{2m^2} \left(\eta^{\mu\rho}\partial^\nu\partial^\sigma + \eta^{\mu\sigma}\partial^\nu\partial^\rho + \eta^{\nu\rho}\partial^\mu\partial^\sigma + \eta^{\nu\sigma}\partial^\mu\partial^\rho \right) \\ & + \frac{2}{3} \left(\frac{1}{2}\eta^{\mu\nu} + \frac{\partial^\mu\partial^\nu}{m^2} \right) \left(\frac{1}{2}\eta^{\rho\sigma} + \frac{\partial^\rho\partial^\sigma}{m^2} \right) \Bigg\} \delta \end{aligned} \tag{30}$$

show the fourth-order massless singularities. In the case of $a = \frac{1}{2}$, on the other hand, field equations reduce to

$$\begin{cases} \left(\Box - m^2 \right) h^{\mu\nu} = 0, \\ \partial_\nu h^{\mu\nu} - \frac{1}{2} \partial^\mu h = 0. \end{cases} \tag{31}$$

The number of physical degrees of freedom of this model is six; five coresoponds to a massive tensor, and one is to an additional scalar ghost field. The two-point functions

$$\langle h^{\mu\nu} h^{\rho\sigma} \rangle = \frac{1}{\Box - m^2} \left\{ \frac{1}{2} \left(\eta^{\mu\rho}\eta^{\nu\sigma} + \eta^{\mu\sigma}\eta^{\nu\rho} - \eta^{\mu\nu}\eta^{\rho\sigma} \right) \right.$$
$$\left. - \frac{1}{2m^2} \left(\eta^{\mu\rho}\partial^\nu\partial^\sigma + \eta^{\mu\sigma}\partial^\nu\partial^\rho + \eta^{\nu\rho}\partial^\mu\partial^\sigma + \eta^{\nu\sigma}\partial^\mu\partial^\rho \right) \right\} \delta \tag{32}$$

contain only the second-order massless singularities in this case.

3.3. ASG-type massive tensor: BRS model

In order to promote the ASG model to a BRS invariant one, introduce an auxiliary vector field θ^μ, replace $h^{\mu\nu}$ with the combination $h^{\mu\nu} - \frac{1}{m}(\partial^\mu\theta^\nu - \partial^\nu\theta^\mu)$ in the Lagrangian (28) with $a = \frac{1}{2}$, and append a certain GF+FP Lagrangian. The total Lagrangian is

$$L_{\mathrm{BRS}}^{a=\frac{1}{2}} = L_m^{a=\frac{1}{2}} \left[h^{\mu\nu} - \frac{1}{m} \left(\partial^\mu\theta^\nu + \partial^\nu\theta^\mu \right) \right] + L_{\mathrm{GF+FP}}^{m\alpha}$$
$$= L_m^{a=\frac{1}{2}} \left[h^{\mu\nu} \right] - 2m\theta_\mu \left(\partial_\nu h^{\mu\nu} - \frac{1}{2}\partial^\mu h \right) - \partial_\mu\theta_\nu\partial^\mu\theta^\nu + L_{\mathrm{GF+FP}}^{m\alpha}. \tag{33}$$

Following the BRS procedure for the massive vector in section 2.4, we adopt the following GF+FP Lagrangian of massive type:

$$L_{\mathrm{GF+FP}}^{m\alpha} = b_\mu \left(\partial_\nu h^{\mu\nu} - \frac{1}{2}\partial^\mu h - m\theta^\mu + \frac{\alpha}{2} b^\mu \right) + \mathrm{i}\bar{c}_\mu \left(\Box - m^2 \right) c^\mu$$
$$= -\mathrm{i}\delta \left[\bar{c}_\mu \left(\partial_\nu h^{\mu\nu} - \frac{1}{2}\partial^\mu h - m\theta^\mu + \frac{\alpha}{2} b^\mu \right) \right]. \tag{34}$$

The BRS transformation

$$\delta h^{\mu\nu} = \partial^\mu c^\nu + \partial^\nu c^\mu, \quad \delta\theta^\mu = mc^\mu, \quad \delta\bar{c}_\mu = \mathrm{i}b_\mu \tag{35}$$

keeps the system invariant. For $\alpha = \frac{1}{2}$, two-point functions are calculated as

$$\langle h^{\mu\nu} h^{\rho\sigma} \rangle = \frac{1}{\Box - m^2} \frac{1}{2} \left(\eta^{\mu\rho}\eta^{\nu\sigma} + \eta^{\mu\sigma}\eta^{\nu\rho} - \eta^{\mu\nu}\eta^{\rho\sigma} \right) \delta,$$
$$\langle h^{\mu\nu} b^\rho \rangle = \frac{1}{\Box - m^2} \left(\eta^{\mu\rho}\partial^\nu + \eta^{\nu\rho}\partial^\mu \right) \delta, \quad \langle b^\mu b^\nu \rangle = 0, \quad \langle c^\mu \bar{c}_\nu \rangle = -\mathrm{i}\frac{1}{\Box - m^2} \delta^\mu_\nu \delta, \tag{36}$$
$$\langle h^{\mu\nu} \theta^\rho \rangle = 0, \quad \langle b^\mu \theta^\nu \rangle = \frac{m}{\Box - m^2}\eta^{\mu\nu}\delta, \quad \langle \theta^\mu \theta^\nu \rangle = \frac{1}{2}\frac{1}{\Box - m^2}\eta^{\mu\nu}\delta.$$

Physical states are defined by

$$Q_{\mathrm{B}}|\mathrm{phys}\rangle = 0. \tag{37}$$

Let us introduce a field $\tilde{\theta}^\mu(p)$ as the combination

$$\tilde{\theta}^\mu(p) \stackrel{\mathrm{d}}{\equiv} \theta^\mu(p) + \mathrm{i}m\chi^\mu(p) + m\omega^{\mu\nu}b_\nu(p) \tag{38}$$

with the matrix $\omega^{\mu\nu}$ having the components

$$\begin{cases} \omega^{00} = \dfrac{1}{8(p^0)^2}, \quad \omega^{03} = \omega^{30} = \dfrac{1}{8p^0p^3}, \quad \omega^{33} = \dfrac{1}{8(p^3)^2}, \\ \omega^{11} = \omega^{22} = -\dfrac{1}{2(p^0)^2}, \quad \text{the others} = 0. \end{cases} \tag{39}$$

Then we find the following particle contents: $\{\phi^i(p)\ (i=1,2),\ \tilde{\theta}^\mu(p)\}$ are BRS singlets and physical; $\{\chi^\mu(p), b_\mu(p), c^\mu(p), \bar{c}_\mu(p)\}$ make up BRS quartets. Note that there remains a ghost $\tilde{\theta}^0(p)$ in the physical states. Nonlinear completion of this model including a possible mechanism of killing the ghost was reported at QTS-4 [6] (See also [5].).

3.4. PT-type massive tensor: BRS+Nakanishi model

We have seen in section 3.2 that the PT model of massive tensor shows fourth-order massless singularities in two-point functions. Those singularities cannot be removed by such simple application of the BRS procedure as done in section 3.3. So we invoke the Nakanishi procedure in addition to the BRS one. For the Lagrangian, we adopt the following form:

$$\begin{aligned} L^{a=1}_{\text{BRS+N}} &= L^{a=1}_m\left[h^{\mu\nu} - \frac{1}{m}\left(\partial^\mu\theta^\nu + \partial^\nu\theta^\mu\right)\right] + L^{m\alpha}_{\text{GF+FP}} + L^{\beta}_{\text{`GF'}} \\ &= L^{a=1}_m\left[h^{\mu\nu}\right] - 2m\theta_\mu\left(\partial_\nu h^{\mu\nu} - \partial^\mu h\right) - \frac{1}{2}\left(\partial^\mu\theta^\nu - \partial^\nu\theta^\mu\right)^2 + L^{m\alpha}_{\text{GF+FP}} + L^{\beta}_{\text{`GF'}}. \end{aligned} \tag{40}$$

Here the first and the second terms on the right side of the first line are from the BRS procedure, and the third term $L^{\beta}_{\text{`GF'}}$ represents the gauge-fixing-like term in the Nakanishi procedure. For $L^{\beta}_{\text{`GF'}}$, we also choose massive type of the following form:

$$L^{\beta}_{\text{`GF'}} = b\left(\partial_\mu\theta^\mu - \frac{m}{2}h + \frac{\beta}{2}b\right) \tag{41}$$

with the second parameter β. Assuming the Nakanishi-Lautrup field b is BRS invariant, the total Lagrangian is invariant under the BRS transformation

$$\delta h^{\mu\nu} = \partial^\mu c^\nu + \partial^\nu c^\mu, \quad \delta\theta^\mu = mc^\mu, \quad \delta\bar{c}_\mu = \mathrm{i}b_\mu, \quad \delta b = 0. \tag{42}$$

Two-point functions show simple pole structure for $\alpha = \beta = \frac{1}{2}$ as follows:

$$\begin{gathered} \langle h^{\mu\nu}h^{\rho\sigma}\rangle = \frac{1}{\Box - m^2}\frac{1}{2}\left(\eta^{\mu\rho}\eta^{\nu\sigma} + \eta^{\mu\sigma}\eta^{\nu\rho} - \eta^{\mu\nu}\eta^{\rho\sigma}\right)\delta, \\ \langle h^{\mu\nu}b^\rho\rangle = \frac{1}{\Box - m^2}\left(\eta^{\mu\rho}\partial^\nu + \eta^{\nu\rho}\partial^\mu\right)\delta, \quad \langle b^\mu b^\nu\rangle = 0, \quad \langle c^\mu\bar{c}_\nu\rangle = -\mathrm{i}\frac{1}{\Box - m^2}\delta^\mu_\nu\delta, \\ \langle h^{\mu\nu}\theta^\rho\rangle = 0, \quad \langle b^\mu\theta^\nu\rangle = \frac{m}{\Box - m^2}\eta^{\mu\nu}\delta, \quad \langle\theta^\mu\theta^\nu\rangle = \frac{1}{2}\frac{1}{\Box - m^2}\eta^{\mu\nu}\delta, \\ \langle h^{\mu\nu}b\rangle = -\frac{m}{\Box - m^2}\eta^{\mu\nu}\delta, \quad \langle\theta^\mu b\rangle = \frac{\partial^\mu}{\Box - m^2}\delta, \quad \langle b^\mu b\rangle = 0, \quad \langle bb\rangle = -\frac{6m^2}{\Box - m^2}\delta. \end{gathered} \tag{43}$$

The fourth-order massless singularities have been driven away indeed. Note that, as seen from the last equation of (43), the field b is a ghost for $m \neq 0$. This is the same situation as in the case of massive vector in section 2.3. Physical states are picked out by two conditions of the BRS type and the Nakanishi type:

$$Q_{\text{B}}|\text{phys}\rangle = 0, \quad b^{(+)}(x)|\text{phys}\rangle = 0. \tag{44}$$

In order to investigate the particle contents, we introduce the following quantities:

$$\varphi^1(p) \stackrel{\text{d}}{\equiv} \tilde{\theta}^1(p), \quad \varphi^2(p) \stackrel{\text{d}}{\equiv} \tilde{\theta}^2(p), \quad \chi(p) \stackrel{\text{d}}{\equiv} \frac{1}{p^3}\tilde{\theta}^3(p). \tag{45}$$

Journal of Physics: Conference Series **128** (2008) 012006 doi:10.1088/1742-6596/128/1/012006

For the massive case, we can introduce the combination

$$\tilde{\chi}(p) \stackrel{\mathrm{d}}{\equiv} \chi(p) + \mathrm{i}\,\frac{1}{6}\left(\frac{1}{m^2} + \frac{1}{2(p^3)^2}\right) b(p). \tag{46}$$

Particle contents are then: $\{\phi^i(p),\ \varphi^i(p)\ (i=1,2),\ \tilde{\chi}(p)\}$ are BRS singlets and physical; $b(p)$ is a BRS singlet but unphysical (*ghost*); $\{\chi^\mu(p), b_\mu(p), c^\mu(p), \bar{c}_\mu(p)\}$ constitute BRS quartets. For the massless case, on the other hand, we cannot define a field like $\tilde{\chi}$. In this case, we find the following particle contents: $\{\phi^i(p),\ \varphi^i(p)\ (i=1,2),\ b(p)\}$ are BRS singlets and physical; $\chi(p)$ is a BRS singlet but unphysical; $\{\chi^\mu(p), b_\mu(p), c^\mu(p), \bar{c}_\mu(p)\}$ form BRS quartets. Note again that in the massless case, b is physical but unobservable because it is zero-normed. From now on, we focus on the model described by the Lagrangian (40) with $\alpha = \beta = \frac{1}{2}$.

4. Massive tensor: nonlinear completion

4.1. Nonlinear BRS transformation

To study nonlinear theories we introduce the metric $g_{\mu\nu}$ and the tetrad $e_k{}^\mu$ through

$$g_{\mu\nu} \stackrel{\mathrm{d}}{\equiv} \eta_{\mu\nu} - \kappa h_{\mu\nu}, \qquad e_k{}^\mu e^{k\nu} = g^{\mu\nu} \tag{47}$$

with the gravitational constant κ. The linear BRS transformation (42) is extended to its nonlinear form:

$$\begin{cases} \delta e_k{}^\mu = \kappa\left(\partial_\rho c^\mu \cdot e_k{}^\rho - c^\rho \partial_\rho e_k{}^\mu\right), \\ \delta\theta^\mu = mc^\mu - \kappa c^\rho \partial_\rho \theta^\mu, \\ \delta c^\mu = -\kappa c^\rho \partial_\rho c^\mu, \\ \delta\bar{c}_\mu = \mathrm{i} b_\mu, \\ \delta b = -\kappa c^\rho \partial_\rho b. \end{cases} \tag{48}$$

Basic quantities invariant under the nonlinear BRS transformation can be constructed as

$$E_k{}^\mu \stackrel{\mathrm{d}}{\equiv} e_k{}^\mu - \frac{\kappa}{m} e_k{}^\rho \partial_\rho \theta^\mu, \tag{49}$$

$$G^{\mu\nu} \stackrel{\mathrm{d}}{\equiv} E_k{}^\mu E^{k\nu} = g^{\mu\nu} - \frac{\kappa}{m}\left(g^{\rho\mu}\partial_\rho\theta^\nu + g^{\rho\nu}\partial_\rho\theta^\mu\right) + \left(\frac{\kappa}{m}\right)^2 g^{\rho\sigma}\partial_\rho\theta^\mu\partial_\sigma\theta^\nu. \tag{50}$$

In fact they behave as scalars under the transformation (48):

$$\delta E_k{}^\mu = -\kappa c^\rho \partial_\rho E_k{}^\mu, \qquad \delta G^{\mu\nu} = -\kappa c^\rho \partial_\rho G^{\mu\nu}. \tag{51}$$

Possible Lagrangians are therefore of the form

$$L = \sqrt{-g}\, F\left(E_k{}^\mu, b\right), \tag{52}$$

where F is an arbitrary function. The action is indeed invariant, because such Lagrangian as (52) is transformed as

$$\delta L = -\kappa \partial_\mu\left(c^\mu L\right). \tag{53}$$

Journal of Physics: Conference Series **128** (2008) 012006 doi:10.1088/1742-6596/128/1/012006

4.2. Nonlinear Lagrangian

We require for the Lagrangian to be at most quadratic in $E_k{}^\mu$ *and* to reduce to $L^{a=1}_{\mathrm{BRS+N}}$ in the flat-spacetime limit ($\kappa \to 0$). Then we have the following form consisting of four terms:

$$L = \tilde{L}_m + \gamma \tilde{L}_{\mathrm{R}} + \tilde{L}^{\alpha}_{\mathrm{GF+FP}} + \tilde{L}^{\beta}_{\mathrm{GF'}}, \tag{54}$$

with an arbitrary real number γ. These terms are given by

$$\tilde{L}_m = \frac{1}{2\kappa^2}\sqrt{-g}\left\{R + \frac{m^2}{2}\left[6 - G^{\mu\nu}\eta_{\mu\nu} - \left(E_k{}^\mu \delta^k_\mu\right)^2 + 2E_k{}^\mu \delta^l_\mu E_l{}^\nu \delta^k_\nu\right]\right\}, \tag{55}$$

$$\tilde{L}_{\mathrm{R}} = \frac{m^2}{\kappa^2}\sqrt{-g}\left[-3 + 2E_k{}^\mu \delta^k_\mu - \frac{1}{2}\left(E_k{}^\mu \delta^k_\mu\right)^2 + \frac{1}{2}E_k{}^\mu \delta^l_\mu E_l{}^\nu \delta^k_\nu\right], \tag{56}$$

$$\begin{aligned}\tilde{L}^{\alpha}_{\mathrm{GF+FP}} &= -\mathrm{i}\delta\left[\bar{c}_\mu\left(\frac{1}{\kappa}\partial_\nu \tilde{g}^{\mu\nu} - m\theta^\mu + \frac{\alpha}{2}\eta^{\mu\nu}b_\nu\right)\right] \\ &= b_\mu\left(\frac{1}{\kappa}\partial_\nu \tilde{g}^{\mu\nu} - m\theta^\mu + \frac{\alpha}{2}\eta^{\mu\nu}b_\nu\right) + \mathrm{i}\bar{c}_\mu\left(\partial_\nu D^{\mu\nu}{}_\rho - m^2\delta^\mu_\rho\right)c^\rho + \mathrm{i}\kappa m\bar{c}_\mu c^\rho \partial_\rho \theta^\mu,\end{aligned} \tag{57}$$

and

$$\tilde{L}^{\beta}_{\mathrm{GF'}} = \sqrt{-g}\,b\left[\frac{m}{\kappa}\left(\delta^\mu_k - E_k{}^\mu\right)\delta^k_\mu + \frac{\beta}{2}b\right], \tag{58}$$

where we have used the definitions

$$\tilde{g}^{\mu\nu} \stackrel{\mathrm{d}}{\equiv} \sqrt{-g}\,g^{\mu\nu}, \tag{59}$$

$$D^{\mu\nu}{}_\rho \stackrel{\mathrm{d}}{\equiv} \tilde{g}^{\mu\sigma}\delta^\nu_\rho\partial_\sigma + \tilde{g}^{\nu\sigma}\delta^\mu_\rho\partial_\sigma - \tilde{g}^{\mu\nu}\partial_\rho - \left(\partial_\rho \tilde{g}^{\mu\nu}\right). \tag{60}$$

We can easily verify that the main part of the Lagrangian $\tilde{L}_m + \tilde{L}^{\alpha}_{\mathrm{GF+FP}} + \tilde{L}^{\beta}_{\mathrm{GF'}}$ goes to $L^{a=1}_{\mathrm{BRS+N}}$ and the redundant part $\tilde{L}_{\mathrm{R}}$ becomes null in the flat spacetime limit, $\kappa \to 0$.

4.3. Physical states

In the linear theory, physical states are picked out by the two conditions, the BRS-type one $Q_{\mathrm{B}}|\mathrm{phys}\rangle = 0$ and the Nakanishi-type one $b^{(+)}(x)|\mathrm{phys}\rangle = 0$, as stated in section 3.4. Going to the nonlinear theory, the BRS-type condition takes over the same form:

$$Q_{\mathrm{B}}|\mathrm{phys}\rangle = 0. \tag{61}$$

However, it is not an easy task to find a nonlinear version of the Nakanishi-type condition:

$$\text{“}b^{(+)}(x)\text{”}\,|\mathrm{phys}\rangle = 0. \tag{62}$$

The problem is how to define "$b^{(+)}(x)$" in the nonlinear case. In the linear case, $b(x)$ satisfies the free field equation $\left(\Box - m^2\right)b(x) = 0$. This fact allows to impose the physical-state condition of the Nakanishi type $b^{(+)}(x)|\mathrm{phys}\rangle = 0$. In the nonlinear case, however, $b(x)$ obeys some nonlinear equation. Setting up an auxiliary condition consistently in that case is still unsolved.

5. Summary and discussion

We have presented a possible nonlinear completion of massive gravity of the Fierz-Pauli type. Physical implications of this model are under study.

This model has the smooth massless ($m \to 0$) as well as the smooth flat-spacetime ($\kappa \to 0$) limits. In the flat-spacetime limit, it reduces to the BRS+Nakanishi extension of the PT (Fierz-Pauli) model.

Finding a nonlinear version of the Nakanishi-type physical-state condition is still unsettled.

Journal of Physics: Conference Series **128** (2008) 012006 doi:10.1088/1742-6596/128/1/012006

References

[1] Hamamoto S 1996 *Prog. Theor. Phys.* **95** 441
[2] Hamamoto S 1996 *Prog. Theor. Phys.* **95** 639
[3] Hamamoto S 1997 *Prog. Theor. Phys.* **97** 141
[4] Hamamoto S 1997 *Prog. Theor. Phys.* **97** 327
[5] Hamamoto S 2005 *Prog. Theor. Phys.* **114** 1261
[6] Hamamoto S 2006 *Quantum Theory and Symmetries IV* (Sofia: Heron Press) p 337
[7] Nakanishi N 1972 *Prog. Theor. Phys. Suppl.* no 51 1

Journal of Physics: Conference Series **128** (2008) 012007 doi:10.1088/1742-6596/128/1/012007

Noncanonical embedded rational map soliton in quantum SU(3) Skyrme model

D Jurčiukonis and E Norvaišas

Vilnius University Institute of Theoretical Physics and Astronomy, Goštauto 12, Vilnius 01108, Lithuania

E-mail: `darius@itpa.lt` and `norvaisas@itpa.lt`

Abstract. The model is considered in non canonical bases SU(3) $\supset$ SO(3) for the state vectors. The rational map ansatz are used to describe the soliton with topological number bigger than one. The canonical quantization generates five different momenta of inertia in Hamiltonian and quantum mass corrections, which can stabilize the quantum soliton solution. The explicit expressions of the Lagrangian and Hamiltonian are derived for this model.

1. Introduction

The Skyrme model was introduced as an effective theory of baryons [1]. Recently the topological soliton solutions – skyrmions find the application in quantum Hall effect, Bose-Einstein condensate and black hole physics. The general exact solutions of the model are unknown even in a classical case and the approximate solitonic solutions are under intensive consideration.

The original model was defined for a unitary field $U(\mathbf{x},t)$ that belongs to fundamental representation of the SU(2) group. Semiclassical quantization suggests that the skyrmion rotates as a "rigid body" [2]. Collective coordinates approach separate the variables which depend on the time and spatial coordinates. The structure of the ansatz which depend on spatial coordinates determine the solitonic solutions. Constructive realization of canonical quantization provides in Hamiltonian a term which may be interpreted as an effective pion mass term [3]. The extension of the model to SU(N) group [4] represents the common structure of the Skyrme Lagrangian.

The aim of this work is to discuss the group-theoretical aspects of the canonical quantization of the SU(3) Skyrme model in rational map antsatz approximation with baryon number $B \geq 2$. The ansatz is defined in noncanonical SU(3) $\supset$ SO(3) bases as SO(3) solitonic solution. The canonical quantization generates five different momenta of inertia. The proposed ansatz can be used to determine light nuclei as special skyrmions.

2. Noncanonical embedding of rational map soliton

The Skyrme model is a Lagrangian density for unitary field $U(\mathbf{x},t)$ that can be defined to the general representation of SU(3) group [5]. Here we consider the unitary field in fundamental representation (1,0) of SU(3) group. The chiraly symmetric Lagrangian density has the form:

$$\mathcal{L} = -\frac{f_\pi^2}{4}\mathrm{Tr}\{\mathbf{R}_\mu \mathbf{R}^\mu\} + \frac{1}{32e^2}\mathrm{Tr}\{[\mathbf{R}_\mu, \mathbf{R}_\nu][\mathbf{R}^\mu, \mathbf{R}^\nu]\}, \tag{1}$$

where the "right" and "left" chiral currents are defined as

$$R_\mu = (\partial_\mu U)\overset{+}{U} = \partial_\mu \alpha^i C_i^{(B)}(\alpha)\left\langle \;\left| J_{(B)}^{(1,1)} \right|\; \right\rangle, \tag{2}$$

$$L_\mu = \overset{+}{U}(\partial_\mu U) = \partial_\mu \alpha^i C_i'^{(B)}(\alpha)\left\langle \;\left| J_{(B)}^{(1,1)} \right|\; \right\rangle \tag{3}$$

and have the values on the SU(3) algebra. The f_π and e in (1) are the phenomenological parameters of the model. The functions $C_i^{(B)}(\alpha)$ and $C_i'^{(B)}(\alpha)$ depend on parameters α^i of the group. $J_{(B)}^{(1,1)}$ are the generators of the group.

The system of noncanonical SU(3) generator in terms of canonical generators $J_{(Z,I,M)}^{(1,1)}$ which are defined in [6] can be expressed as follows:

$$\begin{aligned}
J_{(1,1)} &= \sqrt{2}\left(J^{(1,1)}_{(\frac{1}{2},\frac{1}{2},\frac{1}{2})} - J^{(1,1)}_{(-\frac{1}{2},\frac{1}{2},\frac{1}{2})}\right), & J_{(1,0)} &= 2J^{(1,1)}_{(0,1,0)},\\
J_{(1,-1)} &= \sqrt{2}\left(J^{(1,1)}_{(-\frac{1}{2},\frac{1}{2},-\frac{1}{2})} + J^{(1,1)}_{(\frac{1}{2},\frac{1}{2},-\frac{1}{2})}\right), & J_{(2,2)} &= -2J^{(1,1)}_{(0,1,1)},\\
J_{(2,1)} &= -\sqrt{2}\left(J^{(1,1)}_{(\frac{1}{2},\frac{1}{2},\frac{1}{2})} + J^{(1,1)}_{(-\frac{1}{2},\frac{1}{2},\frac{1}{2})}\right), & J_{(2,0)} &= -2J^{(1,1)}_{(0,0,0)},\\
J_{(2,-1)} &= -\sqrt{2}\left(J^{(1,1)}_{(-\frac{1}{2},\frac{1}{2},-\frac{1}{2})} - J^{(1,1)}_{(\frac{1}{2},\frac{1}{2},-\frac{1}{2})}\right), & J_{(2,-2)} &= 2J^{(1,1)}_{(0,1,-1)}.
\end{aligned} \tag{4}$$

They satisfy the commutation relations

$$\left[J_{(L',M')}, J_{(L'',M'')}\right] = -2\sqrt{3}\begin{bmatrix} (1,1) & (1,1) & (1,1)_a \\ L' & L'' & L \end{bmatrix}\begin{bmatrix} L' & L'' & L \\ M' & M'' & M \end{bmatrix} J_{(L,M)}. \tag{5}$$

The coefficient on the rhs of (5) is SU(3) noncanonical isofactor and SO(3) Clebsch-Gordan coefficient. The state vectors for canonical bases SU(3) $\supset$ SU(2) and non-canonical bases SU(3) $\supset$ SO(3) in fundamental representation are equivalent. The rational map antsatz in SO(3) case we suggest as the matrix

$$\begin{aligned}
(U_R)_{a,a'} &= \left(\exp(2i\hat{n}_m J_{(m)} F(r))\right)_{a,a'} = 2\sin^2 F(-1)^a \hat{n}_{-a}\hat{n}_{a'}\\
&\quad + i\sqrt{2}\sin 2F\begin{bmatrix} 1 & 1 & 1 \\ a & u & a' \end{bmatrix}\hat{n}_u + \cos 2F\,\delta_{a,a'}.
\end{aligned} \tag{6}$$

The unit vector $\hat{\mathbf{n}}$ are defined in [7] in terms of a rational complex function $R(z)$ as

$$\hat{\mathbf{n}}_R = \frac{1}{1+|R|^2}\{2\mathrm{Re}(R), 2\mathrm{Im}(R), 1-|R|^2\}, \tag{7}$$

for every baryon number

$$B = \frac{1}{24\pi^2}\int \mathrm{d}^3 r\,\mathcal{I}(\theta,\varphi)\frac{F'(r)\sin^2 F}{r^2}. \tag{8}$$

The symbol $\mathcal{I}(\theta,\varphi)$ denotes the function

$$\mathcal{I}(\theta,\varphi) = \left(\frac{1+|z|^2}{1+|R|^2}\left|\frac{\mathrm{d}R}{\mathrm{d}z}\right|\right)^2. \tag{9}$$

which depends on angles θ, φ only.

3. Canonical quantization

The quantization of the model can be carried out by means of collective coordinates that separate the variables, which depend on the time and spatial coordinates

$$U(\vec{r}, q(t)) = A(q(t))U_R(\vec{r})A^{\dagger}(q(t)). \tag{10}$$

Here eight SU(3) group parameters $q^i(t)$ are quantum variables. The Skyrme Lagrangian is considered quantum mechanically *ab initio* in contrast to the conventional semiclassical quantization of the soliton as a rigid body. Generalized coordinates $q^i(t)$ and the corresponding velocities $\dot{q}^i\,(t)$ satisfy the following commutation relations

$$\left[\dot{q}^{\alpha}, q^{\beta}\right] = -if^{\alpha\beta}(q), \tag{11}$$

where $f^{\alpha\beta}(q)$ are functions of q only. Weyl ordering of the operators has been employed for arbitrary function $G(q)$

$$\partial_0 G(q) = \frac{1}{2}\left\{\dot{q}^{\alpha}, \frac{\partial}{\partial q^{\alpha}} G(q)\right\}. \tag{12}$$

The curly brackets denote an anticommutator. With thus choice of operator ordering (12) no further ordering ambiguity appears in Lagrangian or Hamiltonian.

After substitution of the ansatz (10) into the model Lagrangian density (1) and integration over spatial coordinates the Lagrangian has this form:

$$L = \frac{1}{2}\dot{q}^{\alpha} g_{\alpha\beta}(q, F)\dot{q}^{\beta} + a^0 \frac{1}{2}\left\{\dot{q}^{\alpha}, C'^{(2,0)}_{\alpha}(q)\right\} + \left[(\dot{q})^0 \text{ order terms}\right], \tag{13}$$

where the metric tensor is

$$g_{\alpha\beta}(q, F) = C'^{(L,M)}_{\alpha}(q) E_{(L,M)(L',M')}(F) C'^{(L',M')}_{\beta}(q), \tag{14}$$

and

$$E_{(L,M)(L',M')}(F) = -(-1)^M a_{L,M}(F)\delta_{L,L'}\delta_{M,-M'}. \tag{15}$$

Note that the exact expression of the coefficient a^0 is not important for the calculation of $g_{\alpha\beta}$. In this case we have five different quantum momenta of inertia – integrals beside the terms of the "physical" SU(3) Casimir operator:

$$a_{(1,0)}(F) = \frac{1}{e^3 f_\pi} \int d\theta d\varphi d\tilde{r}\tilde{r}^2 \sin^2 F \left(n_0^2 - 1\right) \left[1 + F'^2 + \frac{\mathcal{I}}{r^2} \sin^2 F\right]; \tag{16a}$$

$$a_{(1,1)}(F) = \frac{1}{2e^3 f_\pi} \int d\theta d\varphi d\tilde{r}\tilde{r}^2 \sin^2 F \left(n_0^2 + 1\right) \left[1 + F'^2 + \frac{\mathcal{I}}{r^2} \sin^2 F\right]; \tag{16b}$$

$$\begin{aligned} a_{(2,0)}(F) = \frac{1}{e^3 f_\pi} \int d\theta d\varphi d\tilde{r}\tilde{r}^2 \sin^2 F \left(n_0^2 - 1\right) \Big(\cos^2 F + n_0^2 \sin^2 F \\ - (n_0^2 - 4\cos^2 F + 2n_0^2 \cos 2F) F'^2 \\ + \frac{\mathcal{I}}{r^2} \sin^2 F (2\cos^2 F + n_0^2(4 - \cos 2F))\Big); \end{aligned} \tag{16c}$$

$$\begin{aligned} a_{(2,1)}(F) = \frac{1}{2e^3 f_\pi} \int d\theta d\varphi d\tilde{r}\tilde{r}^2 \sin^2 F \Big(3 + 2\cos 2F - 3n_0^2 + 4n_0^4 \sin^2 F \\ + (9 + 8\cos 2F - 3n_0^2 - 4n_0^4(1 + 2\cos 2F)) F'^2 \\ + \frac{\mathcal{I}}{r^2} \sin^2 F (9 + 4\cos 2F - 15n_0^2 + 4n_0^4(4 - \cos 2F))\Big); \end{aligned} \tag{16d}$$

$$\begin{aligned} a_{(2,2)}(F) = \frac{1}{4e^3 f_\pi} \int d\theta d\varphi d\tilde{r}\tilde{r}^2 \sin^2 F \Big(-3 - \cos 2F - 12n_0^2 \cos^2 F + 2n_0^4 \sin^2 F \\ - 2(3 + 2\cos 2F - 24n_0^2 \cos^2 F + n_0^4(1 + 2\cos 2F)) F'^2 \\ - \frac{2\mathcal{I}}{r^2} \sin^2 F (6 + \cos 2F - 12n_0^2 \cos^2 F - n_0^4(4 - \cos 2F))\Big); \end{aligned} \tag{16e}$$

which depend on chiral angle function $F(r)$, one component n_0 of rational map vector and function $\mathcal{I}(\theta, \varphi)$.

The canonical momenta are defined as

$$p_\beta = \frac{\partial L}{\partial \dot{q}^\beta} = \frac{1}{2} \{\dot{q}^\alpha, g_{\alpha\beta}\} + a^0 C_\beta^{\prime(2,0)}(q). \tag{17}$$

Is notable what the momenta have a term which do not contains velocity. The canonical momenta and conjugate coordinates satisfy the commutation relations $[p_\beta, q^\alpha] = -i\delta_{\alpha\beta}$. The commutation relations fix the explicit expressions of the functions

$$f^{\alpha\beta}(q) = (g_{\alpha\beta}(q))^{-1}. \tag{18}$$

It is possible to choose the parametrization on the SU(3) group manifold so that the eight operators

$$\hat{R}_{(LM)} = \frac{i}{2} \left\{p_\beta, C'^\beta_{(L,M)}(q)\right\} \tag{19}$$

are defined as the group generators satisfying the commutation relations (5). Here the functions $C'^\alpha_{(L,M)}(q)$ are dual to the functions defined in (3)

$$\sum_\alpha C'^\alpha_{(L,M)}(q) C'^{(L',M')}_\alpha(q) = \delta_{(L,M)(L',M')}, \tag{20a}$$

$$\sum_{L,M} C'^\alpha_{(L,M)}(q) C'^{(L,M)}_{\alpha'}(q) = \delta_{\alpha\alpha'}. \tag{20b}$$

The generators (19) act on the Wigner matrix of the SU(3) irreducible representation as right transformation generators:

$$\left[\hat{R}_{(LM)}, D^{(\lambda,\mu)}_{(\alpha_1 L_1 M_1)(\alpha_2 L_2 M_2)}(q)\right] = D^{(\lambda,\mu)}_{(\alpha_1 L_1 M_1)(\alpha_2' L_2' M_2')}(q) \left\langle \begin{matrix} (\lambda,\mu) \\ (\alpha_2' L_2' M_2') \end{matrix} \middle| J_{(LM)} \middle| \begin{matrix} (\lambda,\mu) \\ (\alpha_2 L_2 M_2) \end{matrix} \right\rangle. \tag{21}$$

The indices α_1 and α_2 label the multiplets of (L, M).

R. Sugano and collaborators [8] developed the q-number variational method to formulate a theory that has the consistency between the Lagrangian and the Hamiltonian formalisms on the curved space of generalized coordinates. In terms of (19) generators we construct the Hamiltonian of the model:

$$\begin{aligned} H &= \frac{1}{8}\left\{\dot{q}^{\alpha}, C_{\alpha}^{\prime(L_1,M_1)}(q)\right\}E_{(L_1M_1)(L_2M_2)}\left\{\dot{q}^{\beta}, C_{\beta}^{\prime(L_2,M_2')}(q)\right\} - V_3 + M_{\mathrm{cl}} \\ &= -\frac{1}{2}\hat{R}_{(L_1M_1)}E^{(L_1M_1)(L_2M_2)}\hat{R}_{(L_2M_2)} - \frac{2V_2}{a_{(2,0)}}\hat{R}_{(2,0)} - 2\left(\frac{V_2}{a_{(2,0)}}\right)^2 - V_3 + M_{\mathrm{cl}}. \end{aligned} \tag{22}$$

where $E^{(L_1M_1)(L_2M_2)}$ is inverse matrix of eq. (15), M_{cl} is classical mass. The functions V_2 and V_3 results from the trace containing three and four group generators respectively.

$$\begin{aligned} V_2 &= \int \mathrm{d}^3x \frac{f_\pi^2}{4}\left(\frac{(-1)^{M_2+M_1}}{a_{(L,M_2)}}\frac{\sqrt{2}}{\sqrt{3}}\sqrt{L^2+L+1}\begin{bmatrix} L & L & 2 \\ M_2 & -M_2' & M_1 \end{bmatrix}\left(D^L_{M_2',M_2}(\varkappa) - \overset{+}{D}{}^L_{M_2',M_2}(\varkappa)\right)\right. \\ &\quad \left. - \frac{1}{a_2}\left(D^2_{0,M_1}(\varkappa) - \overset{+}{D}{}^2_{0,M_1}(\varkappa)\right)\right) \\ &\quad + \frac{1}{4\mathrm{e}^2}(-1)^{M_1'+M_2}\frac{\sqrt{2\cdot 3}}{a_{(L,M_2)}}\frac{\sqrt{L^2+L+1}}{\sqrt{5-2L}}B_{m,m'}(\varkappa)\begin{bmatrix} 2 & 1 & 2 \\ M_1'' & m & M_1' \end{bmatrix}\begin{bmatrix} L & 1 & L \\ M_2'' & m' & M_2' \end{bmatrix} \\ &\quad \times \begin{bmatrix} L & L & 2 \\ M_2' & M_2''' & -M_1' \end{bmatrix}\left(\delta_{M_1M_1''}\delta_{M_2M_2''}\overset{+}{D}{}^L_{M_2''',-M_2'}(\varkappa) - \delta_{M_1M_1''}\delta_{M_2M_2''}\delta_{-M_2M_2'''}\right. \\ &\quad \left. - \delta_{M_1M_1''}\delta_{-M_2M_2''}D^L_{M_2'',M_2}(\varkappa) + \delta_{M_2M_2''}\delta_{-M_2M_2'''}\overset{+}{D}{}^2_{M_1'',M_1}(\varkappa)\right); \end{aligned} \tag{23}$$

$$\begin{aligned} V_3 &= \int \mathrm{d}^3x \frac{f_\pi^2}{4}\left(\frac{4(2L_1+1)(2L_2+1)}{a_{(L_1,M_1)}a_{(L_2,M_2)}}\begin{Bmatrix} L_1 & L_2 & k \\ 1 & 1 & 1 \end{Bmatrix}^2\begin{bmatrix} L_1 & L_2 & k \\ M_1 & M_2 & u \end{bmatrix}^2 D^k_{u,u}(\varkappa) + \frac{3}{a_0^2}\right. \\ &\quad + \frac{1}{a_2^2}\left(1 + \overset{+}{D}{}^2_{0,0}(\varkappa)\right) - \frac{4}{a_{(L,M)}}\left(\frac{1}{a_0}D^L_{M,M}(\varkappa) - (-1)^M\frac{1}{a_2}\frac{\sqrt{L^2+L+1}}{\sqrt{2\cdot 3}}\right. \\ &\quad \left. \times \begin{bmatrix} L & L & 2 \\ M & -M & 0 \end{bmatrix}D^L_{M,M}(\varkappa)\right) \\ &\quad - \frac{3}{2\mathrm{e}^2}\frac{(2L_1+1)(2L_2+1)}{\sqrt{(5-2L_1)(5-2L_2)}}(-1)^{M_1+M_2+u}B_{m,m'}(\varkappa)\begin{Bmatrix} L_1 & L_1 & k \\ 1 & 1 & 1 \end{Bmatrix}\begin{Bmatrix} L_2 & L_2 & k \\ 1 & 1 & 1 \end{Bmatrix} \\ &\quad \times \begin{bmatrix} L_1 & 1 & L_1 \\ M_1' & m & M_1'' \end{bmatrix}\begin{bmatrix} L_2 & 1 & L_2 \\ M_2' & m' & M_2'' \end{bmatrix}\begin{bmatrix} L_1 & L_1 & k \\ M_1 & -M_1'' & u \end{bmatrix}\begin{bmatrix} L_2 & L_2 & k \\ -M_2 & M_2'' & u \end{bmatrix} \\ &\quad \times \left(\frac{1}{a_{(L_1,M_1)}a_{(L_2,M_2)}}\left(\delta_{M_2M_2'}D^{L_1}_{M_1',M_1}(\varkappa)\left(1-(-1)^k\right) - \delta_{M_1M_1'}\delta_{M_2M_2'}\right.\right. \\ &\quad \left.\left. - D^{L_1}_{M_1',M_1}(\varkappa)D^{L_2}_{M_2',M_2}(\varkappa)\right) + \frac{1}{a_{(L_1,M_1')}a_{(L_2,M_2)}}\delta_{M_2M_2'}D^{L_1}_{M_1',M_1}(\varkappa)\left(1+(-1)^k\right)\right). \end{aligned} \tag{24}$$

where $B_{m,m'}(\varkappa)$ are:

$$B_{m,m'}(\varkappa) = 8(-1)^{m+m'}\hat{n}_{-m}\hat{n}_{-m'}\left(\frac{1}{r^2}\mathcal{I}\sin^2 F - F'^2\right) - (-1)^m\delta_{m,-m'}\frac{8}{r^2}\mathcal{I}\sin^2 F. \tag{25}$$

We define the state vectors as the complex conjugate Wigner matrix elements of the (Λ, Θ) representation depending on eight quantum variables q^α:

$$\left| \begin{matrix} (\Lambda, \Theta) \\ \alpha, S, N; \beta, S', N' \end{matrix} \right\rangle = \sqrt{\dim(\Lambda, \Theta)} D^{*(\Lambda,\Theta)}_{(\alpha,S,N)(\beta,S',N')}(q) \left|0\right\rangle , \tag{26}$$

The indices α and β label the multiplets of the SO(3) group. $|0\rangle$ denotes the vacuum state. Because of five different moments of inertia the vectors (26) are not the eigenstates of the Hamiltonian (22). The action of the Hamiltonian on vectors (26) following (21) can be expressed in terms of the moments of inertia $a_{(L,M)}$ and the SU(3) group Clebsch-Gordan coefficients.

4. Conclusions

We considered a new rational map approximation ansatz for Skyrme model which is noncanonical embedded $SU(3) \supset SO(3)$ soliton with baryon number $B \geq 2$. The canonical quantization leads to five different quantum momenta of inertia in Hamiltonian and quantum corrections V_2 and V_3. The state vectors defined as SU(3) group representation (Λ, Θ) matrix which depend on quantum variables q. Because the momenta $a_{(2,1)}$ and $a_{(1,1)}$ are not equivalent the state vectors (26), they are not eigenstate vectors of the Hamiltonian for the higher representations. The mixing is small but to find eigenstates we must diagonolize the Hamiltonian matrix in every (Λ, Θ) representation. For baryon number $B = 1$ case $\hat{n} = \hat{x}$ and we get soliton with two different momenta of inertia which was considered in [6]. The SU(2) rational map ansatzes are not spherically symmetric and "rotation" in SU(3) manifold leads to five momenta of inertia in Hamiltonian.

References

[1] Skyrme T H R 1961 *Proc. Roy. Soc.* A **260**, 127
[2] Adkins G S, Nappi C R, Witten E 1983 *Nucl. Phys.* B **228**, 552
[3] Acus A, Norvaišas E, Riska D O 1998 *Phys. Rev.* C **57**, 2597
[4] Walliser H 1992 *Nucl. Phys.* A **548**, 649
[5] Jurčiukonis D, Norvaišas E, Riska D O 2005 *J. Math. Phys.* **46**, 072103
[6] Jurčiukonis D, Norvaišas E, 2007 *J. Math. Phys.* **48**, 052101
[7] Houghton C, Manton N and Sutcliff P 1998 *Nucl. Phys.* B **510**, 507
[8] Ohtami T, Sugano R 1973 *Prog. Theor. Phys.* **50**, 1715

Moyal-Weyl Star product and quasiconformal mapping relationship

M. Kachkachi

Département de Mathématiques & Informatique Faculté des Sciences & Techniques, Université Hassan 1er, Morocco

E-mail: mkachkachi@caramail.com

Abstract. The deformation quantization procedure can introduce the quantum effect by including the deformation of complex structures on a bi-dimensional Riemann surface Σ without boundary. This is the platform of bi-dimensional conformal models like string theory and two-dimensional gravity. Then, the star product can become a quasiconformal mapping on this surface. This makes the road toward a Teichmuller theory of the Moyal-Weyl star product through the quasiconformal mappings as lifting of universal covering surfaces of any Riemann surface.

1. Introduction

Deformation quantization was born with the paper [1] and grew out of the work of Weyl, Moyal and Vey. It has become a large research area covering several algebraic theories like the formal deformation of associative algebras and the more recent theory of operades, as well as geometric theories like the theory of symplectic and (more generally) Poisson manifolds, and physical theories like string theory and non commutative gauge theory.

The theory of deformation quantization has culminated with the recent work of Kontsevitch [7] proving that every Poisson manifold admits a non-trivial deformation quantization.

Two-dimensional conformal field theories on Riemann surfaces without boundaries are relevant models in string theory [11]. The dependence of these bidimensional conformal models on the background geometry turns out to be useful for the construction of effective actions for two-dimensional gravity [10].This geometrical dependence is well exhibited using the Beltrami parametrization of the bidimensional world sheet metric of the bosonic string [8].

In the Riemannian manifold approach, Beltrami coefficients parametrize conformal classes of the metrics [2]. However, in the Riemannian surface formalism, they parametrize complex structures of the Riemann surface [8] and satisfy the ellipticity condition:

$$Sup_{\Sigma}\left|\mu\right| \prec 1,\ \ \mu \in C^{\infty}\left(\Sigma\right) \tag{1}$$

where, Σ is a Riemann surface.

2. Beltrami parametrization's formalism

2.1. Riemann Surface

A Riemann surface Σ of genus $g \geq 1$ is a connected topological manifold of real dimension two which is equipped with a reference complex structure say C_0. this is a maximal atlas of local

Journal of Physics: Conference Series **128** (2008) 012008 doi:10.1088/1742-6596/128/1/012008

coordinates with a holomorphic change of coordinates:

$$(z, \overline{z}) \mapsto \left(z' = z'(z), \overline{z'} = \overline{z'}(\overline{z})\right) \tag{2}$$

The natural tangent space basis is $\partial \equiv \frac{\partial}{\partial z}, \frac{\partial}{\partial \overline{z}}$.

Then, if the coordinates z and z' are related by a holomorphic (local) diffeomorphism they are said to belong to the same complex structure. however, if the transition functions $z' = f(z, \overline{z})$ are not holomorphic, the two complex structures (z) and $(\overline{z})$ belong to different atlas.

2.2. Quasiconformal mapping

Now, let us consider a Belrami differential $\mu(z, \overline{z})$ that is $C^{\infty}(\Sigma)$ and satisfies

$$|\mu(z, \overline{z})| \leq 1. \tag{3}$$

The set of Beltrami differentials on the Riemann surface Σ has a structure of topologically contractible complex Banach manifold. It parametrizes the set of all complex structures on Σ. Indeed, to any Beltrami differential μ there is associated a conformal structure C_{μ} on Σ whose generic coodinate $Z(z, \overline{z})$ is a C^{∞}-diffeomorphism of the reference coodinates and satisfy the Beltrami equation:

$$\left(\overline{\partial} - \mu\partial\right) Z = 0. \tag{4}$$

The diffeomorphism

$$z \to Z(z, \overline{z}), \ \mu \neq 0 \tag{5}$$

is called a quasi-conformal transformation, that one can verify, becomes conformal for $\mu = 0$. It can be seen as the transition function from the reference complex structure C_0 to the complex structure C_{μ} $\left(\mu = \frac{\overline{\partial} Z}{\partial Z}\right)$.

On the other hand, let us consider a sense-preserving diffeomorphism f.

The function

$$\partial_{\alpha} f = \partial f + e^{-2i\alpha} \overline{\partial} f \tag{6}$$

is its derivative in the direction α.

The dilatation quotient of the diffeomorphism f is finite and is defined by

$$D_f = \frac{\max_{\alpha} |\partial_{\alpha} f|}{\min_{\alpha} |\partial_{\alpha} f} \tag{7}$$

where

$$\max_{\alpha} |\partial_{\alpha} f| = |\partial f| + |\overline{\partial} f| \tag{8}$$

and

$$\min_{\alpha} |\partial_{\alpha} f = |\partial f| - |\overline{\partial} f|. \tag{9}$$

2.3. Theorem

Let $f : D \to D'$ be a sense-preserving diffeomorphism satisfying the condition

$$D_f(z) \leq K \tag{10}$$

for every $z \in D$ (a domain of Σ). Then, f is a K-quasiconformal mapping.

The jacobian of f :

$$J_f = |\partial f|^2 - \left|\overline{\partial} f\right|^2 \tag{11}$$

is positive.

Then, we can define the quotient

$$\mu(z,\overline{z}) = \frac{\overline{\partial} f}{\partial f} \tag{12}$$

Since f is continuous, the function μ is Borel-measurable and then we have

$$|\mu(z,\overline{z})| \leq \frac{K-1}{K+1} \prec 1. \tag{13}$$

Then, a sense-preserving diffeomorphism f satisfying the Beltrami equation

$$\left(\overline{\partial} - \mu\partial\right) f = 0 \tag{14}$$

is a quasiconformal mapping.

3. Moyal Star product on a Riemann surface

3.1. Symplectic structure

An even dimensional manifold M^{2n} endowed with a closed 2-form ω; $d\omega = 0$ is called a symplectic manifold and the structure given by ω si a symplectic structure.

Locally we have

$$\omega = \omega_{ij} dx^i dx^j, \quad i,j = 1,2,...,2n \tag{15}$$

where $\det(\omega_{ij}) \neq 0$ and $(x^i)_{i=1,2,...2n}$ is a local coordinates system on M^{2n}.
The Poisson bracket of two observables f and g is defined by:

$$\{f,g\} = \omega^{ij}\partial_i f \partial_j g \tag{16}$$

Consider the following Beltrami equations

$$\overline{\partial} f = \mu_f \, \partial f \tag{17}$$

$$\overline{\partial} g = \mu_g \, \partial g. \tag{18}$$

Then, the P. B. is given by

$$\{f,g\}_{(z,\overline{z})} = \partial f \overline{\partial} g - \overline{\partial} f \partial g \tag{19}$$

One can see that the P. B. vanishes at the region $\mu_f = \mu_g$.

3.2. Moyal-Weyl Star product on Σ

Let f and g be two functions defined on a Riemann surface Σ which is equipped with a reference complex strcuture that is parametrized by a system of coordinates $(z,\overline{z})$.

The star product of g and f is a function:

$$w(z,\overline{z}) = f(z,\overline{z}) * g(z,\overline{z}) \tag{20}$$

$$\overline{w}(z,\overline{z}) = \overline{f}(z,\overline{z}) * \overline{g}(z,\overline{z}) \tag{21}$$

such that

$$* \equiv \exp\left[i\hbar\left(\overleftarrow{\partial}\overleftarrow{\overline{\partial}} - \overleftarrow{\overline{\partial}}\overleftarrow{\partial}\right)\right]. \tag{22}$$

The associativity of the star product is satisfied while it does not commute:

Journal of Physics: Conference Series **128** (2008) 012008 doi:10.1088/1742-6596/128/1/012008

$$(f_1 * f_2 * f_3) = f_1 * (f_2 * f_3) \tag{23}$$

$$f_1 * f_2 * f_3 \neq f_2 * f_3 * f_1 \tag{24}$$

Here we choose the operator of the star product to generate the Poisson bracket in the canonical form.

Usually, the star product is interpreted as an expansion of power series of $\hbar$. Our star product may be expanded by $\hbar$ or by the Beltrami coefficients:

$$w = w^{(0)} + \hbar w^{(1)} + \hbar^2 w^{(2)} + ... \tag{25}$$

$$= \widetilde{w}^{(0)} + \mu_f \; \widetilde{w_f}^{(1)} + \mu_g \widetilde{w_g}^{(1)} + \mu_f^2 \widetilde{w_f}^{(2)} + \mu_g^2 \widetilde{w_g}^{(2)} + \mu_f \mu_g \widetilde{w_{fg}}^{(2)} + ... \tag{26}$$

and

$$\mu_w = \mu_w^{(0)} + \hbar \mu_w^{(1)} + \hbar^2 \mu_w^{(2)} + ... \tag{27}$$

Hence, the star product is given as a superposition of $w^{(n)}$ or of that $\widetilde{w}^{(n)}$ $(n \in \mathbb{Z},\ 0 \leq n \leq +\infty)$.

This indicates a deep relationship between the Beltrami coefficients and $\hbar$ in the star product.

4. Some quasiconformal mapping's star product

The Moyal-Weyl star product can become a quasiconformal mapping. Indeed, let us consider the affine mappings defined by:

$$f(z, \overline{z}) = a_1 z + b_1 \overline{z} + c_1 \tag{28}$$

$$g(z, \overline{z}) = a_2 z + b_2 \overline{z} + c_2 \tag{29}$$

$$a_i,\ b_i \ \in \mathcal{C}$$

$$|a_i| \succ |b_i|, \quad i = 1, 2$$

Then, the star product for these two functions is given by:

$$w = f * g \tag{30}$$

$$= fg + i\hbar (a_1 b_2 - a_2 b_1) \tag{31}$$

$$= fg - i\hbar (\mu_f - \mu_g) a_1 a_2 \tag{32}$$

where the criterion for the quasiconformal mapping restricts the Beltrami differentials to satisfy the following conditions

$$0 \leq |\mu_f| \leq 1, \quad 0 \leq |\mu_g| \leq 1. \tag{33}$$

Moreover, one can verify that w is convergent for $\mathcal{C}^2 = \mathcal{C}_{\mu_f} \otimes \mathcal{C}_{\mu_g}$.

Now, using the Moyal-Weyl star product, the deformation quantization can be introduced as a phase factor to the normal product of two diffeomorphisms defined on the Riemann surface Σ. Indeed, let us consider the two functions:

$$f(z, \overline{z}) = e^{i\alpha_1 z} \, e^{i\beta_1 \overline{z}} \tag{34}$$

$$g(z, \overline{z}) = e^{i\alpha_2 z} \, e^{i\beta_2 \overline{z}} \tag{35}$$

$$\alpha_i,\ \beta_i \ \in \ \mathcal{C}, \ \ \alpha_i \neq 0 \tag{36}$$

Journal of Physics: Conference Series **128** (2008) 012008 doi:10.1088/1742-6596/128/1/012008

Then, the first terms of the power series of the star product in terms of $\hbar$ are given by:

$$w^{(0)} = fg = e^{i(\alpha_1+\alpha_2)z}\ e^{i(\beta_1+\beta_2)\overline{z}} \tag{37}$$

$$\begin{aligned} w^{(1)} &= i\{f,g\} = -i(\alpha_1\beta_2 - \alpha_2\beta_1) \\ &= i(\mu_f - \mu_g) \end{aligned}$$

and then

$$w = e^{-i\hbar(\alpha_1\beta_2-\alpha_2\beta_1)} fg \tag{38}$$

with

$$\mu_w = \frac{\alpha_1\mu_f + \alpha_2\mu_g}{\alpha_1+\alpha_2} \tag{39}$$

Hence, the effect of the deformation quantization is introduced as the phase factor to the product fg. w is absolutly convergent at $0 \leq \hbar \leq \infty$.

More generally, let us consider a series of quasiconformal mappings:

$$f_j = e^{i\alpha_j z} e^{i\beta_j \overline{z}}, \quad j = 1, 2, ... \tag{40}$$

Then, the star product is given by

$$w = f_1 * f_2 * ... * f_n \tag{41}$$

where the Beltrami equations for such quasiconformal mapping are given by

$$\overline{\partial} f_j = \mu_j \partial f_j \tag{42}$$

Then, one can find that the Beltrami differential of the star product is

$$\mu_w = \frac{\sum_{j=1}^n \beta_j}{\sum_{j=1}^n \alpha_j} \tag{43}$$

and this later is holomorphic with respect to these Beltrami differential:

$$\frac{\partial w}{\partial \mu_j} = 0 \tag{44}$$

It is an iteresting fact that this effect disappears and the P. B. vanishes identically when this P.B. is defined on the same complex structure $C_{\mu_f=\mu_g}$.

In another point of view, if the star product is considered as a functional of μ_f and μ_g, it can be expressed with the help of the Cauchy theorem of the case of several complex variables as follows:

$$w(\mu_f, \mu_g) = \int_{C_1} \frac{d\eta_1}{2i\pi} \int_{C_2} \frac{d\eta_2}{2i\pi} \frac{w(\eta_1, \eta_2)}{(\eta_1 - \mu_f)(\eta_2 - \mu_g)} \tag{45}$$

The C_i are the appropriate integration paths inside $\mathcal{C}$.

Hence, we can write it as a functional of the Beltrami differentials as follows:

$$w(\mu_{f_1}, \mu_{f_2}, ..., \mu_{f_n}) = \prod_{j=1}^n \left(\int_{C_j} \frac{d\eta_j}{2\pi i} \right) \frac{w(\eta_1, \eta_2, ...\eta_n)}{(\eta_1 - \mu_{f_1})...(\eta_n - \mu_{f_n})} \tag{46}$$

and

$$\frac{\partial^{m_1+...+m_n} w(\mu_1, ..., \mu_n)}{\partial^{m_1}\mu_{f_1}\partial^{m_2}\mu_{f_2}...\partial^{m_n}\mu_{f_n}} = (m_1!...m_n!) \prod_{j=1}^n \left(\int_{C_j} \frac{d\eta_j}{2\pi i} \right) \frac{w(\eta_1, \eta_2, ...\eta_n)}{(\eta_1 - \mu_{f_1})^{m_1+1} ... (\eta_n - \mu_{f_n})^{m_n+1}}. \tag{47}$$

5. Conclusion.

It is an interesting fact that the deformation of complex structures disappears when the poisson bracket is defined on the same complex structure.

This can shed some ligth on the deep relationship between the deformation of complex structures and the deformation quantization on a bi-dimensional Riemann surface.

Acknowledgments

I would like to thank the organizing committee of the International Symposium on Quantum theory and Symmetries "QTS5" that was held in Valladolid for their invitation, their hospitality and their financial support.

References

[1] Bayen F, Flato M, Fronsdal C, Lichnerowicz A and Sternheimer D 1978 *Ann. Phys.* **111** 61
[2] Becchi C 1988 *Nucl. Phys.* B **304** 513
[3] Kachkachi M 2000 *Phys. Lett.* B **490** 242
[4] Kachkachi M 2001 *J. Geom. Phys.* **40** 56
[5] Kachkachi M and Kessabi M 2001 *J. Phys. A: Math. Gen.* **34** 6157
[6] Kachkachi M 1997 *Preprint* hep-th/9707140
[7] Kontsevitch M 1997 *Preprint* q-algebra/9709040
[8] Lazzarini S 1990 Doctoral thesis, LAPP Annecy-Le-Vieux, France
[9] Ohasaku T 2006 *Preprint* math-ph/0610032
[10] Zucchini R 1993 *Comm. Math. Phys.* **152** 269
[11] Witten E 1987 *Superstring Theory* vol 2 (Cambridge: Cambridge University Press)

Journal of Physics: Conference Series **128** (2008) 012009 doi:10.1088/1742-6596/128/1/012009

An analytical vertex operator

Rémi Léandre

Institut de Mathématiques de Bourgogne. Universié de Bourgogne. 21000. Dijon. FRANCE

E-mail: `Remi.leandre@u-bourgogne.fr`

Abstract. We give an analyical meaning to a vertex operator by using an analog of the Hida Fock space

1. Introduction

Vertex operator algebras play a big role in conformal field theory as well as in string theory ([1]). We refer to the book of Kac ([2]) for an introduction. We are interested in this paper by the simplest example.

Let us consider the space of polynomial $C(x_1, .., x_n, ...)$ depending of an infinite number of variables. We consider for $n > 0$ the operator $a_n = \frac{\partial}{\partial x_n}$ and for $n \leq 0$ $a_n = nx_n$ the multiplication operator by nx_n. We introduce the field

$$a(z) = \sum a_n z^{-n-1} \tag{1}$$

The normalized vertex operator is given by [2], chapter 5.2:

$$\Gamma(z) = \exp[-\sum x_n z^n] \exp[-\sum z^{-n} 1/n \frac{\partial}{\partial x_n}] \tag{2}$$

It is a formal series of operators acting on the polynomial algebra:

$$\Gamma(z) = \sum A_r z^{-r-1} \tag{3}$$

A_r is clearly combinatorially defined because we consider finite sums in the polynomial algebra $C(x_1, .., x_n, ..)$. Our goal is to do a suitable completion $L_{\infty-}$ of $C(x_1, .., x_n, ..)$ such that each A_r acts continuously on $L_{\infty-}$.

This follows the works of Léandre-Rogers ([3]), Léandre-Ouerdiane ([4]), Léandre ([5], [6], [7], [8], [9], [10], [11], [12],[13]) where the tools of white-noise analysis ([2]) were used outside their original context. But instead of using Hida weights as in [14] for instance, our weights are closely related to the weights of [4].

2. Statement of the result

We consider $A = (i_1 \leq i_2 \leq .. \leq i_{|A|})$ where $i_1 > 0$. We put

$$\|A\| = \sum_{i_j \in A} |i_j| \tag{4}$$

Journal of Physics: Conference Series **128** (2008) 012009 doi:10.1088/1742-6596/128/1/012009

We have clearly $|A| \leq \|A\|$. We put

$$x^A = \prod_{i_j \in A} x_{i_j} \tag{5}$$

We introduce the space L_k of series $X = \sum \lambda_A x^A$ such that

$$\|X\|_k^2 = \sum (\|A\|!)^k |\lambda_A|^2 < \infty \tag{6}$$

where $k > 0$. The Hida space $L_{\infty-}$ is the intersection of all L_k, $k > 0$. We have:

TheoremA_r *is continuous on* $L_{\infty-}$.

This means that for all k, there exists $k'(r)$ and $C(r,k)$ such that

$$\|A_r X\|_k \leq C(r,k)\|X\|_{k'(r)} \tag{7}$$

3. Proof of the theorem

We remark if $\cup$ denotes the concatenation of multi-indices that

$$\|A \cup B\|! \leq C\|A\|!^k \|B\|!^k \tag{8}$$

for any $k > 1$.

Moreover

$$A_r X = \sum \mu_B x^B \tag{9}$$

where

$$|\mu_B| \leq \sum_{I \subseteq A; (A-I)\cup J = B; \|J\| - \|I\| = -r-1} |\lambda_A| \|A\|! \tag{10}$$

There are at most $(\|B\|!)^C$ terms in the sum. This comes from the fact that

$$\|A\| - \|I\| + \|J\| = \|B\| \tag{11}$$

and therefore in the sum

$$\|A\| = \|B\| + r + 1 \tag{12}$$

and from the Stirling formula. By Jensen inequality, we deduce that

$$|\mu_B|^2 \leq (\|B\|!)^C \sum |\lambda_A|^2 (\|A\|!)^{k'} \tag{13}$$

where we sum on the same set. By using (8), we deduce that

$$|\mu_b|^2 (\|B\|!)^k \leq C \sum |\lambda_A|^2 (\|J\|!)^{k"} (\|A\|!)^{k"} \tag{14}$$

where we sum on the same set. But

$$\|J\| = \|I\| - r - 1 \leq \|A\| + |r| + 1 \tag{15}$$

Therefore

$$|\mu_B|^2 (\|B\|!)^k \leq C \sum |\lambda_A|^2 (\|A\|!)^{k_1} = A_B \tag{16}$$

Due to (11) in

$$\sum_B |\mu_B|^2 (\|B\|!)^k \leq \sum_B A_B \tag{17}$$

there at most $(C\|A\|)^{C\|A\|}$ the same $|\lambda_A|^2 (\|A\|!)^{k_1}$ which appear. By the Stirling formula, we deduce that

$$\sum_B |\mu_B|^2 (\|B\|!)^k \leq C \sum |\lambda_A|^2 (\|A\|!)^{k_2} \tag{18}$$

Therefore the result.

References

[1] Di Francesco P, Mathieu P, and Sénéchal D 1997 *Conformal field theory.* (Berlin: Springer).
[2] Kac V 1998 *Vertex Algebras for Beginners.*(Providence: A.M.S.)
[3] Léandre R and Rogers A 2006 *J.Phys. A: Math. Gen.* **39** 11929
[4] Léandre R and Ouerdiane H 2005 *Stochastic Analysis, Classical and Quantum* ed T Hida (Singapore: World Scientific) pp 72–86
[5] Léandre R 2003 *I.D.A.Q.P.* **6** 505
[6] Léandre R 2004 *Stochastic analysis and Mathematical Physics* ed R Rebolledo et al (Singapore: World Scientific) pp 115–129
[7] Léandre R 2005 *Infinite Dimensional Harmonic Analysis III.* ed H Heyer et al (Singapore: World Scientific) pp 177–187
[8] Léandre R 2005 *XIVth Int. Con. Math. Physics* ed J C Zambrini (Singapore: World Scientific) pp 493–498
[9] Léandre R 2006 *Encyclopedia of Math. Physics* ed J P Françoise et al (Oxford: Elsevier) pp 8–12
[10] Léandre R 2006 *Quantum Probability* ed M Bozejko (Warsaw: Banach Center Publications) vol 73 p. 299
[11] Léandre R 2007 *Geometric Aspects of Integrable Systems*, ed F Calogero and al *S.I.G.M.A.* vol. 3 p. 27
[12] Léandre R 2006 *Fedosov quantization in white-noise analysis* Preprint 484 I.M.B. Univ. Dijon.
[13] Léandre R 2007 it Deformation quantization in infinite dimensional analysis To appear *Festchrift in honour of H.v. Weizsaecker* ed M Scheutzow
[14] Hida T, Kuo H H, Potthoff J and Streit L 1993 *White noise: an infinite dimensional Calculus* (Dordrecht: Kluwer)

Journal of Physics: Conference Series **128** (2008) 012010 doi:10.1088/1742-6596/128/1/012010

The Casimir effect for a massless scalar field in the n dimensional Einstein universe with Dirichlet boundary conditions on a sphere[1]

Patrick Moylan[2]

Institute of Nuclear Research and Nuclear Energy, Bulgarian Academy of Sciences, 72 Tsarigradsko Chaussee, 1784 Sofia, Bulgaria

E-mail: pjm11@psu.edu

Abstract. We use the representation theory of $SO(2,n)$ to determine the renormalized vacuum energy for a massless scalar field in the n-dimensional Einstein universe subject to Dirichlet boundary conditions on a sphere of maximum radius. The problem is an exactly solvable one. This is in remarkable contrast to the analogous problem in flat n dimensional Minkowski space where, except for the lowest dimensional case (n = 2), there is no known exactly solvable method of solution for any radius of the spherical boundary. For $n = 4$ our results agree with those of Bayen and Özcan, *Class. Quant. Grav.*, **10** (1993) L115-L121. We use our results to obtain some qualitative information about the Casimir effect for spherical boundaries of smaller radii, and we comment on how one may apply these results to obtain information about the corresponding problem in Minkowski space.

1. The conformal group in n dimensions and preliminaries

Consider the quadratic form $Q(x)$ defined on $\mathbb{R}^{n+2}$ by $Q(x) = x_{-1}^2 + x_0^2 - x_1^2 - x_2^2 - \ldots - x_n^2$ where $x = (x_{-1}, x_0, x_1, \ldots, x_n) \in \mathbb{R}^{n+2}$. Thus $\mathbb{R}^{n+2}$ equipped with the metric defined by $Q(x)$ is $n+2$ dimensional Minkowski space. We denote n dimensional Minkowski space by M_0. Let $G = SO_0(2,n)$ denote the connected component of the group of linear transformations of $\mathbb{R}^{n+2}$ preserving the symmetric bilinear form which is associated to $Q(x)$ by polarization. We shall call G the n-dimensional conformal group, and we denote the universal cover of G by $G^{\sim} = SO_0(2,n)^{\sim}$. Let $\mathcal{G}$ be the Lie algebra of G. $\mathcal{G}$ is identified with the set of all matrices (a_{ij}) $(-1 \le i,j \le n)$ such that $a_{ii} = 0$ $(0 \le i \le n)$, $a_{ij} = -a_{ji}$ $(1 \le i \le j \le n)$, $a_{0,j} = a_{j,0}$ $(1 \le j \le 4)$, $a_{-1,j} = a_{j,-1}$ $(1 \le j \le 4)$ and $a_{-1,0} = -a_{0,-1}$. We define subalgebras $\mathcal{K}$, $\mathcal{H}$, $\mathcal{A}$, $\mathcal{N}_+$ and $\mathcal{N}_-$ as follows. Let E_{ij} be the matrix such that the (i,j) component is equal to 1 and the other components are all equal to 0. Let $\mathbf{L}_{ij} = E_{ij} - E_{ji}$ $(1 \le i \le j \le n)$, $\mathbf{L}_{0i} = E_{i,o} + E_{o,i}$ $(1 \le i \le n)$, $\mathbf{L}_{-1,i} = E_{i,-1} + E_{-1,i}$ $(1 \le i \le n)$ and $\mathbf{L}_{-10} = E_{-1,0} - E_{0,-1}$. Let $\mathcal{K}$ be the subalgebra spanned by: $\mathbf{L}_{ij}$ $(1 \le i,j \le n)$ and $\mathbf{L}_{-1,0}$; $\mathcal{H}$ be the subalgebra spanned

[1] Dedicated to Professor Arno R. Böhm on the occasion of his 70th birthday. (Paper presented at the 5th International Symposium on Quantum Theory and Symmetries, Valladolid, Spain, July 22-28, 2007)

[2] Permanent address: Physics Dept., The Pennsylvania State University, Abington College, Abington, PA 19001 USA

by: $\mathbf{L}_{ij}$ $(1 \leq i,j \leq n-1)$, $\mathbf{L}_{0i}$ $(1 \leq i \leq n-1)$, $\mathbf{L}_{-1,i}$ $(1 \leq i \leq n-1)$ and $\mathbf{L}_{-1,0}$; $\mathcal{A}$ be the subalgebra spanned by $\mathbf{L}_{-1,n}$; and $\mathcal{N}_+$ $(\mathcal{N}_-)$ be the subalgebra spanned by $\mathbf{P}_i = \mathbf{L}_{n,i} + \mathbf{L}_{-1,i}$ $(0 \leq i \leq n-1)$ ($\tilde{\mathbf{P}}_i = \mathbf{L}_{n,i} - \mathbf{L}_{-1,i}$ $(0 \leq i \leq n-1)$). Denote the analytic subgroups of G corresponding to $\mathcal{K}$, $\mathcal{H}$, $\mathcal{A}$, $\mathcal{N}_+$ and $\mathcal{N}_-$ by K, H, A, N_+ and N_-, respectively. The group elements corresponding to the subalgebras $\mathcal{N}_+$ and $\mathcal{N}_-$ spanned by the $\mathbf{P}_i$ and $\tilde{\mathbf{P}}_i$ are in $(n+2)\times(n+2)$ matrix form:

$$\exp\{x^i(\mathbf{L}_{n,i} \pm \mathbf{L}_{-1,i})\} = \begin{bmatrix} 1 - \frac{1}{2}q(x) & x & \pm\frac{1}{2}q(x) \\ -x^\dagger & I_n & \pm x^\dagger \\ \mp\frac{1}{2}q(x) & \pm x & 1 + \frac{1}{2}q(x) \end{bmatrix}, \tag{1.1}$$

where $x = (x_0, x_1, , , , x_n)$, $x^\dagger = (x_0, -x_1, , , , -x_n)$, $q(x) = x_0^2 - x_1^2 - \ldots - x_{n-1}^2$ and I_n is the $n \times n$ identity matrix. We note that $H = SO_0(2, n-1)$, and we have an Iwasawa like decomposition of G i.e. the map $H' \times A \times N_+ \longrightarrow G$ is an injective diffeomorphism onto an open, dense subset of G, where H' is $SO(2, n-1)$ [1].

Consider the $n+1$ dimensional isotropic cone in $\mathbb{R}^{n+2}$ defined by

$$C = \{x \in \mathbb{R}^{n+2} | Q(x) = 0\}. \tag{1.2}$$

Let $\mathbb{R}^{n+2*}$ and C^* be the sets of nonzero elements in $\mathbb{R}^{n+2}$ and C, respectively. Let $P \subset G$ be the stabilizer subgroup of $e = e_{-1} + e_n$ where $e_i = (0, 0, , , , 0, 1, 0, , , , 0) \in \mathbb{R}^{n+2}$ (i.e. e_i is the vector with 1 in the ith slot and zeros elsewhere). $P \cong SO_0(1, n-1) \times_s N_+$, where $\times_s$ denotes semi-direct product, i.e. P is the n dimensional Poincaré group [2]. The orbit of e under G is C^* [2]. Hence $C^* \cong G/P$. We let S^{n-1} denote the $(n-1)$-sphere:

$$S^{n-1} = \left\{u = (u_1, u_2, \ldots, u_n) \in \mathbb{R}^n | u_1^2 + \ldots u_n^2 = 1\right\} \tag{1.3}$$

We define the upper and lower hemispheres of S^{n-1} as

$$S_\pm^{n-1} = \{s \in S^{n-1} \mid u_n \gtrless 0\}. \tag{1.4}$$

and the Σ^{n-1} is:

$$\Sigma^{n-1} = \partial\, S_+^{n-1} = \partial\, S_-^{n-1} \tag{1.5}$$

Spherical coordinates on S^{n-1} are as follows:

$$u = (sin(\rho)\,\omega, cos(\rho)) \in S^{n-1} \text{ with } \omega \in S^{n-2}. \tag{1.6}$$

Let $Proj(\mathbb{R}^{n+2})$ be the real projective variety of all one dimensional subspaces in $\mathbb{R}^{n+2}$. We have the map

$$X^* \longrightarrow Proj(\mathbb{R}^{n+2}) \qquad x \longrightarrow \overline{x}. \tag{1.7}$$

Let $M \subset Proj(\mathbb{R}^{n+2})$ be the image of C^* under (1.7). The above defined action of $SO_0(2,n)^\sim$ on $\mathbb{R}^{n+2}$ induces an action of $SO_0(2,n)^\sim$ on $Proj(\mathbb{R}^{n+2})$. Since C is stable under the action of $SO_0(2,n)^\sim$, M is stable under the action of $SO_0(2,n)^\sim$ on $Proj(\mathbb{R}^{n+2})$. $SO_0(2,n)$ and therefore $SO_0(2,n)^\sim$ are transitive on C^*, and hence $SO_0(2,n)$ and $SO_0(2,n)^\sim$ are transitive on M. M is naturally diffeomorphic to $(S^1 \times S^{n-1})/\mathbb{Z}_2$ where the $\mathbb{Z}_2$ action is the product of antipodal maps on S^1 and S^{n-1}. Denote $S^1 \times S^{n-1}$ by $\bar{M}$. $K = SO(2) \times SO(n)$ acts transitively on $\bar{M}$, and $\bar{M}$ is the homogeneous space $\bar{M} \cong K/K_0$ where $K_0 = SO(1) \times SO(n-1)$. The universal cover $\bar{M}^\sim$ of $\bar{M}$ is the (n dimensional) Einstein universe. (See [4] for the definition in four dimensions.) Since M is the conformal compactification of M_0 [2], the Einstein universe is the universal cover of the conformal compactification of Minkowski space.

Journal of Physics: Conference Series **128** (2008) 012010 doi:10.1088/1742-6596/128/1/012010

2. Line bundles and invariant differential operators

The line bundle $L^s(M)$ over M associated with the character $\lambda \longrightarrow |\lambda|^{-s}$ of $I\!R^*$ is the bundle whose fibre over $\bar{x}$ is the set of all pairs $(\lambda x,\ |\lambda|^s)\ \in\ C\ \times\ \mathbb{C},\ (s\ \in\ \mathbb{C})$. Denote by $\Gamma^s(M)$ the space of smooth sections of $L^s(M)$. There is a unique isomorphism between $\Gamma^s(M)$ and the space of smooth functions $f\ :\ C^*\ \longrightarrow\ \mathbb{C}$ which satisfy the homogeneity condition $f(\lambda x)\ =\ |\lambda|^{-s}\ f(x)$. $\Gamma^s(M)$ is an $SO_0(2,n)^{\sim}$ module with respect to the representation π_s defined by $(\pi_s(g)\ f)(x)\ =\ f(g^{-1}x)$ ($f \in \Gamma^s(M)$, $g \in SO_0(2,n)$, $x \in C^*$, and $g^{-1}\ x$ denotes the action of g^{-1} on $x\ \in\ C^*$). We denote the associated representation of the Lie algebra $so(2,n)$ by $d\pi_s$.

Let $D \subset I\!R^{n+2}$ be open and be stable under multiplication by $I\!R$ and $C \subset D$. Define

$$\Gamma^s(D)\ =\ \{f\ \in\ C^\infty(D)|\ f(\lambda x)\ =\ |\lambda|^{-s}\ f(x);\ x\ \in\ D^*\ ,\ \ \lambda \in I\!R\ \}\ . \qquad (2.1)$$

Let x_i be the natural linear coordinates on $I\!R^{n+2}$ introduced above. Important operators on $\Gamma^s(D)$ are:

$$S\ =\ \Sigma_{i=-1}^n\ x_i\ \frac{\partial}{\partial x_i}\ \ ,Q\ =\ \Sigma_{i=-1}^n\ \epsilon_i x_i^2\ I\ \ ,\Delta\ =\ \Sigma_{i=-1}^n\ \epsilon_i \frac{\partial^2}{\partial x_i^2}\ \ . \qquad (2.2)$$

ϵ_i is $+1$ for $i = -1$ or 0 and -1 for $i = 1,\ .\ .\ n$. (I = Identity operator on $\Gamma^s(D)$.) We have:

$$[\Delta,\ Q]\ =\ 4\{S\ +\ \frac{n+2}{2}\ I\}\ \ ,[S,\ \Delta]\ =\ -\,2\Delta\ \ ,[S,\ Q]\ =\ 2\,Q\ \ , \qquad (2.3)$$

and we also have Euler's theorem on homogeneous functions:

$$S\ \phi\ =\ \ -\,s\ \phi\ \ \text{for}\ \ \phi\ \in\ \Gamma^s(D)\ \ . \qquad (2.4)$$

Now, we explicitly let $D\ =\ \{(x,y)|x = (x_{-1}, x_0)\ \in\ I\!R^2,\ y = (y_1\ .\ .\ .y_n)\ \in\ I\!R^n,\ (x,y) = (0,0)$ or $x \neq 0,\ y \neq 0\}$, and we have the following results [3]:

Lemma: (i) $\Delta : \Gamma^s(D) \longrightarrow \Gamma^{(s+2)}(D)$. (ii) Let $s = -2\ +\ \frac{n+2}{2}$, then, if $\phi \in \Gamma^s(D)$ vanishes on C^*, so does $\Delta\ \phi$.

Corollary: Let $\phi \in \Gamma^s(M)$ with $s\ =\ \ -2\ +\ \frac{n+2}{2}$ and let ϕ_e be an extension of ϕ to $\Gamma^s(D)$, then $\Delta\phi_e$ is independent of the extension.

Since $\Delta\phi_e$ is independent of the extension, we obtain:

Proposition: For $s\ =\ \ -2\ +\ \frac{n+2}{2}$ we have that Δ induces a map $\Delta_M : \Gamma^s(M) \rightarrow \Gamma^{s+2}(M)$ which intertwines for the G actions on $\Gamma^s(M)$ and $\Gamma^{s+2}(M)$.

To define Δ_M we choose $\phi \in \Gamma^s(M)$ and extend to $\Gamma^s(D)$, we then apply Δ and restrict $\Delta\phi_e$ to C^*. It is clear from the above that only for

$$s\ =\ \ -2\ +\ \frac{n+2}{2} \qquad (2.5)$$

do we obtain a well-defined operator Δ_M on $\Gamma^s(M)$ from this procedure, and we shall henceforth only consider such values of s.

3. The harmonic component of $\Gamma^s(M)$

Let $C_K^s(\bar{M})$ be the space of all K finite elements of $C^\infty(\bar{M})$ for which $\phi(w)\ =\ (-1)^s\phi(-w)$ ($w \in \bar{M}$), and define extensions to D by $(\phi_e \otimes \psi_e)(rv,\ \tilde{r}u)\ =\ \phi_e(rv)\ \psi_e(\tilde{r}u)$ ($v \in S^1$, $u \in S^{n-1}$) and r,$\tilde{r}\ \in\ I\!R^+$ with $\phi_e(ru)\ \ =\ \ r^{-s/2}\phi(v)$, and $\psi_e(\tilde{r}u)\ \ =\ \ \tilde{r}^{-s/2}\psi(u)$ where $\phi \otimes \psi\ \in\ C_K^s(\bar{M})$. Now $(r,\ v)\ \in\ I\!R^+ \times S^1\ =\ I\!R^2\backslash\{0\}$ and $(\tilde{r},\ u)\ \in\ I\!R^+ \times S^{n-1}\ =\ I\!R^n\backslash\{0\}$ and, in terms of $(r,\ v,\ \tilde{r},\ u)$, Δ may be expressed as:

$$\Delta\ =\ \frac{\partial^2}{\partial r^2}\ +\ \frac{1}{r}\frac{\partial}{\partial r}\ +\ \frac{1}{r^2}\Delta_{S^1}\ -\ (\frac{\partial^2}{\partial \tilde{r}^2}\ +\ \frac{n-1}{\tilde{r}}\frac{\partial}{\partial \tilde{r}}\ +\ \frac{1}{\tilde{r}^2}\Delta_{S^{n-1}}) \qquad (3.1)$$

Journal of Physics: Conference Series **128** (2008) 012010 doi:10.1088/1742-6596/128/1/012010

where Δ_{S^1} and $\Delta_{S^{n-1}}$ are the Laplacians on S^1 and S^{n-1}, respectively. Thus, using $s = -2 + \frac{n+2}{2}$ we have

$$(\Delta\ \phi_e(rv)\psi_e(\tilde{r}u))\,|_{C^*} = \frac{1}{r^2}\left\{\left(\frac{n-2}{2}\right)^2 + \Delta_{S^1} - \Delta_{S^{n-1}}\right\}\phi(r\ v)\psi(r\ u)\ . \tag{3.2}$$

Now let $H \subset \Gamma^s(M)$ be the kernel of Δ_M and let H' be the cokernel, then we have an exact sequence

$$0 \to H \longrightarrow \Gamma^s(M) \overset{\Delta_M}{\longrightarrow} \Gamma^{s+2}(M) \longrightarrow H' \to 0$$

and the subspace H_K of elements of H in $C^s_K(\bar{M})$ is an $so(2,n)$ module, with the representation of $so(2,n)$ on H_K being $d\pi_s$.

4. Quantization of a real massless scalar field on $\tilde{\mathrm{M}}$

We have shown above that there is a representation of $SO_0(2,n)^\sim$ on $C^\infty(\tilde{M})$ and Ker Δ_{M} (the harmonic component of $\Gamma^s(M)$) defines a subrepresentation. This harmonic component splits into two irreducible components, and Kobayashi and Orsted [5] show that, for $n \geq 3$, they are spaces of positive and negative energy massless, spin zero fields on $\tilde{M}^n$ (the minimal representations) [3]), and, indeed for $n = 4$, they are described in the paper by Barut and Böhm [6] as representations for massless, spin zero particles. They are just the spin zero, singleton representations of $SO(2,n)$ considered by Angelopoulos and Laoues [7], and, for $n = 3$ they reduce to the positive and negative energy *Rac* representations of $SO_0(2,3)$.[7]

Let Δ_j denote the covariant derivative determined by the semi-Riemannian metric g on $\tilde{M}$, then the energy operator for a massless, scalar field on $\tilde{M}$ is: [8]

$$H = \int_{S^{n-1}} \frac{1}{2}\left\{\sum_{j=1}^{n}(\Delta_j\phi)^*(\Delta_j\phi) + \left(\frac{n-2}{2R}\right)^2|\phi|^2\right\} du \tag{4.1}$$

where $du = R^{n-1}\ \sin^{n-2}(\rho)d\rho\wedge d\omega$ is the volume form on S^{n-1}, where $d\omega$ denotes the volume form on S^{n-2}. (R is the radius of the Einstein universe, which, in the above, we had set equal to one.)

Recall the space of spherical harmonics of degree σ on S^{n-1} is defined to be

$$\mathcal{H}^\sigma = \left\{f \in C^\infty(S^{n-1}) | \Delta_{S^{n-1}}\ f = -\sigma(\sigma+n-2)\ f\right\}\ . \tag{4.2}$$

The spherical harmonics are

$$Y_{\sigma\ \ell\ \{m\}}(v) = N(k,\ell,\{m\})\ \sin^\ell\!\rho\ C^{\ell+\frac{n-2}{2}}_{\sigma-\ell}(\cos\rho)\ Y^{\{m\}}_\ell(\theta_1,\theta_2,\ \ldots\ ,\theta_{n-2}) \tag{4.3}$$

where $\theta_1,\theta_2,\ \ldots\ ,\theta_{n-2},\ \rho$ are spherical coordinates of the point $v \in S^{n-1}(\omega = (\theta_1,\theta_2,\ \ldots\ ,\theta_{n-2}) \in S^{n-2})$, $\{m\}$ is used for the other labels and $N(k,\ell,\{m\})$ is the normalization factor for the spherical harmonics. For $\phi \in \mathrm{Ker}\Delta_{\mathrm{M}}$ we have from (3.2) that

$$\Delta_M\phi = \left\{\frac{\partial^2}{\partial\tau^2} - \Delta_{S^{n-1}} + \left(\frac{n-2}{2}\right)^2\right\}\phi = 0\ . \tag{4.4}$$

For ϕ a K-finite function of the form $\phi_{\nu\sigma\ \ell\ \{m\}}(u,v) = e^{i\nu\tau}Y_{\sigma\ \ell\ \{m\}}(v)$ which are in $\mathrm{Ker}\Delta_{\mathrm{M}}$ we obtain

$$\nu^2 - \sigma(\sigma+n-2) - \left(\frac{n-2}{2}\right)^2 = 0\ . \tag{4.5}$$

The standard quantization of a real, massless scalar field on $\tilde{M}$ is carried out in a similar way to the quantization of a real, massless scalar field on M_0 [8]. Let c be the speed of light and $\hbar$ Planck's constant, then, for a point $x \in \tilde{M}$ with coordinates $(t = \frac{R}{c}\tau, \omega, \rho)$, the quantum field $\phi(x)$ is given by [8]:

$$\phi(t,\omega,\rho) = \sum_{\sigma,\ell,\{m\}} \left(\frac{\hbar}{E_\sigma}\right)^{1/2} \Big\{a_{\sigma\ell\{m\}} e^{-i\nu_\sigma t} Y_{\sigma\ \ell\ \{m\}}(\rho\ ,\omega) + a^\dagger_{\sigma\ell\{m\}} e^{i\nu_\sigma t} Y^*_{\sigma\ \ell\ \{m\}}(\rho\ ,\omega)\Big\} , \tag{4.6}$$

where ν_σ is specified by eqn. (4.5), and the $a_{\sigma\ell\{m\}}$ and $a^\dagger_{\sigma\ell\{m\}}$ satisfy the following relations on the Fock space:

$$[a_{\sigma\ell\{m\}}\ ,\ a^\dagger_{\sigma'\ell'\{m'\}}] = \delta_{\sigma,\sigma'}\delta_{\ell,\ell'}\delta_{\{m\},\{m'\}} \tag{4.7a}$$

and

$$[a_{\sigma\ell\{m\}}\ ,\ a_{\sigma'\ell'\{m'\}}] = [a^\dagger_{\sigma\ell\{m\}}\ ,\ a^\dagger_{\sigma'\ell'\{m'\}}] = 0 . \tag{4.7b}$$

By a calculation, which uses eqn. (4.4), we may rewrite the Hamiltonian as

$$H = \int_{S^{n-1}} \frac{1}{2}\Big\{(\partial_t\phi)^*(\partial_t\phi) - \phi^*\partial_t^2\phi\Big\}\, du . \tag{4.8}$$

5. The Casimir energy of a massless scalar field on $\tilde{M}$ which vanishes on Σ^{n-1}

From eqns. (1.4), (1.5) and (1.6) we obtain: $S_+^{n-1} = \Big\{v \in S^{n-1} \mid \rho > \frac{\pi}{2}\Big\}$, $S_-^{n-1} = \Big\{v \in S^{n-1} \mid \rho < \frac{\pi}{2}\Big\}$, and $\Sigma^{n-1} = \Big\{v \in S^{n-1} \mid \rho = \frac{\pi}{2}\Big\}$. We want to determine the Casimir energy of a quantized massless, scalar field on $\tilde{M}$ subject to vanishing of the field on the equator Σ^{n-1}. In order to do this we need the following result [9]:

Proposition: For $\nu > 0$

$$C_k^\nu(\cos(\frac{\pi}{2})) = 0 \iff k \text{ is odd} . \tag{5.1}$$

Using eqn. (4.6) and this result in eqn. (4.8), we are able to compute the vacuum expectation value of the energy operator (eqn. (4.8)), to obtain the following for the total zero point energy of the problem:

$$E = \frac{\hbar c}{2R} \sum_{\substack{\sigma = 0 \\ \sigma - \ell \text{ odd} \\ \ell, \{m\}}}^{\infty} \sqrt{\sigma(\sigma + n - 2) + (\frac{n-2}{2})^2} , \tag{5.2}$$

where the sum over ℓ goes from zero to σ, and the ranges of summations for the $\{m\}$'s follow from the theory of spherical harmonics of degree ℓ on S^{n-2}.

In order to perform the (finite) sums over ℓ and the $\{m\}$'s we need to recall a little about the representation theory of $SO(n-1)$ on S^{n-2} and its relationship to spherical harmonics in $n-1$ dimensions. Let τ be the left regular representation of $SO(n-1)$ on $C^\infty(S^{n-2}) \cong C^\infty(SO(n-1)/SO(n-2))$. Denote by Λ the set of all integers for $n = 3$, and the set of all nonnegative integers for $n > 3$. The zonal spherical function ω_ℓ with height ℓ ($\ell \in \Lambda$) is given by $C_\ell^{\frac{n-3}{2}}(\theta_{n-2})$ for $n > 3$ and $(x_1 + ix_2)^\ell$ for $n = 3$ [10]. For any $\ell \in \Lambda$ we denote by $\mathcal{H}_\ell$ the subspace of $C^\infty(SO(n-1)/SO(n-2))$, which is spanned by the elements $\tau(k)\omega_\ell$

$(k \in K)$. Then, as is well-known, τ decomposes into the direct sum of inequivalent, irreducible representations of $SO(n-1)$ i.e. $C^\infty(S^{n-2}) \cong \Sigma_\ell^\oplus \mathcal{H}_\ell$, and the spaces $\mathcal{H}_\ell$ are identified as spaces of harmonic, homogeneous polynomials on $\mathbb{R}^{n-1}$ of degree ℓ i.e the spaces of spherical harmonics of degree ℓ. Since an arbitrary homogeneous polynomial of degree ℓ in $n-1$ variables depends on

$$\binom{\ell + n - 2}{\ell}$$

arbitrary constants and the condition that the polynomial be harmonic gives

$$\binom{\ell + n - 4}{\ell - 2}$$

restrictions, the number of linearly independent spherical harmonics $Y_\ell^{\{m\}}(\theta_1, \theta_2, \ldots, \theta_{n-2})$ of degree ℓ on S^{n-2} (i.e. the dimension of $\mathcal{H}_\ell$) is:

$$N(\ell, n) = \binom{\ell + n - 2}{\ell} - \binom{\ell + n - 4}{\ell - 2} = \frac{\Gamma(\ell + n - 3)(2\ell + n - 3)}{\Gamma(n-2)\Gamma(\ell + 1)} \tag{5.3}$$

for $n > 3$ and one for $n = 3$. (It follows from the representation theory of $SO(2)$ that for $n = 3$ the space $\mathcal{H}_\ell$ is one dimensional.) Substitution of eqn. (5.3) into (5.2) gives (for $n > 3$):

$$E = \frac{\hbar c}{2R} \sum_{\substack{\sigma = 0 \\ \sigma - \ell \text{ odd}}}^{\infty} \sqrt{\sigma(\sigma + n - 2) + (\frac{n-2}{2})^2} \sum_{\ell=0}^{\sigma} \frac{\Gamma(\ell + n - 3)(2\ell + n - 3)}{\Gamma(n-2)\Gamma(\ell + 1)} \ . \tag{5.4}$$

We may perform the finite sums over ℓ for ℓ even (σ odd) and ℓ odd (σ even), respectively, and we obtain

$$\frac{\Gamma(\sigma - 2 + n)}{\Gamma(\sigma)\Gamma(n-1)}$$

in both cases. Thus, after a little algebra, eqn. (5.4) becomes

$$E = \frac{\hbar c}{4R} \sum_{\sigma=0}^{\infty} (n - 2 + 2\,\sigma) \frac{\Gamma(\sigma - 2 + n)}{\Gamma(\sigma)\Gamma(n-1)} \ . \tag{5.5}$$

Due to the difference in multiplicities of μ's for $n = 3$ and $n > 3$ the much simpler $n = 3$ case must be treated, with minor alterations, separately. This is an easy exercise which we leave to the reader. Our final result for the Casimir energy in this case ($n = 3$) is recorded below in the Table. In the Table we have also recorded the well known case of one spatial dimension for completeness [11].

We must regularize the above divergent series (eqn. (5.5)). For this purpose we introduce the exponential cutoff $e^{-\frac{\alpha}{R}(\frac{2\sigma+n-2}{2})}$ in eqn. (5.5) to obtain:

$$E^{reg}(\alpha, n) = \frac{\hbar c}{4R} \sum_{\sigma=0}^{\infty} (n - 2 + 2\,\sigma) \frac{\Gamma(\sigma - 2 + n)}{\Gamma(\sigma)\Gamma(n-1)} \, e^{-\frac{\alpha}{R}(\frac{2\sigma+n-2}{2})} =$$

$$= \frac{1}{4R} \, n \, e^{-\frac{\alpha\, n}{2R}} \left(\left\{ 1 - e^{(-\frac{\alpha}{R})} \right\}^{-n+1} + 2\,(\frac{n-1}{n})\,(1 - e^{-\frac{\alpha}{R}})^{-n} \, e^{-\frac{\alpha}{R}} \right) . \tag{5.6}$$

By using the Euler-Maclaurin sum formula we may extract out of this formula the finite part of $E^{reg}(\alpha, n)$ which does not depend upon α. This is the Casimir energy. We have tabulated our results for various n in Table 1.

Table 1. Casimir energy of a massless scalar field in the n dimensional Einstein universe with Dirichlet boundary conditions on Σ^{n-1}

spatial dimension (n - 1)	Casimir Energy / ($\frac{\hbar\, c}{R}$)
1	$-\frac{1}{12}$
2	$-\frac{1}{98}$
3	$+\frac{1}{480}$
4	$+\frac{17}{7680}$
5	$-\frac{31}{120960}$
6	$-\frac{367}{774144}$
7	$+\frac{289}{7257600}$
8	$+\frac{27859}{265420800}$
9	$-\frac{317}{45619200}$
10	$-\frac{1295803}{54499737600}$

We also may extract the Casimir energy out of (5.5) by using the zeta regularization technique [12]. For example, if $n = 4$, we obtain from eqn. (5.5)

$$E = \frac{\hbar c}{4R} \left(\sum_{\sigma=0}^{\infty} (\sigma + 1)^2 \, \sigma \right) . \tag{5.7}$$

We rewrite this as

$$E = \frac{\hbar c}{4R} \left(\sum_{\sigma=0}^{\infty} (\sigma + 1)^3 - \sum_{\sigma=0}^{\infty} (\sigma + 1)^2 \right) . \tag{5.8}$$

Now we use

$$\zeta(-3) = \sum_{\sigma=0}^{\infty} (\sigma + 1)^3 = \frac{1}{120} \tag{5.9a}$$

and

$$\zeta(-2) = \sum_{\sigma=0}^{\infty} (\sigma + 1)^2 = 0 \tag{5.9b}$$

in eqn. (5.8) to obtain

$$E = + \frac{1}{480} \frac{\hbar c}{R} \tag{5.10}$$

which is the result obtained in ref. [13]

6. Conclusions

Our results may be compared with the the corresponding results for the Casimir energy of a massless scalar field in n dimensional Minkowski space with Dirichlet boundary conditions on a sphere S^{n-2} of radius R in $I\!R^{n-1}$ ($n-1$ spatial dimensions) (c.f. ref. [14]). Notice that we obtain the same type of behavior for alteration in signs in steps of two, and in odd spatial dimensions our results agree qualitatively with the corresponding results for the corresponding problem in Minkowski space (Casimir energy of a massless scalar field in n dimensional Minkowski space with Dirichlet boundary conditions on a sphere). However, in contrast to our results for the Einstein universe, calculations for the Casimir energy in Minkowski spaces of odd dimensions (even spatial dimensions) give infinite answers [14], [15]! This is a clear manifestation of the greater regularity properties obtained by working with quantum field theory in the Einstein universe [16].

For the physically important case of three spatial dimensions we obtain from the Table that the Casimir energy is $+\frac{1}{480}\frac{\hbar c}{R}$. Notice the all important sign: the energy is positive, as it is for the corresponding problem in Minkowski space. We can explain why both results (our result and the corresponding result for the Casimir energy of a scalar field with Dirichlet boundary conditions on a sphere in Minkowski space) are positive as follows. First, we argue that the Casimir energy with Dirichlet boundary conditions on spheres of slightly smaller radii in the Einstein universe must also be positive and greater than $+\frac{1}{480}\frac{\hbar c}{R}$. This is because the equator Σ^3 is the sphere of largest radius in the Einstein universe, and, therefore, the Casimir energy must be a minimum for Σ^3. If the energy were not a minimum, then, on the one hand, there would be an outward force on Σ^3 centered at the "north pole" i.e. the point $\hat{u} = (0,\ 0,\ 0, R) \in S^3$. On the other hand, by invariance of conformal inversion, we can equally view the problem from the diametrically opposite point $\hat{u}' = (0,\ 0,\ 0, -R) \in S^3$, and from this point of view the Casimir force will be an inward force towards the point $\hat{u}$. The only way out of this paradox is to insist that the Casimir energy be a minimum for Σ^3. (Arguments of this nature, which exploit the symmetry of the Einstein universe under conformal inversion, have been used before; in particular, Landau and Lifschitz ([17]) use this symmetry together with Gauss' law to show that the total charge of the Einstein universe must be zero.) Hence, by the physical requirement that the observable energy must be a continuous function, we conclude that the Casimir energy must be positive for a sphere of cosmic size in the four dimensional Einstein universe and smaller in radius than Σ^3. (We have also obtained the same conclusion in a much more rigorous way, namely by applying second order perturbation theory to the image of the Einstein energy under scaling [18].) If we relate the Einstein universe and the Minkowski space via a Lorentzian version of stereographic projection (the flat or "Minkowski space" parallelization in the terminology of ref. [19], we can conclude from this result that also for a sphere of cosmic size in Minkowski space-time the Casimir energy must be positive. Then, using a scaling argument [18], we conclude that the Casimir energy of a massless scalar field in four dimensional Minkowski space with Dirichlet boundary conditions on a sphere of radius R is positive.

References

[1] Sekiguchi J 1980 *Nagoya Math. J.* **79** 151-185.
[2] Angelopoulos E and Laoues M 1998 *Review in Math. Phys.* **10** 3 271-299.
[3] Kostant B 1992 The vanishing of scalar curvature on 6-manifolds, Einstein's equation, and representation theory *Prog. Math.*, **105**, Birhäuser, Boston 67-86.
[4] Birrell N. and Davies P 1982 *Quantum Fields in Curved Space* (Cambridge, UK: Cambridge University Press).
[5] Kobayashi T and Orsted B 2003 *Adv. Math.* **180** 486-595.
[6] Barut A and Böhm A 1970 *Jour. Math. Phys.* **11**, (10) 2938-2945.
[7] Angelopoulos E and Laoues M 2000 Singletons on AdS$_n$ *Conférence Moshé flato 1999, Quantization, Deformations and Symmetries*, Vol. II ed Giuseppi Dito and Daniel Sternheimer *Math. Phys. Studies* **22** 3-23
[8] Baez J 1991 *Lett. Math. Phys.*, **21** 117-121.

[9] Gradshteyn I C and Rhyshik I M 2000 *Table of Integrals, Sums, Series and Products* (New York: Academic Press).
[10] Molčanov V F 1970 *Math. Sbornik*, **81** 3 333-347.
[11] Moylan P 2006 $U_q(so(1,2))$ Representations at Roots of Unity and Their Realizations on the Finite Circle *Bulg. J. Phys.* **33** s2 121-132.
[12] Elizalde E 1994 *J. Phys. A* **27** L299-304.
[13] Bayen S and Özcan M 1993 *Class. Quant. Grav.* **10** L115-L121.
[14] Bordgag M et al 2001 *Physics Reports*, **353** 1-205.
[15] Sen S 1981 *Jour. Math. Phys.* **22** 2968.
[16] Segal I E and Zhou Z 1992 *Ann. Phys.* (N.Y.) **218** 279-292.
[17] Landau L D and Lifschitz E M 1973 The Classical Theory of Fields (Moscow: Nauka).
[18] Moylan P 2007 The Casimir effect in the Einstein Universe (work in progress).
[19] Segal I E and Paneitz S M 1982 *Jour. Funct. Anal.* **47** 78-142.

Effective action methods on manifolds with branes and boundaries

Dmitry Nesterov and Andrei Barvinsky
Theory Department, Lebedev Physics Institute, Leninsky prospect 53, Moscow, Russia
E-mail: nesterov@lpi.ru, barvin@lpi.ru

Abstract. We review the developing effective action framework for field-theoretical models in the presence of branes: describe general scheme based on Dirichlet-to-Neumann reduction and present some applications and results.

1. Introduction. Braneworld setup

The field theoretical setup for models with branes motivated from string theory can be realized by the fundamental action which is the sum of bulk and brane actions

$$\mathbf{S}[\Psi,\varphi] = \mathbf{S}_{Bulk}[\Psi] + \mathrm{S}_{brane}[\varphi] \qquad (1)$$

Bulk action is the integral over bulk manifold of the local invariants of fields, propagating in the bulk. Set of fields which are dynamical in the bulk we shall denote as Ψ. Brane action is the integral over brane d-dimensional manifold of invariants built of boundary values of fundamental bulk fields $\Psi\big|_{brane} = \varphi$. The latter are called *induced* or *brane* fields.

$$\begin{array}{llll} \Psi: & G_{AB}, \dots & - & \text{Bulk fields} \\ \varphi \equiv \Psi\big|_{brane}: & g_{\alpha\beta}, \dots & - & \text{induced (brane) fields} \end{array}$$

In braneworld models various additional matter and gauge fields which are binded to branes can be present. But they do not play crucial role in what follows, so we omit them and concentrate on fields, which propagate not only on branes, but in the bulk as well. The main goal is to discuss the effective dynamics of such fields as it seen by the observers living on branes.

Branes can be considered as boundaries of the bulk, no matter they really bound the bulk manifold or they are inside the bulk manifold (so-called fixed points). The only thing one has to do is to select appropriate boundary or junction conditions [1].

In what follows we restrict ourselves to the case when branes are d-dimensional manifolds embedded in the bulk of dimension $(d+1)$ – case of codimension 1 branes, which is motivated by interest in applications to braneworld scenarios such as Randall-Sundrum (RS) or Dvali-Gabadadze-Porratti (DGP) with one large extra dimension [2],[3]. Also for the sake of simplicity in what follows we skip analysis of the gauge aspects of the formalism [4],[5].

The brane effective action approach allows us to introduce the brane functional of only induced fields φ, which incorporate the dynamics of correspondent fields Ψ in the bulk [1],[6]

$$\mathbf{S}[\Psi,\varphi] \rightarrow S^{eff}[\varphi] \tag{2}$$

The main advantage of this approach, when one is allowed to deal with functional of only brane fields, compared for example with Kaluza-Klein approach, is that these induced fields can be "measured" directly by observers on the brane due to local interaction of brane gauge and matter fields with these induced fields [1],[7].

The price one pays for this is that effective action for induced fields appears to be nonlocal.

$$e^{-S^{eff}[\varphi]} = e^{-\mathrm{S}_{brane}[\varphi]} \int\limits_{\Psi\big|_{brane}=\varphi} d\Psi\, e^{-\mathbf{S}_{Bulk}[\Psi]} \tag{3}$$

This means that from the viewpoint of brane observer the dynamics of induced fields along branes is essentially nonlocal. On the other hand such properties as local causality, unitarity to a large extent are preserved by construction that justifies and suggests new ideas for search of consistent modifications of Einstein gravity since one accepts the possibility that our 4-dimensional universe can be the brane embedded in multidimensional bulk.

Passing on to working with brane effective action one does not completely lose the information about bulk dynamics of fundamental fields – the geometry, topology of the bulk and information of brane embedding is incorporated in nonlocality of induced fields dynamics, particularly in operator of small fluctuations. Thus brane observers are able to obtain some information about the bulk by measuring deviations from local d-dimensional behavior of fundamental fields, including gravity.

For example one can study the deviations from the Newton law on very large and very small distances. Alternative way is to observe the deformations of gravitational waves and signals from distant astrophysical objects. For the class of Randall-Sundrum type models that was analyzed in [8].

Another argument in favor of efficiency of the effective action approach is that nonlocalities governing the propagation of induced field perturbations are calculable, at least in models possessing symmetric bulk backgrounds. So one has a handy tool for explicit analysis of induced fields behavior on arbitrary energy scales. It is very helpful that by construction all objects appear to be d-dimensionally covariant.

In order to take into account quantum properties in full (i.e. to quantize brane configurations as well) one needs *braneworld quantum effective action* $\Gamma[\bar{\varphi}]$. It is by definition the functional of the mean brane field emergent as the response to arbitrary sources concentrated on branes.

$$e^{-\Gamma[\bar{\varphi}]} \equiv \int d\Psi\, e^{-\mathbf{S}[\Psi,\varphi] - \int\limits_{brane} (\varphi-\bar{\varphi})\, j}\Bigg|_{j=j[\bar{\varphi}]} = \int d\varphi\, e^{-S^{eff}[\varphi] - \int\limits_{brane} (\varphi-\bar{\varphi})\, j}\Bigg|_{j=j[\bar{\varphi}]} \tag{4}$$

Its construction differs from construction of quantum effective action of the whole theory (1) in the bulk by the fact that one introduces sources (j) only on branes. Owing to that it becomes the functional only of mean values of induced fields $\bar{\varphi} \equiv <\varphi>$. Relevance of that object is again justified by the braneworld philosophy that observers living on branes are able to investigate the world only by means of local interactions on branes.

2. One-loop brane effective action and duality relations

Journal of Physics: Conference Series **128** (2008) 012011 doi:10.1088/1742-6596/128/1/012011

For realistic models which are nonlinear in fields, especially for models such as gravity which is inevitably present in braneworld scenarios, one is unable to calculate explicitly functionals S^{eff} or $\Gamma[\bar{\varphi}]$. So, one has to use some perturbation technique. When weak field approximation is relevant it is convenient to perform loop expansion

$$\Gamma[\bar{\varphi}] = S^{tree}[\bar{\varphi}] + \Gamma^{1-loop}[\bar{\varphi}] + ... \tag{5}$$

Interesting and nontrivial observations can be found already at one-loop order, when in path integral formalism one deals with calculable gaussian functional integrals. In one-loop approximation the effective action is given by the following path integral:

$$e^{-\Gamma^{1-loop}[\bar{\varphi}]} \equiv \int d\Psi \; e^{-\frac{1}{2}\int\limits_{Bulk} \Psi \overleftrightarrow{\mathbf{F}} \Psi - \frac{1}{2}\int\limits_{brane} \phi f \phi} \tag{6}$$

where we perform integration over fluctuations Ψ: $\Psi\big|_{brane} = \phi$, over background configuration $\Psi_0 = \Psi[\bar{\varphi}]$, which is the solution of:

$$\begin{cases} \frac{\delta \mathbf{S}_{Bulk}[\Psi]}{\delta\Psi} = 0, \\ \Psi|_{brane} = \bar{\varphi}. \end{cases} \tag{7}$$

i.e. the on-shell continuation of brane background to the bulk.

In r.h.s of (6) background $\bar{\varphi}$ shows itself through the operators of small fluctuations in the bulk $\mathbf{F}$ and on branes f:

$$\mathbf{F}\delta^{d+1}(X,X') \equiv \frac{\delta^2 \mathbf{S}_{Bulk}[\Psi]}{\delta\Psi(X)\delta\Psi(X')}\bigg|_{\Psi=\Psi[\bar{\varphi}]}, \qquad \mathfrak{f}\,\delta^d(x,x') \equiv \frac{\delta^2 \mathrm{S}_{brane}[\varphi]}{\delta\varphi(x)\delta\varphi(x')}\bigg|_{\varphi=\bar{\varphi}}. \tag{8}$$

In braneworld scenarios under consideration $\mathbf{F}$ is a second order local differential nondegenerate bulk operator $\mathbf{F} = {}^{d+1}\Box + ...$ and $\mathfrak{f}$ – local brane operator.

It can be proved [6] that r.h.s of (6) equals square root of functional determinant of Neumann Green function for bulk operator $\mathbf{F}$ of fluctuations Ψ over background $\Psi_0 \equiv \Psi[\bar{\varphi}]$. It is the same as determinant of the operator $\mathbf{F}$ to the power minus one half on space of functions with Neumann boundary conditions $(-\nabla_n + \mathfrak{f})\Psi|_{brane} = 0$ (∇_n– covariant derivative normal to brane):

$$\int d\Psi \; e^{-\frac{1}{2}\int\limits_{Bulk} \Psi \overleftrightarrow{\mathbf{F}} \Psi - \frac{1}{2}\int\limits_{brane} \phi \mathfrak{f} \phi + \int\limits_{brane} \phi j} = \mathbf{Det}^{1/2}\mathbf{G}_N \cdot e^{\frac{1}{2}\int\limits_{brane} j\,|\mathbf{G}_N|\,j}. \tag{9}$$

We introduced auxiliary sources j in path integral to pick up another useful (*tree-level*) contribution [1] leading to so called Dirichlet-to-Neumann map. Notation $|\mathbf{G}_N|$ stands for boundary operator whose kernel is the restriction of bulk Neumann Green function $\mathbf{G}_N(X, X')$

$$\begin{cases} \mathbf{F}\,\mathbf{G}_N(X,X') = \delta^{d+1}(X,X'), \\ (-\nabla_n + \mathfrak{f})\,\mathbf{G}_N(X,X')|_{X\in\, brane} = 0, \end{cases} \tag{10}$$

to boundary (brane):

$$|\mathbf{G}_N|(x,x') \equiv \mathbf{G}_N(X,X')\Big|_{X=x,\, X'=x'}.$$

Note, that there were made *no* special assumptions about boundary operator $\mathfrak{f}$ but it selfadjointness (and preserving strong ellipticity). So our approach covers most of possible *generalized* Neumann boundary conditions such as Robin, oblique or even more involved ones, arising for example in Dvali-Gabadadze-Porratti model (where $\mathfrak{f}$ is brane d-dimensional D'Alembertian). More detailed discussion of this subject can be found in the next section.

Journal of Physics: Conference Series **128** (2008) 012011 doi:10.1088/1742-6596/128/1/012011

The second way to calculate path integral above is to split integration

$$\int d\Psi ... \rightarrow \int d\phi \int\limits_{\Psi\big|_{brane}=\phi} d\Psi.$$

Performing first integration over bulk fields with fixed boundary values generate determinant of Dirichlet operator in the bulk. Finally this way leads to [6]

$$\int d\Psi\, e^{-\frac{1}{2}\int\limits_{Bulk}\Psi\overleftrightarrow{\mathbf{F}}\Psi-\frac{1}{2}\int\limits_{brane}\phi\mathfrak{f}\phi+\int\limits_{brane}\phi j} = \mathbf{Det}^{1/2}\mathbf{G}_D\cdot\mathrm{Det}^{-1/2}(F^{ind}+\mathfrak{f})\cdot e^{\frac{1}{2}\int\limits_{brane} j\,|F^{ind}+\mathfrak{f}|^{-1}j}. \quad (11)$$

where $\mathbf{G}_D$ is the Dirichlet Green function:

$$\begin{cases} \mathbf{F}\,\mathbf{G}_D(X,X') = \delta^{d+1}(X,X'), \\ \mathbf{G}_D(X,X')|_{X\in brane} = 0, \end{cases} \quad (12)$$

and F^{ind} is the (nonlocal) effective inverse propagator for field fluctuations on the boundary generated by bulk dynamics of Ψ, which kernel is defined in terms of Dirichlet Green function as follows:

$$F^{ind}\delta^d(x,x') \equiv \big(-\overrightarrow{\nabla}_n\mathbf{G}_D(X,X')\overleftarrow{\nabla}_{n'}\big)\Big|_{X=x,\,X'=x'}. \quad (13)$$

Comparing (9) and (11) one can notice that this gives justification for the following duality relations between Green functions for different boundary value problems which is known as Dirichlet-to-Neumann map and between functional determinants of operators with different boundary conditions on manifolds with boundaries [1],[6]:

$$-\Big|\overrightarrow{\nabla}_n\mathbf{G}_D\overleftarrow{\nabla}_n\Big| + \mathfrak{f} = \Big|\mathbf{G}_N\Big|\ , \quad (14)$$

$$\mathbf{Det}\,\mathbf{G}_D\cdot\mathrm{Det}^{-1}(F^{ind}+\mathfrak{f}) = \mathbf{Det}\,\mathbf{G}_N\ . \quad (15)$$

Surely such formal path integral manipulations can not serve as rigorous prove for relations above. In fact more pedantic considerations show that (14,15) are correct for rather general setup [1],[6].

3. Braneworld applications

What can this approach suggest for braneworld physics?

First of all duality relations suggest handy scheme for exact calculation of one-loop effective action depending on background parameters (such as internal mass scales, interbrane distances and brane embeddings, background curvature scales, etc. which in what follows we denote as m_i) and covariant low-energy effective action expressions. The key point is that due to (15) calculation of one-loop contribution which for dynamical sector of the model is nothing but logarithm of the Neumann determinant decouples into calculation of the Dirichlet bulk determinant and brane determinant of some brane nonlocal operator.

First part (Dirichlet determinant) is well-studied problem. New and intriguing is the problem of brane determinant calculation. The approach to this problem depends on what one aims at[1]. In braneworld physics one-loop effective action is interesting in the context of moduli stabilization problem, for cosmological applications it can give the preferred vacuum. In both cases one is

[1] Generic properties of brane determinant based on Schwinger-DeWitt expansion are discussed in the next section.

Journal of Physics: Conference Series **128** (2008) 012011 doi:10.1088/1742-6596/128/1/012011

interested in dependence on background moduli m_i and hence is forced to calculate quantum determinants over parameterized families of highly-symmetric backgrounds.

When bulk background metric appears to be of the form $ds^2 = dy^2 + a^2(y)g_{\mu\nu}(x)dx^\mu dx^\nu$ with brane(s) located at some $y = const$ then induced boundary operator (13) can be *explicitly* expressed as some function of brane d'Alembertian (Laplacian in euclidean signature calculations) $F^{ind} = F^{ind}(\Box)$ parametrically dependent on background moduli (interbrane distances, brane embeddings, bulk curvature scale, etc.).

For example in DGP-like background [3]($(d+1)$-dimensional flat space, flat d-dimensional brane)

$$F^{ind} = m_{DGP}\sqrt{M^2 - \Box}; \qquad \mathfrak{f} = -\Box. \tag{16}$$

In Randall-Sundrum background [2] ($(d+1)$-dimensional AdS in the bulk and flat d-dimensional brane) F^{ind} is some combination of Bessel functions of $\sqrt{-\Box}$, $\mathfrak{f} = 0$ [1],[7].

When boundary operators are expressed as functions of brane d'Alembertian (Laplacian) $\Box$ then expressions for effective propagator and brane contribution to one-loop effective action can be obtained using Fourier transform or proper time technique (Laplace transform). Using latter representation needed quantities can be expressed in terms of weighted proper time integrals

$$\frac{1}{F^{ind} + \mathfrak{f}} = \int\limits_0^\infty ds\, e^{s\Box}\, u(s, m_i), \tag{17}$$

$$\ln \mathrm{Det}\Big(F^{ind} + \mathfrak{f}\Big) = -\mathrm{Tr} \int\limits_0^\infty \frac{ds}{s}\, e^{s\Box}\, v(s, m_i), \tag{18}$$

The advantage of the latter representation is that it is applicable also in the case of the curved brane, so that generally covariant expansion of (17)-(18) in curvatures can be directly obtained by using a well-known Schwinger-DeWitt expansion for $e^{s\Box}$. Thus the lowest order approximation for the exact brane-to-brane operator in models with a curved bulk and curved branes can be considered by means of manifestly covariant technique which can be systematically extended to higher orders. Combined with the method of fixing the background covariant gauge for diffeomorphism invariance in brane models, developed in [4, 5] this will ultimately give the universal background field method of the Schwinger-DeWitt type in gravitational brane systems.

For instance in DGP-type model these weights are given in terms of the error function [9]:

$$u(s, m_{DGP}) = e^{s\, m_{DGP}^2} \cdot \mathrm{erfc}\,(m_{DGP}\sqrt{s}); \qquad v(s, m_{DGP}) = \frac{1}{2}\Big(1 + u(s, m_{DGP})\Big) \tag{19}$$

Another test-application of this scheme is renormalized (using dimensional regularization) effective potential of the usual Coleman-Weinberg structure for DGP case [9]:

$$V_{\mathrm{eff}}(m_{DGP}) = \frac{1}{2}\,\frac{(-1)^d}{\Gamma(\frac{d}{2}+1)}\left(\frac{m_{DGP}^2}{4\pi}\right)^{d/2} \ln\frac{m_{DGP}}{\mu}. \tag{20}$$

Here μ is the parameter reflecting the renormalization ambiguity, and the role of the field (argument of the potential) is played by the DGP-scale m_{DGP} of the brane term in the classical action. This result confirms the boundary effective action calculation of [10].

4. New heat kernel technique

One of interesting byproducts of the brane effective action formalism briefly discussed above is a new technique for calculating the heat kernel expansion on manifolds with boundaries.

Journal of Physics: Conference Series **128** (2008) 012011 doi:10.1088/1742-6596/128/1/012011

In quantum field theory heat kernel $K(s|X,Y) \equiv e^{-s\mathbf{F}}\delta(X,Y)$ for differential operator $\mathbf{F}$ of quantum perturbations is a well-known and convenient tool for studying one-loop divergences, counterterms, quantum anomalies, Casimir effect and various asymptotics of effective actions and propagators [11]. Convenience and efficiency of heat kernel in many respects is based on the possibility of expanding it and its trace in asymptotic series in (integer) powers of proper time s in the (ultraviolet) limit $s \to 0$:

$$e^{-s\mathbf{F}}\delta(x,y) = \frac{e^{-\frac{\sigma(x,y)}{2s}}}{(4\pi s)^{d/2}} \sum_{n=0}^{\infty} s^n a_n(x,y) \tag{21}$$

where $\sigma(x,y)$ is half of the geodesic distance squared. Coefficients $a_n(x,y)$ referred as Schwinger-DeWitt coefficients are universal and can be explicitly found in the coincidence limit ($x \to y$) as local invariants of space-time curvature, fibre-bundle connection and background external fields [11] by solving recurrent equations. In particular this expansion gives rise to local low-energy expansion of the effective action in inverse powers of (auxiliary) large mass parameter M^2 when the inverse propagator is supplied with mass term $\mathbf{F} \to \mathbf{F} + M^2$.

In spacetimes with boundaries the heat kernel theory becomes complicated. The apparent modification consists in that while bulk coefficients of expansion remain unchangeable, the additional series of half-integer powers of proper time (and possible logarithmic terms) with coefficients built up of boundary invariants arise

$$\mathbf{Tr}\, e^{-s\mathbf{F}-sM^2} = \frac{1}{(4\pi s)^{D/2}}\, e^{-sM^2} \sum_{n=0}^{\infty} \Big[s^n\, \mathbf{A}_n + s^{n/2}\, B_{n/2} + \ln\left(s\mu^2\right) s^{n/2}\, C_{n/2}\Big] \tag{22}$$

where

$$\mathbf{A}_n = \int\limits_{Bulk} a_n(X,X); \qquad B_n, C_n = \int\limits_{brane} b_n, c_n(x,x). \tag{23}$$

The crucial difficulty consists in the absence of recurrent local scheme for boundary coefficients B_n (and C_n). In contrast to the bulk coefficients A_n which are universal and independent on boundary conditions, the surface (boundary) coefficients essentially depend on the latter. Surface coefficients can be somehow regularly found only for Dirichlet and homogeneous Neumann boundary conditions [11, 12]. For more complicated boundary conditions such as Robin and oblique cases the problem of calculating the surface contributions in (22) becomes highly involved or needs the use of methods (e.g. based on conformal properties of geometric invariants) which are not universally applicable [11].

One-loop duality (15) suggests new elegant, covariant and universal way for explicit calculation of Schwinger-DeWitt boundary coefficient for generalized Neumann boundary conditions. The latter are parameterized by the boundary operator $W = -\nabla_n + \mathfrak{f}$, which determine selfadjoint extension of bulk symmetric operator $\mathbf{F}$ of the Neumann type. In our approach, beside pure Neumann case $\mathfrak{f} = 0$, which arise for example in Randall-Sundrum type models, one can deal with the following generalized Neumann boundary conditions:

$$\begin{array}{ll} \mathfrak{f} = v(x) & \text{– inhomogeneous Neumann (Robin) case;} \\ \mathfrak{f} = \frac{1}{2}(\Gamma^\mu \nabla_\mu + \nabla_\mu \Gamma^\mu) & \text{– oblique boundary conditions (open string, Chern-Simons);} \\ \mathfrak{f} = -\frac{1}{m}\Box \equiv -\frac{1}{m}\, g^{\alpha\beta}\nabla_\alpha\nabla_\beta & \text{– Dvali-Gabadadze-Porratti (DGP) model.} \end{array} \tag{24}$$

The key point of new technique is that one-loop duality relation (15) being rewritten in the following form

$$\mathbf{Tr}\, \ln(\mathbf{F}_N + M^2) - \mathbf{Tr}\, \ln(\mathbf{F}_D + M^2) = \mathrm{Tr}\, \ln\left(F^{ind}_M + \mathfrak{f}\right) \tag{25}$$

and being expanded in inverse powers of M gives the difference between Schwinger-DeWitt coefficients of heat kernel trace of generalized Neumann operator $\mathbf{F}_N$ in the bulk and trace of Dirichlet operator $\mathbf{F}_D$ in terms of expansion coefficients of *boundary* operator $\left(F_M^{ind} + \mathfrak{f}\right)$.

$$\begin{aligned}
&\mathbf{Tr}\,\ln(\mathbf{F}_{N,D} + M^2) \\
&\sim \int\limits_0^\infty \frac{ds}{s}\, s^{-\frac{d+1}{2}}\, e^{-sM^2} \sum_{n=0}^{\infty} \left[s^n\,\mathbf{A}_n + s^{n/2}\, B_{n/2}^{N,D} + \ln\left(s\mu^2\right) s^{n/2}\, C_{n/2}^N\right] \\
&\sim M^{d+1} \sum_{n=0}^{\infty} \left[M^{-2n}\,\mathbf{A}_n + M^{-n}\left(B_{n/2}^{N,D} + C_{n/2}^N\right) - \ln\left(M^2/\mu^2\right) M^{-n}\, C_{n/2}^N\right]
\end{aligned} \tag{26}$$

Note that in l.h.s. of (25) bulk contributions $(\mathbf{A}_n)$ from above expansion of generalized Neumann and Dirichlet operators are identical and therefore cancel out.

Finally one comes to the following identification

$$\mathrm{Tr}\,\ln\left(F_M^{ind} + \mathfrak{f}\right) \sim M^{d+1} \sum_{n=0}^{\infty} M^{-n} \int\limits_{brane} \left[\left(b_{n/2}^N + c_{n/2}^N\right) - \ln\left(M^2/\mu^2\right) c_{n/2}^N - b_{n/2}^D\right] \tag{27}$$

Boundary coefficients $b_{n/2}^D$ for Dirichlet case can be considered as reference ones since there exist explicit calculational schemes and they were calculated to rather high orders n (see e.g. [11, 12]). The Dirichlet case is a special one since it is unique for a given manifold – it does not depend of boundary operator $\mathfrak{f}$. Mainly for the same reason Dirichlet case is the simplest one.

The only thing one needs in our scheme to obtain generalized Neumann Schwinger-DeWitt coefficients is the handy tool to calculate the large M expansion for trace of logarithm of boundary operator $\left(F_M^{ind} + \mathfrak{f}\right)$ in l.h.s of (27). Despite the nonlocality of boundary operator F^{ind} great simplification comes from the fact that boundary (brane) is itself the d-dimensional manifold without boundary.

As was mentioned in previous section in braneworld scenarios where branes are symmetric manifolds F^{ind} can be expressed explicitely as the function of intrinsic and embedding geometric brane quantities and tangential boundary covariant derivatives ∇_α. This allows one to use straightforwardly the theory of asymptotic expansion of integrals with weak pequliarity to obtain the expansion in inverse powers of M. Finally, identifying terms at the same powers of M in both sides of (27) one attains the aim.

For general case of bounded manifold a recurrent scheme for calculation of l.h.s. in (27) also can be suggested. One possible way is as follows (for simplicity consider scalar case with operator $\mathbf{F}_M = -\,{}^{d+1}\Box + M^2 = -\partial_y^2 - k(x,y)\partial_y - \Box_x(y) + M^2$ in gauss normal coordinates in the bulk). With the help of proper time representation for one-dimensional formal operator $-\partial_y^2 + \hat{V}(y)$ and subsequent restriction to $y = 0$ (where the brane is placed) and integration over proper time one can expand operator (13) $F_M^{ind} = \left|\overrightarrow{\partial_y}\, \frac{-1}{-\partial_y^2 - k(x,y)\partial_y - \Box_x(y) + M^2}\, \overleftarrow{\partial_y}\right|_{y=0}$ either in inverse powers of M or in inverse (odd) powers of $\sqrt{M^2 - \Box(0)}$. Coefficients of this expansion are d-dimensional tangential to brane differential operators expressed in terms of extrinsic curvature, brane Riemann curvature, their normal and tangential derivatives and tangential derivatives themselves [13]. Then scheme depends on what type of operator $\mathfrak{f}$. When $\mathfrak{f} = -\Box$ as in DGP case or when it is a potential term (case of Robin boundary value problem) then it is possible to express coefficients of inverse M expansion of l.h.s. recurrently using known Schwinger-DeWitt coefficients of $\Box$ on brane. The case when $\mathfrak{f}$ is first order in tangential derivatives (oblique boundary conditions case) with dimensionless coefficients should be treated in slightly different way [13].

Examples of implementing this scheme for some nontrivial but particular cases already exist. For instance special (toy) DGP case is considered in [9] where for flat background inverse M expansion coefficients are found and a nonperturbative hypergeometric representation of l.h.s. (27) dependence on DGP scale is obtained. Some insight on how to treat the oblique boundary value problem case can be found in [6].

5. Conclusions

To summarize, we conclude that the technique for quantum effects in brane models is more complicated than in systems without boundaries. Moreover, it does not reduce to a simple bookkeeping of surface terms in the heat kernel expansion of [11, 12], and so on, because of the complicated square-root structure of the brane propagator F^{ind} mediating the effect of the generalized Neumann boundary conditions. The proper time method that was fundamental and efficient in models without boundaries in our approach appear as a derivative one in an alternative calculational scheme. Namely, the surface terms in the heat kernel expansion can be recovered from the $1/M$-expansion of the action obtained by a different method.

Acknowledgements

D.N. is partially supported by Russian Foundation for Basic Research under the grant No 05-01-01049. The work of A.B. was supported by the Russian Foundation for Basic Research under the grant No 08-02-00725. D.N. is also grateful to the Center of Science and Education of Lebedev Institute and target funding program of Presidium of Russian Academy of Sciences for support. This work was also supported by the LSS grant No 4401.2006.2.

References

[1] Barvinsky A O and Nesterov D V 2003 *Nucl. Phys.* B **654** 225 (*Preprint* hep-th/0210005)
[2] Randall L and Sundrum S 1999 *Phys. Rev. Lett.* **83** 4690 (*Preprint* hep-th/9906064)
[3] Dvali G R, Gabadadze G and Porrati M 2000 *Phys. Lett.* B **485** 208 (*Preprint* hep-th/0005016)
[4] Barvinsky A O 2005 The Gospel according to DeWitt revisited: quantum effective action in braneworld models *Preprint* hep-th/0504205
[5] Barvinsky A O 2006 *Phys. Rev.* D **74** 084033 (*Preprint* hep-th/0608004)
[6] Barvinsky A O and Nesterov D V 2006 *Phys.Rev.* D **73** 066012 (*Preprint* hep-th/0512291)
[7] Barvinsky A O, Kamenshchik A Yu, Rathke A and Kiefer C 2003 *Phys.Rev.* D **67** 023513 (*Preprint* hep-th/0206188)
[8] Barvinsky A O and Solodukhin S N 2003 *Nucl.Phys.* B **675** 159-178 (*Preprint* hep-th/0307011)
[9] Barvinsky A O, Kamenshchik A Yu, Kiefer C and Nesterov D V 2007 *Phys.Rev.* D **75** 044010 (*Preprint* hep-th/0611326)
[10] Pujolas O 2006 *JCAP* **0610** 004 (*Preprint* hep-th/0605257)
[11] Vassilevich D V 2003 Heat kernel expansion: user's manual *Preprint* hep-th/0306138
[12] McAvity D M and Osborn H 1991 *Class.Quantum Grav* **8** 603-638;
McKean H P and Singer I M 1967 *J. Diff. Geom.* **1** 43
[13] Barvinsky A O and Nesterov D V, *in preparation*

Journal of Physics: Conference Series **128** (2008) 012012 doi:10.1088/1742-6596/128/1/012012

Application of Noether's theorem to extended particles

A Smida, M Hachemane and A H Hamici
Faculté de Physique, USTHB B.P.32 El-Alia Bab-Ezzouar 16111 Alger Algeria.
E-mail: hachemane@wissal.dz

Abstract. We consider an approach similar to field theory for the application of Noether theorem to extended paticles. We obtain Euler-Lagrange equations for the extended particles as well as an equation binding the internal and external currents. The concrete case of spin 1/2 is considered in detail.

1. Introduction

Previously, we have constructed a model of extended particles based on a fiber bundle structure ([1] and references therein). The extended particle is represented by a bilocal field $\psi(x, \xi)$ where x belongs to the base manifold and ξ to a fiber. The base has been interpreted as the usual (external) space-time which may be Riemannian while the fiber corresponds to an internal space-time accounting for the extension of the particles and having a definite symmetry (Galilei, Minkowki, or de Sitter symmetry). The bilocal field $\psi(x, \xi)$ describes an external quantum mode located at x and an internal one located at ξ. Both modes compose the particle as in the two-body problem (x being the center of mass and ξ the relative motion). The basic foundations of the model have been drawn from a functional theory [2], a semi-classical geometric model [3], and a geometro-stochastic theory [4, 5].

This led us to the conclusion that it may be worth constructing a Lagrangian variant of our model. The present work constitutes one aspect of this construction program. We consider flat external and internal space-times and a vanishing connection and carry out a simple Lagrangian calculation [6]. In section 2, we determine the equations of motion for a set of fields $u_i(x, \xi)$ which are no more than the usual Euler-Lagrange equations with an extra term corresponding to the internal motion. In section 3, Noether's theorem is used to determine the currents of the theory. The latter will be split into external and internal components (whose expressions are analogous to the well known ones [6]) and to additional components representing a mixing between the external and internal spaces. In section 4, we apply the formalism to a specific Lagrangian of bispinor fields before we conclude the work.

2. Euler Lagrange equations

We consider a set of bilocal fields $u_i(x, \xi)$ where $x = \left(x^k\right)$ corresponds to the external Minkowski space-time and $\xi = (\xi^\alpha)$ corresponds to the internal Minkowski space-time so that both superscripts k and α run from 0 to 3. The Lagrangian density is chosen as usual with a distinction between the external and internal derivatives

$$\begin{aligned} L &= L\left(u_i(x,\xi), u_{i;k}(x,\xi), u_{i;\alpha}(x,\xi)\right) \\ u_{i;k} &= \partial_k u_i = \partial u_i/\partial x^k \ ; \qquad u_{i;\alpha} = \partial_\alpha u_i = \partial u_i/\partial \xi^\alpha \end{aligned} \tag{1}$$

Euler-Lagrange equations correspond to an extremum of the action

$$A = \int L dx d\xi \tag{2}$$

$$\delta A = \int \bar{\delta} L dx d\xi = 0 \tag{3}$$

with respect to variations of the fields at fixed space-time points

$$u_i'(x,\xi) = u_i(x,\xi) + \bar{\delta} u_i \tag{4}$$

where $\bar{\delta} u_i$ is the variation of the form of the function u_i. The Lagrangian density variation is

$$\bar{\delta} L = \bar{\delta} u_i \partial L/\partial u_i + \bar{\delta} u_{i;k} \partial L/\partial u_{i;k} + \bar{\delta} u_{i;\alpha} \partial L/\partial u_{i;\alpha} \tag{5}$$

Using the expression of $\bar{\delta} L$ in $\delta A = 0$ and the well known integration by parts with vanishing variations of the fields at the boundaries, the Euler-Lagrange Equations are derived

$$\partial L/\partial u_i - \partial_k \left[\partial L/\partial u_{i;k}\right] - \partial_\alpha \left[\partial L/\partial u_{i;\alpha}\right] = 0 \tag{6}$$

Up to this point, nothing new is obtained for these obvious equations of motion which are ordinary Euler-Lagrange equations with an additional term $\partial_\alpha \left[\partial L/\partial u_{i;\alpha}\right]$ due to the internal degrees of freedom. However, when the Lagrangian density is suitably chosen, these equations will contain influences between the external and internal motions. This will be shown after calculating the currents in a general form.

3. Noether's theorem

Noether's theorem states that the invariance of the action with respect to a n parameter group leads to the existence of n conserved quantities provided the equations of motion hold true.

Hence, we consider a transformation of the external and internal space-time variables

$$x' = x + \delta x \tag{7}$$

$$\xi' = \xi + \delta \xi \tag{8}$$

The fields undergo the usual transformations

$$u_i'(x',\xi') = u_i(x,\xi) + \delta u_i \tag{9}$$

$$\delta u_i = \bar{\delta} u_i + \delta x^k u_{i;k} + \delta \xi^\alpha u_{i;\alpha} \tag{10}$$

so that the lagrangian density variation becomes

$$\delta L = \bar{\delta} L + \delta x^k \partial L/\partial x^k + \delta \xi^\alpha \partial L/\partial \xi^\alpha \tag{11}$$

Invariance of the action means that

$$\delta A = \int \delta L dx d\xi + \int L \delta (dx d\xi) = 0 \tag{12}$$

where the variation of the integration measure

$$\delta\left(dxd\xi\right)=\left(1-|J|\right)dxd\xi \tag{13}$$

is given by the Jacobian determinant

$$|J|\approx 1+\frac{\partial\delta x^{k}}{\partial x^{k}}+\frac{\partial\delta\xi^{\alpha}}{\partial\xi^{\alpha}} \tag{14}$$

Replacing all the above values in the invariance condition without using the equations of motion, one finds the following equation

$$\frac{\delta L}{\delta u_i}\bar{\delta}u_i=\partial_k\left(\delta\omega^n\Theta_n^k\right)+\partial_\alpha\left(\delta\omega^n\Theta_n^\alpha\right) \tag{15}$$

where $\delta\omega^n$ are the parameters of the symmetry. The Lagrangian derivatives are given by

$$\frac{\delta L}{\delta u_i}=\partial L/\partial u_i-\partial_k\left[\partial L/\partial u_{i;k}\right]-\partial_\alpha\left[\partial L/\partial u_{i;\alpha}\right] \tag{16}$$

and the currents are

$$\Theta_n^k = \frac{-1}{\delta\omega^n}\left[\bar{\delta}u_i\partial L/\partial u_{i;k}+\delta x^k L\right] \tag{17}$$

$$\Theta_n^\alpha = \frac{-1}{\delta\omega^n}\left[\bar{\delta}u_i\partial L/\partial u_{i;\alpha}+\delta \xi^\alpha L\right] \tag{18}$$

It is at this stage that the symmetry of the extended particle comes into play. In the general case, we have

$$\delta x^k = \delta\omega^n\left(\left[I_n\right]_l^k x^l+\left[I_n\right]_\alpha^k\xi^\alpha\right) \tag{19}$$

$$\delta \xi^\alpha = \delta\omega^n\left(\left[I_n\right]_l^\alpha x^l+\left[I_n\right]_\beta^\alpha\xi^\beta\right) \tag{20}$$

The generators I of the symmetry are denoted T in the fields representation

$$\delta u_i = \delta\omega^n\left[T_n\right]u_i \tag{21}$$

$$\bar{\delta}u_i = \delta\omega^n\left\{\left[T_n\right]u_i-u_{i;k}\left(\left[I_n\right]_l^k x^l+\left[I_n\right]_\alpha^k\xi^\alpha\right)-u_{i;\alpha}\left(\left[I_n\right]_l^\alpha x^l+\left[I_n\right]_\beta^\alpha\xi^\beta\right)\right\} \tag{22}$$

The currents can be decomposed

$$\Theta_n^k = \theta_n^k+\sigma_n^k \tag{23}$$

$$\Theta_n^\alpha = \theta_n^\alpha+\sigma_n^\alpha \tag{24}$$

The components

$$\theta_n^k = -\partial L/\partial u_{i;k}\left(\left[T_n\right]u_i-u_{i;j}\left[I_n\right]_l^j x^l\right)-L\left[I_n\right]_l^k x^l \tag{25}$$

$$\theta_n^\alpha = -\partial L/\partial u_{i;\alpha}\left(\left[T_n\right]u_i-u_{i;\lambda}\left[I_n\right]_\beta^\lambda \xi^\beta\right)-L\left[I_n\right]_\beta^\alpha \xi^\beta \tag{26}$$

have expressions which are identical to those corresponding to point-like external or internal particles. Depending on the subscript n, they may correspond to external or internal components

Journal of Physics: Conference Series **128** (2008) 012012 doi:10.1088/1742-6596/128/1/012012

(note that this not really true since L, $\partial L/\partial u_{i;k}$, and $\partial L/\partial u_{i;\alpha}$ contain both external and internal parts), or they may contain mixing terms. The other components

$$\sigma_n^k = -\partial L/\partial u_{i;k}\left(-u_{i;j}\,[I_n]^j_\alpha\,\xi^\alpha - u_{i;\alpha}\left([I_n]^\alpha_l\,x^l + [I_n]^\alpha_\beta\,\xi^\beta\right)\right) - L\,[I_n]^k_\alpha\,\xi^\alpha \quad (27)$$

$$\sigma_n^\alpha = -\partial L/\partial u_{i;\alpha}\left(-u_{i;k}\left([I_n]^k_l\,x^l + [I_n]^k_\beta\,\xi^\beta\right) - u_{i;\beta}\,[I_n]^\beta_l\,x^l\right) - L\,[I_n]^\alpha_l\,x^l \quad (28)$$

contain explicit mixing between external and internal space-time components.

If we consider the case where the external and internal symmetries act separately ($[I_n]^k_\alpha = [I_n]^\alpha_l = 0$), then the latter expressions simplify extremely

$$\sigma_n^k = \partial L/\partial u_{i;k}\left(u_{i;\alpha}\,[I_n]^\alpha_\beta\,\xi^\beta\right) \quad (29)$$

$$\sigma_n^\alpha = \partial L/\partial u_{i;\alpha}\left(u_{i;k}\,[I_n]^k_l\,x^l\right) \quad (30)$$

4. Bispinor fields

Let us construct a Lagrangian density assuming that the fields $u(x,\xi)$ are bispinors $u(x,\xi) = u^{AA'}(x,\xi)$ where the first index A transforms by an external Lorentz group representation and the second one (A') transforms with an internal Lorentz group. The conjugate fields $\bar{u} = u^*\left(\gamma^0 \otimes \Gamma^0\right)$ are expressed with external and internal Dirac matrices γ and Γ, respectively. We chose the following expression for the Lagrangian

$$\begin{aligned} L = {} & \frac{i}{2}\left[\bar{u}\gamma^k\partial_k u - \partial_k\bar{u}\gamma^k u\right] \\ & + \frac{i}{2}\left[\bar{u}\Gamma^\alpha\partial_\alpha u - \partial_\alpha\bar{u}\Gamma^\alpha u\right] \\ & - \frac{1}{2}\left[\partial_k\bar{u}\gamma^k\Gamma^\alpha\partial_\alpha u + \partial_\alpha\bar{u}\gamma^k\Gamma^\alpha\partial_k u\right] \\ & - m\bar{u}u \end{aligned} \quad (31)$$

Obvious simplifying notations have been used such that $\gamma^k = \gamma^k \otimes 1$, $\Gamma^\alpha = 1 \otimes \Gamma^\alpha$, and $\gamma^k\Gamma^\alpha = \gamma^k \otimes \Gamma^\alpha$. The equations of motion

$$\left[i\left(\gamma^k\partial_k + \Gamma^\alpha\partial_\alpha\right) - m + \left(\gamma^k\partial_k \otimes \Gamma^\alpha\partial_\alpha\right)\right] u = 0 \quad (32)$$

$$\bar{u}\left[-i\left(\overleftarrow{\partial}_k\gamma^k + \overleftarrow{\partial}_\alpha\Gamma^\alpha\right) - m + \left(\overleftarrow{\partial}_k\gamma^k \otimes \overleftarrow{\partial}_\alpha\Gamma^\alpha\right)\right] = 0 \quad (33)$$

contain, in addition to the external and internal kinetic terms, a mixing term $\left(\gamma^k\partial_k \otimes \Gamma^\alpha\partial_\alpha\right)$ due to the chosen form of the Lagrangian.

As to the currents, let us begin with translations in external and internal spaces. In this case, n is either l or β. The generators are such that $[I_l]^k_j\,x^j$ and $[I_\beta]^\alpha_\lambda\,\xi^\lambda$ should be replaced by δ^k_l and δ^α_β, and $T_n = 0$. The external and internal components of the energy-momentum tensor retain their usual expressions

$$\theta^k_l = \partial L/\partial u_{i;k}\left(u_{i;l}\right) - L\delta^k_l \quad (34)$$

$$\theta^\alpha_\beta = \partial L/\partial u_{i;\alpha}\left(u_{i;\beta}\right) - L\delta^\alpha_\beta \quad (35)$$

The mixing components read

$$\theta^k_\beta = \theta^\alpha_k = 0 \quad (36)$$

$$\sigma^k_\beta = \frac{i}{2}\left[\bar{u}\gamma^k + i\partial_\alpha\bar{u}\left(\gamma^k \otimes \Gamma^\alpha\right)\right]\partial_\beta u \quad (37)$$

$$\sigma^\alpha_k = \frac{i}{2}\left[\bar{u}\Gamma^\alpha + i\partial_k\bar{u}\left(\gamma^k \otimes \Gamma^\alpha\right)\right]\partial_k u \quad (38)$$

Journal of Physics: Conference Series **128** (2008) 012012 doi:10.1088/1742-6596/128/1/012012

For Lorentz rotation in the external plane (m,n) and the internal plane (μ,ν), the generators act in the following form

$$\left[I_{(n,m)}\right]^k_j x^j = x_m\delta^k_n - x_n\delta^k_m \tag{39}$$

$$\left[I_{(\nu,\mu)}\right]^\alpha_\beta \xi^\beta = \xi_\mu\delta^\alpha_\nu - \xi_\nu\delta^\alpha_\mu \tag{40}$$

The mixing components are

$$\theta^k_{(\nu,\mu)} = \frac{i}{2}\left[\bar{u}\gamma^k + i\partial_\alpha\bar{u}\left(\gamma^k\otimes\Gamma^\alpha\right)\right]\left[1\otimes(\Gamma_\nu\Gamma_\mu/2)\right]u \tag{41}$$

$$\theta^\alpha_{(n,m)} = \frac{i}{2}\left[\bar{u}\Gamma^\alpha + i\partial_k\bar{u}\left(\gamma^k\otimes\Gamma^\alpha\right)\right]\left[(\gamma_n\gamma_m/2)\otimes 1\right]u \tag{42}$$

$$\sigma^k_{(\nu,\mu)} = \frac{i}{2}\left[\bar{u}\gamma^k + i\partial_\alpha\bar{u}\left(\gamma^k\otimes\Gamma^\alpha\right)\right](\xi_\mu\partial_\nu - \xi_\nu\partial_\mu)\,u \tag{43}$$

$$\sigma^\alpha_{(n,m)} = \frac{i}{2}\left[\bar{u}\Gamma^\alpha + i\partial_k\bar{u}\left(\gamma^k\otimes\Gamma^\alpha\right)\right](x_m\partial_n - x_n\partial_m)\,u \tag{44}$$

5. Conclusion

We considered a direct product symmetry for a lagrangian of bilocal fields representing extended particles. Very simple calculations lead to complicated currents with external, internal and mixed components. The chosen Lagrangian has thrown some light on these expressions but the results are not conclusive. We hope that forthcoming works will sharpen the physical interpretation of each component.

References

[1] Hachemane M Smida A and Djelid R 1999 *Found. Phys* **29** 1479
[2] Destouches J L 1956 *La Quantification en Théorie Fonctionnelle des Corpuscules* (Gauthier-Villars: Paris)
[3] Drechsler W 1975 *Fortschr. Phys.* **23** 607
[4] Prugovečki E 1992 *Quantum Geometry* (Dordrecht: Kluwer)
[5] Prugovečki E 1995 *Principles of Quantum General Relativity* (Singapore and London: World Scientific)
[6] Bogoliubov N N and Shirkov D V 1980 *Introduction to the Theory of Quantized Fields* (Wiley)

Journal of Physics: Conference Series **128** (2008) 012013 doi:10.1088/1742-6596/128/1/012013

A quantum architecture for multiplying signed integers

J J Álvarez-Sánchez[1], **J V Álvarez-Bravo**[1] **and L M Nieto**[2]

[1] Departamento de Informática, E.U. de Informática, Universidad de Valladolid, 40005 Segovia, Spain

[2] Departamento de Física Teórica, Atómica y Óptica, Universidad de Valladolid, 47071 Valladolid, Spain

E-mail: jjalvarez@infor.uva.es, jvalvarez@infor.uva.es, luismi@metodos.fam.cie.uva.es

Abstract. A new quantum architecture for multiplying signed integers is presented based on Booth's algorithm, which is well known in classical computation. It is shown how a quantum binary chain might be encoded by its flank changes, giving the final product in 2's-complement representation.

1. Introduction and motivation

Factoring big numbers in polynomial time, searching for an item in an unstructured data base faster than any classical (not quantum) algorithm, or achieving quantum cryptographic protocols for a quantum public key distribution that gives us better evesdroping detection techniques, are some of the spectacular theoretical results [1, 2, 3] that have supported the idea of building a quantum computer or, at least, a quantum microprocessor for performing some specific tasks. In this context, a fundamental issue that has to be considered is how could we work out arithmetic calculus by means of quantum architectures.

Indeed, the first step for performing complex quantum arithmetic calculus is to build a quantum architecture which will be able to add two positive integer numbers (a 'quantum adder'). This quantum algorithm has been the target of many quantum arithmetic researchers, who can be divided in two groups: those who carry out the quantization of classical arithmetic algorithms [4, 5, 6] (who, indeed, rewrite classical arithmetical algorithms, that are irreversible, as classical reversible ones), and those who use the Quantum Fourier Transform for building adders [7].

In this paper the first approach is chosen in order to design a quantum circuit that will be able to reproduce the outcomes of a classical arithmetic circuit for multiplying signed integers. The keystone is to avoid the construction of different circuits for different arithmetic tasks. We could consider, for instance, the subtraction of two positive numbers to illustrate the point of view we are assuming. In principle, we could tackle the problem by just applying the corresponding subtraction truth table. Nevertheless, if we adopt this point of view, hardware different from the one used for addition will necessary to perform the subtraction operation.

This work is organized as follows. In section 2 some relevant aspects of classical and quantum computer arithmetics are exposed. In section 3, a brief and basic description of the classical

Journal of Physics: Conference Series **128** (2008) 012013 doi:10.1088/1742-6596/128/1/012013

Booth algorithm for multiplying signed integers is given. Then, the quantum counterpart of Booth's algorithm is explicitly constructed in section 4, which is the central part of this work. We end with a discussion of the results and some indications of future work.

2. Classical and quantum computer arithmetics

Let us start by remembering that, in a classical computer, the arithmetic calculus is performed by the arithmetic logic unit, that is a part of the *brain* of the computer: the central processing unit (CPU). Within the arithmetic logic unit the fundamental arithmetic operations (addition, subtraction, multiplication and division) are carried out over the numbers, which are represented using Boolean algebra. We can assume that the same will be true in the quantum context, using quantum bits ('qubits'). Either in the classical or in the quantum case, any integer Boolean number can be written as

$$p = x_{n-1}2^{n-1} + x_{n-2}2^{n-2} + \ldots + x_0 2^0 \tag{1}$$

with $x_i = 0, 1$, and the algebraic internal operations $(+, \cdot)$, usually called **OR** and **AND**, are defined by the truth tables shown in table 1.

Table 1. True tables of the operations **AND** $(\cdot)$, **OR** $(+)$ and **XOR** $(\oplus)$.

x_i	y_i	$x_i \cdot y_i$	$x_i + y_i$	$x_i \oplus y_i$	c_i
0	0	0	0	0	0
0	1	0	1	1	0
1	0	0	1	1	0
1	1	1	1	0	1

Thus, one can introduce, in an obvious way, the addition of two integers just adding the bits that have the same power of two (x_i and y_i) modulo two, that we will represent by $x_i \oplus y_i$ (usually called **XOR**), and taking into account the carry bit (c_i), as shown in table 1. If we take into account the carry bit c_{i-1} due to the addition of the two previous bits x_{i-1} and y_{i-1}, the adder will have three inputs and a full addition (*full adder*) will be carried out, as shown in table 2.

Table 2. True table of the full addition of two bits x_i and y_i, taking into account the existence of the carry bit c_{i-1} of a previous addition.

x_i	y_i	c_{i-1}	$x_i \oplus y_i \oplus c_{i-1}$	c_i
0	0	0	0	0
0	0	1	1	0
0	1	0	1	0
0	1	1	0	1
1	0	0	1	0
1	0	1	0	1
1	1	0	0	1
1	1	1	1	1

The efficient way of working out operations with positive and negative numbers in classical Boolean arithmetics is to use the 2's-complement representation of the number, which allows us to work with the sign bit in the same way we do with any other bit. Given a Boolean number p of the form (1), its 2's-complement representation [8, 9] is

$$C_2(p) = 2^n - p. \tag{2}$$

That information representation will allow us to design simpler digital quantum circuits and will also avoid fundamental problems, such as having different 'positive' and 'negative' zeros. Signed integer multiplication is usually (classically) performed encoding the multiplier and performing additions, subtractions and bit displacements using the so-called Booth's algorithm [8]. The main idea of this algorithm, which will be analyzed later on, is to represent binary data as unsigned integers in order to perform additions. This can be achieved by the 2's-complement representation, that makes possible performing subtraction by means of addition [9]. With this idea in mind, we can work out additions and subtractions in a very easy way, and tackle the multiplication of signed integers. For this purpose, which is not at all a trivial arithmetic operation from the point of view of computer science, a quantum Booth multiplier (QBM) is presented *based* on the corresponding classical Booth multiplication algorithm, which is briefly described in the next section.

3. The classical Booth algorithm

The classical Booth multiplication algorithm, as a first step, encodes binary chains by means of their transitions between 0's and 1's as shown in Figure 1. Once these transitions are detected, the necessary displacements and additions are performed. This scheme extends to any number of blocks of 1's in a multiplier, including the case in which a single 1 is treated as a block. Indeed, we might say that the classical Booth's algorithm detects 'islands' of 1's.

0 0 1 0 1 1 1 0 1 1 0 1 0 0 1 0   0 1 $\bar{1}$ 1 0 0 $\bar{1}$ 1 0 $\bar{1}$ 1 $\bar{1}$ 0 1 $\bar{1}$

Figure 1. On the right part of the figure is Booth's encoding of the 16 bits chain shown on the left. Transitions $0 \to 1$ and $1 \to 0$ are represented by 1 and $\bar{1}$, respectively.

It is well known that multiplication can be achieved by adding appropriately shifted copies of the multiplicand and that the number of such operations in a block of 1's can be reduced to two if we observe that

$$2^n + 2^{n-1} + \cdots + 2^{n-k} = 2^{n+1} - 2^{n-k}. \tag{3}$$

Therefore, the algorithm encodes the binary alphabet $\{1, 0\}$ in another one, $\{0, 1, -1\}$, that shows the transitions among the binary logic states within a binary chain. Thus, when a -1 appears, a subtraction (using the 2's-complement representation) and a shift are performed, and when a 1 appears an addition and a shift are performed. In the other two cases, just an arithmetic shift is performed for each situation. The algorithm conforms to this scheme by performing a subtraction when the first 1 of the block is encountered ($1 \to 0$) and an addition when the end of the block is encountered ($0 \to 1$). It is straightforward to show that the same scheme works for a negative multiplier using the 2's-complement notation.

The architecture of Booth's algorithm, just described, is shown in Fig. 2, where the multiplier and the multiplicand are placed in the Q and M registers, respectively; there is also a 1-bit register placed logically to the right of the Less Significant Bit (LSB) Q_0 of the Q register and designated Q_{-1}; finally, the result of the multiplication is loaded in the A and Q registers, the first one initialized to 0, and is read from the Most Significant Bit (MSB) on the left to the LSB on the right in the juxtaposition of the registers A and Q.

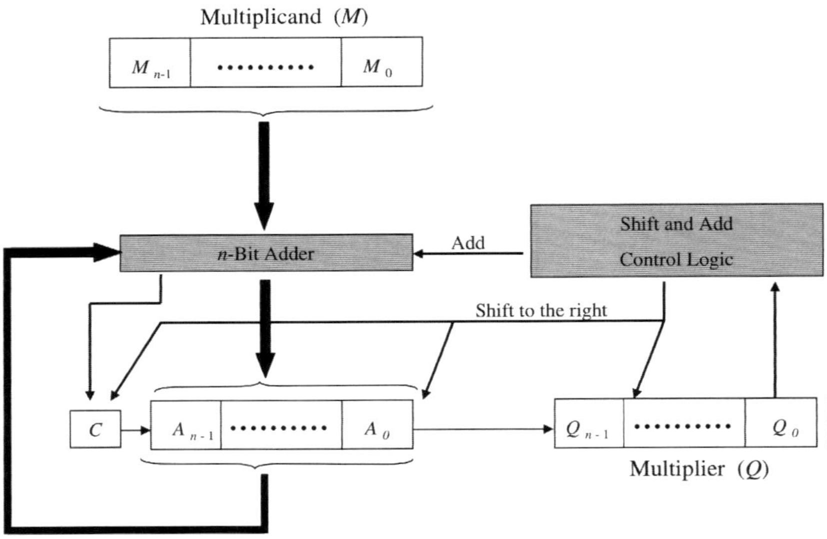

Figure 2. Classical multiplier architecture.

Control logic scans the bits of Q_0 and Q_{-1} one at a time. According to Fig. 3, if the two bits are the same, then all the bits of A, Q, and Q_{-1} are shifted to the right one bit. If the two bits differ, then the multiplicand is added or subtracted from the A register, according as the two bits are changing $0 \rightarrow 1$ or $1 \rightarrow 0$. Following the addition or subtraction, the right shift occurs. In either case, the right shift is such that the leftmost bit of A, namely A_{n-1}, not only is shifted into A_{n-2}, but also remains in A_{n-1}. This is required to preserve the sign of the number in A and Q and it is named *arithmetic shift* [8], since it preserves the sign bit.

4. The quantum Booth algorithm

Now that the classical Booth algorithm is well understood, our version of the quantum Booth's algorithm will be presented. First of all, we have to represent the new alphabet $\{0, 1, -1\}$ by three orthonormal quantum states belonging to the basis of $\mathcal{H}^2 \bigotimes \mathcal{H}^2$ as follows:

$$|00\rangle \equiv 0, \quad |01\rangle \equiv 1, \quad |10\rangle \equiv -1. \tag{4}$$

They have been chosen in this way because they just differ from each other in one binary digit, a useful property that will simplify the control stage in the QBM architecture. To encode two adjacent qubits $|x_{i+1}\rangle$ and $|x_i\rangle$ an ancillary $|0\rangle$ is needed, and therefore the number of qubits required for that is three, which will be denoted as

$$|x_{i+1}\rangle_1, \quad |0\rangle_2 \quad \text{and} \quad |x_i\rangle_3. \tag{5}$$

They are transformed into

$$|x_{i+1}\rangle_1, \quad |x'_{i+1,i}\rangle_2 \quad \text{and} \quad |x''_{i+1,i}\rangle_3 \tag{6}$$

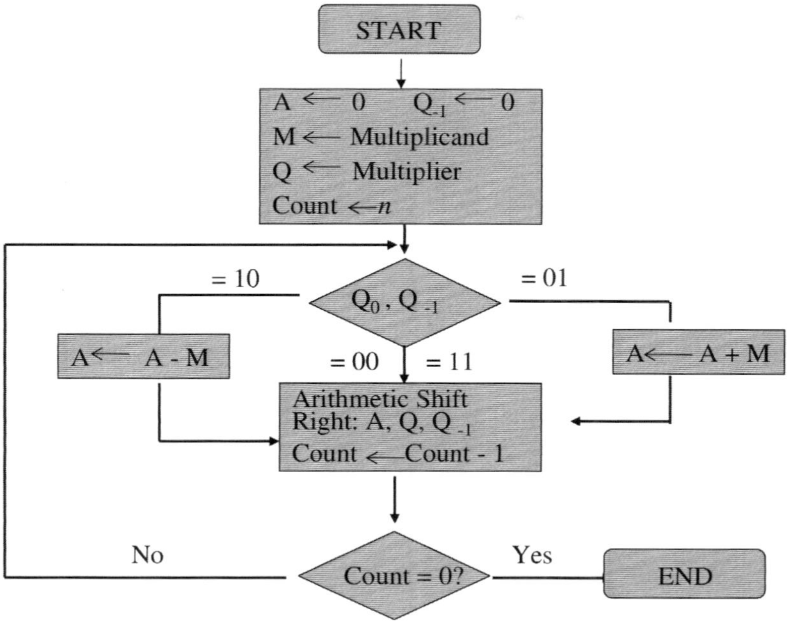

Figure 3. Flow chart for classical Booth's algorithm.

respectively, by means of the circuit shown in Figure 4, that encodes all the input possibilities according to the rules shown in Figure 5, which involves **CNOT** and **CSWAP** gates. From now on, we will denote by $\mathcal{BE}$ this Booth encoding procedure.

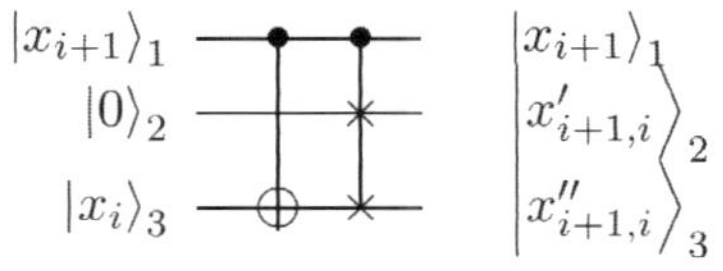

Figure 4. Booth's encoding of two adjacent qubits.

$$
\begin{array}{ccccc}
 & \textbf{CNOT} & & \textbf{CSWAP} & \\
|0_1, 0_2, 0_3\rangle & \longrightarrow & |0_1, 0_2, 0_3\rangle & \longrightarrow & |0_1, 0_2, 0_3\rangle \\
|0_1, 0_2, 1_3\rangle & \longrightarrow & |0_1, 0_2, 1_3\rangle & \longrightarrow & |0_1, 0_2, 1_3\rangle \\
|1_1, 0_2, 0_3\rangle & \longrightarrow & |1_1, 0_2, 1_3\rangle & \longrightarrow & |1_1, 1_2, 0_3\rangle \\
|1_1, 0_2, 1_3\rangle & \longrightarrow & |1_1, 0_2, 0_3\rangle & \longrightarrow & |1_1, 0_2, 0_3\rangle
\end{array}
$$

Figure 5. Transforming the state of two adjacent qubits.

An example showing how a 4-qubit $\mathcal{BE}$ encodes is presented in Figure 6. Notice that for reversibility purposes the inverse of the circuit has to be applied, $\mathcal{BE}^{-1}$, which is the same circuit as $\mathcal{BE}$ but uses the outputs as inputs and viceversa.

Following this 4-qubit example, let us consider now the multiplication of two 4-qubit numbers represented by the multiplier ket $|x_3, x_2, x_1, x_0\rangle$ and the multiplicand ket $|y_3, y_2, y_1, y_0\rangle$. The

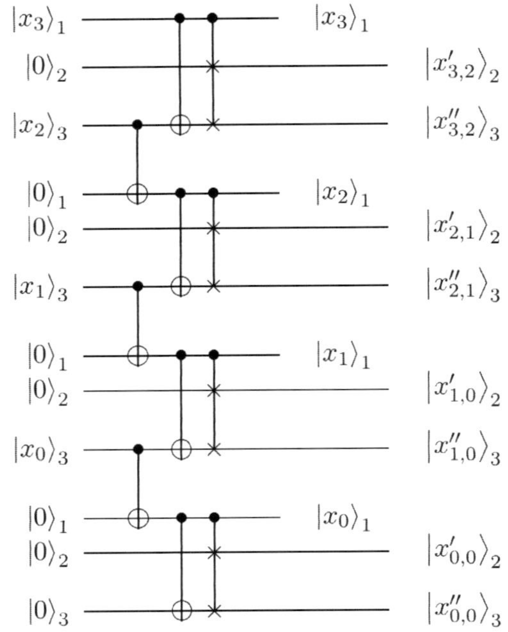

Figure 6. 4-qubits quantum Booth encoder

multiplier is encoded by the Booth encoder circuit and controls the logic operations on the multiplicand that load the proper partial products in their registers. The product (the final output) will have 8 qubits (the same length that will have the partial product registers) and is given in 2's-complement representation. As it is shown in Fig. 7 the partial products are arranged in order to reproduce the multiplication inherent displacements. In summary, the quantum circuit shown in Fig. 6, encodes the binary transitions $(i+1, i)$ and controls, by Toffoli gates, what will be stored at the quantum registers, where the partial products have been loaded. Hence, depending on the transitions in the quantum binary chain, we will perform one of the three following operations over the multiplicand:

- If there is no transition, $|00\rangle$, then 0's are loaded in the partial product registers.
- If there are up-transitions, $|01\rangle$, then the multiplicand is loaded in the partial product registers.
- If there are down-transitions, $|10\rangle$, then the multiplicand 1's-complement representation is loaded in the partial product registers.

Finally, we have to sum all the partial products that we have generated. If the multiplicand and the multiplier have n qubits, each partial product will have $2n$ qubits and they will be shifted, from the encoded multiplier LSB to the MSB, depending on the ordinal position of the multipliers bits. These displacements are implemented within the quantum $2n$-registers hardware.

After the $\mathcal{O}(\log n)$ sums have been performed we still have to do something more because we have used 1's-complement representation instead of 2's-complement representation. Indeed, we have saved the "carry" bit bearing in mind that 2's-complement representation is equal to 1's-complement representation plus one. Therefore, we have to add, to the whole partial products sum, another register where the 1's have been stored in their proper ordinal way, as one can

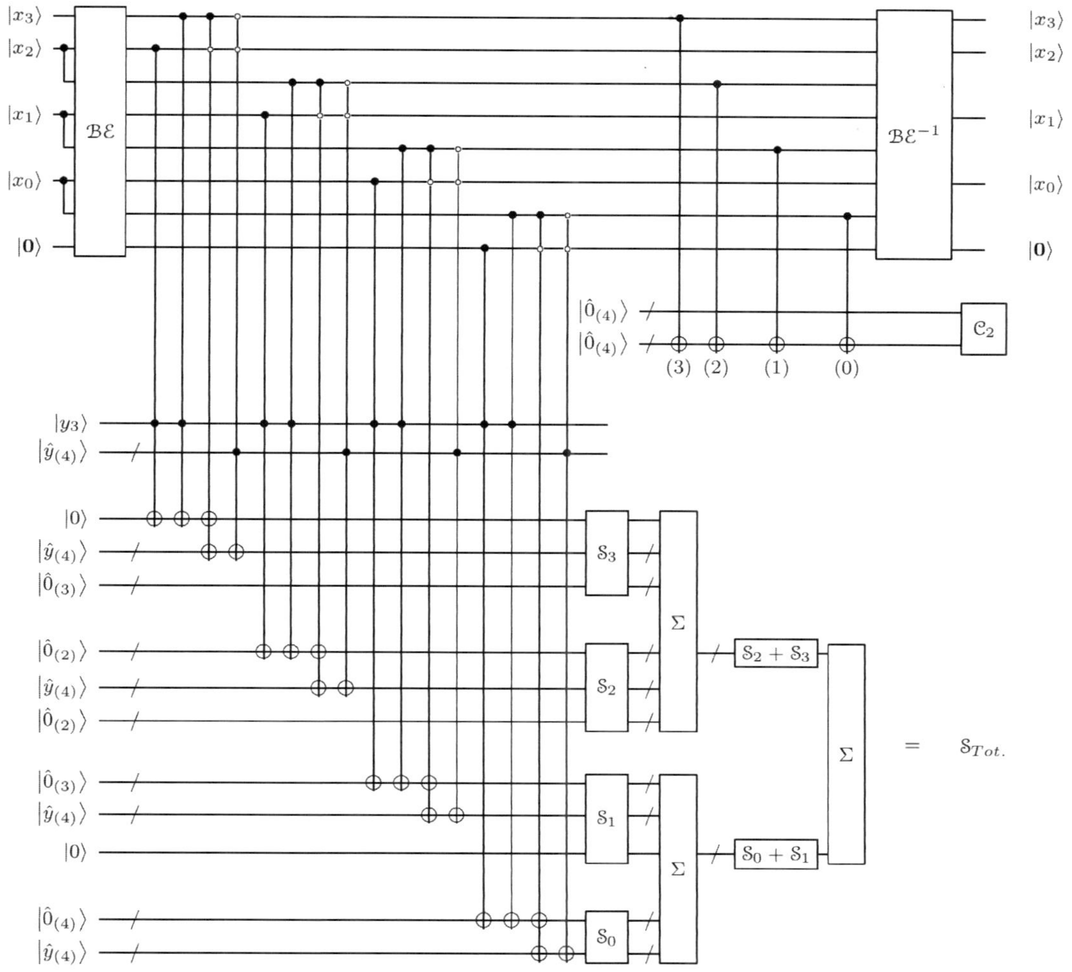

Figure 7. 4-qubits Quantum Booth's array multiplier circuit

see in the example displayed in Fig. 7. Therefore, the final output is the desired input product represented in 2's-complement notation.

As an example of how the QBM circuit shown in Fig. 7 works, let us trace now the fourth partial product $\mathcal{S}_3$ in Figure 7 of the multiplication of $M = 0010\,(2_{2)})$ by $Q = 0110\,(6_{2)})$ and $M' = 1110\,(-2_{2)})$ by $Q' = 0110\,(6_{2)})$, where $6_{B)} \equiv 1\,0 - 1\,0$ is the codification of $6_{2)} = 0110$ by Booth's circuit encoder. For the sake of simplicity the multiplier is the same for both examples, which are explicitly shown in Figure 8.

From the quantum circuit point of view the fourth partial product (for both examples) is the instantiation of the quantum register shown in Fig. 9 with the notation

$$|y_3, y_2, y_1, y_0\rangle \equiv \left|\hat{y}_{(4)}\right\rangle . \tag{7}$$

Then, the circuit adds, using the most efficient quantum adders that already exist [5], all the partial products that, eventually, give us the result called $\mathcal{S}_{Tot.}$. For achieving the final result we still have to add the content of the quantum register $\mathcal{C}_2$ where the proper value for working out the 2's-complement operation has been stored. In that example the value loaded in $\mathcal{C}_2$ is

$2_{2)}$					0	0	1	0
$6_{B)}$				$\times$	1	0	-1	0
	0	0	0	0	0	0	0	0
	1	1	1	1	1	0	1	0
	0	0	0	0	0	0	0	0
	0	0	0	1	0	0	0	0
	0	0	0	0	1	0	1	0

$-2_{2)}$					1	1	1	0
$6_{B)}$				$\times$	1	0	-1	0
	0	0	0	0	0	0	0	0
	0	0	0	0	0	0	1	0
	0	0	0	0	0	0	0	0
	1	1	1	1	0	0	0	0
	1	1	1	1	0	0	1	0

Figure 8. Two examples of multiplication using Booth's encoding.

00000010, giving us, as a final and correct result, 00001100 for the multiplication of $2_{2)} \times 6_{2)}$ and 11110100 (the 2's complement of 00001100) for the multiplication of $-2_{2)} \times 6_{2)}$.

$$
\begin{array}{rl}
|0\rangle & \leftarrow 0 \\
|\hat{y}_{(4)}\rangle & \leftarrow 0010 \\
|\hat{0}_{(3)}\rangle & \leftarrow 000 \\
\\
|\hat{0}_{(2)}\rangle & \leftarrow 00 \\
|\hat{y}_{(4)}\rangle & \leftarrow 0000 \\
|\hat{0}_{(2)}\rangle & \leftarrow 00 \\
\\
|\hat{0}_{(3)}\rangle & \leftarrow 111 \\
|\hat{y}_{(4)}\rangle & \leftarrow 1101 \\
|0\rangle & \leftarrow 0 \\
\\
|\hat{0}_{(4)}\rangle & \leftarrow 0000 \\
|\hat{y}_{(4)}\rangle & \leftarrow 0000
\end{array}
\qquad\qquad
\begin{array}{rl}
|0\rangle & \leftarrow 1 \\
|\hat{y}_{(4)}\rangle & \leftarrow 1110 \\
|\hat{0}_{(3)}\rangle & \leftarrow 000 \\
\\
|\hat{0}_{(2)}\rangle & \leftarrow 00 \\
|\hat{y}_{(4)}\rangle & \leftarrow 0000 \\
|\hat{0}_{(2)}\rangle & \leftarrow 00 \\
\\
|\hat{0}_{(3)}\rangle & \leftarrow 000 \\
|\hat{y}_{(4)}\rangle & \leftarrow 0001 \\
|0\rangle & \leftarrow 0 \\
\\
|\hat{0}_{(4)}\rangle & \leftarrow 0000 \\
|\hat{y}_{(4)}\rangle & \leftarrow 0000
\end{array}
$$

Figure 9. Quantum registers instantiation

5. Discussion and proposed future directions

It is important to notice that the QBM has a very regular structure. It can be easily implemented in order to build bigger circuits for bigger inputs than the one shown in Fig. 7. Its delay is $\mathcal{O}((\log n)^2)$ due to the fact that the adders structure is $\mathcal{O}(\log n)$ and that each adder has a logarithmic depth. One could reduce it by means of a CSA scheme [10], but it would then destroy the circuit's regular structure, which is an undesireble compensation. Thus, the bottlenecks are the adders and our future research efforts will go in the direction of clarifying whether $\mathcal{O}(\log n)$ is a fundamental bound, from the quantum physics point of view, or not.

Note that adder information processing [6] is classical because the input information is also classical; there is neither entanglement nor quantum superposition. Therefore, the Quantum Booth Algorithm architecture presented here is also a classical reversible one, as a consequence of following the research direction pointed by [4, 5, 6], and using only **TOFFOLI**, **SWAP** and **CNOT**, and no Hadamard gates. Indeed, what we have accomplished here is interesting by itself (to our knowledge, there are no quantum circuits for multiplying signed integers), but if we want to go a step further towards designing faster quantum circuits, it is necessary to reconsider the way in which the quantum arithmetic circuits have been designed. It is essential to take into account the intrinsic quantum ingredients: entanglement and superposition of states. Work in this direction is in progress.

Acknowledgments

The authors thank Profs. C Gómez, G Brassard, J M Fernández and M L Glasser for valuable discussions and advice. This work has been partially supported by Spanish Ministerio de Educación y Ciencia (Project MTM2005-09183) and Junta de Castilla y León (Excellence Project VA013C05). The Q-circuit package was used to produce the quantum circuits included in this paper.

References

[1] Shor P W 1997 *SIAM J. Comput.* **26** 1484 (*Preprint* quant-ph/9508027)
[2] Grover L K 1997 *Phys. Rev. Lett.* **79** 4709
[3] Bennett C H and Brassard G 1984 *Proc. of IEEE Int. Conf. on Computers Systems and Signal Porcessing (Bangalore)* p 175–179
[4] Vedral V, Barenco A, and Ekert A 1996 *Phys. Rev. A* **54** 147 (*Preprint* quant-ph/9511018)
[5] Cuccaro S A, Draper T G, Kutin S A, and Moulton D P 2004 A new quantum ripple-carry addition circuit *Preprint* quant-ph/0410184.
[6] Draper T G, Kutin S A, Rains E M, and Svore K M 2004 A logarithmic-deph quantum carry-lookahead adder *Preprint* quant-ph/0406142.
[7] Draper T G 2000 Addition on a quantum computer *Preprint* quant-ph/0008033.
[8] Stallings W 1996 *Computer Organization and Architecture* (New York: Prentice-Hall. Inc.)
[9] Mano M M 1979 *Digital logic and computer design* (New York: Prentice-Hall, Inc.)
[10] Sunar B 2006 Computer Arithmetic Circuits *Preprint* `http://ece.wpi.edu/~sunar/courses/ee579v/`

Journal of Physics: Conference Series **128** (2008) 012014 doi:10.1088/1742-6596/128/1/012014

The role of the Hamiltonian in the interpretation of quantum mechanics

Mario Castagnino[1] **and Olimpia Lombardi**[2]

[1] CONICET-Institutos de Fisica de Rosario y de Astronomia y Fisica del Espacio. Casilla de Correos 67, Sucursal 28, 1428, Buenos Aires, Argentina

[2] CONICET-Unversidad de Buenos Aires. Crisologo Larralde 3440, 1430, Buenos Aires, Argentina

E-mail: `mariocastagnino@citynet.net.ar, olimpiafilo@arnet.com.ar`

Abstract. In this paper we propose a new realist, non-collapse interpretation of quantum mechanics, which moves away from the prevailing trend in the subject by paying special attention to the physical relevance of the interpretation. In particular, our proposal endows the Hamiltonian of the system, systematically ignored in the traditional interpretations, with a central role: it distinguishes between systems and subsystems and is the main ingredient in the selection of the definite-valued observables. We show how this interpretation solves the measurement problem, both in the ideal and in the non-ideal version, and we argue for the physical relevance of the new definite-value assignment.

1. Introduction

In spite of the impressive empirical success of quantum mechanics, its interpretation is still an open problem. During many years since the birth of the theory, instrumentalist interpretations prevailed: the meaning of quantum mechanics was expressed in terms of measurements results. But in the last decades, several attempts to interpret quantum mechanics from a realist viewpoint have been proposed. David Mermin's "Ithaca interpretation" ([1]) is an example of this new trend. From this realist perspective, Mermin proposes six *desiderata* that any interpretation of quantum mechanics should satisfy:

(i) *The theory should describe an objective reality independent of the observers and their knowledge.*

(ii) *The concept of measurement should play no fundamental role.*

(iii) *The theory should describe individual systems, not just ensembles.*

(iv) *The theory should describe small isolated systems without having to invoke interaction with anything external.*

(v) *Objectively real internal properties of an isolated individual system should not change when something is done in another non-interacting system.*

(vi) *It suffices to base the interpretation of quantum mechanics on the interpretation of objective probabilities.*

Journal of Physics: Conference Series **128** (2008) 012014 doi:10.1088/1742-6596/128/1/012014

As it is well known, the Kochen-Specker theorem ([2]) is an insurmountable barrier to any classical-like interpretation of quantum mechanics, since it shows the contextuality of the theory by proving the impossibility of ascribing precise values to all the observables of a quantum system in a non-contradictory manner. Therefore, any realist, non-collapse interpretation is committed to select a *preferred context*, which defines the observables that will acquire definite values. This point is so relevant that Bub ([3]) classifies the best known realist non-collapse interpretations in terms of the observable R chosen for defining the preferred context in each one of them. For instance, in Bohr's complementarity interpretation ([4]), R is defined by the experimental arrangement; in the interpretation of Bohm ([5]), R is simply the position observable; for some modal interpretations ([6], [7], [8]), R is given by the biorthonormal decomposition (Schmidt) theorem.

When the many interpretations of quantum mechanics proposed through the years are reviewed from this general viewpoint, it is easy to realize that the Hamiltonian has been systematically ignored in the discussions. In this paper our purpose is to present a new realist interpretation, where the Hamiltonian of the system acquires a central relevance, besides its well known role in representing a constant of motion and in governing the dynamics of the system. As we will see, in this interpretation the Hamiltonian is decisive in the definition of systems and subsystems (Section 2), and in the selection of the definite-valued observables (Section 3). On this basis, the measurement problem receives an adequate answer (Section 4), and several traditional interpretative difficulties can be solved (Section 5). Moreover, the physical relevance of this interpretation can be argued for (Section 6). Finally, we will show that this new interpretation satisfies the six *desiderata* proposed by Mermin (Conclusions).

2. Systems and subsystems

In order to study the physical world, we have to identify the systems populating that world. We can cut out the physical reality in many different and arbitrary ways, but only when a portion of reality does not interact with others, we have a non-arbitrary, objective criterion to identify that portion as a system. For this reason, we will design our interpretation to account for those pieces of reality non-interacting with other pieces and, so, we will conceive only closed systems as quantum systems. On this basis, and by adopting an algebraic perspective, we define a *quantum system* S as a pair $(\mathcal{O}, H)$ such that (i) $\mathcal{O}$ is a space of self-adjoint operators on a Hilbert space $\mathcal{H}$, representing the *observables* of the system, (ii) $H \in \mathcal{O}$ is the time-independent *Hamiltonian* of the system, and (iii) being $\rho_0 \in \mathcal{O}'$ (where $\mathcal{O}'$ is the dual space of $\mathcal{O}$) the *initial state* of S, it evolves according to the Schrödinger equation in its von Neumann version.

Of course, any quantum system can be decomposed in parts in many ways; however, not any decomposition will lead to parts which are, in turn, quantum systems. This will be the case only when the components' behaviors are *dynamically independent* to each other, that is, the time-evolution of each component can be described independently to the remaining ones (see [9]). In other words, a quantum system can be split into subsystems when there is no interaction among the subsystems. On this basis, we will say that a quantum system $S : (\mathcal{O}, H)$ with initial state $\rho_0 \in \mathcal{O}'$ is *composite* when it can be partitioned into two quantum systems $S^1 : (\mathcal{O}^1, H^1)$ and $S^2 : (\mathcal{O}^2, H^2)$ such that (i) $\mathcal{O} = \mathcal{O}^1 \otimes \mathcal{O}^2$, and (ii) $H = H^1 \otimes I^2 + I^2 \otimes H^1$ (where I^2 and I^2 are the identity operators in the corresponding tensor product spaces). In this case, the initial states of S^1 and S^2 are obtained as the partial traces $\rho_0^1 = Tr_{(2)}\rho_0 \in \mathcal{O}^{1\prime}$ and $\rho_0^2 = Tr_{(1)}\rho_0 \in \mathcal{O}^{2\prime}$, and we will say that S^1 with initial state ρ_0^1 and S^2 with initial state ρ_0^2 are *subsystems* of the composite system $S = S^1 \cup S^2$.

It is worth stressing that this definition of composite system does not imply that the initial state ρ_0 of S is the tensor product $\rho_0^1 \otimes \rho_0^2$: this factored or uncorrelated state is a very special kind of state corresponding to independent preparations of the component systems (see

Journal of Physics: Conference Series **128** (2008) 012014 doi:10.1088/1742-6596/128/1/012014

[10]). On the contrary, in the general case the initial state is a correlated or *entangled state* $\rho_0 \in (\mathcal{O}^1 \otimes \mathcal{O}^2)'$; nevertheless, in spite of correlations, each subsystem evolves independently according to the Schrödinger equation. In fact, since there is no interaction between S^1 and S^2,

$$[H^1 \otimes I^2, I^2 \otimes H^1] = 0 \tag{1}$$

and, as a consequence,

$$\exp[-iHt] = \exp\left[-iH^1t\right] \exp\left[-iH^2t\right] \tag{2}$$

Therefore,

$$\rho^1(t) = Tr_{(2)}\rho(t) = Tr_{(2)}(e^{-iHt}\rho_0\, e^{iHt}) = e^{-iH^1t}(Tr_{(2)}\rho_0)\, e^{iH^1t} = e^{-iH^1t}\rho_0^1\, e^{iH^1t} \tag{3}$$

$$\rho^2(t) = Tr_{(1)}\rho(t) = Tr_{(1)}(e^{-iHt}\rho_0\, e^{iHt}) = e^{-iH^2t}(Tr_{(1)}\rho_0)\, e^{iH^2t} = e^{-iH^2t}\rho_0^2\, e^{iH^2t} \tag{4}$$

This means that S^1 and S^2 are dynamically independent: each one of them will evolve under the action of its own Hamiltonian.

It has to be emphasized that, although this definition of composite quantum system is completely general, the decomposition of a quantum system into subsystems is not always possible: it may happen that there is no partition of the whole S such that the total Hamiltonian can be expressed as a sum of component Hamiltonians. In this case, the quantum system is not composite and we will call it *elemental*. Therefore, this interpretation supplies a precise criterion to distinguish between elemental and composite systems, and such a criterion is *based on the system's Hamiltonian*.

On the other hand, given any two quantum systems $S^1 : (\mathcal{O}^1, H^1)$ and $S^2 : (\mathcal{O}^2, H^2)$ with initial states $\rho_0^1 \in \mathcal{O}^{1\prime}$ and $\rho_0^2 \in \mathcal{O}^{2\prime}$ respectively, we can always define a composite system $S : (\mathcal{O}, H)$ such that $\mathcal{O} = \mathcal{O}^1 \otimes \mathcal{O}^2$, $H = H^1 \otimes I^2 + I^1 \otimes H^2 + H_{int}$, where H_{int} is the interaction Hamiltonian, and $\rho_0 = \rho_0^1 \otimes \rho_0^2 \in \mathcal{O}'$. In this case, the initial state ρ_0 of S and the initial states ρ_0^1 of S^1 and ρ_0^2 of S^2 are still related by a partial trace, since

$$Tr_{(2)}\rho_0 = Tr_{(2)}(\rho_0^1 \otimes \rho_0^2) = \rho_0^1 \tag{5}$$

$$Tr_{(1)}\rho_0 = Tr_{(1)}(\rho_0^1 \otimes \rho_0^2) = \rho_0^2 \tag{6}$$

However, when the two systems S^1 and S^2 interact to each other, $H_{int} \neq 0$ and, therefore, ρ_0^1 and ρ_0^2 do not evolve according to the Schrödinger equation. This means that, strictly speaking, S^1 and S^2 are not subsystems of S but should be considered as mere "parts" of S. Only in the particular case that $H_{int} = 0$, S^1 and S^2 will evolve unitarily as in eqs. (3) and (4), and they will properly be subsystems of S.

3. The selection of the preferred context

As we have pointed out in the introduction, the subtler point in any realist interpretation of quantum mechanics is the choice of the preferred context, which will define the definite-valued observables of the system. In this section we will introduce the rule for selecting such a context. However, before explaining the rule, two remarks are in order.

As it is well known, quantum mechanics is a *probabilistic* theory: by contrast with classical mechanics, it does not ascribe precise values to the observables of the system, but only probabilities to each possible value (see [11]). Moreover, quantum mechanics is *intrinsically* probabilistic: any attempt to ascribe precise values to all the observables of the system, in such a way that probabilities be interpreted in terms of ignorance about an underlying classical-like state, runs into contradiction, as proved by the Kochen-Specker theorem (see [3], [12]). This means that the rule of definite-value ascription, which selects the preferred context, cannot be

Journal of Physics: Conference Series **128** (2008) 012014 doi:10.1088/1742-6596/128/1/012014

inferred from the formalism, but has to be introduced as an interpretational postulate. Therefore, the adequacy of such a rule has to be assessed in the light of its physical relevance and its ability to solve the interpretation problems of the theory.

The second remark concerns to the interpretation of probability. If, as Mermin proposes, quantum mechanics is a fundamental theory describing individual systems and not ensembles, the Kochen-Specker theorem teaches us that there are not hidden variables explaining an underlying level of reality. As Giere ([13]) points out, only the propensity interpretation of probability takes seriously the "no-hidden variables" results, since it requires that quantum probabilities describe ultimate and irreducible features of physical systems. On this basis, our interpretation conceives probabilities as *ontological irreducible propensities*, and the time evolution of the quantum state as the time evolution of those propensities. In this interpretational framework, the rule that selects the preferred context is the interpretational postulate that defines, among all the observables of the system, which of them will acquire actual, and not merely possible, values. For this reason, we will call it *Actualization Rule*.

We have seen that each realist interpretation appeals to a different criterion to select the preferred context: the classicality of the experimental arrangement (Bohr), the assumption of the priority of position (Bohm), the mathematical result of the Schmidt theorem (Dieks-Vermaas). However, the Hamiltonian of the system has never been taken into account. Our Actualization Rule, on the contrary, endows the Hamiltonian, with its particular symmetries, with a central role.

It is widely accepted that the laws of physics are covariant under certain transformations represented by a definite group; in particular, quantum mechanics is covariant under the Galilean transformations, represented by the Galilean group. As it is well known, there are ten symmetry generators K_i of the Galilean group: one time-displacement, three space-displacements, three boost-velocity components and three space-rotations. These generators represent the fundamental dynamical magnitudes of the theory (see [10]): the energy H (time-displacement), the momentum P (space-displacement), the position Q (boost-transformation: boost generator=mQ), the total angular momentum J (space-rotation).[1]

Since we have defined a quantum system as a closed system, its energy is constant in time and, then, the Hamiltonian H is time-independent: H is always invariant under time displacement. Nevertheless, in a given quantum system, H may have the remaining symmetries or not.[2] To say that the Hamiltonian is symmetric or invariant under a certain continuous transformation means that

$$e^{iK\alpha} H e^{-iK\alpha} = H \qquad \textit{then} \quad [H, K] = 0 \tag{7}$$

where α is the parameter of the transformation and K is the corresponding generator. This means that, when H is invariant under a certain transformation, the generator of that transformation is a *constant of motion*: each symmetry of H defines a conserved quantity. For instance, invariance of H under space displacement in the direction α implies that the component P_α of the momentum P is a constant of motion; invariance of H under rotation about an axis α implies that the component J_α of the total angular momentum J is a constant of motion.

Moreover, we know that each symmetry of the Hamiltonian leads to an energy degeneracy. In fact, if H is invariant under a continuous transformation with generator K, we can write

$$K\, H\, |n\rangle = K\, \omega_n\, |n\rangle \quad \Rightarrow \quad HK\, |n\rangle = \omega_n K\, |n\rangle \tag{8}$$

[1] Actually, the generators are proportional to these dynamical magnitudes; for instance, the time-displacement generator is $H/\hbar$. For simplicity, here we are taking $\hbar = 1$.

[2] Other kinds of symmetry transformations, like those of the q-integrable systems (see [14], [15]) will be considered elsewhere.

Journal of Physics: Conference Series **128** (2008) 012014 doi:10.1088/1742-6596/128/1/012014

This means that any vector $K\,|n\rangle$, obtained by applying the operator K on the eigenvector $|n\rangle$, is also an eigenvector of H with the same eigenvalue (see [16]). As a consequence, H can be expressed as

$$H = \sum_n \omega_n P_n \tag{9}$$

where P_n is the projector operator onto the subspace spanned by the degenerate eigenvectors corresponding to ω_n. Then, we can write explicitly the index k_n corresponding to the degeneracy of the eigenvalue ω_n, in such a way that

$$H\,|n, k_n\rangle = \omega_n\,|n, k_n\rangle \quad \Rightarrow \quad H = \sum_n \omega_n \sum_{k_n} |n, k_n\rangle\langle n, k_n| \tag{10}$$

$$K\,|n, k_n\rangle = \kappa_{k_n}\,|n, k_n\rangle \quad \Rightarrow \quad K = \sum_{n,k_n} \kappa_{k_n}\,|n, k_n\rangle\langle n, k_n| \tag{11}$$

The degeneracies with origin in symmetries are called "normal" ([16]) or "systematic" ([17]). On the contrary, degeneracies that have no obvious origin in symmetries are called "accidental" (see [16], [17]). However, deeper study usually shows either that the accidental degeneracy is not exact, or else that a hidden symmetry in the Hamiltonian can be found which explains the degeneracy (see [16]). For this reason it is assumed that, once all the symmetries of the Hamiltonian have been considered, a basis for the Hilbert space $\mathcal{H}$ is obtained and the "good quantum numbers" are well defined.

Now we have all the conceptual elements necessary to present our Actualization Rule. The basic idea is that the Hamiltonian of the system defines actualization; therefore, any observable that does not have the symmetries of the Hamiltonian cannot acquire an actual value, since this actualization would break the symmetry of the system in an arbitrary way. Precisely, given an elemental quantum system $S : (\mathcal{O}, H)$, (i) if the Hamiltonian H is zero, there is no actualization, that is, none observable acquires definite value, and (ii) if the Hamiltonian H is not zero, the definite-valued observables are H and the observables commuting with H and having, at least, the same symmetries as H. Let us see how the rule works in different cases:

- The Hamiltonian H does not have symmetries; this means that it is non-degenerate. In this case,

$$H\,|n\rangle = \omega_n\,|n\rangle \qquad with\ \omega_n \neq \omega_{n'} \tag{12}$$

  where $\{|n\rangle\}$ is a basis of the Hilbert space $\mathcal{H}$. Therefore, n is the only good quantum number: the definite-valued observables of the system are H and all the observables commuting with H.

- The Hamiltonian H has certain symmetries that lead to energy degeneracy. In this case, H can be written as in eq. (9). Therefore, for any index i_n expressing the degeneracy of the eigenvalue ω_n,

$$H\,|n, i_n\rangle = \omega_n\,|n, i_n\rangle \quad \Rightarrow \quad H = \sum_{n,i_n} \omega_n\,|n, i_n\rangle\langle n, i_n| = \sum_n \omega_n \sum_{i_n} |n, i_n\rangle\langle n, i_n| = \sum_n \omega_n P_n \tag{13}$$

  Let us consider an observable of the form

$$A = \sum_{n,i_n} a_n\,|n, i_n\rangle\langle n, i_n| = \sum_n a_n \sum_{i_n} |n, i_n\rangle\langle n, i_n| = \sum_n a_n P_n \tag{14}$$

  It is clear that $[H, A] = 0$. Moreover, A has, at least, the same degeneracy of H since the subspace spanned by the degenerate eigenvectors corresponding to a_n is the same as that

Journal of Physics: Conference Series **128** (2008) 012014 doi:10.1088/1742-6596/128/1/012014

spanned by the degenerate eigenvectors corresponding to ω_n. In other words, A has the same symmetries as H. Therefore, H and all the observables A commuting with H and having the form of eq. (14) are definite-valued. On the contrary, e.g. observables of the form

$$B = \sum_{n,i_n} b_{n,i_n} |n, i_n\rangle\langle n, i_n| \tag{15}$$

in spite of commuting with H, do not acquire definite values, since the actualization of a particular eigenvalue of B would discriminate among the degenerate eigenvectors corresponding to a single eigenvalue of H and, in this way, would introduce in the system an asymmetry not contained in the Hamiltonian.

As we have said, Bub ([3]) classifies the realist non-collapse interpretations in terms of the observable R by means of which the preferred basis is defined. In particular, each interpretation selects a sublattice $\mathcal{D}(|e\rangle, R)$ of the complete lattice of quantum propositions, where $|e\rangle$ is the instantaneous state of the system: in this sublattice, truth values can be assigned and standard Kolmogorov probabilities can be defined. But since, in general, the sublattice depends on the instantaneous state $|e\rangle$, it changes with time. This means that the set of definite-valued observables, that is, the set of properties that acquire actual values, is different at each time as the instantaneous state of the system evolves dynamically. This result not only defies intuitions (a system having, say, position but not momentum at t, and momentum but not position at an infinitesimal time later), but also leads to the need of accounting for the dynamics of actual properties (see [18], [19]). In our interpretation, on the contrary, this step is unnecessary because the dynamics of actual properties is trivial. In fact, since in any case the definite-valued observables commute with the Hamiltonian, they are constants of motion of the system: in spite of the fact that probabilities are continuously evolving, the set of definite-valued observables is time-independent and, thus, completely robust. This supplies a more intuitive picture of the behavior of a quantum system: actualization occurs only once, with the constitution of the system as such, and since then the definite properties are the same at any time, up to the time when the system "disappears" as that particular system by interacting with another system. This picture is consistent with the well known fact that, if the energy of a quantum system is completely definite, time is completely indefinite; therefore, the search of a definite time when actualization occurs in a closed, constant-energy system is senseless.

4. The measurement problem

As we have stressed in the previous section, since the Actualization Rule is an interpretational postulate that cannot be inferred from the formalism, it has to be assessed by its ability to solve the interpretation difficulties of the theory. Among them, the measurement problem is one of the main challenges.

In the standard model (von Neumann), a quantum measurement is conceived as an interaction between a system S and a measuring apparatus M. Before the interaction, M is prepared in a ready-to-measure state $|r_0\rangle$, eigenvector of the pointer observable R of M, and the state of S is a superposition of the eigenstates $|a_i\rangle$ of an observable A of S. The interaction introduces a correlation between the eigenstates $|a_i\rangle$ of A and the eigenstates $|r_i\rangle$ of R (see [20]):

$$|\Psi_0\rangle = \sum_i c_i |a_i\rangle \otimes |r_0\rangle \quad \rightarrow \quad |\Psi\rangle = \sum_i c_i |a_i\rangle \otimes |r_i\rangle \tag{16}$$

The problem consists in explaining why, being the state $|\Psi\rangle$ a superposition of the $|a_i\rangle \otimes |r_i\rangle$, the pointer R acquires a definite value.

In the orthodox collapse interpretation, the pure state $|\Psi\rangle$ is assumed to "collapse" to one of the components of the superposition, say $|a_k\rangle \otimes |r_k\rangle$, with probability $|c_k|^2$ (see [21]):

$$|\Psi\rangle = \sum_i c_i \, |a_i\rangle \otimes |r_i\rangle \Longrightarrow |a_k\rangle \otimes |r_k\rangle| \tag{17}$$

Then, it is supposed that, after detection, the state of the measuring apparatus is the eigenvector $|r_k\rangle$ of the pointer R and, as a consequence, R acquires the definite value r_k, that is, the eigenvalue corresponding to $|r_k\rangle$. In the realist non-collapse interpretations, the problem is to explain the definite reading of the pointer (with its associated probability) without assuming the collapse hypothesis. In our case, the Actualization Rule is what must accomplish this task.

Let us suppose that we want to obtain the coefficients of the state $|\Psi_S\rangle$ of the elemental quantum system $S : (\mathcal{O}_S, H_S)$,

$$|\Psi_S\rangle = \sum_i c_i \, |a_i\rangle \in \mathcal{H}_S \tag{18}$$

where $\{|a_i\rangle\}$ is a basis of $\mathcal{H}_S$ and $A = \sum_i a_i |a_i\rangle\langle a_i|$. For simplicity, we will assume that the Hamiltonian H_S is non-degenerate: $H_S = \sum_i \omega_{Si} |\omega_{Si}\rangle\langle\omega_{Si}|$. The measuring apparatus is an elemental quantum system $M : (\mathcal{O}_M, H_M)$ with an observable $R \in \mathcal{O}_M$, which must possess different and macroscopically distinguishable eigenvalues r_i in order to play the role of the pointer: $R = \sum_i r_i |r_i\rangle\langle r_i|$. Since the apparatus is a macroscopic system, its Hamiltonian is the result of the interaction among a huge number of degrees of freedom. It is quite clear that symmetry decreases with complexity: the probability that the Hamiltonian of a macroscopic system have some symmetry is vanishing small. Therefore, it can be assumed, with no loss of generality, that the Hamiltonian H_M of M is non-degenerate: $H_M = \sum_i \omega_{Mi} |\omega_{Mi}\rangle\langle\omega_{Mi}|$. Moreover, by construction of the apparatus, the pointer R commutes with H_M for the eigenvectors $|r_i\rangle$ to be stationary and, therefore, the reading of R to be possible.[3] M is prepared in a ready-to-measure state $|r_0\rangle$, eigenvector of R:

$$|\Psi_M\rangle = |r_0\rangle \in \mathcal{H}_M \tag{19}$$

Then, at time $t_0 = 0$, the Hamiltonian of the composite system $S \cup M$ is $H = H_S \otimes I_M + I_S \otimes H_M$ and its state is

$$|\Psi_0\rangle = |\Psi_S\rangle \otimes |\Psi_M\rangle = \sum_i c_i \, |a_i\rangle \otimes |r_0\rangle \in \mathcal{H}_S \otimes \mathcal{H}_M \tag{20}$$

In a second, interaction stage, S and M interact through an interaction Hamiltonian H_{int}. The total Hamiltonian of $S \cup M$ is

$$H = H_S \otimes I_M + I_S \otimes H_M + H_{int} \tag{21}$$

H introduces a correlation between the $|a_i\rangle$ and the $|r_i\rangle$ in such a way that, when the interaction ends, the state is

$$|\Psi\rangle = \sum_i c_i \, |a_i\rangle \otimes |r_i\rangle \tag{22}$$

In the third stage, the interaction ends and the Hamiltonian is again $H = H_S \otimes I_M + I_S \otimes H_M$. Therefore, in spite of the fact that the initial state of this stage is a correlated, entangled state $|\Psi\rangle$, S and M turn out to be again subsystems of the composite system $S \cup M$ (see Section II); therefore, we can apply the Actualization Rule to each one of them.

- In the apparatus M, the initial state of the third stage, $\rho_M = Tr_{(S)} |\Psi\rangle\langle\Psi|$, evolves unitarily under the action of H_M. Nevertheless, the preferred context is time-invariant: since H_M is the non-degenerate Hamiltonian of a macroscopic system and $[H_M, R] = 0$, both H_M and R are definite-valued.

[3] In the highly improbable case that H_M were degenerate, $H_M = \sum_i \omega_{Mi} |\omega_{Mi}, \mu\rangle\langle\omega_{Mi}, \mu|$, it would be sufficient that the pointer have the same symmetries (the same degeneracy) as H: $R = \sum_i r_i \, |r_i, \mu\rangle\langle r_i, \mu|$.

- The system S also evolves unitarily under the action of its Hamiltonian H_S, and the preferred context is defined by H_S. However, two cases have to be distinguished. If $[H_S, A] = 0$, then both H_S and A may have definite values. But if $[H_S, A] \neq 0$, the observable A is not definite-valued.

The fact that in certain situations the observable A of the system S may have no definite value turns out to be non-problematic when we recall that the goal of a quantum system is not to "discover" the value of the observable A, but to know the coefficients $|c_i|^2$ of the systems's state $|\Psi_S\rangle$. In fact, the $|c_i|^2$ can be obtained by registering the frequencies of detection of each eigenvalue of R, since

$$\rho_M = Tr_{(S)} |\Psi\rangle\langle\Psi| = \sum_{ij} c_i c_j^* |r_i\rangle\langle r_i| \tag{23}$$

and

$$\Pr(r_i) = \langle r_i | \rho_M | r_i\rangle = |c_i|^2 \tag{24}$$

In summary, according to our interpretation, no matter whether the system's observable acquires a definite value or not, the apparatus' pointer is always definite-valued, and the frequencies of those definite values provide us the correct coefficients of the system's state.[4]

5. Solving old problems

Having rejected the collapse hypothesis to solve the measurement problem, many non-collapse interpretations were specifically designed to supply an alternative answer to the problem. Usually, those attempts work with no difficulties in the case of ideal measurements, but run into troubles when faced to non-ideal measurements. This fact cannot be ignored, since ideal measurement is a situation that can never be achieved in practice: the interaction between the measured system and the measuring apparatus never introduces an absolutely perfect correlation. In spite of this, successful measurements are commonly performed in real experiments. So, we are committed to show that our interpretation is able to account for quantum measurements even in non-ideal situations.

Let us consider the general case where the correlation introduced by H_{int} is not perfect; then, the initial state of M at the third stage of the measurement process is not given by eq. (22), but by

$$|\Psi\rangle = \sum_{ij} d_{ij} |a_i\rangle \otimes |r_j\rangle \tag{25}$$

In the interpretations where the selection of the preferred basis depends on the instantaneous state of the system, the imperfect correlation may represent a very serious challenge. For instance, the rule of property ascription based on the Schmidt theorem ([8], [18]), when applied to non-ideal measurements, leads to results that disagree with those obtained in the orthodox collapse interpretation (see [22], [23]). This fact has been considered as a "silver bullet" for killing the interpretations based on the Schmidt decomposition (see [24]).

In the case of our interpretation, on the contrary, the imperfect correlation is not an obstacle for the definite value of the pointer R. In fact, the initial state of M at the third stage of the process reads

$$\rho_M = Tr_{(S)} |\Psi\rangle\langle\Psi| = \sum_{ij} \rho_{Mij} |r_i\rangle\langle r_j| \tag{26}$$

[4] When the initial state of S is not pure but mixed, $\rho_S = \sum_{ij} \rho_{ij} |a_i\rangle\langle a_j|$, this procedure will supply only the diagonal coeficcients $|\rho_{ii}|^2$. If we want to know the remaining coefficients ρ_{ij} with $i \neq j$, we have to perform further measurements with different experimental arrangements (see [10]).

Journal of Physics: Conference Series **128** (2008) 012014 doi:10.1088/1742-6596/128/1/012014

where

$$\rho_{Mij} = \sum_n d_{ni}\, d^*_{nj} \tag{27}$$

Nevertheless, since H_M and R are non-degenerate and $[H_M, R] = 0$, both H_M and R are definite-valued. In this case, the probability of each eigenvalue r_i of R is computed as

$$\Pr(r_i) = \langle r_i \,|\, \rho_M \,|r_i\rangle = \rho_{Mii} = \sum_n |d_{ni}|^2 = |d_{ii}|^2 + \sum_{n\neq i} |d_{ni}|^2 \tag{28}$$

which reduces to the ideal measurement case when $d_{ni} = 0$, for $n \neq i$. In the non-ideal case, two situations can be distinguished. When the d_{ni}, with $n \neq i$, are small ($\sum_{n\neq i} |d_{ni}|^2 << |d_{ii}|^2$), then $\rho_{Mii} \simeq |d_{ii}|^2$ and the diagonal coefficients of the system's state can be *approximately* obtained. But when the d_{ni}, with $n \neq i$, are not small, the frequencies of the pointer's values do not supply the desired coefficients and the measurement is *non-reliable.*

Summing up, our interpretation can account for the fact that perfect correlation is not a necessary condition for "good" measurements: if the reliability condition of small cross-terms is satisfied, the coefficients of the system's state at the beginning of the process can be approximately computed even when the correlation is not perfect. Nevertheless, both in the reliable and in the non-reliable case, in each measurement a definite reading of the pointer is obtained.

Another argument that stresses the difficulties introduced by non-ideal measurements is that posed by Elby ([25]) in the context of the Stern-Gerlach experiment. This argument points to the fact that the wavefunctions in z variable typically have infinite "tails" that introduce non-zero cross-terms; therefore, the "tail" of the wavefunction of the "down" beam may produce detection in the upper detector, prepared to detect p_+, and vice versa.

Let us consider this new argument in detail by supposing that the imperfection is due to a non-perfect collimation of the incoming beam. In this case, with the magnetic field still turned off, we would obtain a diffuse spot instead of a definite point on the screen. Therefore, the perfect ready-to-measure state $|r_0\rangle$ has to be replaced with a narrow gaussian $|\varphi_0(z)\rangle$. As a consequence, the measurement process turns out to be expressed as

$$|\Psi_0\rangle = (c_1 \,|\uparrow\rangle + c_2 \,|\downarrow\rangle) \otimes |\varphi_0(z)\rangle \longrightarrow |\Psi\rangle = c_1 \,|\uparrow\rangle \otimes |\varphi_+(z)\rangle + c_2 \,|\downarrow\rangle \otimes |\varphi_-(z)\rangle \tag{29}$$

where now $|\varphi_+(z)\rangle$ and $|\varphi_-(z)\rangle$ are gaussians that do not need to be as narrow as the initial one. Let us call the widths of the upper and the lower detectors Δz_+ and Δz_- respectively. Thus, the long tail of the gaussian $|\varphi_+(z)\rangle$ arrives to Δz_- and the long tail of the gaussian $|\varphi_-(z)\rangle$ arrives to Δz_+. We can compute the probabilities corresponding to the four possible cases:

$$\Pr(\ \uparrow\ , +) = |(\langle\uparrow| \otimes \langle\varphi_+(z)\,|)|\Psi\rangle|^2 = |c_1|^2 \int_{\Delta z_+} ||\varphi_+(z)\rangle|^2\, dz = |c_{11}|^2 \tag{30}$$

$$\Pr(\ \uparrow\ , -) = |(\langle\uparrow| \otimes \langle\varphi_-(z)\,|)|\Psi\rangle|^2 = |c_1|^2 \int_{\Delta z_-} |\langle\varphi_-(z)\,|\varphi_+(z)\rangle|^2\, dz = |c_{12}|^2 \tag{31}$$

$$\Pr(\ \downarrow\ , +) = |(\langle\downarrow| \otimes \langle\varphi_+(z)\,|)|\Psi\rangle|^2 = |c_2|^2 \int_{\Delta z_+} |\langle\varphi_+(z)\,|\varphi_-(z)\rangle|^2\, dz = |c_{21}|^2 \tag{32}$$

$$\Pr(\ \downarrow\ , -) = |(\langle\downarrow| \otimes \langle\varphi_-(z)\,|)|\Psi\rangle|^2 = |c_2|^2 \int_{\Delta z_-} ||\varphi_-(z)\rangle|^2\, dz = |c_{22}|^2 \tag{33}$$

These cases can be read as follows:

$|c_{11}|^2$ is the probability that $|\uparrow\rangle$ be detected by Δz_+

$|c_{12}|^2$ is the probability that $|\uparrow\rangle$ be detected by Δz_- (tail)
$|c_{21}|^2$ is the probability that $|\downarrow\rangle$ be detected by Δz_+ (tail)
$|c_{22}|^2$ is the probability that $|\downarrow\rangle$ be detected by Δz_-

If the reliability condition $|c_{21}|^2 << |c_{11}|^2$ and $|c_{12}|^2 << |c_{22}|^2$ holds, then the collimation, even if not perfect, is good enough for the measurement since $|c_{11}|^2 \simeq |c_1|^2$ and $|c_{22}|^2 \simeq |c_2|^2$. If the original gaussian is not very narrow or the screen is placed too far from the magnet, the measurement will be non-reliable since the c_{ij}, with $i \neq j$, are not small enough. Nevertheless, in our interpretation we always obtain a definite reading of the pointer (a definite detection in Δz_+ or Δz_-) because the preferred basis is always the eigenbasis of H_M, and it commutes with the pointer.

6. Physical relevance

During the last decades, the research on the formal properties of the mathematical structure of quantum mechanics has shown a great advance: many results, unknown by the founding fathers of the theory, have been obtained, and this work has greatly improved the understanding of the deep obstacles that any interpretation must face. However, this interest in the features of the formalism has led to forget the physical content of quantum mechanics. In fact, in the last times, the realist interpretations usually rely on mathematical results and focus their attention mainly on the formal model of the measurement problem. But quantum mechanics is a *physical* theory that has been applied to many well known systems and by means of which an impressive amount of experimental evidence has been accounted for. Therefore, a "good" interpretation of quantum mechanics should not only face the traditional interpretational challenges of the theory, but also show its agreement with the orthodox practice of physics.

In this sense, our position moves away from the present trend in the research on the interpretation of quantum mechanics, since it places an element with a clear physical meaning, the Hamiltonian of the system, at the heart of the interpretation. This is what will allow us to argue for the physical relevance of our approach, by showing that it is consistent with many experimental results accounted for by the teory. In this section we will show how our Actualization Rule agrees with the best known results obtained in quantum experiments.

6.1. Free pointlike particle

The Hamiltonian H of the free particle reads

$$H = \frac{P^2}{2m} = \frac{P_x^2 + P_y^2 + P_z^2}{2m} \tag{34}$$

where P is the momentum operator and m is the mass of the particle. This system is invariant under the full Galilean group and, in particular, under space-displacement: the three components P_x, P_y and P_z of the momentum are the generators of the symmetry and, as a consequence, constants of motion of the system. Therefore, the Hamiltonian is degenerate.

According to our Actualization Rule, H acquires a definite value, and also P^2 since it is proportional to H and, then, has the same space-displacement symmetry. Nevertheless, P_x, P_y and P_z are not definite-valued because, being the generators of the symmetry, the actualization of their values would break the symmetry of the free particle.

Of course, the three components of P can be used for the theoretical description of the free particle; in fact, usually any two of them are added to H to constitute a CSCO$=\{H, P_x, P_y\}$, $\{H, P_x, P_z\}$ or $\{H, P_y, P_z\}$ that defines a basis of the Hilbert space (given the functional dependence among the four magnitudes, the CSCO$=\{P_x, P_y, P_z\}$ can be equivalently used). But if we want to know the value of those observables, we have to perform a measurement on the particle. As we have seen, measurement always involves an interaction with the measured object, which breaks the symmetry of the original system by modifying its Hamiltonian. This means that, under measurement, the particle is not free anymore, and the symmetry breaking introduced by the interaction with the measuring apparatus is what allows us to have empirical access to an observable that was a symmetry generator of the original free system.

6.2. Free particle with spin

Spin is an internal contribution to the total angular momentum and, therefore, adds further degrees of freedom to the particle (see [10], [17]): the Hilbert space is now $\mathcal{H} = \mathcal{H}_f \otimes \mathcal{H}_s$, where $\mathcal{H}_f$ is the Hilbert space of the free particle and $\mathcal{H}_s$ is the Hilbert space of the spin. In this case, the Hamiltonian is

$$H = \frac{P^2}{2m} + E_0 \tag{35}$$

where E_0 can only be a multiple of S^2 and, then, may be conceived as an internal contribution to the energy (see [10]). But in this case the system is not elemental, because it can be decomposed into two subsystems which, by definition of subsystem (see Section 2), do not interact to each other: a free particle with Hamiltonian $H_f = \frac{P^2}{2m}$, and a spin system with Hamiltonian $H_s = kS^2$, with $k = const$. Then, the Actualization Rule has to be applied independently to each elemental subsystem.

The rule applies to the free particle subsystem, with its symmetries, as explained above. In the spin subsystem, on the contrary, H_s is invariant under space-rotation: the generators of this symmetry are the three components J_x, J_y and J_z of the total angular momentum J. But since in this case the orbital angular momentum L is zero, the total angular momentum $J = L + S$ turns out to be simply $J = S$, and the three components S_x, S_y and S_z of the spin S are the generators of the space-rotation symmetry. Analogously to the case of the free particle, according to our Actualization Rule in this case H_s acquires a definite value, and also S^2 since it is proportional to H_s; nevertheless, S_x, S_y and S_z are not definite-valued since they are the generators of the symmetry of the Hamiltonian.

Again, we have no direct access to the spin of a free particle. If we want to know the value of the spin, we have to introduce a magnetic field $\mathbf{B}$ of modulus B in some direction, say z, that breaks the isotropy of space and, then, the space-rotation symmetry of the system: the Hamiltonian H_s is not invariant under space-rotation anymore because now it includes the interaction $-\gamma B S_z$ that privileges a particular direction of space. In other words, we only have experimental access to the spin component S_z by means of a measurement that breaks the space-rotation symmetry of the original free system: this is the usual way in which a spin component is measured in a Stern-Gerlach experiment.

6.3. Harmonic oscillator

By definition, an harmonic oscillator is a system with a quadratic potential energy, which produces a restoring force against any displacement from equilibrium that is proportional to the displacement (see [10], [17]). In one dimension, the Hamiltonian of this system is

$$H = \frac{P^2}{2m} + \frac{m\Omega^2 Q^2}{2} \tag{36}$$

where Q is the position observable and Ω is the frequency of oscillation. If the dimensionless position and momentum operators $q = (\frac{m\Omega}{\hbar})^{1/2} Q$ and $p = (\frac{1}{m\Omega\hbar})^{1/2} P$ are introduced in eq. (36), the Hamiltonian reads

$$H = \frac{1}{2}\hbar\Omega\,(q^2 + p^2) \tag{37}$$

In turn, if the observable number of modes $N = a^\dagger a = (\frac{q-ip}{\sqrt{2}})(\frac{q+ip}{\sqrt{2}})$ is used,

$$H = \hbar\Omega\left(N + \frac{1}{2}\right) \tag{38}$$

As it is well known, in this case the energy spectrum of the system and the spectrum of the observable N can be obtained algebraically:

$$N\,|n\rangle = n\,|n\rangle \qquad with\;\; n = 0, 1, 2, ... \tag{39}$$

$$H\,|n\rangle = \omega_n\,|n\rangle \qquad with\;\; \omega_n = \hbar\Omega\left(n + \frac{1}{2}\right) \tag{40}$$

In fact, since H has no symmetries, it is non-degenerate: the CSCO=$\{H\}$ defines a basis of the Hilbert space of the system. According to our Actualization Rule, H acquires a definite value and, due to its non-degeneracy, the preferred context that it defines is the preferred basis $\{|n\rangle\}$: any observable commuting with H (that is, whose eigenvectors are vectors of the preferred basis) are also definite-valued. In particular, the number of particles N acquires a definite value since $[N, H] = 0$.

The harmonic oscillator has a central relevance in quantum mechanics because it provides a model for may kinds of vibrating systems. In particular, the electromagnetic field can be decomposed in terms of linearly independent modes, each one of which behaves as an harmonic oscillator usually associated to a particle; in this case, N is conceived as the observable number of particles. But the point to stress here is that, in all of those vibrating phenomena, the energy of the system is the relevant physical magnitude, whose values are experimentally accessible, and our Actualization Rule accounts for this fact.

6.4. Free Hydrogen atom

The Hydrogen atom is a two-particles system consisting of an electron and a proton interacting to each other by means of a Coulombian interaction. In this case, the Hamiltonian reads

$$H = \frac{P_e^2}{2m_e} + \frac{P_p^2}{2m_p} - \frac{e^2}{|Q_e - Q_p|} \tag{41}$$

where the subindexes e and p refer to the electron and the proton respectively. The usual strategy for solving the energy eigenvalue equation in coordinate representation is to refer the Hamiltonian to the center of mass of the system by means of a canonical transformation, and to write the resulting equation in spherical coordinates (r, θ, ϕ). As it is well known, with this strategy the solution of the equation can be expressed as the product of two functions, one only dependent on r and the other only dependent on the angular coordinates (see [26]):

$$\Psi(r, \theta, \phi) = R(r)\; Y(\theta, \phi) \tag{42}$$

By solving the radial and the angular equations, three "good" quantum numbers are obtained: the principal quantum number n, the orbital angular momentum quantum number l and the magnetic quantum number m_l. These quantum numbers correspond to the eigenvalues of the observables H, L^2 and L_z repectively, where L is the orbital angular momentum, L_x, L_y, L_z are its components, and L^2 is the Casimir operator of the Lie group generated by these components:

$$H\,|n, l, m_l\rangle = \omega_n\,|n, l, m_l\rangle \tag{43}$$

$$L^2\,|n, l, m_l\rangle = l(l+1)\,\hbar^2\;|n, l, m_l\rangle \tag{44}$$

$$L_z\,|n, l, m_l\rangle = m_l\,\hbar\;|n, l, m_l\rangle \tag{45}$$

Journal of Physics: Conference Series **128** (2008) 012014 doi:10.1088/1742-6596/128/1/012014

with $n = 0, 1, 2, ...$, $l < n$ and $-l \leq m_l \leq l$. In particular, the energy eigenvalues are computed as

$$\omega_n = -\frac{\mu\, e^4}{2\hbar^2 n^2} \tag{46}$$

where $\mu = (1/m_e + 1/m_p)^{-1}$ is the reduced mass of the atom. Therefore, the Hydrogen atom is described in terms of the basis $\{|n, l, m_l\rangle\}$ defined by the CSCO=$\{H, L^2, L_z\}$, and the quantum numbers n, l and m_l label the solutions Ψ_{nlm_l} of the energy eigenvalue equation.

It is clear that L^2 and L_z are constants of motion since they commute with H. However, the Hamiltonian is degenerate due to its space-rotation invariance: since in this case $J = L$, then L_x, L_y and L_z are the generators of the symmetry, and L^2 is the Casimir operator of this group. In other words, although l and m_l are good quantum numbers, the eigenvalues ω_n of the Hamiltonian do not depend on them due to the symmetry of H: the values of L^2 and L_z have no manifestations in the energy spectrum. According to our Actualization Rule, this means that L^2 and L_z do not acquire definite values: the only definite-valued observables of the system are H and the observables with the same space-rotation symmetry (at least, the same degeneracy) as H.

In quantum chemistry, the states Ψ_{nlm_l} of the atom (orbitals) are labeled as $X\alpha$, where X is the principal quantum number n, and α is replaced with s, p, d, f, etc., that is, with letters corresponding to the value of the angular momentum quantum number l: $1s$, $2s$, $2p$, $3s$, $3p$, $3d$, etc. The number m_l is not included in those labels because there is no experimental evidence for it in the free Hydrogen atom. In fact, although Ψ_{nlm_l} depends on the three quantum numbers, the space-rotation symmetry of the system makes the selection of L_z for completing the basis a completely arbitrary decision: we can choose L_x or L_y for obtaining an equally legitimate description of the atom. This fact justifies the prescription, derived from our Actualization Rule, of not selecting L_z as a definite-valued observable.

Nevertheless, the value of the orbital angular momentum quantum number l can be inferred from the observed energy spectrum of the Hydrogen atom, and this might be interpreted as a symptom of the definite-valuedness of L^2. However, the manifestation of the value of l requires the interaction between the atom and an electromagnetic field. The usual explanation runs as follows. Since energy transitions involve the absorption or emission of a photon (spin 1), conservation of the angular momentum requires that the atom experiences a change of 1 in its orbital angular momentum L. For this reason, when a photon is absorbed by an atom in an s orbital, the atom acquires orbital momentum and makes a transition to a p orbital; when absorbed by an atom in a p orbital, the orbital momentum increases ($p \longrightarrow d$ transition) or decreases ($p \longrightarrow s$ transition), depending on the relative orientations of the photon and the atom angular momenta. But transitions $s \longrightarrow d$ or $p \longrightarrow f$ are forbidden. It is clear, then, that the value of l is obtained as a consequence of an interaction; then, the system is not the free Hydrogen atom anymore. The new system has a Hamiltonian of the form

$$H = H_{at} + H_{em} + H_{int} \tag{47}$$

where H_{at} is the Hamiltonian of the free Hydrogen atom, H_{em} is the Hamiltonian of the electromagnetic field (which can be computed as the infinite sum of the Hamiltonians of the independent harmonic oscillators corresponding to the infinite modes of the field), and H_{int} is the interaction Hamiltonian (see [10], p. 548)

$$H_{int} \simeq H_D = -D\,E \tag{48}$$

where D is the dipole moment observable of the atom, and E is the electric field observable. The interaction removes the energy degeneracy in the quantum number l of the free atom: this fact is what leads to the manifestation of the value of l in the energy spectrum.

Journal of Physics: Conference Series **128** (2008) 012014 doi:10.1088/1742-6596/128/1/012014

This does not mean that the eigenvalues of the observable L^2 never become actual in a free atom. The particular features of the Hydrogen atom strongly depend on the Coulombian potential generated by its one-proton nucleus. In more complex atoms, the potential is not perfectly Coulombian, and this fact removes the degeneracy of the Hamiltonian in l: the energy eigenvalues ω_{nl} turn out to be functions of both the quantum numbers n and l. This means that L^2 does not discriminate among the degenerate eigenvalues corresponding to a single energy eigenvalue anymore, but rather it removes the degeneracy of the Coulombian case. According to our Actualization Rule, this means that in this case L^2 is a definite-valued observable of the system.

6.5. Zeeman effect

When an external magnetic field is applied to the atom, the spectral lines split into multiple closely spaced lines. First observed by P. Zeeman in 1896, this phenomenon is known as Zeeman effect.

As we have seen, in a free atom the energy eigenvalues are not functions of the magnetic quantum number m_l, either in the Coulombian or in the non-Coulombian potential case: in both situations, m_l corresponds to a degeneracy of the Hamiltonian resulting from its rotation-symmetry. It is precisely this symmetry what makes the selection of L_z for completing the basis of the Hilbert space, instead of L_x or L_y, an arbitrary decision. The arbitrariness in the selection of the z-direction agrees with the fact that there is no experimental evidence of m_l in the energy spectrum.

A magnetic field B along the z-axis breaks the isotropy of space and, as a consequence, the complete space-rotation symmetry of the Hamiltonian. In turn, the breaking of the symmetry removes the degeneracy of H in m_l: now L_z is not arbitrarily chosen but it is selected by the direction of the magnetic field. But now the atom is not free anymore; the Hamiltonian of the new system is approximately (see [10], [27])

$$H = H_{at} + \frac{e}{2m_e c} BL \tag{49}$$

The original degeneracy of the $(2l+1)$-fold multiplet of fixed n and l is now removed: the energy levels turn out to be displaced by an amount

$$\Delta\omega_{nlm_l} = \frac{e\hbar B}{2m_e c} m_l \tag{50}$$

This means that the Hamiltonian, with eigenvalues ω_{nlm_l}, is non-degenerate and, therefore, it constitutes by itself the CSCO=$\{H\}$ that defines the preferred basis $\{|n, l, m_l\rangle\}$. According to our Actualization Rule, in this case H and all the observables commuting with H are definite-valued: since this is the case for L^2 and L_z, both observables will acquire definite values, in agreement with the experimental evidence of the Zeeman effect.

6.6. Fine structure

When the spectral lines of the Hydrogen atom corresponding to $n > 1$ are examined at a very high resolution, they are found to be closely spaced doublets. This splitting was one of the first experimental evidences of electron spin. In fact, the energy levels of the atom are affected by the interaction between the electron spin S and the orbital angular momentum L. Now the Hamiltonian of the system reads

$$H = H_{at} + H_s + H_{s-o} \tag{51}$$

Journal of Physics: Conference Series **128** (2008) 012014 doi:10.1088/1742-6596/128/1/012014

where H_{at} is again the Hamiltonian of the free Hydrogen atom, H_s is the Hamiltonian of the spin, and H_{s-o} is the Hamiltonian representing the spin-orbit interaction, function of the product $L.S$. When the spin-orbit interaction is taken into account, the observables L_z and S_z no longer commute with H and, therefore, they are not constants of motion: it is usually said that m_l and m_s are not good quantum numbers anymore (see [27], [28]). Nevertheless, the system is still invariant under spatial rotation: the components J_x, J_y and J_z of the total angular momentum J are the generators of this symmetry, and J^2 is the Casimir operator of this group. In turn, J is the vectorial sum of the orbital angular momentum L and the spin angular momentum S:

$$J = L + S \qquad\qquad m_j = m_l + m_s \tag{52}$$

So, now m_j is a good quantum number. But we also know that

$$J^2 = (L+S)^2 \quad \Rightarrow \quad L.S = \frac{J^2 - L^2 - S^2}{2} \tag{53}$$

Therefore, H_{s-o} is a function of J^2, L^2 and S^2, and the corresponding quantum numbers j, l and s are also good quantum numbers. As a consequence, the eigenvalues of the total Hamiltonian have the general form (see [16], pp. 181-183)

$$\omega_{nljs} = \omega_{nl} + \xi(nl)\ [j(j+1) - l(l+1) - s(s+1)] \tag{54}$$

where the ω_{nl} represent the energy eigenvalues with no spin-orbit coupling, and ξ is a function of nl. Then, the basis of the Hilbert space of the system is defined by the CSCO=$\{H, L^2, J^2, S^2, J_z\}$, where

$$H\,|n,l,j,s,m_j\rangle = \omega_{nljs}\,|n,l,j,s,m_j\rangle \tag{55}$$
$$L^2\,|n,l,j,s,m_j\rangle = l(l+1)\,\hbar^2\,|n,l,j,s,m_j\rangle \tag{56}$$
$$J^2\,|n,l,j,s,m_j\rangle = j(j+1)\,\hbar^2\,|n,l,j,s,m_j\rangle \tag{57}$$
$$S^2\,|n,l,j,s,m_j\rangle = s(s+1)\,\hbar^2\,|n,l,j,s,m_j\rangle \tag{58}$$
$$J_z\,|n,l,j,s,m_j\rangle = m_j\,\hbar\,|n,l,j,s,m_j\rangle \tag{59}$$

It is quite clear that the spin-orbit coupling removes the original degeneracy of the eigenvalues ω_{nl} of the atom with no coupling. Therefore, in this case our Actualization Rule selects L^2, J^2 and S^2 as definite-valued observables, because all of them commute with H and have the same symmetry in m_j as H. But the rotation symmetry, still present in the system, leads to a degeneracy manifested by the fact that the energy eigenvalues ω_{nljs} do not depend on m_j. Then, according to our Actualization Rule, although in this case m_j is a good quantum number, J_z does not acquire a definite value, and this result agrees with the arbitrariness in the selection of the z-direction for J_z.

As we have seen, when a magnetic field is applied to the atom, the spectral lines split in different ways. The "normal" Zeeman effect explained above is observed in spin 0 states where, obviously, the spin-orbit coupling has no effect. In the states where the spin-orbit coupling is effective, the action of the magnetic field produces a further splitting of the energy levels known as "anomalous" Zeeman effect. Nevertheless, the explanation of the "anomalous" effect is the same as that of the "normal" effect: the action of the magnetic field breaks the full space-rotation symmetry of the system by selecting a z-direction, and this leads to the removal of the original degeneracy of the Hamiltonian in the quantum number m_j (instead of the quantum number m_l as in the "normal" effect). In this case, our Actualization Rule prescribe that J_z will be also definite-valued.

Journal of Physics: Conference Series **128** (2008) 012014 doi:10.1088/1742-6596/128/1/012014

6.7. Definite position

Our Actualization Rule endows the Hamiltonian of the system with the role of selecting the preferred context and, therefore, the energy of the system always acquires an actual definite value. But this does not mean that the momentum becomes actual in any case, since it does not always commute with the Hamiltonian. In fact, when a system is affected by a scalar field, its Hamiltonian has the general form

$$H = \frac{P^2}{2m} + V(Q) \tag{60}$$

When the mass of the system is small, the kinetic term prevails over the potential term, and the Hamiltonian approximately commutes with the momentum operator. In turn, for very large masses, the kinetic term can be neglected and H approximately commutes with the position operator. So, according to our rule, "small" systems actualize in momentum and "large" systems actualize in position. In this sense, our interpretation supports the usual physical assumption that electrons have definite momentum but not position, and nucleus have definite position but not momentum. In general, our interpretational postulate explains the fact that macroscopic systems always posses a definite value of position.

This point has a particular relevance in molecular chemistry, where the description of molecules is based on the adiabatic separation of electron and nuclear motions. As it is well known, the Born-Oppenheimer approximation conceives the nuclei as classical-like particles, that is, as precisely localized objects. This approximation strategy of holding the nucleus at rest in a definite position can be thought off as formally arising from making the masses of the nuclei infinite. However, from a strictly quantum mechanics viewpoint, without a rule for selecting the definite-valued observables of the system, the assumption of infinite nuclear masses does not explain yet why the nucleus can be treated as having a definite value of position. As Primas says ([29], p. 13), "we hardly understand why the Born-Oppenheimer *picture* is compatible with the *concepts* of quantum mechanics" (see also [30], [31]).

Our interpretation provides a simple answer to this conceptual problem. For large masses, the Hamiltonian is approximately invariant under boost transformation and, therefore, it approximately commutes with position. As a consequence, according to the Actualization Rule the position observable acquires a definite value: this provides a conceptual justification to the Born-Oppenheimer assumption. Of course, masses are never infinite: this is what makes the Born-Oppenheimer strategy an approximation and not a precise method. But also in this sense our interpretation agrees with the usual assumptions: since the Hamiltonian perfectly commutes with position only in the infinite mass limit, only in this limit we can say with absolute precision that position actualizes; in real situations, the observable that acquires definite-value will be an observable very similar to position, but which becomes indistinguishable from position for increasing masses.

6.8. Decoherence

As it is widely accepted, quantum correlations are the main obstacle to classicality. Therefore, if the classical world has to be explained in terms of quantum mechanics, the first step is to find a process leading to the stable diagonalization of the operator describing the system: this process is known as decoherence. The subtler point of the theory of decoherence is to find a precise definition of the so-called "pointer basis", that is, the basis of diagonalization, since the observables defined by such a basis are the candidates for classical observables.

In previous papers ([32], [33], [34], [35], [36], [37]) we have developed the *self-induced approach to decoherence* (SID), according to which a system with continuous or quasi-continuous energy spectrum decoheres by destructive interference. Here we will not develop the complete formalism

Journal of Physics: Conference Series **128** (2008) 012014 doi:10.1088/1742-6596/128/1/012014

of our approach: we only want to stress the compatibility between SID and the proposed interpretation of quantum mechanics. In fact, SID is a process that leads the expectation value of any physically relevant observable O of the system to a value that can be computed as if the system were in a state represented by the diagonal functional ρ_*:

$$\lim_{t\to\infty} \langle O\rangle_{\rho(t)} = \langle O\rangle_{\rho_*} \tag{61}$$

Decoherence is the first step towards the classical limit when it is applied to macroscopic systems where, as we have argued above (see Section 4), we have good reasons to assume the non-degeneracy of the Hamiltonian. Under this assumption, ρ_* reads in the discrete case

$$H\,|n\rangle = \omega_n\,|n\rangle \quad \rightarrow \quad \rho_* = \sum_n \rho_n\,|n\rangle\langle n| \tag{62}$$

and in the continuous case

$$H\,|\omega\rangle = \omega\,|\omega\rangle \quad \rightarrow \quad \rho_* = \int_\omega \rho(\omega)\;(\omega|\;d\omega \tag{63}$$

where $\{(\omega|\}$ is the cobasis of $\{\,|\omega\rangle\langle\omega\,|\}$. The pointer basis is, then, the preferred basis $\{\,|\omega\rangle\}$, and it is always the eigenbasis of the Hamiltonian. This result agrees with the prescription of our Actualization Rule which, in the non-degenerate case, selects H and all the observables commuting with H as the definite-valued observables of the system.

6.9. Classical limit

If quantum mechanics is a fundamental theory, it should account for the behavior of any kind of systems, not only microscopic systems but also macroscopic bodies governed by classical mechanics. The problem of explaining the classical description of macroscopic systems in terms of quantum mechanics is known as "the problem of the classical limit", and it has been one of the challenges for the interpretation of quantum mechanics since its birth.

Let us begin with considering the classical description of a system in phase space. If the system is integrable, its trajectory is defined; but when the system is non-integrable, the only observables we can guarantee to be definite-valued at any time are the global constants of motion. As we have shown in previous works ([38], [39]), the quantum correlates of the global constants of motion of classical mechanics are the observables of a CSCO containing the Hamiltonian and, therefore, they are also quantum constants of motion. In fact, given a quantum system endowed with a CSCO $= \{O_0, ..., O_N\}$, with $O_0 = H$, by definition $[O_i, O_j] = 0$, for $i, j = 0$ to N. But we know that the Wigner transformation *symb* turns the quantum commutator into the Moyal bracket $\{\}_{mb}$, where (see [40])

$$\{A(\phi), B(\phi)\}_{mb} = \{A(\phi), B(\phi)\}_{pb} + 0(\hbar^2) \tag{64}$$

If we call $O_i(\phi) = symb\, O_i$, with $\phi \in \mathbb{R}^{2(N+1)}$, the Moyal brackets corresponding to $[O_i, O_j] = 0$ are

$$symb\,[O_i, O_j] = i\hbar\,\{O_i(\phi), O_j(\phi)\}_{mb} = i\hbar\left[\{O_i(\phi), O_j(\phi)\}_{pb} + 0(\hbar^2)\right] = 0 \tag{65}$$

Therefore, in the limit $\hbar \to 0$, $\{O_i(\phi)\}$ is a complete set of constants of motion in involution, globally defined all over the phase space $\mathbb{R}^{2(N+1)}$.

Nevertheless, as we have argued above, in general a macroscopic system has no symmetries and the only constant of motion is the non-degenerate Hamiltonian. In previous papers ([41], [42], [43]) we have proved that, in the non-degenerate case, when the diagonal ρ_* resulting from

Journal of Physics: Conference Series **128** (2008) 012014 doi:10.1088/1742-6596/128/1/012014

decoherence (see eqs. (62) and (63)) is Wigner transformed and the classical limit $\hbar \to 0$ is applied, we obtain a classical distribution $\rho_c(\phi)$

$$\rho_c(\phi) = \int_\omega \rho(\omega)\ \delta(H(\phi) - \omega)\, d\omega \tag{66}$$

where $H(\phi) = symbH$ is the Wigner transformation of the Hamiltonian. The distribution $\rho_c(\phi)$ can be interpreted as an infinite sum of classical densities infinitely peaked in the classical trajectories defined by the values ω of the classical constant of motion $H(\phi)$, and weighted by their corresponding probabilities $\rho(\omega)$.

Of course, from the classical viewpoint we can expect that the classical Hamiltonian $H(\phi)$ involved in the definition of $\rho_c(\phi)$ be definite-valued. But from the quantum viewpoint, without a rule of definite-value assignment, there are no grounds for supposing that the quantum observable H acquires a definite value. Our Actualization Rule provides the desired grounds by positing that the quantum Hamiltonian is always a definite-valued observable. Therefore, our interpretation may explain the fact that the quantum Hamiltonian, after decoherence and in the macroscopic limit, manifests itself as a classical constant of motion in the phase space representation of the system.

7. Conclusions

In this paper we have proposed a new realist, non-collapse interpretation of quantum mechanics, which moves away from the prevailing trend in the subject by paying special attention to the physical relevance of the interpretation. In particular, our proposal endows the Hamiltonian of the system, systematically ignored in the traditional interpretations, with a central role: it distinguishes between systems and subsystems, and is the main ingredient in the selection of the definite-valued observables. We have shown how this interpretation, through its Actualization Rule, solves the measurement problem, both in its ideal and in its non-ideal version. Moreover, we have argued for the physical relevance of the new interpretational rule.

Finally, we want to stress the realist nature of our interpretation by showing that it satisfies the six *desiderata* posed by Mermin:

(i) *The theory should describe an objective reality independent of the observers and their knowledge.* In fact, our interpretation is based on the definition of quantum system, with no reference to observers.

(ii) *The concept of measurement should play no fundamental role.* In our interpretation, measurement is not a departing point, but it is explained in terms of an interpretational postulate.

(iii) *The theory should describe individual systems, not just ensembles.* In our interpretation, states and probabilities apply to individual systems.

(iv) *The theory should describe small isolated systems without having to invoke interaction with anything external.* In our interpretation, the basic entity is the quantum system, closed with respect to any interaction with other systems.

(v) *Objectively real internal properties of an isolated individual system should not change when something is done in another non-interacting system.* According to our interpretation, this is the case even in subsystems with entangled states.

(vi) *It suffices to base the interpretation of quantum mechanics on the interpretation of objective probabilities.* In fact, in our interpretation probabilities measure ontological propensities, non-reducible in terms of ignorance.

Of course, this presentation does not exhaust all the questions involved in the interpretation of quantum mechanics. Nevertheless, we consider that the advantages shown by this perspective open up a new field of research that deserves to be studied.

Acknowledgments

This paper was partially supported by grants of the Buenos Aires University (UBA), the National Research Council (CONICET) and the National Research Agency (FONCYT) of Argentina,

References

[1] Mermin N D 1998 *Pramana* **51** 549-65 (*Preprint* quant-ph/9609013v1)
[2] Kochen S and Specker E 1967 *Jour. Math. Mech.* **17** 59-87
[3] Bub J 1997 *Interpreting the Quantum Wold* (Cambridge: Cambridge University Press)
[4] Bohr N 1948 *Dialectica* **2** 312-9
[5] Bohm D 1952 *Phys. Rev.* **85** 166-79 180-93
[6] Kochen S 1988 *Symp. on the Foundations of Modern Physics*, ed P J Lahti and P Mittelsteadt (Singapore: World Scientific) pp. 151-69
[7] Dieks D 1988 *Ann. Phys.* **7** 174-90
[8] Vermaas P E and Dieks D 1995 *Found. Phys.* **25** 145-58
[9] Harshman N L and Wickramasekara S 2007 *Phys. Rev. Lett.* **98** 080406
[10] Ballentine L 1998 *Quantum Mechanics, a Modern Development* (Singapore: World Sientific)
[11] Dieks D 2007 *Stud. Hist. Phil. Mod. Phys.* **38** 292-310
[12] Hughes R I G 1989 *The Structure and Interpretation of Quantum Mechanics* (Cambridge MA: Harvard University Press)
[13] Giere R N 1976 *Jour. Phil. Logic* **5** 321-53
[14] Olshanetsky M A and Perelomov A M 1983 *Phys. Rep.* **94** 313-404
[15] Zhang W M, Feng D H, Yuan J M and Wang S J 1989 *Phys. Rev.* A **40** 438-47
[16] Tinkham M 1992 *Group Theory and Quantum Mechanics* (New York: Dover)
[17] Cohen-Tannoudji C, Diu B and Laloë F 1977 *Quantum Mechanics* (New York: John Wiley & Sons)
[18] Dieks D and Vermaas P E 1998 *The Modal Interpretation of Quantum Mechanics* (Dordrecht: Kluwer Academic Publishers)
[19] Bacciagaluppi G and Dickson W M 1999 *Found. Phys.* **29** 1165-201
[20] Mittelstaedt P 1998 *The Interpretation of Quantum Mechanics and the Measurement Process* (Cambridge: Cambridge University Press)
[21] Jammer M 1974 *The Philosophy of Quantum Mechanics* (New York: John Wiley & Sons)
[22] Albert D and Loewer B 1990 *Proc. of the 1990 Biennial Meeting of the Philosophy of Science Association* Vol. 1, ed A Fine, M Forbes and L Wessels (East Lansing: Philosophy of Science Association) pp. 277-85
[23] Albert D and Loewer B 1993 *Found. Phys. Lett.* **6** 297-305
[24] Bacciagaluppi G and Hemmo M 1996 *Stud. Hist. Phil. Mod. Phys.* **27** 239-77
[25] Elby A 1993 *Found. Phys. Lett.* **6** 5-19
[26] Bransden B H and Joachain C J 1983 *Physics of Atoms and Molecules* (London: Longman)
[27] Sobelman I I 2006 *Theory of Atomic Spectra* (Oxford: Alpha Science International)
[28] Gil V 2000 *Orbitals in Chemistry* (Cambridge: Cambridge University Press)
[29] Primas H 1983 *Chemistry, Quantum Mechanics and Reductionism* (Berlin: Springer)
[30] Woolley R G 1978 *Jour. Am. Chem. Soc.* **100** 1073-8
[31] Amann A 1992 *South Afr. Jour. Chem.* **45** 29-38
[32] Castagnino M and Laura R 2000 *Phys. Rev.* A **62** 022107
[33] Castagnino M and Laura R 2000 *Int. Jour. Theo Phys.* **39** 1737-65
[34] Castagnino M and Lombardi O 2003 *Int. Jour. Theo. Phys.* **42** 1281-99
[35] Castagnino M and Lombardi O 2004 *Stud. Phil. Hist. Mod. Phys.* **35** 73-107
[36] Castagnino M and Ordoñez A 2004 *Int. Jour. Theo. Phys.* **43** 695-719
[37] Castagnino M and Lombardi O 2005 *Phys. Rev.* A **72** 012102
[38] Castagnino M and Lombardi O 2006 *Chaos Solit. Fract.* **28** 879-98
[39] Castagnino M and Lombardi O 2007 *Stud. Phil. Hist. Mod. Phys.* **38** 482-513
[40] Hillery M, O'Connell R F, Scully M O and Wigner E P 1984 *Phys. Rep.* **106** 121-67
[41] Castagnino M 2004 *Physica* A **335** 511-7
[42] Castagnino M and Lombardi O 2005 *Phil. Scie.* **72** 764-76
[43] Castagnino M and Gadella M 2006 *Found. Phys.* **36** 920-52

Journal of Physics: Conference Series **128** (2008) 012015 doi:10.1088/1742-6596/128/1/012015

Zeros and points of discontinuity of the commutator function of the free Dirac field

Ekatherina A. Karatsuba
Dorodnicyn Computing Center of RAS, Vavilova 40, Moscow, 119991 RUSSIA
E-mail: ekar@ccas.ru

"Evidently it can be asked how far the theory of Dirac goes in reconciling and fusing quantum conceptions and relativity conceptions, the former implying essentially discontinuity, the latter being imbued with continuity."
LOUIS DE BROGLIE
"The Revolution in Physics"

Abstract. The properties of the commutator function of the free electron Dirac field, zeros of which respond to the independent measurements of the electron charge density in two different points of the space-time, are discussed. The established properties of one-space dimensional commutator function $D(x, t)$ in the discrete representation, show the dependency of the number and location of its zeros and points of discontinuity on the value of normalizing factor. In this connection, the following new property of the commutator function is found: the length of the time interval containing at least one zero of $D(x, t)$ depends on the relation between the value of the normalizing factor and the Compton wavelength of the electron.

1. Introduction

In the relativistic theory of Dirac the (anti-)commutation relations determine the causal connections between events that happen at different times and places, and hence the relativistic properties of the theory as a whole.

In what follows, we shall call both commutation and anticommutation for brevity *commutation*.

Suppose we consider in the framework of Dirac's theory a certain physically observable quantity, for example, the electron charge density. The measurements of the values of this observable at two different space-time points can be made without interference of one measurement with the other if the two operators that represent the observable at the two points commute with each other. In this case, the corresponding matrices can be diagonalized simultaneously and hence, exact results (eigenvalues) can be obtained from both measurements.

It is well known (see, for example [10, 11, 13]), that the charge density $\rho(\bar{r'}, t')$ will commute with $\rho(\bar{r''}, t'')$ only when the commutator function $D(\bar{r''} - \bar{r'}, t'' - t')$ is zero. Here the commutator function $D(\bar{r}, t)$ is defined by the expression (see [10, 11, 13])

Journal of Physics: Conference Series **128** (2008) 012015 doi:10.1088/1742-6596/128/1/012015

$$D(\bar{r},t) = \sum_{\bar{k}} L^{-3} e^{i\bar{k}\bar{r}} \frac{\sin\left(ct\sqrt{k^2+k_0^2}\right)}{c\sqrt{k^2+k_0^2}}, \tag{1}$$

where $\bar{k}$ is the wave vector of the length $|\bar{k}| = k = \frac{2\pi}{L}\sqrt{k_x^2+k_y^2+k_z^2}$; k_x, k_y, k_z — any integers,

$$k_0 = 2\pi\frac{mc}{h}; \tag{2}$$

m is the mass of the electron, c is the speed of light, h is the Planck constant. In addition, L is the normalizing factor, representing the length of the edge of the cube, in which the eigenfunctions of the operators of the energy-impulse, subject to the periodicity conditions at the walls of this cube, are defined.

In early works on quantum field theory ([3, 4, 5, 6, 10] see also [1]), the field functions and together with them the commutator functions were introduced in accordance with the concept of the quantization in the form of sums over the states, defined by wave vectors, and only after that for the convenience of work with them, they are substituted by the corresponding integrals. So, from the sum (1) it is obtained the integral which is called the Pauli-Jordan function (see [1])

$$D_{PJ}(\bar{r},t) = \frac{1}{(2\pi)^3 c}\int_{-\infty}^{+\infty} e^{i\bar{k}\bar{r}} \frac{\sin\left(ct\sqrt{k^2+k_0^2}\right)}{c\sqrt{k^2+k_0^2}} d\bar{k}. \tag{3}$$

Later on, the commutator functions are considered merely in the form of the integrals, since this representation is relativistically-invariant and sufficiently simple for studying (see [1, 2, 9, 12, 14]). The space one-dimensional version of the function (3) can be represented in the following form [1, 4, 10, 11, 13]

$$D_{PJ}(x,t) = \frac{1}{4\pi xc}\left(\delta\left(t-\frac{x}{c}\right) - \delta\left(t+\frac{x}{c}\right)\right) - \widetilde{D}_{PJ}(x,t), \tag{4}$$

where

$$\widetilde{D}_{PJ}(x,t) = \begin{cases} \frac{k_0}{4\pi}\frac{J_1(k_0\sqrt{c^2t^2-x^2})}{\sqrt{c^2t^2-x^2}}, & \text{if} \quad ct > x \\ 0, & \text{if} \quad -x < ct < x \\ -\frac{k_0}{4\pi}\frac{J_1(k_0\sqrt{c^2t^2-x^2})}{\sqrt{c^2t^2-x^2}}, & \text{if} \quad ct < -x. \end{cases} \tag{5}$$

Here $J_1(z)$ is a well-known Bessel function of the first kind [15], $\delta(z)$ is the Dirac delta-function:

$$\delta(z) = 0\ , \quad z \neq 0\ , \quad \int_{-\infty}^{+\infty} \delta(z)dz = 1.$$

Thus, the Pauli-Jordan function has the following important property: it vanishes exterior to the light cone, for the space one-dimensional version, for $-x < ct < x$. In physical understanding it is connected with a more general postulate that no signals can be transmitted with velocities greater than that of light, and therefore the measurements, for example, of the electron charge density, at two space points with a space-like distance can never disturb each other.

The commutator function in the discrete representation hasn't aforementioned properties.

The question arises: whether the passage from the discrete representation to the continuous one has any deficiencies? Whether all properties of the commutator function in the discrete representation are reflected in its integral analogue?

Journal of Physics: Conference Series **128** (2008) 012015 doi:10.1088/1742-6596/128/1/012015

The aim of the present paper is to show that the properties of the commutator function in the discrete representation are different from the properties of the commutator function in the integral representation, and that in the problems of quantum theory, in which the quantum effects are essential, the use of the commutator function in the discrete representation can help to get more precise results.

We remark that in accordance with (4), (5) the Pauli-Jordan function has infinitely many zeros (together with the Bessel function $J_1(k_0\sqrt{c^2t^2-x^2})$) interior of the domain $ct=|x|$.

Now we consider the commutator function in the discrete representation. For simplicity, we shall study the space one-dimensional version of the function (1)

$$D(x,t) = L^{-1}\sum_{k_x=-\infty}^{+\infty} e^{i\frac{2\pi}{L}xk_x}\frac{\sin\left(ct\sqrt{\left(\frac{2\pi}{L}\right)^2k_x^2+k_0^2}\right)}{c\sqrt{\left(\frac{2\pi}{L}\right)^2k_x^2+k_0^2}}. \tag{6}$$

In what follows, for the real x, the function *fractional part of the number* x, $y=\{x\}$, is defined by the equality

$$y=\{x\}=x-[x],$$

where $[x]$ is *the integral part of* x, that is an integer such that $x-1<[x]\le x$.

2. The points of discontinuity of the function $D(x,t)$

Rewrite (6) in the form

$$D(x,t) = L^{-1}\left(\frac{\sin(ctk_0)}{k_0}+2\sum_{k=1}^{+\infty}\cos\left(2\pi\frac{x}{L}k\right)\frac{\sin\left(ct\sqrt{\left(\frac{2\pi}{L}\right)^2k^2+k_0^2}\right)}{c\sqrt{\left(\frac{2\pi}{L}\right)^2k^2+k_0^2}}\right). \tag{7}$$

We remark that (7) is a conditionally convergent series. The function $D(x,t)$ is an odd function in t and an even function in x. Besides, $D(x+L,t)=D(x,t)$ and $D(x,0)=0$. Further we shall consider $D(x,t)$ for

$$t>0, \quad 0\le x\le L. \tag{8}$$

From (7) we obtain (see for details [8]):

$$D(x,t) = (2\pi c)^{-1}\left(k_1^{-1}\sin\left(2\pi\frac{ct}{L}k_1\right)+2f(x,t)+2g(x,t)\right), \tag{9}$$

where

$$k_1 = k_0\frac{L}{2\pi}. \tag{10}$$

Here

$$g(x,t)=\sum_{k=1}^{\infty}\cos\left(2\pi\frac{x}{L}k\right)\left(\frac{\sin\left(2\pi\frac{ct}{L}\sqrt{k^2+k_1^2}\right)}{\sqrt{k^2+k_1^2}}-\frac{\sin\left(2\pi\frac{ct}{L}k\right)}{k}\right). \tag{11}$$

is a continuous function in x and t in the domain $|x|\le X$, $|t|\le\tau$, where X and τ are arbitrary numbers. In addition, the function

$$f(x,t)=\sum_{k=1}^{\infty}k^{-1}\cos\left(2\pi\frac{x}{L}k\right)\sin\left(2\pi\frac{ct}{L}k\right),$$

is representable in the form

$$f(x,t)=\frac{\pi}{2}\left(\varrho_0\left(\frac{x+ct}{L}\right)-\varrho_0\left(\frac{x-ct}{L}\right)\right), \tag{12}$$

where

$$\varrho_0(y)=\frac{1}{\pi}\sum_{k=1}^{\infty}\frac{\sin(2\pi yk)}{k}$$

is the Fourier series for the Euler ϱ-function, $\varrho(y)=\frac{1}{2}-\{y\}$ (see, for example, [7]). The function $\varrho_0(y)$ is a piecewise smooth function with the points of discontinuity of the first kind at the integer points

$$y=n;\quad n=0;\pm1;\pm2;\dots;$$

and also

$$\varrho_0(y)=\begin{cases}\varrho(y)=\frac{1}{2}-\{y\}, \text{ if } & \{y\}\neq 0\\ 0, \text{ if } & \{y\}=0.\end{cases} \tag{13}$$

It is easy to see from (9) –(13) that the commutator function $D(x,t)$ is a piecewise smooth function and can be represented in the form

$$D(x,t)=(2\pi c)^{-1}\left(k_1^{-1}\sin\left(2\pi\frac{ct}{L}k_1\right)+\pi\varrho_0\left(\frac{x+ct}{L}\right)-\pi\varrho_0\left(\frac{x-ct}{L}\right)+2g(x,t)\right),$$

where $g(x,t)$ is the continuous function (11), and the points of discontinuity of $D(x,t)$ coincide with the points of discontinuity of $f(x,t)$. We remark that the points of discontinuity of the functions $\varrho_0\left(\frac{x+ct}{L}\right)$ and $\varrho_0\left(\frac{x-ct}{L}\right)$ are correspondingly the points
$x+ct=Ln;\quad n=0;\pm1;\pm2;\dots;$
$x-ct=Ln;\quad n=0;\pm1;\pm2;\dots.$
Since $g(x,t)$ is the continuous function of x and t, and $g(x,0)=0$, then for certain values of x and t, for example, for small values of t with any x, the behavior of the function $D(x,t)$ is determined by the behavior of $f(x,t)$. According to (12)–(13), this function has the following properties in the domain of interest (8):

I. $f(x+L,t)=f(x,t)\ ;\quad f\left(x,t+\frac{L}{c}\right)=f(x,t)$.

II. for the fixed $x=x_1;\ 0\leq x_1<L$;

1) for $x_1=0$,

$$\frac{2}{\pi}f(0,t)=\frac{2}{\pi}f(t)=\begin{cases}0, & \text{if } t=0\\ 1-2\frac{ct}{L}, & \text{if } 0<t<\frac{L}{c}\end{cases}$$

2) for $0<x_1<\frac{L}{2}$,

$$\frac{2}{\pi}f(x_1,t)=\frac{2}{\pi}f(t)=\begin{cases}-2\frac{ct}{L}, & \text{if } 0\leq t<\frac{x_1}{c}\\ \frac{1}{2}-2\frac{x_1}{L}, & \text{if } t=\frac{x_1}{c}\\ 1-2\frac{ct}{L}, & \text{if } \frac{x_1}{c}<t<\frac{L-x_1}{c}\\ -\frac{1}{2}+2\frac{x_1}{L}, & \text{if } t=\frac{L-x_1}{c}\\ 2-2\frac{ct}{L}, & \text{if } \frac{L-x_1}{c}<t\leq\frac{L}{c}\end{cases} \tag{14}$$

3) for $x_1 = \frac{L}{2}$,

$$\frac{2}{\pi} f(x_1, t) = \frac{2}{\pi} f(t) = \begin{cases} -2\frac{ct}{L}, & \text{if} \quad 0 \le t < \frac{L}{2c} \\ 0, & \text{if} \quad t = \frac{L}{2c} \\ 2 - 2\frac{ct}{L}, & \text{if} \quad \frac{L}{2c} < t \le \frac{L}{c} \end{cases}$$

The case, when $\frac{L}{2} < x_1 < L$, is obtained from (14), if we replace x_1 by $L - x_1$.

III. for the fixed $t = t_1$; $0 < t_1 < \frac{L}{c}$;
1) for $0 < t_1 < \frac{L}{2c}$,

$$\frac{2}{\pi} f(x, t_1) = \frac{2}{\pi} f(x) = \begin{cases} 1 - 2\frac{ct_1}{L}, & \text{if} \quad 0 \le x < ct_1 \\ \frac{1}{2} - 2\frac{ct_1}{L}, & \text{if} \quad x = ct_1 \\ -2\frac{ct_1}{L}, & \text{if} \quad ct_1 < x < L - ct_1 \\ \frac{1}{2} - 2\frac{ct_1}{L}, & \text{if} \quad x = L - ct_1 \\ 1 - 2\frac{ct_1}{L}, & \text{if} \quad L - ct_1 < x \le L \end{cases}$$

2) for $t_1 = \frac{L}{2c}$, for $0 \le x \le L$,

$$\frac{2}{\pi} f(x, t_1) = \frac{2}{\pi} f(x) = 0.$$

3) for $\frac{L}{2c} < t_1 < \frac{L}{c}$

$$\frac{2}{\pi} f(x, t_1) = \frac{2}{\pi} f(x) = \begin{cases} 1 - 2\frac{ct_1}{L}, & \text{if} \quad 0 \le x < L - ct_1 \\ \frac{3}{2} - 2\frac{ct_1}{L}, & \text{if} \quad x = L - ct_1 \\ 2 - 2\frac{ct_1}{L}, & \text{if} \quad L - ct_1 < x < ct_1 \\ \frac{3}{2} - 2\frac{ct_1}{L}, & \text{if} \quad x = ct_1 \\ 1 - 2\frac{ct_1}{L}, & \text{if} \quad ct_1 < x \le L \end{cases}$$

From the foregoing properties of the function $f(x, t)$, the following statements are valid on its discontinuities, and therefore, on the discontinuities of the function $D(x, t)$:

Proposition 2.1 *For any fixed $x = x_1$, $0 \le x_1 < L$, the function $D(x_1, t)$ has at most one point of discontinuity*
— *on the segment* $\frac{L}{2c} n \le t \le \frac{L}{2c}(n+1)$; $n = 0, 1, 2, \ldots$; *if* $x_1 \ne \frac{L}{2}$;
— *on the segment* $\frac{L}{c} n \le t \le \frac{L}{c}(n+1)$; $n = 0, 1, 2, \ldots$; *if* $x_1 = \frac{L}{2}$.
On every segment $0 \le t \le \frac{L}{c} n$; $n = 1, 2, 3, \ldots$; *the function $D(x_1, t)$ has at most $2n$ points of discontinuity, if $x_1 \ne \frac{L}{2}$; and at most n points of discontinuity, if $x_1 = \frac{L}{2}$.*

Proposition 2.2 *For any fixed $t = t_1$, $0 < t_1 < \frac{L}{c}$, the function $D(x, t_1)$*
— *has at most one point of discontinuity on the segment* $\frac{L}{2} n \le x \le \frac{L}{2}(n+1)$; $n = 0, 1, 2, \ldots$; *if* $t_1 \ne \frac{L}{2c}$;
— *is continuous,* *if* $t_1 = \frac{L}{2c}$.
On every segment $0 \le x \le Ln$; $n = 1, 2, 3, \ldots$; *the function $D(x, t_1)$ has at most $2n$ points of discontinuity, if $t_1 \ne \frac{L}{2c}$; and $D(x, t_1)$ is continuous, if $t_1 = \frac{L}{2c}$.*

We remark that from (4), (5) the Pauli-Jordan function has one point of discontinuity for any fixed $x = x_1$ (when $t = \frac{x_1}{c}$) and one point of discontinuity for any fixed $t = t_1$ (when $x = ct_1$).

Journal of Physics: Conference Series **128** (2008) 012015 doi:10.1088/1742-6596/128/1/012015

3. Zeros of the function $D(x,t)$ in time variable

The following theorem on the number of the changes of sign of the function $D(x,t)$ in t takes place.

Theorem 3.1 *Let λ_c be the Compton wavelength of the electron. Then for any x, $x \geq 0$, on any interval (t_a, t_b), $t_b > t_a \geq 0$, the function $D(x,t)$ changes sign at least j times, where*

$$j \geq 2\left(\frac{(t_b - t_a)c}{\lambda_c} - \frac{L^2}{\lambda_c^2} - \frac{3}{2}\right).$$

For the proof we introduce the notations

$$\alpha_k = \frac{c}{L}\sqrt{k^2 + k_1^2}; \quad \beta_k = \frac{k}{L}; \quad k = 0, 1, 2, \ldots.$$

Then $D(x,t)$ takes the form:

$$D(x,t) = L^{-1}\left(\frac{\sin(2\pi\alpha_0 t)}{2\pi\alpha_0} + 2\sum_{k=1}^{\infty}\cos(2\pi\beta_k x)\,\frac{\sin(2\pi\alpha_k t)}{2\pi\alpha_k}\right). \tag{15}$$

Let N be an arbitrary natural number, and

$$\widetilde{D}_{2N}(x,t) = \frac{(-1)^N}{L}\left(\frac{\sin(2\pi\alpha_0 t)}{(2\pi\alpha_0)^{2N+1}} + 2\sum_{k=1}^{\infty}\cos(2\pi\beta_k x)\,\frac{\sin(2\pi\alpha_k t)}{(2\pi\alpha_k)^{2N+1}}\right).$$

It is obvious that

$$\frac{d^{2N}}{dt^{2N}}\widetilde{D}_{2N}(x,t) = D(x,t).$$

To find a lower bound of the number of zeros of the function $\widetilde{D}_{2N}(x,t)$ on the interval $t_a < t < t_b$, we represent $\widetilde{D}_{2N}(x,t)$ in the following form

$$\widetilde{D}_{2N}(x,t) = \frac{(-1)^N}{L}(2\pi\alpha_0)^{-2N-1}\left(\sin(2\pi\alpha_0 t) + \widetilde{F}(t)\right), \tag{16}$$

where

$$\widetilde{F}(t) = 2\alpha_0^{2N+1}\sum_{k=1}^{\infty}\alpha_k^{-2N-1}\cos(2\pi\beta_k x)\,\sin(2\pi\alpha_k t).$$

It is not difficult to prove (see for details [8]) that if

$$N = \max([k_1^2] + 1; 5), \tag{17}$$

then

$$\left|\widetilde{F}(t)\right| < \frac{124}{125} < 1.$$

From here and from (16) it follows that the function $\widetilde{D}_{2N}(x,t)$ has on the interval (t_a, t_b) no fewer zeros than the function $y = \sin(2\pi\alpha_0 t)$, which has at least

$$2\alpha_0(t_b - t_a) - 1$$

zeros. Consequently, $D(x,t)$ changes sign on the interval (t_a, t_b) at least j_1 times, where

$$j_1 \geq 2\alpha_0(t_b - t_a) - 2N - 1. \tag{18}$$

Journal of Physics: Conference Series **128** (2008) 012015 doi:10.1088/1742-6596/128/1/012015

Taking into account that the Compton wavelength of the electron is $\lambda_c = h/(mc)$, and hence from (2), (10)

$$k_1 = \frac{L}{\lambda_c},$$

the statement of the theorem follows from (17), (18).

We have the following important corollaries from the theorem.

Corollary 3.2 *On any interval* (t_a, t_b) *of its continuity, the function* $D(x,t)$ *has at least*

$$2\left(\frac{(t_b - t_a)c}{\lambda_c} - \frac{L^2}{\lambda_c^2} - \frac{3}{2}\right)$$

zeros.

Corollary 3.3 *On any interval* (t_a, t_b) *such that* $t_a \geq 0$, $t_b \leq \frac{L}{c}n$, *the function* $D(x,t)$ *has at least*

$$2\left(\frac{(t_b - t_a)c}{\lambda_c} - \frac{L^2}{\lambda_c^2} - n - \frac{3}{2}\right)$$

zeros.

Corollary 3.4 *For any natural number* n *such that*

$$L^2 + \left(n + \frac{3}{2}\right)\lambda_c^2 < nL\lambda_c,$$

the function $D(x,t)$ *has at least one zero on the interval*

$$\frac{L^2}{c\lambda_c} + \left(n + \frac{3}{2}\right)\frac{\lambda_c}{c} < t_b - t_a < n\frac{L}{c}. \tag{19}$$

The proofs of the Corollaries 3.2–3.4 follow immediately from the Theorem 3.1 and the Proposition 2.1.

4. Zeros of the function $D(x,t)$ in space variable

The problem of existence of zeros of $D(x,t)$ in x is more complicated. We shall follow the arguments of the Section 3 with certain modification. We denote by

$$J = J(x_a, x_b)$$

the number of the changes of sign of the function $D(x,t)$ represented by (15) on the interval $x_a < x < x_b$. If $D(x,t)$ is continuous on the interval (x_a, x_b), then J is the number of zeros of $D(x,t)$ on (x_a, x_b). Let N be an arbitrary natural number, and

$$\tilde{\tilde{D}}_{2N}(x,t) = \frac{(x - x_a)^{2N}}{(2N)!}\frac{\sin(2\pi\alpha_0 t)}{2\pi\alpha_0} + (-1)^N 2\sum_{k=1}^{\infty}\frac{\cos(2\pi\beta_k x)}{(2\pi\beta_k)^{2N}}\frac{\sin(2\pi\alpha_k t)}{(2\pi\alpha_k)}. \tag{20}$$

It is obvious that

$$\frac{1}{L}\frac{d^{2N}}{dx^{2N}}\tilde{\tilde{D}}_{2N}(x,t) = D(x,t).$$

Besides, if J_{2N} is equal to the number of zeros of the function $\tilde{\tilde{D}}_{2N}(x,t)$ on (x_a, x_b), then

$$J \geq J_{2N} - 2N.$$

Let $n \geq 2$, n be an integer. Isolating in the sum over k from (20) the summand with $k = n$ and assuming that $\sin(2\pi\alpha_n t) \neq 0$, we rewrite $\tilde{\tilde{D}}_{2N}(x,t)$ in the form

$$\tilde{\tilde{D}}_{2N}(x,t) = (-1)^N \frac{2}{(2\pi\beta_n)^{2N}} \frac{\sin(2\pi\alpha_n t)}{2\pi\alpha_n} \left(\cos(2\pi\beta_n x) + R_1(x,t) + R_2(x,t)\right),$$

where

$$R_1(x,t) = \frac{(-1)^N}{2} \frac{2\pi\alpha_n}{\sin(2\pi\alpha_n t)} \frac{(2\pi\beta_n(x-x_a))^{2N}}{(2N)!} \frac{\sin(2\pi\alpha_0 t)}{2\pi\alpha_0}$$
$$+ \sum_{k=1}^{n-1} \left(\frac{\beta_n}{\beta_k}\right)^{2N} \frac{\alpha_n}{\alpha_k} \frac{\sin(2\pi\alpha_k t)}{\sin(2\pi\alpha_n t)} \cos(2\pi\beta_k x)$$
$$+ \sum_{k=n+1}^{2n} \left(\frac{\beta_n}{\beta_k}\right)^{2N} \frac{\alpha_n}{\alpha_k} \frac{\sin(2\pi\alpha_k t)}{\sin(2\pi\alpha_n t)} \cos(2\pi\beta_k x).$$

$$R_2(x,t) = \sum_{k=2n+1}^{\infty} \left(\frac{\beta_n}{\beta_k}\right)^{2N} \frac{\alpha_n}{\alpha_k} \frac{\sin(2\pi\alpha_k t)}{\sin(2\pi\alpha_n t)} \cos(2\pi\beta_k x).$$

Assume that $t = t_\nu$ is such that for every x from (x_a, x_b) the inequality

$$|R_1(x,t_\nu) + R_2(x,t_\nu)| < 1 \tag{21}$$

is valid. Then the number of zeros of the function $\tilde{\tilde{D}}_{2N}(x,t_\nu)$ coincides on (x_a, x_b) with the number of zeros of the function $\cos(2\pi\beta_n x)$, that is

$$J_{2N} \geq \frac{2n}{L}(x_a - x_b) - 1.$$

In [8] it is proved that in order for the value of $R_1(x,t)$ to be small, we need that there should exist such $t = t_\nu$ that the value $|\sin(2\pi\alpha_n t_\nu)|$ would be sufficiently great, for example,

$$|\sin(2\pi\alpha_n t_\nu)| \geq \frac{1}{A}, \tag{22}$$

A is a constant, $A > 1$, and at the same time all the values $|\sin(2\pi\alpha_k t_\nu)|$; $k = 0, 1, \ldots, n-1, n+1, \ldots, 2n$; would be sufficiently small, for example,

$$|\sin(2\pi\alpha_k t_\nu)| < \frac{1}{B_k}, \tag{23}$$

$$B_k = \max(PA2^k\rho_k; 2); \qquad k = 0, 1, \ldots, n-1, n+1, \ldots, 2n;$$

where P is a constant, $P \geq 1$, and

$$\rho_0 = \frac{(2\pi\beta_n(x-x_a))^{2N}}{(2N)!} \frac{\alpha_n}{\alpha_0},$$

$$\rho_k = 2\left(\frac{\beta_n}{\beta_k}\right)^{2N} \frac{\alpha_n}{\alpha_k}; \qquad k = 1, 2, \ldots, 2n.$$

If the conditions (22), (23) are satisfied, then

$$|R_1(x,t_\nu)| \le \frac{1}{P}\left(1 - \frac{1}{2^{n+1}} - \frac{1}{2^{2n+1}}\right).$$

At the same time, the following bound for $R_2(x,t)$ is proved in [8]:

$$|R_2(x,t)| \le 2A\frac{N+n}{2N-1}\left(\frac{n}{2n+1}\right)^{2N}.$$

Selecting the suitable parameters P, A, n and N, it is possible to refine the estimates for the values $R_1(x,t), R_2(x,t)$, and at the same time to make more precise the estimate of the amount of zeros of $D(x,t)$ in one or another interval (x_a, x_b).

Thus we have the following statement on the condition of the existence of at least one zero of the function $D(x,t)$ on the interval, the length of which doesn't exceed L.

Proposition 4.1 *If the parameters A, P, n, N such that the bound (21) is realized and besides, $N + \frac{3}{2} < n$, then on the interval $x_a < x < x_b$, $x_a \ge 0$, $x_b \le L$, such that*

$$x_a - x_b > L\frac{2N+3}{2n}, \tag{24}$$

the function $D(x,t)$ has at least one zero.

The main problem is, whether it is possible to find $t = t_\nu$, such that for all x in (x_a, x_b) the conditions (22),(23), were satisfied, which for a suitable choice of the parameters, would have provided the estimate (21). The answer gives the following

Lemma 4.2 *Let $n \ge 1$, let real numbers $\alpha_0, \alpha_1, \ldots, \alpha_{2n}$ be linearly independent over the field of rational numbers Q. Let $1 < A \le 2$; $B_k \ge 2$ for $k = 0, 1, 2, \ldots, n-1, n+1, \ldots, 2n$; $B_n = 1$.*

Then on any interval $[T, 2T)$, $T \ge 1$, there exists a set $E \subset [T, 2T)$, the measure of which $\mu(E)$ is defined by the relation

$$\mu(E) = VT + o(T),$$

where

$$V = (B_0 B_1 \ldots B_{2n})^{-1} (2\pi)^{-2n-1}\sqrt{2}\sqrt{1 - \frac{1}{A}},$$

and such that for any $t \in E$ the inequalities

$$|\sin(2\pi\alpha_n t)| \ge \frac{1}{A},$$

$$|\sin(2\pi\alpha_k t)| \le \frac{1}{B_k}, \quad k = 0, 1, 2, \ldots, n-1, n+1, \ldots, 2n,$$

are fulfilled.

The proof of the lemma (see for details [8]) is based on application of the Kronecker theorem (see [7]) in the following statement.

Let $s \ge 1$; let $\alpha_0, \alpha_1, \ldots, \alpha_s$ be real numbers which are linearly independent over Q; let $\theta_0, \theta_1, \ldots, \theta_s$ be arbitrary real numbers; let $b_0, b_1, \ldots, b_s$; $\Delta_0, \Delta_1, \ldots, \Delta_s$ be arbitrary nonnegative numbers such that

$$b_0 + \Delta_0 < 1, \;\; b_1 + \Delta_1 < 1, \;\ldots\;, \;\; b_s + \Delta_s < 1.$$

Let γ be a domain of $s+1$- dimensional Euclidean space $(x_0, x_1, \ldots, x_s)$ of the form

$$b_0 < x_0 \le b_0 + \Delta_0, \quad b_1 < x_1 \le b_1 + \Delta_1, \quad \ldots, \quad b_s < x_s \le b_s + \Delta_s.$$

Then on the interval $[0, T)$, $T \ge 1$, there exists a set E of the numbers t, $E \subset [0, T)$, the measure of which $\mu(E)$ is equal

$$\mu(E) = VT + o(T),$$

where

$$V = \Delta_0 \Delta_1 \ldots \Delta_s,$$

and such that for any $t \in E$ the relation

$$(\{\alpha_0 t + \theta_0\}, \{\alpha_1 t + \theta_1\}, \ldots, \{\alpha_s t + \theta_s\}) \in \gamma$$

is satisfied.

In the considered problem the satisfaction of all conditions of the Kronecker theorem provides the proof of the lemma.

Thus, for any fixed $x = x_1$ the commutator function $D(x,t)$ has at least one zero on the interval (t_a, t_b) which is determined by (19). From the other side, there are infinitely many intervals (T_a, T_b) such that for certain $t_1 \in (T_a, T_b)$ the commutator function in the discrete representation has at least one zero on the interval (x_a, x_b) which is determined by (24).

5. Conclusion

The amount of points of discontinuity and zeros and their location in one or another time or space interval for the commutator function in the discrete representation $D(x,t)$ and its integral analogue, the Pauli-Jordan function $D_{PJ}(x,t)$, are not the same.

The function $D(x,t)$ detects the following interesting inherent property: the number and location of its zeros in x and in t depend on the normalizing factor L, moreover the length of the time interval containing at least one zero of $D(x,t)$ depends on the relation between the value of the normalizing factor and the Compton wavelength of the electron.

A particular role in the study of zeros on space intervals plays the Kronecker theorem.

The described method of the studying $D(x,t)$ is applicable also to $D(\bar{r},t)$, taking into account the effects, connected with the summing the multiple series.

At the same time, for both functions, $D(x,t)$ and $D_{PJ}(x,t)$, the same result on the zeros in the domain $-x > ct > x$ is obtained, and it is possible to generalize it also to three space dimensional case. Thus, the Dirac theory establishes theoretical possibility of the exact measurements of the electron charge density interior of the light cone $|ct| = \bar{r}$.

Acknowledgments. I thank A A Karatsuba for helpful discussions of the Kronecker theorem and the ways of its application.

[1] Bogolubov N N and Shirkov D V 1984 *Introduction to theory of quantized fields* (Moscow: Nauka)
[2] Bjorken J D and Drell S D 1978 *Relativistic quantum fields. Vol.2* (Moscow: Nauka)
[3] Dirac P A M 1926 On the theory of quantum mechanics *Proc. Roy. Soc. London, Ser. A* **112** 661
[4] P. A. M. Dirac 1934 Discussion of the infinite distribution of electrons in the theory of the positron *Proc. Cambr. Phil. Soc.* **30** 150
[5] Jordan P and Pauli W 1928 Zur Quantenelektrodynamik ladungsfreier Felder *Z. Phys.* **47** 151
[6] Heisenberg W and Euler H 1935 Folgerungen aus der Diracschen Theorie des Positrons *Z. Phys.* **98** 714

[7] Karatsuba A A and Voronin S M 1992 *The Riemann Zeta-Function* (W. de Gruyter: Berlin).
[8] Karatsuba E A 2008 The Commutator Function of the Free Dirac Field in the Discrete Representation and its Zeros *Pacif. J. Appl. Math.* **1** 37
[9] Nakawaki Yu and McCartor G 1999 The Indispensability of Ghost Fields in the Light-Cone Gauge Quantization of Gauge Fields *Progr. Theor. Phys.* **102** 149
[10] Pauli W 1941 Relativistic Field Theories of Elementary Particles *Rev. Mod. Phys.* **13** 203
[11] Schiff L I 1955 *Quantum Mechanics. Second ed.* (New York: McGraw-Hill)
[12] Schwinger J 1949 Quantum Electrodynamics II. Vacuum Polarization and Self-Energy. *Phys. Rev.* **75** 651
[13] Sokolov A and Ivanenko D 1952 *The quantum field theory: The selected problems.* (Moscow: Techn.-Theor. Lit. State Press)
[14] Ullrich P 2004 On the restriction of quantum fields to a lightlike surface *J. Math. Phys.* **45** 3109
[15] Watson G N 1966 *A Treatise on the Theory of Bessel Functions. Second ed.* (Cambridge:Cambridge University Press)

Journal of Physics: Conference Series **128** (2008) 012016 doi:10.1088/1742-6596/128/1/012016

Generalized gauge transformations in phase space picture of quantum mechanics: Kirkwood representation

Siamak Khademi[1] **and Sadollah Nasiri**[1,2]

[1]Department of Physics, Zanjan University, ZNU, Zanjan, Iran.

[2]Institute for Advanced Studies in Basic Sciences, IASBS, Zanjan, Iran

E-mail: `khademi@znu.ac.ir` and `siamakkhademi@yahoo.com`

Abstract. Here, the ordinary gauge transformations in configuration space are generalized to the phase space. The corresponding gauge functions are separable in terms of those of the configuration and the momentum spaces. Furthermore, we consider the gauge transformations in phase space for which their gauge functions are not, in general, separable. It is shown that, the assumption of the general global gauge symmetry in phase space immediately yields the quantization of the electric charge. We obtain this result without assuming the concept of the magnetic monopole that Dirac did in his approach to the problem of electric charge quantization and which is not detected yet.

1. Introduction

In 1932 the phase space approach to quantum mechanics is proposed by Wigner [1]. Since then, the phase space picture of quantum mechanics is extended and formulated as different representations [2]. Recent years have seen a surge of interest in applying the representations known as the Husimi [1], the Glauber-Sudarshan [1, 5], Q- [6], generalized P- [7], and Kirkwood representations [8], to the problems encountered in quantum mechanics, quantum statistical mechanics, quantum optics, condensed matter physics, etc. [9, 10, 11]. Different classes of symmetry transformations in phase space picture of quantum mechanics including the rotation, translation, squeezing and boost are extensively investigated by Kim et al [12]. As an important symmetry, the gauge symmetry is widely studied in configuration space by different authors [13, 14, 15]. Here we study the gauge symmetry using the Kirkwood representation of the phase space. We introduce the gauge transformations in phase space corresponding to the ordinary gauges in the configuration space. We show that the gauge functions in the phase space are separable as an additive form in terms of those in configuration and momentum spaces.

We further generalized the gauge transformations in phase space to include the cases that are not, in general, separable. We show that, the Kirkwood equation giving the temporal evolution of Kirkwood distribution functions is invariant under the general gauge symmetry. It can be easily shown that the general gauge transformations are a subset of the canonical transformations in the extended phase space [16]. We show that an immediate result of the general global gauge symmetry, as a special case of the general gauge symmetry in phase space, is the quantization of the electric charge.

Journal of Physics: Conference Series **128** (2008) 012016 doi:10.1088/1742-6596/128/1/012016

In 1932 P.A.M. Dirac formulated the quantization of the electric charge by assuming the dual symmetry in Maxwell's equation as well as the existence of the magnetic monopole [17]. Despite the elaborate and mathematically sophisticated approach of Dirac to the problem of the charge quantization, unfortunately, the magnetic monopole has not already been detected. However, in our approach, the existence of the magnetic monopole is no longer necessary to obtain the quantization of electric charge. In the next section a review of the ordinary gauge transformations in quantum mechanics is given. In Section 3 the generalization of the gauge symmetry in phase space quantum mechanics is presented. This is followed by introducing the relations between gauge functions in configuration, momentum and phase space. In section 4 the quantization of electric charge is obtained by assuming the global gauge symmetry in phase space. Section 5 is devoted to the conclusions and remarks.

2. Ordinary gauge transformations

In quantum mechanics, the gauge transformations in configuration representation are defined as unitary transformations on the wave functions as follows

$$\psi'(q,t) = e^{-i\alpha\Lambda}\psi(q,t), \tag{1}$$

where $\alpha = e/\hbar c$. Here one can distinguish between three kinds of gauges: 1) the global gauge defined as $\Lambda = Const.$; 2) the local defined as $\Lambda = \Lambda(q,t)$ and finally 3) the nonlinear gauges defined as $\Lambda = \Lambda(\psi, \psi^*, q, t)$ [18]. Total electric charge is denoted by "e". Similarly, in momentum representation, one may assume that the wave function $\phi(p,t)$ evolves by a gauge transformation as

$$\phi'(p,t) = e^{-i\alpha\Gamma}\phi(p,t), \tag{2}$$

where $\Gamma = \Gamma(p,t)$ is the gauge function in momentum space. In momentum representation one still has three kinds of gauges corresponding to those of configuration representation mentioned above. One may emphasize that the Schrödinger equation

$$H(-i\hbar\frac{\partial}{\partial q}, q)\psi(q,t) = i\hbar\frac{\partial\psi(q,t)}{\partial t}, \tag{3}$$

is invariant under the gauge transformations, i.e.,

$$H'(-i\hbar\frac{\partial}{\partial q}, q)\psi'(q,t) = i\hbar\frac{\partial\psi'(q,t)}{\partial t}. \tag{4}$$

Therefore, Hamiltonian transforms as

$$H \to H' = e^{i\alpha\Lambda}He^{-i\alpha\Lambda} - i\hbar e^{i\alpha\Lambda}\partial(e^{-i\alpha\Lambda})/\partial t. \tag{5}$$

The minimal coupling Hamiltonian of a system of charged particles interacting with an electromagnetic field is given by

$$H = \frac{(P - \frac{e}{c}A(q,t))^2}{2m} + V(q) - \frac{e}{c}\phi(q,t). \tag{6}$$

One may look to the gauge transformations of minimally coupled Hamiltonian

$$H' = \frac{(P - \frac{e}{c}A'(q,t))^2}{2m} + V(q) - \frac{e}{c}\phi'(q,t), \tag{7}$$

from two equivalent points of view: 1) In the conventional approach, the effect of a gauge transformation on the vector and scalar potentials are as $A' = A + \nabla\Lambda$ and $\phi' = \phi - \frac{1}{c}\frac{\partial\Lambda}{\partial t}$. One

Journal of Physics: Conference Series **128** (2008) 012016 doi:10.1088/1742-6596/128/1/012016

can show that the Maxwell's equations are invariant using the new vector and scalar potentials. In other words, the electromagnetic fields are invariant under a gauge transformation. However, except in a special gauge, the Hamiltonian and conjugate momentum, are not, in general, the conventional energy and kinetic momentum, respectively.

2) An alternative look to the gauge transformations is to consider them as canonical transformations on the phase space coordinates, i.e.,

$$\begin{aligned} Q &= q, \\ P &= p + \nabla\Lambda, \\ H' &= H + \frac{1}{c}\frac{\partial\Lambda}{\partial t}. \end{aligned} \tag{8}$$

For an arbitrary $\Lambda(q,t)$ this transformation is canonical, where the corresponding transformation matrix is

$$M = \begin{pmatrix} 1 & 0 \\ \nabla^2\Lambda & 1 \end{pmatrix}, \tag{9}$$

and satisfies $MJ\widetilde{M} = J$. The gauge symmetry of quantum electrodynamics in configuration space is responsible for conservation of the electric charge.

3. Gauge Transformations in Phase Space

Among infinitely many representations of Phase space quantum mechanics, the Kirkwood representation is a possible one [8]. In this representation the distribution function is defined as

$$\chi = \psi\phi^* e^{-ipq/\hbar}, \tag{10}$$

and satisfies a Schrödinger like equation

$$h\chi = i\hbar\partial\chi/\partial t. \tag{11}$$

In Eq. (11), the extended Hamiltonian h is defied as [8]

$$h = H(p + \pi_q, q) - H(p, q + \pi_p). \tag{12}$$

One can use Eqs. (1), (2) and (10), to obtain the gauge transformed distribution function

$$\begin{aligned} \chi' &= \psi'\phi'^* e^{-ipq/\hbar} = e^{-ie\Lambda/\hbar c}\psi e^{ie\Gamma/\hbar c}\phi^* e^{-ipq/\hbar} \\ &= e^{-ie(\Lambda-\Gamma)/\hbar c}\chi = e^{-ieg(p,q,t)/\hbar c}\chi. \end{aligned} \tag{13}$$

Therefore, the distribution functions as well as the wave functions evolve by unitary transformations and the gauge function in phase space, $g(p,q,t)$, is separable in terms of gauge functions in configuration, Λ , and momentum, Γ, spaces, respectively

$$g(p,q,t) = \Lambda(q,t) - \Gamma(p,t). \tag{14}$$

The gauge functions Λ and Γ, in general, are not independent. To understand the relation between gauge functions in configuration and momentum space, lets have a look to the integral form of the Kirkwood distribution function of Eq. (10), i.e., [19]

$$\chi = \frac{1}{2\pi\hbar}\int \psi(q)\psi^*(q+\lambda)e^{ip\lambda/\hbar}d\lambda. \tag{15}$$

Applying the same gauge transformation on Eq. (15), one gets

$$\begin{aligned}\chi' &= e^{-ie(\Lambda-\Gamma)/\hbar c}\chi \\ &= \frac{1}{2\pi\hbar}\int e^{-ie\Lambda(q)/\hbar c}\psi(q)e^{ie\Lambda(q+\lambda)/\hbar c}\psi^*(q+\lambda)e^{ip\lambda/\hbar}d\lambda, \\ \rightarrow \quad e^{ie\Lambda/\hbar c} &= \frac{\int e^{ie\Lambda(q')/\hbar c}\psi^*(q')e^{ipq'/\hbar}dq'}{\int \psi^*(q')e^{ipq'/\hbar}dq'}. \end{aligned} \tag{16}$$

Equation (16) shows the relation between the gauge functions in configuration and momentum space and could be used to construct the general gauge function in phase space, for a given ordinary gauge in the configuration space. One may release the separable condition on Eq. (14), and look after a larger class of gauge functions that may include that of Eq. (14) as a special case. Then, a general gauge transformation on the extended Hamiltonian for a particle in a potential $V(q)$, is given by

$$h' = \frac{(p+\pi_q-\kappa\nabla_q g)^2}{2m} - \frac{p^2}{2m} + V(q+\pi_p-\kappa\nabla_p g) - \hbar\kappa\frac{\partial g}{\partial t}. \tag{17}$$

Applying the separation condition, Eqs. (14), one gets

$$h' = \{\frac{(\mathbf{p}+\pi_{\mathbf{q}}-\kappa\nabla_q\Lambda)^2}{2m} + V(q)\} - \{\frac{\mathbf{p}^2}{2m} + V(q+\pi_{\mathbf{p}}-\kappa\nabla_p\Gamma)\} - \hbar\kappa\frac{\partial(\Lambda-\Gamma)}{\partial t}. \tag{18}$$

It can be easily shown that for arbitrary gauge functions $\Lambda(q,t)$ and $\Gamma(p,t)$, the Hamiltonian (17) or (18) could be obtained by the following extended canonical transformation on h

$$\begin{aligned} q &\rightarrow q, & \pi_q &\rightarrow \pi_q - \kappa\nabla g, \\ p &\rightarrow p, & \pi_p &\rightarrow \pi_p - \kappa\nabla_p g. \end{aligned} \tag{19}$$

Therefore, the dynamical equation for the distribution function, Eq. (11), is invariant under the canonical transformations (19) in phase space. Due to the arbitrariness of general gauge function $g(p,q,t)$, the general gauge transformation is the gauge symmetry transformation of our system in phase space. The set of ordinary gauge operators, G' , is a subset of general gauge operators G, i.e.,

$$\forall g = \Lambda - \Gamma \in G' | G' \subset G. \tag{20}$$

By extending the Hamiltonian (6) by the extension rule given in [16] one finds

$$\begin{aligned} h' &= \{\frac{(p+\pi_q-\frac{e}{c}A(q,t))^2}{2m} + v(q) + e\phi(q,t)\} \\ &- \{\frac{(p-\frac{e}{c}A(q+\pi_p,t))^2}{2m} + v(q+\pi_p) + e\phi(q+\pi_p,t)\}, \end{aligned} \tag{21}$$

and applying a general gauge symmetry (19) one gets

$$\begin{aligned} h' &= \{\frac{(p+\pi_q-\frac{e}{c}A(q,t)-\alpha\nabla_q g(p,q,t))^2}{2m} + v(q) + e\phi(q,t)\} \\ &- \{\frac{(p-\frac{e}{c}A(q+\pi_p-\alpha\nabla_p g(p,q,t),t))^2}{2m} + v(q+\pi_p-\alpha\nabla_p g(p,q,t)) \\ &+ e\phi(q+\pi_p-\alpha\nabla_p g(p,q,t),t)\} - \hbar\alpha\frac{\partial g(p,q,t)}{\partial t}. \end{aligned} \tag{22}$$

Journal of Physics: Conference Series **128** (2008) 012016 doi:10.1088/1742-6596/128/1/012016

Clearly the operations of extension and general gauge transformations do not commute. If one applies a general gauge transformation before extension of Hamiltonian, the gauge functions in configuration space will contain the conjugate momenta as well as the coordinates, while, the theory of gauge symmetry emphasize that the gauge functions should be in terms of the coordinates and not the conjugate momenta. Therefore, the extended Hamiltonian (22), which follows a gauge transformation after extension, is conventionally preferred with respect to one made by the extension of a gauge transformed Hamiltonian. Note that for the global gauges in phase space, where the gauge functions are constant, or at most are a function of time, $g = g(t)$, the operation of extension and gauge transformations do commute. In contrast to the configuration space, one cannot interpret a general gauge transformation of the extended Hamiltonian (22), as a transformation on vector and scalar potentials (6). As mentioned after Eq. (7), the ordinary gauge transformations could be interpreted either as the change in the vector and the scalar electromagnetic potentials or as a canonical transformations in the configuration space. However, in the case of the general gauge transformation it is clear from Hamiltonian (22) that it can only be interpreted as a canonical transformation in the extended phase space.

In the general gauge transformed Hamiltonian (22), the changes in the vector potential appears in two different ways: i) in the first half of Eq. (22) the vector potential changes as $A'(q,t) = A(q,t) + \frac{\alpha e}{c}\nabla g(p,q,t)$, ii) where in the second half one has $A'(q,t) = A(q + \pi_p - \alpha\nabla_p g(p,q,t),t)$. The same is true for scalar potential. Therefore, the change in the vector and scalar potentials has no unique definition in the general gauge transformed extended Hamiltonian (23), although, both form of general gauge transformed vector and scalar potentials satisfy the Maxwell equations.

4. Charge Quantization

From Eq. (16) it can be easily shown that a gauge function in momentum space corresponding to a global gauge in configuration space, is itself a global gauge, too. If one substitute $\Lambda = Const.$ in Eq. (16) then

$$e^{ie\Gamma/\hbar c} = e^{ie\Lambda/\hbar c} \rightarrow \frac{e}{\hbar c}\Lambda = \frac{e}{\hbar c}\Gamma + 2n\pi, \qquad n = 0, \pm1, \pm2, \cdots. \tag{23}$$

All ordinary global gauges in configuration space are mapped to the gauges in phase space with discrete values, where

$$g = \Lambda - \Gamma = (\frac{2\pi c}{e})n\hbar. \tag{24}$$

Equation (24) is a additional quantization condition in phase space representation of quantum mechanics that gives the quantization of electric charge. In Eq. (24) one may set $n = 1$ to obtain the quanta of electric charge e_1, then

$$g = 2\pi\hbar/e_1, \tag{25}$$

where quanta of charge e_1 is the charge of electron. Substituting Eq. (25) in Eq. (24) gives the quantization of electric charge

$$e = ne_1, \tag{26}$$

as a special property of the general gauge invariance in phase space quantum mechanics. The charge quantization first proposed in 1931 by P.A.M. Dirac was as follows

$$ee_m = 2\pi cn\hbar. \tag{27}$$

He obtained it by assuming the dual symmetry of Maxwells equations and the concept of magnetic monopole with the "magnetic charge" e_m. The assumption of existence of magnetic monopole is essential in the derivation of charge quantization in the Dirac approach. Nevertheless a great effort has been done during the past 76 years, no experimental evidence has been observed for detecting the magnetic monopoles. In our method, however, the assumption of the magnetic monopole is not necessary.

5. Conclusions

Symmetries and the corresponding conservation laws, in general, play an important role in the study of physical problems. Here, the Kirkwood representation of phase space quantum mechanics is considered and its gauge symmetry is investigated. The relation between the gauge functions in configuration and momentum space and their corresponding general gauge function in the phase space are obtained by extending the ordinary gauge transformations in configuration space into phase space. An immediate result of the general global gauge symmetry is the phase space is shown to be the quantization of the electric charge. In contrast to the Dirac's method, which is based on the existence of the magnetic monopole that has never been identified, we obtain the quantization of electric charge on the basis of the global gauge symmetry property inherent in the phase space.

References

[1] Wigner E P 1932 *Phys. Rev.* **40** 749.
[2] Lee H W 1995 *Physics Reports* **259** 147.
[3] Husimi K 1940 *Proc. Phys. Math. Soc. Jpn.* **22** 264.
[4] Glauber R J 1963 *Physical Review* **130** 2529.
[5] Sudarshan E C G 1963 *Physical Review Letter* **10** 277.
[6] Mehta C L 1964 *J. Math. Phys.* **5** 677.
[7] Drummond P D and Gardiner C W 1980 *J. Phys. A: Math. Gen.* **13** 2353.
[8] Nasiri S, Sobouti Y and Taati F 2006 *Journal of Math. Physics* **47** 092106.
[9] Sirera M and Prez A 1999 *Physical Review* D **52** 125011.
[10] Nouri S 1998 *Physical Review* A **57** 1526.
[11] Khademi S and Nasiri S 2002 *Iranian Journal of Science and Technology* **20** 1.
[12] Kim Y S and Noz M E 1991 *Phase Space Picture of Quantum Mechanics* (New York: World Scientific).
[13] Kobe D H and Wen E C T 1980 *Phys. Letter* **80A** 121.
[14] Kobe D H 1982 *Am. J. Phys.* **50** 128.
[15] Baxter C 1991 *Phys. Review* A **44** 3178.
[16] Sobouti Y and Nasiri S 1993 *Int. J. Mod. Phys.* B **7** 3255.
[17] Dirac P A M 1932 *Proc. R. Soc.* A **133** 60.
[18] Doebner H D and Goldin G A 1994 *J. Phys. A: Math. Gen.* **27** 1771.
[19] Hillery M, OConnell R F, Scully M O and Wigner E P 1984 *Phys. Rep.* **106** 121.

Journal of Physics: Conference Series **128** (2008) 012017 doi:10.1088/1742-6596/128/1/012017

Is Bohm's quantum theory equivalent to standard quantum theory?

Boon Leong Lan[1] **and Shiuan-Ni Liang**[2]

[1] School of Engineering, Monash University, Selangor, Malaysia
[2] Faculty of Information Science and Technology, Multimedia University, Malacca, Malaysia

E-mail: `lan.boon.leong@eng.monash.edu.my`

Abstract. It is conventionally believed that Bohm's quantum theory yields all the predictions of standard quantum theory. However, in this paper, we show that the transmission and reflection coefficients (i.e., probabilities) predicted by Bohm's quantum theory can be different from those predicted by standard quantum theory for energy eigenstates of the step barrier.

1. Introduction

In 1952, David Bohm [1, 2] showed that quantum theory, besides predicting probabilities, could also provide a deterministic account of the actual unobserved motion of a particle. In Bohm's quantum theory, the actual motion of a particle is governed by an equation that has the same form as Newton's second law. But in addition to the classical force(s), the particle also experiences what Bohm called a quantum force that is determined by the amplitude of the quantum wave function.

It is conventionally believed [1, 3] that Bohm's quantum theory yields all the predictions of standard quantum theory. According to Bohm [1] himself: "This suggested interpretation leads to precisely the same results for all physical processes as does the usual interpretation." According to Bell [3]: "It [Bohm's quantum theory] is experimentally equivalent to the usual version insofar as the latter is unambiguous." However, in this paper, we show that the transmission and reflection coefficients (i.e., probabilities) predicted by Bohm's quantum theory can be different from those predicted by standard quantum theory for energy eigenstates of the step barrier.

2. Reflection and transmission coefficients from Bohm's quantum theory

Consider a particle in a one-dimensional potential barrier whose initial wavefunction is one of the energy eigenstates $\phi_E(x)$ of the potential barrier with energy E. To predict the particle trajectory using Bohm's theory [1, 2], the particle initial position has to be specified. Assuming that the source of the particle extends from $x = a$ to $x = b$ on the left of the barrier, the initial position x_0 of the particle is in the range $[a, b]$.

The normalized probability distribution for the particle initial position x_0 is given by

$$P_E(x_0) = \frac{|\phi_E(x_0)|^2}{\int_a^b |\phi_E(x_0)|^2 \mathrm{d}x_0} . \tag{1}$$

Journal of Physics: Conference Series **128** (2008) 012017 doi:10.1088/1742-6596/128/1/012017

The reflection and transmission coefficients are [4]-[6], respectively,

$$R(E) = \int_a^b P_E(x_{0\mathrm{R}})\mathrm{d}x_{0\mathrm{R}} \tag{2}$$

$$T(E) = \int_a^b P_E(x_{0\mathrm{T}})\mathrm{d}x_{0\mathrm{T}} \tag{3}$$

where the integrations are, respectively, over the set of initial positions of trajectories that are reflected (not transmitted) and transmitted through the barrier. The reflection coefficient and transmission coefficient above clearly sum to unity

$$R(E) + T(E) = 1\,. \tag{4}$$

3. Step barrier

Consider a one-dimensional step potential defined by

$$\begin{aligned} V\,(x) \quad &= 0 \quad x > 0 \\ &= V \quad x \leq 0, \end{aligned} \tag{5}$$

where V is the height of the step. We will consider the cases $E < V$ and $E > V$ in turn.

3.1. $E < V$

For any energy eigenstate with energy E less than the step height V, the particle is, according to Bohm's theory [7], always at rest, independent of its initial position x_0 on the left of the step barrier. Thus, Bohm's quantum theory predicts $R(E) = 1$ and $T(E) = 0$.

According to standard quantum theory [8], $R(E) = 1$ and $T(E) = 0$.

In this case, the two theories yield the same predictions for the transmission and reflection coefficients.

3.2. $E > V$

For any energy eigenstate with energy E greater than the step height V, the particle, according to Bohm's theory [7], always move to the right, independent of its initial position x_0 on the left of the step barrier. Therefore, Bohm's quantum theory predicts $R(E) = 0$ and $T(E) = 1$.

However, according to standard quantum theory [8]

$$T\,(E) = \frac{4\sqrt{1-\frac{V}{E}}}{\left[1+\sqrt{1-\frac{V}{E}}\right]^2}\,. \tag{6}$$

The transmission coefficient given by equation (6), see figure 1, approaches 1 as V/E approaches 0, i.e., as E approaches infinity for fixed V. For $0 < V/E < 1$, the transmission coefficient from standard quantum theory is not equal to 1 (the value predicted by Bohm's theory). The difference between the transmission coefficients from the two theories grows with increasing V/E, the difference is close to 1 when V/E is close to 1.

4. Conclusion

Contrary to conventional belief, we have shown with the step barrier that the transmission and reflection coefficients predicted by Bohm's quantum theory can be different from those predicted by standard quantum theory. Differences between the predictions of the two theories for the coefficients, if they exist for experimentally realizable barrier systems, allow another possibility of testing Bohm's quantum theory, which has not been conclusively tested so far [9]-[13].

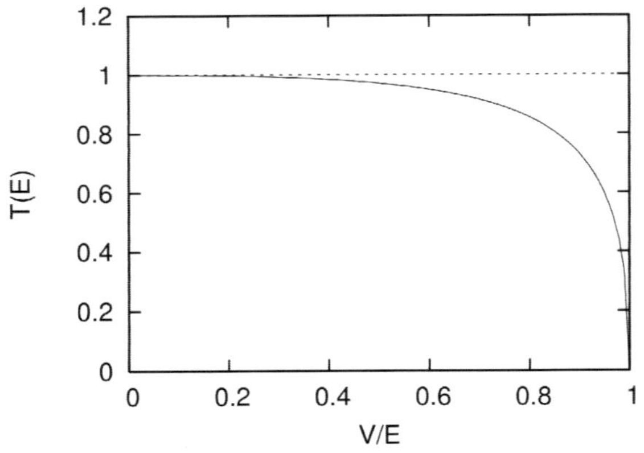

Figure 1. Transmission coefficients (from standard quantum theory —— and from Bohm's quantum theory - - - -) versus V/E for the case $E > V$.

References

[1] Bohm D 1952 *Phys. Rev.* B **85** 166–179
[2] Bohm D 1952 *Phys. Rev.* B **85** 180–193
[3] Bell J S 1987 *Speakable and Unspeakable in Quantum Mechanics* (Cambridge: Cambridge University Press) p 111
[4] Cushing J T 1995 *Found. Phys.* **25** 269–280
[5] Leavens C R 1990 *Solid State Commun.* **74** 923–928
[6] Leavens C R 1990 *Solid State Commun.* **76** 253–261
[7] Holland P 1993 *The Quantum Theory of Motion* (Cambridge: Cambridge University Press) pp 146–148
[8] Liboff R L 1998 *Introductory Quantum Mechanics* (Massachusetts: Addison Wesley) p 246
[9] Brida G, Cagliero E, Falzetta G, Gienovese M, Gramegna M and Predazzi E 2003 *Phys. Rev.* A **68** 033803
[10] Brida G, Cagliero E, Falzetta G, Gienovese M, Gramegna M and Novero C 2002 *J. Phys.* B: *At. Mol. Opt. Phys.* **35** 4751–56 (*Preprint* quant-ph/0206196)
[11] Oriols X 2005 *Phys. Rev.* A **71** 017801
[12] Akhavan O and Golshani M 2004 *J. Phys.* B: *At. Mol. Opt. Phys.* **37** 3777–79
[13] Brida G, Cagliero E, Gienovese M and Gramegna M 2004 *J. Phys.* B: *At. Mol. Opt. Phys.* **37** 3781–83

Journal of Physics: Conference Series **128** (2008) 012018 doi:10.1088/1742-6596/128/1/012018

Semigroups of tomographic probabilities and quantum correlations

Vladimir I. Man'ko

P. N. Lebedev Physical Institute, Leninskii Prospect 53, Moscow 119991, Russia

E-mail: manko@sci.lebedev.ru, manko@na.infn.it

Abstract. Semigroups of stochastic and bistochastic matrices constructed by means of spin tomograms or tomographic probabilities and their relations to the problem of Bell's inequalities and entanglement are reviewed. The probability determining the quantum state of spins and the probability densities determining the quantum states of particles with continuous variables are considered. Entropies for semigroups of stochastic and bisctochastic matrices are studied, in view of both the Shannon information entropy and its generalization like Rényi entropy. Qubit portraits of qudit states are discussed in the connection with the problem of Bell's inequality violation for entangled states.

1. Introduction

The quantum state of a particle with spin j is determined by the probability distribution function $w(m,u)$, where $-j \leq m \leq j$ is the spin projection and u is unitary group $(2j+1)\times(2j+1)$-matrix which can be also considered as matrix of irreducible representation of $SU(2)$ group with matrix elements depending on two Euler angles φ and θ identified with coordinates on a sphere S^2 [1, 2].

For two spins j_1 and j_2, the quantum state is determined by the joint probability distribution $w(m_1, m_2, u)$ of random spin projections m_1 and m_2 depending also on the unitary $N\otimes N$-matrix u, where $N = (2j_1+1)(2j_2+1)$. The matrix u can be taken as tensor product $u = u_1 \times u_2$ of two unitary matrices which can be considered as generic unitary transform matrices or as matrices of two irreducible representations of the $SU(2)$ group depending on two points on two spheres $S^2\otimes S^2$ or two directions given by unit vectors $\vec{n}_1$ and $\vec{n}_2$ [3].

The physical meaning of the probability distribution $w(m,u)$ consists in the fact that it provides the probability to get the spin projection on the direction $\vec{n}$ to be equal to m. The function $w(m_1, m_2, u)$ provides the probability to get the spin projections j_1 and j_2 on the directions $\vec{n}_1$ and $\vec{n}_2$ to be equal, respectively, to m_1 and m_2.

The joint probability $w(m_1, m_2, \ldots, m_N, u)$ of N qudit states, where $m_1, m_2, \ldots, m_N$ are spin j_k projections on corresponding directions $\vec{n}_k$ $(k = 1, 2, \ldots, N)$ determines the multiparticle quantum state of the system. The approach using probabilities to determine quantum states is called the probability representation of quantum mechanics. It is equivalent to all other representations of quantum mechanics reviewed, for example, in [4].

Since in the probability representation the quantum states are identified with probability distributions, such representation provides natural consideration of the aspects related to such probability characteristics of quantum correlations as the entanglement phenomenon [5] and Bell's inequalities [6–8]. The application of the approach in this context was suggestred in [9–15]. One should point out that some probabilities associated to quantum states [16–18] were applied to study Bell's inequalities [6]. Understanding that Bell's inequalities are characteristics not only of quantum picture but also characteristics of standard classical probability distributions was emphasized in [12, 19, 20].

The stochastic and bistochastic matrices have columns which can be considered as probability vectors with nonnegative vector components equal to a probability of some random event. The properties of such matrices which form semigroup are given, for example, in [21]. In view of this, the probability representation of quantum states is naturally connected with stochastic maps realized by semigroups of stochastic matrices. It is worthy noting that semigroups are well-known ingredients in quantum theory of states of decaying particles [22].

The aim of this work is to review the probability approach to qudit states, both separable and entangled, and to relate Bell's inequalities [7, 8] to the properties of stochastic matrix semigroup following [11–13].

The paper is organized as follows.

In section 2, the probabilities determining the quantum states of qudits called tomographic probabilities (or tomograms) are constructed. In section 3, the properties of semigrops of stochastic matrices, including Shannon [23] and Rényi [24] entropies of the matrices, are presented. In section 4, Bell's inequalities [7, 8] are discussed and their connection with separability and entanglement of multipartite qudit states is considered. Conclusions and perspertives are given in section 5.

2. Spin-tomogram construction

The tomographic probability density $w(X,\mu,\nu) \geq 0$ determining the quantum state of a particle with density operator $\hat{\rho}$ reads [25]

$$w(X,\mu,\nu) = \mathrm{Tr}\left[\hat{\rho}\,(X-\mu\hat{q}-\nu\hat{p})\right], \tag{1}$$

where X is a random position, μ and ν are real parameters, and $\hat{q}$ and $\hat{p}$ are position and momentum operators. The tomographic probability of one qudit pure state $\mid\psi\rangle$ reads [1, 2, 26]

$$w(m,u) = \langle m \mid u^{\dagger} \mid \psi\rangle\langle\psi \mid u \mid m\rangle = \mathrm{Tr}\,\hat{\rho}_{\psi} u \mid m\rangle\langle m \mid u^{\dagger}. \tag{2}$$

The tomographic probability of the pure state of two qudits is constructed analogously

$$w(m_1,m_2,u) = \left|\langle m_1 m_2 \mid u^{\dagger} \mid \psi\rangle\right|^2, \tag{3}$$

where matrix u belonging to unitary group $U(N)$ can be taken as matrix of the subgroup of unitary group given as tensor product of local transforms

$$U(N) = U(N_1)\otimes U(N_2), \tag{4}$$

with $N_1 = 2j_1+1$ and $N_2 = 2j_2+1$.

For arbitrary number $\mathcal{K}$ of qudits, i.e., $N=\prod_{k=1}^{\mathcal{K}} N_k$ with $N_k = 2j_k+1$, the state tomogram reads

$$w_{\rho}(m_1,m_2,\ldots,m_{\mathcal{K}},u) = \langle m_1,m_2,\ldots,m_{\mathcal{K}} \mid u^{\dagger}\rho u \mid m_1,m_2,\ldots,m_{\mathcal{K}}\rangle, \tag{5}$$

which for the pure state gives an analog of (3)

$$w_{\psi}(\vec{m},u) = \left|\langle\vec{m} \mid u^{\dagger} \mid \psi\rangle\right|^2, \tag{6}$$

where we introduced the vector $\vec{m} = (m_1, m_2, \ldots, m_K)$.

The probability is normalized for any matrix u

$$\sum_{\vec{m}} w_\rho(\vec{m}, u) = 1. \tag{7}$$

For example, the qubit state $|\,\psi\rangle = \begin{pmatrix} 1 \\ 0 \end{pmatrix}$ has the tomogram, which we associate with the probability vector

$$\vec{w}_+(u) = \begin{pmatrix} |u_{11}|^2 \\ |u_{22}|^2 \end{pmatrix}. \tag{8}$$

This means that the probability of the spin projection on the direction $\vec{n}$ reads

$$w\,(+1/2, \vec{n}) = \cos^2\theta/2, \qquad w\,(-1/2, \vec{n}) = \sin^2\theta/2. \tag{9}$$

The two qubit entangled state

$$|\,\psi\rangle = \frac{1}{\sqrt{2}}\Big(\,|+1/2, +1/2\rangle + |\,-1/2, -1/2\rangle\Big) \tag{10}$$

is determined by tomogram which can be presented as the probability four-vector depending on matrix elements of 4×4-unitary matrix

$$\vec{w}_\psi(u) = \frac{1}{2}\begin{pmatrix} |u_{11} + u_{41}|^2 \\ |u_{12} + u_{42}|^2 \\ |u_{13} + u_{43}|^2 \\ |u_{14} + u_{44}|^2 \end{pmatrix}. \tag{11}$$

For separable state, in the case of $u = u_1 \otimes u_2$, the tomograpgic probability vector can be presented in the form of convex sum of tensor-product vectors

$$\vec{w} = \sum_k p_k \vec{w}^{(k)}(u_1) \otimes \vec{w}^{(k)}(u_2). \tag{12}$$

If one takes the matrix u in the form of tensor-product of two 2×2-unitary matrices u_1 and u_2, vector (11) will be expressed in terms of probability of spin projections on the directions $\vec{n}_1$ and $\vec{n}_2$ given as follows:

$$\begin{aligned} w\,(+1/2, +1/2, u_1, u_2) &= \frac{1}{2}\,|u_{11} + u_{41}|^2, \qquad & w\,(+1/2, -1/2, u_1, u_2) &= \frac{1}{2}\,|u_{12} + u_{42}|^2, \\ w\,(-1/2, +1/2, u_1, u_2) &= \frac{1}{2}\,|u_{13} + u_{43}|^2, \qquad & w\,(-1/2, -1/2, u_1, u_2) &= \frac{1}{2}\,|u_{14} + u_{44}|^2, \end{aligned} \tag{13}$$

In the case of tensor product of 2×2-matrices $u^{(1)}$ and $u^{(2)}$. one has the matrix elements of 4×4-matrix u as follows:

$$\begin{aligned} u_{11} &= u^{(1)}_{11}u^{(2)}_{11}, & u_{12} &= u^{(1)}_{11}u^{(2)}_{12}, & u_{13} &= u^{(1)}_{12}u^{(2)}_{11}, & u_{14} &= u^{(1)}_{12}u^{(2)}_{12}, \\ u_{41} &= u^{(1)}_{21}u^{(2)}_{21}, & u_{42} &= u^{(1)}_{21}u^{(2)}_{22}, & u_{43} &= u^{(1)}_{22}u^{(2)}_{21}, & u_{44} &= u^{(1)}_{22}u^{(2)}_{22}. \end{aligned} \tag{14}$$

The tomographic-probability vectors can be transformed by means of matrices. For example, the qubit state (9) can evolve due to some interaction corresponding to a Hamiltonian which is Hermite 2×2-matrix H providing the evolution of the state vector $\mid \psi\rangle$ as follows:

$$\mid \psi, t\rangle = \exp(-iHt) \mid \psi, 0\rangle. \tag{15}$$

The unitary matrix $u_H(t) = \exp(-iHt)$ gives the change of the tomographic-probability vector

$$w(m,u,0) = \left|\langle m \mid u^\dagger \mid \psi, 0\rangle\right|^2 \longrightarrow \left|\langle m \mid \left(u^\dagger u_H(t)\right) \mid \psi, 0\rangle\right|^2 = w(m,u,t). \tag{16}$$

The time evolution of a quantum state provides a change of the tomographic probability vector of the quantum state. If the initial density matrix is diagonal, the evolution is realized by the matrix

$$M_{jk} = \left|\left(e^{-itH}\right)_{jk}\right|^2. \tag{17}$$

3. Stochastic and bistochastic matrices

Let us consider matrices M_{jk} with nonnegative matrix elements satisfying the condition

$$\sum_{j=1}^{N} M_{jk} = 1. \tag{18}$$

Such stochastic matrices form semigroup. The matrices M_{jk} with the property

$$\sum_{j=1}^{N} B_{jk} = \sum_{k=1}^{N} B_{jk} = 1 \tag{19}$$

also form semigroup. The matrix elements in columns of stochastic matrices provide probability distributions. The matrix elements of bistochastic matrices B_{jk} give the probability distributions in both columns and rows.

In infinite-dimensional Hilbert space with discrete basis like Fock states $\mid n\rangle$, the definitions (18) and (19) hold with the replacement $N \to \infty$. For example, in the oscillator case, one can use the formalism of semigroups transforming the tomograms of oscillator states. If one has a probability vector $\vec{w}$, the stochastic and bistochastic matrices give the linear transform of the vector providing new probability vectors

$$\vec{w}_M = M\vec{w}, \qquad \vec{w}_B = B\vec{w}. \tag{20}$$

One can consider entropy of stochastic and bistochastic matrices

$$H_M = -\sum_{jk=1}^{N} M_{jk} \ln M_{jk}, \qquad H_B = -\sum_{jk=1}^{N} B_{jk} \ln B_{jk}. \tag{21}$$

The entropies are larger than zero. The minima of the entropies are zero and the maxima of the entropies are $N \ln N$. The qudit states are also characterized by entropies. For example, the state with tomogram $w(m,u)$ has entropy, which is a function on unitary group [27, 14]

$$H(u) = -\sum_{m} w(m,u) \ln w(m,u). \tag{22}$$

Journal of Physics: Conference Series **128** (2008) 012018 doi:10.1088/1742-6596/128/1/012018

This is the Shannon entropy of quantum qudit state. The von Neumann entropy of the state

$$S = -\mathrm{Tr}\,\rho \ln \rho \tag{23}$$

is the minimun of the tomographic Shannon entropy [27, 14]

$$S = \min\, H(u). \tag{24}$$

One can construct the stochastic matrices using as their columns the tomographic-probability vectors taken at different unitary group elements u. For example, the qubit state $w(m,u)$ creates 2×2 stochastic matrix

$$M_q(u_1,u_2) = \begin{pmatrix} w(+1/2,u_1) & w(+1/2,u_2) \\ w(-1/2,u_2) & w(-1/2,u_2) \end{pmatrix}, \tag{25}$$

which depends on two unitary group 2×2-matrices u_1 and u_2.

For $u_1 = u_2 = u$, the matrix constructed has the property

$$M_q(u,u)M = M_q(u,u), \tag{26}$$

where M is an arbitrary stochastic matrix. Any tomogram under the action of matrix M_q becomes the tomogram which is the vector in column of the matrix $M_q(u,u)$.

For given tomogram of two qubits $w(m_1,m_2,u)$, one can construct some probability distributions. For example, one can consider the probability to find spin projection of the first particle to be equal to $+1/2$ $(-1/2)$. One has for these probabilities

$$w(+1/2,u) = \sum_{m_2} w(+1/2,m_2,u), \qquad w(-1/2,u) = \sum_{m_2} w(-1/2,m_2,u). \tag{27}$$

Analogous probabilities can be found for the second particle. These probabilities are the marginals of joint probability distributions

$$w(m_1,u) = \sum_{m_2} w(m_1,m_2,u), \qquad w(m_2,u) = \sum_{m_1} w(m_1,m_2,u). \tag{28}$$

Other probabilities can be found answering questions like "what is the probability that both spin projections are equal? and "what is the probability that both spin projections are different?

The answer to these questions provides the probability distributions

$$P_{\mathrm{eq}}(u) = \sum_{m} w(m,m,u), \qquad P_{\mathrm{d}}(u) = \sum_{m} w(m,m+1,u), \tag{29}$$

where the sum $(m+1)$ means summation modulo 1. This means that here $-1/2+1 = 1/2$ and $1/2+1 = -1/2$. The marginals (29) provide the qubit portrait of the initial probability distribution [13, 10].

Let us study a qubit state with density matrix ρ_1 and the corresponding tomographic-probability vector

$$\vec{w}_1(u_1) = \begin{pmatrix} w_1(+1/2,u_1) \\ w_1(-1/2,u_1) \end{pmatrix} \tag{30}$$

and construct stochastic matrix

$$M_{q1} = \begin{pmatrix} w_1(+1/2,u_a) & w_1(+1/2,u_d) \\ w_1(-1/2,u_a) & w_1(-1/2,u_d) \end{pmatrix}, \tag{31}$$

Journal of Physics: Conference Series **128** (2008) 012018 doi:10.1088/1742-6596/128/1/012018

taking two values of unitary 2×2-matrix u_1 denoted u_a and u_d corresponding to two directions $\vec{n}_a$ and $\vec{n}_d$, respectively.

Then we consider the second qubit with density matrix ρ_2, which is determined by the tomographic-probability vector

$$\vec{w}_2(u_2) = \begin{pmatrix} w_2(+1/2, u_2) \\ w_2(-1/2, u_2) \end{pmatrix}, \tag{32}$$

and construct stochastic matrix

$$M_{q2} = \begin{pmatrix} w_2(+1/2, u_b) & w_2(+1/2, u_c) \\ w_2(-1/2, u_b) & w_2(-1/2, u_c) \end{pmatrix}, \tag{33}$$

taking two values of unitary 2×2-matrix u_2 denoted u_b and u_c corresponding to two directions $\vec{n}_b$ and $\vec{n}_c$, respectively.

The tensor product of two stochastic matrices

$$M = M_{q1} \otimes M_{q2} \tag{34}$$

is 4×4-matrix with the column being tomographic-probability distributions of a two-qubit quantum state with density matrix $\rho_{12} = \rho_1 \otimes \rho_2$ given by (13), in view of the corresponding choice of matrices u_1 and u_2.

The stochastic matrix of the form M has a specific property. Let us consider 4×4-matrix [10, 12, 13]

$$I = \begin{pmatrix} 1 & -1 & -1 & 1 \\ 1 & -1 & -1 & 1 \\ 1 & -1 & -1 & 1 \\ -1 & 1 & 1 & -1 \end{pmatrix} \tag{35}$$

and calculate the trace

$$B = \mathrm{Tr}\, IM. \tag{36}$$

One can show that for any stochastic 4×4-matrix M of the form (34) the following inequality holds: $|B| \leq 2$. The Shannon entropy for matrices (29), (33), and (34) read

$$H_{q1} = H(u_a) + H(u_d), \qquad H_{q2} = H(u_b) + H(u_c), \tag{37}$$

where the Shannon entropies depending on unitary 2×2-matrices are given by (22).

One can consider Rényi entropy [24] of the stochastic matrix. For matrix M_q (25), the Rényi entropy is

$$\begin{aligned} R(u_1, u_2) &= \frac{1}{1-q} \ln \Big\{ \big[w^q(+1/2, u_1) + w^q(-1/2, u_1) \big] \big[w^q(+1/2, u_2) + w^q(-1/2, u_2) \big] \Big\} \\ &= R_1(u_1) + R_2(u_2), \end{aligned} \tag{38}$$

where the number q provides parametric dependence of the Rényi entropy and in the limit $q \to 1$ the Rényi entropy becomes the Shannon entropy. In (38)

$$R_j(u_j) = \frac{1}{1-q} \ln \big[w^q(+1/2, u_j) + w^q(-1/2, u_j) \big], \qquad j = 1, 2 \tag{39}$$

is the Rényi entropy associated to the probability vector in jth column of matrix (25).

Some new inequalities for the Rényi entropy of symplectic tomogram describing the quantum state of a system with continuous variables were obtained in [28–33].

Journal of Physics: Conference Series **128** (2008) 012018 doi:10.1088/1742-6596/128/1/012018

4. Bell's inequalities

The inequality properties for stochastic matrices constructed by means of tomographic probability vectors correspond to Bell's inequalities [6–8]. The inequality given in [7] is written for correlations of spin projections onto different directions. The inequality [8] generalizing the inequality [7] is written for the following probabilities.

Given two qudits with equal j. We produce a shift labelling the spin projections as follows:

$$-j \to 0, -j+1 \to 1, \ldots, j \to d, \qquad d = 2j+1.$$

This means that in the tomographic probability distribution all spin projections are denoted by integer nonnegative numbers. Let us have for two-qudit state the joint tomographic probability distribution $w(m_1, m_2, u)$, where now $m_1, m_2 = 0, 1, \ldots, 2j$. Now we consider the following probability of the event that the results of measuring m_1 and m_2 differe by integer k, i.e.,

$$P(m_1 = m_2 + k, u) = \sum_{m=0}^{2j} w(m, m-k, u), \tag{40}$$

where the difference $m-k$ means the nonnegative number equal to $m-k$ modulo d. For example, if $d=2$, one has $0-1=1$, and if $d=3$, one has $2-3=2$, $0-2=1$.

If one uses the matrix $u = u_1 \otimes u_2$ the probability can be rewritten in the form

$$P(m_1 = m_2 + k, u_1, u_2) = \sum_{m=0}^{2j} w(m, m-k, u_1, u_2), \tag{41}$$

where now u_1 and u_2 are local unitary transforms.

Other notation can be used for these probabilities and for the tomogram [8, 11, 34]

$$P(A_a = B_b + k) = \sum_{m=0}^{d-1} P(A_a = m, B_b = m-k), \tag{42}$$

where $d = 2j+1$ and subscripts a and b correspond to u_1 and u_2. There exists the inequality formulated in this notation as follows:

$$\begin{aligned} I_d = &\sum_{k=0}^{[\frac{d}{2}]-1} \left(1 - \frac{2k}{d-1}\right) \\ &\times\Big\{\Big[P(A_1 = B_1 + k) + P(B_1 = A_2 + k + 1) + P(A_2 = B_2 + k) + P(B_2 = A_1 + k)\Big] \\ &-\Big[P(A_1 = B_1 - k - 1) + P(B_1 = A_2 - k) + P(A_2 = B_2 - k - 1) + P(B_2 = A_1 - k - 1)\Big]\Big\} \leq 2. \end{aligned} \tag{43}$$

Using notation (41) in terms of qudit-state tomogram for the probabilities, one can check that for $d=2$ inequality (43) is identical to the inequality for Bell's number (38).

The trace of product of real $N \times N$-matrices can be constructed as scalar product of N^2-vectors in N^2-dimensional Hilbert space. Thus the inequality $|\mathrm{Tr}\, IM| \leq 2$ can be considered as the inequality $|\vec{I}\vec{M}| \leq 2$, where vectors $\vec{I}$ and $\vec{M}$ have matrix elements of the matrices I and M as their components.

The inequality [7] has the form

$$\left|\sum_{m_1 m_2} \sum_k w(m_1, m_2, u_k) c(m_1, m_2, k)\right| \leq 2, \tag{44}$$

where the coefficients of the linear form $c(m_1, m_2, k)$ depend on the index of unitary matrix u_k and the spin projections m_1 and m_2.

For Bell's inequality [7], $c(m_1, m_2, k)$ are just matrix elements of the matrix I (37). For two qutrits, the coefficients can be obtained too [11]. For entangled states, Bell's inequalities can be violated. Thus, for two qubits, the number $|B|$ (38) can reach value $2\sqrt{2}$ which is the Cirelson bound [35]. One can construct the stochastic matrix M for two qudit states with the probability vector $\vec{w}(u)$ using these vectors as columns $\vec{w}(u)$ of the matrix. The index $k = 1, 2, \ldots, d^2$ can be considered as a pair of indices $k = (\alpha, \beta)$ where $\alpha, \beta = 1, 2, \ldots, d$. For the matrices $u_{(\alpha,\beta)} = u_\alpha \otimes u_\beta$, where α and β are unitary $d \otimes d$-matrices. the simply separable two-qudit state has the stochastic matrix M of the tensor-product form $M = M_1(u_\alpha) \otimes M_2(u_\beta)$, where the stochastic matrices M_1 and M_2 are constructed from the tomographic-probability vectors of qudits $\vec{w}(u_\alpha)$ and $\vec{w}(u_\beta)$ used as columns of these matrices. The separable state of two qudits can be associated with convex sum of the stochastic matrices of the given tensor-product form. Both the tensor product of such matrices and convex sum of the tensor products satisfy the same inequality of the form $\left|\sum_{i,j} c_{ij} M_{ij}\right| \leq L$. The bound L is different for the matrix M corresponding to entangled state. The coefficients c_{ij} have some geometrical meaning which one needs to clarify.

5. Conclusions

To conclude, we formulate our main results.

We reviewed the probability representation of quantum mechanics in which quantum states are described by standard ptobability distributions. The probability distribution (or tomographic probability distribution) can be considered as the probability vector. For qudits, the probability vector depends on random spin projections and unitary group. Using the tomographic probability vectors we constructed the stochastic matrices. The columns of the stochastic matrices were taken as tomographic probability vectors at different values of the unitary-group elements. The properties of the constructed stochastic matrices associated with quantum states and forming semigroups provide the possibility to understand mathematical mechanism of Bell's inequality violations for entangled quantum states of multipartite systems.

We pointed out that the joint probability distributions used to study Bell's inequality violations in [8] are just spin tomograms introduced to describe quantum states of bipartite systems in [3] (see, also [9]).

The example of two qubit states considered in this paper demonstrates that Bell's inequality violations is a specific property of joint probability distributions which can be used either in classical statistics or in quantum mechanics. This observation is coherent with considerations presented in recent works [19, 20] where Bell's inequalities were discussed for classical probability distributions.

Acknowledgments

The author thanks the Russian Foundation for Basic Research for support under Projects Nos. 07-02-00598 and 07-02-08534 and the Organizers of the 5th International Conference "Quantum Theory and Symmetries" (July 2007, Valladolid, Spain) for invitation and kind hospitality.

References

[1] Dodonov V V and Man'ko V I 1997 *Phys. Lett.* A **229** 335
[2] Man'ko V I and Man'ko O V 1997 *Zh. Éksp. Teor. Fiz.* **112** 796 [1997 JETP **85** 430]

[3] Man'ko V I and Safonov S S 1998 *Yad. Fiz.* **4** 658
[4] Stayer D F et al. 2002 *Amer. J. Phys.* **70** 288
[5] Schrödinger E 1935 *Proc. Cambridge Philos. Soc.* **31** 555
[6] Bell J S 1964 *Physics* (Long Island City, N.Y.) **1** 195
[7] Clauser J F, Horne M A, Shimony A and Holt R A 1969 *Phys. Rev. Lett.* **23** 880
[8] Collins D, Gisin N, Linden N, Massar S and Popescu S 2002 *Phys. Rev. Lett.* **88** 040404
[9] Man'ko V I, Marmo G, Sudarshan E C G and Zaccaria F 2004 *Phys. Lett.* A **327** 353
[10] Lupo C, Man'ko V I and Marmo G and Sudarshan E C G 2005 *J. Phys. A: Math. Gen.* **38** 10377
[11] Lupo C, Man'ko V I and Marmo G 2007 *J. Phys. A: Math. Gen.* **40** 13091
[12] Andreev V A, Man'ko V I, Man'ko O V and Shchukin E V 2006 *Teor. Mat. Fiz.* **146** 140 [*Theor. Math. Phys.* 146 140]
[13] Chernega V N and Man'ko V I 2007 *J. Russ. Laser Res.* **28** 103
[14] Man'ko M A, Man'ko V I and Mendes R V 2006 *J. Russ. Laser Res.* **27** 507 [quant-ph/0602129]
[15] Man'ko V I 2007 *Probability instead of wave function and Bell inequalities as entanglement criterion* 2007 Talk at IV Intern. Conf. "Quantum Theory: Reconsideration of Foundations-4" (Vaxjo, Sweden, June 2007), in: G. Adenier, A. Yu. Khrennikov, P. Lahti, V. I. Man'ko and T. Nieuwenhuizen (eds.), AIP Conference Proceedings Series, Vol. 962, p. 140
[16] Brunner N, Branciard C and Gisin N 2008 *Can one see entanglement?* quant-ph/0802.0472v1
[17] Gisin N 2008 quant-ph/0702.021v2
[18] Acin A, Chen J L, Gisin N et al. 2004 *Phys. Rev. Lett.* **92** 250404
[19] Matzkin A 2007 *Classical statistical distribution can violate Bell's inequalities* quant-ph/0703251
[20] Matzkin A 2007 *Bell's theorem as signature of nonlocality: a classical counterexample* quant-ph/07019.2114
[21] Aliski R and Lendi K 1987 *Dynamical Semigroups and Applications* (Springer: Berlin)
[22] Bohm A, Kaldass H and Komy S 2007 *Time–asymmetry in quantum physics* quant-ph/0703162
[23] Shannon C E 1948 *Bell Systems Tech. J.* **27** 379
[24] Rènyi A 1970 *Probability Theory* (North-Holland: Amsterdam)
[25] Man'ko M A, Man'ko V and Mendes R V 2001 *J. Phys. A: Math. Gen.* **34** 8321
[26] Klimov A B, Man'ko O V, Man'ko V I, Smirnov Y F and Tolstoy V N 2002 *J. Phys. A: Math. Gen.* **35** 6101
[27] Man'ko O V and Man'ko V I 2004 *J. Russ. Laser Res.* **25** 115 [quant-ph/041131]
[28] Man'ko M A, Man'ko O V and Man'ko V I 2006 *Intern. J. Mod. Phys.* B **20** 1399
[29] De Nicola S, Fedele R, Man'ko M A and Man'ko V I 2006 *Eur. Phys. J.* B **52** 191 [quant-ph/0607200v1]
[30] Man'ko M A *New inequalities for tomograms in the probability representation of quantum states* 2006 Talk at the 3d Int Workshop "Nonlinear Physics. Theory and Experiment. III (Gallipoli, Lecce, Italy, June 2006) De Nicola S, Fedele R, Man'ko M A and Man'ko V I 2007 *Teor. Mat. Fiz.* **152** 241 [*Theor. Math. Phys.* **152** 1081] [quant-ph/0611114v1]
[31] Man'ko M A 2006 *J. Russ. Laser Res.* **27** 405
[32] Man'ko M A 2006 *Symplectic entropy* Talk at the 3d Feynman Festival (25–29 August 2006, University of Maryland, USA)
De Nicola S, Fedele R, Man'ko M A and Man'ko V I 2007 *J. Phys.: Conf. Ser.* **70** 012007
[33] Man'ko M A 2007 *Tomographic entropy and new entropic uncertainty relations* 2007 Talk at IV Intern. Conf. "Quantum Theory: Reconsideration of Foundations-4" (Vaxjo, Sweden, June 2007), in: G. Adenier, A. Yu. Khrennikov, P. Lahti, V. I. Man'ko and T. Nieuwenhuizen (eds.), AIP Conference Proceedings Series, Vol. 962, p. 132
[34] Chen J L, Deng D L and Hu M G 2008 *Analytic proof of Gisin's theorem for two d-dimensional systems* quant-ph/0802.0125v1
[35] Cirel'son B S 1980 *Lett. Math. Phys.* **4** 93

Journal of Physics: Conference Series **128** (2008) 012019 doi:10.1088/1742-6596/128/1/012019

Dirac Equation for Electrodynamic Model Particles

J.X. Zheng-Johansson

Institute of Fundamental Physics Research, 611 93 Nyköping, Sweden; in affiliation with the Swedish Institute of Space Physics, Kiruna, Sweden

E-mail: jxzj@iofpr.org

Abstract. We set up the Maxwell's equations and subsequently the classical wave equations for the electromagnetic waves which together with their generating source, an oscillatory charge of zero rest mass in general travelling, make up a particle travelling similarly as the source at velocity v in the field of an external scalar and vector potentials. The direct solutions in constant external field are Doppler-displaced plane waves propagating at the velocity of light c; at the de Broglie wavelength scale and expressed in terms of the dynamically equivalent and appropriate geometric mean wave variables, these render as functions identical to the space-time functions of a corresponding Dirac spinor, and in turn identical to de Broglie phase waves previously obtained from explicit superposition. For two spin-half particles of a common set of space-time functions constrained with antisymmetric spin functions as follows the Pauli principle for same charges and as separately indirectly induced based on experiment for opposite charges, the complete wave functions are identical to the Dirac spinor. The back-substitution of the so explicitly determined complete wave functions in the corresponding classical wave equations of the two particles, subjected further to reductions appropriate for the stationary-state particle motion and to rotation invariance when in three dimensions, give a Dirac equation set; the procedure and conclusion are directly extendible to arbitrarily varying potentials by use of the Furious theorem and to particle motions in three dimensions by virtue of the characteristics of de Broglie particle motion. Through the derivation of the Dirac equation, the study hopes to lend insight into the connections between the Dirac wave functions and the electrodynamic components of simple particles under the government by the well established basic laws of electrodynamics.

1. Introduction

P.A.M. Dirac established in [1] a relativistic quantum mechanical wave equation, Dirac equation, for a point electron based on the relativistic energy-momentum relation subjected to Lorentz transformation under rotation. In [1] P.A.M. Dirac also theoretically predicted for the electron the existence of an internal oscillation state, a magnetic moment, and by interpretation of the negative energy solution, an anti-particle state known today as the positron. The Dirac equation has proven to be an accurate equation of motion for (two) spin-half quantum particles at high velocities; most notably, Dirac predicted based on his equation the relativistic intensities of Zeeman components of spectral lines and the frequency differences [1 (1928b)] in exact agreement with experiment. Up to the present however it has remained an open question that, what is waving with the Dirac wave functions, or Dirac spinor, a similar question as for the Schrödinger wave functions and the de Broglie waves? In addition, the Dirac theory meets with a few its own open questions. What is the nature of a Dirac internal oscillation? How are the Dirac space-time functions explicitly connected with the spin orientations, the signs of charges, the signs of the

Journal of Physics: Conference Series **128** (2008) 012019 doi:10.1088/1742-6596/128/1/012019

energies, and in the extreme situation when an electron and position annihilate, the emitted two gamma rays and conversely? What is the symmetry of the total spin of an electron and positron? Also, in the case of an isolated single electron or positron in zero external field where the spin orientation is of no consequence, it would be desirable to have a way to directly write down the corresponding Dirac equation without involving the Pauli matrices. These as well as various other not fully addressed questions relating to fundamental physics seem to consistently point to the inadequacy of the point particle picture of today and the need for a representation of the internal processes of the particles.

Recently, using overall experimental observations as input data the author proposed an internally electrodynamic (IED) particle model [2a,l] or sometimes termed a basic particle formation (BPF) scheme (with coauthor P.-I. Johansson), which states that *a simple (basic) particle like an electron and positron, etc., briefly, is constituted of an oscillatory point-like (elementary) charge with a specified sign and a zero rest mass, and the resulting electromagnetic waves in the vacuum.* As a broad test of the IED particle model and also as an endeavour of understanding the various puzzles relating to fundamental physics, in terms of solutions for the electrodynamic processes of the model particle with its charge's sign and total energy as two sole input data, the author has further achieved with coauthor(s) derivations/predictions of a range of basic properties and relations of the simple particles [2a-k] including the relativistic mass, de Broglie wave, de Broglie relations, Schrödinger equation, Einstein energy-mass relation, Newton's law of gravity and Doebner-Goldin equation, among others. As to the Schrödinger wave function specifically of relevance here, the solution[2a,c] showed that it is the (envelope of the) standing beat wave resulting from superposition of the Doppler-differentiated electromagnetic waves generated by the particle's travelling charge, that is waving.

As shown in [2a,c-d], the direct solutions for the classical wave equations, derivable from the Maxwell's equations, for the electromagnetic waves comprising a free particle consist of Doppler-displaced plane waves; these superpose to two opposite-travelling beat waves that represent directly the de Broglie phase waves and in turn the Dirac space-time functions. It therefore is foreseeable that the classical wave equations for the electromagnetic waves here of two given particles would more naturally lead to wave equations of the particles corresponding directly to the Dirac equation in comparison to the Schrödinger equation. We elucidate in this paper a formal procedure which transforms the classical wave equations for the electromagnetic waves of two spin-half particles, of identical space-time functions and tending to approach one another, to the Dirac equation. Through the procedure we show that the Dirac internal oscillation corresponds to the oscillation of the electromagnetic waves at a geometric mean of the respective wave frequencies which in general are Doppler-displaced owing to source motion. And we elucidate the explicit relationships between the internal electromagnetic waves, charges, spins, the centre-of-mass and total wave motions and the associated energies of the particles, pertaining to entities and motions at a sub-quantum mechanical level, under the government of a few established elementary laws of electrodynamics.

2. Wave equations for the electromagnetic waves of particle. Solutions

We consider an IED particle, here an electron or positron, is as its source charge q ($= e$ or $-e$) travelling at a velocity $\boldsymbol{v}$ in $+z$-direction for the present along a one-dimensional box of side L in the vacuum. The charge q of the particle has an oscillation associated with a total energy ε_q, which is minimum at $v = 0$, denoted by $\mathcal{E}_q$; $\mathcal{E}_q$ may be endowed e.g. in a pair production in the vacuum. In virtue that it describes the ground state, $\mathcal{E}_q$ cannot be dissipated or detached from the charge except in a pair annihilation.

The charge q of the particle generates owing to its oscillation electromagnetic waves of radiation electric fields $\mathbf{E}^j$'s and magnetic fields $\mathbf{B}^j$'s described in zero applied potential field by

Journal of Physics: Conference Series **128** (2008) 012019 doi:10.1088/1742-6596/128/1/012019

the Maxwell's equations as:

$$\nabla \cdot \mathbf{E}^j = \rho_q^j/\epsilon_0,\ \nabla \cdot \mathbf{B}^j = 0,\ \nabla \times \mathbf{B}^j = \mu_0 \mathbf{j}_q^j + (1/c^2)\partial_t \mathbf{E}^j,\ \nabla \times \mathbf{E}^j = -\partial_t \mathbf{B}^j. \tag{1}$$

Where ρ_q^j is the density and $\mathbf{j}_q^j$ the current of the particle's charge, assuming no other charges and currents present; ϵ_0 is the permittivity and μ_0 the permeability of the vacuum, and c is the velocity of light; $\partial_t \equiv \partial/\partial t$. Expressing the jth fields generally by a dimensionless displacement φ^j, $E^j = D\varphi^j$, thus $B^j = E^j/c = D\varphi^j/c$, with D a conversion constant, considering regions sufficiently away from the source only so that $\rho_q^j = j_q^j = 0$, and with some otherwise standard algebra of the Maxwell's equations (1), we obtain the corresponding classical wave equations for the electromagnetic waves φ^j's

$$c^2\nabla^2\varphi^j = \partial_t^2\varphi^j, \tag{2}$$

with $\partial_t^2 \equiv \partial^2/\partial t^2$. In the above, $j = \dagger$ labels the component wave generated in the direction parallel with $+v$, and $j = \ddagger$ the wave parallel with $-v$; within walls there prevail also their reflected components generated by the reflected charge at an earlier time and being at the present time as if generated by a virtual charge travelling in the $-z$-direction, labelled by $j =$ vir$\dagger$ and $j =$ vir$\ddagger$. j is to distinguish a Doppler effect owing to the source motion to be expressed in (7) below. In Appendix A we outline in relevance to the particle model a few further standard relations of classical and quantum electrodynamics for the electromagnetic waves, and a derivation of the particle's mass given by the author previously[2a,e] (with P.-I. Johansson).

To the particle we now apply an electromagnetic force $\mathbf{F} = \mathbf{F}_e + \mathbf{F}_m$, with $\mathbf{F}_e = -q\nabla\phi_a$ the Coulomb force in z- direction and $\mathbf{F}_m = -q\mathbf{v} \times \nabla \times \mathbf{A}_a$ the Lorentz force due to an external scalar potential ϕ_a and vector potential $\mathbf{A}_a$, each expressed in SI units as for all other quantities in this paper. $\mathbf{F}_m$ may be simplified using the BAC-CAB rule as $\mathbf{F}_m = -q[\nabla(\boldsymbol{v}\cdot\mathbf{A}_a) - \mathbf{A}_a(\boldsymbol{v}\cdot\nabla)]$. In the applications below $\mathbf{A}_a$ is constant in L or in each small division in question (see end of Sec. 3), and v is constant in L for the particle being in stationary state and also is parallel with ∇ and $\mathbf{z}$, so $\nabla(\boldsymbol{v}\cdot\mathbf{A}_a) = 0$ and $\mathbf{F}_m = q\mathbf{A}_a(\boldsymbol{v}\cdot\nabla) = q\mathbf{A}_a v\nabla$. Thus, $\mathbf{F} = -q\nabla\phi_a + q\mathbf{A}_a v\nabla$. The formula of $\mathbf{F}$ is in the usual usage established for a point particle; so when extending to the extensive IED particle here, $\mathbf{F}$ apparently directly acts on the particle's point charge.

We need to map the $\mathbf{F}$ to a force directly interacting with the internal fields E^j, B^j, or φ^j of the particle. We observe that, in virtue of its form, (2) after multiplying a conversion factor a of length dimension represents just a classical wave equation for a mechanical wave of a transverse displacement $a\varphi^j$ propagated in an apparent *elastic medium*. On grounds of this direct correspondence, but taken as a heuristic means only in this paper (so that we here need not involve the details of this elastic medium), $\mathbf{F}$ therefore interacts with the internal fields through a force $\mathbf{F}_{\text{med}}^j$ directly acting on this apparent medium. We can think of the medium to be composed of coupled dipole charges which do not move along the z-axis but the $\mathbf{F}_{\text{med}}^j$ propagates across the dipoles at the wave speed c. Viewed in the frame where $\mathbf{F}_{\text{med}}^j$ is at rest, we then find the dipole charges are travelling at the speed c; so as a first step of the mapping, the Lorentz force on the medium ought to scale as $\mathbf{F}^{j\prime}_m = (\pm c/v)\mathbf{F}_m = \pm q\mathbf{A}_a c\nabla$ with $+, -$ for the $j = \dagger, \ddagger$ waves; thus $\mathbf{F}^{j\prime} = \mathbf{F}_e + \mathbf{F}^{j\prime}_m$. Under the actions of the respective forces, the acceleration $F^{j\prime}/m^j$ of the particle's charge of a dynamical mass m^j (due to the charge's total motion and equivalently the φ^j motion, see further Appendix A), and that of the medium of a dynamical mass $\mathfrak{M}_\varphi^j$, $F_{\text{med}}^j/\mathfrak{M}_\varphi^j$ must equal, i.e. $\mathbf{F}^{j\prime}/m^j = \mathbf{F}_{\text{med}}^j/\mathfrak{M}_\varphi^j$. Thus $\mathbf{F}_{\text{med}}^j = \frac{\mathfrak{M}_\varphi^j}{m^j}\mathbf{F}^{j\prime} = \frac{\mathfrak{M}_\varphi^j q}{m^j L_\varphi^j}(-\nabla\phi_a \pm \mathbf{A}_a c\nabla)$.

By its pure mechanical virtue the force $\mathbf{F}_{\text{med}}^j$ acting on the continuous medium is nonlocal and will be transmitted uniformly across the medium here along the z-axis of an effective length L_φ^j, $j = \dagger, \ddagger$ (L_φ^j winds J^j loops about L; thus, using the geometric mean $L_\varphi = \sqrt{L_\varphi^\dagger L_\varphi^\ddagger}$, we have $\nabla\phi_a = \pm(\phi_a/L_\varphi)\hat{z}$ and $\nabla = \pm 1/L_\varphi$. With these, putting $\mathfrak{M}^j_\varphi = \rho_l L^j_\varphi$ where ρ_l is the mean linear mass density of the medium, writing for conciseness $\nabla\phi_a$ and also the final $\mathbf{F}^j_{\rm med}$ in scalar forms and keeping the generally arbitrarily oriented $\mathbf{A}_a$ in vector form only, $\mathbf{F}^j_{\rm med}$ becomes

$$F^j_{\rm med} = -\rho_l V^j/m^j, \quad V^\dagger = q\phi_a - q\mathbf{A}_a c, \quad V^\ddagger = -q\phi_a - q\mathbf{A}_a c. \tag{3}$$

$F^j_{\rm med}$ can be implemented in (2) by directly establishing the corresponding wave equation for the apparent elastic medium acted by $F^j_{\rm med}$. If without $F^j_{\rm med}$, the elastic medium would be deformed owing to the disturbance of the oscillation of the source charge alone, by a total displacement $u = a\sum_j \varphi^j$ with φ^j given for $\phi_a = A_a = 0$, and be thus subject to a tensile force $F_R = \rho_l c^2$. The applied $F^j_{\rm med}$ and F_R add up to a total force acting on the particle through acting directly on the medium

$$F^{j\prime}_R = F_R - F^j_{\rm med} = \rho_l\left[c^2 + V^j/m^j\right]. \tag{4}$$

Where, the minus sign of $F^j_{\rm med}$ is because this force tends to contract the chain. Assuming $\sum\varphi^j$, with φ^j now given for a finite ϕ_a, $\mathbf{A}_a$, is relatively small which in general is the case in practical applications, $F^{j\prime}_R$ is thus uniform across the L. A segment ΔL of the medium along the box, of mass $\Delta\mathfrak{M}_\varphi = \Delta\mathfrak{M}^j_\varphi/J^j = \rho_l\Delta L \simeq \rho_l\Delta z$, will upon deformation be tilted from its equilibrium position z-axis an angle ϑ^j and $\vartheta^j + \Delta\vartheta^j$ at z and $z + \Delta z$. The transverse (y-) component force acting on $\Delta\mathfrak{M}_\varphi$ is $\Delta F^{j\prime}_{Rt} = F^{j\prime}_R[\sin(\vartheta^j + \Delta\vartheta^j) - \sin\vartheta^j] = F^{j\prime}_R\nabla^2(a\varphi^j)\Delta z = a\rho_l\left[c^2 + \frac{V^j}{m^j}\right]\nabla^2\varphi^j\Delta z$. Newton's second law for the mass $\Delta\mathfrak{M}_\varphi$ writes $\rho_l\Delta z\partial_t^2(a\varphi^j) = \Delta F^{j\prime}_{Rt}$. The two last equations give the equations of motion, on dividing $a\rho_l\Delta z$, for per unit length per unit linear mass density of the medium at z or equivalently the classical wave equations for the electromagnetic waves φ^j's in the fields of the applied potentials ϕ_a, $\mathbf{A}_a$:

$$\left[c^2 + q(\phi_a - \mathbf{A}_a c)/m^\dagger\right]\nabla^2\varphi^\dagger = \partial_t^2\varphi^\dagger, \quad [c^2 - q(\phi_a + \mathbf{A}_a c)/m^\ddagger]\nabla^2\varphi^\ddagger = \partial_t^2\varphi^\ddagger. \tag{5}$$

This for $\phi_a = A_a = 0$ reduces to (2) given directly from the Maxwell's equations earlier.

Assuming for the present ϕ_a, $\mathbf{A}_a$ are constant and also $\mathbf{A}_a$ is small such that the particle motion effectively deviates not from the linear path, so the solution of (5) consists of plane waves $\varphi^\dagger = \mathfrak{C}f^\dagger$ and $\varphi^\ddagger = \mathfrak{C}f^\ddagger$ (Figure 1a, solid and dotted curves) generated in $+z$- and $-z$-directions and initially also travelling in these directions at speed $\omega^j/k^j = c$, where

$$f^\dagger = Ce^{i[k^\dagger_d z - \omega^\dagger t + \alpha_0]}, \quad f^\ddagger = -Ce^{i[-k^\ddagger_d z + \omega^\ddagger t - \alpha_0]} \tag{6}$$

(Figure 1 a-b, single-dot-dashed and triple-dot-dashed curves), $\mathfrak{C} = e^{iKz}$, and C $(= 4C_1/\sqrt{L})$ a normalisation constant. And,

$$k^\dagger = K/(1 - v/c) = \gamma^\dagger K, \ k^\ddagger = K/(1 + v/c) = \gamma^\ddagger K \quad \text{and} \quad \omega^\dagger = \gamma^\dagger\Omega, \ \omega^\ddagger = \gamma^\ddagger\Omega \tag{7}$$

are the source-motion resultant Doppler-displaced wavevectors and angular frequencies; $\gamma^\dagger = \frac{1}{1-v/c}$, $\gamma^\ddagger = \frac{1}{1+v/c}$; $K, \Omega = Kc$ are values of k^j, ω^j at $v = 0$. (7) further gives

$$k^\dagger_d = k^\dagger - K = \gamma^\dagger K_d, \ k^\ddagger_d = K - k^\ddagger = \gamma^\ddagger K_d \quad \text{where } K_d = (v/c)K. \tag{8}$$

Suppose $v << c$ (yet v^2/c^2 may be large so that dynamically the factor γ in (9) below can be different from 1) and accordingly the de Broglie wavelength ($\lambda_d = 2\pi/(\gamma K_d)$) later will be much greater than the electromagnetic wavelength ($\Lambda = \frac{2\pi}{K}$). So at the scale of λ_d the rapid variation of $\mathfrak{C}$ is to an external observer no different from the constant 1, that is $\lim_{v<<c}\mathfrak{C} = 1$. Thus

$\varphi^\dagger|_{\mathcal{C}=1} = f^\dagger$, $\varphi^\ddagger|_{\mathcal{C}=1} = f^\ddagger$ and the $f^\dagger, f^\ddagger$ as given in (6) represent external-effective space-time functions. We see that $f^\dagger$ and $f^\ddagger$ are two new plane waves travelling each to the right at equal phase velocities, $\frac{\omega^\dagger}{k_d^\dagger} = \frac{-\omega^\ddagger}{-k_d^\ddagger} = W = \frac{c^2}{v}$ which is c/v times the velocity of light c. It can be checked (Appendix B) that the exact solutions φ^j's and in turn the external effective f^j's given above placed in the respective wave equations (5) above and (12) below yield exactly the expected relativistic energy-momentum relation.

The Doppler-displaced variables k_d^j's,ω^j's in the functions f^j's,φ^j's are not single valued and thus are not good dynamical variables of the particle. The respective geometric means

$$k_d = \sqrt{k_d^\dagger k_d^\ddagger} = \gamma K_d, \quad \omega = \sqrt{\omega^\dagger \omega^\ddagger} = \gamma \Omega, \quad \text{with } \gamma = \sqrt{\gamma^\dagger \gamma^\ddagger} = 1/\sqrt{1 - v^2/c^2}, \tag{9}$$

are evidently good dynamical variables of particle and also are appropriate in view of the stochastic virtue of the electromagnetic waves (these being distributed functions and having in general random initial phases over a macroscopic time interval. These are also the natural independent variables of the superposed wave functions (for a detailed elucidation see [2c-d]): $\widetilde{\psi} = \varphi^\dagger + \varphi^\ddagger = \mathcal{C}_d e^{i[k_d z - \omega t + \alpha_0]}$ and $\widetilde{\psi}^{\text{vir}} = \varphi^{\text{vir}\dagger} + \varphi^{\text{vir}\ddagger} = -\mathcal{C}_d^{\text{vir}} e^{i[k_d z + \omega t + \alpha_0]}$ which are two beat waves of a wavelength $\lambda_d = 2\pi/k_d$, travelling each at the phase speed $W = c^2/v$ to the right and to the left as due to the actual and reflected (virtual) charges travelling to the right and left respectively; $\mathcal{C}_d = 2C_1 e^{i[(K + \frac{v}{c}k_d)z - \frac{v}{c}\omega t]}$, $\mathcal{C}_d^{\text{vir}} = 2C_1 e^{i[(K + \frac{v}{c}k_d)z + \frac{v}{c}\omega t]}$ and $\mathcal{C}_d \doteq \mathcal{C}_d^{\text{vir}} \doteq 2C_1 e^{iKz} = 2C_1$ for $v << c$. An inspection will show that clearly $\widetilde{\varphi}, \widetilde{\varphi}^{\text{vir}}$ resemble directly the de Broglie phase waves of the particle in the constant $\phi_a, \mathbf{A}_a$ fields here, λ_d and k_d are the de Broglie wavelength and wavevector, and accordingly $\hbar k_d = p_v$ the linear momentum. If $\phi_a, \mathbf{A}_a$ are arbitrarily varying in L and well behaved, we can divide L in a large N number of small divisions in each of which the plane waves hold true and their sum gives according to the Fourier theorem the total wave. Further, in three dimensions the de Broglie particle motion is a straightforward extension of a locally one-dimensional motion. The wave equations to be given will formally be otherwise the same (for a formal treatment in the case of a Schrödinger system in a varying potential and in three dimensions see [2c,j]). We shall thus for simplicity proceed the remainder of treatment for constant $\phi_a, \mathbf{A}_a$ and, until the discussion regarding spin rotation in Sec. 5, for a particle motion in one-dimension.

3. Wave equation for total motion of particle

For a single particle or for many particles without regarding the spins, the functions $\widetilde{\psi}$ and $\widetilde{\psi}^{\text{vir}}$, or equivalently the f_r and f_ℓ of (15) below, are seen to be identical to the usual solutions to the Dirac equation, c.f. Appendix C. So their wave equations, originally the (5), evidently must have a direct correspondence with the Dirac equation. The remainder of the task mainly will be to identify a physically justifiable procedure to transform (5) to a form of the Dirac equation under corresponding considerations.

First, similarly as Dirac (or as alternatively but compatibly argued in [2c]) we want the eventual wave functions of the particle, and thus immediately the f^j or the original φ^j to be linear in $\hbar\omega$ and thus in $\partial_t f^j$ here, so that f^j at any initial time determines its value at any future time; and we want similarly for the linear momentum here. We shall thus transform the second order differential equations (5) to first order ones and in the end take the limit for $c >> v$ as follows. For the $c^2\nabla^2\varphi^j$ terms of (5), starting with the full wave functions $\varphi^j = \mathcal{C} f^j$ we first lower the spatial derivative one order as $\nabla^2\varphi^j = \nabla[i\gamma^j K \mathcal{C} f^j] = i\gamma^j K[(\nabla\mathcal{C}) f^j + \mathcal{C}\nabla f^j]$. Restricting to $v << c$, we thus can replace $\nabla\mathcal{C}$ by its computed value $iK\mathcal{C}$ and in turn put $\mathcal{C} \doteq 1$ for each term; this gives

$$\nabla^2\varphi^j|_{\mathcal{C}\doteq 1} = [-\gamma^j K^2 \mathcal{C} f^j + i\gamma^j K \mathcal{C}\nabla f^j]_{\mathcal{C}\doteq 1} = -\gamma^j K^2 f^j + i\gamma^j K \nabla f^j, \quad j = \dagger, \ddagger. \tag{10}$$

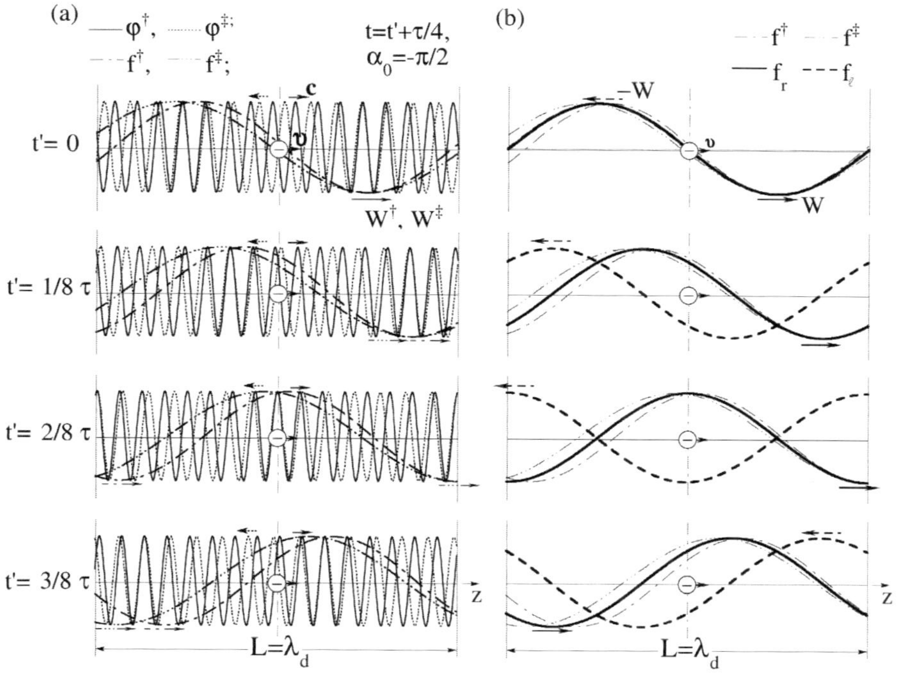

Figure 1. (a) shows an IED model electron constituted of an oscillatory charge $-e$ of zero rest mass ($\ominus$), travelling at velocity v, and the resulting Doppler-differentiated electromagnetic waves $\varphi^\dagger$ and $\varphi^\ddagger$ (solid and dotted curves) of a angular frequencies $\omega^\dagger, \omega^\ddagger$ generated in $+z$- and $-z$- directions, plotted for the real parts in a time interval $(3/4)\tau$ in a one-dimensional box of side $L = \lambda_d$; $\tau = 2\pi/\omega$, $\omega = \sqrt{\omega^\dagger \omega^\ddagger}$, $\lambda_d = 2\pi((v/c)\omega)$ being the de Broglie wavelength. $f^\dagger$ and $f^\ddagger$ (single-dot-dashed and triple-dot-dashed curves) are the corresponding external-effective waves, shown in both (a) and (b). f_r and f_ℓ (solid and dashed curves) in (b) are the dynamically equivalent mean-variable wave functions; these resemble directly the (opposite-travelling) de Broglie phase waves and are equivalent to the space-time functions of Dirac spinor.

Next, assume ϕ_a, $\mathbf{A}_a$ relatively small as typically is true in practical applications, so the resulting force constant (i.e. force per unit displacement) on the particle does not vary across L. We can thus replace the $\nabla^2\varphi^j$ in the ϕ_a, $\mathbf{A}_a$ terms of (5) by its computed value as $\nabla^2\varphi^j|_{\mathcal{C}=1} = -{\gamma^j}^2 K^2 f^j$; accordingly

$$\frac{q(\phi_a - \mathbf{A}_a c)}{m^\dagger}\nabla^2\varphi^\dagger|_{\mathcal{C}=1} = q(-\phi_a + \mathbf{A}_a c)\frac{\gamma^\dagger \Omega f^\dagger}{\hbar}, \quad \frac{-q(\phi_a + \mathbf{A}_a c)}{m^\ddagger}\nabla^2\varphi^\ddagger|_{\mathcal{C}=1} = q(\phi_a + \mathbf{A}_a c)\frac{\gamma^\ddagger \Omega f^\dagger}{\hbar} \quad (11)$$

For the final expressions we used $Kc = \Omega$ as earlier, and $m^j = \gamma^j M$ and $Mc^2 = \hbar\Omega$ given after (A.1) and (9). Finally, the $\partial_t^2\varphi^j$'s of (5) lower one order as $\partial_t^2\varphi^\dagger|_{\mathcal{C}=1} = -i\gamma^\dagger\Omega\partial_t f^\dagger$, $\partial_t^2\varphi^\ddagger|_{\mathcal{C}=1} = i\gamma^\ddagger\Omega\partial_t f^\ddagger$. Substituting these and equations (10)–(11) in wave equations (5), multiplying the first resulting equation by $-\frac{\hbar c}{K\gamma^\dagger}$ and the second by $\frac{\hbar c}{K\gamma^\ddagger}$, with $cK = \Omega$ and $\hbar K = Mc$ as before and as given after (A.1), we eventually obtain the wave equations for the electromagnetic waves of the particle expressed by $f^\dagger, f^\ddagger$:

$$[Mc^2 + q\phi_a - c(i\hbar\nabla + q\mathbf{A}_a)]f^\dagger = i\hbar\partial_t f^\dagger, \quad [-Mc^2 + q\phi_a + c(i\hbar\nabla + q\mathbf{A}_a)]f^\ddagger = i\hbar\partial_t f^\ddagger. \quad (12)$$

For the particle dynamics in question we want to further transform (12) to be expressed by the particle wave variables, i.e. the k_d and ω defined in (9), and the corresponding wave functions, the f_r, f_ℓ to be obtained below. We shall below obtain the sought-for functions f_r, f_ℓ through a dynamic equivalence transformation directly from the $f^\dagger, f^\ddagger$; these ought to be and will show to be identical to the $\widetilde{\psi}, \widetilde{\psi}^{\text{vir}}$ obtained in a physically more transparent way earlier. The present approach below will advantageously preserve a direct tractable connection with the original $f^\dagger, f^\ddagger$, thus also $\varphi^\dagger, \varphi^\ddagger$, whose wave equations (12) or (5) give the relativistic energy-momentum relation exactly based on the Doppler equations (7), see (B.2) of Appendix B. What accordingly is in question in the transformation mainly is to maintain an equivalence to the quadratic equation (B.2); this corresponds to the equations $\frac{\partial f^j}{\partial z^\nu}\frac{\partial f^{j'}}{\partial z^\nu} = \frac{\partial f_\mu}{\partial z^{\nu n}}\frac{\partial f_{\mu'}}{\partial z^\nu}$, etc., with $j, j' = \dagger, \ddagger$, $\mu, \mu' = r, l$, $z^\nu = t, z$ $(\nu = 0, 3)$; and $f_\mu \frac{\partial f_{\mu'}}{\partial z} = 0$. The equivalence condition requires in particular the transformed quadratic to be $\frac{\partial f_r}{\partial z}\frac{\partial f_l}{\partial z} = k_d^2$, that is, it has a plus sign in front and the cross-term product of $\frac{\partial f_\mu}{\partial z}$ with Mc^2 (i.e. the $\mathcal{O}$ discussed after 17) is absent. This can be achieved if we introduce a wavevector being the imaginary of (and thus orthogonal to) k_d:

$$\bar{k}_d = (\gamma/i\gamma^\dagger)k_d^\dagger, \quad \bar{k}_d = (\gamma/i\gamma^\ddagger)k_d^\ddagger; \quad \text{thus } k_d^\dagger k_d^\ddagger = (1/i^2)\bar{k}_d^2 \tag{13}$$

(compare $\bar{k}_d$ with the operator $p_{v.op} = \frac{\hbar}{i}\nabla$ later). (13) alternatively can be expressed by

$$(\text{a}):\ k_d^\dagger(-k_d^\ddagger) = \bar{k}_d\bar{k}_d \quad \text{or} \quad (\text{b}):\ (-k_d^\dagger)k_d^\ddagger = \bar{k}_d\bar{k}_d. \tag{14}$$

We now first transform the Doppler-differentiated $f^j(;k_d^j, \omega^j)$'s (as the short hand notations of $f^j(z,t;k_d^j,\omega^j)$'s which in turn are expanded notations of $f^j(z,t)$'s) to a pair of mean (wave)-variable functions $f_\mu(;\bar{k}_d,\omega)$'s (denoting $f_\mu(z,t;\bar{k}_d,\omega)$'s) by, say, satisfying (a) of (14) and ordinarily $\omega = \sqrt{\omega^\dagger\omega^\ddagger}$ of (9):

$$f^\dagger(;k_d^\dagger,\omega^\dagger) \to f_r(;\bar{k}_d,\omega) = C_r e^{i[\bar{k}_d z - \omega t + \alpha_0]}, \quad f^\ddagger(;k_d^\ddagger,\omega^\ddagger) \to f_\ell(;\bar{k}_d,\omega) = C_\ell e^{i[\bar{k}_d z + \omega t + \alpha_0]}; \tag{15}$$

see these functions plotted in Figure 1b. The transformed f_r, f_ℓ indeed are desirably identical functions to the original $f^\dagger, f^\ddagger$ if disregarding the high-order differences between the coefficients $\gamma^\dagger, \gamma^\ddagger$ and γ in the wave variables and the reversed travel direction of f_ℓ from $f^\ddagger$. The f_r, f_ℓ, being identical functions to the $\widetilde{\psi}, \widetilde{\psi}^{\text{vir}}$ earlier, indeed are therefore the pertinent space-time functions of the particle; these are each functions of the source motion and the total (electromagnetic) wave oscillation and are formally equivalent to de Broglie phase waves; and these in turn are equivalent to Dirac's space-time functions.

To entail that in the matrix representation later (Sec. 5) a cross-term product $\mathcal{O}$ discussed after (17) is similarly absent, for transformation of the first derivatives we are compelled to satisfy the alternative condition (b) of (14). The use of (14b) and the common $\omega = \sqrt{\omega^\dagger\omega^\ddagger}$ of (9) first directly leads to the intermediate transformations for the f^j's given in the left column below:

$$\begin{aligned} f^\dagger \xrightarrow[\substack{k_d^\dagger \to -\bar{k}_d \\ \alpha_0 \to -\alpha_0'}]{} f_\ell^* \xrightarrow[(\text{a1})]{} f_\ell, \quad \nabla f^\dagger = ik_d^\dagger f^\dagger \xrightarrow[\substack{k_d^\dagger \to -\bar{k}_d, \\ f^\dagger \to f_\ell}]{} i(-\bar{k}_d)f_\ell = -\nabla f_\ell, \\ f^\ddagger \xrightarrow[\substack{k_d^\ddagger \to \bar{k}_d \\ \alpha_0 \to \alpha_0'}]{} -f_r^* \xrightarrow[(\text{a2})]{} -f_r, \quad \nabla f^\ddagger = -ik_d^\ddagger f^\ddagger \xrightarrow[\substack{k_d^\ddagger \to \bar{k}_d, \\ f^\ddagger \to -f_r}]{} -i\bar{k}_d(-f_r) = \nabla f_r \end{aligned} \tag{16}$$

where $f_\ell^* = C_\ell e^{i[-\bar{k}_d z - \omega t - \alpha_0']}$, $f_r^* = C_r e^{i[-\bar{k}_d z + \omega t + \alpha_0']}$, $\alpha_0' = -\alpha_0$. The transformations (a1) and (a2) in (16) to the same f_ℓ, f_r as in (15) are on the basis that the latter indeed represent the original f_ℓ^*, f_r^* in all aspects (having the same phase velocities and wave forms) except the opposite rotating phases on the complex plane and the opposite signs of α_0' and α_0 that altogether are dynamically inconsequential. The relations in the left column and the use of (14b) again then lead to the results in the right column which is actually in question in respect to dynamical equivalence here.

Substituting in wave equations (12) the transformation relations (15) and (16) gives

$$(Mc^2 + q\phi_a)f_r + c(i\hbar\nabla - q\mathbf{A}_a)f_\ell = i\hbar\partial_t f_r, \quad (-Mc^2 + q\phi_a)f_\ell + c(i\hbar\nabla + q\mathbf{A}_a)f_r = i\hbar\partial_t f_\ell. \quad (17)$$

As a check, placing in (17) the f_r, f_ℓ of (15), multiplying the first and the negative of the second resulting equations, dividing $f_r f_\ell$, putting for simplicity $\phi_a = A_a = 0$ for the problem mainly of concern here, we correctly obtain the same result as (B.2): $M^2c^4 - \hbar^2\bar{k}_d^2c^2 + \mathcal{O} = \hbar^2\omega^2$ where $\bar{k}_d^2 = -k_d^2$ following (13) and (9); $\mathcal{O} = 0$.

4. Two-particle system: spins, charges, and time-arrows

Consider two spin-half particles 1,2 having identical sets of $\{f_r, f_\ell\}$'s tend to occupy the same location z or more precisely the same region $(0, L)$; suppose these are noninteracting (a finite particle-particle interaction can in principle be included in V^j and will not affect the general conclusions below). In virtue of the statistical nature of the electromagnetic displacements

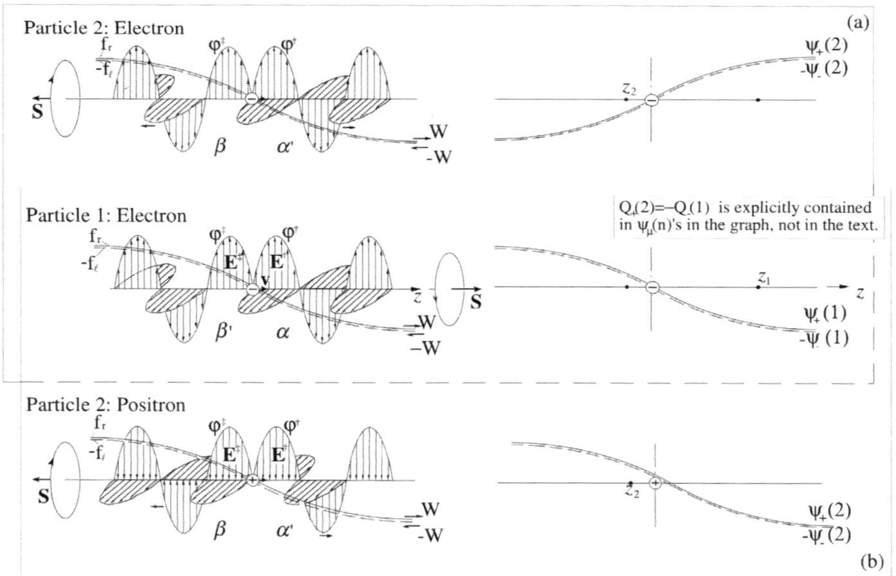

Figure 2. Two spin-half IED particles 1 and 2 described by identical, thus mutually symmetric sets of Doppler-displaced space-time functions $\{\varphi^\dagger, \varphi^\ddagger\}$'s or effectively $\{f_r, f_\ell\}$'s tend to occupy the same location z; panel (a) shows two electrons and (b) an electron and positron. Particle 1 has a spin $\mathbf{S}$ in $+z$- direction, denoted as spin up, $\alpha(1)$, and is parallel with the generating direction of the f_r wave; particle 2 has spin $\mathbf{S}$ in $-z$-direction, denoted as spin down, $\beta(2)$, and is parallel with f_ℓ. Their opposite counterparts (termed virtual spins) $\beta'(1)$ and $\alpha'(2)$ are parallel with f_ℓ and f_r. Right graphs show the complete wave functions $\{\psi_+(1), \psi_-(1)\}$ and $\{\psi_+(2), \psi_-(2)\}$ of particles 1 and 2. In panel (a), $\psi_\nu(1)$ and $\psi_\nu(2)$ ($\nu = +$,-) are antisymmetric due to the antisymmetric spin functions and same charges. In (b), these are symmetric (as explicitly shown in the graph) due to antisymmetric spins and also opposite charges ($\mathcal{Q}_+(2) = -\mathcal{Q}_-(1)$), the latter of which leads to the radiation electric fields $\mathbf{E}^j$'s are opposite in direction (in the text the $\mathcal{Q}_+(2), \mathcal{Q}_-(1)$ are not explicitly regarded, rendering the $\psi_\nu(1)$ and $\psi_\nu(2)$ for the electron and positron to be antisymmetric, thus the same as for two electrons).

and combinatorially of the particles' center-of mass motions (see an elaborate discussion about this in [2k]), the probability of finding a portion of particles 1 and 2 at locations z_1 and z_2 is proportional to the product $f_\mu(z_1,t)f_{\mu'}(z_2,t)$. Since these have identical space-time function sets, the corresponding total space-time function is evidently symmetric, thus f_s, here in the only form $f_s(z_1,z_2) = \frac{1}{\sqrt{2}}[f_r(z_1,t)f_\ell(z_2,t)+f_\ell(z_2,t)f_r(z_1,t)]$ to be compatible with the antisymmetric total spin function later.

In the case of two identical electrons (Figure 2, panel a), their spins then need according to Pauli principle be opposite (left graph in the figure) to avoid both particles occupying the same quantum state; the total spin function for this is antisymmetric ($\chi_a = \frac{1}{\sqrt{2}}[\alpha(1)\beta(2)-\alpha'(2)\beta'(1)]$). Apparently it is in general also relevant that we introduce charge functions $\mathcal{Q}_-(1), \mathcal{Q}_-(2)$ to reflect the sings of charges 1,2; these for the two like charges are trivial identities and lead to a trivial symmetric total charge function, $\mathcal{Q}_s$. The above functions together define an antisymmetric two-electron function (Figure 2a, right graph), $\psi_a = f_s\mathcal{Q}_s\chi_a$, yielding as expected a total probability that is unchanged when interchanging the two stationary-state identical, indistinguishable (as result of being identically extensively distributed in L) particles.

In the case of an electron and positron (Figure 2, panel b), the total two-particle function is symmetric (right graph), thus ψ_s, as follows from the observational fact that two such particles can approach each other arbitrarily close and, in an extreme case annihilate into "one point" in the vacuum. The total charge function for their two opposite signed charges evidently is antisymmetric, thus $\mathcal{Q}_a$. (—If imagine the wave displacement φ^j in the medium is executed by a chain of dipole charges then it is immediately clear that the corresponding radiation electric field E^j, with $|E^j| \propto \varphi^j$, produced by the positive charge is reversed from that by the negative charge, see left graph in Figure 2b.) Placing the two known functions in $\psi_s = \mathcal{Q}_a f_s \chi_{\rm Sym}$ gives therefore Sym=antisymmetric and thus $\chi_{\rm Sym} = \chi_a$ for the electron-positron system.

Our particles contain each the electromagnetic waves $\varphi^\dagger, \varphi^\ddagger$, or effectively f_r, f_ℓ, generated in the specified $+z$- and $-z$- directions which assume definite relationships with the spin orientations that turn out in a measurement: Given, say, particle 1 is measured to be spin up, $\alpha(1)$, with f_r being parallel with it, then its other internal process, f_ℓ, is parallel with the opposite of $\alpha(1)$, denoted by a primed, virtual spin-down state, $\beta'(1)$. Similarly for particle 2, being then actually spin down, thus its f_ℓ is parallel with $\beta(2)$, and its f_r is parallel with the opposite of $\beta(2)$, a primed virtual spin-up state, $\alpha'(2)$. From the foregoing antisymmetric spin requirement follow the relations for the spin functions:

$$\alpha'(2) = -\alpha(1), \quad \beta'(1) = -\beta(2); \quad \text{with } \beta'(1) = -\alpha(1), \quad \alpha'(2) = -\beta(2) \tag{18}$$

for the opposite signs of each particle as is meant by the "virtual" spins. Through (18), the virtual spin vector of particle 1, $\mathbf{S}'(1)$ (virtual spin down) and the actual spin vector of particle 2, $\mathbf{S}(2)$ (actual spin down), pointed each in the $-z$- direction, are now each represented as scalar quantities with minus signs.

If disregarding the signs of charges explicitly, the electron-electron and the electron-positron are two equivalent systems of identical spin-half particles, each described by the space-time functions $f_\mu(;\bar{k}_d,\omega)$'s with $\mu = r,\ell$. These joined together with the spin functions give the complete wave functions: $\psi_\nu(n;\bar{k}_d,\omega) = f_\mu(;\bar{k}_d,\omega)\alpha_\nu(n)$ with $\nu = $+,-, $n = 1,2$, $\alpha_\nu = \alpha,\beta,\alpha',\beta'$. Now as a further step to conform our wave equation later to matrix form, we hereafter require that the $\psi_\nu(n)$ functions are elements of a matrix of one column, $\boldsymbol{\psi}$. The matrix wave equation itself will entail the two desired features discussed after equation (17), that is, (i) the product of $-k_d$ and k_d in the quadratic equation is positive: k_d^2 (entailed by the situation that these in the matrix form are offdiagonal elements, see (22) or (C.1), and (ii) the total cross-term product $\mathcal{O} = 0$ (entailed by the characteristics that in matrix equation the $\psi_\nu(n)$'s are explicitly mutually orthogonal). The first of these two features which we have up to now enforced by use of $\bar{k}_d$ for k_d, which should no longer be used in the matrix form to avoid a dual accounting. The space-time functions accordingly write $f_r(;k_d,\omega) = C_r e^{i[k_d z-\omega t]}$, $f_\ell(;k_d,\omega) = C_\ell e^{i[k_d z+\omega t]}$ with k_d the ordinary scalar quantity and related with $k_d^\dagger, k_d^\ddagger$ through (9). Accordingly,

$$\psi_+(1) = \alpha(1) f_r(;k_d,\omega) = \alpha(1) C_r e^{i[k_d z-\omega t]}, \quad \psi_-(2) = \beta(2) f_r(;k_d,\omega) = \beta(2) C_r e^{i[k_d z-\omega t]}, \tag{19a}$$
$$\psi_-(1) = \beta'(1) f_\ell(;k_d,\omega) = \beta'(1) C_\ell e^{i[k_d z+\omega t]}; \quad \psi_+(2) = \alpha'(2) f_\ell(;k_d,\omega) = \alpha'(2) C_\ell e^{i[k_d z+\omega t]}. \tag{19b}$$

The complete wave functions (19a)–(b) (Figure 2, right graphs) describe two identical particles of opposite oriented actual spins and accordingly opposite virtual spins, and these are identical to the solutions (see equation C.3) for Dirac equation. We can readily check that placing the foregoing relations in the antisymmetric total function for two spin-half (like charge) particles, $\psi_a = \frac{1}{\sqrt{2}}[\psi_+(1)\psi_-(2) - \psi_-(1)\psi_+(2)]$ correctly leads to that the probability of finding two identical particles at any location z in L is not altered by interchanging the locations of the particles (the indistinguishability). We can also check that the same ψ_a is given by the product of the separate total functions: $\psi_a(z_1,z_2) = f_s(z_1,z_2)\chi_a(1,2)$; notice that once we specified say particle 1 is spin up and 2 spin down, then $f_r(z_1,t)f_\ell(z_2,t)\beta'(1)\alpha'(2)$ and $f_\ell(z_1,t)f_r(z_2,t)\alpha(1)\beta(2)$ are zero since these do not describe the present reality.

Lastly, the spin-up state of particle 1, $\alpha(1)$, is associated with an effective electromagnetic wave f_r travelling to the right, thus $\partial_t f_r/f_r = -i\omega$, while the spin-up state of particle 2 with f_ℓ travelling to the left, thus $\partial_t f_\ell/f_\ell = i\omega$; the latter has as if a reversed time arrow relative to the former. We may introduce the time arrow functions defined for particles 1 and 2 as $\mathcal{T}(1) = 1, \mathcal{T}(2) = -1$, such that the action of these on the time derivatives project the wave propagations to be both in the $+z$-direction: $\mathcal{T}(1)\partial_t\psi_\nu(1) = \partial_t\psi_\nu(1)$, $\mathcal{T}(2)\partial_t\psi_\nu(2) = -\partial_t\psi_\nu(2)$.

5. Dirac equation

For two identical, spin-half particles of identical sets of space-time functions $\{f_r, f_\ell\}'s$ each described by wave equations (17) tending to occupy the same location z, we shall now express the corresponding wave equations in terms of the complete wave functions of Sec. 4. For particle 1, we thus multiply the first equation of (17) by $\alpha(1)$ and the second by $\beta'(1)$, act $\mathcal{T}(1)$ in front of the time derivatives, denote its charge by q_1, and get

$$\begin{aligned}(Mc^2 + q_1\phi_a) f_r\alpha(1) + c(i\hbar\nabla - q_1\mathbf{A}_a) f_\ell(-\beta'(1)) &= i\hbar\mathcal{T}(1)\partial_t f_r\alpha(1),\\ (-Mc^2 + q_1\phi_a) f_\ell\beta'(1) + c(i\hbar\nabla + q_1\mathbf{A}_a) f_r(-\alpha(1)) &= i\hbar\mathcal{T}(1)\partial_t(f_\ell\beta'(1)).\end{aligned} \tag{20}$$

For particle 2, instead we multiply the first equation of (17) by $-\beta(2)$ and the second by $-\alpha'(2)$, act both equations by $\mathcal{T}(2)$, denote its charge by q_2, and get

$$\begin{aligned}(Mc^2 + q_2\phi_a) f_r(-\beta(2)) + c(i\hbar\nabla - q_2\mathbf{A}_a) f_\ell(+\alpha'(2)) &= i\hbar\mathcal{T}(2)\partial_t f_r(-\beta(2)),\\ (-Mc^2 + q_2\phi_a) f_\ell(-\alpha'(2)) + c(i\hbar\nabla + q_2\mathbf{A}_a) f_r(+\beta(2)) &= i\hbar\mathcal{T}(2)\partial_t f_\ell(-\alpha'(2)).\end{aligned} \tag{21}$$

In the second terms in equations (20)–(21) we made the replacements $\alpha(1) \to -\beta'(1)$ and $\beta'(1) \to -\alpha(1)$, $\beta(2) \to -\alpha'(2)$, $\alpha'(2) \to -\beta(2)$ based on (18), to conform to the transformed space-time functions earlier.

Substituting in (20)–(21) with (19a)–(b) for the $\psi_\nu(n)$'s and the $\mathcal{T}(n)$'s expressed earlier, and, to form a direct contrast between the actual spin directions of the two particles, re-arranging the resulting four equations in the order of spin-up states of particles 1 and 2 first and then spin-down states of particles 1 and 2, we finally obtain a set of four coupled linear first order partial differential equations governing the motions of the two particles in terms of $\psi_\nu(n)'s$:

$$(Mc^2 + q_1\phi_a)\psi_+(1) - c(i\hbar\nabla - q_1\mathbf{A}_a)\psi_-(1) = i\hbar\partial_t\psi_+(1) \quad \text{(particle 1, spin up)}$$

$$\begin{aligned}
(Mc^2 - q_2\phi_a)\psi_+(2) + c(i\hbar\nabla + q_2\mathbf{A}_a)\psi_-(2) &= i\hbar\partial_t\psi_+(2) \quad \text{(particle 2, spin up)}\\
(-Mc^2 + q_1\phi_a)\psi_-(1) - c(i\hbar\nabla + q_1\mathbf{A}_a)\psi_+(1) &= i\hbar\partial_t\psi_-(1) \quad \text{(particle 1, spin down)}\\
(-Mc^2 - q_2\phi_a)\psi_-(2) + c(i\hbar\nabla - q_2\mathbf{A}_a)\psi_+(2) &= i\hbar\partial_t\psi_-(2) \quad \text{(particle 2, spin down)}
\end{aligned} \tag{22}$$

From the discussion of Sec. 4 that the $\psi_\nu(n)$'s and accordingly also their first derivatives are mutually orthogonal, it follows that the linear equations (22) are equivalent to a matrix equation. Supposing specifically the two particles are a positron and an electron and therefore $q_1 = q$, $q_2 = -q$, the matrix form of (22) is thus

$$H_{op}\boldsymbol{\psi} = i\hbar\partial_t\boldsymbol{\psi}, \quad \text{where } H_{op} = \mathbf{b}Mc^2 + q\phi_a + c\boldsymbol{\alpha}(\mathbf{p}_{\upsilon.op} - q\mathbf{A}_a) \text{ and } \mathbf{p}_{\upsilon.op} = -i\hbar\nabla \tag{23}$$

are the relativistic total Hamiltonian and linear momentum operators. And,

$$\mathbf{b} = \begin{pmatrix} I & 0 \\ 0 & -I \end{pmatrix}, \quad \boldsymbol{\alpha} = \begin{pmatrix} 0 & \boldsymbol{\sigma}_z \\ \boldsymbol{\sigma}_z & 0 \end{pmatrix}, \quad \boldsymbol{\psi} = \begin{pmatrix} \boldsymbol{\psi}_+ \\ \boldsymbol{\psi}_- \end{pmatrix} \quad \text{and}$$
$$I = \begin{pmatrix} 1 & 0 \\ 0 & 1 \end{pmatrix}; \quad \boldsymbol{\sigma}_z = \begin{pmatrix} 1 & 0 \\ 0 & -1 \end{pmatrix} \quad \boldsymbol{\psi}_+ = \begin{pmatrix} \psi_+(1) \\ \psi_+(2) \end{pmatrix}; \quad \boldsymbol{\psi}_- = \begin{pmatrix} \psi_-(1) \\ \psi_-(2) \end{pmatrix}. \tag{24}$$

The off-diagonal elements of the matrix $\boldsymbol{\sigma}_z$, $\sigma_{z11}(=1)$ and $\sigma_{z22}(=-1)$ here correspond to the $\alpha(1) = 1$ and $\alpha'(2) = -1$ earlier. We see that, $\boldsymbol{\psi}$ is equivalent to a Dirac spinor, σ_z the z-component of Pauli matrices, and as a whole, equation (23) is identical to the Dirac equation for an electron-positron system equivalent to here. For the present case rotation transformation is trivial, so $\vec{\sigma} = \sigma_z\hat{x}$.

Suppose more generally the two particles' spin angular momenta, $\mathbf{S}$ $(= \frac{\hbar}{2}\boldsymbol{\sigma})$'s, are along an axis $\mathbf{n}$ executing in general a precession about the z-axis at a fixed angle ($\arccos(\frac{S_z}{S})$). For each particle being in stationary state, its $\mathbf{S}$ (similarly its magnetic moment $\boldsymbol{\mu}_s(= -e\mathbf{S}/m)$) as a vector quantity when in small rotations about the z-axis must maintain invariant with respect to its projection on the z-axis, $\mathbf{n}\cdot\mathbf{S} = \pm\frac{1}{2}\hbar$, and is Hermitian. In addition to the antisymmetric condition given by the σ_z of (24) above, an infinitesimal rotation transformation as such needs be unitary. A specific set of transformation matrices having these properties are known to be the Pauli matrices, σ_x, σ_y and σ_z of the standard expressions and σ_z as expressed in (24), $\boldsymbol{\sigma} = \boldsymbol{\sigma}_x\hat{x} + \boldsymbol{\sigma}_y\hat{y} + \boldsymbol{\sigma}_z\hat{z}$. And, the unitary matrix $I\hat{z}$ about the z-axis naturally extends to a unitary matrix about the new $\mathbf{n}$- axis in three dimensions, given by $\vec{I} = I\hat{x} + I\hat{y} + I\hat{z}$. Substituting in (23) with $\boldsymbol{\sigma}$ and I for $\boldsymbol{\sigma}_z$ and I gives a Dirac equation of the same form, now for spins in arbitrary directions.

Appendix

Appendix A. Total energy and inertia of particle wave

As a general result of classical electrodynamics based on solution to the Maxwell's equations combined with Lorentz force law, an electromagnetic wave j transmits at the speed of light c a wave energy ε^j and a linear momentum $p^j = \varepsilon^j/c$. Here, the amplitudes of ε^j, accordingly of p^j, E^j, B^j and φ^j, etc., are continuous values. Following M. Planck's discovery of quantum theory in 1901, it has been additionally understood that these quantities are *by nature* quantized in amplitudes; an electromagnetic wave of frequency $\omega/2\pi$ has an energy $\varepsilon = n\hbar\omega$, consisting in general of n momentum-space quanta, or photons, each of an energy $\hbar\omega$; and the classical continuous amplitude solutions to these are only approximations when n is large. In the present problem, in conformity with experiments, especially the pair processes, the electromagnetic wave comprising our electron or positron has a "single energy quantum", $n = 1$; so $\varepsilon = \hbar\omega$. It has been further proven especially through quantum electrodynamics that the Maxwell's equations, and the subsequent classical wave equation (2) or more generally (5), continue to hold, and the quantisation of the fields and wave energy etc. is the result of subjecting the canonical displacement and momentum, the $u(=a\varphi)$ and $\dot{u}$ here, to the quantum commutation relation $[u, \dot{u}] = i\hbar$.

The total wave of our particle of a single energy quantum $\hbar\omega$ in a one-dimensional box has, following the solution to the Maxwell's equations earlier [see after (2)], two components, $\varphi^\dagger$ and $\varphi^\ddagger$, with their frequencies being Doppler-displaced to $\omega^\dagger$ and $\omega^\ddagger$ as a result of the source motion as given in (7), which are related to ω through (9). For the total wave comprising the particle, the total wave energy ε represents therefore a dynamical variable of the particle, here the total energy of the particle.

The electromagnetic waves, E^j, B^j's or φ^j's, rapidly oscillating at frequencies $(\omega^j/2\pi)$'s, of a geometric mean frequency $\omega/2\pi$ and wavelength $\lambda = c/(\omega/2\pi)$ will, when overlooking the oscillation details and speaking of the total dynamical variables ε, p only, appear as if two rigid objects, wavetrains, travelling at the speed of light c. In view that their speed of travel, c, is *finite* as contrasted to infinite, the wavetrains have inevitably each *finite* inertial masses, m^j's, thus an inertial mass $m = \sqrt{m^\dagger m^\ddagger}$ for the total wavetrain and hence its resulting particle. This mechanical depiction of the total wave, as a rigid "wavetrain", permits us at once to express according to Newtonian mechanics the linear momentum of the wavetrain to be $p = mc$. Combining this with the classical electrodynamic result $\varepsilon = pc$ above gives the kinetic energy of the wavetrain $\varepsilon = mc^2$, being equivalent to the Einstein's mass-energy relation. This energy and the Planck energy earlier ought to equal, thus

$$m = \hbar\omega/c^2; \qquad \text{or at } v = 0\text{: } M = \hbar\Omega/c^2 \tag{A.1}$$

with M the rest mass of the particle. Combining (A.1) with (9) gives $m = \gamma M$. Combining (A.1) with $p = mc$ further gives $mc = (\hbar\omega/c^2)c = \hbar k$ and accordingly $Mc = \hbar K$, with $\omega = kc$, $\Omega = Kc$ and $k = \gamma K$ as earlier.

Appendix B. Relativistic energy–momentum relation for the electromagnetic waves of particle

Consider first the simpler case of $A_a = 0$. Placing in wave equations (12) with $f^\dagger, f^\ddagger$ of (6), dividing the resulting first and second equations by $f^\dagger$ and $-f^\ddagger$ and sorting give $Mc^2 + \hbar k_d^\dagger c = \hbar\omega^\dagger - q\phi_a$, $Mc^2 - \hbar k_d^\ddagger c = \hbar\omega^\ddagger + q\phi_a$. Multiplying gives

$$M^2c^4 - \hbar^2 k_d^\dagger k_d^\ddagger c^2 + Q = \hbar^2\omega^\dagger\omega^\ddagger - q^2\phi_a^2 \tag{B.1}$$

Where $k_d^\dagger k_d^\ddagger = k_d^2$ and $\omega^\dagger\omega^\ddagger = \omega^2$ following (9); $Q = Mc^2c\hbar(k_d^\dagger - k_d^\ddagger)$, with $k_d^\dagger - k_d^\ddagger = 2k_d(\frac{v}{c})\gamma$ and $Mc^2 = \hbar Kc$, so $Q = 2\hbar^2 k_d{}^2 c^2$. With these, putting $\hbar k_d = \pm p_v$, $\hbar\omega = \pm\varepsilon$ where p_v,ε are here variables having positive and negative solutions and thus the right hand side of (B.1) reduces as $\sqrt{[(\hbar\omega - q\phi_a)(\hbar\omega + q\phi_a)]^2} = \sqrt{[-(\varepsilon - q\phi_a)(\varepsilon - q\phi_a)]^2} = (\varepsilon - q\phi_a)^2$, then (B.1) reduces exactly to $M^2c^4 + p_v^2c^2 = (\varepsilon - q\phi_a)^2$. This, or this in the more familiar form for $\phi_a = 0$,

$$M^2c^4 + c^2p_v^2 = \varepsilon^2, \tag{B.2}$$

gives just the experimentally widely corroborated relativistic energy-momentum relation. For the more general case of $\mathbf{A}_a$ finite, denoting $k_d^{\dagger\prime} = \bar{k}_d^\dagger - \frac{q\mathbf{A}_a}{\hbar}$, $k_d^{\ddagger\prime} = k_d^\ddagger + \frac{q\mathbf{A}_a}{\hbar}$, the particular feature that (the effective portion of) $\mathbf{A}_a$ is always perpendicular to $k_d\mathbf{z}^0$ leads to $k_d{}'^2 = k_d^\dagger{}' k_d^{\ddagger\prime} = k_d^2 - q^2\mathbf{A}_a^2/\hbar^2$, or, $(\pm p_v')^2 \equiv (\pm\hbar k_d')^2 = \mp(\hbar k_d - q\mathbf{A}_a)(-\hbar k_d - q\mathbf{A}_a) = [\mp(\mathbf{p}_v - q\mathbf{A}_a)]^2$. (B.2) thus generalises to $M^2c^4 + c^2p_v{}'^2 = (\varepsilon - q\phi_a)^2$.

Journal of Physics: Conference Series **128** (2008) 012019 doi:10.1088/1742-6596/128/1/012019

Appendix C. Solution of Dirac equation from the standpoint of particle internal process

We shall here mainly discuss the choice of the solution forms of the Dirac equation from the standpoint of internal processes of the IED particle model for simplicity for spins along z-axis, in an otherwise basically standard procedure. The two equations of (20) or (21) for particle $n = 1$ or 2 are coupled in $\psi_+(n)$ and $\psi_-(n)$ and can not be solved separately as in Appendix B. We need to solve each two, or more generally the four equations of the Dirac equation (20) together. Let the trial functions be: $\psi_\nu(n) = C_{sn}e^{\frac{i}{\hbar}[p_\upsilon z - \varepsilon t]}$, $s = +, -$, $n = 1, 2$. Placing these in (23) and rearranging give

$$\begin{pmatrix} \varepsilon - Mc^2 - q\phi_a & 0 & -[\mathbf{p}_\upsilon - q\mathbf{A}_a]c & 0 \\ 0 & \varepsilon - Mc^2 - q\phi_a & 0 & [\mathbf{p}_\upsilon - q\mathbf{A}_a]c \\ -[\mathbf{p}_\upsilon - q\mathbf{A}_a]c & 0 & \varepsilon + Mc^2 - q\phi_a & 0 \\ 0 & [\mathbf{p}_\upsilon - q\mathbf{A}_a]c & 0 & \varepsilon + Mc^2 - q\phi_a \end{pmatrix} \begin{pmatrix} \psi_+(1) \\ \psi_+(2) \\ \psi_-(1) \\ \psi_-(2) \end{pmatrix} = 0 \qquad (C.1)$$

(C.1) corresponds to the four linear, homogeneous algebraic equations (C.2) for the $\psi_{s,j}$'s as four unknowns below; for these to have nontrivial solutions, the determinant for the matrix of the coefficients of (C.1) needs be zero. This is det= $[(\varepsilon - q\phi_a)^2 - M^2c^4]^2 - c^4\mathbf{p}'^4 = 0$, with $\mathbf{p}'_\upsilon = \mathbf{p}_\upsilon - q\mathbf{A}_a$. This has two degenerate sets of square roots solutions: $\varepsilon - q\phi_a = \pm\sqrt{M^2c^4 + p_\upsilon'^2c^2}$, which being identical to (B.2). In view that each particle has internal processes, we thus naturally assign symmetrically two of the four solutions to particle 1, as $\varepsilon - q\phi_a = \pm\sqrt{M^2c^4 + p_\upsilon'^2c^2}$, and the other two for particle 2 as $\varepsilon - q\phi_a = \mp\sqrt{M^2c^4 + p_\upsilon'^2c^2}$. These two distinct sets of square-roots solutions to the algebraic equation above represent two identical (in the manner said earlier) yet distinct particles, like an electron and a positron, which do not transit from one to the other, a point agreeing with reality and having been stressed by P.A.M. Dirac from the very beginning in [1]. These algebraic solutions are in contrast to the usual problem of eigen values arising generally from boundary conditions and being each possible states of same particle between which transitions generally can occur.

With p_υ and ε as known parameters, we further solve the four algebraic equations

$$(\varepsilon - (Mc^2 + q\phi_a))\psi_+(1) = p'_\upsilon c\psi_-(1) \quad (a), \quad (\varepsilon - (Mc^2 + q\phi_a))\psi_+(2) = -p'_\upsilon c\psi_-(2) \quad (b)$$
$$(\varepsilon + (Mc^2 - q\phi_a))\psi_-(1) = p'_\upsilon c\psi_+(1) \quad (c), \quad (\varepsilon + (Mc^2 - q\phi_a))\psi_-(2) = -p'_\upsilon c\psi_+(2) \quad (d) \quad (C.2)$$

corresponding to (C.1), for the wave functions. Taking the imaginary of equations (a) and (b) first, multiplying the resulting equation with equation (c) and the second with (d) respectively on opposite sides, substituting with $\psi_-^*(1)\psi_-(1) = C_{-1}^2$, $\psi_+^*(1)\psi_+(1) = C_{+1}^2$, $\psi_-^*(2)\psi_-(2) = C_{-2}^2$ and $\psi_+^*(2)\psi_+(2) = C_{+2}^2$, we get
$(\varepsilon - (Mc^2 + q\phi_a))C_{+1}^2 = (\varepsilon + Mc^2 - q\phi_a)C_{-1}^2$, $(\varepsilon - (Mc^2 + q\phi_a))C_{+2}^2 = (\varepsilon + Mc^2 - q\phi_a)C_{-2}^2$ These have two independent solutions, and in mathematical terms two of the four wave functions can thus be arbitrarily chosen. In view of the IED particle model by which each particle has internal, wave processes consisting of two components in the one-dimensional box, it is natural here that we choose the values for C_{+1} and C_{+2} symmetrically, in the sense also $C_{+2} = -C_{+1}$, with these in the two equations above the values for C_{-1} and C_{-2} then follow to be uniquely given as

$$\frac{C_{+1}}{C_{+2}} = \pm\left(\sqrt{\frac{\varepsilon+(Mc^2-q\phi_a)}{\varepsilon-(Mc^2+q\phi_a)}}\right)^{1/2} C, \quad \frac{C_{-1}}{C_{-2}} = \mp\left(\sqrt{\frac{\varepsilon-(Mc^2+q\phi_a)}{\varepsilon+(Mc^2-q\phi_a)}}\right)^{1/2} C$$

where $|C_{+1}C_{-1}| = C^2$, $|C_{+2}C_{-2}| = C^2$. With the above in the trial functions, we get the complete solution for Dirac equation

$$\boldsymbol{\psi} = \begin{pmatrix} \psi_+(1) \\ \psi_+(2) \\ \psi_-(1) \\ \psi_-(2) \end{pmatrix} = \begin{pmatrix} \left(\sqrt{\frac{\varepsilon+(Mc^2-q\phi_a)}{\varepsilon-(Mc^2+q\phi_a)}}\right)^{1/2} Ce^{i(k_d z-\omega t)} \\ -\left(\sqrt{\frac{\varepsilon+(Mc^2-q\phi_a)}{\varepsilon-(Mc^2+q\phi_a)}}\right)^{1/2} Ce^{i(k_d z+\omega t)} \\ -\left(\sqrt{\frac{\varepsilon-(Mc^2+q\phi_a)}{\varepsilon+(Mc^2-q\phi_a)}}\right)^{1/2} Ce^{i(k_d z+\omega t)} \\ \left(\sqrt{\frac{\varepsilon-(Mc^2+q\phi_a)}{\varepsilon+(Mc^2-q\phi_a)}}\right)^{1/2} Ce^{i(k_d z-\omega t)} \end{pmatrix}. \qquad (C.3)$$

Agreeing with the wave functions directly based on IED particle model, $\psi_+(1), \psi_-(1)$ are two opposite travelling component waves of particle 1, and $\psi_+(2), \psi_-(2)$ of particle 2; in the meantime, the spin-up component waves of particles 1 and 2, $\psi_+(1)$ and $\psi_+(2)$ travel in opposite directions and similarly the spin-down component waves.

The author would like to thank scientist P.-I. Johansson for his moral and funding support of the research, the Committee of the 5th International Symposium on Quantum Theory and Symmetries (QTS 5) for a grant covering the conference fee, and the Swedish Research Council for a travel grant enabling the author to present this work at the QTS 5, Valladolid and the Swedish Institute of Space Physics for administrating the travel grant. The author would like to thank Professor H.-D. Doebner, Professor J. Goldin and Professor V. Dobrev for valuable reading of this and the related papers, for their valuable discussion and suggestions for a more informative and adaptable introduction to the particle model, and thank several distinguished Professors in Sweden for valuable reading of this and the related papers. The author would also like to acknowledge the interesting discussion of Professors J. Patera, M. Berry, J. Gazeau, D. Schuch, L. Boyle, R. Picken, A. Bohm, M. Olmo and others at the QTS 5.

References

[1] Dirac, P. A. M. 1928a The quantum theory of the electron *Proc. Roy. Soc.* **A117** 610—624; 1928b The quantum theory of the electron. Pt ii *Proc. Roy. Soc.* **A118** 351–361.

[2] Zheng-Johansson J. X. and Johansson P.-I. (a) 2006 *Unification of Classical, Quantum and Relativistic Mechanics and of the Four Forces*, Foreword by R. Lundin (New York: Nova Science Publishers. Inc., 2nd printing) (see a very early sketch of the ideas in *Preprint* physics/0412168); (b) 2006 *Inference of Basic Laws of Classical, Quantum and Relativistic Mechanics from First-Principles Classical-Mechanics Solutions* (New York: Nova Science Publishers Inc.); (c) 2006 Inference of Schrödinger equation from classical mechanics solution *Quantum Theory and Symmetries IV* vol. 2 (*Suppl. Bulg. J. Phys.* **33**), ed. V.K. Dobrev (Sofia: Heron Press) pp. 763-770 (*Preprint* phyiscs/0411134 v5); 2007 Schrödinger equation for electrodynamic model particle, submitted for publication; (d) 2006 Developing de Broglie wave *Prog. Phys.* **4** 32-35 (*Preprint* phyiscs/0608265); (e) 2006 Mass and mass–energy equation from classical-mechanics solution, *Phys. Essays* **19**, nr. 4 (*Preprint* phyiscs/0501037); (f) Zheng-Johansson J.X. 2006 Spectral emission of moving atom *Prog. Phys.* **3** 78-81 (*Preprint* phyiscs/060616); (g) Zheng-Johansson J.X. 2007 Vacuum structure and potential *Preprint* physics/0704.0131; (h) Zheng-Johansson J.X. 2006 Dielectric theory of the vacuum *Preprint* physics/0612096; (i) 2006 Depolarisation radiation force in a dielectric medium. Its analogy with gravity, with R. Lundin *Quantum Theory and Symmetries IV* vol. 2 (*Suppl. Bulg. J. Phys.* **33**), ed. V.K. Dobrev (Sofia: Heron Press) pp. 771-779 (*Preprint* phyiscs/0411245); (j) Zheng-Johansson J.X. 2007 Doebner-Goldin equation for electrodynamic model particle. The implied applications *7th Int. Conf. Symmetry in Nonlinear Mathematical Physics*, June 24-30, 2007, Inst. of Math., Kyiv (*Preprint* phyiscs/0801.4279); 2008 submitted for publication; (k) Zheng-Johansson J.X. 2008 Quantum mechanics for many identical electrodynamic particles (urged from discussions of Professor G.A Goldin and Professor H-D Doebner), in preparation; (l) Zheng-Johansson J.X. 2008 A review of the experimental basis for the electrodynamic particle model (motivated from discussion of Professor H-D Doebner), in preparation.

V International Symposium on Quantum Theory and Symmetries IOP Publishing
Journal of Physics: Conference Series **128** (2008) 012020 doi:10.1088/1742-6596/128/1/012020

Multi-photon coherent states via a Lie algebra method

D Afshar and M Jafarpour

Physics Department, Shahid Chamran University, Ahvaz, Iran

E-mail: da_afshar@yahoo.com and mojtaba_jafarpour@hotmail.com

Abstract. Using Lie algebra methods, we find a new class of Bogoliubov transformations which generalize the notion of squeezed states. The Hamiltonians for the simple harmonic and anharmonic oscillators, turn out to be the generators of a Lie group, whose other generators may be found exactly, or up to any desired order of the perturbation parameter. An element of this Lie group, which is realized as the multi-photon operator, transforms the anharmonic Hamiltonian to the harmonic one. The transformation of the ordinary annihilation and creation operators under this unitary transformation leads to the introduction of multi-photon coherent states. We specifically consider four-photon coherent states in detail and study the time dependent position and momentum uncertainties in these states.

1. Introduction

When fields interact with matter, that is characterized by nonlinear properties and can be described by anharmonic oscillators, nonlinear optical processes occur [1, 2]. Such processes correspond to the production of two or k-photon coherent states [3, 4]. One expects to generalize squeezed states or two-photon coherent states to k-photon ones, by means of straightforward generalization of the two-photon unitary operator; however, this task encounters divergence difficulties [5]and several approaches [6, 7, 8, 9, 10, 11, 12]have been taken toward the alleviation of this problem.

In this work, we present a simple method to generalize squeezed states. It also bridges between the notion of the latter and the anharmonic oscillators. We use a Lie algebra method to relate the Hamiltonian of an anharmonic oscillator to that of a harmonic one, in a perturbation sense. This is achieved via a canonical transformation, which is unique to the structure of the specific Hamiltonian under study. We introduce our method in section 2, where we apply it to three examples: a harmonic oscillator with a linear perturbation term, one with a quadratic perturbation term that just brings about a frequency shift and also the quartic anharmonic oscillator. Section 3 will be devoted to the study of four-photon coherent states that emerge from our method. Time dependent position and momentum uncertainties are obtained in these states and their oscillating nature are studied. Finally, we deal with discussion and conclusions in section 4.

2. Our method

The anharmonic oscillator Hamiltonian with a perturbation term of order n may be expressed by

$$H_n = a^\dagger a + \frac{1}{2} + \lambda(\frac{a+a^\dagger}{\sqrt{2}})^n = H_0 + \lambda(\frac{a+a^\dagger}{\sqrt{2}})^n, \tag{1}$$

where, H_0 is the Hamiltonian for a simple harmonic oscillator. If N hermitian generators L_i, including $L_1 = H_0$ and $L_2 = H_n$, form a closed commutator algebra, the operators L_i specify a Lie group, whose generators they are. The elements of this group are given by

$$U = exp(-\mathrm{i}\sum_{i=1}^{N} \alpha_i L_i), \tag{2}$$

where, α_i's are real parameters.

The case $n = 1$ in equation (1) may be considered as a harmonic oscillator in an external electric field. Considering the commutator $[H_0, H_1] = \lambda(a^\dagger - a)/\sqrt{2}$, we realize that the operators H_0, H_1, $\mathrm{i}(a^\dagger - a)$ and the identity operator I, form a closed commutator algebra and are the generators of a Lie group. The Glauber operator $U_1 = exp(\alpha a^\dagger - \alpha^* a)$, with real α, is an element of this group which transforms H_0 to H_1 in the following manner

$$H_1 = U_1^\dagger H_0 U_1 - \frac{\lambda^2}{2}, \tag{3}$$

where, α has been replaced by $\lambda/\sqrt{2}$ in U_1. Using Eq. (3), the eigenfunctions of H_1 are given by $U_1^\dagger|n\rangle_a$ where $\{|n\rangle_a\}$ are a-mode or the displaced number states [13].

Inserting $n = 2$ in Eq.(1)gives the Hamiltonian of a shifted frequency harmonic oscillator. In this case the set of hermitian operators $\{H_0, H_2, \mathrm{i}(a^{\dagger 2} - a^2), I\}$ are the generators of a Lie group. The squeezing operator, $U_2 = exp(\beta a^{\dagger 2} - \beta^* a^2)$ with β assumed real, is an element of this group. The transformation of the fundamental mode operators a and $a^\dagger$ into the new ones, b and $b^\dagger$ under U_2, is carried out by the following Bogoliubov transformations

$$\begin{aligned} b &= U_2^\dagger a U_2 = \mu a + \nu a^\dagger, \\ b^\dagger &= U_2^\dagger a^\dagger U_2 = \mu a^\dagger + \nu a, \end{aligned}$$

where, $\mu = \cosh(2\beta)$ and $\nu = \sinh(2\beta)$. As $\mu^2 - \nu^2 = 1$, we have also $[b, b^\dagger] = 1$. The squeezing operator transforms H_0 to H_2 in the following manner

$$H_2 = \sqrt{1+2\lambda}U_2^\dagger H_0 U_2 = \sqrt{1+2\lambda}(b^\dagger b + \frac{1}{2}), \tag{4}$$

where, $\lambda = 2\mu\nu/(\mu - \nu)^2$ has been assumed. Using equation (4), the eigenfunctions of H_2, called b-mode or squeezed number states [13, 14] are given by $|n\rangle_b = U_2^\dagger|n\rangle_a$.

We now deal with the quartic anharmonic oscillator by inserting $n = 4$ in Eq. (1). For small values of λ, the set of operators $\{H_0, H_4, \mathrm{i}\lambda(a^{\dagger 4} - a^4), \mathrm{i}\lambda(a^{\dagger 2} - a^2), \mathrm{i}\lambda(a^{\dagger 3}a - a^\dagger a^3), \lambda(a^{\dagger 4} + a^4), \lambda(a^{\dagger 2} + a^2), \lambda(a^{\dagger 3}a + a^\dagger a^3), I\}$ form a closed commutator algebra, up to first order, and therefore they are the generators of a Lie group up to that order. Let's consider the following element of this Lie group, known as the four photon operator [5]

$$U_4 = exp[A\lambda(a^{\dagger 4} - a^4) + B\lambda(a^{\dagger 2} - a^2) + C\lambda(a^{\dagger 3}a - a^\dagger a^3)], \tag{5}$$

where A, B and C are real parameters. This operator transforms the fundamental mode annihilation and creation operators, a and $a^\dagger$ to the new ones c and $c^\dagger$, which may be expressed up to the first order in the parameter λ, by the following expressions

$$c = U_4^\dagger a U_4 = a + \lambda(4Aa^{\dagger 3} + 2Ba^\dagger + 3Ca^{\dagger 2}a - Ca^3), \tag{6}$$

$$c^\dagger = U_4^\dagger a^\dagger U_4 = a^\dagger + \lambda(4Aa^3 + 2Ba + 3Ca^\dagger a^2 - Ca^{\dagger 3}). \tag{7}$$

It should be emphasized that c and $c^\dagger$ also obey the canonical commutation relation $[c, c^\dagger] = 1$; therefore, we have introduced a new class of Bogoliubov transformations by (6) and (7). Assuming $A = 1/16$, $B = 3/4$ and $C = 1/2$, the four photon operator transforms H_0 to H_4, up to the first order in the parameter λ, as follows

$$U_4^\dagger H_0 U_4 = c^\dagger c + \frac{1}{2} = H_4 - \frac{3\lambda}{2}(N_a^2 + N_a) - \frac{3\lambda}{4}\,, \tag{8}$$

where, $N_a = a^\dagger a$ is the normal number operator. Using equation (8), the eigenstates of the quartic anharmonic oscillator, our c-mode number states, are given by

$$|n\rangle_c \equiv U_4^\dagger |n\rangle_a = \mathrm{e}^{-\lambda[\frac{1}{16}(a^{\dagger 4} - a^4) + \frac{3}{4}(a^{\dagger 2} - a^2) + \frac{1}{2}(a^{\dagger 3}a - a^\dagger a^3)]}|n\rangle_a, \tag{9}$$

and the first order perturbed energies are expressed by

$$E_n = n + \frac{1}{2} + \frac{3\lambda}{4} + \frac{3\lambda}{2}n(n+1)\,.$$

The above result is in complete agreement with that obtained, using first order Rayleigh-Schrödinger perturbation theory [15].

3. Four photon coherent states

Ordinary coherent states, or so called one photon coherent states, are eigenstates of the ordinary annihilation operator a

$$a|\alpha\rangle = \alpha|\alpha\rangle,\ \ |\alpha\rangle = U_1(\alpha)|0\rangle_a = \mathrm{e}^{\alpha a^\dagger - \alpha^* a}|0\rangle_a,$$

entailing a Poisson distribution, in terms of the ordinary number states. Similarly, defining b- mode vacuum state $|0\rangle_b$, by the relation $b|0\rangle_b = 0$, we may construct the so called b-mode number states by repeated application of b-mode creation operator $b^\dagger$ on the b-mode vacuum state as follows

$$|n\rangle_b = \frac{b^{\dagger n}}{\sqrt{n!}}|0\rangle_b,\ \ |n\rangle_b \equiv U_2^\dagger |n\rangle_a.$$

Eigenstates of the operator b, $|\beta\rangle$ are called the b- mode coherent states, two photon coherent states or squeezed states

$$b|\beta\rangle = \beta|\beta\rangle,\ \ |\beta\rangle = U_2^\dagger U_1(\beta)|0\rangle_a = \mathrm{e}^{\beta b^\dagger - \beta^* b}|0\rangle_b.$$

In a similar way, we may introduce the c-mode number states, by the relation

$$|n\rangle_c = \frac{c^{\dagger n}}{\sqrt{n!}}|0\rangle_c,\ \ |n\rangle_c \equiv U_4^\dagger |n\rangle_a,$$

where, $|0\rangle_c$ is the c-mode vacuum state and it is defined by $c|0\rangle_c = 0$. The four photon or c-mode coherent states, or the generalized squeezed states $|\gamma\rangle$, are defined as the eigenstates of the c-mode annihilation operator as follows

$$c|\gamma\rangle = \gamma|\gamma\rangle, \tag{10}$$

where, γ is a complex number. Using (6), we have

$$U_4^\dagger a U_4|\gamma\rangle = \gamma|\gamma\rangle \ ,$$
$$a(U_4|\gamma\rangle) = \gamma(U_4|\gamma\rangle) \ .$$

Thus, $U_4|\gamma\rangle$ is an a-mode coherent state, and we have

$$U_4|\gamma\rangle = U_1(\gamma)|0\rangle_a \ . \tag{11}$$

Therefore, the four photon coherent states are given by

$$|\gamma\rangle = U_4^\dagger(\lambda)U_1(\gamma)|0\rangle_a = \mathrm{e}^{\gamma c^\dagger - \gamma^* c}|0\rangle_c \ . \tag{12}$$

Now we embark upon the derivation of uncertainties. The time dependent position operator for a quartic anharmonic oscillator, up to the first order in the parameter λ, has been derived in [16] as follows

$$\begin{aligned} x(t) &= \frac{1}{\sqrt{2}}\{(\cos t - \mathrm{i}\sin t)a(0) - \lambda[A_1 a(0) + A_2 a^3(0) \\ &\quad + A_1^* a^{\dagger 2}(0)a(0)]\} + h.c. \ , \end{aligned} \tag{13}$$

where h.c. stands for the Hermitian conjugate of the other terms, and the parameters A_1 and A_2 are defined as follows

$$A_1 = \frac{3}{4}[t\sin t + \mathrm{i}(t\cos t - \sin t)],$$
$$A_2 = \frac{1}{16}[(\cos t - \cos 3t) + \mathrm{i}(\sin 3t - 3\sin t)].$$

We also write down the momentum operator

$$\begin{aligned} p(t) &= -\frac{1}{\sqrt{2}}\{(\sin t + \mathrm{i}\cos t)a(0) + \lambda[\dot{A}_1 a(0) + \dot{A}_2 a^3(0) \\ &\quad + \dot{A}_1^* a^{\dagger 2}(0)a(0)]\} + h.c. \ , \end{aligned} \tag{14}$$

where $\dot{A}_1$ and $\dot{A}_2$ are the time derivatives of A_1 and A_2 respectively.

Using (13) and (14), we obtain the position and momentum uncertainties in the four photon coherent state $|\gamma\rangle$ as follows

$$\begin{aligned} \Delta^2 x(t) &= \frac{1}{2} - \frac{\lambda}{2}\{(\frac{3}{2} + 3|\gamma|^2)\sin^2 t + 2Re[(\frac{3}{4}\gamma^2 + 3|\gamma|^2 + \frac{3}{2})\mathrm{e}^{2\mathrm{i}t} \\ &\quad + A_1\gamma^2\mathrm{e}^{-\mathrm{i}t} + 3A_2\gamma^2\mathrm{e}^{\mathrm{i}t}]\} \ , \end{aligned} \tag{15}$$

$$\begin{aligned} \Delta^2 p(t) &= \frac{1}{2} + \frac{\lambda}{2}\{(\frac{3}{2} + 3|\gamma|^2)\sin^2 t + 2Re[(\frac{3}{4}\gamma^2 + 3|\gamma|^2 + \frac{3}{2})\mathrm{e}^{2\mathrm{i}t} \\ &\quad + \mathrm{i}\dot{A}_1\gamma^2\mathrm{e}^{-\mathrm{i}t} - 3\mathrm{i}\dot{A}_2\gamma^2\mathrm{e}^{\mathrm{i}t}]\} \ , \end{aligned} \tag{16}$$

where, all the expressions of order two and higher in terms of λ are neglected. The product of uncertainties (15) and (16) reduces to $\frac{1}{4}$, up to the first order in λ at $t \to 0$ limit. Thus we conclude that the four photon coherent state (12) is also a minimum uncertainty state at $t = 0$. As time passes, one of the uncertainties dips well more and more, bellow the value $\frac{1}{2}$ in successive periods; thus revealing the squeezing properties of such states. We have illustrated the uncertainties of $\hat{x}$ and $\hat{p}$ versus time in figure 1. These quantities depend on γ and λ, oscillate in opposite directions and their overall range of variation increase with time.

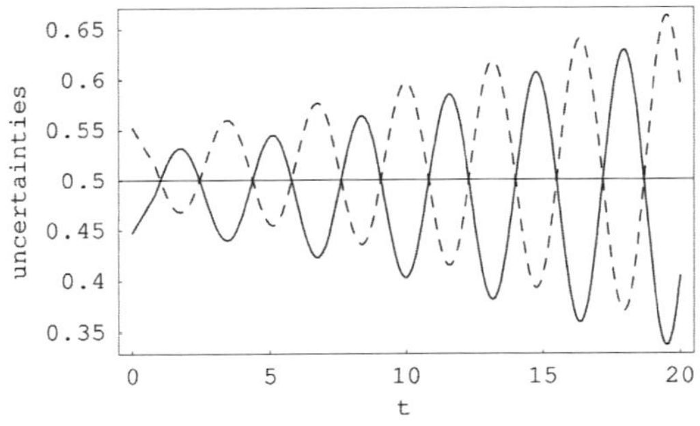

Figure 1. Plot of the $\hat{x}$-uncertainty (solid line) and $\hat{p}$-uncertainty (dashed line) for four-photon coherent state versus time. $\lambda = 0.01$, $\gamma = 1$.

4. Discussion

We have presented multi-photon coherent states via anharmonic oscillators. We have constructed a Lie group, whose generators include H_0 and H_n. The generator of the group H_0 may be transformed to the generator H_n by a unitary transformation, endowed by an element of the group. In the cases of H_1 and H_2, the familiar Glauber and squeezed operators, being canonical transformations, emerge respectively; while in the case of H_4, the four photon operators are generated. The transformation of the annihilation operator a, under the squeezed operator U_2, leads to the Bogoliubov transformation, and the associated squeezed states. Keeping terms up to the first order, the transformation of the ordinary annihilation operator under four photon operator, leads to a new class of Bogoliubov transformations, in which the associated eigenstates are four photon coherent states. The uncertainties for the four photon coherent state $|\gamma\rangle$, oscillate around $\frac{1}{2}$ in time, similar to the squeezed states, but the overall range of their variation grows with time. Finally, one may generalize this procedure to six photon coherent states, via sextic anharmonic oscillators, eight photon coherent states, via octic anharmonic oscillator and so on.

References

[1] Bondurant R S, Kumar P, Shapiro J H and Maeda M 1984 *Phys. Rev. A* **30** 343-53
[2] Levenson M D, Shelby R M, Aspect A, Reid M and Walls D F 1985 *Phys. Rev. A* **32** 1550-62
[3] Slusher R E, Hollberg L W, Yurke B, Mertz J C and Valley J F 1985 *Phys. Rev. A* **31** 3512-5
[4] Slusher R E, Hollberg L W, Yurke B, Mertz J C and Valley J F 1985 *Phys. Rev. Lett.* **55** 2409-12
[5] Fisher R A, Nieto M M and Sandberg V D 1984 *Phys. Rev. D* **29** 1107-10
[6] Braunstein S L and McLachlan R I 1987 *Phys. Rev. A* **35** 1659-67
[7] Brandt R A and Greenberg O W 1969 *J. Math. Phys.* **10** 1168-76
[8] D'Ariano G, Rasetti M and Vadacchino M 1985 *Phys. Rev. D* **32** 1034-37
[9] D'Ariano G M and Sterpi N 1989 *Phys. Rev. A* **39** 1860-68
[10] Bužek V and Jex I 1990 *Phys. Rev. A* **41** 4079-82
[11] DeSiena S, DiLisi A and Illuminati F 2001 *Phys. Rev. A* **64** 063803
[12] Wu Y and Côté R 2002 *Phys. Rev. A* **66** 025801
[13] Nieto M M 1997 *Phys. Lett. A* **229** 135-43
[14] Jafarpour M and Afshar D 2002 *J. Phys. A:Math. Gen.* **35** 87-92
[15] McRae S M and Vrscay E R 1997 *J. Math. Phys.* **38** 2899-921
[16] Pathak A and Mandal S 2001 *Phys. Lett. A* **286** 261-76

Journal of Physics: Conference Series **128** (2008) 012021 doi:10.1088/1742-6596/128/1/012021

Spin squeezing properties of an initial coherent state via a two-axis countertwisting Hamiltonian in the presence and absence of an external field

A Akhound and M Jafarpour

Physics Department, Shahid Chamran University, Ahwaz, Iran.

E-mail: aakhond@pnu.ac.ir and mojtaba_jafarpour@hotmail.com

Abstract. We consider a four-qubit system, initially in a coherent state along the z-axis, and study its time dependence under the influence of the well-known two-axis countertwisting Hamiltonian. Analyzing the time dependence of the spin operators, we observe its squeezing properties, according to the criterion given by Kitagawa et al. We also perform similar calculations for the same Hamiltonian in the presence of an external field and compare its squeezing properties with that of the former. It is observed that the details of the squeezing properties are different in the two cases. Plots of the squeezing parameters as a function of time and the rotation angle for both cases are presented.

1. Introduction

It is well known that spin squeezing may be achieved via nonlinear interactions [1, 2, 3, 4, 5, 6]. Kitagawa and Ueda have introduced two fundamental types of nonlinear Hamiltonians that lead to spin squeezing; namely, one-axis twisting and two-axis countertwisting Hamiltonians [7]. In this paper we focus on the two-axis countertwisting Hamiltonian

$$H = \frac{\chi}{2i}(\hat{S}_+^2 - \hat{S}_-^2), \tag{1}$$

where, $\hat{S}_+$ and $\hat{S}_-$ are the spin ladder operators. We shall also consider the Hamiltonian model

$$H_f = \frac{\chi}{2i}(\hat{S}_+^2 - \hat{S}_-^2) + \Omega \hat{S}_z, \tag{2}$$

where, the second linear term in H_f, may be realized as an additional external field. The main purpose of this paper is to compare the squeezing property of a 4-qubit system, which is dynamically evolved via the Hamiltonians H and H_f. We shall use the criterion of spin squeezing as has been advanced by Kitagawa and Ueda. They define the parameter ξ_K^2, as a measure of squeezing,

$$\xi_K^2 = \frac{2(\Delta S_\perp)^2}{S}, \tag{3}$$

where, $\Delta S_\perp$ is the uncertainty of an angular momentum component perpendicular to the mean angular momentum direction $\langle \vec{\hat{S}} \rangle$. A spin state is said to be squeezed according to this criterion if $\xi_K^2 < 1$.

Journal of Physics: Conference Series **128** (2008) 012021 doi:10.1088/1742-6596/128/1/012021

2. Squeezing in the absence of the field

Our initial coherent state is a collection of four spin-half systems, with all of them in the upward direction denoted by $|2,2\rangle_z$ [8].

We calculate the following quantities in our initial state $|2,2\rangle_z$

$$|\langle \vec{\hat{S}} \rangle| = \langle \hat{S}_z \rangle \quad = \quad {}_z\langle 2,2|\hat{S}_z|2,2\rangle_z = 2, \tag{4}$$

$$\Delta S_x = \Delta S_y = \frac{S}{2} = 1. \tag{5}$$

We note that the uncertainty relation

$$(\Delta S_x)^2(\Delta S_y)^2 \geq \frac{1}{4}|\langle \hat{S}_z \rangle|^2, \tag{6}$$

is satisfied here with equal sign, as we expect for a coherent state. We also have $\xi_K^2 = 1$; an indication of the lack of squeezing in a coherent state.

Now, it is assumed that the spin system is evolved via the countertwisting Hamiltonian H. To find the optimal squeezing direction, we also rotate the reference frame about the z-axis by an angle φ and thus replace x and y by $\acute{x}$ and $\acute{y}$, respectively [9]. it is trivial to show that

$$_z\langle 2,2|\hat{S}_{\acute{x}}|2,2\rangle_z \quad = \quad {}_z\langle 2,2|\hat{S}_{\acute{y}}|2,2\rangle_z = 0. \tag{7}$$

We work in the Schrödinger picture and the evolution is enforced by the unitary transformation operator $U(t) = e^{-iHt}$. Thus, the wave function at time $t = 0$, is evolved in time according to the transformation $\Psi(t) = e^{-iHt}\Psi(0)$. We can also show that

$$\langle \Psi(t)|\hat{S}_{\acute{x}}|\Psi(t)\rangle = \langle \Psi(t)|\hat{S}_{\acute{y}}|\Psi(t)\rangle = 0. \tag{8}$$

Therefore, the average spin direction remains along the z-direction and we find

$$|\langle \vec{\hat{S}} \rangle| = \langle \hat{S}_z \rangle = \langle \Psi(t)|\hat{S}_z|\Psi(t)\rangle = 2\cos(2\sqrt{3}\chi t), \tag{9}$$

we also obtain ξ_K^2 as follows

$$\xi_K^2 = (\Delta \hat{S}_{\acute{x}})^2 = \frac{3}{2} - \frac{1}{2}\cos(4\sqrt{3}\chi t) + \sqrt{3}\cos(2\varphi)\sin(2\sqrt{3}\chi t); \tag{10}$$

it is a function of the rotation angle φ, time t and the Hamiltonian parameter χ. Minimizing (10) with respect to these variables we find $\chi = -0.3781$, $t = 0.7994$ and $\varphi = 0$, which leads to $(\xi_K^2)_{min} = 0.25$.

We now study the squeezing property of the system along the $\varphi = 0$ direction implemented by H at $\chi = -0.3781$. Substituting the above values for χ and φ in (10), we have plotted ξ_K^2 versus t in figure 1a. It is observed that the squeezing condition $\xi_K^2 < 1$ is satisfied during alternate time intervals. Also assuming $\chi = -0.3781$ and $t = 0.7994$ we have plotted ξ_K^2 versus φ in figure 1b. We observe that the system is squeezed through alternate angle ranges at these conditions.

3. Squeezing in the presence of the field

We now consider the evolution of our initial coherent state via H_f. Again the average spin is in the z direction and we may write

$$|\langle \vec{\hat{S}} \rangle| = \langle \hat{S}_z \rangle = \langle \Psi(t)|\hat{S}_z|\Psi(t)\rangle = -\frac{2[\Omega^2 + 3\chi^2\cosh(2At)]}{A^2}, \tag{11}$$

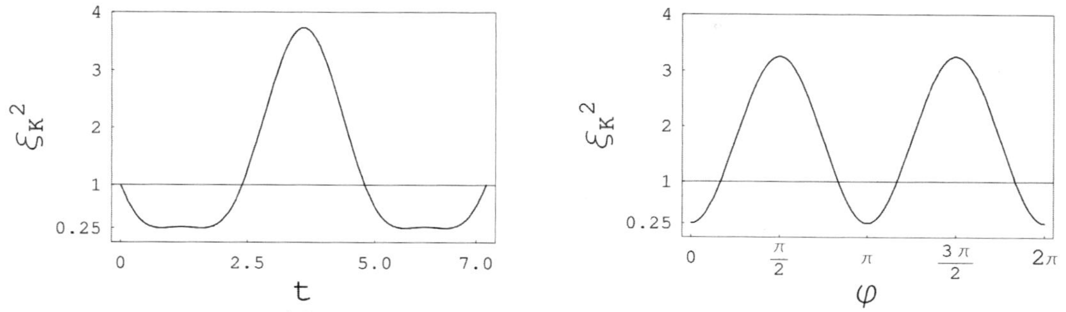

(*a*) (*b*)

Figure 1. Plots of ξ_K^2(a) versus time at $\varphi = 0$ (b) versus the rotation angle at $t = 0.7994$; $H = (-0.3781/2i)(\hat{S}_+^2 - \hat{S}_-^2)$.

where we have defined $A = \sqrt{-3\chi^2 - \Omega^2}$. We also find

$$\begin{aligned}
\xi_{K,f}^2 &= (\Delta\hat{S}_{\acute{x}})^2 = \langle\Psi(t)|\hat{S}_{\acute{x}}^2|\Psi(t)\rangle \\
&= \frac{1}{4A^5}\exp(-4At)\{6\exp(2At)[-1+\exp(4At)]\chi A^4\cos(2\varphi) \\
&- 2A\exp(4At)[-27\chi^4 - 24\chi^2\Omega^2 - 2\Omega^4 + 3\chi\{4\chi\Omega^2\cosh(2At) \\
&+ 3\chi^3\cosh(4At) - 4A^2\Omega\sin(2\varphi)\sinh^2(At)\}]\}.
\end{aligned} \tag{12}$$

Minimizing $\xi_{K,f}^2$ with respect to all its parameters, we find $\chi = 0.3445$, $t = 1.3162$, $\Omega = -1.0335$ and $\varphi = \pi/4$, leading to $\xi_{K,f}^2 = 0.25$.

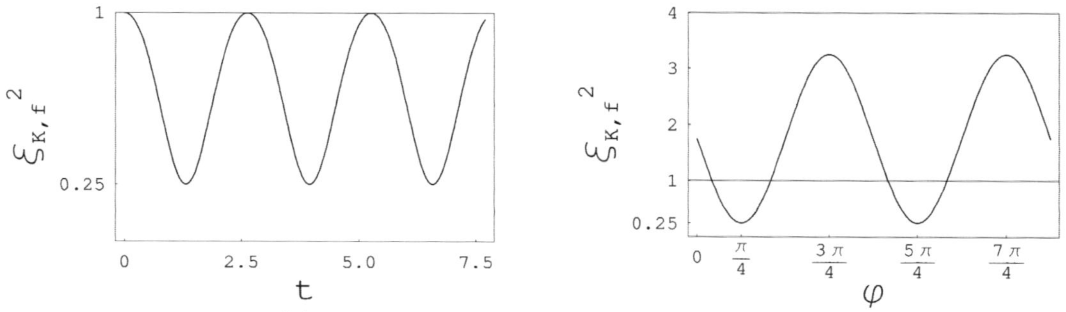

(*a*) (*b*)

Figure 2. Plots of $\xi_{K,f}^2$(a) versus time at $\varphi = \pi/4$ (b) versus the rotation angle at $t = 1.3162$; $H_f = (0.3445/2i)(\hat{S}_+^2 - \hat{S}_-^2) - 1.0335\hat{S}_z$.

Inspecting the $\xi_{K,f}^2$ plots in figure 2a, discloses an interesting phenomenon; the addition of the external field has increased the time span of the squeezing, in the deepest squeezing direction, substantially. In fact, barring some separate instances of time, the inequality $\xi_{K,f}^2 < 1$, at the given values of the parameters χ, φ and Ω, is always satisfied and the system is always squeezed in the presence of the field.

References

[1] Law C K, Ng H T and Leung P T 2001 *Phys. Rev. A* **63** 055601
[2] Song L, Wang X, Yan D and Zong Z 2006 *J. Phys. B: At. Mol. Opt. Phys.* **39** 559

[3] Vidal J, Palacios G and Aslangul C 2004 *Phys. Rev. A* **70** 062304
[4] Müstecaplıoğlu Ö E, Zhang M and You L 2002 *Phys. Rev. A* **66** 033611
[5] Prakash H and Kumar R 2005 *J. Opt. B: Quantum Semiclass. Opt.* **7** S757
[6] Choi S and Bigelow N P 2005 *Phys. Rev. A* **72** 033612
[7] Kitagawa M and Ueda M 1993 *Phys. Rev. A* **47** 5138
[8] Radcliffe J M 1971 *J. Phys. A: Gen. Phys.* **4** 313
[9] Wang X and Sanders B C 2003 *Phys. Rev. A* **68** 012101

Journal of Physics: Conference Series **128** (2008) 012022 doi:10.1088/1742-6596/128/1/012022

Temporal evolution of two quantum radiators at interaction with electromagnetic cavity field

Nellu Ciobanu and Nicolae Enaki

Institute of Applied Physics, Academy of Sciences, Chisinau,MD 2028, Republic of Moldova

E-mail: nellu_ciobanu@yahoo.com

Abstract. In this paper, we study a new quantum reversible effect for a three-level atom which interacts with electromagnetic cavity field. We investigate the reversible conditions for a three-level atom that is in resonance with two cavity modes and in off resonance with a single mode of cavity. The possibility of the reversible factorization between the atom and cavity radiation is the subject of this work. It is found the reversible condition for which these quantum subsystems can (or not) restore their diagonal moments.

1. Introduction

Recently, considerable attention in quantum physics has been devoted to interaction between the atoms and electromagnetic cavity field$[1-3]$. One of this, which is very popular and in great demand for realization of quantum computer represents the quantum time separation of two quantum subsystems. The quantum collapse and revival realization of two radiators is an interesting problem due to the possibilities of restoration of initially separated states of atom and electromagnetic field. There are a lot of works in the literature devoted to the revival problem of two quantum subsystems in interaction$[4-6]$.

In the first part of this paper, it is investigated the reversible condition for a non-equidistant three-level atom in interaction with two cavity modes. We analyze a general case in that the atom is prepared in a superposition of all three states and the cavity field is regarded as a finite superposition of Fock states. Taking into account the various possibilities to restore the initial state, we put the reversible problem of this system in general case. The recursion relation between the coefficients of bimodal cavity field is obtained. From this relation it is found the particular cases for that the coherent trapping conditions are realized in squeezed field proposed in [4]. We propose a *new reversible* condition, taking into account the coherent trapping effect [4] and dynamic trapping effect [5] as particular cases. This condition has direct connection with entangled states and is reduced to quantum regular nutation between the excited and ground states of three-level atom. The temporal behavior and the properties of these new states are investigated.

In the second part, it is analyzed the quantum mixing, when we have a detunig from resonance between the radiator and single mode cavity field. From quantum mechanical points of view we put the problem of reversibility in such interaction. In order to obtain this condition it is established the recursion relation between the decomposition coefficients of wave functions on the Fock states of atom and cavity field. We analytically analyze the possibility of states restoration of two quantum oscillators in off resonance interaction. As follows from our results,

Journal of Physics: Conference Series **128** (2008) 012022 doi:10.1088/1742-6596/128/1/012022

in the off-resonace case the full reversibility is impossible. We find the conditions for which the atom and cavity field restore all diagonal elements of density operator. This is possible only for the case in which the transit time through the micro-cavity strongly satisfies the reversible condition.

2. Quantum reversible condition for a three-level atom in resonance with bimodal cavity field

In this section we analyzed the reversible condition for a non-equidistant three-level atom in interaction with two cavity modes. We consider a cavity with a high quality factor, into which a flux of three-level atoms with a very low rate is pumped, so that during the flying time, τ, only one atom in the cavity is considered and the field damping time is larger than the atomic transit time through the cavity. Preparing in the superpositions of all three-level states (see *Figure* 1) in classical Ramsay zone[6], at the time moment t the atom enters into the cavity. Let us consider the situation when the states of the atomic and the field are considered totally separated, so that the wave function can be represented as

$$|\Psi(t)\rangle = (\alpha\,|e\rangle + \beta\,|i\rangle + \gamma\,|g\rangle) \otimes \sum_{n,m=N_d M_d}^{N_u,M_u} S_{n,m}\,|n\rangle_a\,|m\rangle_b\,. \tag{1}$$

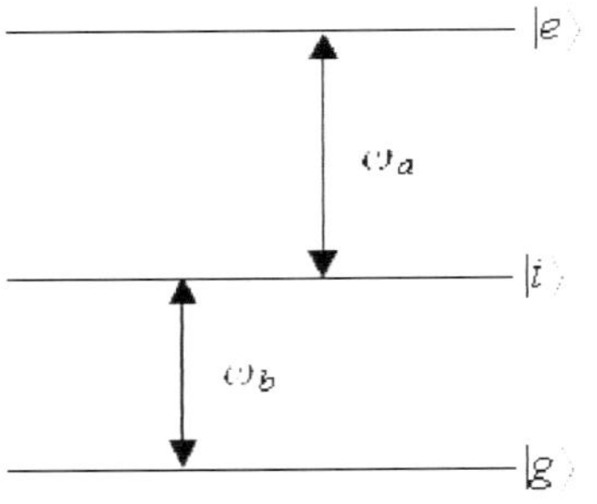

Figure 1. Configuration of the energy levels of a three-level atom.

In the expression (1), N_d, M_d and N_u, M_u represent the bottom and upper limits. We emphasize that the energies of excited and ground states are measured relatively to the position of the intermediate state, so that $E_e = \hbar\omega_a$, $E_g = \hbar\omega_b$ and $E_i = 0$. The frequencies ω_a and ω_b correspond to the resonance cavity modes described by annihilation a, b and creation a^+, b^+ field operators.

In the time moment, $t + \tau$, the atom enters into an interaction with the cavity field and the system is described by the Hamiltonian

$$H = H_0 + H_i, \tag{2}$$

with the following free and interaction parts

$$\begin{aligned} H_0 &= \hbar\omega_a |e\rangle\langle e| - \hbar\omega_b |g\rangle\langle g| + \hbar\omega_a a^+ a + \hbar\omega_b b^+ b, \\ H_i &= \hbar\lambda_a \left(a^\dagger |i\rangle\langle e| + a|e\rangle\langle i|\right) + \hbar\lambda_b \left(b^\dagger |g\rangle\langle i| + b|i\rangle\langle g|\right). \end{aligned}$$

The coefficients $\hbar\lambda_a$ and $\hbar\lambda_b$ describe the coupling energies between the modes a , b and level transitions $i \longleftrightarrow e$, $i \longleftrightarrow g$.

Taking into account the commutative propriety between the free and interaction parts of the Hamiltonian, $[H_0, H_I] = 0$, the wave function solution in the interaction picture $\left|\tilde{\Psi}(t+\tau)\right\rangle = \exp(\frac{i}{\hbar}H_0\tau)\left|\Psi(t+\tau)\right\rangle$ of the Schrodinger equation $i\hbar\partial_\tau \left|\tilde{\Psi}(t+\tau)\right\rangle = H_i \left|\tilde{\Psi}(t+\tau)\right\rangle$ can be exactly solved

$$\left|\tilde{\Psi}(t+\tau)\right\rangle = \left|\tilde{\Psi}_e(t+\tau)\right\rangle + \left|\tilde{\Psi}_g(t+\tau)\right\rangle + \left|\tilde{\Psi}_i(t+\tau)\right\rangle, \tag{3}$$

where

$$\begin{aligned} \left|\tilde{\Psi}_e(t+\tau)\right\rangle &= \sum_{n,m=N_d M_d}^{N_u,M_u} \{S_{n,m}\left[-2\lambda_a^2\alpha\frac{n+1}{\Omega^2_{n+1,m+1}}\sin^2(\Omega_{n+1,m+1}\tau/2)+\alpha\right] \\ &\quad - S_{n+1,m+1}2\gamma\lambda_a\lambda_b\frac{[(n+1)(m+1)]^{1/2}}{\Omega^2_{n+1,m+1}}\sin^2(\Omega_{n+1,m+1}\tau/2) \\ &\quad - S_{n+1,m}i\beta\lambda_a\frac{(n+1)^{1/2}}{\Omega_{n+1,m+1}}\sin(\Omega_{n+1,m+1}\tau)\}\,|e\rangle|n\rangle|m\rangle, \end{aligned}$$

$$\begin{aligned} \left|\tilde{\Psi}_g(t+\tau)\right\rangle &= \sum_{n,m=N_d M_d}^{N_u,M_u} \{-S_{n-1,m-1}2\alpha\lambda_a\lambda_b\frac{(nm)^{1/2}}{\Omega^2_{n,m}}\sin^2(\Omega_{n,m}\tau/2) \\ &+ S_{n,m}\left\{\gamma - 2\gamma\lambda_b^2\frac{m}{\Omega^2_{n,m}}\sin^2(\Omega_{n,m}\tau/2)\right\} \\ &- S_{n,m-1}i\beta\lambda_b\frac{m^{1/2}}{\Omega_{n,m}}\sin(\Omega_{n,m}\tau)\}\,|g\rangle\}\,|n\rangle|m\rangle, \end{aligned}$$

$$\begin{aligned} \left|\tilde{\Psi}_i(t+\tau)\right\rangle &= \sum_{n,m=N_d M_d}^{N_u,M_u} \{-S_{n-1,m}i\alpha\lambda_a\frac{n^{1/2}}{\Omega_{n,m+1}}\sin(\Omega_{n,m+1}\tau) + S_{n,m}\beta\cos(\Omega_{n,m+1}\tau) \\ &- S_{n,m+1}i\gamma\lambda_b\frac{(m+1)^{1/2}}{\Omega_{n,m+1}}\sin(\Omega_{n,m+1}\tau)\}\,|i\rangle|n\rangle|m\rangle. \end{aligned}$$

In order to obtain the recursion relation between the coefficients, $S_{n,m}$, we must require the probability amplitudes for the ground, $\gamma S_{n,m}|g\rangle|n\rangle|m\rangle$, excited $\alpha S_{n,m}|n\rangle|m\rangle|e\rangle$, and intermediate $-\beta S_{n,m}|i\rangle|n\rangle|m\rangle$ states to remain unchanged after the flying time, τ. As follows, all three reversible conditions are satisfied by the same recursion relation

$$S_{n,m} = -\frac{\alpha}{\gamma}\frac{\lambda_a}{\lambda_b}\left(\frac{n}{m}\right)^{1/2} S_{n-1,m-1} - i\frac{\beta}{\gamma}\frac{1}{\lambda_b}\frac{\Omega_{n,m}}{m^{1/2}}\cot(\Omega_{n,m}\frac{\tau}{2})S_{n,m-1}, \tag{4}$$

where $\Omega_{n,m} = \sqrt{\lambda_a^2 n + \lambda_b^2 m}$.

In the case when, $\lambda_a = \lambda_b = \lambda$, and $\alpha = 0$ (or $\gamma = 0$), three-level system is reduced to a two-level quantum oscillator relative to the ground and intermediate states (or intermediate and excited states), so that the recursion relation (4) passes into the same recursion relation proposed in [5].

For simplicity, considering the intermediate level is not populated, $\beta = 0$, we can obtain a similarity between expression (4) and the coherent trapping effect studied in the literature. In this case, the upper and bottom limits, are considered finite and the class of the biharmonic cavity field which satisfies the recursion relation is

$$\sin\left(\left(\lambda_a^2 n + \lambda_b^2 m\right)^{1/2} \frac{\tau}{2}\right)\left(S_{n,m} + \frac{\alpha}{\gamma}\frac{\lambda_a}{\lambda_b}\left(\frac{n}{m}\right)^{1/2} S_{n-1,m-1}\right) = 0. \tag{5}$$

In the case when the bottom limits for both modes are the vacuum states $N_d \to 0$, $M_d \to 0$, and the upper limits are infinite $N_u \to \infty$, $M_u \to \infty$. If the number of photons in both modes are equal, $m = n$, and $S_{n,m} = \delta_{n,m} S_{n,m}$, from relation (5) we can use only the second factor, $S_{n,m} - AS_{n-1,m-1} = 0$, for which the inversion is not changed (here $A = -\alpha\lambda_a/\gamma\lambda_b$). In such situation, the EM field state in which this effect is realized can be regarded as squeezed states of two cavity modes

$$|\Psi\rangle = \sqrt{1 - A^2} \exp\left(Aa^\dagger b^\dagger\right) |0\rangle_a |0\rangle_b . \tag{6}$$

Let us consider the situation for which the number of photons in both modes are equal, $m = n$, and the upper states are finite $n \le N_u$, $m \le M_u$. In this case when, $S_{N_u+1,N_u+1} = 0$, the second part of relation (5) is truncated, $AS_{N_u,N_u} \neq S_{N_u+1,N_u+1}$, from which follows

$$\begin{aligned} &\sin\left(\tfrac{\tau_k}{2}\left[(N_u + 1)(\lambda_a^2 + \lambda_b^2)\right]^{1/2}\right) = 0, \\ &\tau_k = 2k\pi / \left[(N_u + 1)(\lambda_a^2 + \lambda_b^2)\right]^{1/2}, \quad (k \text{ integer}). \end{aligned} \tag{7}$$

The relation (7) represents the reversible condition which describes the atom flying time, τ_k, through the cavity.

From a physical point of view, the reversible condition (7) becomes more clear, if we analyze the following partial cases:

a. *Reversible condition on the vacuum state of EM cavity field, when,* $N_d = N_u$, $M_d = M_u$ *and* $N_u = M_u = 0$. Considering, $\sin\left(\frac{\tau}{2}(\lambda_a^2 + \lambda_b^2)^{1/2}\right) = 0$, we can drop the above recursion relation, $S_{1,1} - AS_{0,0} \neq 0$, so that only one element is non-equal with zero, $S_{00} = 1$. From this condition we obtain the following discrete values for reversible time, τ_k

$$\tau_k = \frac{2k\pi}{\left(\lambda_a^2 + \lambda_b^2\right)^{1/2}}. \tag{8}$$

In accord with flying time (8), the populations on the excited and intermediate states oscillate with vacuum Rabi frequency, $\bar{n}_e = |\alpha|^2 \cos^4\left[\frac{\tau}{2}\left(\lambda_a^2 + \lambda_b^2\right)^{1/2}\right]$, and $\bar{n}_i = |\alpha|^2 \sin^2\left[\tau\left(\lambda_a^2 + \lambda_b^2\right)^{1/2}\right]/2$.

b. *Entangled bimodal field, when* $N_u = M_u = 1$, *and* $N_d = M_d = 0$. As follows, the recursion relation (5) is realized only between the coefficients, S_{22}, and S_{11}. Choosing, $S_{2,2} - AS_{1,1} \neq 0$, and $\sin\left(\frac{\tau}{2}\left[2(\lambda_a^2 + \lambda_b^2)\right]^{1/2}\right) = 0$, we obtain the following expression for flying time $\tau_k = \sqrt{2}k\pi/\sqrt{\lambda_a^2 + \lambda_b^2}$.

When, $\lambda_a = \lambda_b = \lambda$, the EM wave function in the time moment, $\tau = \tau_k$, corresponds to the entangled state of photons

$$\left|\tilde{\Psi}_{EM}(t+\tau_k)\right\rangle = \left(1+\frac{\alpha^2}{\gamma^2}\right)^{-1/2}\left[|0\rangle_a|0\rangle_b - \frac{\alpha}{\gamma}|1\rangle_a|1\rangle_b\right]. \tag{9}$$

3. Quantum mixing for off resonance case

In this section, we investigate the quantum mixing between a three-level atom and one cavity mod. It is consider the case when these oscillators are not in resonance each other (see *Figure* 2).

Assuming the energy is measured from the intermediate level of the atom, we study the influence of the detuning from resonance of its state. The general Hamiltonian of the system can be written

$$H = H_0 + H_i, \tag{10}$$

where

$$H_0 = \hbar\omega a^+ a + \hbar\omega_e |e\rangle\langle e| - \hbar\omega_g |g\rangle\langle g|,$$

$$H_i = \hbar\Delta\left(|e\rangle\langle e| + |g\rangle\langle g|\right) + \hbar\lambda_1\left(|i\rangle\langle g| a + a^+|g\rangle\langle i|\right) + \hbar\lambda_2\left(|e\rangle\langle i| a + a^+|i\rangle\langle e|\right)$$

are the free and interaction parts of the Hamiltonian; $E_g = -\hbar\omega_g$, $E_e = \hbar\omega_e$ represent the energetically positions of ground and excited levels of the atom. In the Hamiltonian (10) the frequency ω corresponds to the resonance cavity mode described by annihilation a, and creation a^+ field operators. The coefficients $\hbar\lambda_1 = \left(\overrightarrow{d}_{i,g}, \overrightarrow{g}(\omega)\right)$ and $\hbar\lambda_2 = \left(\overrightarrow{d}_{i,e}, \overrightarrow{g}(\omega)\right)$ represent the coupling energies between the cavity mode, level excitations $|i\rangle\langle g|$, $|e\rangle\langle i|$ and relaxation $|g\rangle\langle i|$, $|i\rangle\langle e|$ transitions. $\overrightarrow{d}_{i,g}$and $\overrightarrow{d}_{i,e}$ are the transitions of dipole matrix elements; $\overrightarrow{g}(\omega)$ is the strength constant.

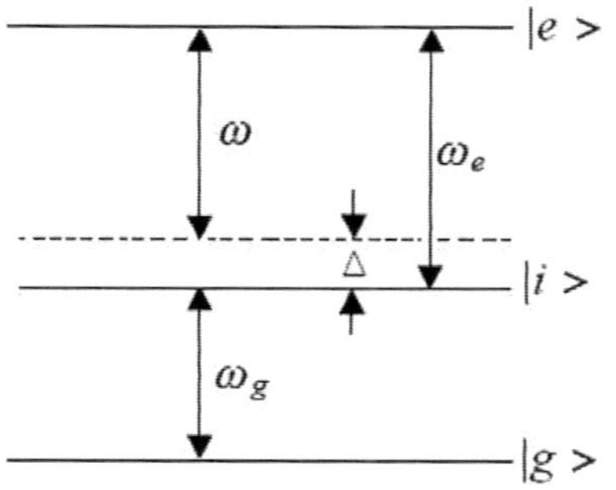

Figure 2. Energy levels relevant to the two photons laser with atomic polarization.

Let us considering that at the time, $\tau = 0$, the atom enters in the cavity and the atomic and cavity field states are totally factorized

$$|\Psi(0)\rangle = \sum_{n=N_d}^{N_u}\left(\alpha|e\rangle + \beta|i\rangle + \gamma|g\rangle\right)\otimes S_n|n\rangle. \tag{11}$$

After interaction, at the time moment, τ, the wave function of these two oscillators can be written in the following way

$$
\begin{aligned}
|\Psi(\tau)\rangle = & \sum_{n=N_d}^{N_u} \{\{\alpha\Bigg[\frac{\lambda_2^2\,(n+1)}{\Theta_{n+1}}\left[e^{-i\Delta\tau/2}\cos\left(\Omega_{n+1}\frac{\tau}{2}\right)-i\frac{\Delta}{\Omega_{n+1}}e^{-i\Delta\tau/2}\sin\left(\Omega_{n+1}\frac{\tau}{2}\right)-e^{-i\Delta\tau}\right] \\
& + e^{-i\Delta\tau}\Bigg]S_n - 2\beta i\lambda_2\frac{\sqrt{n+1}}{\Omega_{n+1}}e^{-i\Delta\tau/2}\sin\left(\Omega_{n+1}\frac{\tau}{2}\right)S_{n+1} \\
& + \gamma\frac{\lambda_1\lambda_2\sqrt{(n+1)\,(n+2)}}{\Theta_{n+1}}\left[e^{-i\Delta\tau/2}\cos\left(\Omega_{n+1}\frac{\tau}{2}\right)-i\frac{\Delta e^{-i\Delta\tau/2}}{\Omega_{n+1}}\sin\left(\Omega_{n+1}\frac{\tau}{2}\right)-e^{-i\Delta\tau}\right]S_{n+2}\}\,|e\rangle \\
& + \{-2i\alpha\frac{\lambda_2\sqrt{n}}{\Omega_n}e^{-i\Delta\tau/2}\sin\left(\Omega_n\frac{\tau}{2}\right)S_{n-1}+\beta e^{-i\Delta\tau/2}\left[\cos\left(\Omega_n\frac{\tau}{2}\right)+i\frac{\Delta}{\Omega_n}\sin\left(\Omega_n\frac{\tau}{2}\right)\right]S_n \\
& - 2i\gamma\frac{\lambda_1\sqrt{n+1}}{2\Omega_n}e^{-i\Delta\tau/2}\sin\left(\Omega_n\frac{\tau}{2}\right)S_{n+1}\}\,|i\rangle \\
& + \{\alpha\frac{\lambda_1\lambda_2\sqrt{n(n-1)}}{\Theta_{n-1}}\left[e^{-i\Delta\tau/2}\cos\left(\Omega_{n-1}\frac{\tau}{2}\right)-i\frac{\Delta}{\Omega_{n-1}}e^{-i\Delta\tau/2}\sin\left(\Omega_{n-1}\frac{\tau}{2}\right)-e^{-i\Delta\tau}\right]S_{n-2} \\
& - 2\beta i\lambda_1\frac{\sqrt{n}}{\Omega_{n-1}}e^{-i\Delta\tau/2}\sin\left(\Omega_{n-1}\frac{\tau}{2}\right)S_{n-1} \\
& + \gamma\left\{\frac{\lambda_1^2 n}{\Theta_{n-1}}\left[e^{-i\Delta\tau/2}\cos\left(\Omega_{n-1}\frac{\tau}{2}\right)-i\frac{\Delta e^{-i\Delta\tau/2}}{\Omega_{n-1}}\sin\left(\Omega_{n-1}\frac{\tau}{2}\right)-e^{-i\Delta\tau}\right]+e^{-i\Delta\tau}\right\}S_n\}\,|g\rangle\}\,|n\rangle\,,
\end{aligned}
\tag{12}
$$

where $\Omega_n = \sqrt{\Delta^2 + 4\left(\lambda_1^2\,(n+1)+\lambda_2^2 n\right)}$ is the Rabi frequency and $\Theta_n = \left(\Omega_n^2 - \Delta^2\right)/4$.

During the interaction process the atom and cavity field are coupled in according with the solution (12). The aim of this paper consists in the finding of quantum field proprieties for which the atomic population and photon number distributions on the Fock states after the transit time, τ, restore their initial proprieties (11). In other words the population probabilities of the field on the Fock states $|n>$, $|\langle\Psi(\tau)|\,|n\rangle|^2 = |S_n|^2$, atom in the ground, $|\langle\Psi(\tau)|\,|g\rangle|^2 = |\alpha_g|^2$, intermediate, $|\langle\Psi(\tau)|\,|i\rangle|^2 = |\alpha_i|^2$ and excited $|\langle\Psi(\tau)|\,|e\rangle|^2 = |\alpha_e|^2$ states become reversible. Such field must have special coefficients S_n, for which the reversibility is possible. Indeed, according to the wave function (12) this requirement can be satisfied for the evolution of the ground state amplitude during the transit time, τ. Thus, after the interaction the amplitude on the ground and excited states remain as in the initial state (11), but with other phase : $\gamma'(\tau)S_n'(\tau)\to\gamma S_n e^{-i\Delta\tau}$, $\alpha'(\tau)S_n'(\tau)\to\alpha S_n e^{-i\Delta\tau}$.

As follows, we obtain the same recursion relation between the coefficients S_n

$$
S_n = \frac{1}{2\beta\Theta_n}\left[i\Omega_n\frac{e^{-i\Delta\tau/2}-\cos\left(\Omega_n\frac{\tau}{2}\right)}{\sin\left(\Omega_n\frac{\tau}{2}\right)}-\Delta\right]\left[\alpha\lambda_2\sqrt{n}S_{n-1}+\gamma\lambda_1\sqrt{n+1}S_{n+1}\right]. \tag{13}
$$

The recursion relation (13) must be satisfied for all three-states of the atom at the same time. Introducing the coefficient, S_n, from this relation in the amplitude of the intermediate atomic state, $|i>$, from solution (12) we obtain that the amplitude of the intermediate state also pass in the initial amplitude, but with a phase which depends on the field state numbers

$$
\beta'(\tau)S_n'(\tau)\,|i\rangle\,|n\rangle = -\beta e^{-i\Delta\tau}\frac{Z\,(n)}{Z^*\,(n)}S_n\,|i\rangle\,|n\rangle = -\beta e^{-i\Delta\tau+2i\arg Z}S_n\,|i\rangle\,|n\rangle\,. \tag{14}
$$

Here the notation, $Z\,(n) = \cos\left(\Omega_n\tau/2\right) + i\Delta/\Omega_n\sin\left(\Omega_n\tau/2\right)$, represents the complex number. The module $|Z(n)|$ and argument $\phi(n) = \arctan\left(\Delta\tan\Omega_n/\Omega_n\right)$ depend on the Fock state number.

As a conclusion, we observe that the recursion relation (13) will be satisfied for all atomic states and in the time moment, τ, the wave function of the system evaluates in a quasifactorised state with the same probability amplitude as in the initial state

$$|\Psi(\tau)\rangle = e^{-i\Delta\tau} \sum_{n=N_d}^{N_u} S_n \left[\alpha|e\rangle + \gamma|g\rangle - \beta\frac{Z(n)}{Z^*(n)}|i\rangle\right]|n\rangle . \quad (15)$$

The reversibility between two quantum oscillators can be analyzed in two cases:

a. Full separation, that corresponds to exact resonance case, $\Delta = 0$. In this situation the wave function which describes the atomic and field subsystem is separated $\Psi(\tau) = \Psi_A(\tau) \otimes \Psi_{EM}(\tau)$. In according with the expression (15), the coefficient $Z(n)/Z^*(n) = 1$ and the wave function are totally factorized $|\Psi(\tau)\rangle = \sum_{n=N_d}^{N_u} S_n|n\rangle \otimes [\alpha_e|e\rangle + \alpha_g|g\rangle - \alpha_i|i\rangle]$.

b. Partial separation, that corresponds to $\Delta \neq 0$. In this case, the coefficient, $|a_k|^2\,|S_n|^2$, represents the probability to find the atom and field in the same initial Fock state $|k\rangle\,|n\rangle$, where $k \equiv e, i, g$. Using the definition (14), we observe that the phase factor $\phi(n)$ does not change the probability to find the atom and field in the Fock states $|k\rangle$ and $|n\rangle$: $\left|a_k S_n e^{i\phi(n)}\right|^2 = |a_k|^2\,|S_n|^2$.

As follows from both definitions the amplitude phases do not affect the diagonal matrix elements of the system after the interaction process. According to second case, we observe that the wave function remains unfactorized. This effect can be regarded as a partial factorization and becomes more clear, introducing the reduced density operator for atomic states $W_A = Sp\{|\Psi\rangle\langle\Psi|\}_{ph}$:

$$\begin{aligned} W_A = {} & [\alpha|e\rangle + \gamma|g\rangle][\alpha^*\langle e| + \gamma^*\langle g|] + |\beta|^2|i\rangle\langle i| \\ & - \beta^*(\alpha|e\rangle + \gamma|g\rangle)\langle i| \sum_{n=N_d}^{N_u} |S_n|^2\frac{Z^*(n)}{Z(n)} \\ & - \beta[\alpha^* < e| + \gamma^* < g|]\,|i\rangle \sum_{n=N_d}^{N_u} |S_n|^2\frac{Z(n)}{Z^*(n)}. \end{aligned} \quad (16)$$

In analogical way, we can write the reduced density matrix for the cavity field

$$W_{EM} = Sp_A|\Psi(\tau)\rangle\langle\Psi(\tau)| = \sum_{n=N_d}^{N_u}\sum_{m=M_d}^{M_u}\left(1 + |\beta|^2\left(\frac{Z_m^* Z_n}{Z_m Z_n^*} - 1\right)\right) S_n S_m^*|n\rangle\langle m| . \quad (17)$$

As follows from definitions (16) and (17) after the transit time, τ, the diagonal matrix elements coincide with their initial values, but the mixing phases between the atom and field of non-diagonal matrix elements remain coupled.

4. Conclusions

In this paper we have found the reversible conditions for a three-level atom flying through the cavity in the resonance case and in the absence of it. In the case of non-vacuum state the decompose coefficients of Fock states, S_{nm}, must comply with the reversible conditions described by recursion relations (4) and (7). Here, we proposed a new reversible condition which not coincides with earlier discussed trapping conditions [$4-6$], but in particular case this general condition contains the reversibilities discussed in above papers. We consider that this reversibility open new experimental [8] and theoretical realizations in quantum mechanics.

In the off resonance case is demonstrated that the phases coupling between the cavity field and atomic inversion remain unchanged. As follows from reduced density operator, after the interaction process the atom and cavity subsystem can restore only their initial diagonal matrix elements (atomic population, mean number of photons and its fluctuations), but the nondiagonal elements remain coupled. This effect was named partial restoration of two quantum interaction subsystems.

The effect discussed here gives us the possibility to use the prepared atomic inversion of three-level atom flying through the cavity in quantum computing. Choosing the flying time accordingly to the reversible condition (7) and (8) one can obtain the factorized wave function of two subsystems: atom and cavity vacuum after interaction. Quantum computer describes the evolution of two or more subsystems in which the reversible transition from non-entangled to the entangled state and vice versa represents an essential feature [9]. This manipulation with quantum phase can be used in logical gates of information processing. Quantum logic gates are the basic units of quantum control. For example, if the system remains in the same factorized state the quantum control is realized (or not realized). It is equivalent to the controlled-NOT gate operation proposed in paper [10].

5. References

[1] Weidinger M, Varcoe B T H, Heerlein R and Walther H 1999 *Phys. Rev. Lett.* **82** 3795
[2] Meschede D. Walther H and Muller G 1985 *Phys. Rev. Lett.* **54** 551
[3] Rempe G, Schmidt-Kaler F and Walther H 1990 *Phys. Rev. Lett.* **64** 2783
[4] Agarwal G S 1993 *Phys. Rev. Lett.* **71** 1351; Orszag M,Ramirez R, Retamal J C and Roa L 1992 *Phys. Rev.* A **45** 6717
[5] Slosser J J and Meystre P 1990 *Phys. Rev.* A **41** 3867
[6] Enaki N A and Koroli V I 1999 *J. Opt. Soc. Am.* A **16** 2973; Enaki N A and Ciobanu N *Journal of Modern Optics* **272097** (*In press*)
[7] Bergou J and Hillery M 1997 *Phys. Rev.* A **55** 4585
[8] Chaneliere T, Matsukevich D N, Jenkins S D, Lan S Y, Zhao R, Kennedy T A B and Kuzmich A 2007 *Phys. Rev. Lett.* **98** 113602; Laurat J, Chou C, Deng H, Choi K S, Felinto D, Riedmatten H and Kimble H J 2007 *New Journal of Phys.* **9** 207
[9] Bennett C 1973 *IBM Jl Res.Dev.* **6** 525; Deutsch D. 1989 *Proc. R. Soc. Lond.* A **425** 73
[10] Ekert A and Jozsa R 1996 *Modern Physics* **68** (3) 733

Journal of Physics: Conference Series **128** (2008) 012023 doi:10.1088/1742-6596/128/1/012023

Coherent states for SUSY partner Hamiltonians

David J. Fernández C.[1], **Oscar Rosas-Ortiz**[2]
Departamento de Física, Cinvestav, A.P. 14-740, 07000 México D.F., Mexico

Véronique Hussin[3]
Département de Mathématiques et Centre de Recherches Mathématiques, Université de Montréal, C.P. 6128, Succ. Centre-Ville, Montréal (Québec), H3C 3J7, Canada

E-mail: [1]`david@fis.cinvestav.mx`, [2]`orosas@fis.cinvestav.mx`, [3]`hussin@DMS.UMontreal.CA`

Abstract. Coherent states are constructed for systems generated by supersymmetry from an initial Hamiltonian with a purely discrete spectrum such that the levels depend analytically on their subindex. The technique is illustrated by means of the trigonometric Pöschl-Teller potentials.

1. Introduction

The beautiful attributes of the standard coherent states (CS) stems from the Heisenberg-Weyl algebra, characteristic of the harmonic oscillator. It is well known that some of these properties play the role of definitions, which can be used to built up CS for systems different from the oscillator [1–8]. In particular, if the CS are to be constructed through algebraic techniques, e.g. as eigenstates of the annihilation operator, it is fundamental to identify the algebra characterizing the system under study. Surprisingly, for Hamiltonians whose spectra consist of an infinite set of discrete energy levels $E_n = E(n)$, where $E(n)$ is a function specifying the analyzed system, the appropriate algebraic structure was just recently identified, and this allowed at the same time to perform the CS construction [9].

On the other hand, supersymmetric quantum mechanics (SUSY QM) has proved successful to generate exactly solvable potentials from an initial solvable one [10–22]. If the Hamiltonian of departure is ruled by a given algebra, it is natural to ask how much of this algebraic structure will be inherited by its SUSY partner potentials and what will be the similarities and/or differences of the corresponding CS.

Some partial answers to these question were found for the first time in 1994 [23] (see however the seminal work of Mielnik [24]). There, third order differential annihilation and creation operators were identified as the generators of the *natural algebra* (which turned out to be of polynomial type) for the simplest family of first-order SUSY partners of the oscillator (the so called Abraham-Moses-Mielnik potentials [24, 25]). The CS were also built as eigenstates of the *natural annihilation operator*, and its decomposition in terms of energy eigenstates turned out to be more involved than the standard one. In further works it was shown that simpler CS exist for the same family of potentials (working with deformations of the natural annihilation and creation operators), and it was shown that their expansion in terms of eigenstates of the Hamiltonian was essentially the same as the standard one [26, 27]. In 1999 it was identified the *natural algebra* characterizing a quite general family of SUSY partner potentials of the oscillator. The order $k \in \mathbb{N}$ of the intertwining operator was finite, and the positions of the new levels created by the transformation were arbitrary, although all of them below the ground state energy of the oscillator [28]. Once again, it was found that the *natural CS* were more involved than the standard ones. In the same paper it was shown that the standard expression for the CS can be as well recovered working with a *deformation* of the natural algebra, which mimics the Heisenberg-Weyl algebra in the subspace associated to the isospectral part of the spectrum.

In this paper we are going to perform a similar study for the SUSY partners of potentials different from the oscillator, which are not longer described by the Heisenberg-Weyl algebra. The main results to be presented here involve an initial Hamiltonians whose spectrum is composed of an infinite set of energy levels such that $E_n = E(n),\ n = 0, 1, \ldots$ For a detailed discussion of these and some other results we recommend [29].

The organization of the paper is as follows. In the next section the algebraic structures of the initial Hamiltonian H_0 will be explored, while in section 3 we are going to derive the corresponding coherent states. The SUSY partner Hamiltonians of H_0 will be discussed in section 4, while in section 5 the corresponding algebraic structures will be analyzed. In section 6 the coherent states of H_k shall be constructed. The complete procedure is going to be applied to the trigonometric Pöschl-Teller potential in section 7 while in section 8 we will finish the paper with our conclusions.

2. Algebraic structures of H_0

Let us start with an initial Schrödinger Hamiltonian

$$H_0 = -\frac{1}{2}\frac{d^2}{dx^2} + V_0(x), \tag{1}$$

whose eigenfunctions and eigenvalues satisfy

$$H_0|\psi_n\rangle = E_n|\psi_n\rangle, \qquad E_0 < E_1 < E_2 < \ldots \tag{2}$$

where

$$E_n \equiv E(n), \tag{3}$$

i.e., we are assuming that there is an analytical dependence, characterized by the function $E(x)$, of the energy eigenvalues with the subindex labeling them. Let us introduce now the number operator N_0 in the way:

$$N_0|\psi_n\rangle = n|\psi_n\rangle. \tag{4}$$

The *intrinsic nonlinear algebra* of H_0 is defined through the following expressions:

$$a_0^-|\psi_n\rangle = r_{\mathcal{I}}(n)|\psi_{n-1}\rangle, \quad a_0^+|\psi_n\rangle = \bar{r}_{\mathcal{I}}(n+1)|\psi_{n+1}\rangle, \tag{5}$$

$$r_{\mathcal{I}}(n) = e^{i\alpha(E_n - E_{n-1})}\sqrt{E_n - E_0}, \quad \alpha \in \mathbb{R}. \tag{6}$$

At the operator level the previous relationships lead to:

$$a_0^+ a_0^- = E(N_0) - E_0, \quad a_0^- a_0^+ = E(N_0 + 1) - E_0. \tag{7}$$

Thus, we arrive at the following set of non-null commutation relationships characteristic of the *intrinsic algebra* of H_0:

$$[N_0, a_0^\pm] = \pm a_0^\pm, \qquad [a_0^-, a_0^+] = E(N_0+1) - E(N_0) \equiv f(N_0), \tag{8}$$

$$[H_0, a_0^\pm] = \pm a_0^\pm f\left(N_0 - \frac{1}{2} \pm \frac{1}{2}\right) = \pm f\left(N_0 - \frac{1}{2} \mp \frac{1}{2}\right) a_0^\pm. \tag{9}$$

The Hubbard representation of $a_0^\pm$ is given by:

$$a_0^- = r_\mathcal{I}(N_0+1) \sum_{m=0}^{\infty} |\psi_m\rangle\langle\psi_{m+1}|, \qquad a_0^+ = \bar{r}_\mathcal{I}(N_0) \sum_{m=0}^{\infty} |\psi_{m+1}\rangle\langle\psi_m|. \tag{10}$$

On the other hand, the *linear algebra* of H_0 is obtained from the following *deformation* of the intrinsic annihilation and creation operators $a_0^\pm$:

$$a_{0_\mathcal{L}}^- = b(N_0)\, a_0^-, \quad a_{0_\mathcal{L}}^+ = a_0^+\, b(N_0), \tag{11}$$

with

$$b(n) = \frac{r_\mathcal{L}(n+1)}{r_\mathcal{I}(n+1)} = \sqrt{\frac{n+1}{E(n+1)-E_0}}, \qquad r_\mathcal{L}(n) = e^{i\alpha f(n-1)}\sqrt{n}. \tag{12}$$

It turns out that:

$$a_{0_\mathcal{L}}^-|\psi_n\rangle = r_\mathcal{L}(n)|\psi_{n-1}\rangle, \qquad a_{0_\mathcal{L}}^+|\psi_n\rangle = \bar{r}_\mathcal{L}(n+1)|\psi_{n+1}\rangle, \tag{13}$$

which is equivalent to the following commutation relationships:

$$[N_0, a_{0_\mathcal{L}}^\pm] = \pm a_{0_\mathcal{L}}^\pm, \qquad [a_{0_\mathcal{L}}^-, a_{0_\mathcal{L}}^+] = 1, \tag{14}$$

i.e., the *linear annihilation-creation operators* $a_{0_\mathcal{L}}^\pm$ (together with N_0) satisfy the standard Heisenberg-Weyl algebra.

3. Coherent states of H_0

Along this paper we will derive the coherent states as eigenstates of an involved annihilation operator. As regards to the *intrinsic* CS of H_0, they obey:

$$a_0^-|z,\alpha\rangle_0 = z|z,\alpha\rangle_0, \qquad z \in \mathbb{C}. \tag{15}$$

An standard procedure leads to the following expression:

$$|z,\alpha\rangle_0 = \left(\sum_{m=0}^{\infty} \frac{|z|^{2m}}{\rho_m}\right)^{-\frac{1}{2}} \sum_{m=0}^{\infty} e^{-i\alpha(E_m-E_0)\frac{z^m}{\sqrt{\rho_m}}|\psi_m\rangle}, \tag{16}$$

$$\rho_m = \begin{cases} 1 & \text{if } m=0, \\ (E_m-E_0)\dots(E_1-E_0) & \text{if } m>0. \end{cases} \tag{17}$$

The completeness relationship of the *intrinsic* CS

$$\int |z,\alpha\rangle_0\ {}_0\langle z,\alpha\,| d\mu(z) = 1, \tag{18}$$

Journal of Physics: Conference Series **128** (2008) 012023 doi:10.1088/1742-6596/128/1/012023

with a measure given by

$$d\mu(z) = \frac{1}{\pi}\left(\sum_{m=0}^{\infty}\frac{|z|^{2m}}{\rho_m}\right)\rho(|z|^2)\, d^2z, \tag{19}$$

is valid provided that ρ_m and $\rho(x)$ satisfy the following moment problem:

$$\int_0^{\infty} y^m \rho(y)\, dy = \rho_m, \quad m = 0, 1, \ldots \tag{20}$$

The reproducing kernel is given by

$${}_0\langle z, \alpha | z', \alpha\rangle_0 = \left(\sum_{m=0}^{\infty}\frac{|z|^{2m}}{\rho_m}\right)^{-\frac{1}{2}}\left(\sum_{m=0}^{\infty}\frac{|z'|^{2m}}{\rho_m}\right)^{-\frac{1}{2}}\left(\sum_{m=0}^{\infty}\frac{(\bar{z}z')^m}{\rho_m}\right), \tag{21}$$

which in particular means that two *intrinsic* CS associated to different complex eigenvalues $z \neq z'$ are non-orthogonal. It is worth to notice that a CS evolves always as a CS since

$$U_0(t)|z, \alpha\rangle_0 = \exp(-itH_0)|z, \alpha\rangle_0 = e^{-itE_0}|z, \alpha + t\rangle_0. \tag{22}$$

This property is a consequence of the phase choice of (5-6). That is why these CS sometimes are called *adapted.* Note that the eigenvalue $z = 0$ is non-degenerated:

$$|z = 0, \alpha\rangle_0 = |\psi_0\rangle. \tag{23}$$

Concerning the *linear* CS of H_0, which are eigenstates of $a_{0_{\mathcal{L}}}^-$, they turn out to be:

$$|z, \alpha\rangle_{0_{\mathcal{L}}} = e^{-\frac{|z|^2}{2}}\sum_{m=0}^{\infty} e^{-i\alpha(E_m - E_0)}\frac{z^m}{\sqrt{m!}}|\psi_m\rangle, \tag{24}$$

which, up to the phase factors, have the form of the standard CS. This implies that the completeness relationship is automatically satisfied:

$$\frac{1}{\pi}\int |z, \alpha\rangle_{0_{\mathcal{L}}}\ {}_{0_{\mathcal{L}}}\langle z, \alpha|\, d^2z = 1. \tag{25}$$

The reproducing kernel is the standard one:

$${}_{0_{\mathcal{L}}}\langle z, \alpha | z', \alpha\rangle_{0_{\mathcal{L}}} = \exp\left(-\frac{|z|^2}{2} + \bar{z}z' - \frac{|z'|^2}{2}\right), \tag{26}$$

and, since the phase choice for the linear is the same as for the intrinsic annihilation and creation operators, it turns out that once again a CS evolves in time into a CS.

It is worth to notice that for the linear CS it is valid

$$|z, \alpha\rangle_{0_{\mathcal{L}}} = D_{\mathcal{L}}(z)|\psi_0\rangle = \exp(z a_{0_{\mathcal{L}}}^+ - \bar{z} a_{0_{\mathcal{L}}}^-)|\psi_0\rangle, \tag{27}$$

meaning that the *linear* CS can be obtained through the action of the displacement operator $D_{\mathcal{L}}(z)$ onto the ground state of the system, as for the harmonic oscillator.

4. SUSY partner Hamiltonians H_k

The SUSY partner Hamiltonians of H_0, denoted by H_k, are linked to each other through the following intertwining relationships involving two k-th order differential operators B, B^+:

$$H_k B_k^+ = B_k^+ H_0, \qquad H_0 B_k = B_k H_k, \tag{28}$$

where

$$H_k = -\frac{1}{2}\frac{d^2}{dx^2} + V_k(x). \tag{29}$$

The potential $V_k(x)$, in terms of the initial one and the solutions $\alpha_i(x, \epsilon_i)$ of a sequence of Riccati equations, is given by:

$$V_k(x) = V_0(x) - \sum_{i=1}^{k} \alpha_i'(x, \epsilon_i). \tag{30}$$

In case that all the $\epsilon_j, j = 1, \dots, k$ are different, it turns out that the i-th Riccati solution at ϵ_i can be expressed algebraically in terms of two previous ones at ϵ_i, ϵ_{i-1} through the following finite-difference (Bäcklund) formula:

$$\alpha_i(x, \epsilon_i) = -\alpha_{i-1}(x, \epsilon_{i-1}) - \frac{2(\epsilon_i - \epsilon_{i-1})}{\alpha_{i-1}(x, \epsilon_i) - \alpha_{i-1}(x, \epsilon_{i-1})}, \qquad i = 2, \dots k. \tag{31}$$

By iterating down this formula, it is straightforward to realize that all the α_i can be expressed in terms of the k solutions $\alpha_1(x, \epsilon_i)$ of the first Riccati equation:

$$\alpha_1'(x, \epsilon_i) + \alpha_1^2(x, \epsilon_i) = 2[V_0(x) - \epsilon_i], \quad i = 1, \dots, k. \tag{32}$$

The above formulae can be expressed in terms of k solutions $u_i(x) \propto \exp[\int_0^x \alpha_1(y, \epsilon_i) dy]$ of the associated Schrödinger equation, in particular the potential $V_k(x)$:

$$V_k(x) = V_0(x) - \{\ln[W(u_1, \dots, u_k)]\}''. \tag{33}$$

Two relevant equations resulting from this procedure are:

$$B_k^+ B_k = \prod_{i=1}^{k} (H_k - \epsilon_i), \qquad B_k B_k^+ = \prod_{i=1}^{k} (H_0 - \epsilon_i). \tag{34}$$

They show the connection between the k-th order SUSY QM and the *factorization method* [18].

Let us denote by $|\theta_{\epsilon_i}\rangle$, $i = 1, \dots, s$, the s physical states belonging to the kernel of B_k which are at the same time eigenstates of H_k, namely, $B_k|\theta_{\epsilon_i}\rangle = 0$, $H_k|\theta_{\epsilon_i}\rangle = \epsilon_i|\theta_{\epsilon_i}\rangle$, $s \leq k$. Suppose that $\mathrm{Sp}(H_k) = \{\epsilon_1, \dots, \epsilon_q, E_0, E_1, \dots\}$, $q \leq s$, i.e., $\epsilon_{q+j} = E_{m_j}$, $j = 1, \dots, p = s - q$, $m_j < m_{j+1}$. We thus have $B_k^+|\psi_{m_j}\rangle = 0$.

5. Algebraic structures of H_k

Let us define in the first place the number operator N_k by its action onto the eigenstates of H_k:

$$N_k|\theta_n\rangle = n|\theta_n\rangle, \quad N_k|\theta_{\epsilon_i}\rangle = 0, \quad n = 0, 1, \dots, \quad i = 1, \dots, q. \tag{35}$$

Following [23, 24, 26–28], let us introduce now a *natural algebra* for the SUSY partner Hamiltonians H_k, whose generators are constructed from the intertwining and intrinsic annihilation and creation operators of H_0:

$$a_{k\mathcal{N}}^{\pm} = B_k^{+} a_0^{\pm} B_k. \tag{36}$$

It turns out that the action of the *natural annihilation and creation operators* $a_{k\mathcal{N}}^{\pm}$ onto the eigenstates of H_k is given by:

$$a_{k\mathcal{N}}^{\pm}|\theta_{\epsilon_i}\rangle = 0, \qquad i = 1, \dots, q, \tag{37}$$

$$a_{k\mathcal{N}}^{-}|\theta_n\rangle = r_{\mathcal{N}}(n)\ |\theta_{n-1}\rangle, \qquad a_{k\mathcal{N}}^{+}|\theta_n\rangle = \bar{r}_{\mathcal{N}}(n+1)\ |\theta_{n+1}\rangle, \tag{38}$$

where

$$r_{\mathcal{N}}(n) = \left\{\prod_{i=1}^{k}[E(n)-\epsilon_i][E(n-1)-\epsilon_i]\right\}^{\frac{1}{2}} r_{\mathcal{I}}(n). \tag{39}$$

An alternative expression for $a_{k\mathcal{N}}^{\pm}$ is provided by the corresponding Hubbard representation:

$$a_{k\mathcal{N}}^{-} = r_{\mathcal{N}}(N_k+1)\sum_{m=0}^{\infty}|\theta_m\rangle\langle\theta_{m+1}|, \qquad a_{k\mathcal{N}}^{+} = \bar{r}_{\mathcal{N}}(N_k)\sum_{m=0}^{\infty}|\theta_{m+1}\rangle\langle\theta_m|. \tag{40}$$

On the subspace associated to the levels $\{E_n, n = 0, 1, \dots\}$ it is valid:

$$[a_{k\mathcal{N}}^{-}, a_{k\mathcal{N}}^{+}] = [\bar{r}_{\mathcal{N}}(N_k+1)r_{\mathcal{N}}(N_k+1) - \bar{r}_{\mathcal{N}}(N_k)r_{\mathcal{N}}(N_k)]\sum_{m=0}^{\infty}|\theta_m\rangle\langle\theta_m|. \tag{41}$$

Let us analyze now the *intrinsic algebra* of H_k, which is generated by the following annihilation and creation operators:

$$a_k^{-} = r_{\mathcal{I}}(N_k+1)\sum_{m=0}^{\infty}|\theta_m\rangle\langle\theta_{m+1}|, \qquad a_k^{+} = \bar{r}_{\mathcal{I}}(N_k)\sum_{m=0}^{\infty}|\theta_{m+1}\rangle\langle\theta_m|. \tag{42}$$

Their action onto the energy eigenstates of H_k becomes:

$$a_k^{\pm}|\theta_{\epsilon_i}\rangle = 0, \qquad i = 1, \dots, q, \tag{43}$$

$$a_k^{-}|\theta_n\rangle = r_{\mathcal{I}}(n)|\theta_{n-1}\rangle, \qquad a_k^{+}|\theta_n\rangle = \bar{r}_{\mathcal{I}}(n+1)|\theta_{n+1}\rangle. \tag{44}$$

On the subspace associated to the levels $\{E_n,\ n = 0, 1, \dots\}$, this algebra is equal to the corresponding one for H_0:

$$[a_k^{-}, a_k^{+}] = f(N_k)\sum_{m=0}^{\infty}|\theta_m\rangle\langle\theta_m|. \tag{45}$$

Note that the natural and intrinsic algebras of H_k can be seen as deformations from each other, since:

$$a_{k\mathcal{N}}^{-} = \frac{r_{\mathcal{N}}(N_k+1)}{r_{\mathcal{I}}(N_k+1)}\,a_k^{-}, \quad a_{k\mathcal{N}}^{+} = \frac{r_{\mathcal{N}}(N_k)}{r_{\mathcal{I}}(N_k)}\,a_k^{+}, \quad a_{k\mathcal{N}}^{+}a_{k\mathcal{N}}^{-} = [E(N_k)-E_0]\left[\frac{r_{\mathcal{N}}(N_k)}{r_{\mathcal{I}}(N_k)}\right]^2. \tag{46}$$

Finally, let us analyze the *linear algebra* of H_k, which is generated by:

$$a^-_{k_\mathcal{L}} = r_\mathcal{L}(N_k+1)\sum_{m=0}^{\infty}|\theta_m\rangle\langle\theta_{m+1}|, \qquad a^+_{k_\mathcal{L}} = \bar{r}_\mathcal{L}(N_k)\sum_{m=0}^{\infty}|\theta_{m+1}\rangle\langle\theta_m|. \tag{47}$$

The corresponding action on the eigenstates of H_k reads:

$$a^{\pm}_{k_\mathcal{L}}|\theta_{\epsilon_i}\rangle = 0, \qquad i=1,\ldots,q, \tag{48}$$

$$a^-_{k_\mathcal{L}}|\theta_n\rangle = r_\mathcal{L}(n)|\theta_{n-1}\rangle, \qquad a^+_{k_\mathcal{L}}|\theta_n\rangle = \bar{r}_\mathcal{L}(n+1)|\theta_{n+1}\rangle. \tag{49}$$

At the operator level, this algebra is described by the following commutation relationships:

$$[N_k, a^{\pm}_{k_\mathcal{L}}] = \pm a^{\pm}_{k_\mathcal{L}}, \qquad [a^-_{k_\mathcal{L}}, a^+_{k_\mathcal{L}}] = \sum_{m=0}^{\infty}|\theta_m\rangle\langle\theta_m|. \tag{50}$$

Once again, the intrinsic and linear algebra are deformations from each other due to:

$$a^-_{k_\mathcal{L}} = \frac{r_\mathcal{L}(N_k+1)}{r_\mathcal{I}(N_k+1)}\,a^-_k, \qquad a^+_{k_\mathcal{L}} = \frac{r_\mathcal{L}(N_k)}{r_\mathcal{I}(N_k)}\,a^+_k, \qquad a^+_{k_\mathcal{L}}a^-_{k_\mathcal{L}} = N_k. \tag{51}$$

The previous results lead us to conclude that the algebraic structures associated to H_0, intrinsic and linear, characterize as well its SUSY partner Hamiltonians H_k on the subspace associated to the initial levels $\{E_n,\ n=0,1,\ldots\}$. It will be interesting to see what happens with the corresponding coherent states.

6. Coherent states of H_k

The *natural nonlinear coherent states* of H_k, which are eigenstates of $a^-_{k_\mathcal{N}}$, are given by:

$$|z,\alpha\rangle_{k_\mathcal{N}} = \left[\sum_{m=0}^{\infty}\frac{|z|^{2m}}{\widetilde{\rho}_m}\right]^{-\frac{1}{2}}\sum_{m=0}^{\infty}e^{-i\alpha(E_{m+m_p+1}-E_{m_p+1})}\frac{z^m}{\sqrt{\widetilde{\rho}_m}}|\theta_{m+m_p+1}\rangle, \tag{52}$$

where $\widetilde{\rho}_0 = 1$ and, for $m>0$,

$$\widetilde{\rho}_m = \frac{\rho_{m+m_p+1}}{\rho_{m_p+1}}\prod_{i=1}^{k}(E_{m+m_p+1}-\epsilon_i)(E_{m+m_p}-\epsilon_i)^2\ldots(E_{m_p+2}-\epsilon_i)^2(E_{m_p+1}-\epsilon_i). \tag{53}$$

Since the CS decomposition (52) does not involve the eigenstates $\{|\theta_{\epsilon_i}\rangle,\ |\theta_n\rangle,\ i=1,\ldots,q,\ n=0,\ldots,m_p\}$ of H_k, the completeness relationship has to be modified, namely,

$$\sum_{i=1}^{q}|\theta_{\epsilon_i}\rangle\langle\theta_{\epsilon_i}| + \sum_{m=0}^{m_p}|\theta_m\rangle\langle\theta_m| + \int |z,\alpha\rangle_{k_\mathcal{N}}\,{}_{k_\mathcal{N}}\langle z,\alpha|\,d\widetilde{\mu}(z) = 1, \tag{54}$$

where the measure reads:

$$d\widetilde{\mu}(z) = \frac{1}{\pi}\left(\sum_{m=0}^{\infty}\frac{|z|^{2m}}{\widetilde{\rho}_m}\right)\widetilde{\rho}(|z|^2)\,d^2z. \tag{55}$$

The associated moment problem becomes now:

$$\int_0^{\infty} y^m\widetilde{\rho}(y)\,dy = \widetilde{\rho}_m, \quad m\geq 0. \tag{56}$$

Note that here the degeneracy of the eigenvalue $z = 0$ can be any of the integers $\{s+1, \ldots, s+p+1\}$. The time evolution of the natural CS become:

$$U_k(t)|z,\alpha\rangle_{k_\mathcal{N}} = \exp(-itH_k)|z,\alpha\rangle_{k_\mathcal{N}} = e^{-itE_{m_p+1}}|z,\alpha+t\rangle_{k_\mathcal{N}}. \tag{57}$$

As regards the *intrinsic nonlinear coherent states* of H_k, it turns out that:

$$|z,\alpha\rangle_k = \left(\sum_{m=0}^{\infty}\frac{|z|^{2m}}{\rho_m}\right)^{-\frac{1}{2}}\sum_{m=0}^{\infty}e^{-i\alpha(E_m-E_0)}\frac{z^m}{\sqrt{\rho_m}}|\theta_m\rangle. \tag{58}$$

Note that, if equation (18) is satisfied, then the completeness relationship,

$$\sum_{i=1}^{q}|\theta_{\epsilon_i}\rangle\langle\theta_{\epsilon_i}| + \int |z,\alpha\rangle_k\, {}_k\langle z,\alpha|\, d\mu(z) = 1, \tag{59}$$

is also valid. Let us remark that the eigenvalue $z = 0$ is now (q+1)th degenerated.

For the *linear coherent states* of H_k we get:

$$|z,\alpha\rangle_{k_\mathcal{L}} = e^{-\frac{|z|^2}{2}}\sum_{m=0}^{\infty}e^{-i\alpha(E_m-E_0)}\frac{z^m}{\sqrt{m!}}|\theta_m\rangle. \tag{60}$$

Now the completeness relationship is simply:

$$\sum_{i=1}^{q}|\theta_{\epsilon_i}\rangle\langle\theta_{\epsilon_i}| + \frac{1}{\pi}\int |z,\alpha\rangle_{k_\mathcal{L}}\, {}_{k_\mathcal{L}}\langle z,\alpha|\, d^2z = 1. \tag{61}$$

Once again, the eigenvalue $z = 0$ is $(q+1)$-th degenerated. Notice finally that:

$$|z,\alpha\rangle_{k_\mathcal{L}} = D_{k_\mathcal{L}}|\theta_0\rangle = \exp(za^+_{k_\mathcal{L}} - \bar{z}a^-_{k_\mathcal{L}})|\theta_0\rangle. \tag{62}$$

It is worth to point out that the decompositions for the intrinsic and linear CS of H_k, which involve just energy eigenstates belonging to the subspace associated to the initial levels $\{E_n,\ n = 0, 1, \ldots\}$, are the same as the corresponding ones for H_0. This is due to the fact that the linear and intrinsic algebras of H_0 characterize as well the H_k on the subspace associated to the initial levels. Let us illustrate next our treatment by one simple non-trivial example.

7. Example

Let us apply the previous techniques to the trigonometric Pöschl-Teller potentials:

$$V_0(x) = \frac{(\nu-1)\nu}{2\cos^2(x)}, \qquad \nu > 1, \tag{63}$$

which are illustrated in gray in Figure 1. The eigenfunctions of the corresponding Hamiltonian, satisfying the boundary conditions $\psi_n(-\pi/2) = \psi_n(\pi/2) = 0$, as well as the eigenvalues are given by:

$$\psi_n(x) = \left[\frac{n!(n+\nu)\Gamma(\nu)\Gamma(2\nu)}{\sqrt{\pi}\,\Gamma(\nu+\frac{1}{2})\Gamma(n+2\nu)}\right]^{1/2}\cos^\nu(x)\ C_n^\nu(\sin(x)), \tag{64}$$

$$E_n = E(n) = \frac{(n+\nu)^2}{2}, \qquad n = 0, 1, 2, \ldots \tag{65}$$

Journal of Physics: Conference Series **128** (2008) 012023 doi:10.1088/1742-6596/128/1/012023

where $C_n^\nu(y)$ are the Gegenbauer polynomials.

The *intrinsic algebra* of H_0 is characterized by:

$$E(N_0) = \frac{(N_0+\nu)^2}{2}, \tag{66}$$

which leads to:

$$f(N_0) = E(N_0+1) - E(N_0) = N_0 + \nu + \tfrac{1}{2}. \tag{67}$$

Thus, the commutation relationships for the intrinsic algebra of H_0 become:

$$[N_0, a_0^\pm] = \pm a_0^\pm, \qquad [a_0^-, a_0^+] = N_0 + \nu + \tfrac{1}{2}, \qquad [H_0, a_0^\pm] = \pm(N_0 + \nu \mp \tfrac{1}{2}) a_0^\pm. \tag{68}$$

Notice that the first two relations essentially correspond to the su(1,1) algebra.

On the other hand, the *linear annihilation and creation operators* of H_0, which generate the Heisenberg-Weyl algebra, are given by:

$$a_{0_\mathcal{L}}^- = \sqrt{\frac{2}{N_0+2\nu+1}}\, a_0^-, \qquad a_{0_\mathcal{L}}^+ = a_0^+\sqrt{\frac{2}{N_0+2\nu+1}}, \tag{69}$$

such that

$$a_{0_\mathcal{L}}^+ a_{0_\mathcal{L}}^- = N_0, \qquad [a_{0_\mathcal{L}}^-, a_{0_\mathcal{L}}^+] = 1. \tag{70}$$

Concerning the coherent states of H_0, it turns out that the intrinsic and linear ones become respectively:

$$|z,\alpha\rangle_0 = \left[{}_0F_1(2\nu+1; 2|z|^2)\right]^{-\frac{1}{2}} \sum_{m=0}^{\infty} e^{-i\frac{\alpha}{2}m(m+2\nu)} \sqrt{\frac{2^m}{m!(2\nu+1)_m}}\, z^m |\psi_m\rangle, \tag{71}$$

$$|z,\alpha\rangle_{0_\mathcal{L}} = e^{-\frac{|z|^2}{2}} \sum_{m=0}^{\infty} e^{-i\frac{\alpha}{2}m(m+2\nu)} \frac{z^m}{\sqrt{m!}} |\psi_m\rangle. \tag{72}$$

The moment problem for the intrinsic CS involves:

$$\rho_m = \frac{m!\,(2\nu+1)_m}{2^m}, \tag{73}$$

leading to

$$\rho(y) = \frac{2^{\nu+2} y^\nu}{\Gamma(2\nu+1)} K_{2\nu}(2\sqrt{2y}). \tag{74}$$

The invariant measure is thus:

$$d\mu(z) = \frac{2^{\nu+2}|z|^{2\nu}}{\pi\Gamma(2\nu+1)}\, {}_0F_1(2\nu+1; 2|z|^2) K_{2\nu}(2\sqrt{2}\,|z|)\, d^2z, \tag{75}$$

while the reproducing kernel is:

$${}_0\langle z,\alpha|z',\alpha\rangle_0 = \left[{}_0F_1(2\nu+1; 2|z|^2)\ {}_0F_1(2\nu+1; 2|z'|^2)\right]^{-\frac{1}{2}} {}_0F_1(2\nu+1; 2\bar{z}z'). \tag{76}$$

For analyzing the SUSY partner Hamiltonians H_k, let us suppose that we use just seed solutions associated to non-physical eigenvalues of H_0, q of them becoming at the end physical

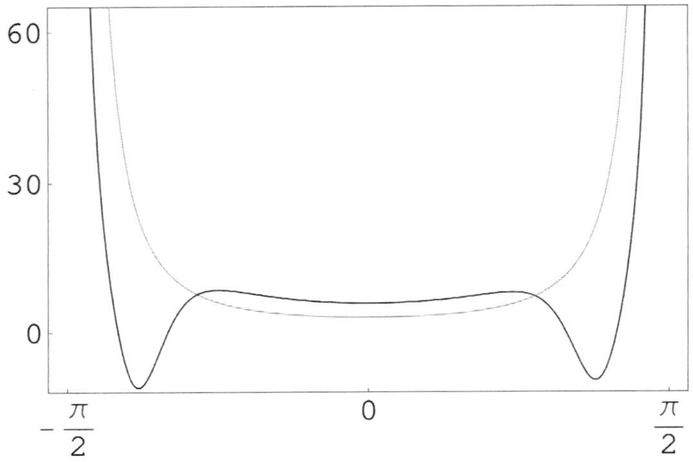

Figure 1. Illustration of the first-order SUSY partner potential $V_1(x)$ (black curve) of the Pöschl-Teller potential with $\nu = 3$ (gray curve). The potential $V_1(x)$ has an additional level, compared with $V_0(x)$, at $\epsilon = 4.4 < E_0 = 9/2$.

levels of H_k. An illustration of the corresponding potentials is plotted in black in Figure 1. The several algebraic structures of H_k are characterized by the following functions:

$$r_{\mathcal{I}}(n) = e^{i\alpha\left(n+\nu-\frac{1}{2}\right)} \sqrt{\frac{n(n+2\nu)}{2}}, \tag{77}$$

$$\frac{r_{\mathcal{N}}(n)}{r_{\mathcal{I}}(n)} = \frac{1}{2^k}\prod_{i=1}^{k} \sqrt{[(n+\nu-1)^2 - 2\epsilon_i][(n+\nu)^2 - 2\epsilon_i]}, \tag{78}$$

$$\frac{r_{\mathcal{L}}(n)}{r_{\mathcal{I}}(n)} = \sqrt{\frac{2}{n+2\nu}}. \tag{79}$$

The *natural CS* of H_k can now be straightforwardly evaluated:

$$|z,\alpha\rangle_{k\mathcal{N}} = \frac{1}{\sqrt{{}_0F_{4k+1}(2\nu+1,\dots,\nu-\sqrt{2\epsilon_i},\nu-\sqrt{2\epsilon_i}+1,\nu+\sqrt{2\epsilon_i},\nu+\sqrt{2\epsilon_i}+1,\dots;2^{2k+1}|z|^2)}} \times \sum_{m=0}^{\infty} \frac{e^{-\frac{i}{2}\alpha m(m+2\nu)}\sqrt{2^{m(2k+1)}}\, z^m|\theta_m\rangle}{\sqrt{m!(2\nu+1)_m}\prod_{i=1}^{k}\sqrt{\left(\nu-\sqrt{2\epsilon_i}\right)_m\left(\nu-\sqrt{2\epsilon_i}+1\right)_m\left(\nu+\sqrt{2\epsilon_i}\right)_m\left(\nu+\sqrt{2\epsilon_i}+1\right)_m}}. \tag{80}$$

The corresponding moment problem involves:

$$\widetilde{\rho}_m = \frac{m!(2\nu+1)_m}{2^{m(2k+1)}}\prod_{i=1}^{k}(\nu-\sqrt{2\epsilon_i})_m(\nu-\sqrt{2\epsilon_i}+1)_m(\nu+\sqrt{2\epsilon_i})_m(\nu+\sqrt{2\epsilon_i}+1)_m. \tag{81}$$

The *intrinsic and linear CS* of H_k can be obtained from the expressions (71-72) by the substitution $|\psi_m\rangle \to |\theta_m\rangle$.

8. Conclusions

In this paper we have shown that the intrinsic and linear algebras of H_0 are inherited by H_k on the subspace associated to the initial levels $\{E_n,\ n = 0, 1, \dots\}$. It turns out that the linear algebra for both SUSY partner Hamiltonians H_0 and H_k simplifies at maximum the intrinsic algebraic structure of our systems. We have also generalized successfully the natural algebra for H_k, introduced previously for the SUSY partners of the harmonic oscillator. Once determined the algebra characterizing our systems, it is straightforward to perform the CS analysis.

Acknowledgments

The authors acknowledge the support of Conacyt, projects 49253-F and 24233-50766-F, and research grants from NSERC of Canada.

References

[1] Glauber R J 2006 *ChemPhysChem* **7** 1618
[2] Klauder J R and Skagerstam B S Eds 1985 *Coherent States. Applications in Physics and Mathematical Physics*, (Singapore: World Scientific)
[3] Perelomov A 1986 *Generalized Coherent States and Their Applications* (Heidelberg: Springer-Verlag)
[4] Zhang W M, Feng D H and Gilmore R 1990 *Rev Mod Phys* **62** 867
[5] Ali S T, Antoine J P and Gazeau J P 2000 *Coherent States, Wavelets and Their Generalizations* (New York: Springer-Verlag)
[6] Dodonov V V 2002 *J Opt* B **4** R1;
Dodonov V V and Manko V I Eds 2003 *Theory of non-classical states of light* (New York: Taylor & Francis)
[7] Nieto L M 2006 *AIP Conf Proc* **809** 3
[8] Fernández D J 2006 *AIP Conf Proc* **809** 80
[9] Daoud M and Hussin V 2002 *J Phys* A: *Math Gen* **35** 7381
[10] Andrianov A A, Ioffe M V and Spiridonov V 1993 *Phys Lett* A **174** 273;
Andrianov A A, Ioffe M V, Cannata F and Dedonder J P 1995 *Int J Mod Phys* A **10** 2683
[11] Fernández D J, Negro J and del Olmo M A 1996 *Ann Phys* **252** 386
[12] Bagrov V G and Samsonov B F 1997 *Phys Part Nucl* **28** 374
[13] Fernández D J 1997 *Int J Mod Phys* A **12** 171;
Fernández D J, Hussin V and Mielnik B 1998 *Phys Lett* A **244** 309;
Cariñena J F, Ramos A and Fernández D J 2001 *Ann Phys* **292** 42
[14] Bagchi B, Ganguly A, Bhaumik D and Mitra A 1999 *Mod Phys Lett* A **14** 27;
Bagchi B K 2001 *Supersymmetry in quantum and classical mechanics* (Boca Raton: Chapman & Hall/CRC)
[15] Aoyama H, Sato M and Tanaka T 2001 *Phys Lett* B **503** 423;
Aoyama H, Sato M and Tanaka T 2001 *Nucl Phys* B **619** 105
[16] Andrianov A A and Sokolov A V 2003 *Nucl Phys* B **660** 25;
Andrianov A A and Cannata F 2004 *J Phys* A: *Math Gen* **37** 10297
[17] Leiva C and Plyushchay M S 2003 *JHEP* **10** 069;
Plyushchay M 2004 *J Phys* A: *Math Gen* **37** 10375
[18] Mielnik B and Rosas-Ortiz O 2004 *J Phys* A: *Math Gen* **37** 10007
[19] Ioffe M V and Nishnianidze D N 2004 *Phys Lett* A **327** 425
[20] González-López A and Tanaka T 2004 *Phys Lett* B **586** 117
[21] Sukumar C V 2005 *AIP Conf Proc* **744** 166
[22] Fernández D J and Fernández-García N 2005 *AIP Conf Proc* **744** 236
[23] Fernández D J, Hussin V and Nieto L M 1994 *J Phys* A: *Math Gen* **27** 3547
[24] Mielnik B 1984 *J Math Phys* **25** 3387
[25] Abraham P B and Moses H E 1980 *Phys Rev* A **22** 1333
[26] Fernández D J, Nieto L M and Rosas-Ortiz O 1995 *J Phys* A: *Math Gen* **28** 2693
[27] Rosas-Ortiz J O 1996 *J Phys* A: *Math Gen* **29** 3281
[28] Fernández D J and Hussin V 1999 *J Phys* A: *Math Gen* **32** 3303
[29] Fernández D J, Hussin V and Rosas-Ortiz O 2007 *J Phys* A: *Math Theor* **40** 6491

Journal of Physics: Conference Series **128** (2008) 012024 doi:10.1088/1742-6596/128/1/012024

The Fedosov deformation quantization with the induced symplectic connection

Jaromir Tosiek

Institute of Physics, Technical University of Lodz, Wolczanska 219, 93-005 Lodz, Poland

E-mail: tosiek@p.lodz.pl

Abstract. A symplectic connection on a cotangent bundle $\mathcal{T}^*\mathcal{M}$ induced by a Levi-Civita connection on a configuration space $\mathcal{M}$ is constructed. General properties of an Abelian connection built from the induced symplectic connection are presented. An example of a finite Abelian connection determined by the induced symplectic connection has been found.

1. Introduction

Quantum mechanics is usually formulated in terms of theory of a Hilbert space and linear operators acting in it. Unfortunately, serious problems appear when one tries to describe quantum effects in systems with curved configuration spaces. On the other side, classical physics works perfectly on arbitrary differentiable manifolds. Hence the formulation of quantum theory based on differential geometry may eliminate obstacles.

An alternative construction of quantum mechanics on a phase space $\mathbb{R}^{2n}$ was proposed Moyal [1]. In his paper ideas of Weyl [2], Wigner [3] and Groenewold [4] were developed. Generalization of Moyal's results for systems with nontrivial phase spaces was presented by Bayen *et al* [5, 6]. Their fresh look at the problem gave birth to deformation quantization.

One of practical realisations of the Bayen quantization programme was invented by Fedosov [7, 8] and is known as the Fedosov quantization. Roughly speaking his algorithm is composed of four steps. We start from a phase space of a system, which is a symplectic manifold and equip it with some symplectic connection. Then the symplectic connection is lifted to the bundle of formal Weyl algebras. After that we introduce an Abelian connection in the Weyl algebra bundle, which is determined by the symplectic connection. Next we identify functions representing observables with smooth sections of the subalgebra of the flat sections of the Weyl algebra bundle. By a "quantum" $*$-product of these functions we understand a projection of a product of smooth sections representing them.

Unlike the Riemannian geometry, a symplectic manifold may be equipped with many symplectic connections. Hence the natural question arises, which of them is the most physical and how for it Fedosov quantization works. In our contribution we propose, following [9], the symplectic connection induced by the Levi- Civita connection on the configuration space $\mathcal{M}$. For such a symplectic connection we analyse properties of the Abelian connection. Another way leading to the induced symplectic connection was indicated by Bordemann, Neumaier and Waldmann [10, 11].

In all formulas in which summation limits are obvious we use the Einstein summation convention.

Journal of Physics: Conference Series **128** (2008) 012024 doi:10.1088/1742-6596/128/1/012024

2. Induced symplectic connection

In this section we present a construction of some symplectic connection determined by the Riemannian structure of a configuration space $\mathcal{M}$. Reader interested in details is pleased to see [9].

Assume that the configuration space $\mathcal{M}$, $\dim \mathcal{M} = n$, is a differentiable manifold endowed with a Riemannian metric g and $\mathcal{T}^*\mathcal{M}$ is a cotangent bundle over it. The bundle $\mathcal{T}^*\mathcal{M}$ is $2n$-D symplectic manifold. We cover it with so called proper Darboux atlas.

Definition 1. *Let $\{(U_\varrho, \phi_\varrho)\}_{\varrho\in I}$ be an atlas on the symplectic manifold $\mathcal{T}^*\mathcal{M}$ such that in every chart the coordinates q^i, $1 \leq i \leq n$ determine points on the basic manifold $\mathcal{M}$ and $q^{i+n} = p_i$, $1 \leq i \leq n$, denote momenta in natural coordinates. Every atlas of this form is called the* **proper Darboux atlas** *and every chart of this atlas the* **proper Darboux chart**. *The transitions functions are the point transformations*

$$Q^k = Q^k(q^i), \quad P_i = \frac{\partial q^k}{\partial Q^i} p_k.$$

We equip the cotangent bundle $\mathcal{T}^*\mathcal{M}$ with a metric structure. Locally in a proper Darboux chart, the metric tensor

$$\tilde{g}_{jk} = \begin{pmatrix} -2p_i\Gamma^i_{jk} & \mathbf{1} \\ \mathbf{1} & \mathbf{0} \end{pmatrix}. \tag{1}$$

By $\mathbf{1}$ and $\mathbf{0}$ we denote the identity and zero matrices $n \times n$ respectively. Coefficients Γ^i_{jk} are components of the Levi- Civita connection of the metric g on $\mathcal{M}$ i.e.

$$\Gamma^i_{jk} = \frac{1}{2} g^{il} \left(\frac{\partial g_{lk}}{\partial q^j} + \frac{\partial g_{lj}}{\partial q^k} - \frac{\partial g_{jk}}{\partial q^l} \right), \quad 1 \leq i, j, k, l \leq n. \tag{2}$$

It is interesting, that the metric tensor (1) appears as a consequence of the operator ordering in Fedosov quantization [12].

Theorem 1. *Let $(\mathcal{M}, g)$ be a Riemannian manifold. Then $(\mathcal{T}^*\mathcal{M}, \tilde{g})$ is also a Riemannian manifold.*

Definition 2. *The* **symplectic connection** *on a symplectic manifold $(\mathcal{W}, \omega)$ is a torsion free connection satisfying locally a set of conditions*

$$\omega_{ij;k} = 0, \quad 1 \leq i, j, k \leq \dim \mathcal{W}.$$

The symbol ω denotes a symplectic form.

In every chart $(\mathcal{U}_z, \phi_z)$ on $\mathcal{W}$

$$\omega_{ij;k} = \frac{\partial \omega_{ij}}{\partial q^k} - \gamma^l_{ik}\omega_{lj} - \gamma^l_{jk}\omega_{il} = 0, \tag{3}$$

$$\gamma^l_{ik} = \gamma^l_{ki}, \quad \text{(torsion free)}.$$

In Darboux coordinates the set of equations (3) takes a simple form

$$\omega_{ij;k} = -\gamma^l_{ik}\omega_{lj} - \gamma^l_{jk}\omega_{il} = -\gamma_{jik} + \gamma_{ijk} = 0,$$

where

$$\gamma_{ijk} \stackrel{\text{def.}}{=} \gamma^l_{jk}\omega_{li}.$$

Hence the connection is symplectic if and only if in every Darboux chart all of components γ^l_{jk} are totally symmetric with respect to indices $\{i, j, k\}$.

Definition 3. *The symplectic manifold* $(\mathcal{W}, \omega)$ *equipped with the symplectic connection* γ *is called* **the Fedosov manifold** $(\mathcal{W}, \omega, \gamma)$.

Every symplectic manifold admits a symplectic connection, but it is not uniquely defined. Locally the difference between two symplectic connections $\gamma_{ijk} - \tilde{\gamma}_{ijk} = r_{ijk}$ is a tensor completely symmetric in indices $\{i, j, k\}$.

From now on the symplectic manifold $(\mathcal{W}, \omega)$ will be the cotangent bundle $\mathcal{T}^*\mathcal{M}$ with its symplectic structure. As we mentioned (compare Theorem 1), the space $\mathcal{T}^*\mathcal{M}$ is also the Riemannian manifold and the Levi-Civita connection $\overline{\Gamma}$ on it is determined by the tensor $\tilde{g}$.

Let us lower the upper index of Christoffel symbols $\overline{\Gamma}^i_{jk}$ so $\overline{\Gamma}_{ijk} \stackrel{\text{def}}{=} \omega_{il}\overline{\Gamma}^l_{jk}$. Then in any Darboux chart

$$\gamma_{ijk} \stackrel{\text{def.}}{=} \frac{1}{3}\left(\overline{\Gamma}_{ijk} + \overline{\Gamma}_{jik} + \overline{\Gamma}_{kij}\right) \tag{4}$$

is the symplectic connection on $\mathcal{T}^*\mathcal{M}$ induced by the Levi-Civita connection on $\mathcal{T}^*\mathcal{M}$. Moreover, since the metric tensor $\tilde{g}$ (see formula (1)) is a function of the Levi- Civita connection on the configuration space $\mathcal{M}$ then in fact the symplectic connection γ_{ijk} is determined by the connection on the base space $\mathcal{M}$.

Indeed, in proper Darboux coordinates the coefficients of the induced symplectic connection on $\mathcal{T}^*\mathcal{M}$

$$\gamma_{\bar{i}jk} = \Gamma^i_{jk}, \quad \gamma_{\bar{i}\bar{j}k} = 0, \quad \gamma_{\bar{i}\bar{j}\bar{k}} = 0,$$

$$\gamma_{ijk} = \frac{1}{3}p_l\left(\frac{\partial \Gamma^l_{jk}}{\partial q^i} + \frac{\partial \Gamma^l_{ij}}{\partial q^k} + \frac{\Gamma^l_{ik}}{\partial q^j} - 2\Gamma^l_{mi}\Gamma^m_{jk} - 2\Gamma^l_{mk}\Gamma^m_{ij} - 2\Gamma^l_{mj}\Gamma^m_{ik}\right). \tag{5}$$

We use the convention $\bar{i} = i + n$.

The coefficients $\gamma_{\bar{i}jk}$ of the induced symplectic connection are functions of spatial coordinates only. Terms γ_{ijk} depend linarly on momenta and are also functions of spatial coordinates. The curvature tensor of the induced symplectic connection contains three kinds of elements: R_{ijkl}, $R_{ijk\bar{l}}$ and $R_{\bar{i}jkl}$. Exact formulas defining them can be seen in [9]. Relations (5) are invariant under proper Darboux transformations.

3. Abelian connection in the Fedosov formalism

In this section we present a brief review of the Fedosov deformation. Its expanded version can be found for example in original Fedosov works [7, 8].

A Fedosov manifold $(\mathcal{W}, \omega, \gamma)$ is given. We deal with formal series

$$a \stackrel{\text{def.}}{=} \sum_{k=0}^{\infty}\sum_{l=0}^{\infty} \hbar^k a_{k,i_1\ldots i_l} X^{i_1}_{\mathrm{p}} \ldots X^{i_l}_{\mathrm{p}}. \tag{6}$$

In the upper expression $\hbar$ denotes some positive parameter identified by physicists with the Dirac constant. $X^1_{\mathrm{p}}, \ldots, X^{2n}_{\mathrm{p}}$ are components of an arbitrary vector $\mathbf{X}_{\mathrm{p}}$ belonging to the tangent space $T_{\mathrm{p}}\mathcal{W}$ at the point p. The components $X^1_{\mathrm{p}}, \ldots, X^{2n}_{\mathrm{p}}$ are written in the natural basis. By $a_{k,i_1\ldots i_l}$ we denote components of a covariant tensor symmetric with respect to indices $\{i_1 \ldots, i_l\}$ taken in the basis $dq^{i_1} \odot \ldots \odot dq^{i_l}$. The part of the series a standing at $\hbar^k$ and containing l components of the vector $\mathbf{X}_{\mathrm{p}}$ will be denoted by $a[k, l]$ so that $a = \sum_{k=0}^{\infty}\sum_{l=0}^{\infty}\hbar^k a[k, l]$. The **degree** $\deg(a[k, l])$ of the component $a[k, l]$ is the sum $2k + l$. The degree of the series a is the maximal degree of its nonzero components $a[k, l]$.

Let $P^*_{\mathrm{p}}\mathcal{W}[[\hbar]]$ be the set of all elements a of the kind (6) at the point p.

Definition 4. *The product* $\circ : P^*_{\mathrm{p}}\mathcal{W}[[\hbar]] \times P^*_{\mathrm{p}}\mathcal{W}[[\hbar]] \to P^*_{\mathrm{p}}\mathcal{W}[[\hbar]]$ *of two elements* $a, b \in P^*_{\mathrm{p}}\mathcal{W}[[\hbar]]$ *is the mapping*

$$a \circ b \stackrel{\text{def.}}{=} \sum_{t=0}^{\infty} \frac{1}{t!}\left(\frac{i\hbar}{2}\right)^t \omega^{i_1 j_1} \cdots \omega^{i_t j_t} \frac{\partial^t a}{\partial X^{i_1}_{\mathrm{p}} \ldots \partial X^{i_t}_{\mathrm{p}}} \frac{\partial^t b}{\partial X^{j_1}_{\mathrm{p}} \ldots \partial X^{j_t}_{\mathrm{p}}}. \tag{7}$$

The tensor ω^{ij} and the symplectic form ω_{jk} are related by $\omega^{ij}\omega_{jk} = \delta^i_k$. The pair $(P^*_{\mathrm{p}}\mathcal{W}[[\hbar]], \circ)$ is a noncommutative associative algebra called the **Weyl algebra**.

A **Weyl bundle** is a triplet $(\mathcal{P}^*\mathcal{W}[[\hbar]], \pi, \mathcal{W})$, where $\mathcal{P}^*\mathcal{W}[[\hbar]] \stackrel{\text{def.}}{=} \bigcup_{\mathrm{p}\in\mathcal{W}}(P^*_{\mathrm{p}}\mathcal{W}[[\hbar]], \circ)$ is a differentiable manifold called the total space, $\mathcal{W}$ is the base space and $\pi : \mathcal{P}^*\mathcal{W}[[\hbar]] \to \mathcal{W}$ the projection. A Weyl bundle is a vector bundle in which the typical fibre is also an algebra.

Definition 5. An m**-differential form** *with value in the Weyl bundle is a form written locally*

$$a = \sum_{k=0}^{\infty}\sum_{l=0}^{\infty} \hbar^k a_{k,i_1\ldots i_l,j_1\ldots j_m}(q^1, \ldots, q^{2n}) X^{i_1} \ldots X^{i_l} dq^{j_1} \wedge \cdots \wedge dq^{j_m} \tag{8}$$

where $0 \leq m \leq 2n$. *Now* $a_{k,i_1\ldots i_l,j_1\ldots j_m}(q^1, \ldots, q^{2n})$ *are components of smooth tensor fields on* $\mathcal{W}$ *and* $C^\infty(T\mathcal{W}) \ni \mathbf{X} \stackrel{\text{locally}}{=} X^i \frac{\partial}{\partial q^i}$ *is a smooth vector field.*

Let Λ^m be a smooth field of m-forms on the symplectic manifold $\mathcal{W}$. Forms of the kind (8) are smooth sections of $\mathcal{P}^*\mathcal{W}[[\hbar]] \otimes \Lambda \stackrel{\text{def.}}{=} \oplus_{m=0}^{2n}(\mathcal{P}^*\mathcal{W}[[\hbar]] \otimes \Lambda^m)$. For simplicity we will omit the variables $(q^1, \ldots, q^{2n})$.

Definition 6. The commutator *of forms* $a \in C^\infty(\mathcal{P}^*\mathcal{W}[[\hbar]] \otimes \Lambda^{m_1})$ *and* $b \in C^\infty(\mathcal{P}^*\mathcal{W}[[\hbar]] \otimes \Lambda^{m_2})$ *is the form* $[a, b] \in C^\infty(\mathcal{P}^*\mathcal{W}[[\hbar]] \otimes \Lambda^{m_1+m_2})$

$$[a, b] \stackrel{\text{def.}}{=} a \circ b - (-1)^{m_1 \cdot m_2} b \circ a. \tag{9}$$

A form $a \in C^\infty(\mathcal{P}^*\mathcal{W}[[\hbar]] \otimes \Lambda)$ is called **central**, if for every $b \in C^\infty(\mathcal{P}^*\mathcal{W}[[\hbar]] \otimes \Lambda)$ the commutator $[a, b]$ vanishes. Only forms not containing X^i's are central.

Definition 7. *The antiderivation operator* $\delta : C^\infty(\mathcal{P}^*\mathcal{M}[[\hbar]] \otimes \Lambda^m) \to C^\infty(\mathcal{P}^*\mathcal{M}[[\hbar]] \otimes \Lambda^{m+1})$ *is defined by*

$$\delta a \stackrel{\text{def.}}{=} dq^k \wedge \frac{\partial a}{\partial X^k}.$$

Definition 8. *The operator* $\delta^{-1} : C^\infty(\mathcal{P}^*\mathcal{W}[[\hbar]] \otimes \Lambda^m) \to C^\infty(\mathcal{P}^*\mathcal{W}[[\hbar]] \otimes \Lambda^{m-1})$ *is*

$$\delta^{-1}a = \begin{cases} \frac{1}{l+m} X^k \frac{\partial}{\partial q^k} \rfloor a & \text{for} \quad l + m > 0 \\ 0 & \text{for} \quad l + m = 0, \end{cases} \tag{10}$$

where l *is the degree of* a *in* X^i*'s, i.e. the number of* X^i*'s.*

Definition 9. The exterior covariant derivative ∂_ξ *of the form* $a \in C^\infty(\mathcal{P}^*\mathcal{W}[[\hbar]] \otimes \Lambda^m)$ *determined by a connection* 1*-form* $\xi \in C^\infty(\mathcal{P}^*\mathcal{W}[[\hbar]] \otimes \Lambda^1)$ *is the linear operator*

$$\partial_\xi : C^\infty(\mathcal{P}^*\mathcal{W}[[\hbar]] \otimes \Lambda^m) \to C^\infty(\mathcal{P}^*\mathcal{W}[[\hbar]] \otimes \Lambda^{m+1})$$

defined in a Darboux chart by the formula

$$\partial_\xi a \stackrel{\text{def.}}{=} da + \frac{1}{i\hbar}[\xi, a]. \tag{11}$$

In the case of a symplectic connection, we use γ instead of ξ and put $\gamma = \frac{1}{2}\gamma_{ijk}X^iX^jdq^k$.

The curvature form R_γ of a connection 1-form γ in a Darboux chart can be expressed by the formula

$$R_\gamma = d\gamma + \frac{1}{2i\hbar}[\gamma,\gamma] = d\gamma + \frac{1}{i\hbar}\gamma\circ\gamma. \tag{12}$$

A crucial role in the Fedosov deformation quantization is played by an Abelian connection $\tilde{\Gamma}$. By the **Abelian** connection we mean a connection $\tilde{\Gamma}$ whose curvature form $R_{\tilde{\Gamma}}$ is central so $\partial_{\tilde{\Gamma}}(\partial_{\tilde{\Gamma}}a) = 0$ for every $a \in C^\infty(\mathcal{P}^*\mathcal{W}[[\hbar]]\otimes\Lambda)$.

The Abelian connection proposed by Fedosov is of the form

$$\tilde{\Gamma} = \omega_{ij}X^idq^j + \gamma + r. \tag{13}$$

Its curvature is

$$R_{\tilde{\Gamma}} = -\frac{1}{2}\omega_{j_1j_2}dq^{j_1}\wedge dq^{j_2} + R_\gamma - \delta r + \partial_\gamma r + \frac{1}{i\hbar}r\circ r. \tag{14}$$

The requirement that the central curvature 2-form $R_{\tilde{\Gamma}} = -\frac{1}{2}\omega_{j_1j_2}dq^{j_1}\wedge dq^{j_2}$ means that r must satisfy the equation

$$\delta r = R_\gamma + \partial_\gamma r + \frac{1}{i\hbar}r\circ r. \tag{15}$$

Theorem 2. *The equation (15) has a unique solution*

$$r = \delta^{-1}R_\gamma + \delta^{-1}\left(\partial_\gamma r + \frac{1}{i\hbar}r\circ r\right) \tag{16}$$

fulfilling the following conditions

$$\delta^{-1}r = 0, \quad 3 \le \deg(r). \tag{17}$$

We work only with the Abelian connection of the form (13) with the correction r defined by (16) and fulfilling (17).

Here some of properties of the Abelian connection are pointed out. More information can be found in [13].

(i) The correction

$$r = \delta^{-1}R_\gamma + \sum_{z=4}^{\infty}\sum_{k=0}^{[\frac{z-1}{4}]}\hbar^{2k}\, r_m[2k, z-4k]dq^m, \tag{18}$$

so r contains only terms with at least one vector component X^i and only even powers of $\hbar$. By $[z]$ we mean the floor of an integer z.

(ii) If $R_\gamma \neq R_{\overline{\gamma}}$ then corrections r determined by connections γ and $\overline{\gamma}$ respectively are different. On the contrary, two different symplectic connections γ and $\overline{\gamma}$ with the same curvature may generate the same Abelian correction r.

(iii) Explicit formulas on the part $r[z]$ of the degree z are [14, 15]

$$r[3] = \delta^{-1}R_\gamma,$$

$$r[z] = \delta^{-1}\left(\partial_\gamma r[z-1] + \frac{1}{i\hbar}\sum_{j=3}^{z-2} r[j]\circ r[z+1-j]\right), \quad 4 \le z. \tag{19}$$

Definition 10. $\mathcal{P}^*\mathcal{W}[[\hbar]]_{\tilde{\gamma}} \subset C^\infty(\mathcal{P}^*\mathcal{W}[[\hbar]]\otimes\Lambda^0)$ *is the subalgebra consisting of flat sections, i.e. sections such that* $\partial_{\tilde{\gamma}}a = 0$.

Theorem 3. *For any* $a_0 \in C^\infty(\mathcal{W})$ *there exists a unique section* $a \in \mathcal{P}^*\mathcal{W}[[\hbar]]_{\tilde{\gamma}}$ *such that* $\sigma(a) = a_0$.

It can be proved [7, 8] that

$$a = a_0 + \delta^{-1}\left(\partial_\gamma a + \frac{1}{i\hbar}[r, a]\right). \tag{20}$$

By a star product '$*$' of functions $a_0, b_0 \in C^\infty(\mathcal{M})$ we mean

$$a_0 * b_0 \stackrel{\text{def.}}{=} \sigma(\sigma^{-1}(a_0) \circ \sigma^{-1}(b_0)). \tag{21}$$

The $*$- product (21) is interpreted as the quantum multiplication of observables.

4. Abelian connection of the induced symplectic connection

In this paragraph we consider the situation when the Abelian connection is determined by the induced symplectic connection. We present some general remarks about a shape of a formal series representing the Abelian connection in that case and look for examples of finite Abelian connections.

After easy but rather long computations it can be checked that the correction r to the Abelian connection does not contain the deformation parameter $\hbar$. Moreover, each term $r[z]$, $3 \leq z$ is a sum of three kinds of elements:

(i) $r_{i_1 \ldots i_z, j}(q^1, \ldots, q^n) X^{i_1} \cdots X^{i_z} dp_j, \ \ 1 \leq i_1, \ldots, i_z \leq n,$
(ii) $r_{i_1 \ldots i_{z-1} j, i_z}(q^1, \ldots, q^n) X^{i_1} \cdots X^{i_{z-1}} X^j dq^{i_z}, \ \ 1 \leq i_1, \ldots, i_z \leq n, \ \ n < j \leq 2n,$
(iii) $\sum_{k=1}^n p_k \cdot r_{k, i_1 \ldots i_z, j}(q^1, \ldots, q^n) X^{i_1} \cdots X^{i_z} dq^j, \ \ 1 \leq i_1, \ldots, i_z, j \leq n.$

Coefficients of the 1st and 2nd kind depend only on spatial coordinates. They are related by a condition $\delta^{-1} r = 0$. Terms of the 3rd type are linear functions in momenta. They also fulfill the requirement $\delta^{-1} r = 0$.

It seems to be extremely difficult to find the general expression of the Abelian connection in the Fedosov formalism. Even in case when the Abelian connection is determined by the induced symplectic connection, this problem has not been completely solved yet.

Hence we look for examples of finite Abelian connections. By definition, we say that r is a **finite formal series** when $\deg(r) = d, \ d \in \mathbb{N}$. The series representing a finite Abelian connection can be easily found even by a computer programme without any additional considerations.

Now we present the only one known example of the finite Abelian connection.

Let $(\mathcal{W}, \omega, \gamma)$ be a Fedosov manifold. Locally the nonvanishing components of a symplectic connection are

$$\gamma_{l_1 l_2 l_3}(q^{l_4}, \ldots, q^{l_r}), \ \ 1 \leq l_1, \ldots, l_r \leq \dim \mathcal{W}$$

such that all of Poisson brackets

$$\{q^{l_i}, q^{l_j}\}_P = 0, \ 1 \leq i, j \leq r.$$

This connection can be curved only if $4 \leq \dim \mathcal{W}$. The Ricci tensor $K_{ij} \stackrel{\text{def.}}{=} \omega^{kl} R_{likj}$ of γ disappears.

Then the correction

$$r[z] = (\delta^{-1}\partial_\gamma)^{z-3} \delta^{-1} R_\gamma. \tag{22}$$

As it was shown [13], the sufficient and necessary condition for r to be a finite series is

$$\exists_{z \geq 4} \ \ (\partial_\gamma \delta^{-1})^{z-3} R_\gamma = 0. \tag{23}$$

It would be interesting to apply this example for the induced symplectic connection. Let us consider first the case when the configuration space is 2-dimensional. As it was proved by Plebański *et al* [9], the Fedosov manifold $(\mathcal{T}^*\mathcal{M},\omega,\gamma)$ with the induced symplectic connection γ is Ricci flat (i.e. $K_{ij}=0$) if and only if the Riemannian manifold $(\mathcal{M},g)$ is Ricci flat (R_{ij}=0).

The only example of the Ricci flat 2-D Riemannian manifold is just a flat space so this is not the case we look for.

Let $\mathcal{M}$ be the 3-D Riemannian manifold covered by an atlas $\{(U_\rho,\phi_\rho)\}_{\rho\in I}$. Assume that in some chart (U_ρ,ϕ_ρ) the nonvanishing Levi-Civita connection components are only $\Gamma^3_{11}(q^1,q^2),\Gamma^3_{22}(q^1,q^2)$ and $\Gamma^3_{12}(q^1,q^2)$. Then from (5)

$$\gamma_{116}=\Gamma^3_{11}(q^1,q^2)\ ,\ \ \gamma_{126}=\Gamma^3_{12}(q^1,q^2)\ ,\ \ \gamma_{226}=\Gamma^3_{22}(q^1,q^2),$$

$$\gamma_{111}=q^6\frac{\partial\Gamma^3_{11}}{\partial q^1}\ ,\ \gamma_{112}=\frac{1}{3}q^6\left(2\frac{\partial\Gamma^3_{12}}{\partial q^1}+\frac{\partial\Gamma^3_{11}}{\partial q^2}\right)\ ,\ \gamma_{122}=\frac{1}{3}q^6\left(2\frac{\partial\Gamma^3_{12}}{\partial q^2}+\frac{\partial\Gamma^3_{22}}{\partial q^1}\right)\ ,\ \gamma_{222}=q^6\frac{\partial\Gamma^3_{22}}{\partial q^2}.$$

In the chart (U_ρ,ϕ_ρ) the symplectic curvature 2-form $R_\gamma=d\Gamma$ and the correction

$$r[z]=(\delta^{-1}d)^{z-3}\delta^{-1}R_\gamma,\ 3\leq z.$$

The term $r[z]$ consists of four kinds of elements:

(i) $q^6\sum_{i=0}^{z}f_{i,z-i,1}(q^1,q^2)(X^1)^i(X^2)^{z-i}dq^1\ ,\ \ q^6\sum_{i=0}^{z}f_{i,z-i,2}(q^1,q^2)(X^1)^i(X^2)^{z-i}dq^2$

(ii) $\sum_{i=0}^{z-1}g_{i,z-i-1,1}(q^1,q^2)(X^1)^i(X^2)^{z-i-1}X^6dq^1\ ,\ \sum_{i=0}^{z-1}g_{i,z-i-1,2}(q^1,q^2)(X^1)^i(X^2)^{z-i-1}X^6dq^2.$

Straightforward calculations lead to the conclusion that functions $f_{i,z-i,1}(q^1,q^2),f_{i,z-i,2}$ are linear combinations of partial derivatives of $\Gamma^3_{11}(q^1,q^2),\Gamma^3_{22}(q^1,q^2)$ and $\Gamma^3_{12}(q^1,q^2)$ of the total degree $z+1$. Moreover, coefficients $g_{i,z-i-1,1}(q^1,q^2),g_{i,z-i-1,2}(q^1,q^2)$ are linear combinations of partial derivatives of $\Gamma^3_{11}(q^1,q^2),\Gamma^3_{22}(q^1,q^2)$ and $\Gamma^3_{12}(q^1,q^2)$ of the total degree z.

Hence we deduce that the Abelian connection $\tilde{\gamma}$ on U_ρ is represented by a finite formal series for example if $\Gamma^3_{11}(q^1,q^2),\Gamma^3_{22}(q^1,q^2)$ and $\Gamma^3_{12}(q^1,q^2)$ are polynomials in q^1,q^2.

5. Results and Perspectives

The fact, that every symplectic manifold may be equipped with many symplectic connections, offers some freedom in frames of the Fedosov deformation quantization. In our opinion the most physically justifiable symplectic connection is the induced symplectic connection built according to the idea presented in [9].

In the current contribution we presented basic properties and the general shape of the Abelian connections generated by the induced symplectic connections. But even in this narrow class of Abelian connections we may search for especially convenient ones namely finite formal series. We found the example of the finite Abelian connection. The correction r determined by the formula (22), for which (23) holds, is the finite series. Unfortunately this example works only on a Ricci flat space and it cannot be considered as an universal solution. Therefore we will search for finite Abelian connections built over non Ricci flat configuration spaces.

The Abelian connection is a step on the way to the $*$-product. It would be valuable to find the explicit form of the $*$- product for observables on a symplectic manifold with the induced symplectic connection.

References

[1] Moyal J E 1949 *Proc. Camb. Phil. Soc.* **45** 99
[2] Weyl H 1931 *The Theory of Groups and Quantum Mechanics* New York: Dover
[3] Wigner E P 1932 *Phys. Rev.* **40** 749
[4] Groenewold H J 1946 *Physica* **12** 405

Journal of Physics: Conference Series **128** (2008) 012024 doi:10.1088/1742-6596/128/1/012024

[5] Bayen F, Flato M, Fronsdal C, Lichnerowicz A and Sternheimer D 1977 *Lett. Math. Phys.* **1** 521
[6] Bayen F, Flato M, Fronsdal C, Lichnerowicz A and Sternheimer D 1978 *Ann. Phys. NY* **11**161; *Ann. Phys. NY* **111** 111
[7] Fedosov B 1994 *J. Diff. Geom.* **40** 213
[8] Fedosov B 1996 *Deformation Quantization and Index Theory* Berlin: Akademie Verlag
[9] Plebański J F, Przanowski M and Turrubiates F 2001 *Acta Phys. Pol.* **B 32** 3
[10] Bordemann M, Neumaier N and Waldmann S 1998 *Commun. Math. Phys.* **198** 363
[11] Bordemann M, Neumaier N and Waldmann S 1999 *J. Geom. Phys.* **29** 199
[12] Przanowski M and Tosiek J 1999 *Acta Phys. Pol.* **B 30** 179
[13] Tosiek J 2007 *Acta Phys. Pol.* **B 38** 3069
[14] Bordemann M, Neumaier N and Waldmann S 1998 *Comm. Math. Phys.* **198** 363
[15] Kravchenko O *Preprint* SG/0008157

Journal of Physics: Conference Series **128** (2008) 012025 doi:10.1088/1742-6596/128/1/012025

Generalized Integrability and Volume-preserving Diffeomorphisms

C. Adam[1], J. Sánchez-Guillén[1] and A. Wereszczyński[2]

[1] Departamento de Fisica de Particulas, Universidad de Santiago and Instituto Galego de Fisica de Altas Enerxias (IGFAE) E-15782 Santiago de Compostela, Spain

[2] Institute of Physics, Jagiellonian University, Reymonta 4, 30-059 Kraków, Poland

E-mail: adam@fpaxp1.usc.es joaquin@fpaxp1.usc.es wereszczynski@th.if.uj.edu.pl

Abstract. The concepts of generalized zero curvature conditions and integrability in higher dimensions are briefly reviewed, where integrability in this context always means the existence of infinitely many conservation laws. It turns out that the conservation laws provided by the generalized integrability are, under certain additional assumptions, generated by the volume-preserving diffeomorphisms on target space. The possible conservation laws for field theories with a three-dimensional target space are classified. Further, an explicit example is discussed in some detail.

1. Introduction

Non-linear field theories are important in many fields of physics, with applications ranging from elementary particle theory to condensed matter physics. One feature of these theories which adds to their relevance is the possibility for the existence of extended static (solitons) or stationary (Q-balls) solutions. On the other hand, non-linear field theories are notoriously difficult to analyse, where the degree of difficulty strongly depends on the dimension of the base space (space-time) on which the fields are defined. In 1+1 dimensions, an ample mathematical apparatus has been developed for the analysis of non-linear theories, among which there are the inverse scattering method, Bäcklund transformations, or the zero curvature representation for integrable systems, which generalizes the Lax pair representation of finite-dimensional integrable systems. Integrability, that is, the existence of infinitely many conserved quantities, is related to all of these methods, and seems to be crucial in the analytical treatment of nonlinear theories, like, e.g., the explicit construction of solutions.

In higher dimensions, much less is known about non-linear field theories. A general concept of integrability has not yet been developed there. One may have, however, theories which contain an integrable subsector like, e.g., in the non-linear sigma model in 2+1 dimensions, where the integrable subsector is formed by the holomorphic solitons of Belavin and Polyakov. One generalization of the zero curvature representation of Shabat and Zakharov to higher dimensions has been proposed in [1], and it was demonstrated there that this proposal leads to non-linear field theories which have either infinitely many conservation laws in the full theory, or which contain integrable subsectors, defined by some additional constraint equations on the fields, such that the solutions belonging to this subsectors have infinitely many conservation laws. This zero

Journal of Physics: Conference Series **128** (2008) 012025 doi:10.1088/1742-6596/128/1/012025

curvature representation, therefore, realizes the concept of integrability in higher-dimensional non-linear field theories in a specific and well-defined manner.

These methods have later been applied to specific models and to the analytic construction of both static and time-dependent solutions. For models with infinitely many conservation laws (the so-called AFZ model and related models), static and time-dependent solutions have been constructed, e.g., in [2], [3], [4], [5], and in [6], [7], respectively. Solutions in integrable subsectors of models which are, themselves, not integrable, have been constructed, e.g., in [8], [9]–[11] (the Nicole model and versions thereof) and in [12], [13] (diverse models on base space S^3). All these models share the property that their target space has dimension two.

A wellknown nonlinear field theory with three-dimensional target space is the Skyrme model [14] with target space SU(2) (or equivalently the three-sphere S^3). Further, this model contains an integrable subsector, and the simplest Skyrmion (i.e., the simplest soliton of the Skyrme model with baryon number equal to one) belongs to this integrable subsector, see [16].

In many cases, it turns out that most of the new conserved currents in models and their subsectors are Noether currents and generalizations thereof, i.e., they are related to transformations of the target space variables (see [17]). So a direct, geometric approach has been succesfully undertaken to find those currents, first for models with two-dimensional target spaces, [18], [19], and later also for three-dimensional target spaces, [20].

It is the purpose of this paper to give a short overview of some of these recent results on higher-dimensional integrability mentioned above, and to present some applications. Concretely, in Section 2 we briefly review the generalized curvature condition which was proposed in Ref. [1] as a possible way to generalize integrability to higher dimensions. In Section 3 we introduce volume-preserving diffeomorphisms on target space. For the case of three-dimensional target space we then classify for a wide class of Lagrangians all possible conservation laws, where the conserved currents in all cases are Noether currents of the volume-preserving target space diffeomorphisms. This section closely follows Ref. [20], but provides a slightly more refined classification, which turns out to be useful for applications. In Section 4 we study as an explicit example the abelian projection of Yang–Mills dilaton theory, which turns out to be integrable. Here we closely follow the results of [21]. Section 5 contains a brief discussion.

2. Generalized zero curvature condition

Here we briefly review the proposal for generalized integrability of Ref. [1], to which we refer for the details. The structure needed consists of a reducible Lie algebra $\tilde{G}$ which is a direct sum of another Lie algebra G and an abelian ideal H,

$$\tilde{G} = G \oplus H \tag{1}$$

together with a flat connection

$$A_\mu \in G\,, \quad \partial_\mu A_\nu - \partial_\nu A_\mu + [A_\mu, A_\nu] = 0, \tag{2}$$

and a covariantly constant vector field

$$B_\mu \in H\,, \quad \partial_\mu B^\mu + [A_\mu, B^\mu] = 0 \tag{3}$$

with

$$A_\mu = A_\mu^a T^a\,, \quad B_\mu = B_\mu^\alpha P^\alpha\,, \quad [P^\alpha, P^\beta] = 0, \tag{4}$$

where T^a and P^α form a basis in G and H, respectively. Further, $\mu = 1 \ldots d$ are base space indices.

To gain some intuition, let us first remark that Eq. (2) is just a generalization to higher dimensions of the zero curvature condition of Zakharov and Shabat in 1+1 dimensions. Further,

Eq. 3 in some sense just generalizes the Lax pair $\dot{L} = [L, M]$ to higher dimensions. Another important point is that Eqs. (2), (3) are not chosen arbitrarily but may, in fact be derived from a generalized curvature condition as follows.

Firstly, the zero curvature condition Eq. (2) of Zakharov and Shabat may be derived as a consequence of the path independence of the Wilson line (or parallel transport) operator

$$W = P\exp\left(\int_0^\sigma d\sigma' A_\mu \frac{dx^\mu}{d\sigma'}\right) \tag{5}$$

where P indicates the path ordering. In an analogous way, Eqs. (2), (3) may be derived from the hypersurface independence of the following hypersurface ordered operator V in d dimensions,

$$V = \tilde{P}\exp\left(\int_{\Sigma_{d-1}} d\sigma^1 \dots d\sigma^{d-1} W^{-1}\tilde{B}_{\mu_1\dots\mu_{d-1}} W \frac{dx^{\mu_1}}{d\sigma^1}\cdots\frac{dx^{\mu_{d-1}}}{d\sigma^{d-1}}\right). \tag{6}$$

Here the $d-1$ form $\tilde{B}$ is the Hodge dual of the vector (one-form) of Eq. (3) (in fact, $\tilde{B}$ is the more natural object from the point of view of generalized integrability). Further, Σ_{d-1} is a based, ordered, closed hypersurface with base point $x_0 \equiv x(\sigma_j = 0)$. $\tilde{P}$ is the hypersurface ordering, which we shall explain a bit more in a moment.

The hypersurface independence of V, in turn, may be derived from the zero curvature condition for a connection $\mathcal{A}$ in higher loop space $\Omega^n(M, x_0)$ where

$$\Omega^n(M, x_0) = \{\gamma : S^n \to M, \gamma(0, \dots, 0) = x_0\} \tag{7}$$

Explicitly, the connection $\mathcal{A}$ reads

$$\mathcal{A} = \int_{\Sigma_{d-1}} d\sigma^1 \dots d\sigma^{d-2} W^{-1}\tilde{B}_{\mu_1\dots\mu_{d-1}} W \frac{dx^{\mu_1}}{d\sigma^1}\cdots\frac{dx^{\mu_{d-2}}}{d\sigma^{d-2}}\delta x^{\mu_{d-1}} \tag{8}$$

where $\delta x^{\mu_{d-1}}$ is the differential on higher loop space which provides an arbitrary infinitesimal variation of the higher loop. A closed ordered based hypersurface Σ^{d-1} may be interpreted as a closed loop in loop space Ω^{d-2}, and this observation allows to understand the hypersurface ordering. It is just ordinary path ordering of the corresponding ordinary loop in higher loop space Ω^{d-2}.

We want to emphasize that the conditions Eqs. (2), (3) are *sufficient, local* conditions for the zero curvature condition on the connection $\mathcal{A}$ of Eq. (8), but certainly they are not the most general ones.

After this brief review of the generalized zero curvature condition, we assume that Eqs. (2), (3) hold and make the following additional simplifying assumptions that

- G is a semisimple Lie algebra (e.g. su(2)) with

$$[T^a, T^b] = f_c^{ab} T^c \tag{9}$$

  a structure constants f_c^{ab}.
- H is a (in general, reducible) representation space of G:

$$[T^a, P^\alpha] = R^{\alpha\beta}(T^a)P^\beta \tag{10}$$

- A_μ is explicitly flat:

$$A_\mu = g^{-1}\partial_\mu g\,, \quad g \in \mathcal{G} \tag{11}$$

  (where e.g. $\mathcal{G} = \mathrm{SU}(2)$) such that only Eq. (3) provides a nontrivial condition ($D_\mu B^\mu = 0$).

Journal of Physics: Conference Series **128** (2008) 012025 doi:10.1088/1742-6596/128/1/012025

- Under these conditions, the currents

$$J_\mu = gB_\mu g^{-1} \tag{12}$$

  are automatically conserved, $\partial^\mu J_\mu = 0$, and therefore the number of the conserved currents equals the dimension of the representation space H, dim H. If dim $H = \infty$, then we say that the corresponding field theory is integrable.

In general, the conserved currents of an integrable theory may be either Noether currents or may be related to hidden symmetries. Under the assumptions Eq. (9) - (12), however, the currents J_μ turn out to be Noether currents of geometric target space transformations, where the target space is spanned by the parameters of $g \in \mathcal{G}$.

A first example for this structure is like follows.

- The Lie group $\mathcal{G}$ is SU(2) where, however, its elements $g \in \mathcal{G}$ are restricted to the equator of SU(2) such that the target space is two-dimensional.
- The representation space H is the space of representations of SU(2) with arbitrary integer angular momentum quantum number l, but magnetic qunatum number m restricted to ± 1,

$$H = \{\text{reps } R_{lm} \text{ of SU}(2),\, m = \pm 1, l = 1, \ldots, \infty\} \tag{13}$$

- Then the conserved currents J_μ of Eq. (12) turn out to generate area preserving diffeomorphisms on the two-dimensional target space (which may be, e.g., the two-sphere S^2, but this depends on the Lagrangian).

A detailed discussion of this case may be found in [17], or in [18], [19].

Another class of theories with three-dimensional target spaces is obtained when the group element g is assumed to take values in the full unrestricted group SU(2). There it turns out that the resulting conservation laws are generated by some subsets of the generators of volume-preserving diffeomorphisms on that target space. This case is discussed in the next section, where also a classification of the conservation laws of these theories is given. More details may be found in Ref. [20]

3. Conservation laws for Skyrme-type models

3.1. *Volume-preserving diffeomorphisms*

Let us start with a three-dimensional manifold (later to be identified with target space) with local coordinates X^i and with a volume form which in local coordinates reads

$$dV = h(X^i)dX^1 \wedge dX^2 \wedge dX^3 \tag{14}$$

where h is the volume density. Further, a diffeomorphism is an infinitesimal transformation

$$X^i \to X^i + \epsilon Y^i(X^j), \tag{15}$$

where ϵ is infinitesimal, and the Y^i are arbitrary functions of the X^j. A volume-preserving diffeomorphism has to obey

$$\partial_i(hY^i) \equiv \frac{\partial}{\partial X^i}(hY^i) = 0, \tag{16}$$

in addition. The corresponding vector fields

$$v^{(Y)} = Y^i \partial_i \tag{17}$$

form a closed Lie algebra, that is

$$[v^{(Y)}, v^{(\tilde{Y})}] = v^{(\tilde{\tilde{Y}})} \tag{18}$$

such that

$$\tilde{\tilde{Y}}^i = (\partial_j Y^i)\tilde{Y}^j - (\partial_j \tilde{Y}^i)Y^j\,, \quad \partial_i(h\tilde{\tilde{Y}}^i) = 0 \tag{19}$$

is again a volume-preserving diffeomorphism. For later convenience we change coordinates according to

$$\begin{aligned} u \equiv X^1 + iX^2 \quad &, \quad \xi \equiv X^3 \\ Y^u \equiv Y^1 + iY^2 \quad &, \quad Y^\xi \equiv Y^3. \end{aligned} \tag{20}$$

These new coordinates are especially useful for a parametrization of $g \in \mathrm{SU}(2)$,

$$g = \exp(i\xi\vec{n}\cdot\vec{\sigma}) = \cos\xi + i\sin\xi\,\vec{n}\cdot\vec{\sigma} \tag{21}$$

provided that we also replace the unit vector field $\vec{n}$ by a complex field u via stereographic projection

$$\vec{n} \to u = \frac{n_1 + in_2}{1+n_3}. \tag{22}$$

Further, we assume from now on the following form of the volume density $h = h(u\bar{u},\xi)$ for simplicity. Finally, we interpret u,ξ as target space variables of a Lagrangian field theory with general Lagrangian $\mathcal{L}(u,\bar{u},\xi,u_\mu,\bar{u}_\mu,\xi_\mu)$, where $u_\mu \equiv \partial_\mu u$, etc., then the Noether currents corresponding to the vector fields $v^{(Y)}$ generating volume-preserving diffeomorphisms on target space are given by

$$J^{(Y)}_\mu = Y^u\Pi_\mu + Y^{\bar{u}}\bar{\Pi}_\mu + Y^\xi P_\mu \tag{23}$$

with the usual canonical four-momenta

$$\Pi_\mu \equiv \partial_{u^\mu}\mathcal{L}\,, \quad P_\mu \equiv \partial_{\xi^\mu}\mathcal{L}. \tag{24}$$

The charges $Q^{(Y)} = \int d^3\mathbf{r} J_0^{(Y)}$ generate a Lie algebra isomorphic to the algebra of the vector fields $v^{(Y)}$ via the Poisson bracket, as usual.

3.2. Classification of conserved currents

We now specialize to the class of Lagrangians

$$\mathcal{L}(a,b,c,\xi,d,e) \tag{25}$$

where

$$a = u\bar{u}\,, \quad b = u^\mu\bar{u}_\mu\,, \quad d = \xi^\mu\xi_\mu \tag{26}$$

$$c = (u^\mu\bar{u}_\mu)^2 - u_\mu^2\bar{u}_\nu^2\,, \quad e = \xi^\mu u_\mu \xi^\nu \bar{u}_\nu. \tag{27}$$

To motivate this choice let us mention that, e.g., the Skyrme model belongs to this class. Indeed, the Skyrme model has Lagrangian $\mathcal{L}_{\mathrm{Sk}} = \frac{m^2}{2}\mathcal{L}_2 - \lambda\mathcal{L}_4$ where

$$\begin{aligned} \mathcal{L}_2 &= \mathrm{tr}(g^{-1}g_\mu g^{-1}g^\mu) = d + 4b\frac{\sin^2\xi}{(1+a)^2} \\ \mathcal{L}_4 &= \mathrm{tr}[g^{-1}g_\mu, g^{-1}g_\nu]^2 = \frac{\sin^2\xi}{(1+a)^2}(bd-e) + \frac{\sin^4\xi}{(1+a)^4}c \end{aligned}$$

and g is the SU(2) group element of Eq. (21).

For the class of Lagrangians (25) we now want to find which subsets of the currents (23) are conserved under which conditions. The calculation is lengthy but straight forward, and we

Table 1. Case a) corresponds to the symmetry $u \to e^{i\alpha}u$.
Case b) implies that the Lagrangian can be expressed by the pullback of a certain target space metric, such that h is the Riemannian volume density of that metric. One example for this case is the Skyrme model.
One example for case d) is provided by the abelian projection of Yang–Mills dilaton theory, which is discussed in Section 4.

	No integrability conditions.
a)	no condition on $\mathcal{L}$. Generically there exists only one vector field Y: $Y^u = iu, \quad Y^{\bar{u}} = -i\bar{u}, \quad Y^\xi = 0.$
b)	$\mathcal{L} = \mathcal{F}(hb, h^2c, d, he)$. There exist finitely many Y generating the isometries of the target space metric.
c)	$\mathcal{L}_b = 0$ and $\mathcal{L}_e = 0$. Y form the abelian subalgebra ($\tilde{G} = \tilde{G}(a)$): $Y^u = iu\tilde{G}_a, \quad Y^{\bar{u}} = -i\bar{u}\tilde{G}_a, \quad Y^\xi = 0.$
d)	conditions b) and c) on $\mathcal{L}$, and factorizing $h = h_1(a)h_2(\xi)$. Y forms the non-abelian subalgebra (for $G = G(u,\bar{u})$): $Y^u = ih_1^{-1}G_{\bar{u}}, \quad Y^{\bar{u}} = -ih_1^{-1}G_u, \quad Y^\xi = 0.$

Table 2. The integrability condition $u^2\bar{u}_\mu^2 - \bar{u}^2u_\mu^2 = 0$ may also be expressed like $\partial^\mu(\mathrm{mod}(u))\partial_\mu(\arg(u)) = 0$, which provides a more geometric interpretation.

	Integrability condition $u^2\bar{u}_\mu^2 - \bar{u}^2u_\mu^2 = 0$
a)	$\mathcal{L}_b = 0$ and $\mathcal{L}_e = 0$. Y forms the abelian subalgebra (for $G = G(a,\xi)$): $Y^u = ih^{-1}uG_a, \quad Y^{\bar{u}} = -ih^{-1}\bar{u}G_a, \quad Y^\xi = 0.$
b)	$\mathcal{L}_e = 0$. Y form the abelian subalgebra ($\tilde{G} = \tilde{G}(a)$): $Y^u = iu\tilde{G}_a, \quad Y^{\bar{u}} = -i\bar{u}\tilde{G}_a, \quad Y^\xi = 0.$

present the resulting classification in the following four tables. In Table 1 the fields just obey the field equations and, therefore, there is a one to one correspondence between symmetries and conservation laws. In Tables 2-4, on the other hand, the fields have to obey certain first order equations ("integrability conditions"), in addition. These integrability conditions are not of the Euler–Lagrange type and, therefore, there is no longer a one to one correspondence between symmetries and conservation laws, see Ref. [22] for a detailed discussion.

Journal of Physics: Conference Series **128** (2008) 012025 doi:10.1088/1742-6596/128/1/012025

Table 3. Case a) is obeyed by many configurations of the Skyrme model. E.g., the simplest Skyrmion with baryon number one as well as many ansaetze for Skyrmion configurations satisfy the conditions of case a).
Case d): the "weight number" $\mathcal{W}$ is defined for monomials of first derivatives of fields as $\mathcal{W} = \text{power}(u_\mu) + \text{power}(\bar{u}_\mu) - 2\text{power}(\xi_\mu)$, which gives e.g. $\mathcal{W}(b) = 2$, $\mathcal{W}(e) = -2$.

	Integrability conditions $u^\mu \xi_\mu = 0$.
a)	no condition on $\mathcal{L}$; or $\mathcal{L} = \mathcal{F}(hb, h^2c, d, he)$. Y forms the abelian subalgebra (for $G = G(a, \xi)$): $Y^u = ih^{-1}uG_a, \quad Y^{\bar{u}} = -ih^{-1}\bar{u}G_a, \quad Y^\xi = 0.$ *And* the further integrability condition $u^2\bar{u}_\mu^2 - \bar{u}^2 u_\mu^2$ holds.
b)	$\mathcal{L}_b = 0$. Y forms the abelian subalgebra (for $G = G(a, \xi)$): $Y^u = ih^{-1}uG_a, \quad Y^{\bar{u}} = -ih^{-1}\bar{u}G_a, \quad Y^\xi = 0.$
c)	$\mathcal{L}_b = 0$ and $\mathcal{L} = \mathcal{F}(hb, h^2c, d, he)$. Y form the subset $Y_\xi^\xi = 0$ (is *not* a subalgebra).
d)	$\mathcal{L}_b = 0$ and $\mathcal{L} = \mathcal{F}(hb, h^2c, d, he)$ and $\mathcal{W}(\mathcal{L}) = 0$. no further condition on Y.

Table 4. Case b) is obeyed by many configurations of the Skyrme model. E.g., the simplest Skyrmion with baryon number one or the rational map ansaetze for Skyrmion configurations satisfy the conditions of case b).

	Integrability conditions $u_\mu^2 = 0$ and $u^\mu \xi_\mu = 0$.
a)	no condition on $\mathcal{L}$. Y forms the abelian subalgebra (for $G = G(a, \xi)$): $Y^u = ih^{-1}uG_a, \quad Y^{\bar{u}} = -ih^{-1}\bar{u}G_a, \quad Y^\xi = 0.$
b)	$\mathcal{L} = \mathcal{F}(hb, h^2c, d, he)$ Y form the subset $Y_\xi^\xi = 0$ (is *not* a subalgebra).
c)	$\mathcal{L} = \mathcal{F}(hb, h^2c, d, he)$ and $\mathcal{W}(\mathcal{L}) = 0$. no further condition on Y.

4. Example: Abelian projection of YM dilaton theory

Here we want to demonstrate that the abelian projection of Yang–Mills dilaton theory is integrable in our sense. It belongs, in fact, to case d) of Table 1. A more detailed discussion can be found in Ref. [21]. The Lagrangian of Yang–Mills dilaton theory is

$$\mathcal{L} = \frac{1}{4}(2\xi^\mu \xi_\mu - e^{-2\kappa\xi} F^{a\mu\nu} F^a_{\mu\nu}) \tag{28}$$

Journal of Physics: Conference Series **128** (2008) 012025 doi:10.1088/1742-6596/128/1/012025

where A^a_μ is an SU(2) Yang–Mills field and $F^a_{\mu\nu}$ is the corresponding field strength. Next, we want to employ the Cho–Faddeev–Niemi decomposition of the gauge field,

$$A^a_\mu = n^a C_\mu + \epsilon^{abc} n^b_\mu n^c + W^a_\mu, \tag{29}$$

where C_μ is an abelian gauge field, n^a is a unit vector in color space, and the so-called "valence field" W^a_μ is perpendicular to n^a in color space, $n^a W^a_\mu = 0$. To be consistent, the decomposition fields have to obey the constraint

$$\partial^\mu W^a_\mu + C_\mu \epsilon^{abc} n^b W^c_\mu + n^a W^b_\mu n^b_\mu = 0. \tag{30}$$

This constraint makes that the number of degrees of freedom of the original gauge field and of the decomposition match, and further, it provides the correct behaviour under gauge transformations for the decompostion fields, which infinitesimally read

$$\begin{aligned} \delta n^a &= \epsilon^{abc} n^b \alpha^c \\ \delta W^a_\mu &= \epsilon^{abc} W^b_\mu \alpha^c \\ \delta C_\mu &= n^a \alpha^a_\mu . \end{aligned} \tag{31}$$

In a next step, we perform the abelian projection, which consists in setting the valence field equal to zero,

$$W^a_\mu = 0. \tag{32}$$

Observe that the abelian projection is gauge invariant and obeys the constraint (30). The resulting gauge field $\hat{A}^a_\mu = n^a C_\mu + \epsilon^{abc} n^b_\mu n^c$ still is a full SU(2) connection, but with abelian field strength. The resulting abelian projected Yang–Mills dilaton theory is already integrable, that is, it has infinitely many conserved currents, see Ref. [21] for details. Here we make the further simplifying assumption $C_\mu \equiv 0$ (which is no longer gauge invariant). The resulting abelian projected Lagrangian is

$$\mathcal{L}_{\rm AP} = \frac{1}{4} \left(2\xi^\mu \xi_\mu - e^{-2\kappa\xi} H^{\mu\nu} H_{\mu\nu} \right) \tag{33}$$

with

$$H_{\mu\nu} = \epsilon^{abc} n^a n^b_\mu n^c_\nu \tag{34}$$

or, after the stereographic projection (22),

$$\begin{aligned} \mathcal{L}_{\rm AP} &= \frac{1}{2}\xi^\mu \xi_\mu - 2e^{-2\kappa\xi} \frac{(u^\mu \bar{u}_\mu)^2 - u^2_\mu \bar{u}^2_\nu}{(1+u\bar{u})^4} \\ &\equiv \frac{1}{2} d - 2h^2 c \end{aligned} \tag{35}$$

where $h = h_1(a) h_2(\xi) \equiv (1+a)^{-2} e^{-\kappa\xi}$. It corresponds to case d) of Table 1 and has, therefore, infinitely many symmetries and infinitely many conservation laws.

We now want to use our explicitly integrable parametrization of the abelian projection of Yang–Mills dilaton theory to discuss the problem of static solutions. For that purpose, we should first review the known results on that issue. It is known that there exist static, sphaleron type solutions in Yang–Mills dilaton theory. For the fully nonabelian theory, solutions both for radially and cylindrically symmetric ansaetze are known numerically, whereas for the abelian subsector solutions for radially and cylindrically symmetric ansaetze are known analytically. In the latter case, it is further known that the energy of the analytic solutions grows linearly with a certain integer m from the ansatz (the magnetic quantum number). The latter fact points to the existence of a Bogomolny bound in the abelian projection, but an ansatz-independent derivation of this Bogomolny bound has not yet been given in the literature.

In our integrable abelian projection of YM dilaton theory, the analytic solutions may be calculated easily by quadratures, and the Bogomolny bound may be derived explicitly. Indeed, upon introducing spherical polar coordinates (r, θ, φ) in three-dimensional base space, the ansatz $\xi = \xi(r)$, $u = v(\theta)\exp(im\varphi)$ turns out to be consistent with the static field equations because of the base space symmetries of the theory. The resulting ordinary differential equations for $\xi(r)$ and $v(\theta)$ turn out to be solvable by quadratures, such that the corresponding exact analytic solutions may be calculated easily. The solvability by quadratures of the field equations might be related to the integrability of the theory. For details we refer to [21].

Finally, the Bogomolny bound may be derived easily within our parametrization. Indeed, we find for the energy corresponding to the Lagrangian (33) for static configurations

$$\begin{aligned} E_{\rm AP} &= \frac{1}{2}\int d^3\mathbf{r}\left((\nabla\xi)^2 + e^{-2\kappa\xi}\vec{H}^2\right) \\ &= \frac{1}{2}\int d^3\mathbf{r}\left(\nabla\xi - e^{-\kappa\xi}\vec{H}\right)^2 + \int d^3\mathbf{r} e^{-\kappa\xi}\nabla\xi\cdot\vec{H} \\ &\geq \int d^3\mathbf{r} e^{-\kappa\xi}\nabla\xi\cdot\vec{H} \equiv E_{\rm Bog.} \end{aligned}$$

(where $\vec{H}$ is the Hodge dual of H_{jk}) and, therefore, the Bogomolny equation

$$\nabla\xi - e^{-\kappa\xi}\vec{H} = 0. \tag{36}$$

All the analytic static solutions mentioned above satisfy this equation and are, therefore, Bogomolny solutions. Further, the Bogomolny energy $E_{\rm Bog.}$ may be expressed by the winding number of a map $S^3 \to S^3$, see again Ref. [21].

To recapitulate, our main results for the abelian projection of YM dilaton theory are that

- there exist infinitely many symmetries and infinitely many conserved currents,
- this fact may explain the infinitely many analytic solutions (this still is a conjecture, which exploits the analogy to the lower dimensional cases),
- there exist both a Bogomolny bound and a Bogomolny equation for static configurations, and the latter is solved by all known analytic solutions.

5. Discussion and Outlook

It was the purpose of this article to briefly review some recent develpoments in the attempts to generalize the concept of integrability to higher-dimensional nonlinear field theories. We gave a brief introduction to the general proposal for higher-dimensional integrability of Ref. [1] and then showed how, under certain additional assumptions, this higher-dimensional integrability is related to certain geometric target space transformations (concretely, volume preserving differomorphisms) which provide infinitely many conservation laws. We discussed in some more detail the case of a three-dimenisonal target space and, as a specific example, the abelian projection of Yang–Mills dilaton theory and its static analytic solutions. Some more applications to specific theories have already been studied (see, e.g., the references quoted in the Introduction), which already demonstrates the usefulness and importance of the concept of generalized integrability for the study of hgher-dimensional nonlinear field theories. There exist, however, many more applications which are still open to further investigation. One obvious application is the search for time-dependent solutions (e.g. Q-balls) in theories where till now only static solutions have been found (e.g. in the integrable submodel of Yang–Mills dilaton theory of the previous section). Another possibility for generalizations consists in the choice of larger groups $\mathcal{G}$ instead of SU(2) in the integrability construction discussed in Section 2. This leads to integrable theories with higher-dimensional target spaces. The search for an integrable submodel of Einstein Yang–Mills dilaton theory and for analytic solutions within this submodel would be an obvious candidate, especially as for this theory only numerical solutions are known so far.

The generalizations and further investigations mentioned here still deal with a connection which is trivially flat, $A_\mu = g^{-1}\partial_\mu g$, such that the zero curvature condition (2) of Section 2 is trivially fulfilled, and Eq. (3) remains the only nontrivial generalized zero curvature condition. A further possible generalization consists in treating Eq. (2) as a nontrivial condition, too, which generates nontrivial constraints on the connection. The resulting modified generalized integrability might then lead to nonlocal conserved currents and to conservation laws which are not generated by geometric transformations, as is well-known to be the case in 1+1 dimensions. This line of investigation is, however, almost completely unexplored, and the above remarks are, therefore, largely tentative at the moment.

Acknowledgments
A.W. gratefully acknowledges support from Adam Krzyżanowski Fund and Jagiellonian University (grant WRBW 41/07). C.A. and J.S.-G. thank MCyT (Spain) and FEDER (FPA2005-01963), and support from Xunta de Galicia (grant PGIDIT06PXIB296182PR and Conselleria de Educacion). Further, C.A. acknowledges support from the Austrian START award project FWF-Y-137-TEC and from the FWF project P161 05 NO 5 of N.J. Mauser.

References

[1] Alvarez O, Ferreira LA and Sanchez-Guillen J 1998 *Nucl. Phys. B* **529**, 689
[2] Aratyn H, Ferreira LA and Zimerman A 1999 *Phys. Lett.* **B456** 162
[3] Aratyn H, Ferreira LA and Zimerman A 1999 *Phys. Rev. Lett.* **83** 1723
[4] De Carli E and Ferreira LA 2005 *J. Math. Phys.* **46** 012703 (*Preprint* hep-th/0406244)
[5] Wereszczynski A. 2004 *Eur. Phys. J.* **C38** 261 (*Preprint* hep-th/0405155)
[6] Ferreira LA 2006 *J. High Energy Phys.* JHEP03(2006)075 (*Preprint* hep-th/0601235)
[7] Riserio do Bonfim AC and Ferreira LA 2006 *J. High Energy Phys.* JHEP03(2006)097 (*Preprint* hep-th/0602234)
[8] Nicole DA 1978 *J. Phys.* **G4** 1363
[9] Wereszczynski A 2005 *Eur. Phys. J.* **C41** 265 (*Preprint* math-ph/0504008)
[10] Wereszczynski A 2005 *Phys. Lett.* **B621** 201 (*Preprint* hep-th/0508121)
[11] Adam C, Sanchez-Guillen J, Vazquez RA and Wereszczynski A 2006 *J. Math. Phys.* **47** 052302 (*Preprint* hep-th/0602152)
[12] Ward RS 1999 *Nonlinearity* **12** 241
[13] Adam C, Sanchez-Guillen J and Wereszczynski A 2006 *Eur. Phys. J.* **C47** 513 (*Preprint* hep-th/0602008)
[14] Skyrme THR 1961 *Proc. R. Soc. Lond.* **A260** 127
[15] Skyrme THR 1962 *Nucl. Phys.* **31** 556
[16] Ferreira LA and Sanchez-Guillen J 2001 *Phys. Lett.* **B504** 195 (*Preprint* hep-th/0010168)
[17] Babelon O and Ferreira LA 2002 *J. High Energy Phys.* JHEP11(2002)020
[18] Adam C and Sanchez-Guillen J 2005 *Phys. Lett.* **B626** 235 (*Preprint* hep-th/0508011)
[19] Adam C, Sanchez-Guillen J and Wereszczynski A 2006 *J. Math. Phys.* **47** 022303 (*Preprint* hep-th/0511277)
[20] Adam C, Sanchez-Guillen J and Wereszczynski A *J. Math. Phys.* **48** 032302 (*Preprint* hep-th/0610227)
[21] Adam C, Sanchez-Guillen J and Wereszczynski A 2007 *Preprint* hep-th/0703224
[22] Adam C and Sanchez-Guillen J 2005 *J. High Energy Phys.* JHEP01(2005)004 (*Preprint* hep-th/0412028)

Journal of Physics: Conference Series **128** (2008) 012026 doi:10.1088/1742-6596/128/1/012026

Intertwining algebras of quantum superintegrable systems on the hyperboloid

J A Calzada[1], **Ş Kuru**[2], **J Negro**[3] **and M A del Olmo**[3]

[1]Departamento de Matemática Aplicada, Escuela Superior de Ingenieros Industriales, Universidad de Valladolid, 47011 Valladolid, Spain

[2]Department of Physics, Faculty of Science, Ankara University, 06100 Ankara, Turkey

[3]Departamento de Física Teórica, Atómica y Optica, Facultad de Ciencias, Universidad de Valladolid, 47011 Valladolid, Spain

E-mail: juacal@eis.uva.es, kuru@science.ankara.edu.tr, jnegro@fta.uva.es, olmo@fta.uva.es

Abstract. A class of quantum superintegrable Hamiltonians defined on a two-dimensional hyperboloid is considered together with a set of intertwining operators connecting all of them. It is shown that such intertwining operators close a $su(2,1)$ Lie algebra and determine the Hamiltonians through the Casimir operators. The physical states are characterized as unitary representations of $su(2,1)$.

1. Introduction

In this work we will consider a quantum system superintegrable living in a two-dimensional hyperboloid of two-sheets. Although this system is well known in the literature [1]-[6] and can be dealt with standard procedures [7]-[9], it will be studied here under a different point of view based on the properties of intertwining operators (IO), a form of Darboux transformations [10]. We will see how this approach can give a simple explanation of the main features of this physical system. The intertwining operators and integrable Hamiltonians have been studied in previous references [11]-[14], but we will supply here a thorough non-trivial application by means of this example. Besides, there are several points of interest for the specific case here considered.

The intertwining operators are first order differential operators connecting different Hamiltonians in the same class (called hierarchy) and they are associated to separable coordinates of the Hamiltonians. We will obtain just a complete set of such intertwining operators, in the sense that any of the Hamiltonians can be expressed in terms of these operators.

In our case the IO's close an algebraic structure which is the non-compact Lie algebra $su(2,1)$ (see [15] for a compact case). This structure allows us to characterize the discrete spectrum and the corresponding eigenfunctions of the system that constitute a (infinite dimensional) unitary representation. The construction of such representations is not so standard as for compact Lie algebras. We will compute the ground state and characterize the representation space of the wave-functions which share the same energy. Notice that this system includes also a continuum spectrum, but we will not go into this point here.

2. Parametrizations of the two-sheet hyperboloid

Let us consider the two-dimensional two-sheet hyperboloid $s_0^2 + s_1^2 - s_2^2 = -1$, with positive metric $ds^2 = -ds_0^2 - ds_1^2 + ds_2^2$. On this surface, we define the following Hamiltonian

$$H_\ell = J_2^2 - J_1^2 - J_0^2 - \frac{l_2^2 - \frac{1}{4}}{s_2^2} + \frac{l_1^2 - \frac{1}{4}}{s_1^2} + \frac{l_0^2 - \frac{1}{4}}{s_0^2}, \tag{1}$$

where $\ell = (l_0, l_1, l_2) \in R^3$, and the differential operators

$$J_0 = s_1\partial_2 + s_2\partial_1, \qquad J_1 = s_2\partial_0 + s_0\partial_2, \qquad J_2 = s_0\partial_1 - s_1\partial_0, \tag{2}$$

constitute a realization of the $so(2,1)$ Lie algebra with Lie commutators

$$[J_0, J_1] = -J_2, \qquad [J_2, J_0] = J_1, \qquad [J_1, J_2] = J_0.$$

The generator J_2 corresponds to a rotation around the axis s_2, while the generators J_0 and J_1 give pseudo-rotations (i.e., non-compact rotations) around the axes s_0 and s_1, respectively. The Casimir operator

$$C = J_0^2 + J_1^2 - J_2^2$$

is the 'kinetic' part of the Hamiltonian.

We can parametrize the hyperbolic surface by means of the 'analogue' of the spherical coordinates

$$s_0 = \sinh\xi\,\cos\theta, \qquad s_1 = \sinh\xi\,\sin\theta, \qquad s_2 = \cosh\xi, \tag{3}$$

where $0 \le \theta < 2\pi$, $0 \le \xi < \infty$ and the invariant measure is $d\mu(\xi,\theta) = \sinh\xi\, d\xi\, d\theta$. In this coordinate system, the infinitesimal generators (2) have the following expressions

$$J_0 = \sin\theta\,\partial_\xi + \cos\theta\coth\xi\,\partial_\theta, \qquad J_1 = \cos\theta\,\partial_\xi - \sin\theta\coth\xi\,\partial_\theta, \qquad J_2 = \partial_\theta\,. \tag{4}$$

Also using these coordinates (3), the Hamiltonian (1) is rewritten as

$$H_\ell = -\partial_\xi^2 - \coth\xi\,\partial_\xi - \frac{l_2^2 - \frac{1}{4}}{\cosh^2\xi} + \frac{1}{\sinh^2\xi}\left[-\partial_\theta^2 + \frac{l_1^2 - \frac{1}{4}}{\sin^2\theta} + \frac{l_0^2 - \frac{1}{4}}{\cos^2\theta}\right]. \tag{5}$$

Therefore, H_ℓ can be separated in the variables ξ and θ. Choosing its eigenfunctions Φ ($H_\ell\Phi = E\Phi$) in the form

$$\Phi(\theta,\xi) = f(\theta)\,g(\xi), \tag{6}$$

we get the separated equations

$$H^\theta_{l_0,l_1} f(\theta) \equiv \left[-\partial_\theta^2 + \frac{l_1^2 - \frac{1}{4}}{\sin^2\theta} + \frac{l_0^2 - \frac{1}{4}}{\cos^2\theta}\right] f(\theta) = \alpha\, f(\theta) \tag{7}$$

and

$$\left[-\partial_\xi^2 - \coth\xi\,\partial_\xi - \frac{l_2^2 - \frac{1}{4}}{\cosh^2\xi} + \frac{\alpha}{\sinh^2\xi}\right] g(\xi) = E\,g(\xi),$$

where $\alpha > 0$ is a separation constant.

3. A complete set of intertwining operators

The second order operator at the l.h.s. of (7) can be factorized in terms of first order operators [16, 17]

$$H^{\theta}_{l_0,l_1} = A^{+}_{l_0,l_1} A^{-}_{l_0,l_1} + \lambda_{l_0,l_1},$$

being

$$A^{\pm}_{l_0,l_1} = \pm\partial_\theta - (l_0 + 1/2)\tan\theta + (l_1 + 1/2)\cot\theta, \qquad \lambda_{l_0,l_1} = (1 + l_0 + l_1)^2.$$

The Hamiltonian can be also rewritten in terms of the triplet $(A^{\pm}_{l_0-1,l_1-1}, \lambda_{l_0-1,l_1-1})$

$$H^{\theta}_{l_0,l_1} = A^{-}_{l_0-1,l_1-1} A^{+}_{l_0-1,l_1-1} + \lambda_{l_0-1,l_1-1} = A^{+}_{l_0,l_1} A^{-}_{l_0,l_1} + \lambda_{l_0,l_1}. \tag{8}$$

In this way we can determine a hierarchy of Hamiltonians

$$\dots, H^{\theta}_{l_0-n,l_1-n}, \dots, H^{\theta}_{l_0-1,l_1-1}, H^{\theta}_{l_0,l_1}, H^{\theta}_{l_0+1,l_1+1}, \dots, H^{\theta}_{l_0+n,l_1+n}, \dots \tag{9}$$

satisfying the following recurrence relations

$$A^{-}_{l_0-1,l_1-1} H^{\theta}_{l_0-1,l_1-1} = H^{\theta}_{l_0,l_1} A^{-}_{l_0-1,l_1-1}, \qquad A^{+}_{l_0-1,l_1-1} H^{\theta}_{l_0,l_1} = H^{\theta}_{l_0-1,l_1-1} A^{+}_{l_0-1,l_1-1}.$$

Hence, the operators $\{A^{\pm}_{l_0+n,l_1+n}\}_{n\in\mathbb{Z}}$ are intertwining operators and act as transformations between the eigenfunctions of the hierarchy of Hamiltonians (9),

$$A^{-}_{l_0-1,l_1-1} : f_{l_0-1,l_1-1} \longrightarrow f_{l_0,l_1}, \qquad A^{+}_{l_0-1,l_1-1} : f_{l_0,l_1} \longrightarrow f_{l_0-1,l_1-1},$$

where the subindex refers to the corresponding Hamiltonian of the hierarchy.

We can define new operators in terms of $A^{\pm}_{l_0,l_1}$ and a diagonal operator $A_{l_0,l_1} = (l_0 + l_1)\mathbb{I}$ acting in the following way in the space of eigenfunctions

$$\hat{A}^{-} f_{l_0,l_1} := \frac{1}{2} A^{-}_{l_0,l_1} f_{l_0,l_1}, \qquad \hat{A}^{+} f_{l_0,l_1} := \frac{1}{2} A^{+}_{l_0,l_1} f_{l_0,l_1}, \qquad \hat{A} f_{l_0,l_1} := -\frac{1}{2}(l_0 + l_1) f_{l_0,l_1}.$$

It can be shown from (8) that $\hat{A}^{-}$, $\hat{A}^{+}$ and $\hat{A}$ satisfy the commutation relations of a $su(2)$ Lie algebra, i.e.

$$[\hat{A}^{-}, \hat{A}^{+}] = -2\hat{A}, \qquad [\hat{A}, \hat{A}^{\pm}] = \pm\hat{A}^{\pm}.$$

The 'fundamental' states, $f^{0}_{l_0,l_1}$, of the $su(2)$ representations are determined by the relation $A^{-}_{l_0,l_1} f^{0}_{l_0,l_1}(\theta) = 0$. They are

$$f^{0}_{l_0,l_1}(\theta) = N(\cos\theta)^{l_0+1/2}(\sin\theta)^{l_1+1/2},$$

where N is a normalization constant. These functions are regular and square-integrable when $l_0, l_1 \geq -1/2$. Since $\hat{A} f_{l_0,l_1} \equiv -\frac{1}{2}(l_0 + l_1) f_{l_0,l_1}$, then $j = \frac{1}{2}(l_0 + l_1)$ and the dimension of the unitary representation will be $2j + 1 = l_0 + l_1 + 1$.

Now, observe that because the IO's $A^{\pm}_{l_0,l_1}$ depend only on the θ-variable, they can act also as IO's of the total Hamiltonian H_ℓ (5) and its global eigenfunctions Φ_ℓ (6), leaving the parameter l_2 unchanged (in this framework we will use three-fold indexes)

$$A^{-}_{\ell'} H_{\ell'} = H_\ell A^{-}_{\ell'}, \qquad A^{+}_{\ell'} H_\ell = H_{\ell'} A^{+}_{\ell'},$$

where $\ell = (l_0, l_1, l_2)$ and $\ell' = (l_0 - 1, l_1 - 1, l_2)$. In this sense, all the above relations can be extended under this global point of view.

3.1. Pseudo-spherical coordinates around s_1

These coordinates are obtained from the non-compact rotations about the axes s_0 and s_1, respectively. In this way we obtain the following parametrization of the hyperboloid

$$s_0 = \cosh\psi \sinh\chi, \qquad s_1 = \sinh\psi, \qquad s_2 = \cosh\psi\cosh\chi\,. \tag{10}$$

The expressions of the $so(2,1)$ generators in these coordinates are

$$J_0 = -\tanh\psi\sinh\chi\partial_\chi + \cosh\chi\,\partial_\psi, \qquad J_1 = \partial_\chi, \qquad J_2 = \sinh\chi\partial_\psi - \tanh\psi\cosh\chi\,\partial_\chi\,.$$

The explicit expression of the Hamiltonian (1) is now

$$H_\ell = -\partial_\psi^2 - \tanh\psi\,\partial_\psi + \frac{l_1^2-\frac14}{\sinh^2\psi} + \frac{1}{\cosh^2\psi}\left[-\partial_\chi^2 + \frac{l_0^2-\frac14}{\sinh^2\chi} - \frac{l_2^2-\frac14}{\cosh^2\chi}\right].$$

This Hamiltonian can be separated in the variables ψ and χ considering the eigenfunctions Φ of H_ℓ ($H_\ell\,\Phi = E\,\Phi$) as $\Phi(\chi,\psi) = f(\chi)\,g(\psi)$. We obtain the two folllowing(separated) equations

$$H^\chi_{l_0,l_2} f(\chi) \equiv \left[-\partial_\chi^2 + \frac{l_0^2-\frac14}{\sinh^2\chi} - \frac{l_2^2-\frac14}{\cosh^2\chi}\right] f(\chi) = \alpha\, f(\chi), \tag{11}$$

$$\left[-\partial_\psi^2 - \tanh\psi\,\partial_\psi + \frac{l_1^2-\frac14}{\sinh^2\psi} + \frac{\alpha}{\cosh^2\psi}\right] g(\psi) = E\,g(\psi),$$

with α a separation constant. The second order operator at the l.h.s. of (11) can be factorized as a product of first order operators

$$H^\chi_{l_0,l_2} = B^+_{l_0,l_2}B^-_{l_0,l_2} + \lambda_{l_0,l_2} = B^-_{l_0-1,l_2-1}B^+_{l_0-1,l_2-1} + \lambda_{l_0-1,l_2-1}, \tag{12}$$

being

$$B^\pm_{l_0,l_2} = \pm\partial_\chi + (l_2+1/2)\,\tanh\chi + (l_0+1/2)\coth\chi, \qquad \lambda_{l_0,l_2} = -(1+l_0+l_2)^2. \tag{13}$$

In this case the intertwining relations take the form

$$B^-_{l_0-1,l_2-1}H^\chi_{l_0-1,l_2-1} = H^\chi_{l_0,l_2}B^-_{l_0-1,l_2-1}, \qquad B^+_{l_0-1,l_2-1}H^\chi_{l_0,l_2} = H^\chi_{l_0-1,l_2-1}B^+_{l_0-1,l_2-1},$$

and imply that these operators $B^\pm$ connect eigenfunctions in the following way

$$B^-_{l_0-1,l_2-1} : f_{l_0-1,l_2-1} \to f_{l_0,l_2}, \qquad B^+_{l_0-1,l_2-1} : f_{l_0,l_2} \to f_{l_0-1,l_2-1}\,.$$

The operators $B^\pm_{l_0,l_2}$ can be expressed in terms of ξ and θ using relations (3) and (10)

$$B^\pm_{l_0,l_2} = \pm J_1 + (l_2+1/2)\tanh\xi\,\cos\theta + (l_0+1/2)\coth\xi\,\sec\theta,$$

where J_1 is given by (4). We define new operators in the following way

$$\hat B^-\,f_{l_0,l_2} := \frac12\,B^-_{l_0,l_2}\,f_{l_0,l_2}, \qquad \hat B^+\,f_{l_0,l_2} := \frac12\,B^+_{l_0,l_2}\,f_{l_0,l_2}, \qquad \hat B\,f_{l_0,l_2} := -\frac12\,(l_0+l_2)\,f_{l_0,l_2},$$

and, having in mind the expressions (12) and (13), we can prove than they close the $su(1,1)$ Lie algebra

$$[\hat B^-,\hat B^+] = 2\,\hat B, \qquad [\hat B,\hat B^\pm] = \pm\hat B^\pm.$$

As the Lie algebra $su(1,1)$ is non-compact its unitary representations are infinite dimensional. In particular, we will be interested in those having a ground state annihilated by the lowering operator.

The IO's $B^\pm$ can be considered as intertwining operators of the Hamiltonians H_ℓ and their eigenfunctions Φ_ℓ, similarly to the IO's $A^\pm$ described before (in this situation we will also use three-fold indexes but now with l_1 remaining unchanged).

Journal of Physics: Conference Series **128** (2008) 012026 doi:10.1088/1742-6596/128/1/012026

3.2. Pseudo-spherical coordinates around s_0

These coordinates are obtained from the noncompact rotations about the axes s_1 and s_0, respectively. They give rise to the parametrization

$$s_0 = \sinh\phi, \qquad s_1 = \cosh\phi \sinh\beta, \qquad s_2 = \cosh\phi\cosh\beta, \tag{14}$$

where the generators take the form

$$J_0 = \partial_\beta, \qquad J_1 = \cosh\beta\,\partial_\phi - \tanh\phi\sinh\beta\,\partial_\beta, \qquad J_2 = -\sinh\beta\,\partial_\phi + \tanh\phi\cosh\beta\,\partial_\beta.$$

Now, the Hamiltonian is

$$H_\ell = -\partial_\phi^2 - \tanh\phi\,\partial_\phi + \frac{l_0^2 - \frac14}{\sinh^2\phi} + \frac{1}{\cosh^2\phi}\left[-\partial_\beta^2 + \frac{l_1^2-\frac14}{\sinh^2\beta} - \frac{l_2^2-\frac14}{\cosh^2\beta}\right],$$

that can be separated in the variables β in terms of the eigenfunctions Φ ($H_\ell\,\Phi = E\,\Phi$) such that $\Phi(\beta,\phi) = f(\beta)\,g(\phi)$ in the following way

$$H^\beta_{l_1,l_2} f(\beta) \equiv \left[-\partial_\beta^2 + \frac{l_1^2-\frac14}{\sinh^2\beta} - \frac{l_2^2-\frac14}{\cosh^2\beta}\right] f(\beta) = \alpha\, f(\beta), \tag{15}$$

$$\left[-\partial_\phi^2 - \tanh\phi\,\partial_\phi + \frac{l_0^2-\frac14}{\sinh^2\phi} + \frac{\alpha}{\cosh^2\phi}\right] g(\phi) = E\,g(\phi),$$

with α a separation constant. The second order operator at the l.h.s. of expression (15) can be factorized as a product of first order operators

$$H^\beta_{l_1,l_2} = C^+_{l_1,l_2} C^-_{l_1,l_2} + \lambda_{l_1,l_2} = C^-_{l_1+1,l_2-1} C^+_{l_1+1,l_2-1} + \lambda_{l_1+1,l_2-1},$$

being

$$C^\pm_{l_1,l_2} = \pm\partial_\beta + (l_2 + 1/2)\tanh\beta + (-l_1 + 1/2)\coth\beta, \qquad \lambda_{l_1,l_2} = -(1 - l_1 + l_2)^2.$$

These operators $C^\pm_{l_1,l_2}$ give rise to the intertwining relations

$$C^+_{l_1+1,l_2-1} H^\beta_{l_1,l_2} = H^\beta_{l_1+1,l_2-1} C^+_{l_1+1,l_2-1} \qquad C^-_{l_1+1,l_2-1} H^\beta_{l_1+1,l_2-1} = H^\beta_{l_1,l_2} C^-_{l_1+1,l_2-1},$$

which imply that

$$C^-_{l_1+1,l_2-1} : f_{l_1+1,l_2-1} \to f_{l_1,l_2}, \qquad C^+_{l_1+1,l_2-1} : f_{l_1,l_2} \to f_{l_1+1,l_2-1}\,.$$

In this case, $C^\pm_{l_1,l_2}$ can also be expressed in terms of ξ and θ using relations (3) and (14)

$$C^\pm_{l_1,l_2} = \pm J_0 + (l_2 + 1/2)\tanh\xi\,\sin\theta + (-l_1 + 1/2)\coth\xi\,\csc\theta,$$

where J_0 is given by (4). Now, the new operators are defined as

$$\hat C^-\, f_{l_1,l_2} := \frac12\, C^-_{l_1,l_2}\, f_{l_1,l_2} \qquad \hat C^+\, f_{l_1,l_2} := \frac12\, C^+_{l_1,l_2}\, f_{l_1,l_2} \qquad \hat C\, f_{l_1,l_2} := -\frac12\,(l_2 - l_1) f_{l_1,l_2}.$$

They satisfy the commutation relations of the $su(1,1)$ algebra

$$[\hat C^-, \hat C^+] = 2\,\hat C, \qquad [\hat C, \hat C^\pm] = \pm\hat C^\pm.$$

As in the previous cases, we can consider the IO's $C^\pm$ as connecting global Hamiltonians H_ℓ and their eigenfunctions, having in mind that now the parameter l_0 is unaltered.

4. Algebraic structure of the intertwining operators

If we consider together all the IO's that have appeared in section 3 ($\hat{A}^{\pm}, \hat{A}, \hat{B}^{\pm}, \hat{B}, \hat{C}^{\pm}, \hat{C}$), in addition to the above algebras $su(2)$ and $su(1,1)$, we find that they close the Lie algebra $su(2,1)$ since they satisfy the following commutation relations

$$
\begin{array}{llll}
[\hat{A}^+, \hat{B}^+] = 0 & [\hat{A}^-, \hat{B}^-] = 0 & [\hat{A}^+, \hat{B}^-] = -\hat{C}^- & [\hat{A}^-, \hat{B}^+] = \hat{C}^+ \\
[\hat{C}^+, \hat{B}^+] = 0 & [\hat{C}^-, \hat{B}^-] = 0 & [\hat{C}^+, \hat{A}^+] = -\hat{B}^+ & [\hat{C}^-, \hat{A}^-] = \hat{B}^- \\
[\hat{C}^+, \hat{B}^-] = -\hat{A}^- & [\hat{C}^-, \hat{B}^+] = \hat{A}^+ & [\hat{C}^+, \hat{A}^-] = 0 & [\hat{C}^-, \hat{A}^+] = 0 \\
[\hat{A}, \hat{B}^+] = \frac{1}{2}\hat{B}^+ & [\hat{A}, \hat{B}^-] = -\frac{1}{2}\hat{B}^- & [\hat{B}, \hat{A}^+] = \frac{1}{2}\hat{A}^+ & [\hat{B}, \hat{A}^-] = -\frac{1}{2}\hat{A}^- \\
[\hat{C}, \hat{B}^+] = \frac{1}{2}\hat{B}^+ & [\hat{C}, \hat{B}^-] = -\frac{1}{2}\hat{B}^- & [\hat{C}, \hat{A}^+] = -\frac{1}{2}\hat{A}^+ & [\hat{C}, \hat{A}^-] = \frac{1}{2}\hat{A}^- \\
[\hat{A}, \hat{C}^-] = \frac{1}{2}\hat{C}^- & [\hat{A}, \hat{C}^+] = -\frac{1}{2}\hat{C}^+ & [\hat{B}, \hat{C}^-] = -\frac{1}{2}\hat{C}^- & [\hat{B}, \hat{C}^+] = \frac{1}{2}\hat{C}^+ \\
[\hat{A}, \hat{B}] = 0 & [\hat{A}, \hat{C}] = 0 & [\hat{B}, \hat{C}] = 0\,. &
\end{array}
$$

Obviously $su(2,1)$ includes as subalgebras the Lie algebras $su(2)$ and $su(1,1)$ defined in the previous section 3. The second order Casimir operator of $su(2,1)$ can be written as follows

$$\mathcal{C} = \hat{A}^+\hat{A}^- - \hat{B}^+\hat{B}^- - \hat{C}^+\hat{C}^- + \frac{2}{3}\left(\hat{A}^2 + \hat{B}^2 + \hat{C}^2\right) - (\hat{A} + \hat{B} + \hat{C})\,.$$

It is worthy noticing that in our differential realization we have $\hat{A} - \hat{B} + \hat{C} = 0$, and that there is another generator

$$\mathcal{C}' = l_1 + l_2 - l_0$$

commuting with the rest of generators. Hence, adding this new generator $\mathcal{C}'$ to the other ones we get the Lie algebra $u(2,1)$.

The eigenfunctions of the Hamiltonians H_ℓ that have the same energy support a unitary representation of $su(2,1)$ and also they are characterized by a value of $\mathcal{C}$ and other of $\mathcal{C}'$. In fact, we can show that

$$H_\ell = -4\,\mathcal{C} + \frac{1}{3}\,\mathcal{C}'^2 - \frac{15}{4}.$$

These representations can be obtained, as usual, starting with a fundamental state annihilated by the lowering operators A^-, C^- and B^-

$$A_\ell^- \Phi_\ell^0 = C_\ell^- \Phi_\ell^0 = B_\ell^- \Phi_\ell^0 = 0\,. \tag{16}$$

Solving equations (16) we find

$$\Phi_\ell^0(\xi, \theta) = N(\cos\theta)^{l_0+1/2}(\sin\theta)^{1/2}(\cosh\xi)^{l_2+1/2}(\sinh\xi)^{l_0+1},$$

where $\ell = (l_0, 0, l_2)$. From Φ_ℓ^0 we can get the rest of eigenfunctions in the representation using the raising operators A^+, B^+ and C^+.

5. Concluding remarks

In this work we have built a set of intertwining operators for a superintegrable system defined on a two-sheet hyperboloid and we have found that they close a non-compact $su(2,1)$ Lie algebra structure.

We have shown how these IO's can be very helpful in the characterization of the physical system by selecting separable coordinates, determining the eigenvalues and building eigenfunctions.

The IO's can also be used to find the second order integrals of motion for a Hamiltonian H_ℓ and their algebraic relations, which is the usual approach to (super) integrable systems. However, we see that it is much easier to deal directly with the IO's, which are more elementary and simpler, than with constants of motion.

Our program in the near future is the application of this method to wider situations. Besides, in principle, we can also adapt the method to classical versions of such systems. On this aspect we must remark that some symmetry procedures usually considered only for quantum systems can be extended in an appropriate way to classical ones [18].

Acknowledgements.
Partial financial support is acknowledged to Junta de Castilla y León (Spain) under project VA013C05 and the Ministerio de Educación y Ciencia of Spain under project FIS2005-03989.

References

[1] del Olmo M A, Rodríguez M A and Winternitz P 1993 *J. Math. Phys.* **34** 5118
[2] Kalnins E G, Miller W and Pogosyan G S 1996 *J. Math. Phys.* **37** 6439
[3] del Olmo M A, Rodríguez M A and Winternitz P 1996 *Fortschritte der Physik*, **44**, 91
[4] Calzada J A, del Olmo M A and Rodríguez M A 1997 *J. Geom. Phys.* **23** 14
[5] Calzada J A, del Olmo M A and Rodriguez M A 1999 *J. Math. Phys.* **40** 88
[6] Calzada J A, Negro J, del Olmo M A and Rodríguez M A 1999 *J. Math. Phys.* **41** 317
[7] Rañada M F 2000 *J. Math. Phys.* **41** 2121
[8] Rañada M F and Santander M 2003 *J. Math. Phys.* **44** 2149
[9] Evans N W 1990 *Phys. Rev.* **41** 5666
1990 *Phys. Lett.* **147A** 483
1991 *J. Math. Phys.* **32** 3369
[10] Matveev V B and Salle M A 1991 *Darboux Transformations and Solitons* (Berlin: Springer)
[11] Kuru Ş, Teğmen A and Verçin A 2001 *J. Math. Phys.* **42** 3344
[12] Demircioğlu B, Kuru Ş, Önder M and Verçin A 2002 *J. Math. Phys.* **43** 2133
[13] Samani K A and Zarei M 2005 *Ann. Phys.* **316** 466
[14] Fernández D J, Negro J and del Olmo M A 1996 *Ann. Phys.* **252** 386
[15] Calzada J A, Negro J and del Olmo M A 2006 J. Math. Phys. **47** 043511
[16] Barut A O, Inomata A and Wilson R 1987 *J. Phys. A* **20** 4075
1987 *J. Phys. A* **20** 4083
[17] del Sol Mesa A, Quesne C and Smirnov Yu F 1998 *J. Phys. A* **31** 321
[18] Kuru Ş and Negro J 2008 *Ann. Phys.* **323** 413.

V International Symposium on Quantum Theory and Symmetries IOP Publishing
Journal of Physics: Conference Series **128** (2008) 012027 doi:10.1088/1742-6596/128/1/012027

Dynamics of the topological structures in inhomogeneous media

LA Ferreira[1], Bernard Piette[2] and WJ Zakrzewski[2]

[1]Instituto de Física de São Carlos, IFSC/USP, São Carlos, SP, Brazil
[2]Department of Mathematical Sciences, University of Durham, Durham DH1 3LE, U.K

E-mail: `laf@ifsc.usp.br`, `B.M.A.G.Piette@durham.ac.uk`, `W.J.Zakrzewski@durham.ac.uk`

Abstract. We present a general review of the dynamics of topological solitons in 1 and 2 dimensions and then discuss some recent work on the scattering of various solitonic objects (such as kinks and breathers etc) on potential obstructions.

1. Introduction

Topological solitons arise in many areas of applied mathematics and in the mathematical description of some processes in physics - for a good review see [1].

They arise in the mathematical description of objects that are localised in space as the energy density of these objects is nonzero only in a finite region; *ie* it is significantly nonzero in a small region and goes to zero, exponentially or as an inverse power, as one moves away from this region.

Their stability is guaranteed by topological considerations, normally associated with the topology of $S^N \to S^N$ maps.

The simplest examples of such maps (in 1+1 and 2+1) dimensions involve, respectively, the Sine-Gordon kinks and the solitons based on sigma models.

As is well known the Lagrangian density of the (1+1) dimensional sine-Gordon model is given by

$$\mathcal{L} = \partial_\mu \phi \cdot \partial^\mu \phi \; - \; \lambda \sin^2(\phi). \tag{1}$$

Solutions of the equations of motion which follow from (1) are well known. They involve kinks and antikinks, which are topological solitons; breathers, which can be thought of as bound states of kinks and antikinks, further bound states of kinks and breathers, and many other solutions, less interesting from our point of view.

The Lagrangian for the (2+1) dimensional sigma model is given by:

$$\mathcal{L} = \partial_\mu \vec{\phi} \cdot \partial^\mu \vec{\phi} - \theta_S \left[(\partial_\mu \vec{\phi} \cdot \partial^\mu \vec{\phi})^2 - (\partial_\mu \vec{\phi} \cdot \partial_\nu \vec{\phi})(\partial^\mu \vec{\phi} \cdot \partial^\nu \vec{\phi}) \right] - V(\vec{\phi}), \tag{2}$$

where for V we can take any 'simple' function of ϕ_3 which vanishes as $\phi_3 = 1$. So, in this work we take

$$V(\vec{\phi}) = \mu(1 - \phi_3^2). \tag{3}$$

Furthermore, we require that the vector $\vec{\phi}$ lies on the unit sphere $\mathcal{S}^2$ hence $\vec{\phi} \cdot \vec{\phi} = 1$. These last two conditions make the static finite energy solutions of the Euler Lagrange equations which follow from (2), the solitons, topological and based on the topology of $S^2 \to S^2$ maps. The requirement that the static energy is finite forces the field ϕ_3 to go to +1 at spatial infinity thus compactifying R^2 and allowing us to consider $\vec{\phi}$ on the extended plane $R^2 \bigcup \infty$ topologically equivalent to $\mathcal{S}^2$.

The three terms in (2) are, from left to right, the pure $\mathcal{S}^2$ sigma model, the Skyrme and the potential term. The last two terms are needed to stabilize the solitons. As long as they are small they have no influence on the topology of the fields.

Incidentally, other potential terms (3) have also been studied [2]. The results, in these cases, are only slightly different but their generic features are the same. Hence in this work we discuss our results obtained for (3). In 3 spatial dimensions we have skyrmions and monopoles but they will not be discussed here; the interested reader can find many interesting details in [1].

2. Dynamics

The dynamics of the Sine Gordon kinks is well known: the kinks reflect from each other and, in fact, not much can happen as the motion is in one dimension.

The dynamics of the two dimensional sigma model solitons can be either, relativistic, (*i.e.* based on the Lagrangian above) or based on the Landau-Lifshitz equation which is given by

$$\frac{\partial \vec{\phi}}{\partial t} = \vec{\phi} \times \frac{\partial L}{\partial \vec{\phi}} \tag{4}$$

where L stands for the spatial part of $\mathcal{L}$, *i.e.* of (2).

The dynamics of both cases is very different. In the relativisitic case we have the familiar 90^0 scattering. Thus, when two solitons are sent towards each other head on (*i.e.* at a zero impact parameter), the system evolves in such a way that after the scattering we have two outgoing solitons which are moving in the direction perpendicular to the motion of the original solitons.

This has been explained in many ways; the most compelling one involves the indistinguishability of solitons [1]. As the system of two solitons is described by a function which is symmetric with respect of the interchange of the positions of these solitons, the phase space of their relative position, is really described by R^2 mod a reflection with respect to the line joining their positions. Hence, effectively, the space is $\frac{R^2}{Z_2}$, where Z_2 describes this reflection, and so it is a cone. A straight line motion in this space through the vertex of the cone corresponds to a 90^0 motion when viewed in the 'opened-up' R^2.

In the nonrelativistic case the situation is different as the equations involve only first order time derivatives. Hence the motion takes place in a lower dimensional phase space.

This has been analysed in some detail by Papanicolaou and Tamaras [3] who showed that when one introduces, for a system of two solitons, $\vec{r} = (x_1, x_2)$ - a 2 dimensional vector describing their relative position, the corresponding relative momentum $\vec{p}$ satisfies

$$p_i \sim \alpha\, \epsilon_{ij} x_j. \tag{5}$$

Hence the equation motion is of the form

$$\frac{d^2 x_i}{dt^2} \sim \alpha \epsilon_{ij} \frac{dx_j}{dt} \tag{6}$$

resulting in a motion along a circle.

The dynamics of solitons in 3 spatial dimensions is even more complicated [1]; however, many aspects of it can be related to the dynamics in 2 dimensions.

All this discussion concerned solitons moving in a free space, *i.e.* in a space with no potential obstructions. In the next section we summarise our recent results on the scattering of individual solitons on a spatial obstruction - either in the shape of a potential bump or a hole.

3. Potential Obstruction

To introduce a potential obstruction we note that the energy of the soliton field, strictly speaking, is never zero, even though it vanishes exponentially as we move away from soliton's position. Hence the obstruction has to be introduced in such a way that it does not change the "tail" of the soliton *i.e.* it has to vanish when, in one dimension, $\phi = 0$ or π or, in two dimensions, when $\phi_3 = 1$.

3.1. Sigma Models

Although a choice of obstruction is highly nonunique, in our studies, we added an extra term to the Lagrangian and we chose it of the form $\alpha(1-\phi_3^2)$ (*i.e.* like (3) with α nonzero in some region of x and y). We chose this term to be independent of y so that it represented an obstruction, located in some finite region of x resembling a trough in the "hole" case or a dam in the "barrier" case. Then sending the soliton from a point far away from this obstruction we can study its scattering on this obstruction.

We have performed many numerical simulations of such systems, varying both the sign and value of α and the velocity of the incoming solitons. In the barrier case we have found that the scattering is very elastic with the soliton behaving like a point particle, depending on its kinetic energy - either going over the barrier or being reflected by it. Hence the velocity of the outgoing soliton is always very close, in magnitude, to the original velocity.

For the hole case the situation was different. Depending on the value of the velocity solitons were either tranmitted or trapped and sometimes were even reflected. In fig 1. we present plots of the solitons' positions (in x as $y = 0$) as a function of time seen in some simulations (*i.e.* started with different values of initial velocities).

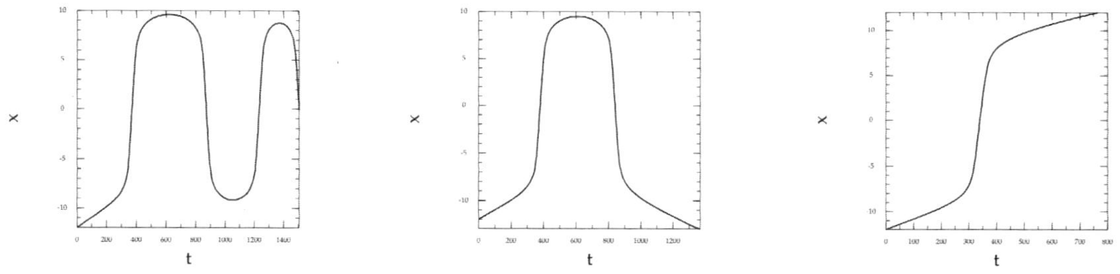

Figure 1. Trajectory of a soliton during the scattering for a well of width $L = 10$ and depth 0.2, (μ reduced from 0.25 to 0.2). Here $\theta_s = 0.25$ and a) $v = 0.0106$, b) $v = 0.0102$, c) $v = 0.012$, respectively

In addition to the trapping, which shows that, this time, the soliton does not behave like a point particle, we have also seen reflections! We have studied this process in great detail and in [4] we argued that the interaction of the soliton with the hole proceeds through the excitement of the vibrational modes of the soliton. This has led us to look at other models and, in particular, at the Sine Gordon model in one dimension, which does not have any genuine vibrational modes. We discuss the results in this model in the next section.

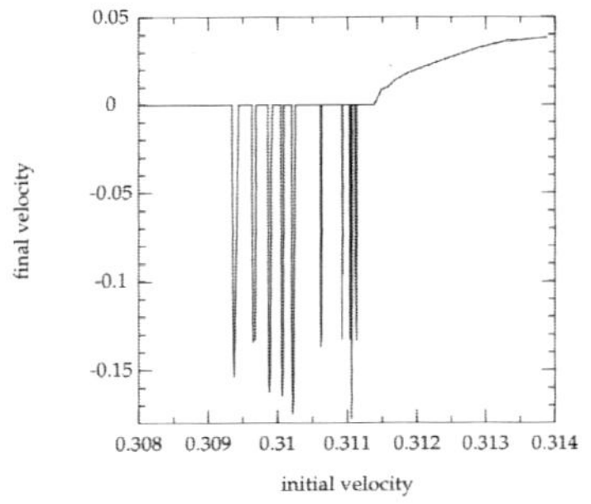

Figure 2. Outgoing velocity of the kink as a function of its initial velocity

3.2. Sine Gordon Model

There are two easy ways of introducing the obstruction into the Sine-Gordon model: either by making λ in (1) position dependent [5] or by altering the basic metric [6]. Here we discuss the results reported in [5]; the results of [6] are qualitatively very similar. We restrict our discussion to the case of the potential hole.

The results of our studies have shown that in the sine Gordon model, as in the sigma model in two dimensions, the solitons can get trapped, be transmitted and bounce back. The process is inelastic and depends on the initial condition and on the size and the depth of the hole. If the initial condition of the soliton corresponds to an exact Sine-Gordon kink moving with a given velocity then there is a well defined critical velocity above which the kink get transmitted (with a certain loss of energy). Below this critical velocity the kink can be trapped or reflected. The ranges of velocities, at which the kink is reflected are very narrow. The value of the critical velocity decreases with the decrease of the depth of the well and as the hole becomes narrower the number of velocity windows corresponding to reflections gets larger. These windows tend to be very narrow indeed. In fig.2 we present a typical plot of the outgoing velocity as a function of the incoming velocity for the case of a relatively narrow well (in this case λ in the well is lowered from 1 down to 0.8) and the hole is relatively narrow - *i.e.* a soliton fits in it about 3 times).

Thus we note that in the Sine-Gordon model the solitons, which have no vibrational modes, can still be reflected by a potential hole. Of course, although the solitons have no vibrational modes they have pseudovibrational ones which can get excited and radiate (similar conclusions, in a different context, were reached by Romanczukiewicz [7]). An example of such a pseudovibrational mode is the mode which describes the variation of the slope of the kink. The usual kink solution is given by

$$\phi(x) = 2\arctan(\exp(\theta(x - x_0)), \quad (7)$$

where x_0 is the kink position and θ is its slope. For (7) to be a solution of the equation of motion, which follows from (1), we need to put θ and λ equal. However, if we put θ different from λ we excite the mode which corresponds to the variation of θ. In fact, looking at the simulations

Journal of Physics: Conference Series **128** (2008) 012027 doi:10.1088/1742-6596/128/1/012027

in detail we have noticed that when the kink enters the hole for which λ is different from θ, it automatically tries to adjust its slope and so it excites this mode. Of course, as soon as this mode is excited the kink begin to radiate and it is the interaction of this radiation with the kink itself which is reponsible for the final outcome of the scattering process.

4. Breathers

The Sine-Gordon model, in addition to the kink, also possesses many other solutions. Amongst them are breathers which are given by

$$\phi(x,t) = 2\,\arctan\left(\frac{\sin(\omega t)}{\omega\,\cosh(\sqrt{1-\omega^2}\,x)}\right). \tag{8}$$

As their energy is $16\sqrt{1-\omega^2}$ they are often thought of as bound states of a kink and an antikink with the binding increasing as $\omega \to 1$. Hence it is interesting to see what happens when a breather is sent towards a hole.

We have performed several numerical experiments of such systems and we are now preparing a long paper with a description of our results [8]. So here we will mention only some of them. As expected our results have shown that the breathers do get trapped, pass with, in general, a different ω (note that increasing ω releases some energy), or split with either a kink or an antikink being ejected from the hole.

In fig 3 we present a couple of pictures showing a breather just before a scattering on a hole, and some time later, in the case when the interaction with the hole leads to the splitting of the breather.

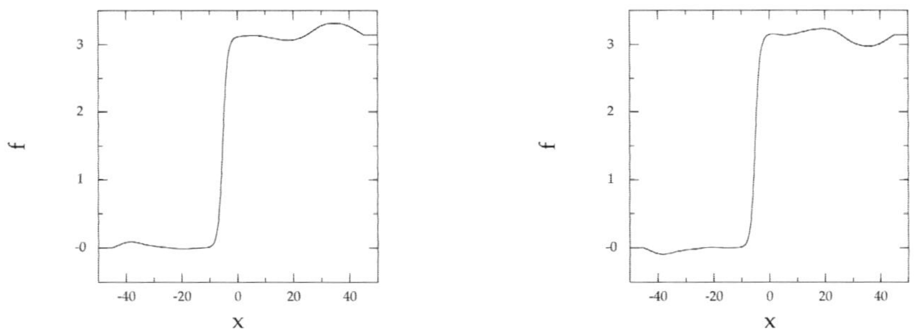

Figure 3. Trajectory of a breather sent with velocity $v = 0.06587375$ towards a well of width $L = 10$ and depth 0.8 (*ie* $\lambda = 0.8$); $t = 600$ and $t = 800$, respectively

5. Further solutions (wobbles etc).

However, the kinks and breathers are not the only finite energy solutions of the sine-Gordon model. In fact, the model possesses also solutions which describe bound states of kinks and breathers. One of such solutions, the 'wobble' was recently discussed in detail by Kälberman [9].

In his paper Kälberman presents an analytic form of this solution, shows that it describes a static kink in interaction with a breather, and then discusses some of its properties.

Recently [11], we have looked at more general solutions of the sine-Gordon model describing kinks and breathers. Our work was based on the Hirota method [10] of deriving such solutions. Using this method we obtained

$$\phi = 2\ \mathrm{ArcTan}\ \frac{\left[2\left\{\cot\mathrm{an}\theta\right\}\ \cos\Gamma_I + e^{\tilde{\Gamma}_3}\{e^{-\Gamma_R} + \rho^2\, e^{\Gamma_R}\}\right]}{\left[\{e^{-\Gamma_R} + e^{\Gamma_R}\} - 2\left\{\cot\mathrm{an}\theta\right\}\rho\, e^{\tilde{\Gamma}_3}\ \cos\{\Gamma_I + \varphi\}\right]}, \tag{9}$$

where

$$\tilde{\Gamma}_3 = \gamma_K\{x - v_K\, t\} + \eta_K \qquad \gamma_K = \cosh\alpha_K \qquad v_K = \tanh\alpha_K, \tag{10}$$

$$\rho = \frac{\cosh\{\alpha_B - \alpha_K\} - \cos\theta}{\cosh\{\alpha_B - \alpha_K\} + \cos\theta} \qquad \text{and} \qquad \varphi = 2\,\mathrm{ArcTan}\,\frac{\varepsilon\,\sin\theta}{\sinh\{\alpha_B - \alpha_K\}}. \tag{11}$$

and

$$\Gamma_R = \frac{1}{\sqrt{1 - v_B^2}}\cos\theta\,\{x - v_B\, t\} + \eta_B$$

$$\Gamma_I = \frac{1}{\sqrt{1 - v_B^2}}\sin\theta\{t - v_B\, x\} + \xi_B.$$

This corresponds to the most general kink/breather field configuration in which the kink and the breather have arbitrary velocities (and so move with respect to each other). If we now take

$$\eta_K = \eta_B = v_B = v_K = 0 \qquad \xi_B = \frac{\pi}{2}$$

and denote

$$\omega = \sin\theta \qquad -\frac{\pi}{2} \le \theta \le \frac{\pi}{2}$$

then

$$\rho = \frac{1 - \sqrt{1 - \omega^2}}{1 + \sqrt{1 - \omega^2}} \qquad \varphi = \pm\,\pi. \tag{12}$$

Hence our expression becomes

$$\phi = 2\ \mathrm{ArcTan}\ \frac{\left[\frac{\sqrt{1-\omega^2}}{\omega}\ \sin\{\,\omega\, t\} + \frac{1}{2}\, e^{\,x}\{e^{-\sqrt{1-\omega^2}\,x} + \rho^2\, e^{\sqrt{1-\omega^2}\,x}\}\right]}{\left[\cosh\{\,\sqrt{1 - \omega^2}\,x\} + \frac{\sqrt{1-\omega^2}}{\omega}\,\rho\, e^{x}\ \sin\{\omega\, t\}\right]} \tag{13}$$

where ω is a frequency varying from -1 to 1. This agrees with the expression given by Kälberman which describes a stationary field configuration in which the kink and the breather sit on 'top of each other' and are not in relative motion.

As is clear from (13) the field configuration depends on one parameter (the frequency of the breather) and so, as we showed in [11], we have studied the stability of this field configuration. To do this we calculated ϕ and its time derivative from (13) and then used them as the initial condition for our simulations. The results of our simulations were in complete agreement with the analytical expression thus showing that the solution is stable with respect to small perturbations (due to the discretisations).

Next we studied the stability of the wobble with respect to larger perturbations. In particular we changed the original slope of the kink: (*i.e.* in the expression (13) we replaced $exp(x)$ by

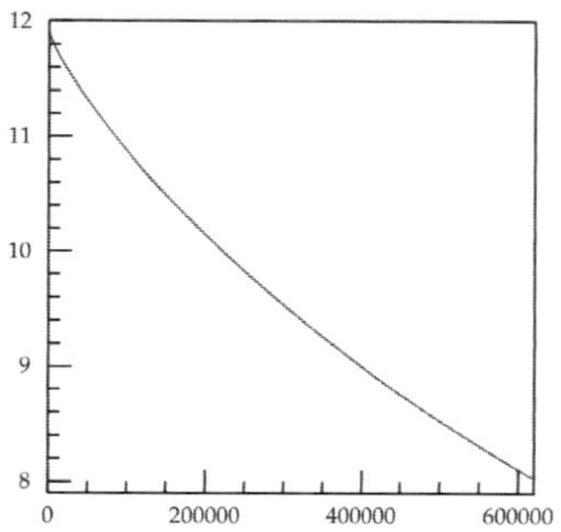

Figure 4. Energy as a function of time as seen in a simulation started with $\lambda = 1.15$

$exp(\lambda x)$ where $\lambda \neq 1$). As discussed in [11] we performed several simulations with λ ranging from 1.05 to 1.3.

Our initial conditions corresponded to incorrect initial field configurations which added some extra energy to the system. The system was then unstable and so it evolved towards a stable wobble emitting some radiation which was sent out towards the boundaries of the grid. For λ close to one - the perturbations were small - hence the system returned to its initial configuration (with $\lambda = 1.0$). For larger values of the perturbation the system was more perturbed and often not only kept on sending out its excess of energy but also, at regular intervals, altered its frequency of oscillation (increasing it) which allowed it to send out even more radiation. In fig 4. we present the plots of the time dependence of the total energy as seen in the simulation in which λ was set at 1.15.

Next we performed a series of simulations in which the initial ('wobble') configuration was sent towards a potential hole. As for a single breather [8] we have found many possibilities. We have also found that the hole can separate the breather from the kink (in one simulation we even saw the kink being trapped in the hole while the breather bounced off the kink and returned to where the wobble originally came from). Our results on this are still preliminary as we have decided to postpone the full study of this problem to some future work.

We have also looked at field configurations involving a kink with more breathers. In particular in [11] we gave an explicit expression for such a configuration corresponding to a kink and two breathers. We have also verified that this configuration is stable.

6. The energy

As we mentioned before the energies of the kink or the breather configurations have very simple forms. This has led us to enquire whether this is also the case for the more general field configurations mentioned above. In fact, we managed to show this: the energies of all field configurations that can be derived by the Hirota method are additive as they are determined entirely by the asymptotic values of the τ_i functions that arise in the construction of these field configurations. The τ_i functions arise when one sets $\phi = 2\log\frac{\tau_1}{\tau_0}$ and then solves the relevant equations (more details in [11]). Any interested reader can find more detail in [11].

In the cases we have discussed here the Hamiltonian density is given by

$$H = (\partial_t \phi)^2 + (\partial_x \phi)^2 + [\sin(\phi)]^2, \tag{14}$$

and so the energy becomes

$$E = \int_{-\infty}^{\infty} dx\, H = 2\partial_x (\ln \tau_0 + \ln \tau_1) |_{x=-\infty}^{x=\infty} \tag{15}$$

and so is determined entirely by the asymptotic values of τ_i functions.

In this way we have managed to show that the energies of solutions we have considered in this talk are:

(i) For the 1-kink:

$$E_{1-\text{soliton}} = 8\frac{1}{\sqrt{1-v^2}}. \tag{16}$$

(ii) For the breather

$$E_{\text{breather}} = 16\frac{\sqrt{1-\omega^2}}{\sqrt{1-v^2}}. \tag{17}$$

(iii) For the wobble

$$E_{\text{wobble}} = 8\frac{1}{\sqrt{1-v_K^2}} + 16\frac{\sqrt{1-\omega^2}}{\sqrt{1-v_B^2}}. \tag{18}$$

(iv) For a solution involving the kink and two breathers mentioned above (with their velocities set to zero)

$$E_{\text{kink+2breathers}} = 8 + 16\sqrt{1-\omega_1^2} + 16\sqrt{1-\omega_2^2}, \tag{19}$$

where $\omega_i = \sin\theta_i$, $i = 1, 2$.

7. Perturbed Field Configurations

Given that the sine-Gordon model possesses many solutions which resemble perturbed kinks (*i.e.* which are given by kinks and breathers) we have also tried to see what happens when one perturbs a kink and lets it evolve in time. Again we have looked at various perturbations, paying particular attention to configurations which were generated by adding to the kink an extra perturbation of the form

$$\delta\phi(t=0) = \frac{B}{\cosh(\mu x)}, \quad \delta\frac{\partial\phi}{\partial t}(t=0) = \frac{A}{\cosh(\nu x)}. \tag{20}$$

In our simulations we used various values of A, B, μ and ν. In all cases the perturbation made the kink move and generated many moving breather-like configurations. To see what the system would finally settled at we absorbed the energy at the boundaries of the grid. This had the effect of slowing down the kink and also of absorbing and/or altering some breather-like structures. The process was very slow and the results were somewhat inconclusive. What we can say at this stage is that a general field configuration gradually splits into moving kinks and breathers, and some radiation, which quickly moves out to the boundaries. However, the resultant field configuration is metastable; it still radiates, albeit very slowly, and gradually evolves towards a field configuration involving mainly a kink. Whether at the end of its evolution we end up with a kink or a kink with some breathers is hard to determine.

8. Simple models

Here we mention briefly our simple models which partially explain what we have seen in our numerical simulations. First we consider the sigma model in 2 dimensions. In this case the soliton possesses many vibrational modes [12] and so we have constructed a model which treats the soliton as a system of four masses (connected to each other by springs) [4]. The system of four masses is sent towards the potential hole and then as it falls into it the masses begin to oscillate. These oscillations then model the soliton vibrations seen in full simulations. The energy is transferred to these oscillations and if the energy of the centre of mass is too low the system is trapped in the hole. Sometimes, when the system reaches one of the edges of the hole it happens to be in a state that allows the energy of the oscillations to get transferred back to the system as a whole (the energy of its centre of mass) and the soliton can come out. Whether this happens or not depends on the flow of the energy between the variational modes and on the kinetic and potential energy of the soliton. Hence, as we showed in [4], the model reproduces quite well the main features of the scattering pattern seen in the full simulations.

In the Sine-Gordon case we have looked at the old results on the scattering of kinks on point inpurities [13] (showing a similar trapping/transmission/reflection pattern) and their recent explanation by Goodman and collaborators [14]. In [14] the authors explained the observed results by invoking an interaction of the kink with the oscillation of the vacuum (around the impurity) which was described by a standing wave whose amplitude was a further degree of freedom (in addition to the position of the kink). The model reproduced all the features of the results of the original simulations reported in [13] and so the two models discussed by us in [5] are based on the adaptation of the ideas of Goodman et al to our case. In both models we introduced degrees of freedom describing various standing waves in the hole (in one model the waves were restricted to the edges of the hole and in the other they described the global standing waves in the hole). The waves in both models were chosen somewhat arbitrarily as the idea was not to reproduce the pattern in any detail but just to check whether the mechanism of Goodman can be applied to our case too.

In fact both models worked surprisingly well. They reproduced the pattern quite well, although the critical value of velocity was a little too high. Given that these models involved only very few (3 or 4) degrees of freedom we were very encouraged by these results; they require further work to understand better which modes are important and which are less so.

9. Final comments and Conclusions.

We started this talk with a very brief review of the topological solitons and of their dynamics. Then we reviewed the results of our studies of the scattering of topological solitons on a potential obstruction, of both a barrier and a hole-type. We finished by reporting some results on breathers and wobbles (bound states of breathers and kinks).

Our results have shown that when a soliton was sent towards the barrier its behaviour resembled that of a point particle. Thus at low energies the soliton was reflected by the barrier and at higher energy it was transmitted. The scattering process was very elastic. During the scattering the kinetic energy of the soliton was gradually converted into the energy needed to 'climb the barrier'. If the soliton had enough energy to get to the 'top' of the barrier then it was transmitted, otherwise it slid back regaining its kinetic energy.

In the hole case, the situation was different. This time, as the soliton entered the hole it gained extra energy. Some of this energy was converted into the kinetic energy of the soliton, some was radiated away. So when the soliton tried to 'get out' of the hole it had less kinetic energy than at its entry and, when this energy was too low, it remained trapped in the hole. During the scattering process, like in the case of a barrier, the soliton's size changed and so it started oscillating. Afterwards, even when the soliton left the hole, its size continued to oscillate. Hence some energy was transferred to the oscillations resulting in the emission of radiation *i.e.*

Journal of Physics: Conference Series **128** (2008) 012027 doi:10.1088/1742-6596/128/1/012027

in an inelastic process.

We have also looked at the scattering of breathers on potential holes. As breathers can be thought of as bound states of kinks and antikinks they can split leaving a trapped kink or an antikink in the hole and allowing its partner to escape either forwards or backwards. In addition, as the energy of the breather depends on the frequency of its oscillation ('breathing') this frequency can change as well. All such phenomena were seen in our simulations and we hope to report on this more fully in the near future [8].

We have also looked at some field configurations, which are solutions of the equations of motion and which describe bound states of kinks and breathers. As the energy of each breather depends on its frequency (and vanishes in the limit of this frequency going to 1) the extra energy, due to these extra breathers, does not have to be very large. The solutions appear to be stable and this stability is guaranteed by the integrability of the model. We have tested this numerically and have found that small perturbations, due to the discretisations, do not alter this stability. To change it we need something more drastic - like the absorption or the space variation of the potential (*i.e.* the coefficient of the sin^2 term in the Lagrangian). But even then the effects are not very large - one sees splitting of breathers *etc* but no 'global annihilation'.

Finally, all our results (on the scattering of topological solitons on obstructions) generalise to other models; such as *i.e.* the Landau-Lifschitz model with a position dependent potential or an external magnetic field, other models in (1+1) dimensions, such as a $\lambda\phi^4$ model or even models describing ferro- and anti-ferro-magnets [15].

Clearly a lot of work still has to be done in this area.

Acknowledgments

This talk was given by WJZ at the QTS5 conference which was held in Valladolid in July 2007. WJZ wants to thank the Organisers for their hospitality.

References

[1] Manton N and Sutcliffe PM 2004 *Topological Solitons* (Cambridge University Press)
[2] Weidig T 1999 *Nonlinearity* **12** 1489–1503
Eslami P, Sarbishaei M and Zakrzewski WJ 2000 *Nonlinearity*, **13**, 1867-1881
[3] Papanicolaou N and Tomaras TN 1991 *Nucl. Phys.* **B 360**, 425
[4] Piette BMAG and Zakrzewski WJ 2006 *J. Phys.* **A 40**, 329-346
[5] Piette BMAG and Zakrzewski WJ 2007 *J. Phys.* **A 40**, 5995-6010
[6] Javidan K 2006 *J.Phys.* **A 39** 10565-10574
[7] Romanczukiewicz T 2006 *J. Phys.* **A 39**, 3479-3494
[8] Piette BMAG and Zakrzewski WJ, in preparation (2007)
[9] Kälberman G 2004 *J. Phys.* **A 37**, 11607
[10] Hirota R 1980 "Direct Methods In Soliton Theory," In *Soliton*, eds. Bullough, R.K., Caudrey, P.J., 157-176
[11] Ferreira LA, Piette BMAG and Zakrzewski WJ 2007 Wobbles and other kink-breather solutions of the Sine Gordon model *Preprint*, th-hep:0708/1088
[12] Piette BMAG and Ward RS 2005 *Physica* **D 201**, 45-55
[13] Fei Z, Kivshar YuS, and Vazquez L 1992 *Phys. Rev.* **A 45**, 6019-6030
[14] Goodman RH and Haberman R 2004 *Physica* **D 195**, 303-323
[15] Martin Speight J 2006 *Nonlinearity* **19**, 1565-79

V International Symposium on Quantum Theory and Symmetries IOP Publishing
Journal of Physics: Conference Series **128** (2008) 012028 doi:10.1088/1742-6596/128/1/012028

A linear chain of interacting harmonic oscillators: solutions as a Wigner Quantum System

S Lievens, N I Stoilova[1] and J Van der Jeugt

Department of Applied Mathematics and Computer Science, Ghent University, Krijgslaan 281-S9, B-9000 Gent, Belgium.

E-mail: `Stijn.Lievens@UGent.be`, `Neli.Stoilova@UGent.be`, `Joris.VanderJeugt@UGent.be`

Abstract. We consider a quantum mechanical system consisting of a linear chain of harmonic oscillators coupled by a nearest neighbor interaction. The system configuration can be closed (periodic boundary conditions) or open (non-periodic case). We show that such systems can be considered as Wigner Quantum Systems (WQS), thus yielding extra solutions apart from the canonical solution. In particular, a class of WQS-solutions is given in terms of unitary representations of the Lie superalgebra $\mathfrak{gl}(1|n)$. In order to determine physical properties of the new solutions, one needs to solve a number of interesting but difficult representation theoretical problems. We present these problems and their solution, and show how the new results yield attractive properties for the quantum system (energy spectrum, position probabilities, spacial properties).

1. Introduction

Coupled systems describing the interaction of oscillating or scattering subsystems and the corresponding operators have been widely used in classical and quantum mechanics [1–5].

Recently, we have taken up the study of such systems as Wigner Quantum Systems (WQS) [6–8]. The particular system under consideration consists of a string of n identical harmonic oscillators, each having the same mass m and frequency ω. The position and momentum operator for the rth oscillator ($r = 1, 2, \ldots, n$) are given by $\hat{q}_r$ and $\hat{p}_r$ respectively; more precisely, $\hat{q}_r$ measures the displacement of the rth mass point with respect to its equilibrium position. The oscillators are coupled by some nearest neighbor coupling, represented by terms of the form $(\hat{q}_r - \hat{q}_{r+1})^2$ in the Hamiltonian. The system configuration can either be closed (periodic boundary conditions), i.e. $\hat{q}_{n+1} = \hat{q}_1$, or open (fixed wall boundary conditions), i.e. $\hat{q}_0 = \hat{q}_{n+1} = 0$.

When treating a system as a WQS, the canonical commutation relations (CCRs) are *not* required to hold. Instead, the compatibility of the Hamilton equations and the Heisenberg equations is imposed. Expressing this compatibility leads to the so-called compatibility conditions (CCs), and these need to be solved (subject to some unitarity conditions). In [9, 10], we have shown that the CCs associated with a system consisting of a coupled oscillator chain has solutions in terms of generators of the $\mathfrak{gl}(1|n)$ Lie superalgebra [11]. We are only going to work with finite-dimensional unitary irreducible representations (unirreps) of $\mathfrak{gl}(1|n)$, so in fact,

[1] Permanent address: Institute for Nuclear Research and Nuclear Energy, Boul. Tsarigradsko Chaussee 72, 1784 Sofia, Bulgaria

all the solutions in this article are non canonical, and fall within the framework of Quantum Systems with a finite Hilbert space [12].

In order to determine physical properties of the system, such as the energy spectrum, position eigenvalues or position probabilities, one is led to considering problems in the representation theory of $\mathfrak{gl}(1|n)$. For the finite-dimensional unirreps of $\mathfrak{gl}(1|n)$, a Gel'fand-Zetlin basis (GZ-basis) is known, as well as the explicit action of the generators on the basis vectors of the representation [13]. These actions, especially the non-diagonal ones, on a general GZ-pattern are, however, quite involved and thus some of the tasks at hand (e.g. determining position eigenvalues and eigenvectors) are far from trivial. Nevertheless, we have been able to solve these problems quite generally [14].

In this article, we first introduce the Hamiltonians of the quantum systems. We then show briefly how to derive and manipulate the CCs such that they are in a form suitable for finding Lie superalgebraic solutions. Next it is indicated how a solution in terms of the odd generators of the Lie superalgebra $\mathfrak{gl}(1|n)$ is found. As a last part of Section 2, the unirreps of $\mathfrak{gl}(1|n)$ are introduced, focusing on one particular class of representations, namely the ladder representations $V(p)$.

Then, in Section 3, we turn to the study of the physical properties of the systems, starting with the energy spectrum, stressing the difference between the systems with open and closed boundary conditions.

Determining the energy spectrum is a relatively easy task compared to determining the spectrum of the position operators. The key point here is to introduce a second set of generators of $\mathfrak{gl}(1|n)$ and recognizing that for this second set of generators the position operators $\hat{q}_r$ are in fact elements of a $\mathfrak{gl}(1|1)$ subalgebra. The eigenvalue problem is thus solved by considering the branching $\mathfrak{gl}(1|n) \to \mathfrak{gl}(1|1) \oplus \mathfrak{gl}(n-1)$, the multiplicity of the different eigenvalues being determined by the dimension of the various $\mathfrak{gl}(n-1)$ representations. The eigenvector problem is even harder as this involves expressing the highest weight vector w.r.t. the second set of generators in terms of the original basis vectors of the representation. For this article, we restrict ourselves completely to the ladder representations.

Finally, as an application, we give some numerical results concerning the position probabilities when the system is in a stationary state. We remark how the system behaves differently according to whether it is in the ground state or in the most excited state. Also, this behavior is in line with physical intuition about such systems. This is the subject of Section 5.

In Section 6 a short conclusion and outlook are given.

2. A Lie superalgebraic solution to the compatibility conditions

2.1. System with periodic boundary conditions

We start with a quick derivation of the compatibility conditions for a linear system consisting of n interacting but otherwise identical harmonic oscillators. We assume that each oscillator only interacts explicitly with its direct neighbors. The Hamiltonian of such a system is given by:

$$\hat{H}_P = \sum_{r=1}^{n} \left(\frac{\hat{p}_r^2}{2m} + \frac{m\omega^2}{2}\hat{q}_r^2\right) + \sum_{r=1}^{n} \frac{cm}{2}(\hat{q}_r - \hat{q}_{r+1})^2, \tag{1}$$

where each oscillator has mass m and frequency ω. The position and momentum operator of each oscillator are given by $\hat{q}_r$ and $\hat{p}_r$ respectively. To be more precise, $\hat{q}_r$ measures the displacement of the rth oscillator with respect to its equilibrium position. Finally, the positive constant c denotes the coupling strength. For now, we assume periodic boundary conditions, i.e. we assume that

$$\hat{q}_{n+1} \equiv \hat{q}_1. \tag{2}$$

When treating the system as a Wigner Quantum System, one imposes the compatibility between the Hamilton equations and the Heisenberg equations. It is known that some systems then also have non canonical solutions, i.e. solutions for which the canonical commutation relations between the position and momentum operators do not hold [6, 9, 15]. For the Hamiltonian under consideration, the compatibility conditions can be computed as:

$$[\hat{H}_P, \hat{q}_r] = -\frac{i\hbar}{m}\hat{p}_r \ (= -i\hbar\dot{\hat{q}}_r), \tag{3}$$

$$[\hat{H}_P, \hat{p}_r] = -i\hbar cm\, \hat{q}_{r-1} + i\hbar m(\omega^2 + 2c)\, \hat{q}_r - i\hbar cm\, \hat{q}_{r+1} \ (= -i\hbar\dot{\hat{p}}_r), \tag{4}$$

with $r \in \{1, 2, \ldots, n\}$, $\hat{q}_{n+1} = \hat{q}_1$ and $\hat{q}_0 = \hat{q}_n$. The goal is thus to find (self-adjoint) operators $\hat{q}_r$ and $\hat{p}_r$ $(r = 1, \ldots, n))$ such that (3) and (4) are satisfied with $\hat{H}_P$ given by (1).

In order to find solutions of the CCs (other than the canonical solution), one introduces the following discrete Fourier transforms of position and momentum operators:

$$\hat{q}_r = \sum_{j=1}^{n} \sqrt{\frac{\hbar}{2mn\omega_j}} \left(e^{-2\pi ijr/n} a_j^+ + e^{2\pi ijr/n} a_j^-\right), \tag{5}$$

$$\hat{p}_r = i \sum_{j=1}^{n} \sqrt{\frac{m\omega_j\hbar}{2n}} \left(e^{-2\pi ijr/n} a_j^+ - e^{2\pi ijr/n} a_j^-\right). \tag{6}$$

Here, the ω_j are positive numbers given by

$$\omega_j^2 = \omega^2 + 2c - 2c\cos(\frac{2\pi j}{n}) = \omega^2 + 4c\sin^2(\frac{\pi j}{n}), \tag{7}$$

while the (yet unknown) operators $a_j^\pm$ satisfy $(a_j^\pm)^\dagger = a_j^\mp$, implying that the operators $\hat{q}_r$ and $\hat{p}_r$ are indeed self-adjoint. In terms of the operators $a_j^\pm$, the Hamiltonian has a simple expression:

$$\hat{H}_P = \sum_{j=1}^{n} \frac{\hbar\omega_j}{2}(a_j^- a_j^+ + a_j^+ a_j^-) = \sum_{j=1}^{n} \frac{\hbar\omega_j}{2}\{a_j^-, a_j^+\}. \tag{8}$$

It is now easy to check that the CCs (3)-(4) are equivalent with finding operators $a_k^\pm$ (subject to $(a_k^\pm)^\dagger = a_k^\mp$) such that

$$\left[\sum_{j=1}^{n} \omega_j\{a_j^-, a_j^+\}, a_k^\pm\right] = \pm 2\omega_k a_k^\pm, \qquad (k = 1, 2, \ldots, n). \tag{9}$$

Note how these relations involve both commutators and anti-commutators.

2.2. System with fixed wall boundary conditions

We now consider the same system of coupled harmonic oscillators, but this time with fixed wall boundary conditions. In this case, the Hamiltonian is

$$\hat{H}_{FW} = \sum_{r=1}^{n} (\frac{\hat{p}_r^2}{2m} + \frac{m\omega^2}{2}\hat{q}_r^2) + \sum_{r=0}^{n} \frac{cm}{2}(\hat{q}_r - \hat{q}_{r+1})^2, \tag{10}$$

with the same data as in (1) but with boundary conditions

$$\hat{q}_0 = \hat{q}_{n+1} \equiv 0 \quad (\text{and } \hat{p}_0 = \hat{p}_{n+1} \equiv 0). \tag{11}$$

Stated otherwise, the first and last oscillator (those numbered 1 and n) are assumed to be attached to a fixed wall rather than being adjacent to one another.

The treatment of this system as a Wigner Quantum system is very similar to that of the system with periodic boundary conditions. In fact, the CCs are identical to (3) and (4) but with boundary conditions given by (11) (and with $\hat{H}_P$ replaced by $\hat{H}_{FW}$). These new boundary conditions, however, imply the use of a different kind of transform (instead of the discrete Fourier transform). In this case, the following transformations are introduced:

$$\hat{q}_r = \sum_{j=1}^{n} \sqrt{\frac{\hbar}{m(n+1)\tilde{\omega}_j}} \sin(\frac{rj\pi}{n+1})(a_j^+ + a_j^-), \tag{12}$$

$$\hat{p}_r = i \sum_{j=1}^{n} \sqrt{\frac{m\tilde{\omega}_j\hbar}{n+1}} \sin(\frac{rj\pi}{n+1})(a_j^+ - a_j^-), \tag{13}$$

where the $\tilde{\omega}_j$ are positive numbers given by

$$\tilde{\omega}_j^2 = \omega^2 + 2c - 2c\cos(\frac{j\pi}{n+1}) = \omega^2 + 4c\sin^2(\frac{j\pi}{2(n+1)}). \tag{14}$$

As before, the operators $a_j^\pm$ satisfy the adjointness conditions $(a_j^\pm)^\dagger = a_j^\mp$. The Hamiltonian $\hat{H}_{FW}$, when expressed in terms of the operators $a_j^\pm$ is given by

$$\hat{H}_{FW} = \sum_{j=1}^{n} \frac{\hbar\tilde{\omega}_j}{2}\{a_j^-, a_j^+\}, \tag{15}$$

which is identical to (8) but with ω_j replaced by $\tilde{\omega}_j$. Also the CCs are identical to (9) up to the same replacement. This implies that the algebraic solutions will be similar, even though the conclusions about physical properties will be different due to the different numerical values of the numbers ω_j and $\tilde{\omega}_j$ (and due to the different transformations used).

2.3. Solution of the CCs in terms of $\mathfrak{gl}(1|n)$ *generators*

The CCs (9) involve both commutators and anti-commutators. Moreover, they resemble certain algebraic relations satisfied by the odd $\mathfrak{gl}(1|n)$ generators. The Lie superalgebra $\mathfrak{gl}(1|n)$ [11] has standard basis elements e_{jk} $(j,k = 0,1\ldots,n)$, where e_{k0} and e_{0k} $(k = 1,\ldots,n)$ are the odd elements (having degree one) while the other elements are the even elements (having degree zero). The bracket is given by:

$$[\![e_{ij}, e_{kl}]\!] = \delta_{jk}e_{il} - (-1)^{\deg(e_{ij})\deg(e_{kl})}\delta_{il}e_{kj}, \tag{16}$$

and we impose the star condition $e_{ij}^\dagger = e_{ji}$. Then, one can give a solution of (9) in terms of the odd elements of this algebra. More explicitly, one has

$$a_j^- = \sqrt{\frac{2\beta_j}{\omega_j}}\, e_{j0}, \quad a_j^+ = \sqrt{\frac{2\beta_j}{\omega_j}}\, e_{0j} \quad (j = 1,\ldots,n) \tag{17}$$

where

$$\beta_j = -\omega_j + \frac{1}{n-1}\sum_{k=1}^{n}\omega_k, \quad (j = 1,\ldots,n). \tag{18}$$

Clearly, all these numbers β_j should be nonnegative. It can be proved [9] that (for $n \geq 4$) this implies that the coupling constant is bounded above by some positive critical value c_0 (dependent upon n). As was to be expected, the same solution is valid for the case of fixed wall boundary conditions but with the constants ω_j replaced by $\tilde{\omega}_j$ (the β_j are then denoted naturally by $\tilde{\beta}_j$). Also in the fixed wall boundary conditions case, the coupling constant c is bounded above by a critical value $\tilde{c}_0$ (different but similar to c_0) [10].

2.4. Unitary irreducible representations of $\mathfrak{gl}(1|n)$

In order to study physical properties of the systems, one has to work with specific unitary irreducible representations of $\mathfrak{gl}(1|n)$. These unitary representations $W = W([m]_{n+1})$ of $\mathfrak{gl}(1|n)$ are well known [16]: they are labeled by some $(n+1)$-tuple $[m]_{n+1}$ subject to certain conditions. Even more: for such representations, a Gel'fand-Zetlin basis has been constructed and the explicit action of the $\mathfrak{gl}(1|n)$ generators on the basis vectors of the representation is also known [13]. Explicit actions of generators on a Gel'fand-Zetlin basis are usually quite involved, and this is also the case for $\mathfrak{gl}(1|n)$. In particular, the action of the odd generators e_{0j} and e_{j0} on a GZ-basis vector is very complicated, see [13, Eqs. (2.25) and (2.26)].

As a special case, we consider a simple, yet interesting and quite rich, class of finite-dimensional representations $W([1, p-1, 0, \ldots, 0]) \equiv V(p)$, characterized by a positive integer p. A notation more convenient than the general GZ-patterns for the basis vectors of $V(p)$ is:

$$w(\theta; s) \equiv w(\theta; s_1, s_2, \ldots, s_n), \qquad \theta \in \{0,1\},\ s_i \in \{0,1,2,\ldots\},\ \text{and } \theta + s_1 + \cdots + s_n = p. \quad (19)$$

The action of the $\mathfrak{gl}(1|n)$ generators on the basis (19) is given by $(1 \leq k \leq n)$ [13, 14]:

$$e_{00} w(\theta; s) = \theta\ w(\theta; s), \quad (20)$$
$$e_{kk} w(\theta; s) = s_k\ w(\theta; s), \quad (21)$$
$$e_{k0} w(\theta; s) = \theta\sqrt{s_k+1}\ w(1-\theta; s_1, \ldots, s_k+1, \ldots, s_n), \quad (22)$$
$$e_{0k} w(\theta; s) = (1-\theta)\sqrt{s_k}\ w(1-\theta; s_1, \ldots, s_k-1, \ldots, s_n). \quad (23)$$

From these one deduces the action of other elements e_{kl}. The basis $w(\theta; s)$ of $V(p)$ is orthonormal, i.e. $\langle w(\theta; s), w(\theta'; s')\rangle = \delta_{\theta,\theta'}\delta_{s,s'}$, and with respect to this inner product the action of the generating elements satisfies the conjugacy relations $e_{k0}^\dagger = e_{0k}$ and $e_{0k}^\dagger = e_{k0}$.

3. Energy spectrum of the systems

As can be seen from (8) and (17), the Hamiltonian for both systems is a linear combination of elements e_{kk} $(k = 0, 1, \ldots, n)$. Since such elements act diagonally on the basis vectors of any representation $W([m]_{n+1})$, [13], the basis vectors of such a representation will be eigenvectors of the Hamiltonian. Stated otherwise, the basis vectors of the representation are the stationary states of the system.

We will now restrict our attention to the ladder representations $V(p)$ introduced in the previous section. In particular, for fixed wall boundary conditions we have:

$$\hat{H}_{FW} = \hbar(\tilde{\beta}\, e_{00} + \sum_{k=1}^{n} \tilde{\beta}_k\, e_{kk}),$$

with $\tilde{\beta} = \sum_{k=1}^{n} \tilde{\beta}_k = \sum_{k=1}^{n} \tilde{\omega}_k$. From (20) and (21) it follows immediately that:

$$\hat{H}_{FW}\, w(\theta; s) = \hbar(\tilde{\beta}\theta + \sum_{k=1}^{n} \tilde{\beta}_k s_k) w(\theta; s) = \hbar \tilde{E}_{\theta,s}\, w(\theta; s).$$

Journal of Physics: Conference Series **128** (2008) 012028 doi:10.1088/1742-6596/128/1/012028

When $c = 0$, one has that $\tilde{\beta}_k = \omega/(n-1)$ and $\tilde{\beta} = \omega n/(n-1)$, so in this case there are only two eigenvalues namely

$$\tilde{E}_{0,s} = \frac{\omega p}{n-1} \quad \text{and} \quad \tilde{E}_{1,s} = \frac{\omega p}{n-1} + \omega$$

with multiplicities

$$\binom{p+n-1}{n-1} \quad \text{and} \quad \binom{p+n-2}{n-1}$$

respectively. For $c > 0$, these two energy levels split into

$$\binom{p+n-1}{n-1} + \binom{p+n-2}{n-1}$$

energy levels, each with multiplicity one. Figure 1(a) gives an example.

In the case of periodic boundary conditions, however, the system has much greater symmetry and hence one expects fewer different energy levels with higher degeneracies. Indeed, in [10] it is shown that the number of energy levels is in general given by

$$\binom{p+\lfloor n/2 \rfloor}{p} + \binom{p+\lfloor n/2 \rfloor - 1}{p-1}.$$

Figure 1(b) illustrated this. Also, the degeneracy of an energy level $E_{\theta,s}$ can in general be calculated from θ and s.

4. On the eigenvalues and eigenvectors for the position operators

Under the solution (17), the position operators $\hat{q}_r$ are quite arbitrary odd elements of the Lie superalgebra $\mathfrak{gl}(1|n)$ as can be seen from (12). Since the action of the odd generators on a GZ-pattern is very complicated, determining the spectrum of the position operators is an involved task. However, in [14] a method was developed to determine the eigenvalues (and eigenvectors) of such an odd operator for a general unirrep $W([m]_{n+1})$ of $\mathfrak{gl}(1|n)$. This method goes as follows: in a first step, a new set of generators $E_{0j}^{(r)}$ and $E_{j0}^{(r)}$ (and $E_{00}^{(r)}$) for $\mathfrak{gl}(1|n)$ is introduced such that the operator $\hat{q}_r$ has an easy form in terms of these new generators. More in particular, the operator $\hat{q}_r$ will be an element of a $\mathfrak{gl}(1|1)$ subalgebra generated by $E_{00}^{(r)}$, $E_{nn}^{(r)}$, $E_{0n}^{(r)}$ and $E_{n0}^{(r)}$:

$$\hat{q}_r \sim E_{0n}^{(r)} + E_{n0}^{(r)}.$$

Hence, it it clear that the decomposition of the representation $W([m]_{n+1})$ as a $\mathfrak{gl}(1|1) \oplus \mathfrak{gl}(n-1)$ representation solves the eigenvalue problem. The eigenvalues are given by solving the $\mathfrak{gl}(1|1)$ eigenvalue problem for the various $\mathfrak{gl}(1|1)$ representations occurring, and their multiplicities by the dimension of the corresponding $\mathfrak{gl}(n-1)$ representation (for which there is a known dimension formula). The eigenvector problem is much harder, and involves expressing the new GZ-patterns in terms of the old GZ-patterns.

Let us now go into a little more detail. First, the position operator $\hat{q}_r$ is written as:

$$\hat{q}_r = \sqrt{\frac{\hbar}{mn}} \sum_{j=1}^{n} \left(\gamma_j \, e^{2\pi i j r/n} e_{j0} + \gamma_j \, e^{-2\pi i j r/n} e_{0j} \right) \tag{24}$$

$$= \sqrt{\frac{\hbar \gamma}{mn}} (E_{n0}^{(r)} + E_{0n}^{(r)}), \tag{25}$$

where we use the notation

$$\gamma_j = \sqrt{\beta_j}/\omega_j \quad (j = 1, \ldots, n) \text{ and } \gamma = \gamma_1^2 + \cdots + \gamma_n^2. \tag{26}$$

(a) Fixed Wall Boundary Conditions (b) Periodic Boundary Conditions

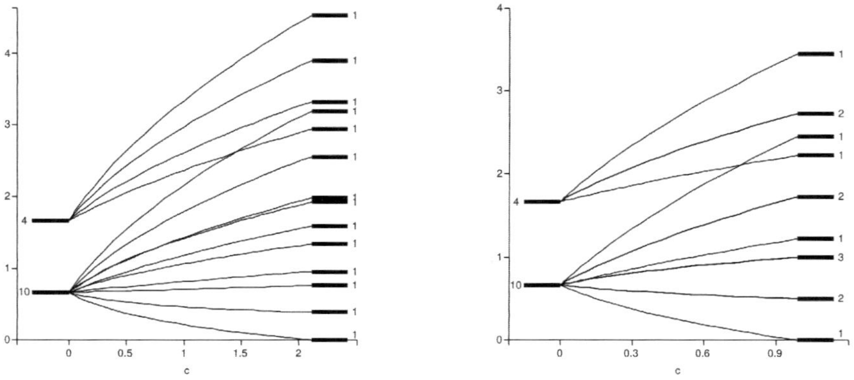

Figure 1. (a) The energy levels of the quantum system with fixed wall boundary conditions for $n = 4$ in the representation $V(p)$ with $p = 2$ and $\hbar = \omega = 1$; c ranges from 0 to $\tilde{c}_0$. The vertical axis gives the energy values and the numbers next to the levels refer to their multiplicity. When $c = 0$, there are only two energy levels with multiplicities 10 and 4 respectively. When $0 < c < \tilde{c}_0$, however, there are (in general) 14 energy levels each with multiplicity 1. (b) The energy levels of the quantum system with periodic boundary conditions for $n = 4$ in the representation $V(p)$ with $p = 2$ and $\hbar = \omega = 1$; c ranges from 0 to c_0. For $c = 0$, one always has the same system irrespective of the boundary conditions in place, so again we find the same two energy levels. When $c > 0$, however, there are in general only 9 different energy levels, some non-degenerate others two-fold and one three-fold degenerate.

This defines the operators $E_{n0}^{(r)}$ and $E_{0n}^{(r)}$. Then, this is extended to $E_{j0}^{(r)}$ (and $E_{0j}^{(r)}$) with $1 \le j \le n-1$ by using a unitary (Hessenberg) matrix. The operators $E_{j0}^{(r)}$ and $E_{0j}^{(r)}$ $(j = 1, \ldots, n)$ then also generate $\mathfrak{sl}(1|n)$ as they also satisfy the defining relations of $\mathfrak{sl}(1|n)$ [14, 17]. (The operator $E_{00}^{(r)}$ is then defined in the appropriate way.)

A unirrep of $\mathfrak{gl}(1|1)$ is characterized by two real numbers a and b, such that either $a + b > 0$ or $a + b = 0$. When $a + b > 0$, the representation $W([a, b])$ is two-dimensional with weights (a, b) and $(a - 1, b + 1)$. Furthermore, calling the basis vectors of the representation v and w one has:

$$e_{01}v = 0, \quad e_{10}v = \sqrt{a+b}\,w, \quad e_{01}w = \sqrt{a+b}\,v, \quad e_{10}w = 0.$$

In this case, it is clear that $(e_{01} + e_{10})(v \pm w) = \pm\sqrt{a+b}(v \pm w)$. When $a + b = 0$, the representation $W([a, -a])$ is one-dimensional and one has that $(e_{01} + e_{10})(v) = 0$. In this case the weight is given by $(a, -a)$.

In the case of the ladder representation $V(p)$ the decomposition $\mathfrak{gl}(1|n) \to \mathfrak{gl}(1|1) \oplus \mathfrak{gl}(n-1)$ is given by

$$V(p) \to W([0,0]) \times V([p, 0, \ldots, 0]) \oplus \bigoplus_{K=0}^{p-1} W([1, p-1-K]) \times V([K, 0, \ldots, 0]), \qquad (27)$$

where the $\mathfrak{gl}(n-1)$ representation $V([K, 0, \ldots, 0])$ has dimension $\binom{n-2+K}{n-2}$. So $\hat{q}_r$ has $2p + 1$

eigenvalues in all, namely $\pm x_K = \pm\sqrt{\frac{\hbar\gamma}{mn}(p-K)}$, where $0 \leq K \leq p-1$, with multiplicities $\binom{n-2+K}{n-2}$, and $x_p = 0$ with multiplicity $\binom{n-2+p}{n-2}$.

One can also explicitly compute the eigenvectors of $\hat{q}_r$ in the ladder representation as follows. First, any basis vector $w(\theta; s_1, \ldots, s_n)$ can be obtained from the highest weight vector $w(1; p-1, 0, \ldots, 0)$ by acting on it with negative root vectors. Explicitly, one has:

$$w(\theta; s_1, \ldots, s_n) = \frac{e_{n,n-1}^{p-\theta-\sum_{j=1}^{n-1} s_j} e_{n-1,n-2}^{p-\theta-\sum_{j=1}^{n-2} s_j} \cdots e_{32}^{p-\theta-\sum_{j=1}^{2} s_j} e_{21}^{p-\theta-\sum_{j=1}^{1} s_j} e_{10}^{1-\theta}}{\sqrt{p^{1-\theta}\prod_{k=1}^{n-1}(p-\theta-\sum_{j=1}^{k} s_j)!(s_k+1)_{p-\theta-\sum_{j=1}^{k} s_j}}} w(1; p-1, 0, \ldots, 0),$$

where $(a)_j = a(a+1)\cdots(a+j-1)$ is the Pochhammer symbol or rising factorial.

We now define a basis $v(\phi; t_1, \ldots, t_n)$ relative to the new generators E_{jk} as follows (for brevity, the superscripts r are temporarily dropped):

$$v(\phi; t_1, \ldots, t_n) = \frac{E_{n,n-1}^{p-\phi-\sum_{j=1}^{n-1} t_j} E_{n-1,n-2}^{p-\phi-\sum_{j=1}^{n-2} t_j} \cdots E_{32}^{p-\phi-\sum_{j=1}^{2} t_j} E_{21}^{p-\phi-\sum_{j=1}^{1} t_j} E_{10}^{1-\phi}}{\sqrt{p^{1-\phi}\prod_{k=1}^{n-1}(p-\phi-\sum_{j=1}^{k} t_j)!(t_k+1)_{p-\phi-\sum_{j=1}^{k} t_j}}} v(1; p-1, 0, \ldots, 0). \tag{28}$$

Of course, in order to make this definition work, we have to express the highest weight vector $v(1; p-1, 0, \ldots, 0)$ in terms of the "old" basis vectors $w(\theta; s)$. So, one has to look for the unique vector that is annihilated by all positive root vectors $E_{jk}^{(r)}$ with $k > j$. For the ladder representation $V(p)$ one has that

$$v(1; p-1, 0, \ldots, 0) = \frac{1}{(\gamma_1^2+\gamma_2^2)^{(p-1)/2}} \sum_{u=0}^{p-1} (-1)^u e^{-2\pi i r u/n} \sqrt{\binom{p-1}{u}} \times \gamma_1^{p-1-u}\gamma_2^u \, w(1; u, p-1-u, 0, \ldots, 0). \tag{29}$$

We remark that Proposition 4 of [14] solves this particular problem for arbitrary unirreps $W([m]_{n+1})$. Using the actions (22) and (23), it is now easy to check that the orthonormal eigenvectors for $\pm x_K \neq 0$ are:

$$\psi_{r,\pm x_K,t} = \frac{1}{\sqrt{2}} v(1; t_1, \ldots, t_{n-1}, p-1-K) \pm \frac{1}{\sqrt{2}} v(0; t_1, \ldots, t_{n-1}, p-K), \tag{30}$$

where $t_1 + \cdots + t_{n-1} = K$. For the eigenvalue 0, the eigenvectors read

$$\psi_{r,0,t} = v(0; t_1, \ldots, t_{n-1}, 0), \qquad t_1 + \cdots + t_{n-1} = p. \tag{31}$$

Note how the multiplicity labels t are in accordance with the previously stated multiplicities of the eigenvalues x_K.

5. Application: position probabilities

For the ladder representations $V(p)$, the expressions (30) (or (31)) together with (29) and (28) give the eigenvectors of the position operators $\hat{q}_r$. Although, admittedly, these expressions do not yield an analytical expression of the eigenvectors in terms of the basis vectors $w(\theta; s)$, they do allow to compute them efficiently using only matrix multiplications. Stated otherwise, the coefficients $C_{r,\pm x_K,t}^{\theta,s}$ in the expansion

$$\psi_{r,\pm x_K,t} = \sum_{\theta,s} C_{r,\pm x_K,t}^{\theta,s} w(\theta; s),$$

Journal of Physics: Conference Series **128** (2008) 012028 doi:10.1088/1742-6596/128/1/012028

may be efficiently computed (numerically).

It is a well known fact (postulate) of quantum mechanics that when measuring an observable, the measurement always yields an eigenvalue of the (self-adjoint) operator associated with that observable. The probability of measuring a certain eigenvalue when the system is in a particular state is determined by the expansion of that state in terms of (orthonormal) eigenvectors of the operator at hand.

So, when the quantum system is in the fixed stationary state $w(\theta, s)$, the probability of measuring for $\hat{q}_r$ the eigenvalue $\pm x_K$ is given by:

$$P(\theta, s, r, \pm x_K) = \sum_{t_1+\cdots+t_{n-1}=K} \left| C^{\theta,s}_{r,\pm x_K,t} \right|^2 . \tag{32}$$

Also, from (30) it is immediately clear that

$$P(\theta, s, r, x_K) = P(\theta, s, r, -x_K). \tag{33}$$

For now, we will restrict our attention to the case of periodic boundary conditions, the case of fixed wall boundary conditions being elaborated upon in [10]. For the case of periodic boundary conditions, it is clear that the position probabilities are independent of the number of the oscillator r, as all oscillators are equivalent. So, one can take $r = 1$. As an explicit example, we choose $n = 4$ (so there are four coupled oscillators) and $p = 10$. This means that each position operator has 21 distinct eigenvalues. We will plot the position probability distributions for the ground state (this is the state $w(0; 0, p, 0, 0) = w(0; 0, 10, 0, 0)$) and for the most excited state (this is the state $w(1; 0, 0, 0, p-1) = w(1; 0, 0, 0, 9)$). These distributions are given in Figure 2, for some values of the coupling constant c.

Let us make a number of observations on these distributions. When the system is in the ground state, the probability distribution function of each position operator is symmetric around its equilibrium position and unimodal. Of course it is also discrete (as we are working in finite-dimensional representations). As the coupling constant c increases, the "peak" around the equilibrium position is sharper. In other words, as the coupling constant becomes larger, the oscillators are more likely to be close to their equilibrium position when the system is in its ground state.

When the system is in its most excited state, the position probabilities are quite different. The probability of finding the oscillator in its equilibrium position is zero. On the other hand, there are certain peaks away from the equilibrium position. As c increases, these peaks are further away from the equilibrium position. In other words, as the coupling constant becomes larger, the oscillators are more likely to be further away from their equilibrium position when the system is in its most excited state.

Note that in this figure one also observes the fact that the range of the spectrum of the position operators depends on the coupling constant c [10].

6. Conclusion

In this article, we have treated a linear chain of harmonic oscillators as a Wigner Quantum System. By suitably manipulating the compatibility conditions, we showed how a solution in terms of odd elements of the Lie superalgebra $\mathfrak{gl}(1|n)$ can be found. The physical properties of the system are then determined by the $\mathfrak{gl}(1|n)$ unitary representation used. Investigating these physical properties leads to quite interesting and challenging problems in the representation theory of $\mathfrak{gl}(1|n)$, and a solution to these problems was presented. The specific examples and figures given used a simple yet interesting class of representations $V(p)$ with p an integer. The properties obtained for such a system, e.g. the position probabilities are in line with physical intuition.

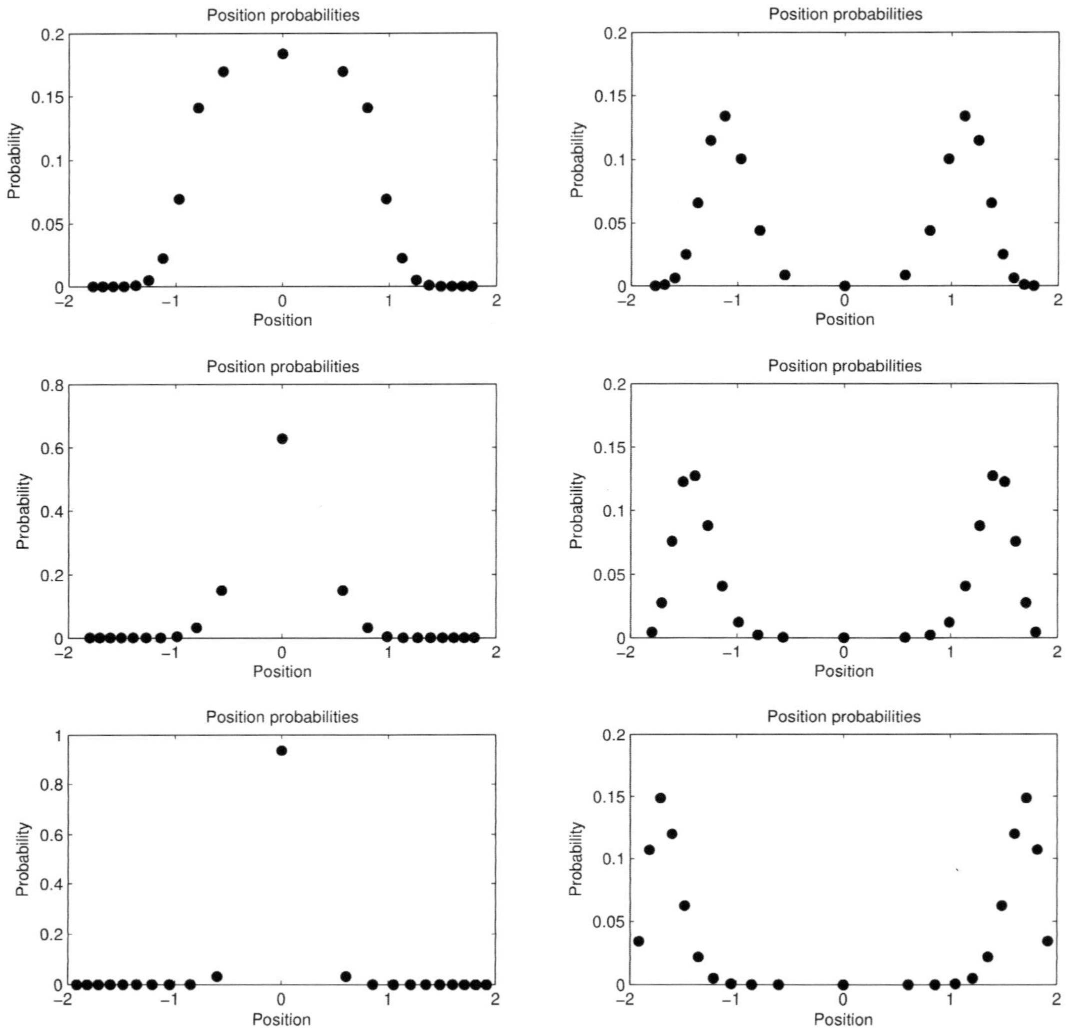

Figure 2. Position probability distribution function for the position operator $\hat{q}_1$, in the periodic boundary case, when $n = 4$ and $p = 10$. In the three rows, $c = 0.1$, $c = 0.4$, $c = 0.8$. The plotted value is $P(\theta, s, 1, \pm x_K)$ for each of the 21 eigenvalues $\pm x_K$ $(K = 0, 1, \ldots, 10)$ of $\hat{q}_1$. This is given for the case when the system is in the stationary state $w(\theta; s)$ corresponding to the ground state (minimum energy) in the left column and in the right column when it is in the stationary state $w(\theta; s)$ corresponding to the most excited state (maximum energy).

Although we restricted ourselves here to the case of $\mathfrak{gl}(1|n)$ solutions, other types of solutions exist, more specifically in terms of $\mathfrak{osp}(1|2n)$ elements. We expect to further investigate this system (and others, such as a multi-dimensional anisotropic oscillator) using a newly constructed class of infinite dimensional unirreps of $\mathfrak{osp}(1|2n)$ [18]. As this class of solutions includes the canonical solution, such a study will be most interesting, although we expect it to be computationally difficult.

Acknowledgments
N.I. Stoilova was supported by a project from the Fund for Scientific Research – Flanders (Belgium) and by project P6/02 of the Interuniversity Attraction Poles Programme (Belgian State – Belgian Science Policy). S. Lievens was also supported by project P6/02.

References

[1] Brun T A and Hartle J B 1999 *Phys. Rev.* D **60** 123503
[2] Audenaert K, Eisert J, Plenio M B and Werner R F 2002 *Phys. Rev.* A **66** 042327
[3] Eisert J and Plenio M B 2003 *Int. J. Quant. Inf.* **1** 479-506
[4] Halliwell J J 2003 *Phys. Rev.* D **68** 025018
[5] Plenio M B, Hartley J and Eisert J 2004 *New J. Phys.* **6** 36
[6] Wigner E P 1950 *Phys. Rev.* **77** 711-712
[7] Kamupingene A H, Palev T D and Tsavena S P 1986 *J. Math. Phys.* **27** 2067-75
[8] Palev T D 1982 *J. Math. Phys.* **23** 1778-84
[9] Lievens S, Stoilova N I and Van der Jeugt J 2006 *J. Math. Phys.* **47** 113504
[10] Lievens S, Stoilova N I and Van der Jeugt J 2007 *Preprint* hep-th/0709.0180
[11] Kac V G 1977 *Adv. Math.* **26** 8-96
[12] Vourdas A 2004 *Rep. Prog. Phys.* **67** 267-320
[13] King R C, Stoilova N I and Van der Jeugt J 2006 *J. Phys.* A **39** 5763-85
[14] Lievens S, Stoilova N I and Van der Jeugt J 2007 *J. Phys.* A **40** 3869-88
[15] Stoilova N I and Van der Jeugt J 2005 *J. Phys.* A **38** 9681-87
[16] Gould M D and Zhang R B 1990 *J. Math. Phys.* **31** 2552-59
[17] King R C, Palev T D, Stoilova N I and Van der Jeugt J 2003 *J. Phys.* A **36** 4337-62
[18] Lievens S, Stoilova N I and Van der Jeugt J 2007 *Preprint* hep-th/0706.4196

Journal of Physics: Conference Series **128** (2008) 012029 doi:10.1088/1742-6596/128/1/012029

Decomposition of scalar potentials of natural Hamiltonians into integrable and perturbative terms: a naive approach

Giovanni Rastelli

Dipartimento di Matematica, Universitá di Torino. Torino, via Carlo Alberto 10, Italia

E-mail: giorast.giorast@alice.it

Abstract. The geometry of Stäckel systems is employed to find, for a given (classical) natural Hamiltonian, another natural Hamiltonian which is integrable by separation of variables and in some sense close to the first one. The study is intended as a first step for a perturbative analysis of a given Hamiltonian system. The method proposed here, still in development, is in large part coordinate independent and effective mainly on manifolds of constant curvature. Examples are given for the quadrupole field and Hénon-Heiles systems. Stäckel systems are associated with quadratic in the momenta first integrals which play a fundamental role in quantization of classical systems.

1. Introduction

In Hamiltonian perturbation theory, Hamiltonian functions are usually prepared to analysis by decomposing them as follows

$$H = L_0(J) + \epsilon H_1(J,\theta) + \epsilon^2 H_2(J,\theta) + \dots \tag{1}$$

where L_0 is an integrable Hamiltonian and (J,θ) its action-angle coordinates, H_i the perturbative terms. We are searching for the best-fitting L_0 among natural Hamilton-Jacobi separable (Stäckel) systems.

Separable systems in that perspective are considered in literature (for example [5, 6, 2] about galactical dynamics) and analytic techniques are employed to find best-fitting separable systems for given problems. However, these approaches assume the separable coordinate systems are given and all computations for the search of the best approximation are performed by using the same coordinates. Here, most of the approximation and comparison work necessary to find the best fitting separable systems is done by considering tensorial objects and therefore in whatever coordinate system. While at a "conjectural stage" the method seems effective, mainly in Riemannian manifolds of constant curvature like Euclidean spaces, n-dimensional spheres and pseudo-spheres and Minkowski spaces, where there is abundance of separable systems and their geometric characterization is very well understood. In two examples on the Euclidean plane, the quadrupole field and Hénon-Heiles systems, numerical evaluations of the method are provided. Stäckel systems are characterized by n independent quadratic in the momenta first integrals in involution and play a basic role in quantization of classical systems, as for example in the Born's model of the hydrogen atom and more recently in [4] and references therein. Therefore, the

Journal of Physics: Conference Series **128** (2008) 012029 doi:10.1088/1742-6596/128/1/012029

present approach naturally concerns quantum systems also. However, such an extension is not yet developed here.

1.1. The objective

Given a Riemannian manifold (Q_n, g), the cotangent bundle T^*Q with canonical coordinates (q^i, p_i) and a natural Hamiltonian function

$$H = \frac{1}{2}g^{ij}p_ip_j + V(q^i) \tag{2}$$

we want to find a function $W_0(q^i)$ such that the natural Hamiltonian

$$L_0 = \frac{1}{2}g^{ij}p_ip_j + W_0 \tag{3}$$

is *Hamilton-Jacobi separable* in some coordinate system and the integral curves associated with L_0 are *close*, in some sense, to those associated with H in some subset of T^*Q.

The perturbative term is defined naturally by $H_p = H - L_0$. By writing $H = L_0 + \epsilon H_p$ in action-angle coordinates for L_0 and expanding in Taylor series for ϵ one obtain exactly the expression (1). The integration of L_0 by the Hamilton-Jacobi method naturally provides canonical coordinates (Jacobi coordinates) that can be transformed into action-angle by standard techniques [10]. Action-angle (or Jacobi) coordinates depend not only on L_0 but also on the other $n-1$ first integrals L_i that make L_0 integrable. Therefore, all the functions (L_0, L_i) must be taken in account in order to provide the best fit in (1). The idea here is to order first the separable systems according to the best fitting given by the first integrals L_i and then consider the associated separated Hamiltonians L_0. The Hamiltonian potentials W_0 can be found, given the separable system associated, by analytic methods [5, 6, 2], here, we will determine these functions just "by inspection" after writing the potential V in the separable coordinates we obtain by applying our method. While rough, this approach provides for the simple systems considered here good results.

2. Separable systems

We recall that a Hamiltonian L_0 is said to be (Stäckel-) separable if the (time independent) Hamilton-Jacobi equation

$$L_0(q^i, p_i = \frac{\partial S}{\partial q^i}) - l = 0$$

$l \in R$, admits a solution $S(q^i, a_i)$ depending on n real parameters (a_i) such that

$$det\left(\frac{\partial^2 S}{\partial q^i \partial a_j}\right) \neq 0$$

and

$$S = S_1(q^1, a_j) + \ldots + S_n(q^n, a_j)$$

In this case S can be found by solving a system of n separated ODE's. Once S has been found, the dynamics of L_0 can be determined by algebraic methods (see for example [1] and references therein).

Journal of Physics: Conference Series **128** (2008) 012029 doi:10.1088/1742-6596/128/1/012029

2.1. Geometric characterization of separable Hamiltonians

The natural Hamiltonian (3) is separable in orthogonal coordinates if and only if [1] there exist n independent symmetric Killing tensors K_i, $i = 0 \ldots n-1$, $K_0 = g$, with common eigenvectors, satisfying

$$[K_i, K_j] = 0, \qquad d(K_i\, dW_0) = 0, \tag{4}$$

where [,] denotes the Schouten bracket. At each point the coordinate curves are the integral curves of the eigenvectors of the tensors K_i. By assuming:

$$L_i = \frac{1}{2} K_i^{jh} p_j p_h + W_i \tag{5}$$

we see that (4) are equivalent to

$$\{L_j, L_i\} = 0 \qquad \{L_0, L_i\} = 0 \tag{6}$$

where { , } denotes the canonical Poisson bracket. Such L_i are then said to be in *involution*. After (5), the second of (6) are equivalent to

$$\nabla W_i - K_i\, dW_0 = 0 \tag{7}$$

2.2. Separable potentials

In orthogonal separable coordinates (q^i) the potential W_0 is separable if and only if [1] there exist n functions $\phi_i(q^i)$ depending each one on a single coordinate, such that we can write (Stäckel)

$$W_0 = g^{11}\phi_1(q^1) + \ldots + g^{nn}\phi_n(q^n) \tag{8}$$

$$W_i = K_i^{jj}\phi_j(q^j) = \lambda_i^1 g^{11}\phi_1(q^1) + \ldots + \lambda_i^n g^{nn}\phi_n(q^n),$$

where (λ_i^j) are the eigenvalues of K_i.

If all the symmetric Killing two-tensors of a manifold are known, it follows that all the possible orthogonal Hamilton-Jacobi separable coordinate systems are (in principle) also known (the nonorthogonal ones are derivable from these ones [1]). It is necessary and sufficient to find the spaces of Killing tensors with the properties described above. In manifolds of constant curvature as Euclidean and Minkowski manifolds or n-Spheres these spaces, together with the associated coordinates, are well known (see for example [8]).

3. A comparison criterion between separable systems

For any Hamiltonian H and any function F on T^*Q,

$$\{H, F\} = \frac{dF}{dt}$$

where t is the natural time parametrizing the integral curves of H.

Given H as in (2) and n independent functions in involution L_i on T^*Q as in (5), (6), we define n functions m_i on T^*Q by

$$m_i := \{H, L_i\}, \qquad i = 0 \ldots n-1.$$

The functions m_i are linear homogeneous polynomials in p_i.

Indeed, since the quantities K_i are Killing tensors, it follows that the cubic term in the momenta of the function $\{H, L_i\}$ vanishes. Then

$$m_i = m_i^j p_j.$$

The n^2 components m_i^j can be interpreted as the components of n vector fields on Q

$$\mu_i = m_i^j \partial_j$$

in the coordinates (q^i) canonically conjugate to the coordinates (p_i). The vector fields μ_i express the 'projection' on each base point of the fibers of T^*Q of the rate of change of each function (L_0, L_i) along the integral curves of H.

We set

$$|\mu_i| = g_{hl} m_i^h m_i^l.$$

By expanding each L_i in some neighborhood of a point $P \in T^*Q$ in Taylor series along the integral curves of H, we get

$$L_i(P_t) = L_i(P) + m_i(P)t + \frac{1}{2}\{\{L_i, H\}, H\}(P)t^2 + \Delta(P, t^3, t^4, \ldots). \tag{9}$$

Then, for t small enough the change of L_i on the integral curves of H is given mainly by the functions m_i, homogeneous polynomial in (p_i), whose coefficients are exactly the m_i^j i.e. the components of the vectors μ_i.

4. The comparison criterion

Let us consider for simplicity that the configuration manifold is properly Riemannian.

Conjecture 1 *Under the hypothesis that the series (9) are well defined in some neighborhood of P for the systems $(L_0, L_i)_1$ and $(L_0, L_i)_2$, for (positive) t small enough, the trajectories of H are better approximated by the system for which the quantity $|\mu| = |\mu_0| + \ldots + |\mu_{n-1}|$ in P is smaller. In other words, the more (L_0, L_i) are close to be constant along the trajectories of H, the more the trajectories of L_0 are close to those of H.*

By considering the moduli $|\mu_i|$ we make the analysis independent from the coordinates we are using.

4.1. A necessary condition?

It is not easy to compute a priori and for all separable systems the functions μ_i. We recall that, for $i \neq 0$,

$$d\mu_i = dK_i dV$$

is independent from (W_0, W_i). Moreover, it is the only obstruction to the separability of H. Therefore, we try to minimize the obstruction by a suitable choice of the (K_i) in the behalf that

Conjecture 2 *Under the same hypothesis of Conjecture 1 and for $i \neq 0$: $d\mu_i$ minimal is a necessary condition for $|\mu_i|$ minimal.*

In dimension 2 we have to evaluate the only non null component of $d\mu_i$. In the examples (Euclidean plane) this will be done according to its degree in Cartesian or Polar coordinates. In higher dimensions some suitable norm should be introduced in the space of the 2-forms (work in progress).

5. The practical strategy in 5 steps

Let H be any natural Hamiltonian with potential V in , for example, the 3-dim Euclidean space.

Step 1: Compute $dKdV$ for the generic Killing Tensor K. It depends on 20 parameters.

Step 2: Find the subset MK of the Killing tensors such that the obstructions $d\mu_i = d(K_i dV)$ are minimal in some suitable neighborhood of Q.

In our 2-dim examples, the degree of the component of $dKdV$ in (x, y) or r will be required maximum if in a neighborhood of the origin, minimum if $(x, y) >> (1, 1)$ or $r >> 1$.

Step 3: Find in MK all possible spaces of 3 (including the metric tensor) independent KT in involution with common eigenvectors. They are associated with separable coordinate systems. If no such subspace exists, return to step 2 after weakening the minimality requirement.

Step 4: If step 3 determines some well defined separable system, order them according to the functions $|\mu_i|$ and find for each system the functions W_0 through (8) such that $|\mu_0| = |\nabla(V - W_0)|$ is minimal, determine ∇W_1 and ∇W_2 through the (K_i) and W_0 by (7) and compute the vectors (μ_0, μ_1, μ_2).

Step 5: Choose for the decomposition the system such that the associated function $|\mu|$ is minimal among the separable systems found in the neighborhood we are interested in. Remember that $(\nabla W_i - K_i dV)$ "minimal" does not assure that $(\nabla W_0 - K_0 dV) = \nabla(W_0 - V)$ also is; Conjecture 2 does not apply to L_0.

6. Examples

The examples we give here are both systems in the Euclidean plane. For this manifold we recall:

In dimension two, one only symmetric Killing two-tensor with distinct eigenvectors is sufficient to define a separable system.

In the Euclidean plane the space of the symmetric Killing two-tensors has dimension six.

In the Euclidean plane there are essentially four distinct separable systems: Cartesian, Parabolic, Polar and Elliptic.

7. The quadrupole field

Let $Q = \mathbf{E}^2$ be the Euclidean plane, (x, y) cartesian coordinates and $r = \sqrt{x^2 + y^2}$. Let us consider the Hamiltonian function

$$H = \frac{1}{2}(p_x^2 + p_y^2) + V$$

with

$$V = \frac{G}{r} + \frac{D}{r^3}\left(3\frac{x^2}{r^2} - 1\right)$$

where G, D are real constants. H is the Hamiltonian function of one charged particle (of mass and electric charge $= 1$) in a gravitational (Coulomb) field centered into the origin of the coordinates plus an electrostatic quadrupole field corresponding to a three-dimensional distribution of static charges with the coordinate axis x as symmetry axis [9]. We assume that the dipole momentum is zero. This system provides also a good approximation of the Keres-Israel potential, the newtonian approximation of the relativistic Kerr solution (see [3] where Stäckel approximations are considered). The motion will be restricted to a plane determined by the axes, the rotation around X being a symmetry of the system. The potential function has also Y as a symmetry axis

Step 1 We consider the space of the symmetric Killing two-tensors of $\mathbf{E}^2$, defined by

$$K^{xx} = a - 2ey + cy^2, \quad K^{xy} = d + ex - fy - cxy, \tag{10}$$

$$K^{yy} = b + 2fx + cx^2, \tag{11}$$

where a, b, c, d, e, f are real numbers.

Steps 2 and 3 We calculate for such a K the function $d(K\ dV)_{xy}$ and obtain a rational function whose numerator is a 6-degree polynomial in x, y with coefficients depending on $a...f$ and whose denominator is $(x^2 + y^2)^{9/2}$. We are then led to choose a neighborhood Ω of $r = \infty$.

By choosing the tensor K in a suitable way it is possible to reduce the degree of the numerator polynomial to 4. Then the components of K take the forms

$$K^{xx} = a + \frac{1}{2}\frac{(a-b)Gy^2}{D}, \qquad K^{xy} = -\frac{1}{2}\frac{(a-b)Gxy}{D}, \tag{12}$$

$$K^{yy} = b + \frac{1}{2}\frac{(a-b)Gx^2}{D}, \qquad a \neq b. \tag{13}$$

The corresponding separable coordinate system is of elliptic type, with foci on the x axis if $D/G > 0$ with equation $x^2 = 2D/G$; with foci on the y axis if $D/G < 0$ with equation $y^2 = -2D/G$.

Let us choose $D/G > 0$ and put $A^2 = 2D/G$. Elliptic coordinates (u, v) are then defined by

$$u = r_1 - r_2, \qquad v = r_1 + r_2,$$

where $r_1 = \sqrt{(x+A)^2 + y^2}$ and $r_2 = \sqrt{(x-A)^2 + y^2}$. We have

$$g^{uu} = -4\frac{u^2 - 4A^2}{v^2 - u^2}, \qquad g^{vv} = 4\frac{v^2 - 4A^2}{v^2 - u^2},$$

and

$$r = \frac{1}{2}\sqrt{u^2 + v^2 - 4A^2}, \qquad x = -\frac{uv}{4A}.$$

Step 4 Now we can try to approximate V by a separable function:

$$W_0 = g^{uu}\phi_u(u) + g^{vv}\phi_v(v)$$

which is the most general separable potential in coordinates (u, v). For $\phi_u = 0$, $\phi_v = \frac{1}{2(v^2-4A^2)}\left(Gv - 4\frac{D}{v}\right)$ we have

$$W_0 = \frac{2}{v^2 - u^2}\left(Gv - 4\frac{D}{v}\right)$$

which is in Ω a good approximation of V. Because K is diagonalized in the elliptic coordinates (u, v), we have $K^{uu} = \lambda_u g^{uu}$ and $K^{vv} = \lambda_v g^{vv}$, with $\lambda_u = \frac{1}{4}(v^2 - 4A^2)$ $\lambda_v = \frac{1}{4}(u^2 - 4A^2)$, then, since $W = K^{ii}\phi_i$,

$$W_1 = \frac{1}{2}\frac{u^2 - 4A^2}{v^2 - u^2}\left(Gv - 4\frac{D}{v}\right), \tag{14}$$

$$L_1 = \frac{2}{v^2 - u^2}(-(u^2 - 4A^2)p_1^2 + (v^2 - 4A^2)p_2^2) + W_1. \tag{15}$$

Let us compare now the previous system with the one associated with the usual polar coordinates (r, θ), where

$$g^{rr} = 1, \qquad g^{\theta\theta} = \frac{1}{r^2}$$

and

$$K^{rr} = 0, \qquad K^{\theta\theta} = 1.$$

By following the same procedure as as above, we find $\phi_r = \frac{G}{r} - \frac{D}{r^3}$, $\phi_\theta = 0$. Therefore

$$Wp_0 = \frac{G}{r} - \frac{D}{r^3}, \qquad Wp_1 = 0.$$

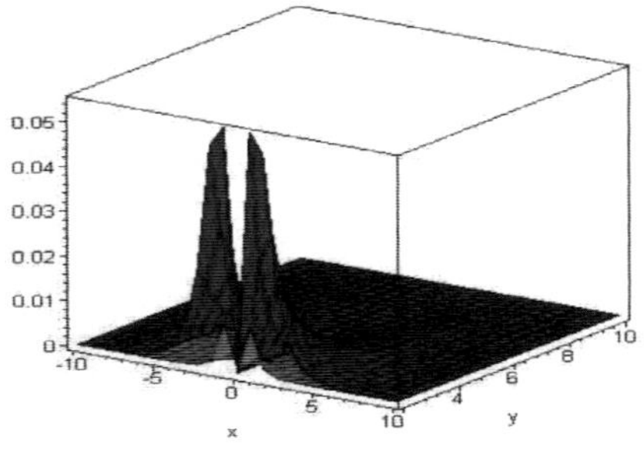

Figure 1. μ for the elliptic (red) and polar (blue) systems.

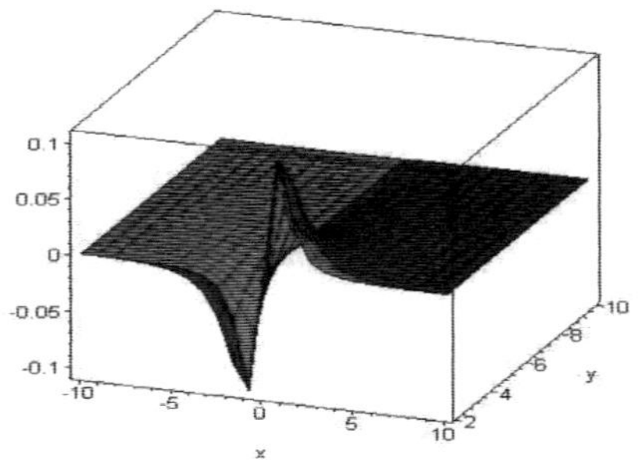

Figure 2. $d\mu_1$ for the elliptic (red) and polar (blue) systems.

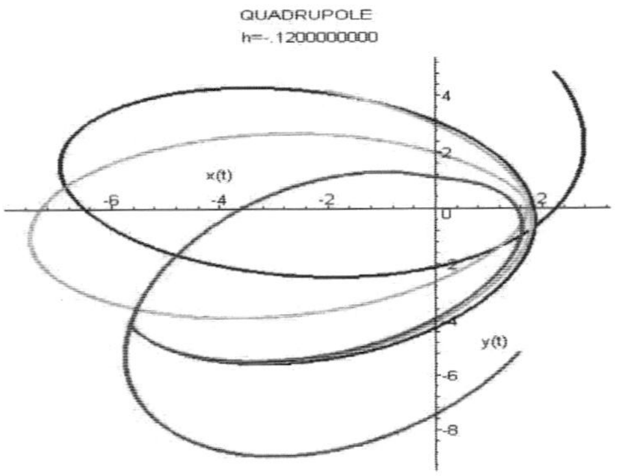

Figure 3.

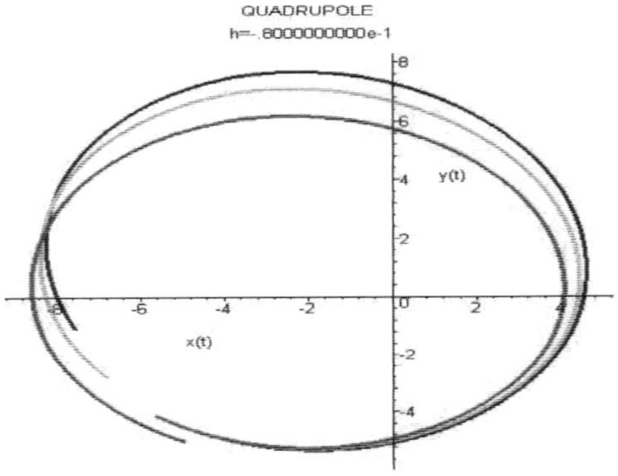

Figure 4.

It is evident that the elliptic system exhibits the same axial symmetry of the original system (axes x and y), while the polar one exhibits a central symmetry.

Step 5 The quantities $\mu = |\mu_0| + |\mu_1|$ are computed and plotted in cartesian coordinates for the elliptic and the polar systems of above, with $G = -1, D = -1/2$ (Figure 1). The non null component of $d\mu_1$ in cartesian coordinates is plotted for the same systems (Figure 2).

For the same systems, trajectories are numerically evaluated for $0 < t < 100$ at several values of the energy h. The original quadrupole in red, the elliptic system in green and the polar in blue (Figures 3 - 6).

For the same systems, the projection of the integral curves on the conjugate plane (x, p_x) is plotted (Figures 7 - 10).

It appears that the value of the energy is not relevant on the level of the approximation given by the Stäckel systems considered, at least for small values of t as considered in our approach.

8. Hénon-Heiles systems

Let the Hénon-Heiles system be defined by

$$H = \frac{1}{2}(p_1^2 + p_2^2) + V$$

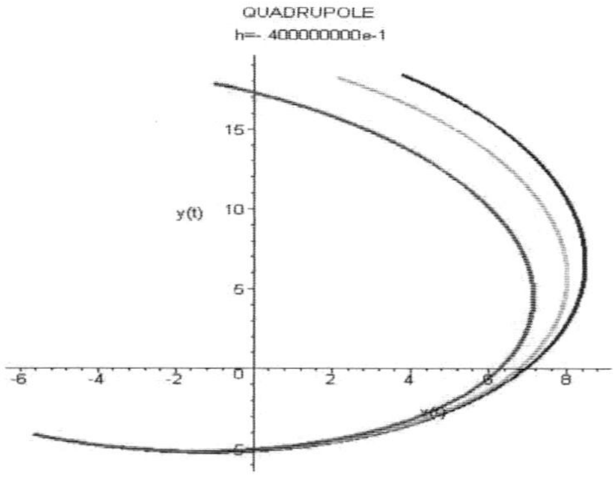

Figure 5.

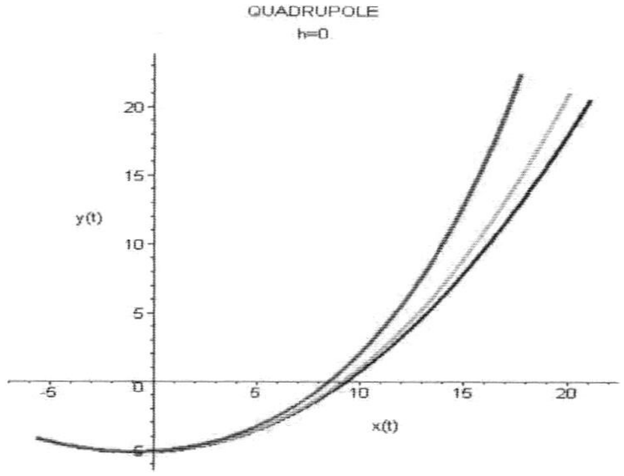

Figure 6.

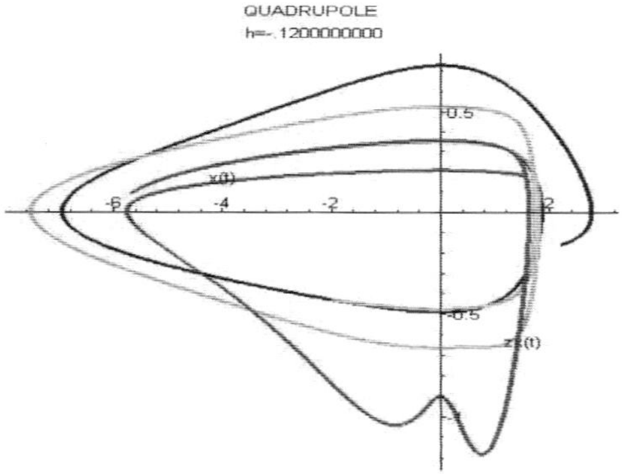

Figure 7.

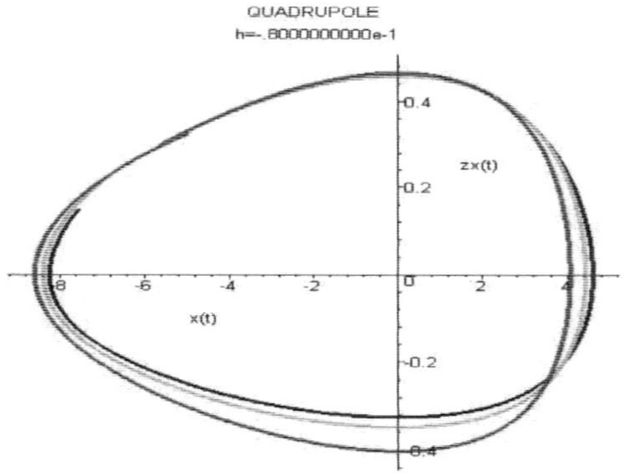

Figure 8.

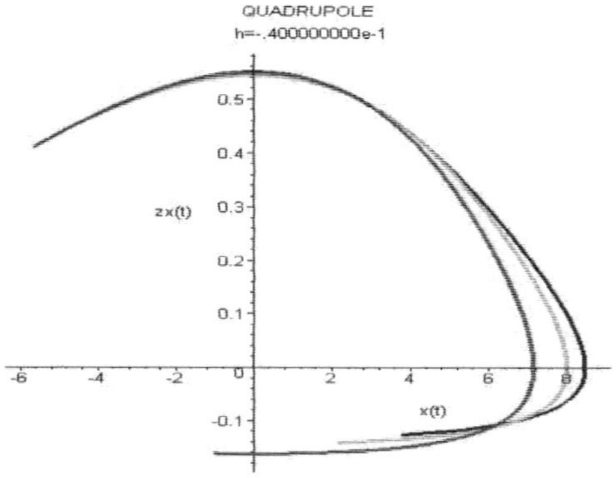

Figure 9.

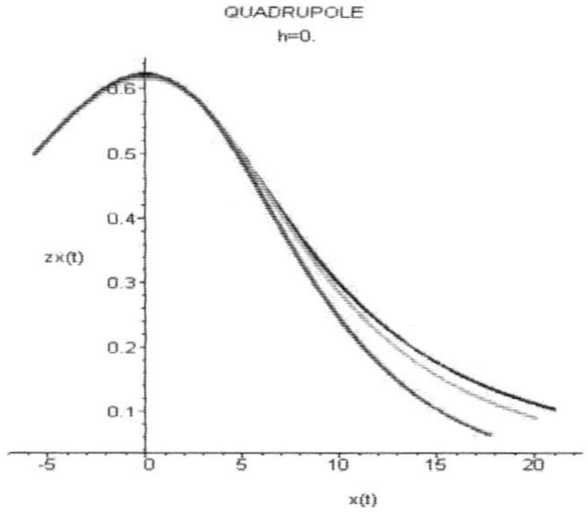

Figure 10.

Journal of Physics: Conference Series **128** (2008) 012029 doi:10.1088/1742-6596/128/1/012029

with

$$V = \frac{1}{2}(Cx^2 + Dy^2) + Axy^2 - \frac{1}{3}Bx^3.$$

Step 1 The generic Killing two-tensor of the Euclidean plane is again (11), the obstruction to separability is now

$$\begin{aligned} dKdV_{12} &= d(D-C) + [2A(a-b) + f(C-4D)]y + [2d(A+B) + \quad (16)\\ &+e(D-4C)]x + e(5B+2A)x^2 - 7eAy^2 + \quad (17)\\ &+[4c(C-D) - 2f(B+6A)]xy + 5cAy^3 - 5c(B+2A)x^2y. \quad (18)\end{aligned}$$

Steps 2 and 3 If we choose to approximate the system for $(x, y) >> (1, 1)$, we have to reduce the degree of $dKdV_{12}$ to be linear in (x, y). Then $c = e = f = 0$, Killing tensors of this kind are associated with Cartesian coordinates. If we consider instead small neighborhoods of the origin, the Polar system centered in the origin, $c = 1$ and the other parameters zero, provides the best $d\mu_1$.

The separable system defined by $c = e = f = a = b = 0, d = 1$ provides one of the best separable approximations to H-H (particularly good if $A+B$ is small). The Cartesian coordinates associated with such a choice of Killing tensor are defined by

$$q^1 = \frac{1}{\sqrt{2}}(x+y), \qquad q^2 = \frac{1}{\sqrt{2}}(y-x).$$

In these coordinates we have

$$dKdV_{12} = D - C + \sqrt{2}(A+B)(q^1 - q^2)$$

and

$$\begin{aligned} V &= \tfrac{\sqrt{2}}{4}(A+B)[(q^1)^2q^2 - q^1(q^2)^2] + \tfrac{1}{4}(C+D)[(q^1)^2 + (q^2)^2] + \quad (19)\\ &+\tfrac{1}{2}(D-C)q^1q^2 + \tfrac{\sqrt{2}}{4}(\tfrac{B}{3} - A)[(q^2)^3 - (q^1)^3]. \quad (20)\end{aligned}$$

Step 4 We can extract from V the separable potential

$$W_0 = \frac{1}{4}(C+D)[(q^1)^2 + (q^2)^2] + \frac{\sqrt{2}}{4}(\frac{B}{3} - A)[(q^2)^3 - (q^1)^3].$$

which in (x, y) is

$$W_0 = \frac{1}{4}[(C+D)(x^2+y^2) + (3A-B)xy^2 + (A - \frac{B}{3})x^3]$$

In (q^1, q^2) we have $K^{11} = 1, K^{22} = -1, K^{12} = 0$ and $g^{11} = g^{22} = 1, g^{12} = 0$, therefore, from $W_0 = g^{ii}\phi_i(q^i)$ and $W_1 = K^{ii}\phi_i(q^i)$ we obtain

$$\phi_1 = \frac{1}{4}[(C+D)(q^1)^2 + \sqrt{2}(A - \frac{B}{3})(q^1)^3], \quad (21)$$

$$\phi_2 = \frac{1}{4}[(C+D)(q^2)^2 - \sqrt{2}(A - \frac{B}{3})(q^2)^3], \quad (22)$$

and

$$W_1 = \phi_1 - \phi_2, \qquad L_1 = \frac{1}{2}(p_1^2 - p_2^2) + W_1.$$

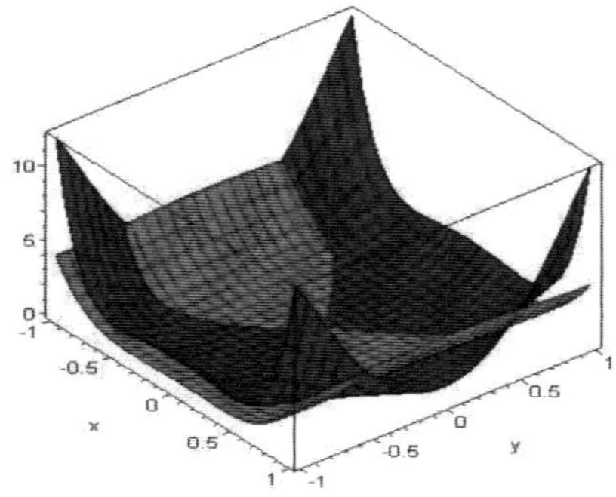

Figure 11. μ for the cartesian (red) and polar (blue) systems.

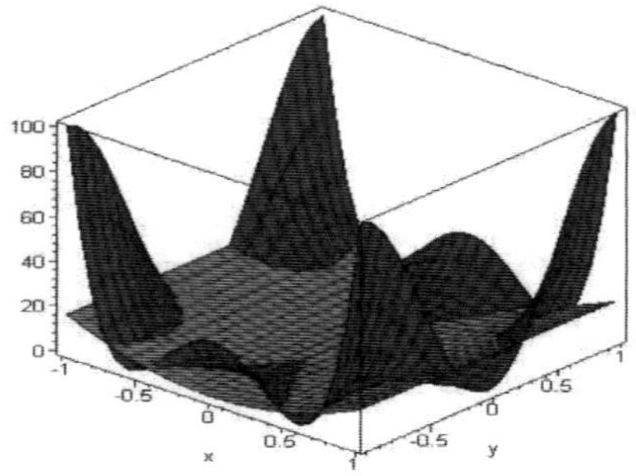

Figure 12. $d\mu_1$ for the cartesian (red) and polar (blue) systems.

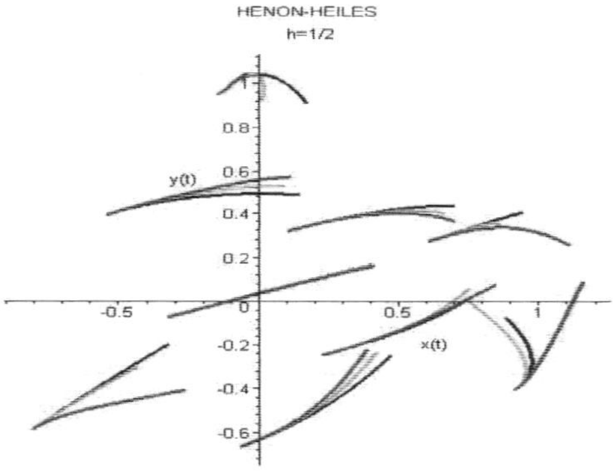

Figure 13.

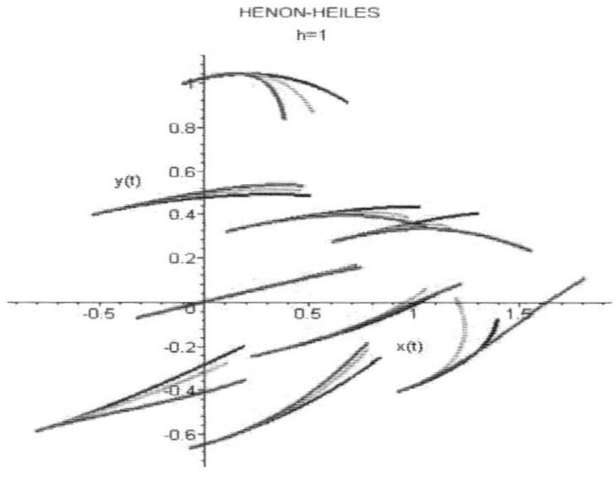

Figure 14.

Step 5 For the Polar system centered in the origin we have

$$K^{11} = y^2, \quad K^{12} = -xy, \quad K^{22} = x^2,$$

and

$$Wp_0 = \frac{1}{2}(x^2 + y^2), \qquad Wp_1 = 0.$$

Therefore, the polar system coincides with the harmonic oscillator. We can now make the comparison between the original H-H, the cartesian and the polar systems. We will find that, after considering the whole μ, the Cartesian system can be a better choice for an approximation also in small neighborhoods of the origin. Indeed, (Figure 11) the quantities $\mu = |\mu_0| + |\mu_1|$ are calculated and plotted for the systems cartesian and polar for $C = D = 1, A = 1, B = 1$ corresponding to the classical chaotic Hénon-Heiles system [7], in (Figure 12) the corresponding non null component of $d\mu_1$ are plotted.

From the (Figure 11) we should expect that, in the regions where the corresponding value for the polar system is smaller than that of the cartesian one, the polar system provide a better approximation, while in the remaining regions the contrary happens.

Journal of Physics: Conference Series **128** (2008) 012029 doi:10.1088/1742-6596/128/1/012029

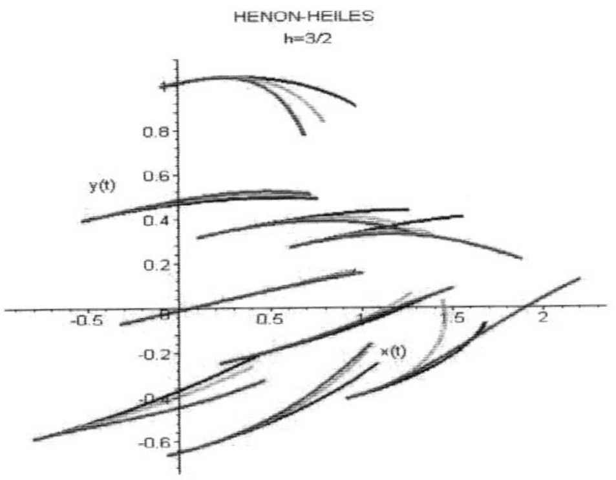

Figure 15.

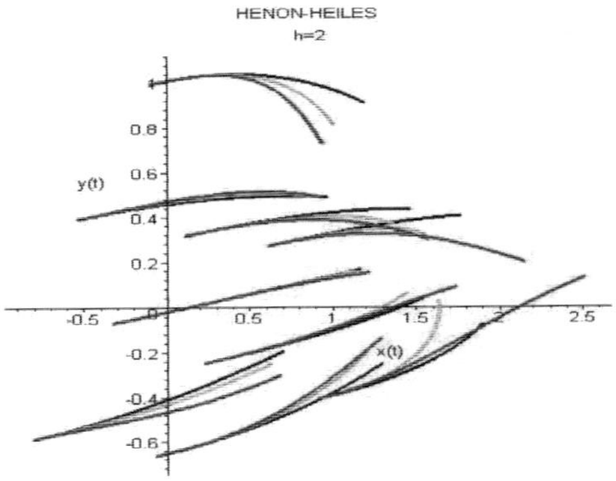

Figure 16.

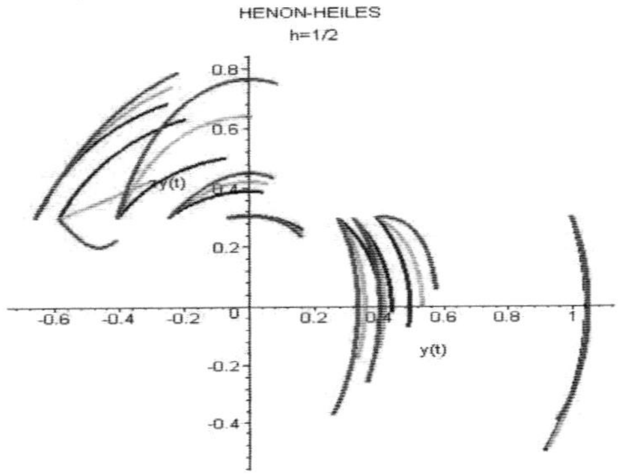

Figure 17.

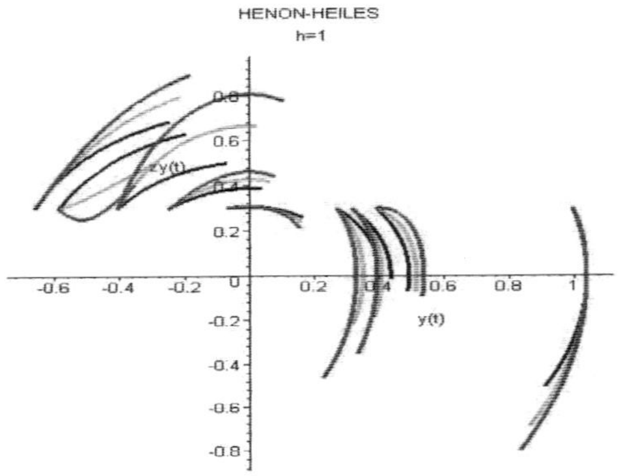

Figure 18.

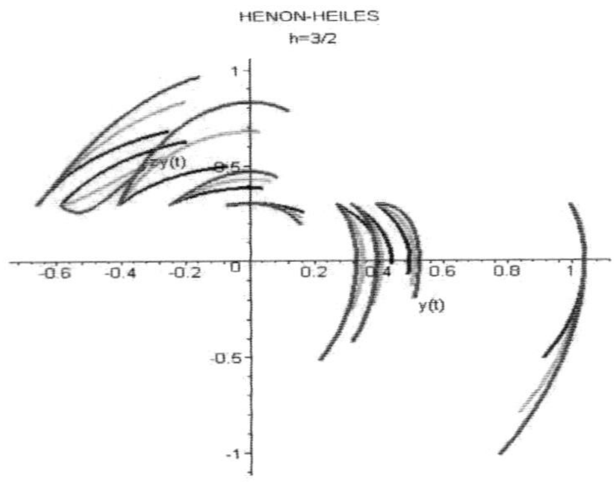

Figure 19.

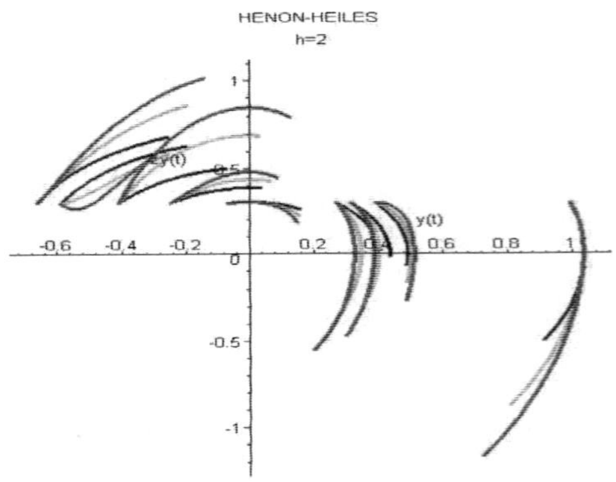

Figure 20.

For the same systems, trajectories are numerically evaluated for several initial points with the same energy, for $0 < t < 0.8$. The original H-H system in red, the cartesian system in green and the polar in blue (Figures 13 - 16).

For the same systems, the projections of the integral curves on the conjugate plane (y, p_y) are plotted (Figures 17 - 20).

It is remarkable that, in neighborhoods of the origin of radius < 1, even if $d\mu_1$ is smaller for polar coordinates, the best μ is given by the cartesian systems again, as appear in the example from the numerical evaluation of μ plotted above. This is due to the fact that , in this case, the relevant part of μ is provided by μ_0 and is not controlled by Conjecture 2, which is concerning μ_i with $i > 0$ only. Remark that trajectories developing closely to the positive x axis show a better approximation provided by the polar system, this is consistent with the evaluation of μ plotted above. Similar comparisons with Parabolic systems have been made elsewhere verifying the two Conjectures.

9. Conclusions and future directions

Given a natural Hamiltonian, we have a strategy to follow for find natural separable systems whose dynamics is locally close to the given one. The method is in large part coordinate-independent and computations are easily done by computer-algebra systems. As future development of the approach, the two conjectures should be transformed, mutatis mutandis, into theorems. Moreover, $n > 2$ dimensional manifolds and quantum mechanical systems can be considered, together with the characterization of those Hamiltonian systems admitting some "Stäckel approximation" of the kind treated here but over larger intervals of time.

Acknowledgments

I am grateful to Roman Smirnov and Giuseppe Pucacco for the fruitful discussions about the subject of this research.

References

[1] Benenti S 1997 *J. Math. Phys.* **38** 6578
[2] De Bruyne V, Leeuwin F and Dejonge H 2000 *MNRAS* **311** 297
[3] de Felice F and Preti G 2000 *J. Phys. A: Math. Gen.* **33** 2767
[4] Duval C and Valent G 2005 *J. Math. Phys.* **46**
[5] de Zeeuw P T 1985 *MNRAS* **216** 273
[6] de Zeeuw P T and Lynden-Bell D 1985 *MNRAS* **215** 713
[7] Hénon M and Heiles C 1964 *Astr. J.* **69** 73
[8] Kalnins E G 1986 *Separation of variables for Riemannian spaces of constant curvature* (Pitman Monographs and Surveys in P. A. Math.) **28**
[9] Landau L D and Lifsits E M 1972 *Field theory* (Moscow: Mir) V 41
[10] Arnold V I 1976 *Les méthodes mathématiques de la mécanique classique* (Moscow: Mir)

Journal of Physics: Conference Series **128** (2008) 012030 doi:10.1088/1742-6596/128/1/012030

Analysis of gene expression: case study for bacteria

Maia Angelova and Chris Myers

Intelligent Modelling Lab, School of CEIS, Northumbria University, Pandon Building, Newcastle NE2 1HE, UK

E-mail: maia.angelova@unn.ac.uk, chris.myers@unn.ac.uk

Abstract. The analysis of gene expression allow to study the functions of genes and their roles in different processes in the cell of a living system, including the cell cycle. Clustering is widely used in the analysis of high-throughput gene expression data to find patterns of similarity that enable related gene groups and functions to be identified. Clustering algorithms are very sensitive to the choice of initial conditions and optimal number of clusters. In this paper, we investigate the impact of metrics and cluster parametrisation for three clustering models and propose a method for optimisation of cluster parameters based on cluster compactness and separation. A case study presents the analysis of gene expression data for *E.coli* bacteria.

1. Introduction

The world of biological sciences has undergone an information revolution in recent years due to the development of rapid DNA sequencing techniques, the progress in computer based technologies and mathematical and computer modelling. The development of DNA microarray technology has made it possible to measure the expression levels of thousand of genes simultaneously under controlled experimental conditions. The completion of the genome sequence for human and many model organisms has provided powerful biological data of richness and scale that require a data driven approach. Moreover, the atomic description of molecular structure and functions, as well as molecular explanation of cellular behaviour, have become increasingly possible. The growing amount of detailed information about biological systems and the complexity of these systems is a challenge that needs modelling and computational tools, adaptable to large data sets, and an integrative modelling approach [1, 2].

The central dogma of molecular biology is that the information is stored in DNA, transcribed to messenger RNA (mRNA) and then translated into proteins (see for example [1]), i.e. once the gene has been activated it is expressed in the cell. This justifies the premise that the information about the functional state of an organism is to a great extent determined by the information on gene expression and is the motivation for the analysis of large-scale gene expression data. Among many powerful automatic techniques for analysing high-throughput gene expression data from microarray experiments, clustering is widely used [2, 3]. It is accomplished by finding similarity patterns within gene expression data thus enabling related gene groups and functions to be identified. Clustering results are used to learn about gene functions, gene regulation, cellular processes and subtypes of cells. The genes with similar expression patterns, known as co-expressed genes, can be clustered together. Co-expressed genes are likely to have similar cellular functions, or to be involved in the same cellular processes [4, 5, 6, 7]. Further, the regulatory modules of biological function of unknown genes can be discovered by associating them with

Journal of Physics: Conference Series **128** (2008) 012030 doi:10.1088/1742-6596/128/1/012030

other genes that have similar expression patterns and known regulatory elements or functions. However, most clustering algorithms are very sensitive to the initial choice of parameters and optimal selection of clusters.

In this paper, we investigate the impact of metrics and cluster parametrisation for three clustering models and propose a method for optimisation of cluster parameters based on cluster compactness, separation and stability. A case study for bacteria is used to illustrate the main concepts and results. The rest of the paper is organised as follows. In Section 2 we consider the effect of similarity metrics and compare the optimisation criteria in three clustering models, K-means, EM and minimal entropy models . In Section 3 we discuss the quality and stability of the clusters and propose a method for optimisation of cluster parameters. In Section 4 the analysis of clustering results for genome-wide expresion of *Escherichia coli (E.coli)* bacteria is presented, the discussion and conclusions are given in Section 5.

2. Effect of similarity metrics in clustering models

Clustering models are very popular in the analysis of high-throughput gene expression data from microarray experiments [2, 3]. Clustering is the exploratory, unsupervised process of grouping data objects into a set of disjoint classes, called clusters, so that objects within one class have high similarity to each other and are dissimilar to objects in another class [8].

Microarray gene expression data can be presented by a real value gene expression matrix,

$$M = \{w_{ij} | 1 \leq i \leq n, 1 \leq j \leq m\} \tag{1}$$

where the rows $G = \{\boldsymbol{g_1}, \boldsymbol{g_2}, ..., \boldsymbol{g_n}\}$ form the expression patterns of the genes, the columns $S = \{\boldsymbol{s_1}, \boldsymbol{s_2}, ..., \boldsymbol{s_m}\}$ represent the expression profiles of the samples (experiments), and each cell of the expression matrix, w_{ij}, is the measured expression level of gene i in sample j, $i = 1, 2, ...n, j = 1, 2, ..., m$. In what follows, the vector $\boldsymbol{r}$ indicates a gene expression data object, which can represent a gene $\boldsymbol{g_i}$ in the n-dimensional gene space or a sample $\boldsymbol{s_j}$ expression profile in the m-dimensional array space. The similarity pattern can be established by comparing genes or experimental conditions (samples). The original gene expression matrix contains noise, missing values, and systematic variations arising from the experimental procedure. In most cases, data preparation and normalization is necessary before any data analysis can be performed.

The choice of similarity measure used within the clustering algorithm is very important as it influences the output and interpretation of the results. It has received much discussion [2, 3, 9] but there is still no agreement over the best metric to use and little work has been done on assessing the impact of different measures in the analysis of gene expression [10, 11].

The similarity is defined as a function, Sim, that measures association, usually distance or correlation, between data objects, representing genes or samples in the expression matrix. The distance measures the proximity between data objects $\boldsymbol{r_i}$ and $\boldsymbol{r_j}$ and represents the dissimilarity or unlikeness between them. A typical distance metric is the Euclidean distance,

$$D(\boldsymbol{r_i}, \boldsymbol{r_j}) = \sqrt{\sum_{d=1}^{m} (w_{id} - w_{jd})^2}, i, j = 1, 2, ..., n \tag{2}$$

Other distance measures used in this work are Manhattan and Minkowski distance.

The correlation measures the similarity or alikeness between two objects $\boldsymbol{r_i}$ and $\boldsymbol{r_j}$. It is different from the distance as a measure of the relationship between gene expression profiles. A typical correlation function is the Pearson correlation coefficient,

$$P(\boldsymbol{r_i}, \boldsymbol{r_j}) = \frac{\sum_{d=1}^{m}(w_{id} - \mu_i)(w_{jd} - \mu_j)}{\sqrt{\sum_{d=1}^{m}(w_{id} - \mu_i)^2}\sqrt{\sum_{d=1}^{m}(w_{jd} - \mu_j)^2}} \tag{3}$$

where μ_i and μ_j are the means for the objects $\boldsymbol{r_i}$ and $\boldsymbol{r_j}$ respectively. The correlation coefficient takes a value between -1 and $+1$. For example, a value of 0 represents perfectly uncorrelated variables, values close to ± 1 represents strong positive (negative) correlation between genes and could point to association of their functions. Cosine correlation coefficient is another example of correlation function. The correlation metric can be converted to a distance memetric [3]. Other examples of similarity metrics, are Jaccard and dice similarities, which take values between 0 and 1. We have investigated the impact of generalised similarity metrics using K-means, EM and entropy based models.

2.1. K-means model

K-means is an iterative partitioning method in which objects are moved among a pre-specified number of clusters, K, until an optimal solution is reached. It is an example of hard clustering algorithm in which each gene is assigned to a single cluster. The basic K-means clustering algorithm can be summarised with the following steps (see for example [8]): the number of clusters K is specified; the initial values for the cluster centres are chosen; each data object $\boldsymbol{r}$ is assigned to a cluster $C_i, i = 1, ..., K$, using the selected Sim function; the cluster centres $\boldsymbol{m_i}, i = 1, ..., K$, are re-calculated using the mean of all objects in each cluster; the objects are re-assigned; the last two steps are repeated until the cluster membership is stable.

The algorithm minimises a global error criterion, known as cost function [2], which depends on the preliminary selection of number of cluster K, cluster centres $\boldsymbol{m_i}$ and similarity function Sim. Although there is no universally accepted definition and the cost function could be tailored to the problem, it is usually defined as "within-cluster" sum of the squared distances between each data object $\boldsymbol{r}$ belonging to the cluster C_i and its cluster centre $\boldsymbol{m_i}$,

$$CF_2 = \sum_{i=1}^{K} \sum_{\boldsymbol{r} \in C_i} | \boldsymbol{r} - \boldsymbol{m_i} |^2 \tag{4}$$

and represents the total error. It is well-known that the algorithm minimises CF_2 but converges to a local rather than a global minimum depending on the choice of initial parameters, such as number of clusters K and cluster centres. Thus, the automatic selection of an optimal number of clusters for the clustering algorithm is a complex task and is a common problem for all partitioning algorithms. An optimisation procedure of these parameters is discussed in Section 3.

The impact of similarity measures has been investigated using microarray data for *E.coli* gene expression. The case study is considered in more details in Section 4. Here we present a summary of the metric's analysis for a reduced data set of 264 genes.

The data set was analysed using seven different metrics such as distance: Euclidean(E), Manhattan (Mh), Minkowski (Mk), correlation: Cosine(C), Pearson correlation coefficients (P) and other similarity: Dice (D) and Jaccard (J). The results are given in Table 1. The table compares the cluster sizes (number of genes per cluster) for different metrics and values of K. The similarity metric is indicated in the first column, the first row gives the number of clusters $K = 4, 5, 6, 7$. The size of each cluster for each value of K is given in the remaining columns for each similarity measure. The table shows that the clustering results are dependent on both, the choice of similarity measure and the value of K. Whilst the cluster sizes naturally decrease as K increases, there is some evidence of stability between the small clusters which tend, in the case of this data set, to contain the most highly expressed genes. The table illustrates the difference, for each value of K, between cluster sizes obtained using distance, correlation or other metrics. However, the results are similar for the three distance measures E, Mh and Mk, same applies for the correlation metrics C and P, and other similarity metrics D and J.

Table 1. Comparison of cluster sizes for different similarity metrics

Number of clusters / Metric	4	5	6	7
E	247,9,4,4	159,89,9,4,3	160,87,9,4,3,1	92,84,71,9,4,3,1
Mh	247,9,4,4	158,90,9,4,3	158,89,8,5,3,1	93,91,63,9,4,3,1
Mk	160,88,13,3	160,88,9,4,3	160,87,8,5,3,1	160,87,7,4,3,2,1
D	159,88,13,4	157,90,10,4,3	91,87,69,10,4,3	96,76,71,7,6,5,3
J	159,88,13,4	157,90,10,4,3	156,86,8,6,5,3	94,75,71,7,7,6,4
C	98,96,59,11	117,69,51,17,10	101,67,50,19,17,10	98,57,40,31,23,8,7
P	87,81,67,29	110,61,38,29,26	88,61,56,28,2,8,3	83,53,43,38,28,16,3

2.2. EM model

In the probabilistic models, data is assumed to be drawn from a series of probability distributions, usually multivariate Gaussian distributions. These models use the Expectation-Maximization (EM) algorithm [12] to produce the best fit between the data and a series of Gaussian distributions. The EM algorithm uses the likelihood as a similarity measure instead of distance or correlation. The algorithm takes into account that each object can belong to each cluster with a certain probability and finds the maximal log-likelihood. The log-likelihood is given by

$$\mathcal{L} = \sum_{i=1}^{n} \log(\sum_{k=1}^{K} \lambda_k p_k(\boldsymbol{r}_i|M_k)) \tag{5}$$

where λ_k is the probability that data object $\boldsymbol{r_i}$ belongs to cluster C_k, $\lambda_k \geq 0$, $\sum_k \lambda_k = 1$, each cluster C_k is represented by a model M_k, $p(\boldsymbol{r_i}|M_k)$ is the probability density of $\boldsymbol{r_i}$ in $M_k, k = 1, ..., K$ and K is the number of clusters. Each model M_k can be represented by a multivariate d-dimensional Gaussian distribution with mean μ_k and covariance Σ_k. The EM algorithm finds the maximal log-likelihood $\mathcal{L}_M$ for a given data model $M\{M_1, ..., M_K\}$.

It can be shown that when the cost function corresponds to an underlying probabilistic model, K-means can be regarded as an approximation of the classical EM algorithm on a spherical Gaussian mixture model [2]. Like K-means, the EM algorithm requires the number of clusters to be specified in advance.

2.3. Minimal entropy model

The entropy-based models use the entropy of the clusters as a similarity metric. The entropy measures the uncertainty of a random variable. In Shannon's information theory [13], the entropy of a random variable X is defined as

$$H(X) = -\sum_{i} p(\boldsymbol{r_i}) \log(p(\boldsymbol{r_i})) \tag{6}$$

In thermodynamics, the entropy is a measure of the disorder in the system. Applied to clustering, the concept of entropy means that each cluster should have a low entropy as objects in the same cluster are similar. Thus, the search for clusters with minimal entropy can be used as a clustering criterium.

The entropy of the clusters can be written as,

$$H = \sum_{j=1}^{K} p_j H(X|C_j) \tag{7}$$

where $H(X|C_j)$ is the entropy of the cluster C_j, p_j is the probability of the cluster C_j such that $\sum_j p_j = 1$ and K is the number of clusters. A clustering algorithm that minimises (7) has been developed in [14]. The entropy of the cluster $H(X|C_j)$ can be measured using the actual relationship between data objects and clusters. Li *et al* [14] have evaluated this relationship by using *posteriori* probabilities $p(C_j|\boldsymbol{r_i})$ of object $\boldsymbol{r_i}$ in cluster C_j. However, the choice of particular data distribution (such as Gaussian distribution) can lead to a poor representation of the data. An alternative method is based on the actual density of data objects using the Parzen density approach [15]. The probability $p(C_j|\boldsymbol{r_i})$ is evaluated using the Parzen density estimation for the clustering problem as [14],

$$p(C_j|\boldsymbol{r_i}) = \frac{n_{ij}}{n_i} \tag{8}$$

where n_{ij} is the number of samples $\boldsymbol{r_i}$ from cluster C_j and n_i is the number of all samples located in a selected region $R(\boldsymbol{r_i})$. The entropy clustering criterion can be written as

$$H = -\sum_{i=1}^{n}\sum_{j=1}^{K} \frac{n_{ij}}{n_i}\log(\frac{n_{ij}}{n_i}). \tag{9}$$

The algorithm minimises (9) to find the minimal entropy of the clusters. Like the other clustering functions discussed in the paper, H has local minima rather then a global minimum. We have used (9) with K-means and EM to optimise and improve the clustering results.

3. Quality, stability and optimisation of clusters

Clustering algorithms, like K-means, EM and minimal entropy, require the number of clusters to be given in advance. This is often very difficult as biologists may not know the exact number of functional categories as some of the genes may have unknown functions or belong to groups with unknown functional categories. The optimal selection of the number of clusters in the clustering algorithm and the stability of the clusters is important as it impacts upon the clustering solution (see Table 1). A good clustering result should produce tightly packed clusters which are stable and well separated [2, 9].

3.1. Quality of clusters

The quality of the clusters can be measured in terms of "intra cluster" homogeneity and "inter cluster" separateness. The term homogeneity [9] is used to represent the sameness of data points within a cluster. The corresponding function can be defined as,

$$Hom(C_i) = \frac{1}{\|C_i\|}\sum_{\boldsymbol{r}\in C_i} Sim(\boldsymbol{r}, \boldsymbol{m_i}) \tag{10}$$

where $\|C_i\|$ is the number of points allocated to cluster C_i and Sim is the similarity function. To illustrate these ideas in K-means, we have chosen Sim as Euclidean distance (2). The term separateness is used to estimate the separability between clusters and can be measured by the function,

$$S(C_i, C_j) = Sim(\boldsymbol{m_i}, \boldsymbol{m_j}), \quad i \neq j, \tag{11}$$

giving the distance between two cluster centres.

The average cluster homogeneity measures the "intra cluster" sameness and represents the average distance between each data object and its cluster centre. The average cluster separation measures "inter cluster" separateness and gives the average (weighted) distance between cluster centres. When considered in terms of distance between objects, our aim is to find solutions based around compact, well separated clusters, with low homogeneity (high density of packing)

Journal of Physics: Conference Series **128** (2008) 012030 doi:10.1088/1742-6596/128/1/012030

and high separation. Based on these concepts, we evaluate the quality of the clusters by defining two cost functions and two quality functions.

The error function CF_1 is the sum of distances between each data object and its cluster centre,

$$CF_1 = \sum_{i=1}^{K} \sum_{\boldsymbol{r} \in C_i} \mid \boldsymbol{r} - \boldsymbol{m_i} \mid \qquad (12)$$

It represents the mean error per data point and measures average cluster homogeneity. The cost function CF_2 (4) is the sum of squared distances between each data point and its cluster centre. The quality function QF_1 is the difference between the mean distance between cluster centres and the mean error per data point,

$$QF_1 = \frac{1}{K(K-1)} \sum_{i,j,i \neq j} \mid \boldsymbol{m_i} - \boldsymbol{m_j} \mid - \frac{CF_1}{n} \qquad (13)$$

It represents the balance between cluster separateness and cluster compactness. QF_2 is the difference between the mean cluster separation and the mean error per data point,

$$QF_2 = \frac{1}{\sum_{i,j,i \neq j} \|C_i\| \|C_j\|} \sum_{i,j,i \neq j} \|C_i\| \|C_j\| S(C_i, C_j) - \frac{CF_1}{n}. \qquad (14)$$

We have optimised the number of clusters within K-means using the functions CF_i, $QF_i, i = 1, 2$. A low cost, high density and high separation solution corresponds to a minimum of CF_i and a maximum of $QF_i, i = 1, 2$, for the same values of the parameters.

Figure 1 represents the cost functions CF_1 and CF_2 plotted against the number of clusters, K, for the reduced set of 264 genes. The quality functions QF_1 and QF_2 are given on Figure 2. The functions CF_1 and CF_2 do not exhibit a global minimum but a series of local minima with respect to K. The function QF_2, however, has a global maximum. Thus, the optimal solution corresponds to a local minimum of CF_2 and a global maximum of QF_1, which for the set of 264 genes is at $K = 6$.

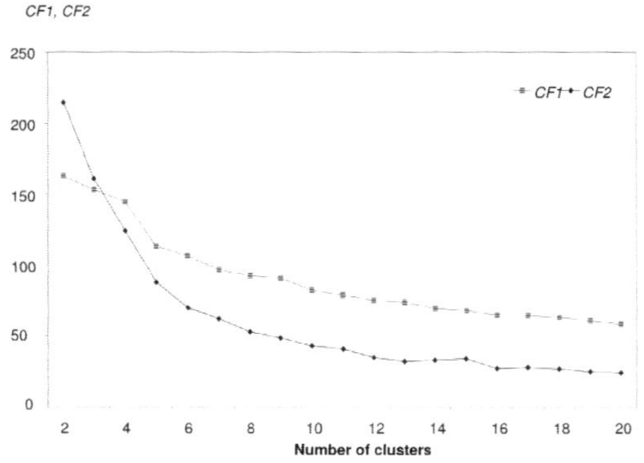

Figure 1. Cost functions CF_1, CF_2 vs number of clusters.

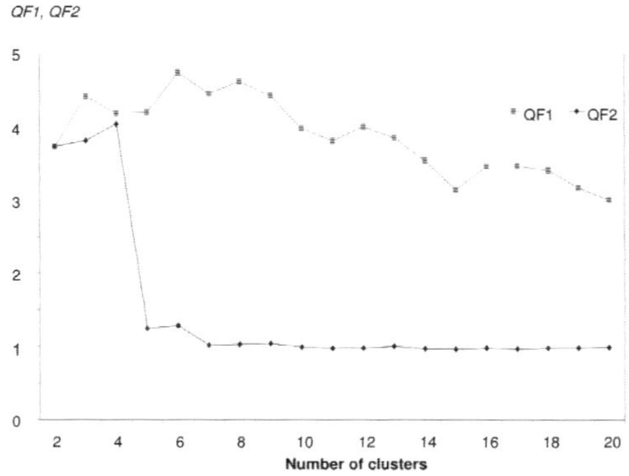

Figure 2. Quality functions QF_1, QF_2 vs number of clusters.

Smet *et al* [16] have used an alternative quality based approach in which clusters are defined sequentially and can only contain genes lying within a specified volume. Other approaches to the optimisation are associated with the statistics of the data [17].

We have analysed the results of the EM model using the Bayesian Information Criterion, BIC (see for example [18] and references therein),

$$BIC = 2\mathcal{L}_M - m_M \log(n) \tag{15}$$

where $\mathcal{L}_M$ is the maximised mixture log likelihood (5) for the data model M, m_M is the number of independent parameters to be estimated and n is the number of data objects. BIC has been used in [18] to estimate the optimal number of components within a mixture model. Whilst the maximised log likelihood will increase as the number of components increases, the second term in (15), which is based on the number of parameters, compensates this increase. Equation (15) can be considered as an alternative of the cluster balance (13). We have used (15) to search for optimal solution in EM with maximised log likelihood and maximal BIC. Figure 3 presents BIC as a function of the number of clusters for the set of 264 genes. The optimal solution is at $K = 5$ where BIC has a global maximum.

We have estimated the entropy of the clusters in K-means and EM-model using the criterium (9). An illustration for the set of 264 genes in K-means is given on Figure 4. We have combined the intra cluster sameness CF_2 and the minimal entropy H to identify the optimal solution. It corresponds to a global minimum of the function $CF_2 + H$, which for the set of 264 genes is at $K = 7$.

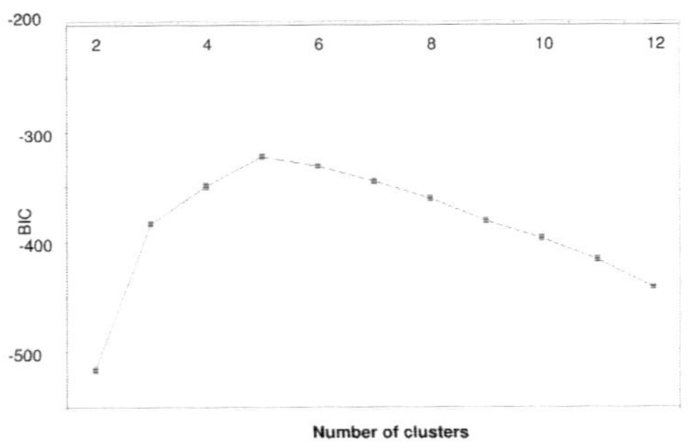

Figure 3. BIC vs number of clusters.

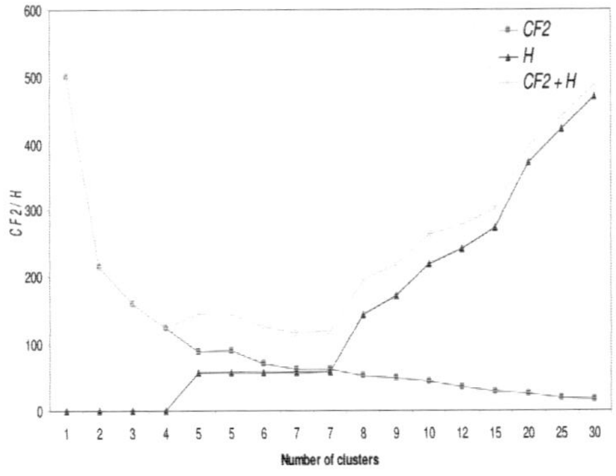

Figure 4. Cost function CF_2 and entropy H vs number of clusters

3.2. Stability of clusters

The Rand index is a useful way of comparing different clustering results from a given dataset. Given a set of data objects $G = \{\boldsymbol{r_1}, \boldsymbol{r_2}, ..., \boldsymbol{r_n}\}$ and two different clustering solutions $C^1 = \{C^1_1, C^1_2, ..., C^1_K\}$ and $C^2 = \{C^2_1, C^2_2, ..., C^2_P\}$, $K \neq P$, the Rand index, R, can be defined as,

$$R = \frac{a+b}{a+b+c+d}, \tag{16}$$

where a is the number of pairs of data objects in G that are in the same cluster in both C^1 and C^2, b is the number of data objects in G that are not in the same cluster in both C^1 and C^2, c is the number of data objects in G that are in the same cluster in C^1 but not in the same cluster in C_2 and d is the number of pairs of data objects in G that are not in the same cluster in C^1 but are in the same cluster in C^2. R has a value between 0 and 1 with 0 indicating that

Journal of Physics: Conference Series **128** (2008) 012030 doi:10.1088/1742-6596/128/1/012030

the two clustering solutions do not agree on any pair of data objects and 1 indicating that the two solutions are identical.

Rand index was used to compare different solutions in one model obtained by varying the number of clusters, a reliable solution would be expected to exhibit some level of stability as the number of clusters is varied. We have also used Rand index calculations to compare the effect of similarity metrics. For the set of 264 genes, the optimal results in K-means are for six clusters. Rand index calculations confirm that distance metrics lead to similar clustering results, R(E/Mh)=0.99, R(E/Mk)=0.99, R(Mh/Mk)=0.97. For correlation metrics, R(P/C)=0.79, R(J/D)=0.74,R(P/D)=0.73; R(P/E)=0.7 indicates some difference in clustering results when Pearson coefficient or Euclidean distance is used as metric (see also Table 1).

Rand index calculations were used to compare results from K-means and EM model. 6 clusters in both K-means (Euclidean metric) and EM give R=0.67, 6 clusters in K-means and 8 clusters in EM give R=0.65. The relatively low value of R indicates that the two models produce to an extent different results, one of the reason being that cluster sizes are different. Table 2 gives the number of genes per cluster in K-means and EM for a solution with 6 clusters for set of 264 genes.

Table 2. Comparison of cluster sizes in K-means and EM

Cluster Model	1	2	3	4	5	6
K-means	160	87	9	4	3	1
EM	67	54	47	43	35	18

A stability based method is developed in [19] by measuring the overlap between clusters as the number of clusters is varied. The optimal number of clusters is determined by selecting the solution with the highest stability. The method is applied to a set of clustering results with the number of clusters varying from 2 to m. Let $C_{c,l}$ be a set of data objects in cluster l resulting from a clustering solution with c clusters, $2 < c < m$. Let k, $0 < k < m - c$, be the threshold at which the stability calculation will stop. The cluster stability [19] of cluster l is,

$$S_{c,l} = Min_{i=c+l}^{c+k}\{Max_{j=1}^{i}\{\frac{||C_{c,j} \bigcap C_{i,j}||}{||C_{c,l}||}\}\}. \tag{17}$$

Equation (17) calculates the values of k for the maximum number of overlapping objects between cluster l and all other clusters in the clustering solution $c, c < i < c + k$. Then it takes the minimum of the k maximum values as the stability of the cluster l. The stability value is normalised to range from 0 to 1 by dividing it with the number of objects in cluster l. The average stability of all clusters in this solution is,

$$< S_c >= \frac{1}{c}\sum_{l=1}^{c} S_{c,l} \tag{18}$$

We have used Rand index (16) together with equations (17) and (18) to evaluate the stability of the clustering results in K-means and the EM model. The values of $< S_c >$ for K-means and EM models are given in Table 3 for $K = 4, .., 7$ for the set of 264 genes. The table indicates that the most stable solution is obtained for 6 clusters. Alternative approaches to the stability of the clustering results are investigated in [20, 21].

Table 3. Average cluster stability for K-means and EM models

Number of clusters Model	4	5	6	7
K-means	0.8484	0.9288	0.9203	0.8968
EM	0.8081	0.9057	0.9471	0.9340

4. Case study: *E.coli*

Ferenc *et al* [22] have studied the effects of knocking out the methionine repressor gene, *metJ*, on the *E.coli* transcriptome. Genome-wide expression data has been obtained where strains of LU106(pFM26) and LU106(pFM20) have been compared using oligonucleotide based array of 4288 *E.coli* genes. The study has confirmed that repression is largely restricted to known genes involved in the biosynthesis and uptake of methionine. The number of additional genes that are up-regulated in the absence of the repressor has been identified. Several other recently characterised genes in the methionine regulon have been identified and previously unknown potentially regulated loci highlighted.

We have used raw data from these experiments, normalised to eliminate background noise and systematic error. The *E.coli* genes have been clustered by hierarchical clustering (with GeneSpring [23]), K-means, EM algorithm and entropy-based algorithms. Data was filtered to select differentially expressed genes and a subset of 265 genes was obtained and analysed. In this subset, *metE* gene has the highest expression level and always forms a cluster by itself. This is due to the biological setting of the experiments. The reduced set of the remaining 264 genes is used in the paper to illustrate the results. The genes in the reduced data set were clustered using different clustering models. For comparison, genes in the complete set were also clustered.

The clustering was performed using our Java-based clustering tool with embedded entropy algorithm to refine the results of K-means and EM algorithms. K-means was executed multiple times with $K = 1, ..., 20$ with a random starting values for selecting the cluster centres. EM-algorithm was executed in a similar way with a number of Gaussians varying from 2 to 10. For each clustering model, the quality and stability of the clusters were examined and optimal solution was chosen based on the criteria described in Section 3. The entropy-based algorithm was executed with K-means and EM to provide a refined set of initial conditions. The solutions of the different models were compared. The optimal solution in K-means was chosen with 6 clusters for the 264 genes. The data was analysed with seven different similarity metrics. Although different metrics can lead to different clustering patterns (as illustrated by Rand index calculations and Table 1), we have established that a number of highly expressed genes are always clustered together. Table 4 shows highly expressed genes consistently clustered together in K-means with different similarity metrics. The genes with the highest expression levels are presented in the first column of the table, distance metrics are given in the second column, other similarity metrics in third column and correlation metrics in the last column. If the genes in the given row belong to the same cluster for the corresponding group of metrics, this is indicated by "yes", otherwise by "no".

K-means clusters of the 20 genes with the highest expression level are given in Table 5. Euclidean distance is used as a metric. Each row gives the genes belonging to the same cluster. genes *lit* and *b*1240 are clustered together if K=5 but split when K=6. The optimal solution in EM model for the 264 genes showed that genes with the highest expression level are distributed in 3 clusters (Table 6). The refinement with the minimal entropy algorithm showed that the two genes *lit* and *bi*240 split in a separate cluster. Genes *metI* and *metN*, grouped in the same EM cluster, are also clustered together in K-means when Pearson correlation coefficients are used

Journal of Physics: Conference Series **128** (2008) 012030 doi:10.1088/1742-6596/128/1/012030

Table 4. Highly expressed genes consistently clustered together

Genes	E, Mh, Mk	D, J	C, P
$yaeS, metI$	yes	yes	yes
$metA, metR$	yes	yes	yes
$narV, yaeO$	yes	no	no
$metK, metF, b0539, yi82_1$	yes	no	no
$polB, ydcN, metN, prpD$	yes	yes	no
$folE, metK$	no	no	yes
$lit, b1240$	no	no	yes
$metF, metN$	yes	yes	yes
$polB, prpD$	yes	yes	yes

Table 5. K-means clusters of genes with the highest expression levels

$metE$
$metA, metR$
$narV, yaeO$
$metI, metB, yeaS, folB$
$metF, metK, metN, cspA, polB, ydcN, prpD, b0539, yi82_1$
$b1240$
lit

as a metric. $metE$ gene forms a cluster by itself in both models and is included in the tables for completeness.

Table 6. EM clusters of genes with the highest expression levels

$metE$
$metA, metR, narV$
$metI, metB, metF, metK, metN, folE, yeaS, cspA, polB, ydcN, prpD, b0539, yi82_1$ $lit, b1240$
$yaeO$

Co-expressed genes in the same cluster could have similar functions or indicate co-regulation. This has to be investigated further by using meaningful biological criteria. We have used a consensus sequence to identify genes belonging to the met box and our results are in agreement with the biological findings [22].

5. Conclusion and further work

Clustering techniques are used frequently in the analysis of gene expression data from microarray experiments. The identification of co-expressed genes allows to infer the function of unknown genes by comparing the co-regulated genes to the genes with known functions. Co-expressed genes in the same cluster are probably involved in the same cellular process and strong expression correlation between those genes indicates co-regulation.

However, clustering models are very sensitive to the choice of initial conditions and optimal selection of clusters. In this paper we have investigated the impact of initial conditions in three clustering models, K-means, EM and entropy-based model. The effect of similarity metrics is investigated in K-means. We have used cluster quality, stability and balance to optimise the clustering solutions. The entropy of the clusters is explored for improvement of the clustering results. We have clustered gene expression data of *E.coli* obtained in experiments investigating the effect of knocking out the methionine repressor gene. Work is in progress to compare automatically clustering results with known gene functionalities using the Gene Ontology (GO) and published literature. The visualisation of high dimensional gene expression data is an essential part of the analysis as it facilitates the discovery of structures, features, patterns and relationships, and enables human exploration and communication of the data. Our recently published method for targeted projection [25] and the tool for exploration and visualisation of high dimensional data [26] can be used to improve classifications of clustering results following the requirements of the user.

Acknowledgements: The authors thank the Astbury Centre for Molecular Biology at Leeds University, UK, for providing raw data for the genome-wide expression of *E.coli*.

References

[1] Higgs P G and Atwood T K 2006 *Bioinformatics and Molecular Evolution* (Oxford: Blackwell Publishing).
[2] Baldi P and Hatfield G W 2003 *DNA Microarrays and Gene Expression* (Cambridge: CUP).
[3] Stekel D 2003 *Microarray Bioinformatics* (Cambridge:CUP).
[4] DeRisi J L, Iyer V R and Brown P O 1997 *Science* **278** 680–686.
[5] M.B. Eisen MB , Spellman P T, Brown P O and Botstein D 1998 *Proc. Nat. Acad. Sci. USA* **95** 14863–14868.
[6] J. Khan,J.S. Wei,M. Ringnr,L.H. Saal,M. Ladanyi, F. Westermann, F.Berthold, M. Schwab, C.R. Antonescu, C. Peterson and P.S. Meltzer 2001 *Nature Medicine*, **7** 673–679.
[7] Golub T R, Slonim D K, Tamayo P, Huard C, Gaasenbeek M, Mesirov J P, Coller H, Loh M L, Downing J R, Caligiuri M A, Bloomfield C D and Lander E S 1999 *Science* **286** 531–5377.
[8] Dunham M 2003 *Data Mining. Introductory and Advanced Topics* (New Jersey: Prentice Hall).
[9] Jiang D, Tang C and Zhang A 2004 *IEEE Trans Knowledge Data Engin***16** 1370–1386.
[10] Brazma A and Vilo J 2000 Fed. Eur. Biochem. Soc. Lett. **480** 17–24.
[11] Verdoucci J S, Mefi V F, Lin S, Wang Z, Roy S and Sen Ch 2006 *Physiol Genomics* **25** 355–363.
[12] Dempster A P, Laird N M andRubin D B 1970 *Proc. Roy. Stat. Soc.* **B39** 1–38.
[13] Shannon C E 1948 *Bell System Tech. J.* **27** 389–423 ibid 623–656.
[14] Li H, Zhang K and Jiang T 2004 *IEEE Trans. Knowledge data Engineering* **16** 1370–1386.
[15] Parzen E 1962 *An. Math. Stat.* **33** 1065–1076.
[16] Smet F D, Mathys J, Marchal K , Thijis G, Moor B D and Moreau Y 2002 *Bioinformatics* **18** 735–746.
[17] L. Kaufman and P.J. Rousseeuw 1990 *Finding Groups in Data* (New York: Wiley).
[18] Fraley C and Raftery A E 1988 *The Computer Journal* **41** 578–588.
[19] Famili A F, Liu G and Liu Z 2004 *Bioinformatics* **20** 1535–1545.
[20] Datta S and Datta S 2002 *Bioinformatics* **19** 459–466.
[21] Bolshakova N, Azuaje F and Cunningham P 2005 *Bioinformatics* **21** 2546–2547.
[22] Marincs F, Manfield I, Stead J, McDowall K and Stockley P 2006 *Biochem. J.* **396** 227–234.
[23] Angelova M 2006 *Bulg. J. Phys.* **33** 876–883.
[24] Misra J, Schmitt W, Hwang D, Hsiao L L, Gullans S, Stephanopoulos Ge and Stephanopoulos Gr 2002 *Genome Research* **12** 1112–1120.
[25] J. Faith, R. Mintram and M. Angelova 2006 *Bioinformatics* **22** 2667–2673.
[26] J. Faith, M.Brockway 2006 *J. Integrative Biology* **3** 43–50.

Journal of Physics: Conference Series **128** (2008) 012031 doi:10.1088/1742-6596/128/1/012031

Particle-number conservation in thermodynamic systems within the Richardson model

N Benhamouda[1], N H Allal[1,2], M Fellah[1,2] and M R Oudih[1]

[1] Laboratoire de Physique Théorique , Faculté de physique, USTHB, BP32, El-Alia, 16111 Bab-Ezzouar, Alger, Algeria

[2] Centre de Recherche Nucléaire d'Alger,
2 Bd. Frantz Fanon, BP. 399 Alger-Gare, Algiers, Algeria

E-mail: Benhamoudan@yahoo.fr

Abstract. To describe thermal pairing systems, an approach that combines the modified BCS method and the Lipkin-Nogami particle-number projection method is proposed. The expressions of the thermal pairing gap and internal energy are derived. The latter are numerically studied as a function of the temperature within the Richardson model.

1. Introduction

In finite systems, such as atomic nuclei, it is well known that the symmetry violations (as the particle-number symmetry or the rotation one) in the BCS theory may imply important effects (generally so-called quantal fluctuations) on the calculation of various physical observables [1] such as the energy [2, 3, 4, 5, 6], the charge square radius and the quadrupole and hexadecapole moments [7, 8], or the moment of inertia [9].

At finite temperature T, statistical fluctuations are added to the quantal ones. These fluctuations arise from another symmetry violation: the unitary relation of the particle-density matrix [11]. Recently, the modified BCS method (MBCS) [10, 11, 12, 13, 14, 15] was suggested to take into account this kind of fluctuations by considering the fluctuations of the quasi-particle number which are neglected by the conventionnal finite temperature BCS approach (FTBCS) [16, 17, 18, 19, 20, 21]. The evaluation of the pairing gap as a function of the temperature within this approach has shown that this quantity decreases monotonously with increasing T. As a consequence, the sharp superfluid-normal phase transition observed using the BCS approach is washed out [10, 11, 12]. It has been shown in ref. [12] that the thermodynamic approach overcomes only the quasiparticle-number fluctuations but does not resolve the problem of the particle-number symmetry violation in the BCS approach. The aim of the present paper is to propose an approach that combines the MBCS method and the Lipkin-Nogami particle-number projection method [22, 23]. The paper is organized as follows. The formalism is presented in section 2, after a brief recall of the MBCS method. The thermal pairing gap and the internal energy are then deduced. They are numerically studied as a function of the temperature within the Richardson model [24], which also provides exact solutions, in section 3. Main conclusions are summarized in section 4.

2. Formalism

The intrinsic motion of N=2P paired particles is described by the Hamiltonian:

$$\hat{H} = \sum_{\nu>0} \varepsilon_\nu \left(a_\nu^\dagger a_\nu + a_{\tilde{\nu}}^\dagger a_{\tilde{\nu}} \right) - G \sum_{\nu\,\mu>0} a_\nu^\dagger a_{\tilde{\nu}}^\dagger a_{\tilde{\mu}} a_\mu, \tag{1}$$

where $a_\nu^\dagger$ and a_ν respectively represent the creation and annihilation operators of the state $\mid \nu\rangle$, of energy ε_ν; and $a_{\tilde{\nu}}^\dagger$ and $a_{\tilde{\nu}}$ those of the state $\mid \tilde{\nu}\rangle$, which is the time reverse of $\mid \nu\rangle$ and has the same energy. G is the pairing strength which is assumed to be constant.
Let us recall that in the case of the BCS approach, the use of the canonical Bogoliubov transformation from the particle operators $a_\nu^\dagger$ and a_ν to the quasiparticle ones $\alpha_\nu^\dagger$ and α_ν [1]:

$$\begin{aligned} a_\nu^\dagger &= u_\nu \alpha_\nu^\dagger + v_\nu \alpha_{\tilde{\nu}} \\ a_{\tilde{\nu}} &= u_\nu \alpha_{\tilde{\nu}} - v_\nu \alpha_\nu^\dagger \end{aligned} \tag{2}$$

allows one to obtain the gap equations in the form [17]:

$$\Delta = G \sum_{\nu>0} u_\nu v_\nu \left(1 - 2\eta_\nu\right) \tag{3}$$

$$N = 2 \sum_{\nu>0} \left[v_\nu^2 + (1 - 2v_\nu^2)\eta_\nu \right] \tag{4}$$

where η_ν is the quasi-particle occupation number given by the Fermi-Dirac distribution:

$$\eta_\nu = \frac{1}{1 + exp(\beta E_\nu)} \tag{5}$$

and E_ν is the quasiparticle energy given by:

$$E_\nu = \sqrt{(\varepsilon_\nu - \lambda - G v_\nu^2)^2 + \Delta^2} \tag{6}$$

The internal energy of the system then reads:

$$E_{FTBCS} = 2 \sum_{\nu>0} [\varepsilon_\nu - 2G \sum_{\mu>0} (v_\mu^2 + (1 - 2v_\mu^2)\eta_\mu)][v_\nu^2 + (1 - 2v_\nu^2)\eta_\nu] - \frac{\Delta^2}{G} \tag{7}$$

Ref.[11, 12] have shown that the quasi-particle number fluctuations are not taken into account in the FTBCS theory. The modified BCS (MBCS) approach allows one to overcome this defect, by introducing a secondary Bogoliubov transformation:

$$\begin{aligned} \bar{\alpha}_\nu^\dagger &= \sqrt{1 - \eta_\nu}\alpha_\nu^\dagger + \sqrt{\eta_\nu}\alpha_{\tilde{\nu}} \\ \bar{\alpha}_{\tilde{\nu}} &= \sqrt{1 - \eta_\nu}\alpha_{\tilde{\nu}} - \sqrt{\eta_\nu}\alpha_\nu^\dagger \end{aligned} \tag{8}$$

that connects between the usual quasi-particle (QP) operators $\alpha_\nu^\dagger$ and α_ν and the modified quasi-particle operators (MQP) $\bar{\alpha}_\nu^\dagger$ and $\bar{\alpha}_\nu$. By combining the two transformations (2) and (8), one obtains:

$$\begin{aligned} a_\nu^\dagger &= \bar{u}_\nu \bar{\alpha}_\nu^\dagger + \bar{v}_\nu \bar{\alpha}_{\tilde{\nu}} \\ a_{\tilde{\nu}} &= \bar{u}_\nu \bar{\alpha}_{\tilde{\nu}} - \bar{v}_\nu \bar{\alpha}_\nu^\dagger \end{aligned} \tag{9}$$

where the variational parameters $\bar{u}_\nu$ and $\bar{v}_\nu$ read:

$$\begin{aligned} \bar{u}_\nu &= \sqrt{1 - \eta_\nu}u_\nu + \sqrt{\eta_\nu}v_\nu \\ \bar{v}_\nu &= \sqrt{1 - \eta_\nu}v_\nu - \sqrt{\eta_\nu}u_\nu \end{aligned} \tag{10}$$

The Hamiltonian in the MQP representation is analogous to that of the QP representation. One has just to replace the $\alpha_\nu^\dagger$ and α_ν operators by $\bar{\alpha}_\nu^\dagger$ and $\bar{\alpha}_\nu$ and the u_ν and v_ν coefficients by $\bar{u}_\nu$ and $\bar{v}_\nu$. One then obtains, after some algebra, the gap equations:

$$\bar{\Delta} = G\sum_{\nu>0}\bar{u}_\nu\bar{v}_\nu = G\sum_{\nu>0}[u_\nu v_\nu + \left(1-2v_\nu^2\right)\delta\eta_\nu] \tag{11}$$

$$N = 2\sum_{\nu>0}\bar{v}_\nu^2 = 2\sum_{\nu>0}\left[v_\nu^2 - 2\left(1-2u_\nu v_\nu\right)\delta\eta_\nu\right] \tag{12}$$

where $\delta\eta_\nu = \sqrt{\eta_\nu(1-\eta_\nu)}$ is the QP fluctuation number. The MBCS internal energy has then an expression similar to the BCS one:

$$E_{MBCS} = 2\sum_{\nu>0}(\varepsilon_\nu - \frac{G}{2}\bar{v}_\nu^2)\bar{v}_\nu^2 - \frac{\bar{\Delta}^2}{G} \tag{13}$$

This approach allows one to establish the gap equations and the physical quantities in a simple way. However, it neglects the fluctuations of the particle number. In order to take them into account, one has to perform the same calculation by starting from the Lipkin-Nogami Hamiltonian [22, 23] instead of the usual BCS auxiliary Hamiltonian, that is:

$$\hat{H}' = \hat{H} - \lambda\hat{N} - \lambda_2\hat{N}^2 \tag{14}$$

and to express it in the MQP representation. One notes that λ is a Lagrange multiplier and λ_2 is kept constant, $\hat{N}$ is the particle number operator defined by:

$$\hat{N} = \sum_{\nu>0}\left(a_\nu^\dagger a_\nu + a_{\tilde{\nu}}^\dagger a_{\tilde{\nu}}\right). \tag{15}$$

The $\bar{u}_\nu$ and $\bar{v}_\nu$ parameters are then chosen such as to minimize the expectation value of the Hamiltonian $\hat{H}'$:

$$\frac{\partial}{\partial\bar{v}_\nu}\langle\psi|\,\hat{H}'\,|\psi\rangle = 0, \qquad \forall\nu. \tag{16}$$

Thus, $\bar{u}_\nu$ and $\bar{v}_\nu$ are given by:

$$\left.\begin{matrix}\bar{u}_\nu^2\\ \bar{v}_\nu^2\end{matrix}\right\} = \frac{1}{2}\left\{1 \pm \frac{\varepsilon_\nu - \lambda + (4\lambda_2 - G)\bar{v}_\nu^2}{\sqrt{(\varepsilon_\nu - \lambda - G\bar{v}_\nu^2 + 4\lambda_2\bar{v}_\nu^2)^2 + \bar{\Delta}^2}}\right\} \tag{17}$$

with:

$$\lambda_2 = \frac{G}{4}\frac{\sum\limits_{\nu>0}\bar{u}_\nu^3\bar{v}_\nu\sum\limits_{\nu>0}\bar{u}_\nu\bar{v}_\nu^3 - \sum\limits_{\nu>0}\bar{u}_\nu^4\bar{v}_\nu^4}{(\sum\limits_{\nu>0}\bar{u}_\nu^2\bar{v}_\nu^2)^2 - \sum\limits_{\nu>0}\bar{u}_\nu^4\bar{v}_\nu^4} \tag{18}$$

It appears that the fact to take into account the particle-number fluctuations induce a renormalization of the single particle energy by a $(4\lambda_2\bar{v}_\nu^2)$ term.
Let us introduce the occupation number:

$$\bar{\eta}_\nu = \frac{1}{1+exp(\beta\bar{E}_\nu)} \tag{19}$$

where $\bar{E}_\nu$ is given by:

$$\bar{E}_\nu = \sqrt{(\varepsilon_\nu - \lambda - G\bar{v}_\nu^2 + 4\lambda_2\bar{v}_\nu^2)^2 + \bar{\Delta}^2} \tag{20}$$

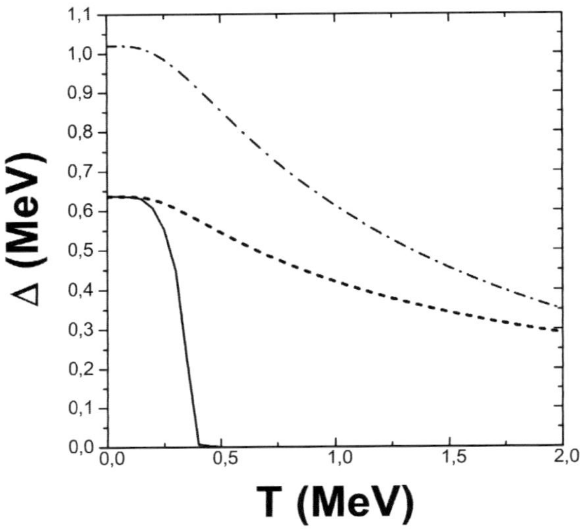

Figure 1. Variation of the thermal pairing gap as a function of temperature obtained within the FTBCS (solid line), MBCS (dashed line) and the present work (dotted-dashed line) calculations .

The calculations are thus formally the same as in the usual Lipkin-Nogami method. One just has to replace u_ν and v_ν and the quasi-particle operators by the modified ones. One deduces then the Modified Lipkin-Nogami internal energy:

$$E_{LN} = 2\sum_{\nu>0}(\varepsilon_\nu - G\bar{v}_\nu^2 + 4\lambda_2\bar{v}_\nu^2)\bar{v}_\nu^2 - \frac{\bar{\Delta}^2}{G} - 4\lambda_2\sum_{\nu>0}\bar{u}_\nu^2\bar{v}_\nu^2 \tag{21}$$

The last term of equation (21) represents the correction to the energy, that is due to the particle-number fluctuations. One notes that when T=0 MeV, equations (17) and (18) reduce to those of the usual Lipkin-Nogami method.

3. Numerical results - Discussion

In order to test the previously described approach, we performed numerical calculations using the Richardson model [24], which provides exact solutions. We considered a system of $N = 10$ particles, with a total degeneracy of pairs $\Omega = 10$ and a pairing-strength $G = 0.4$.
In a first step, we have evaluated the thermal pairing gap as a function of the temperature when $0 \leq T \leq 2MeV$. In figure 1, we reported the results obtained using the conventional FTBCS theory, the MBCS method and the Modified Lipkin-Nogami approach. The figure shows that the pairing gap collapses at a critical temperature $T_c = 0.4MeV$ in the FTBCS theory. This result is a signature of the phase transition from the superfluid state to a normal one. However, the MBCS and Modified Lipkin-Nogami methods predict a monotonous decrease of the pairing gap as a function of the temperature. N. Dinh Dang et al. [10] achieved to similar results by taking into account the particle-number fluctuations using the Lipkin Nogami method in the static path approximation.

Journal of Physics: Conference Series **128** (2008) 012031 doi:10.1088/1742-6596/128/1/012031

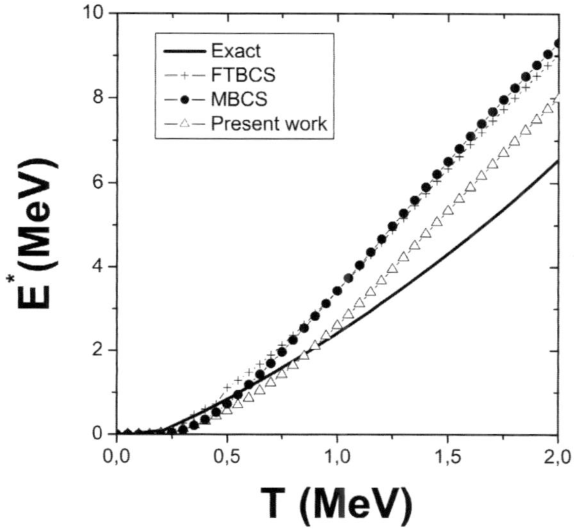

Figure 2. Variation of the excitation energy as a function of temperature obtained within the FTBCS ($+ - +$), MBCS ($\bullet - \bullet$), present work ($\triangle - \triangle$) and exact (solid line) calculations.

Moreover, one notes that the pairing gap in the case of the Modified Lipkin-Nogami method is greater than the predictions of the two other approaches. At $T = 0MeV$ the discrepancy is of about $0.5MeV$, and it decreases with increasing temperatures. This discrepancy is attributed to the particle-number fluctuations effect. Indeed, it was shown in ref.[10] that this effect is important at low temperature i.e. when $T < T_c$. Moreover, their effect vanishes at high temperature while the statistical fluctuations become important. As a consequence, the predictions of the Modified Lipkin-Nogami method converge to the MBCS results when $T > T_c$. As a second step, the excitation energy given by:

$$E^* = E(T) - E(0) \tag{22}$$

has been evaluated as a function of the temperature. The obtained results are reported in figure 2. The latter shows the behavior of this quantity in the case of the FTBCS, MBCS and the Modified Lipkin-Nogami methods. The predictions of these models are also compared to the exact ones. One notes that the exact results are better reproduced when the Modified Lipkin-Nogami method is used. One deduces then that the effect of the particle-number fluctuations is non-negligible. Moreover, one observes an important discrepancy of the order of 1 MeV between the Modified Lipkin-Nogami method and the exact results at high temperature. This effect could be attributed to the fact that the Lipkin-Nogami method remains an approximate projection method.

4. Conclusion

In the present work, we have performed an analysis of the pairing gap and the excitation energy in the case of thermodynamic systems by combining the Lipkin-Nogami method and the MBCS approach. The evaluation of these quantities as a function of the temperature when

Journal of Physics: Conference Series **128** (2008) 012031 doi:10.1088/1742-6596/128/1/012031

$0MeV \leq T \leq 2MeV$ has shown a cancelation of the sharp superfluid-normal phase transition that was observed with the usual FTBCS method. On the other hand, the results confirm that the particle-number fluctuations are important at low temperature, and their effects are non-negligible. Finally, the excitation energy is better reproduced when the modified Lipkin-Nogami method is used instead of the MBCS one.

References

[1] Ring P and Schuck P 1980 *The Nuclear Many-Body Problem*(Berlin: SPRINGER)
[2] Fellah M, Hammann T F and Medjadi D E 1973 *Phys. Rev.C* **8** 1585
[3] Oudih M R, Fellah M and Allal N H 2003 *Int. J. Mod. Phys. E* **12** 109
[4] Oudih M R, Fellah M, Allal N H and Benhamouda N 2006 *Int. J.Mod. Phys. E* **15** 643
[5] Allal N H, Fellah M, Oudih M R and Benhamouda N 2006 *Eur. Phys. J. A* **27** s01, 301
[6] Oudih M R, Fellah M, Allal N H and Benhamouda N 2007 *Phys. Rev. C* **76** 047307
[7] Benhamouda N, Oudih M R, Allal N H and Fellah M 2001 *Nucl. Phys. A* **690** 219
[8] Benhamouda N, Allal N H, Fellah M and Oudih M R 2005 *Int. J. Mod. Phys. E* **14** 197
[9] Allal N H and Fellah M 1991 *Phys. Rev. C* **43** 2648
[10] Dinh Dang N 1990 *Z.Phys. A* **335** 253
[11] Dinh Dang N and Arima A 2003 *Phys. Rev. C* **68** 014318
[12] Dinh Dang N 2006 *Nucl. Phys. A* **784** 147
[13] Ponomarev V Yu and Vdovin A I 2005 *Phys. Rev. C* **72** 034309
[14] Dinh Dang N and Arima A 2006 *Phys. Rev. C* **74** 059801
[15] Ponomarev V Yu and Vdovin A I 2006 *Phys. Rev. C* **74** 059802
[16] Sano M and Yamazaki S 1963 *Prog. Theor. Phys.* **29** 397
[17] Moretto L G 1972 *Nucl. Phys. A* **185** 145
[18] Goodman A 1981 *Nucl. Phys. A* **352** 30
[19] Alasia F, Civitarese O and Reboiro M 1987 *Phys. Rev. C* **35** 812
[20] Esebbag C and Egido J L 1993 *Nucl. Phys. A* **552** 205
[21] Sandulescu N, Civitarese O and Liotta R J 2000 *Phys. Rev. C* **61** 044317
[22] Lipkin H J 1960 *Ann. Phys.* **12** 425
[23] Nogami Y 1964 *Phys. Rev. B* **313** 134
[24] Richardson R W and Sherman N 1964 *Nucl. Phys.* **52** 253
[25] Dinh Dang N, Ring P and Rossignoli R 1993 *Phys. Rev. C* **47** 47

V International Symposium on Quantum Theory and Symmetries IOP Publishing
Journal of Physics: Conference Series **128** (2008) 012032 doi:10.1088/1742-6596/128/1/012032

Unified dynamical symmetries in the symplectic extension of the Interacting Vector Boson Model

A.I. Georgieva[1,2], **H. G. Ganev**[1], **J. P. Draayer**[2] **and V. P. Garistov**[1]
[1]Institute of Nuclear Research and Nuclear Energy,
Bulgarian Academy of Sciences, Sofia 1784, Bulgaria
[2]Department of Physics and Astronomy,
Louisiana State University, Baton Rouge, Louisiana, USA

E-mail: anageorg@inrne.bas.bg

Abstract. The algebraic Interacting Vector Boson Model (IVBM) is extended by exploiting three new subgroup chains in the reduction of its highest symplectic dynamical symmetry group Sp(12,R) to the physical angular momentum subgroup SO(3). The corresponding exactly solvable limiting cases are applied to achieve a description of complex nuclear collective spectra of even-even nuclei in the rare earth and actinide regions up to states of very high angular momentum.

First we exploit two reductions in which collective modes can be mixed, and obtain successful descriptions of both positive and negative parity band configurations. The structure of band-head configurations, whose importance is established in the first two limits, is examined in a third reduction, that also provides important links between the subgroups of the other limits.

1. Introduction

With the intensive development of the experimental facilities a lot of new data on the collective spectra of nuclei throughout the nuclear chart is accumulated [1]. It reveals the complicated character of the nuclear motions, yielding different degrees of mixing of the basic rotational and vibrational collective modes. In turn this requires a corresponding degree of complication of the nuclear structure models, but still they should remain analytically solvable and so providing the tool for the interpretation of the observed data.

The algebraic models like IBM [2], based on the notion of the dynamical symmetries have been widely used for the description of low spin states. An other example of such successful applications [3] for the description of the low-lying collective rotational spectra of the even-even medium and heavy mass nuclei, is the number of bosons preserving version of the phenomenological algebraic Interacting Vector Boson Model /IVBM/ [4].

At the same time in the relatively early stages of the development of the algebraic methods in the nuclear structure theory [5, 6, 7] the separation of the collective degrees of freedom from the intrinsic ones, their development and mixing was attempted in the framework of the symplectic geometry of the nuclear many-body problem. The advantages of the application of symplectic structures, are based on their property to change the number of phonons building the collective states and so to provide larger basis spaces to incorporate the more and more complex nuclear spectra.

With the aim to extend the earlier applications of the IVBM [4], and in order to incorporate the new experimental data on the states with higher spins and on the various new excited bands, we explore the symplectic extension of the model, with the $Sp(12,R)$ as a group of dynamical symmetry. The extension is realized and has its physical interpretation over the basis of its maximal compact subgroup $U(6) \subset Sp(12,R)$, which was interpreted as the rotational limit [3] of the number of boson's preserving model. This naturally leads to the additional description [8] of positive and negative parity states with mixed collective vibrational and rotational modes.

In the larger infinite boson spaces of the dynamical $Sp(12,R)$ there are also new chains of subgroups, starting with noncompact symplectic subalgebras. We investigate the reduction $Sp(2,R) \otimes SO(6) \subset Sp(12,R)$, which allows the inclusion of a 6-dimensional Davidson potential [9], that is known to allow mixing of the rotational and vibrational modes, so in applications it reproduces very well the transitional behavior in nuclear spectra. The other reduction [10] we consider $Sp(4,R) \otimes SO(3) \subset Sp(12,R)$ allows the selection of states with fixed angular momentum L, given by the $SO(3)$ irreducible representations (irreps). The physics behind this dynamical symmetry outlines the importance of the structure of the band-head configurations in the development of the nuclear spectra.

In this work, the above mentioned dynamical symmetries are unified in a generalized reduction scheme for the symplectic extension of the IVBM, which contains a lot of relations between the subgroups from the different chains. These clarify the physical meaning and motivation of the model and lead to rather successful applications for the description of the vast amount of new experimental data on the nuclear structure, while still retaining the advantages of the use of dynamical symmetries, namely the exact analytic solutions for the energy spectra of nuclei.

2. The symplectic extension of the IVBM dynamical symmetry

The algebraic structure of the IVBM is realized in terms of creation (annihilation) operators $u_m^+(\alpha)(u_m(\alpha))$, in a 3- dimensional oscillator potential $m = 0, \pm 1$ of *two* types of bosons differing by the value of the "pseudo-spin" projection $\alpha = 1/2(p)$ or $\alpha = -1/2(n)$. The later are related with the cyclic coordinates $x_{\pm 1}(\alpha) = \mp \frac{1}{\sqrt{2}}(x_1(\alpha) \pm i x_2(\alpha)), x_0(\alpha) = x_3(\alpha)$ and their associated momenta $q_m(\alpha) = -i\partial/\partial x^m(\alpha)$, in the standard way

$$u_m^\dagger(\alpha) = \frac{1}{\sqrt{2}}(x_m(\alpha) - i q_m(\alpha)), \quad u_m(\alpha) = (u_m^\dagger(\alpha))^\dagger), \tag{1}$$

where $x_i(\alpha)$ $i = 1, 2, 3$ are Cartesian coordinates of a quasi-particle vectors with an additional index - the projection of the "pseudo-spin" $\alpha = \pm\frac{1}{2}$. The bilinear products of the creation and annihilation operators of the two vector bosons (1) generate the boson representations of the non-compact symplectic group $Sp(12,R)$ [4] :

$$\begin{aligned} F_M^L(\alpha,\beta) &= \sum\nolimits_{k,m} C_{1k1m}^{LM} u_k^\dagger(\alpha) u_m^\dagger(\beta), \\ G_M^L(\alpha,\beta) &= \sum\nolimits_{k,m} C_{1k1m}^{LM} u_k(\alpha) u_m(\beta), \end{aligned} \tag{2}$$

$$A_M^L(\alpha,\beta) = \sum\nolimits_{k,m} C_{1k1m}^{LM} u_k^\dagger(\alpha) u_m(\beta), \tag{3}$$

where C_{1k1m}^{LM}, the usual Clebsch-Gordon coefficients for $L = 0, 1, 2$ and $M = -L, -L+1, ...L$, define the transformation properties of (2) and (3) under rotations. The commutation relations between the pair creation and annihilation operators (2) and the number preserving operators (3) are calculated in [4].

2.1. Reduction through the compact U(6)

2.1.1. Algebraic structure The set of operators (3) close under commutation the algebra of the maximal compact subgroup of $U(6) \subset Sp(12,R)$ [11]. The linear invariant of $U(6)$ is the number operator:

$$N = \sqrt{3}(A^0(p,p) + A^0(n,n)) = N_+ + N_-, \tag{4}$$

that counts the total number of bosons. Being the first order invariant of $U(6)$, the operator (4) splits the boson representations of $Sp(12,R)$ into a countless number of symmetric unitary irreducible representations /UIR/ of the type $[N,0,0,0,0,0] = [N]_6$, where $N = 0,2,4,...$ for the even UIR and $N = 1,3,5,...$ for the odd ones. The rest of the operators of the physical observables introduced in this limit, which define the algebra of $SU(3)$ are the truncated ("Elliott's") [12] quadrupole operator $Q_M = \sqrt{6}\sum_{M,\alpha} A^2_M(\alpha,\alpha)$, $M = 0,\pm1,\pm2$ and the angular momentum operator with components $L_M = -\sqrt{2}\sum_{M,\alpha} A^1_M(\alpha,\alpha)$, $M = 0,\pm1$ that generate its $SO(3)$ subalgebra. These operators are obtained exactly from the quantization of the classical momenta in the coordinate representation and further motivate the use of the vector bosons (1) as building blocks of the model. The scalar operators:

$$\begin{array}{ll} A^0(p,n) = \sqrt{\frac{2}{3}}T_+ & A^0(n,p) = -\sqrt{\frac{2}{3}}T_- \\ A^0(p,p) = \sqrt{\frac{1}{3}}N_+ & A^0(n,n) = \sqrt{\frac{1}{3}}N_- \end{array} \tag{5}$$

are the Weyl generators of the algebra of $U(2)$. We can use the equivalent set of infinitesimal operators containing in addition to the raising T_+ and lowering T_- components of the pseudospin (see (5)) the Cartan operators N (4) and

$$T_0 = -\sqrt{\frac{3}{2}}[A^0(p,p) - A^0(n,n)] \tag{6}$$

the third projection of the pseudospin operator T. The operators $T_0, T_\pm$ satisfy the commutation relations $[T_0,T_\pm] = \pm T_\pm$, $[T_+,T_-] = 2T_0$, $[N,T_{0,\pm}] = 0$ of the pseudospin algebra $su(2)$. These operators play an important role in the consideration of the nuclear system as composed by two interacting subsystems. Obviously (5) commute with the $SU(3)$ generators $Q_M$and L_M, so that the algebras of the two groups are mutually complementary.

As a result of the above considerations, the rotational limit [3] of the number preserving version of the model defined by the chain:

$$Sp(12,R) \supset U(6) \supset SU(3) \otimes U(2) \supset SO(3) \otimes U(1) \tag{7}$$

$$[N] \quad (\lambda,\mu) \quad (N,T)\; K \quad L \qquad T_0 \tag{8}$$

is extended to the $Sp(12,R)$- group of dynamical symmetry. The labels below the subgroups are the quantum numbers (8) corresponding to their irreducible representations. Since the reduction from $U(6)$ to $SO(3)$ is carried out by the direct product of the groups $SU(3)$ and $U(2)$, their quantum numbers are related in the following way: $T = \frac{\lambda}{2}, N = 2\mu + \lambda$. Making use of the latter we can write the basis as

$$\mid [N]_6; (\lambda,\mu); K, L, M; T_0\rangle = \mid (N,T); K, L, M; T_0\rangle \tag{9}$$

The ground state of the system is the vacuum state $|0\rangle$ with $N = 0, T = 0, K = 0, L = 0, M = 0, T_0 = 0$. The basis states associated with the even irreducible representation of the $Sp(12,R)$ can be constructed by the application on it of powers of raising generators $F^L_M(\alpha,\beta)$ of the same group. Each raising operator will increase the number of bosons N by two. The $Sp(12,R)$

Table 1.

N (T)\ T_0	...	±4	±3	±2	±1	0
0 0						(0, 0)
2 1 0					(2, 0)	(2, 0) (0, 1)
4 2 1 0				(4, 0)	(4, 0) (2, 1)	(4, 0) (2, 1) (0, 2)
6 3 2 1 0			(6, 0)	(6, 0) (4, 1)	(6, 0) (4, 1) (2, 2)	(6, 0) (4, 1) (2, 2) (0, 3)
8 4 3 2 1 0		(8, 0)	(8, 0) (6, 1)	(8, 0) (6, 1) (4, 2)	(8, 0) (6, 1) (4, 2) (2, 3)	(8, 0) (6, 1) (4, 2) (2, 3) (0, 4)
...	...	...	...	...	...	...

classification scheme for the $SU(3)$ boson representations for even value of the number of bosons N is shown on Table 1. Each row (fixed N) of the table corresponds to a given irreducible representation of the $U(6)$. Then the possible values for the pseudospin $T = \frac{N}{2}, \frac{N}{2} - 1, \dots 0$ are given in the column next to the respective value of N. Thus when N and T are fixed, $2T + 1$ equivalent representations of the group $SU(3)$ arise, for which λ and μ are obtained from their relations with N and T. Each of them is labelled by the eigenvalues of the operator $T_0 : -T, -T+1, \dots, T$, defining the columns of Table 1. The same $SU(3)$ representations (λ, μ) arise for the positive and negative eigenvalues of T_0.

Finally in order to obtain the values of the angular momenta contained in a given $SU(3)$ representation (λ, μ) we use the standard reduction rules [12] for the chain $SU(3) \supset SO(3)$. The multiplicity index K appearing in this reduction is related to the projection of L in the body fixed frame and is used with the parity (π) to label the different bands (K^π) in the energy spectra of the nuclei. We define the parity of the states as $\pi = (-1)^T$, which allows us to describe both positive and negative bands.

The Hamiltonian, corresponding to the considered limit of IVBM, is expressed in terms of the first and second order invariant operators of the different subgroups in the chain (7). As a result of the complementarity of the $SU(3)$ and $SU(2)$ groups, the Casimir operator of $SU(3)$ with eigenvalue $(\lambda^2 + \mu^2 + \lambda\mu + 3\lambda + 3\mu)$, is expressed in terms of the operators N and T, so we use the expression:

$$H = aN + bN^2 + \alpha_3 T^2 + \beta_3 \pi_3 + \alpha_1 T_0^2, \tag{10}$$

where π_3 is the $SO(3)$ second order Casimir operator. H (10) is obviously diagonal in the basis (9) labelled by the quantum numbers of the subgroups of the chain (7). Its eigenvalues are the energies of the basis states of the boson representations of $Sp(12, R)$:

$$E((N, T), L, T_0) = aN + bN^2 + \alpha_3 T(T+1) + \beta_3 L(L+1) + \alpha_1 T_0^2 \tag{11}$$

The subspaces $[N, 0, 0, 0, 0, 0] = [N]_6$, where $N = 0, 2, 4, \dots$ given in Table 1 are of the finite dimension, which simplifies the problem of diagonalization. Therefore the complete spectrum of the system is calculated trough the diagonalization of the Hamiltonian in the subspaces of all the UIR of $U(6)$, belonging to a given UIR of $Sp(12, R)$, which further clarifies its role of a group of dynamical symmetry.

Journal of Physics: Conference Series **128** (2008) 012032 doi:10.1088/1742-6596/128/1/012032

2.1.2. Description of the ground and octupole bands energies The most important application of the above limit of the model is the possibility to describe both even and odd parity bands up to very high angular momentum. In order to do this we first have to identify these experimentally observed bands with the sequences of basis states for the even representation of $Sp(12, R)$ given in Table 1. As we deal with the symplectic extension of boson representations of the number preserving $U(6)$ symmetry we are able to consider all even eigenvalues of the number of vector bosons N with the corresponding set of pseudospins T. For the description of the ground band, we choose the $SU(3)$ multiplets $(0, \mu)$ from the different $U(6)$ irreps given by the numbers of bosons, $N = 0, 4, 8, \ldots$ and pseudospin $T = 0$ in the column labelled by $T_0 = 0$ of Table 1. While for the octupole band the $SU(3)$ multiplets $(2, \mu - 1)$ for $N = 8, 12, \ldots$ and $T = 1$ from the same column $T_0 = 0$ are used. In terms of (N, T) this choice corresponds to $(N = 2\mu, T = 0)$ for the positive $(K^\pi = 0^+)$ and $(N = 2\mu + 2, T = 1)$ for the negative $(K^\pi = 0^-)$ parity band, respectively.

Further we define the energies of each state with given L as yrast energy with respect to N in the two considered bands. Hence their minimum values are obtained at $N = 2L$ for the ground band, and $N = 2L + 2$ for the octupole band, respectively. In the so defined $SU(3)$ representations for each N the maximal values of L appear for the first time (see Table 1).

For the above basis states $T_0 = 0$ and the experimental data on the ground and octupole bands, the last term in the energy formula (11) vanishes. The phenomenological model parameters a, b, α_3, and β_3 are evaluated by a fit to the experimental data [13] on the considered bands of the even-even deformed nuclei belonging to light actinides and rare earth region.

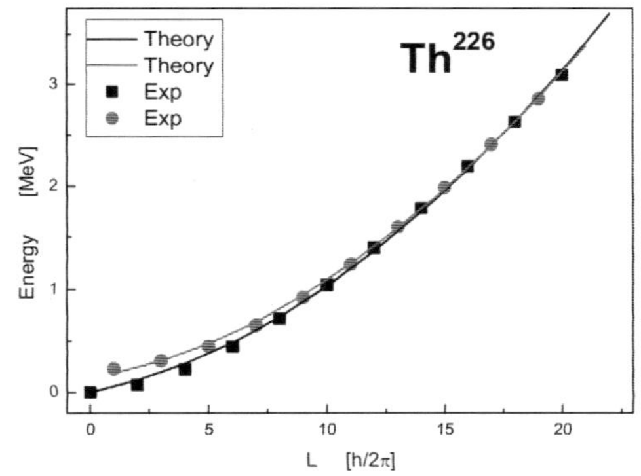

Figure 1. Comparison of the experimental and theoretical energies for ground and octupole bands of ^{228}Th. The later are calculated with the parameters $a = 0.0194, b = -0.0010$, $\alpha_3 = 0,015224$ and $\beta_3 = 0.0094$.

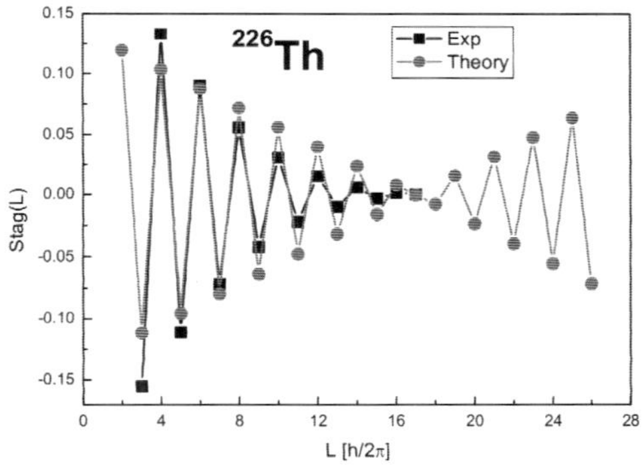

Figure 2. Comparison of the theoretical and experimental staggering functions for the ground and octupole bands of ^{226}Th

The agreement between the theoretical values obtained with only four model parameters and the experimental data for many nuclei considered in [8] is rather good. Applying the yrast conditions relating N and L the energies (11) for the two considered bands can be rewritten as:

$$E(L) = \beta L(L + 1) + (\gamma + \eta)L + \xi. \tag{12}$$

The new free parameters are $\beta = 4b + \beta_3, \gamma = 2a - 4b, \eta = 8b, \xi = 2a + 4b + 2\alpha_3$. The values of the parameters in (11) determine the behavior of the energies of the two bands and their position with respect to each other. The shift between the bands depends on the parameter ξ. When they are very close they interact through the L -dependent interaction with a strength η.

From (12) we can see that eigenstates of the first positive and negative bands consists of rotational $L(L+1)$ and vibrational L modes. The rotational interaction is with equal strength β in both of the bands. In the nuclei, considered in [8], the obtained values of the parameter η are always negative, which means that the negative parity band is less vibrational than the positive one.

Odd-even staggering patterns between ground and octupole bands are a typical feature for this type of collectivity. In order to test further our model we applied on the energies the staggering function defined as [14]:

$$Stg(L) = 6\Delta E(L) - 4\Delta E(L-1) - 4\Delta E(L+1) + \Delta E(L+2) + \Delta E(L-2), \qquad (13)$$

where $\Delta E(L) = E(L) - E(L-1)$. The comparison of the calculated and experimental energies and staggering patterns are illustrated in Figures 1 and 2. One can see a good agreement with experiment, as well as the reproduction of the "beat" patterns of the staggering behavior. This is due to the interaction term ηL in (12) between the positive and negative parity bands, which is a result of the introduced notion of yrast energies in the framework of the symplectic extension of the IVBM.

2.1.3. Description of other excited bands In order to demonstrate further applicability of this dynamical symmetry, we use the theory to describe some additional excited positive and negative parity bands. The next most important bands for determining the collective properties of heavy nuclei are the positive parity β ($K^{\pi} = 0^{+}$) [15] and γ ($K^{\pi} = 2^{+}$) collective bands and low-lying negative parity bands with $K^{\pi} = 1^{-}, 3^{-}$. These excited bands have to be mapped on the basis states as well. In general the appropriate [7] subset of $SU(3)$ states are the so called "stretched" states. Their domination is determined by the important role of the quadrupole-quadrupole interactions in the collective excitations. Thus, the most important $SU(3)$ states will be those with maximal weight, i.e. those which have maximal eigenvalues of the second order $SU(3)$ Casimir operator. For the considered chain we have two types of stretched states: $(\lambda, \mu) = (\lambda_0, \mu_0 + i)$ or $(\lambda, \mu) = (\lambda_0 + 2i, \mu_0)$, where λ_0 and μ_0 fix the starting $SU(3)$ build by $N_0 = \lambda_0 + 2\mu_0$ bosons and i is changing.

In the first case obviously $\lambda_0 = 2T$ is fixed, which fixes the parity of the bands. We require that $\lambda_0 \leq \mu$ and this relates it to the value of K, that labels the bands. For the $K = 0$ bands we have $i = 0, 2, 4\ldots$ and N changing in steps of 4 and for the $K \neq 0$ the values of $i = 0, 1, 2, 3\ldots$ and $\Delta N = 2$. In this case the yrast condition that relates the number of bosons N with the angular momenta is $N = 2L + 6T - K$. We note that the states of ground and octupole bands which were selected above by means of the algebraic notion of yrast bands are of this type.

For the second type of stretched states $(\lambda_0 + 2i, \mu_0)$, obviously $T = (\lambda_0 + 2i)/2$ is changing and in order to preserve the parity of the bands we need to have $i = 0, 2, 4\ldots$ and so $\Delta N = 4$. Now we have $\mu_0 \leq \lambda_0$ fixed and related to the value of K, that labels the bands. In this case the yrast condition is defined by the relation $N = 2L + 4\mu_0 - K$.

If we consider T_0 as fixed in both cases the stretched $SU(3)$ irreps belong to a column in Table 1 and if they change one moves along a diagonal. In the second case T_0 plays a more important role because the value of T is changing with the development of the bands and we have more choices in changing respectively the values of $T_0 = \pm T, \pm(T-1), \ldots, 0$ and hence we can make use of the additional parameter α_1 in (11) when fitting the values of the energies. For the excited $\beta-$bands and $\gamma-$bands, we use the sequences of stretched $SU(3)$ irreps determined by the "diagonals" $(\lambda_0 + 2i, \mu_0)$ with $\mu_0 = 0$ or 2.

In our phenomenological approach, because of the symplectic extension of the number preserving model [3], we have a great flexibility in identifying bands in any of the ways described

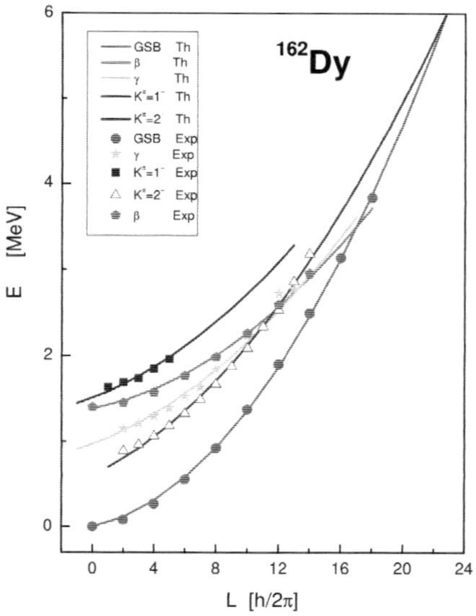

Figure 3. Comparison of the experimental and theoretical energies for the listed excited bands of ^{162}Dy. The later are calculated with the parameters $a = 0.0106$, $b = 0.0014$, $\alpha_3 = -0,0048$, $\alpha_1 = -0,1663$ and $\beta_3 = 0.0052$.

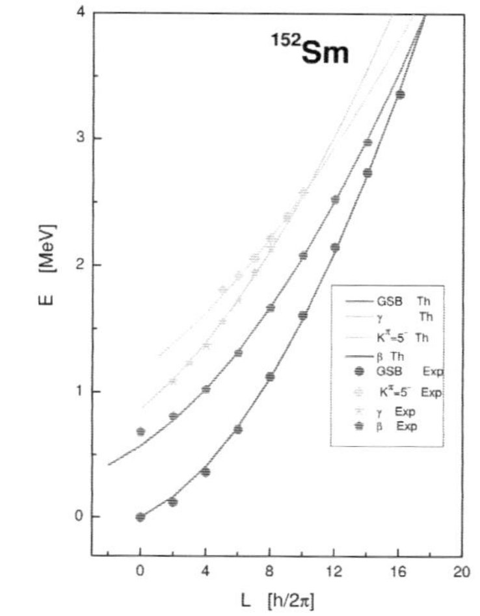

Figure 4. Comparison of the experimental and theoretical energies for the listed excited bands of ^{152}Sm. The later are calculated with the parameters $a = 0.0308$, $b = -0.0012$, $\alpha_3 = -0,0003$, $\alpha_1 = -0,0018$ and $\beta_3 = 0.0044$.

above. As a result of the $N \longleftrightarrow L$ connections we always get a mixing of the rotational and vibrational collective modes, represented by the $L(L+1)$ and L terms, respectively.

We illustrate these advantages for the well deformed nuclei ^{172}Yb with ($R_{4/2} = 3.3$) and ^{152}Sm ($R_{4/2} = 3.025$), which is an example of a nucleus at the critical point symmetry $X(5)$ [16]. The values of the Hamiltonian (10) parameters a, b, β_3, α_3 and α_1, are obtained in a fitting procedure for all the states of the considered collective bands. The comparison of the theoretical results with the experimental data is presented on Figures 3 and 4 respectively, that prove the good reproduction of the different positions of the collective bands in the framework of the model which can be used further for studying finer effects in the structure of the nuclear spectra.

In summary with the symplectic extension of the rotational $U(6)-$limit of the number preserving version of the IVBM, we extend the applications to the description of nuclei with more complex features.

2.2. Reduction through the non-compact Sp(2,R)

2.2.1. Algebraic construction containing the $6-$*dimensional Davidson potential* The need for a description of nuclei in which rotational-vibrational interactions are taken into account has led to a search for algebraically solvable potentials and a meaningful set of basis states that make the transitional nature of these systems more transparent. An algebraically solvable theory that can describe systems with rotational-vibrational interactions and which has known algebraic solutions when applied to diatomic molecules, is the one containing the Davidson potential [17]. In an algebraic approach for either the nuclear many-body problem or the

Bohr-Mottelson collective model, the addition of the Davidson potential to the Hamiltonian requires the consideration of a dynamical subgroup chain that starts with the direct product $Sp(2,R) \otimes SO(n) \subset Sp(2n,R)$, with $n = 3$ and 5, respectively [18].

Indeed if the local isomorphism of the $sp(2,R) \approx su(1,1)$ algebras is taken into account its relation to the spectrum generating algebra of the many body nuclear system with the Davidson interaction becomes explicit. This provides the motivation for considering this reduction in seeking a description of a more complex modes that includes rotational-vibrational interactions. Within the framework of the phenomenological IVBM [4], the more general case of a 6-dimensional Davidson potential naturally appears. The new reduction chain of its dynamical symmetry group [5], [19], [20]:

$$Sp(12,R) \supset Sp(2,R) \otimes SO(6). \tag{14}$$

extends the applicability of the theory to include mixed modes collective interactions.

As can be deduced from the considerations given in [18], this construction obviously survives the addition of Davidson potential. The infinitesimal generators of the $Sp(2,R)$ algebra: $F = \sum_{k,m,\alpha} C^{00}_{1k1m} u^+_k(\alpha) u^+_m(\alpha) = 2S^+$, $G = \sum_{k,m,\alpha} C^{00}_{1k1m} u_k(\alpha) u_m(\alpha) = 2S^-$ and $A = \sum_{k,m,\alpha} C^{00}_{1k1m} u^+_k(\alpha) u_m(\alpha) = \frac{1}{\sqrt{3}} N = 2S^0 - 1$ are obtained from the $Sp(12,R)$ generators (2) and (3) by means of contraction with respect to both the spatial $m = 0, \pm 1$ and the "pseudospin" $\alpha = \pm 1/2$. It is straightforward to show that the operators $S^\tau, \tau = 0, \pm$ commute in a standard way for the $SU(1,1)$ algebra generators [21] $[S^0, S^\pm] = S^\pm$, $[S^+, S^-] = -2S^0$, so the $sp(2,R)$ and the $su(1,1)$ algebras are locally isomorphic with a Casimir operator written as $C_2(SU(1,1)) = S^0(S^0 - 1) - S^+S^-$.

By construction, the generators F, G and A are scalars with respect to 6-dimensional rotations and they commute with the components of the 6-dimensional momentum operators [4],

$$\Lambda^L_M(\alpha,\beta) = A^L_M(\alpha,\beta) - (-1)^L A^L_M(\beta,\alpha), \tag{15}$$

which obey the property $\Lambda^L_M(\alpha,\beta) = (-1)^L \Lambda^L_M(\beta,\alpha)$ and generate the $SO(6) \supset U(6)$ algebra. In this way, the direct product of the two groups (14) is realized. The second order invariant for the $SO(6)$ group is $\Lambda^2 = \sum_{L,\alpha,\beta} (-1)^M \Lambda^L_M(\alpha,\beta) \Lambda^L_{-M}(\beta,\alpha)$, and is related to the second order invariant of the $Sp(2,R)$, as in the direct product (14) the two groups are complementary [19], which means that the irreps of the group $SO(6)$ determine those of $Sp(2,R) \approx SU(1,1)$ and vice versa.

In order to define the basis of the system with (14) as a dynamical symmetry that allows one to include the 6-dimensional Davidson potential, we consider the reduction of the $SO(6)$ algebra to the $SO(3)$ algebra of the angular momentum through the following chain [4], [22]

$$\begin{array}{ccccccc} SO(6) & \supset & \overline{SU(3)} & \otimes & O(2) & \supset & SO(3) \\ \omega & & (\bar{\lambda},\bar{\mu}) & & \nu & & L \end{array}, \tag{16}$$

which could be defined as the γ-unstable limit of the IVBM. The single infinitesimal operator of $O(2)$ is proportional to the scalar operator $\Lambda^0(\alpha,\beta)$ from the $SO(6)$ generators (15),

$$M_{\alpha\beta} = -\sqrt{3}\Lambda^0(\alpha,\beta) = -\sqrt{3}[A^0(p,n) - A^0(n,p)], \tag{17}$$

and the generators of $\overline{SU(3)}$ [4] are

$$X^2_M = i(A^2_M(p,n) - A^2_M(n,p)), \quad M = 0, \pm 1, \pm 2, \tag{18}$$

$$Y^1_M = A^1_M(p,p) + A^1_M(n,n) = -\frac{1}{\sqrt{2}} L_M, \quad M = 0, \pm 1. \tag{19}$$

Journal of Physics: Conference Series **128** (2008) 012032 doi:10.1088/1742-6596/128/1/012032

Note, that in this case the quadrupole moment X (18) is the proton-neutron quadrupole interaction, which makes the difference with the subgroup $SU(3) \subset U(6)$ considered in the previous limit. The second-order Casimir invariants of the two groups in the direct product in (16) are $2C_2(O_2) = M^2 = \sum_{\alpha,\beta} M_{\alpha\beta}M_{\beta\alpha}$ and $C_2(\overline{SU(3)}) = \sum_M (-1)^M (X_M X_{-M} + Y_M Y_{-M})$. Furthermore, the following relation between the quadratic Casimir operators $C_2(\overline{SU(3)})$, M^2 of $O(2)$ and Λ^2 of $SO(6)$ holds [23]:

$$\Lambda^2 = 2C_2(\overline{SU(3)}) - \frac{1}{3}M^2, \tag{20}$$

which means that the reduction from $SO(6)$ to the rotational group $SO(3)$ is carried out through the complementary groups $O(2)$ and $\overline{SU(3)}$ [19].

As the introduced above $SO(6) \subset U(6)$, the obtained symmetric representations $[N]_6$ of $U(6)$ decompose into fully symmetric $(\omega, 0, 0)_6 \equiv (\omega)_6$ irreps of $SO(6)$ according to the rule

$$[N]_6 = \bigoplus_{\omega=N,N-2,\ldots,0(1)} (\omega, 0, 0)_6 = \bigoplus_{i=0}^{<\frac{N}{2}>} (N-2i)_6, \tag{21}$$

where $<\frac{N}{2}>=\frac{N}{2}$ if N is even and $\frac{N-1}{2}$ if N is odd. As a consequence, the Elliott's notation [12] $(\bar{\lambda}, \bar{\mu})$ of $\overline{SU(3)}$ are determined by $(\omega)_6$ of $SO(6)$ and by the integer label $(\nu)_2$ of the associated irrep of $O(2)$ i.e.

$$(\omega)_6 = \bigoplus_{\nu=\pm\omega,\pm\omega-2,\ldots,0(\pm1)} (\bar{\lambda} = \frac{\omega+\nu}{2}, \bar{\mu} = \frac{\omega-\nu}{2}) \otimes (\nu)_2. \tag{22}$$

Finally, the convenience of this reduction can be further enhanced through the use of the standard rules for the reduction of the $SU(3) \supset SO(3)$ chain. Hence the basis, labelled by the quantum numbers classified by the group-subgroup chain (16), can be written as $|N\omega; (\bar{\lambda}, \bar{\mu})\nu; K, L\rangle$, where the reduction rules for obtaining specific values for each state are given earlier. By means of these labels, the basis states can be classified in each of the two irreducible even H_+ with $N = 0, 2, 4, \ldots$, and odd H_- with $N = 1, 3, 5, \ldots$ representations of $Sp(12, R)$. We illustrate this in Table 2 for the even H_+ irreducible representation, where N

Table 2. Classification scheme of the basis states in the even H_+ space of $Sp(12, R)$ according the decompositions given by (14) and (16).

N	ω	$\nu/$ 6	4	2	0	−2	−4	−6
0	0				(0, 0)			
2	2 0			(2, 0)	(1, 1) (0, 0)	(0, 2)		
4	4 2 0		(4, 0)	(3, 1) (2, 0)	(2, 2) (1, 1) (0, 0)	(1, 3) (0, 2)	(0, 4)	
6	6 4 2 0	(6, 0)	(5, 1) (4, 0)	(4, 2) (3, 1) (2, 0)	(3, 3) (2, 2) (1, 1) (0, 0)	(2, 4) (1, 3) (0, 2)	(1, 5) (0, 4)	(0, 6)
⋮		⋮	⋮	⋮	⋮	⋮	⋮	⋮

with the set of ω contained in it (21) label the rows and the values of the quantum number ν label the columns. The $\overline{SU(3)}$ quantum numbers $(\bar{\lambda}, \bar{\mu})$ define the cells of the Table 2 as they are obtained with the help of ω and ν (22).

At this point some of the similarities between the reductions trough the $U(6)$ (7) and $O(6)$ (16) become apparent. The values of the quantum number ω are clearly related with the values of the pseudospin $T = \frac{\omega}{2}$, so we can introduce a parity operator defined as $(-1)^{\frac{\omega}{2}}$ like in the $U(6)-$ limit (7). Respectively the values of ν have similar relation to the values of $T_0 = \frac{\nu}{2}$. The important difference in this case is that there is no degeneracy in the values of the $\overline{SU(3)}$ irreps $(\bar{\lambda}, \bar{\mu})$, that belong to a given row defined by ω, but in the columns (fixed value of ν) the $\overline{SU(3)}$ irreps repeat each other except the ones corresponding to the maximal value of $\omega = N$ (the first row for each N) which is added for each $\nu = \bar{\lambda} - \bar{\mu}$. Also in this case $\bar{\lambda}$ and $\bar{\mu}$ are always both even or odd.

The Hamiltonian with the considered dynamical symmetry (25) is expressed in terms of the first and second order Casimir operators of the different subgroups in its corresponding chain (16):

$$H = aN + bN^2 + \alpha_6 \Lambda^2 + \alpha_2 M^2 + \beta_3 L^2. \tag{23}$$

and it is obviously diagonal in the basis $|N\omega; (\bar{\lambda}, \bar{\mu})\nu; K, L\rangle$. The second order invariant of $\overline{SU(3)}$ is dropped in (23), because of its linear dependence on the Casimir operators of the $SO(6)$ and $O(2)$ (20). Then the eigenvalues of the Hamiltonian (23) that yield the spectrum of a system interacting with 6−dimensional Davidson potential are:

$$E(N, \omega, \nu, L) = aN + bN^2 + \alpha_6 \omega(\omega + 4) + \alpha_2 \nu^2 + \beta_3 L(L + 1). \tag{24}$$

This expression is very similar to the one obtained in the $U(6)$ (7) case (11), with a difference coming only from the $SO(6)$ second order invariant Λ^2, with the parameter α_6. Other differences in the applications to real nuclei follow from the possible choices in mapping the experimentally observed collective states to the basis of this dynamical symmetry.

2.2.2. Application to real nuclei In the applications of this new dynamical symmetry of the IVBM to real nuclear systems that we present here, we exploit again the "algebraic" definition of yrast states as introduced in [8] and in the previous paragraph. Here we use for the states of the ground band, which are the yrast states of the nucleus, a correspondence to the basis states different from the one in [9], where $N = \omega = \mp\nu = 2L = 0, 4, 8, \ldots$ $(\Delta N = 4)$. At $-\nu = 2L$ we obtain the left to right diagonal $(0, 2L)$ of Table 2 with $\bar{\lambda} = 0$ and changing $\bar{\mu} = 2L = 0, 4, 8, \ldots$. At $\nu = 2L$ the respective right to left diagonal of Table 2 is $(2L, 0)$. Here $T = \frac{\omega}{2} = L$ is always even and we consider the states belonging to the $K^\pi = 0^+$ band. Than the ground state band's (GSB) energies are obtained with the expression $E_g(L) = \alpha L + \beta L(L + 1)$ where $\alpha = 2a - 4b + 4\alpha_6 - 4\alpha_2$ and $\beta = 4b + 4\alpha_6 + 4\alpha_2 + \beta_3$. It is obvious that in the GSB we will have a certain degree of mixing of the vibrational and rotational modes depending on the values of the parameters α and β. The later depend on all the Hamiltonian (23) parameters. There is no an additive constant to the energies, like in the $U(6)$ case and we can not describe an octupole $(K^\pi = 0^-)$ band here, because $\bar{\lambda}$ and $\bar{\mu}$ are both always even or odd.

Further, the states of the excited bands are mapped to the theoretical ones, generally in the two ways described in the previous paragraph, namely the stretched states $(\bar{\lambda}, \bar{\mu}) = (\bar{\lambda}_0 + k, \bar{\mu}_0)$ or $(\bar{\lambda}_0, \bar{\mu}_0 + k)$, where k is changing as $k = 1, 2, 3...$ which in turn can belong to the left ($\nu-$possitive) or right ($\nu-$negative) "diagonal" of Table 2. The sitiuation is similar to the case described for the ground state band, where $\bar{\lambda}$ is changing and $\bar{\mu}$ is fixed or vice versa.

The correct placement of the excited bands in the spectrum strongly depends on their band-head configurations and in particular, on the number of bosons $N_0 = \bar{\lambda}_0 + \bar{\mu}_0$, from which they are built [10]. So we define the number of bosons for each state in the band with a given L in

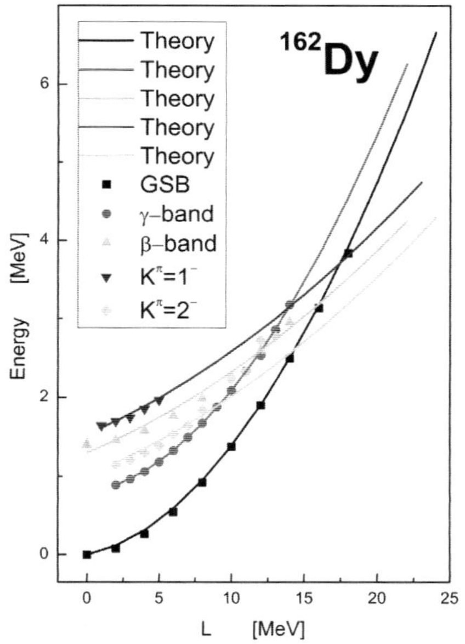

Figure 5. Comparison of the experimental and theoretical energies for the listed excited bands of ^{162}Dy. The later are calculated with the parameters $a = 0.0191$, $b = 0.0002$, $\alpha_6 = -0,00007$, $\alpha_2 = -0,0020$, $\alpha_2' = -0,0002$ and $\beta_3 = 0.0007$.

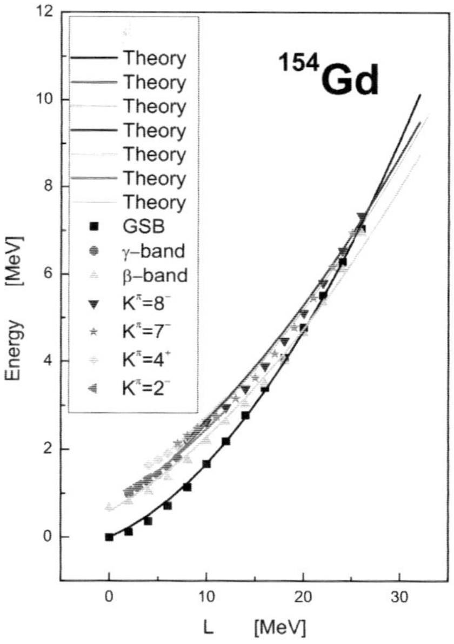

Figure 6. Comparison of the experimental and theoretical energies for the listed excited bands of ^{154}Gd. The later are calculated with the parameters $a = 0.0473$, $b = -0.0004$, $\alpha_3 = -0,0004$, $\alpha_2 = -0,0004$, $\alpha_2' = -0,0003$ and $\beta_3 = 0.0017$.

the following way $N = N_0 + \nu + 2L$ and the respective $\omega = \nu + 2L$. In this case the shifted values of $N = \omega = N_0 + k + 2L$ remain the same but we have different sequences of $(\bar{\lambda}, \bar{\mu})$ multiplets, defining the bands along the diagonals with $\pm\nu = \bar{\lambda} - \bar{\mu} = \pm(2L - k)$.

The variety of possible choices for the correspondence of the excited bands to sequences of states in the symplectic space and the mixing of the rotational and vibrational degrees of freedom like in the $U(6)$-limit allows us to reproduce correctly the behavior of the excited bands with respect to one another, which can change a lot even in neighboring nuclei [25].

The five free parameters of the Hamiltonian (23), are determined by fitting the theoretical predictions for the energies of the ground and few excited bands to the experimental data [1] , using a χ^2-procedure. The bands are developed according to the rules described above for their corresponding K^π values. We choose for N_0 shifting the band-head states, the value that gives the lowest χ^2, after obtaining the parameters mainly from the ground state band. That is why, we usually choose in the applications, nuclei with long spin sequences in their ground bands.

Here in addition to the basic ground, $\beta-$ and $\gamma-$bands, described in [9], we illustrate the possible description of negative parity bands with $K^\pi \neq 0 = 1^-, 2^-, 3^- \ldots$ like in the $U(6)$-reduction chain and use this application to compare the results of both limits. Hence we present here the fit of the Hamiltonian parameters for this limit for the rotational ^{162}Dy nucleus (Figure 6), and for another $X(5)$ symmetry nucleus -^{154}Gd [26]. As shown in Figures 5 and 6, the experimental data is reproduced remarkably well. The energies of the excited bands confirm the correct identification of the experimental states with the basis states. This reveals again the importance of N_0, which vary quite a lot in the different bands.

From the presented results, it could be seen, that both of the above limits of the symplectic extension of the IVBM are equally appropriate for the description of a rather broad range of nuclei, and most importantly nuclei that display different degree of mixing of the rotational and vibrational degrees of freedom.

2.3. The reduction through the noncompact $Sp(4, R)$

2.3.1. The algebraic realization Finally we will introduce the group - theoretical background [10] of the application of the symplectic IVBM, for the description of the energy distributions of collective excited states with fixed angular momenta. This new reduction further elucidates the importance of considering the structure of the band-head's configurations in respect to the number of bosons N that build them, established trough the applications of the first two chains. It also plays the role of a connecting branch in the general reduction scheme of $Sp(12, R)$, clarifying the relations between the subgroups of the above two chains.

In terms of the introduced boson representations of $Sp(12, R)$, the third chain of subgroups starts with the reduction [5, 19, 10]:

$$Sp(12, R) \supset Sp(4, R) \otimes SO(3) \tag{25}$$

The infinitesimal operators of the $Sp(4, R)$ algebra are the $L = 0$ part of the $Sp(12, R)$ generators (2) and (3): $F^0(\alpha, \beta) = \sum_{k,m} C^{00}_{1k1m} u^+_k(\alpha) u^+_m(\beta)$, $G^0(\alpha, \beta) = \sum_{k,m} C^{00}_{1k1m} u_k(\alpha) u_m(\beta)$ and $A^0(\alpha, \beta) = \sum_{k,m} C^{00}_{1k1m} u^+_k(\alpha) u_m(\beta)$ for $\alpha, \beta = \pm 1/2$. Hence by construction, all these operators are scalars in respect to the 3-dimensional rotations. Obviously they commute with the components of the angular momentum L^1_M, that generate the $SO(3)$ algebra, i.e. we have a direct product of the two groups (25). As a result the $Sp(4, R)$ irreps can be labeled by the quantum numbers of their corresponding $SO(3)$ ones, namely the angular momentum L.

The maximal compact subalgebra $u(2)$ of $sp(4, R)$ is generated by the Weyl generators $A^0(\alpha, \beta)$ of (5), and is the same one that corresponds to the pseudospin algebra of $U(2)$ (5) in the direct product with the $SU(3)$ in (7). The operator N (4) generates $u(1)$ and plays role of the first-order invariant of $u(2) \supset su_T(2) \otimes u_N(1)$. Hence the following correspondence between the chains of subalgebras of $sp(12, R)$ – through $u(6)$ and through $sp(4, R)$, exists [5]:

$$\begin{array}{ccccc} sp(12, R) & \supset & sp(4, R) & \otimes & so(3) \\ \cup & & \cup & & \cap \\ u(6) & \supset & u(2) & \otimes & su(3) \end{array} \tag{26}$$

Each of the $sp(4, R)$ irreps that is contained in the $sp(12, R)$ boson representations is of infinite dimension and consists of countless number of $u(2)$ irreps [11]. A basis for the $sp(4, R)$ representations is generated by a consecutive application of the symmetrically coupled products of the operators $F^0(\alpha, \beta)$ to the lowest weight state (lws) with angular momentum L that labels the considered $Sp(4, R)$ irrep [5],[24]. Each starting $u(2)$ configuration is characterized by a totally symmetric representation $[L]_2$ formed by $L = N_{min}$ vector bosons. The procedure to obtain the rest of the $su(2)$ irreps $[k]_2$ that are contained in a given L irrep of $sp(4, R)$ is given in details in [10]. It is illustrated for the cases $L = 0$ and $L = 2$ with the Tables 3 and 4. The columns are defined by the pseudospin quantum number T and the rows by the eigenvalues of $N = N_{min} + n$ for $n = 0, 2, 4, 6, \ldots$.

By means of the correspondence (26) between the two considered chains of subgroups of $Sp(12, R)$ and the relations between the $U(2)$ and $SU(3)$ quantum numbers we were able also to present the respective ($\lambda = k$, $\mu = (N - k)/2$) irreps in the cells of the Tables 3, 4. For a given value of N these could be compared to the classification scheme of the $SU(3)$ irreps contained in the even $U(6)$ irreps of $Sp(12, R)$ given in Table 1. Note that the missing $su(2)$ irreps on the above tables do not contain in their corresponding (λ, μ) representations, states with the considered value of L according to the $SU(3) \supset SO(3)$ reduction rules [3]. In the obtained

Table 3. $L = 0$

...	$T=4$	$T=3$	$T=2$	$T=1$	$T=0$	T/N
					$[0]_2(0,0)$	$N=0$
				$[2]_2(2,0)$		$N=2$
			$[4]_2(4,0)$		$[0]_2(0,2)$	$N=4$
		$[6]_2(6,0)$		$[2]_2(2,2)$		$N=6$
	$[8]_2(8,0)$		$[4]_2(4,2)$		$[0]_2(0,4)$	$N=8$
	$\vdots$	$\vdots$	$\vdots$	$\vdots$	$\vdots$	$\vdots$

Table 4. $L = 2$

...	$T=4$	$T=3$	$T=2$	$T=1$	$T=0$	T/N
				$[2]_2(2,0)$		$N=2$
			$[4]_2(4,0)$	$[2]_2(2,1)$	$[0]_2(0,2)$	$N=4$
		$[6]_2(6,0)$	$[4]_2(4,1)$	$2\times[2]_2(2,2)$		$N=6$
	$[8](8,0)$	$[6]_2(6,1)$	$2\times[4]_2(4,2)$	$2\times[2]_2(2,3)$	$[0]_2(0,4)$	$N=8$
$\vdots$	$\vdots$	$\vdots$	$\vdots$	$\vdots$	$\vdots$	$\vdots$

decomposition of the $sp(4,R)$ representations L into $[k]_2$ - $su(2)$ ones , except for the $L = 0$ case in Table 3, there is a multiplicity, denoted as $\rho\times$ in the appearance of some of the irreps, that shows how many times the respective irrep $[k]_2$ appears for the specified value of N. This multiplicity is exactly equal to the multiplicity of the appearance of the considered value of L in the reduction of the corresponding $su(3)$ irrep (λ, μ) to the $so(3)$ values of L [5](see $N = 6$ and $N = 8$ rows of Table 4).

2.3.2. Energy distribution of states with fixed L As established above, because of the correspondence (26) and the relation between the $SU(3)$ and $SU(2)$ second order Casimir operators [8], the Hamiltonian (10) and the bases in this case are equivalent to the ones in the $U(6)$ limit of the model (7). As a result the eigenvalues of the Hamiltonian (10) for the states with a fixed L are the energies, given by (11). Obviously in (11) the dependence of the energies of the collective states on the number of phonons (vector bosons) N is parabolic. The rest of the quantum numbers T, T_0 defining the states with fixed L in this case are expressed in terms of N by means of the reduction procedure described in [10]. The parity of the states is defined [8] in the same way, as $\pi = (-1)^T$. Although we use the same labels for the basis states, it must be kept in mind, that the states that we consider in this limit are ordered in different from the $U(6)$ case, sequences of $SU(3)$ irreps, reflecting their positions among the other states with the same angular momenta. This not only requires their interpretation in terms of N, but also gives a different physical meaning of the parameters of the Hamiltonian, hence they are fitted again in this limit, but in respect to the variable N.

Here, we first evaluate the parameter of inertia β_3 in front of the term $L(L+1)$ in (11) by fitting the energies of the ground states band (GSB) with $J^\pi = 0_1^+, 2_1^+, 4_1^+, 6_1^+, ...$ to their experimental values in each nuclei. Further, the values of N_{L_i} corresponding to the experimentally observed $E_{L_i}^{\exp}$ and the values of the parameters in (11) are evaluated in a multi-step χ-square fitting procedure making use only of the three sets of states with $J_i^\pi = 0_i^+, 2_i^+, 4_i^+$. We choose nuclei

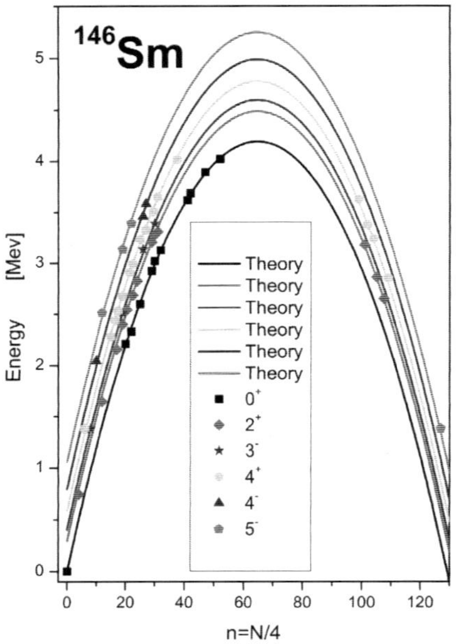

Figure 7. Comparison of the theoretical and experimental energy distributions of states with fixed L in the spectra of ^{146}Sm. The later are calculated with the parameters $a = 0.03243$, $b = -0.00006$, $\alpha_3 = -0,00161$, $\alpha_1 = -0,00553$ and $\beta_3 = 0.04134$.

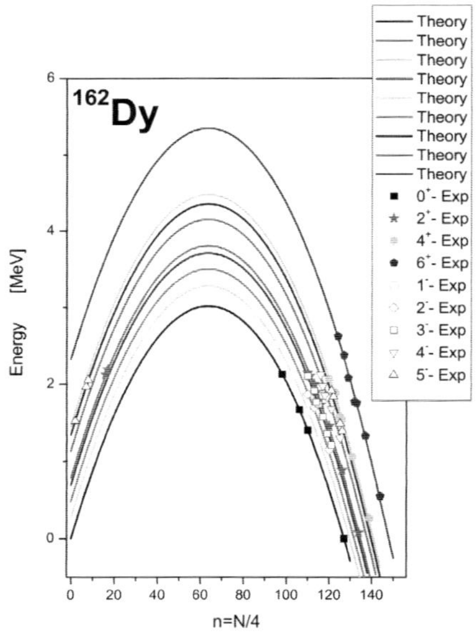

Figure 8. Comparison of the experimental and theoretical energies for the listed excited bands of ^{162}Dy. The later are calculated with the parameters $a = 0.02376$, $b = -0.00005$, $\alpha_3 = 0,03543$, $\alpha_1 = -0,02110$ and $\beta_3 = 0.01288$.

with enough of these states to have good statistics in the fit. Most of these states are band heads (all the 0^+, some of the 2^+ and 4^+) of collective bands and as a result the situation of the whole band depends on them. The set of N_{L_i} with minimal value of χ^2 determines the distribution (the parameters of the Hamiltonian) of the L_i^+ states energies with respect to the number of bosons N_{L_i} that build them. For the set of 0^+ states ($L = 0$) we chose both $T = T_0 = 0$, so the parameters a and b in (11) are evaluated and fixed. Further, for the 2^+ we use even $T > 0$, $T_0 = 0$ and so we get the α_3 parameter and finally for the 4^+ we determine $T > 0$ and also a possible value of $T_0 > 0$ in order to obtain α_1. The first two parameters a and b ($b < 0$) of the Hamiltonian actually determine the form of the parabolas, and they are the same for each set of states with fixed L. The rest of them with the values of the quantum numbers T, T_0 and L, only shift the curves in respect to each other. The predicting power of the model in this limit is related to the possibility to describe any other set of states, once all the parameters of the Hamiltonian are evaluated from the distributions of 0^+,2^+ and 4^+ states, without involving additional parameters.

The examples chosen for the present application are the nuclei ^{146}Sm (Figure 7), which has a typical vibrational spectra and the other one, Dy^{162}, that has typical rotational character (see Figure 8).

For the nucleus with vibrational spectra ^{146}Sm we apply the procedure described above with values of T that differ quite significantly ($\Delta T = 4$) for the sets with $L = 0, 2, 4$, which allows us to place them on the left hand side of the symmetric parabolas. As a result (see Figure 7) the

Journal of Physics: Conference Series **128** (2008) 012032 doi:10.1088/1742-6596/128/1/012032

values of N_{L_i} increase with increasing energy of these states.

The low-lying states of the ground band, $L = 0_1^+, 2_1^+, 4_1^+$, of the typical rotational nuclus [1] ^{162}Dy (Figure 8) require small changes in the number of quanta N_{L_1} that build the corresponding initial states in each sequence. Making use of the latter and the symmetric feature of the second order curves the states with a given L in the rotational spectra are placed on the right-hand-side of the theoretical curves. On a parabola, specified for a fixed L, the number of bosons that build the states will decrease with increasing energies.

With the procedure employed, the ordering of the states into different bands can be recognized. In order to avoid nearly degeneracies of the energies in respect to N_{L_i}, we use the symmetric feature of the second order curves and place some of the states on the right side of the parabolas.

Hence, if the number of quanta that is required to build a collective state is taken as a measure of collectivity, the states from a rotational spectra are much more collective than vibrational ones, which is the expected result.

3. Generalized reduction scheme for the IVBM

The use of symplectic geometry in the investigation of the nuclear collective motion, relates the later to its microscopic structure [27]. A further elaboration in the problem can be achieved, if we consider the nuclear many body system as consisting of two interacting proton and neutron subsystems. This motivation is behind the algebraic construction of the phenomenological IVBM [4], where $Sp(12, R)$ – the group of linear canonical transformation in a 12 -dimensional phase space [19] appears as the group of dynamical symmetry of the model. The three considered above dynamical symmetries can be unified in the reduction scheme (27).

$$\begin{array}{lllll}
U(6) & \supset & U(2) & \otimes & SU(3) \\
[N] & & T & & (\lambda,\mu) \\
\cap & & \cap & & \cup \;\; K \\
Sp(12,R) & \supset & Sp(4,R) & \otimes & SO(3) \\
\cup & & \cup & & \cap \\
SO(6) & \otimes & Sp(2,R) & & \\
\cup \;\; \omega & & (SU(1,1)) & & \\
\overline{SU(3)} & \otimes & O(2) & & \\
\left(\overline{\lambda},\overline{\mu}\right) & & \nu & &
\end{array} \tag{27}$$

In mathematical terms the established relations are based [11] on the appearance of the physically important $U(2)$ [8] group of the pseudospin as the maximal compact subgroup of $Sp(4, R)$ [10], as well as its noncompact counterpart $SU(1, 1)$ [9]. By means of this vertical structure the dynamical symmetries describing the ground and excited bands are connected with the dynamical symmetry describing the sets of states with fixed angular momentum, most of which are band-head configurations. A very important model characteristic leading to the correct description of the experimental energies is the interaction between the proton and neutron subsystems, yielded in the symplectic extension of the model, which still retains the exact analytic solutions in each of the considered applications.

This generalized reduction scheme relates the presented applications of the three dynamical symmetries in their physical interpretations and the conclusions following from each one of them. It was established that the two reduction schemes, that describe the developments of collective bands in various types of nuclear spectra, the one trough $U(6)$ and the one trough $SO(6)$, yield very similar applications for the description of the ground bands and the excited positive and negative parity bands. A common feature of these applications is the possibility to mix with varying strength the two main collective modes - vibrational and rotational, which results in the accurate description even of nuclei at the critical points of phase/shape transitions in the framework of these exactly solvable cases. The success of this approach is due not only to the easy evaluation of the relatively small number of model parameters by means of fitting to the experiment, moreover only of the first lowest bands like the ground band and the immediately following one or two excited bands. The important predicting power of the model is related to the symplectic extension, which allows only by correctly finding the number of bosons that build the band-head configurations of the other observed excited bands, to evaluate the energies of all states that belong to them.

3.1. Acknowledgments

A. Georgieva acknowledges a financial support from the East West Task Force of the European Physical Society and the Institute of Nuclear Research and Nuclear Energy, Sofia, Bulgaria for her participation in QTS-5. The presented work was supported by the U.S. National Science Foundation, Grant Number 0140300 and the Bulgarian National Foundation for Scientific Research under Grant Number $\Phi - 1501$.

References

[1] Sakai M 1984 *Atomic Data and Nuclear Data Tables* **31** 399; Level Retrieval Parameters, http://iaeand.iaea.or.at/nudat/levform.html
[2] Iachello F and Arima A 1978 *The Interacting Boson Model* (Cambridge University Press, Cambridge)
[3] Georgieva A, Raychev P and Roussev R 1983 *J. Phys. G: Nucl. Phys* **9**, 521
[4] Georgieva A, Raychev P and Roussev R 1982 *J. Phys. G: Nucl. Phys* **8** 1377
[5] Vanagas V V, 1988 *Algebraic foundations of microscopic nuclear theory,* (Nauka, Moscow) (in russian)
[6] Moshinsky M 1968 *Group Theory and the Many Body Problem* (Gordon and Breach, New York); Moshinsky M. 1984 *J. Math. Phys.* **25** 1555
[7] Rowe D J 1995 *Prog. Part. Nucl. Phys.* **37** 4711
[8] Ganev H, Garistov V and Georgieva A 2004 *Phys. Rev.* C **69**, 014305
[9] Ganev H G, Georgieva A I and Draayer J P 2005 *Phys. Rev.* C **72**, 014314
[10] Ganev H, Garistov V P, Georgieva A I and Draayer J P 2004 *Phys. Rev.* C **70** 010411
[11] Georgieva A, Ivanov M, Raychev P and Roussev R. 1989 *Int.J. Theor. Phys.* **28** 769
[12] Elliott J P 1958 *Proc. R. Soc. (London)* Ser.A **254** 128; 526
[13] Butler P A and Nazarewicz W 1996 *Rev. of Mod. Phys.* **68** 349
[14] Bonatsos D *et al.* 2000 , *Phys. Rev.* C **62** 024301
[15] Garrett P E 2001 *J. Phys. G: Nucl. Part. Phys.* **27** R1-R22
[16] Iachello F 2000 *Phys.Rev. Lett* **85** 3580; Casten R F and Zamfir N V 2000 *Phys.Rev. Lett* **85** 3584
[17] Davidson P M 1932 *Proc. Roy. Soc.* A **135** 459
[18] Rowe D J 1994 *J. Math. Phys.* **35** 3178; Rowe D J and Bahri C 1998 *J. Phys. A: Math. Gen* **31**, 4947
[19] Moshinsky M and Quesne C 1971 *J. Math. Phys.* **12** 1772
[20] Moshinsky M 1984 *J. Math. Phys.* **25** 1555
[21] Feng Pan and Draayer J P 1998 *Nucl. Phys.* A **636** 156
[22] Chacon E, Castanos O and Frank A 1984 *J. Math. Phys.* **25** 1442
[23] Dragt A J 1965 *J. Math. Phys.* **6** 533
[24] Rosensteel G and Rowe D J 1977 *Phys.Rev.Lett.* **38**10
[25] Popa G, Georgieva A and Draayer J P 2004 *Phys. Rev.* C **69** 064307
[26] Tonev D *et al.*, 2004 *Phys. Rev.* C **69** 034334
[27] Draayer J P and Rosensteel G 1983 *Phys. Lett.* B **124** 281; Rosensteel G and Draayer J P 1985 *Nucl. Phys.* A **436** 445

Dual symmetry in Born-Infeld theory

Siamak Khademi[1] **and Abbas Ayoubi**

Department of Physics, Zanjan University, ZNU, Zanjan, Iran.

E-mail: `khademi@znu.ac.ir` and `siamakkhademi@yahoo.com`

Abstract. Born-Infeld theory is a non-linear formalism which has many applications in string and electromagnetic theories. Although, the existence of magnetic monopoles and dyons are suggested by Born-Infeld theory, but this theory is not invariant under the dual transformations. In this theory electric fields for point charged particles are not singular at origin ($r = 0$), but magnetic fields and vector potentials are still singular. In this paper we show that the vanishing of dual symmetry is responsible for these singularities. Furthermore, we present the dual symmetric Born-Infeld theory, by a symmetric definition of electromagnetic fields in terms of new scalar and vector potentials, as well as the ordinary ones. All singularities of vector potential and magnetic field are removed as an immediate consequence of this symmetry.

1. Introduction

Study of symmetries is very important in all disciplines of science. One of the important symmetries in electrodynamics is dual symmetry. Dual symmetry of Maxwell's equations gives us the existence of magnetic monopoles and quantization of electromagnetic charges [1]. In this paper we investigate the dual symmetry of Born-Infeld (B-I) theory. This theory has many applications in string and electrodynamics theories in presence of magnetic monopoles or dyons which are interacted with electromagnetic fields [2]. In contrast to the ordinary electrodynamics, B-I theory is a non-linear formalism. In this formalism of electrodynamics, electric fields are expressed not only in terms of electric charge but also in terms of magnetic monopole charges. Then the singularity of electric field for a point charge at the origin is removed. But magnetic field for a point magnetic charge at the origin and vector potential are still singular. In this non-linear theory one has non-symmetric definitions for electric and magnetic fields in terms of electromagnetic potentials [3]. Thus dual symmetry between electric and magnetic fields is broken. We show symmetry breaking is a consequence of non-symmetric definitions of electromagnetic fields in terms of vector and scalar potentials. So, the treatment of electric fields, in the non-linear B-I theory, depends on the existence/non-existence of magnetic fields but magnetic field is just explained in terms of the magnetic charges [2, 3]. Thus the treatment of electric and magnetic fields are different. This result is a consequence of dual symmetry breaking of B-I theory. We introduce two additional (new) scalar and vector potentials to have new symmetric definitions of electromagnetic fields. Then we obtain the dual symmetry invariance of B-I theory, by symmetrically explained electromagnetic fields in terms of new scalar and vector potentials as well as the ordinary ones. As an immediate consequence of dual symmetry of B-I theory and new electromagnetic potentials, the singularities in vector potentials and magnetic fields are removed.

[1] Corresponding Author

In the next section a brief review of dual symmetry in electrodynamics is presented. Section three belongs to the description of B-I theory and in section four dual symmetry invariance of B-I theory is presented. Section five is devoted to the conclusions.

2. Dual symmetry in electrodynamics

Dual symmetry of Maxwell's equations predicts the existence of magnetic monopole, although it didn't get still any experimental confirmation. Dual transformations are defined by [4]

$$\begin{aligned}
\mathbf{E} &= \mathbf{B}Sin\theta + \mathbf{E}Cos\theta, \\
\mathbf{B} &= -\mathbf{E}Sin\theta + \mathbf{B}Cos\theta, \\
\rho_e &= \rho_m Sin\theta + \rho_e Cos\theta, \\
\rho_m &= \rho_m Cos\theta - \rho_e Sin\theta, \\
\mathbf{J}_e &= \mathbf{J}_m Sin\theta + \mathbf{J}_e Cos\theta, \\
\mathbf{J}_m &= \mathbf{J}_m Cos\theta - \mathbf{J}_e Sin\theta.
\end{aligned} \tag{1}$$

Symmetrized Maxwell's equations

$$\begin{aligned}
\nabla \cdot \mathbf{E} &= \rho_e, \\
\nabla \cdot \mathbf{B} &= \rho_m, \\
\nabla \times \mathbf{E} &+ \frac{\partial \mathbf{B}}{\partial t} = -\mathbf{J_m}, \\
\nabla \times \mathbf{B} &- \frac{\partial \mathbf{E}}{\partial t} = \mathbf{J_e},
\end{aligned} \tag{2}$$

are invariant under the dual transformation (1). In symmetric Maxwell's equations (2), ρ_m and $\mathbf{J}_m$ are magnetic charge and current density, respectively. The electric and magnetic fields of static point charges are obtained from Maxwell's equations (2), as

$$\mathbf{E} = \frac{q}{4\pi r^2}\hat{\mathbf{r}}, \tag{3}$$

$$\mathbf{B} = \frac{g}{4\pi r^2}\hat{\mathbf{r}}. \tag{4}$$

These electric and magnetic fields are symmetric and proportional to r^{-2} and are singular at $r = 0$. Electromagnetic fields in Maxwell's equations are explained in terms of electromagnetic scalar and vector potentials

$$\mathbf{E} = -\frac{\partial \mathbf{A}}{\partial t} - \nabla\phi, \tag{5}$$

$$\mathbf{B} = \nabla \times \mathbf{A}. \tag{6}$$

Clearly, the electric and magnetic fields (5), (6) are not symmetric (dual symmetry) in terms of potentials. Thus, Maxwell's equations are not invariant under dual transformation (1) in the potential levels, although, they are invariant in terms of electromagnetic fields. Therefore substituting, Eqs. (5) and (6) into Maxwell's equations (2), will break dual symmetry.

To remove this inconsistency and keep dual symmetric Maxwell's equations in potential levels, we introduce two additional (new) scalar and vector potentials $(\psi, \mathbf{G})$. Symmetric electromagnetic fields (5), (6) are explained symmetrically in terms of four electromagnetic (two scalar and two vector) potentials

$$\mathbf{E} = -\frac{\partial \mathbf{A}}{\partial t} - \nabla\phi + \nabla \times \mathbf{G}, \tag{7}$$

$$\mathbf{B} = \frac{\partial \mathbf{G}}{\partial t} - \nabla\psi + \nabla \times \mathbf{A}. \tag{8}$$

Dual transformations for electromagnetic potentials are

$$\begin{aligned} \psi &\rightarrow \phi \\ \phi &\rightarrow -\psi \\ \mathbf{G} &\rightarrow \mathbf{A} \\ \mathbf{A} &\rightarrow -\mathbf{G}. \end{aligned} \tag{9}$$

Now we have a dual symmetric invariant Maxwell's equations in both potential and electromagnetic field levels. To obtain the electric and magnetic fields of a moving magnetic point charge, a singular vector potential is used [5]. But if one applies the new definition of electric and magnetic fields (7) and (8), the singular potentials is unnecessary. One obtains electric and magnetic fields for moving magnetic point chargs by a well defined new scalar and vector potentials. Substituting Eq. (8) in Maxwell's equations give us

$$\nabla \cdot \mathbf{B} = \rho_m \rightarrow \nabla \cdot (\frac{\partial \mathbf{G}}{\partial t} + \nabla \psi + \nabla \times \mathbf{A}) = \rho_m. \tag{10}$$

By a (Coulomb like) gauge freedom for new vector potential $\nabla \cdot \mathbf{G} = 0$, Eq. (10) reduces to $\nabla^2 \psi = \rho_m$, which give us Eq. (6), for a static point magnetic charge, with $\psi = g/4\pi r$. These new scalar and vector potentials are known and sometimes used by other authors. In reference[**?**], the new scalar potential is called "magnetic potential" and denoted by ϕ^*. Recently, new vector potential is utilized by P. Drummond to obtain the dual symmetric Lagrangian in quantum electrodynamics [7]. Here we use new electromagnetic potentials in classical electrodynamics to obtain the equation of motions for classical electric and magnetic charges interacting with electromagnetic fields, by the well defined, and nonsingular, electromagnetic potentials. In section four these new potentials and dual symmetry are investigated for B-I theory.

3. A brief review of B-I theory

Born-Infeld theory is a non-liner method and has many applications in electromagnetic formalism, where magnetic monopoles or dyons are assumed to be existed. We apply B-I theory to study the electrodynamics of dyons and magnetic monopoles which are interacting with external electromagnetic fields. Dyons are particles which have electric and magnetic charges, simultaneously. Dynamical equations in B-I theory are obtained from a nonlinear Lagrangian

$$L_{BI} = \beta^2 \left[1 - \sqrt{1 + \frac{1}{2\beta^2} F_{\mu\nu} F^{\mu\nu} - \frac{1}{16\beta^4} (F_{\mu\nu} \tilde{F}^{\mu\nu})} \right] + J^\mu A_\mu, \tag{11}$$

where$F_{\mu\nu}$,J^μ and A_μ are conventional electromagnetic field strength tensor, four-current density and four-vector potential. If parameter β goes to infinity, B-I Lagrangian (11) reduce to a Lagrangian for the ordinary classical electrodynamics. Dynamical equations in B-I theory are obtained by applying Euler-Lagrange equation, $\partial_\mu \frac{\partial L_{BI}}{\partial A_{\nu,\mu}} - \frac{\partial L_{BI}}{\partial A_\nu} = 0$ as

$$\begin{aligned} &\nabla \cdot \frac{1}{R} \left[\mathbf{E} + \frac{1}{\beta^2} (\mathbf{E} \cdot \mathbf{B}) \mathbf{B} \right] = \rho_e, \\ &\nabla \times \frac{1}{R} \left[\mathbf{B} - \frac{1}{\beta^2} (\mathbf{E} \cdot \mathbf{B}) \mathbf{E} \right] - \frac{\partial}{\partial t} \frac{1}{R} \left[\mathbf{E} + \frac{1}{\beta^2} (\mathbf{E} \cdot \mathbf{B}) \mathbf{B} \right] = \mathbf{J}_e, \\ &\nabla \cdot \mathbf{B} = \rho_m, \\ &\nabla \times \mathbf{E} + \frac{\partial \mathbf{B}}{\partial t} = \mathbf{J}_m, \end{aligned} \tag{12}$$

where

$$R = \sqrt{1 - \frac{1}{\beta^2}(\mathbf{E^2} - \mathbf{B^2}) - \frac{1}{\beta^4}(\mathbf{E} \cdot \mathbf{B})^2}. \tag{13}$$

Equations (12) are Maxwell like equations corresponding to B-I theory. Hereafter these equations are called B-I equations. First part of B-I equations are obtained from Euler-Lagrange equation while the second one are borrowed from dual symmetric Maxwell's equations. Electric field for a point charge, in the absence of magnetic fields, is obtained from B-I equations as

$$\mathbf{E} = \frac{q}{4\pi r^2}\left(1 + \frac{1}{r^4}\frac{q^2}{(4\pi\beta)^2}\right)^{1/2}\hat{\mathbf{r}}, \qquad \mathbf{B} = 0. \tag{14}$$

Clearly, electric field is not singular at $r = 0$. In the presence of magnetic fields the electric field is

$$\mathbf{E} = \frac{q}{4\pi r^2}\left(1 + \frac{1}{r^4}\frac{q^2 + g^2}{(4\pi\beta)^2}\right)^{1/2}\hat{\mathbf{r}}, \qquad \mathbf{B} \neq 0, \tag{15}$$

which describes a dyon. The effect of magnetic charge on electric field is a consequence of nonlinearity in B-I equations. Thus the existence of magnetic fields affects the magnitude of electric field. Magnetic field in Born-Infeld equations is equivalent to the magnetic field in Maxwell's equations (4), either in the presence or in the absence of electric fields. Although, the singularity of electric field is removed in B-I theory, but the magnetic field of a magnetic point charge is still singular. Therefore, the magnetic and electric fields in B-I equations are not invariant under dual transformations. In the next section we use the new electromagnetic potentials to have a dual symmetric B-I equations and remove the singularity of magnetic field and vector potential.

4. Dual symmetry invariant B-I theory

As we explained in the previous section, B-I electromagnetic fields are not invariant under dual transformations. Existence of singularity in magnetic field is due to the dual symmetry breaking. In this section we apply a new definition of field strength tensor

$$F'_{\mu\nu} = \partial_\mu A_\nu - \partial_\nu A_\mu + \frac{1}{2}\epsilon^{\mu\nu\alpha\beta}(\partial_\alpha G_\beta - \partial_\beta G_\alpha), \tag{16}$$

in B-I Lagrangian, where $G_\alpha \equiv (\psi, -\mathbf{G})$ and J^μ_m are new four-potential and four-magnetic current density, respectively. This field strength tensor (16) is symmetric under dual transformations (1) and (9). Thus, nonlinear and dual symmetric Lagrangian of B-I theory is

$$L'_{BI} = \beta^2\left[1 - \sqrt{1 + \frac{1}{2\beta^2}F'_{\mu\nu}F'^{\mu\nu} - \frac{1}{16\beta^4}(F'_{\mu\nu}\widetilde{F}'^{\mu\nu})}\right] + J^\mu_e A_\mu + J^\mu_m G_\mu. \tag{17}$$

By applying the Euler-Lagrange equations, $\partial_\mu \frac{\partial L'_{BI}}{\partial G_{\nu,\mu}} - \frac{\partial L'_{BI}}{\partial G_\nu} = 0$ and $\partial_\mu \frac{\partial L'_{BI}}{\partial A_{\nu,\mu}} - \frac{\partial L'_{BI}}{\partial A_\nu} = 0$, one obtains the symmetric B-I equations

$$\begin{aligned}
&\nabla \cdot \frac{1}{R}\left[\mathbf{E} + \frac{1}{\beta^2}(\mathbf{E}\cdot\mathbf{B})\mathbf{B}\right] = \rho_e,\\
&\nabla \times \frac{1}{R}\left[\mathbf{B} - \frac{1}{\beta^2}(\mathbf{E}\cdot\mathbf{B})\mathbf{E}\right] - \frac{\partial}{\partial t}\frac{1}{R}\left[\mathbf{E} + \frac{1}{\beta^2}(\mathbf{E}\cdot\mathbf{B})\mathbf{B}\right] = \mathbf{J}_e,\\
&\nabla \cdot \frac{1}{R}\left[\mathbf{B} - \frac{1}{\beta^2}(\mathbf{E}\cdot\mathbf{B})\mathbf{E}\right] = \rho_m,\\
&\nabla \times \frac{1}{R}\left[\mathbf{E} + \frac{1}{\beta^2}(\mathbf{E}\cdot\mathbf{B})\mathbf{B}\right] + \frac{\partial}{\partial t}\frac{1}{R}\left[\mathbf{B} - \frac{1}{\beta^2}(\mathbf{E}\cdot\mathbf{B})\mathbf{E}\right] = -\mathbf{J}_m.
\end{aligned} \tag{18}$$

Journal of Physics: Conference Series **128** (2008) 012033 doi:10.1088/1742-6596/128/1/012033

The new Born-Infeld equations (18) are invariant under dual transformations, (1) and (9) where $\beta \rightarrow i\beta$. Electric and magnetic fields for the electric and magnetic point charge (or dyons) are obtained from symmetric B-I equations (18) as:

$$\mathbf{E} = \frac{q}{(4\pi r^2)} \left[\frac{1 - \frac{1}{r^4} \frac{g^2}{(4\pi\beta)^2}}{1 + \frac{1}{r^4} \frac{q^2}{(4\pi\beta)^2}} \right] \hat{\mathbf{r}}, \tag{19}$$

and

$$\mathbf{B} = \frac{g}{(4\pi r^2)} \left[\frac{1 + \frac{1}{r^4} \frac{q^2}{(4\pi\beta)^2}}{1 - \frac{1}{r^4} \frac{g^2}{(4\pi\beta)^2}} \right] \hat{\mathbf{r}}. \tag{20}$$

These symmetric electric and magnetic fields describe the electromagnetic fields of dyons. As an important consequence of nonlinearity in B-I theory, the electric (magnetic) field is not independent of magnetic (electric) point charge. Electric field (15) will be obtained by expansion of (19) in terms of powers of q and g , up to $q^2 g^2$,

$$\frac{1 + \frac{1}{r^4} \frac{q^2}{(4\pi\beta)^2}}{1 - \frac{1}{r^4} \frac{g^2}{(4\pi\beta)^2}} = \left[1 + \frac{1}{r^4} \frac{q^2}{(4\pi\beta)^2} \right] \left[1 - \frac{1}{r^4} \frac{g^2}{(4\pi\beta)^2} \right]^{-1} \left[1 + \frac{1}{r^4} \frac{q^2}{(4\pi\beta)^2} + \frac{1}{r^4} \frac{g^2}{(4\pi\beta)^2} + O(> q^2 p^2) \right]. \tag{21}$$

Ordinary electromagnetic fields for point charges (3),(4), and Maxwell's equations (2) are obtained from symmetric electromagnetic fields (19) and (20) and symmetric B-I equations (18) while β approaches to infinity.

5. Conclusions

In ordinary electrodynamics theory, to describe magnetic monopole, one may use a singular vector potential. In the other hand, ordinary electric and magnetic fields, for electric and magnetic point charges, are also singular at the origin. We found that the origin of the singularities is the absence of new scalar and vector potentials. Using these new electromagnetic potentials , one may remove these singularities not only in vector potential of ordinary Maxwell's equations, but also in electromagnetic fields of dyons in B-I theory. We obtained symmetric electric and magnetic fields as well as symmetric Maxwell's equations (B-I dynamical equations) by introducing the new scalar and vector potentials. They reduced to the ordinary electromagnetic fields and Maxwell's equations, by setting. Although, in this theory we have two scalar and two vector potentials, however, we have just one kind of electric and magnetic fields, which are determined by these electromagnetic potentials.

References

[1] Born M 1934 *Proc. R. Soc. London* **A143** 410.
[2] Born M and Infeld M 1934 *Proc. R. Soc. London* **A144** 425.
[3] Kim H 1999 Genuine Dyons in Born-Infeld Electrodynamics *Preprint hep-th/9910261.*
[4] Jakson J D 1973 *Classical Electrodynamics* (New York: John Wiley & Sons).
[5] Dirac P A M 1962 *Proc. R. Soc. London* **A265** 57.
[6] Reitz J R, Milford F J and Christy R W 1979 *Fundamentals of Electromagnetic Theory* (Reading: Addison-Wesley).
[7] Drummond P D 2006 *J. Phys. B: At. Mol. Opt. Phys.* **39** S573.

Journal of Physics: Conference Series **128** (2008) 012034 doi:10.1088/1742-6596/128/1/012034

Current understanding of the pseudospin symmetry in atomic nuclei

S. Marcos[1], M. López-Quelle[2], R. Niembro[1] and L. N. Savushkin[3]

[1] Departamento de Física Moderna, Universidad de Cantabria, E-39005 Santander, Spain
[2] Departamento de Física Aplicada, Universidad de Cantabria, E-39005 Santander, Spain
[3] Department of Physics, St. Petersburg University for Telecommunications, 191065 St. Petersburg, Russia

E-mail: marcoss@unican.es

Abstract. We use the relativistic mean field framework to analyse the reliability of the explanation of the pseudospin symmetry (PSS) that has been accepted, quite generally, by the scientific community, in the last decade. We make a comparative analysis of the mechanisms responsible for the breaking of the spin and pseudospin symmetries that shows the different nature of these symmetries. We propose an explanation of the PSS, also valid in the non-relativistic limit, in which the effect of the deviation of the single-particle central potential from a harmonic oscillator on the breaking of the degeneracy of pseudospin doublets is partially compensated by the effect of the spin-orbit interaction.

1. Introduction

1.1. The pseudospin symmetry

When one observes the single-particle (SP) energy spectrum of the shell model, one can distinguish couples of states with close energy values. In figure 1, they are represented for neutrons with thicker lines. By analogy with the spin-orbit formalism, each pair of these states is defined as a pseudospin doublet (PSD). Thus, about forty years ago an attempt has been made to introduce the pseudospin (PS) and the pseudo-orbital angular momentum $\tilde{l}$ as new dynamical variables, instead of those used in the conventional shell model, to classify the SP energy levels in nuclei [1, 2]. The main purpose of this classification was to get a small magnitude of the "new" effective pseudospin-orbit force. The quantum number $\tilde{l}$ is defined as

$$\tilde{l} = (2j - l) = \begin{cases} l + 1, & \text{for } j = l + 1/2; \\ l - 1, & \text{for } j = l - 1/2. \end{cases} \tag{1}$$

If we consider the two states "a" and "b" of a PSD, they can be labelled by the quantum numbers n_a, l_a, $j_a = l_a + 1/2$ and $n_b = n_a - 1$, $l_b = l_a + 2$, $j_b = l_a + 3/2$, where n, l, and j are the SP radial, orbital and total angular momentum quantum numbers, respectively. Thus, $\tilde{l}_a = \tilde{l}_b = l_a + 1$. If the pseudospin symmetry (PSS) were exact, the two states a and b would be degenerate. Similarly, we shall consider a spin doublet (SD) shows spin symmetry (SS) when the two respective partners have the same energy.

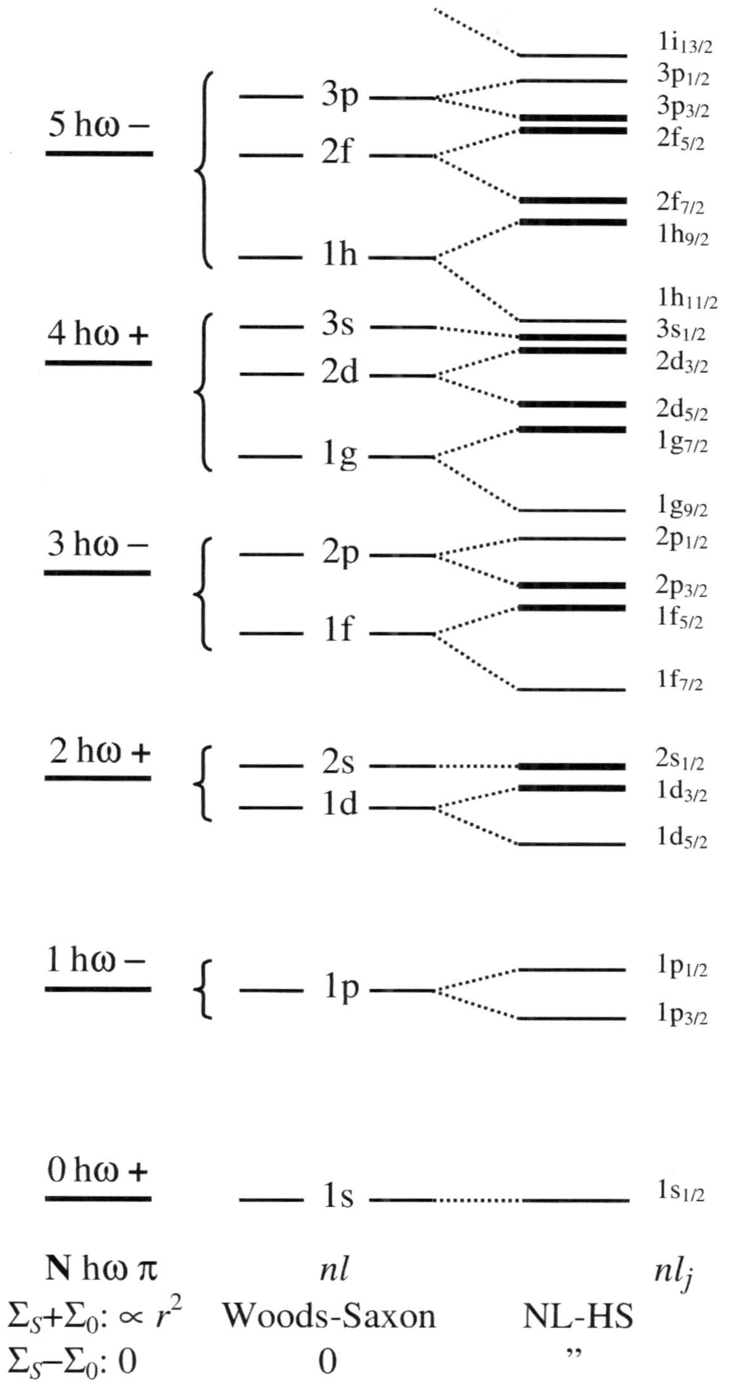

Figure 1. Qualitative evolution of the neutron energy levels from the shell model taking for the central potential, successively, a harmonic oscillator and a Woods-Saxon potential and no spin-orbit interaction (we use standard notation). The NL-SH column indicates qualitative results of the Dirac-Hartree approximation with the NL-SH force [31]. Notice that, in the relativistic case, the single-particle energy does not increase linearly with N for a harmonic oscillator potential.

Journal of Physics: Conference Series **128** (2008) 012034 doi:10.1088/1742-6596/128/1/012034

In the last decade, it has been established that the PSS has its origin in the relativistic symmetry of the Dirac equation [3]-[15], claiming that is not possible to explain this symmetry, properly, in the non-relativistic framework. Thus, in the recent years, many authors have studied also the PSS using the relativistic formalism either based on a Dirac equation with appropriate potentials fitted to reproduce bulk properties of finite nuclei [13]-[15] or through the Hartree [16]-[23] and Hartree-Fock [24]-[26] relativistic approximations.

In spite of the claimed important progress made in understanding the PSS during the last decade, there are still essential points of controversy, mainly related to the effect of the PSS breaking term and the sufficient conditions for the PSS in finite nuclei, between the different groups working in this field [18]-[25]. To give a concrete example of the doubts that some physicists have on the explanation of the PSS based on relativisctic grounds, we reproduce here a few lines from the paper [28] by J. Piekarewicz "notice that while the qualitative nature of the pseudospin symmetry emerges naturally in the relativistic approach, it is not without subtleties. For example, in the limit of exact pseudospin symmetry the nucleons spectrum is unbound. Thus, a realistic description of nuclei necessarily requires the breaking of the symmetry. Moreover, as nucleons become bound, the pseudospin-orbit potential becomes singular. As a result, great care is required in dealing with such a potential and useful insights into this subtle symmetry will undoubtedly continue to emerge."

The aim of this work is to analyse, in the framework of the Dirac Hartree approximation (DHA), the explanation of the PSS given in the last decade. To this end, we study the effect of the PSS breaking term and the reliability of the sufficient conditions proposed to insure an approximate PSS in finite nuclei, making a comparative analysis with the corresponding situation in relation with the SS.

1.2. Dirac Hartree framework and basic equations for the spin and pseudospin symmetries

In the DHA (the tensor contribution of the vector mesons being ignored), the SP states are obtained from a Dirac equation that can be written as

$$[-i\boldsymbol{\alpha}\cdot\boldsymbol{\nabla}+\beta(M+\Sigma_S)+\Sigma_0]\psi(\vec{r})=E\psi(\vec{r}), \tag{2}$$

where, $E=M+\epsilon$ is the relativistic energy, M is the nucleon mass, Σ_S is the scalar self-energy coming from the scalar σ meson and Σ_0 is the time component of the vector self-energy coming from the vector ω and ρ mesons and the Coulomb field [29, 30]. For spherical nuclei (to which we restrict ourselves in this work), the nucleon Dirac spinor $\psi(\vec{r})$ can be written, in standard notation, as

$$\psi(\vec{r})=\frac{1}{r}\begin{pmatrix} iG(r)y^m_{jl}(\theta,\phi) \\ F(r)y^m_{j\tilde{l}}(\theta,\phi)\end{pmatrix}, \tag{3}$$

where $\frac{G(r)}{r}$ and $\frac{F(r)}{r}$ represent the radial parts of its big and small components, respectively. As noticed in [3], $\tilde{l}=l\pm 1$ appears in the small component of the spinor.

By substituting eq. (3) into eq. (2), one can obtain the following Dirac equation for the G and F components:

$$\begin{aligned}\frac{d}{dr}G(r)&=-\frac{\kappa}{r}\,G(r)+W\,F(r),\\ \frac{d}{dr}F(r)&=V\,G(r)+\frac{\kappa}{r}\,F(r).\end{aligned} \tag{4}$$

In this equation,

$$\kappa\equiv(2j+1)(l-j)=j(j+1)-\tilde{l}(\tilde{l}+1)+1/4, \tag{5}$$

Journal of Physics: Conference Series **128** (2008) 012034 doi:10.1088/1742-6596/128/1/012034

whereas

$$V \equiv \Sigma_S + \Sigma_0 - \epsilon, \qquad W \equiv 2M + \Sigma_S - \Sigma_0 + \epsilon \tag{6}$$

are energy dependent potentials.

From the Dirac equation (4), one can get the two following equivalent second order differential equations for the G and F components of the nucleon Dirac spinor:

$$-G'' + \left[\frac{W'}{W}\left(\frac{G'}{G} + \frac{\kappa}{r}\right) + \frac{l(l+1)}{r^2} + VW\right] G = 0, \tag{7}$$

$$-F'' + \left[\frac{V'}{V}\left(\frac{F'}{F} - \frac{\kappa}{r}\right) + \frac{\tilde{l}(\tilde{l}+1)}{r^2} + VW\right] F = 0, \tag{8}$$

where the prime stands for the radial derivative. The following relations

$$l(l+1) = \kappa(\kappa+1) \tag{9}$$

and

$$\tilde{l}(\tilde{l}+1) = \kappa(\kappa-1) \tag{10}$$

hold, in accordance with eq. (5), and the quantity

$$VW = 2MV + 2\epsilon\Sigma_0 + (\Sigma_S^2 - \Sigma_0^2) - \epsilon^2 \tag{11}$$

represents an effective state-dependent central potential (multiplied by $2M$). The quantity $\Sigma_S + \Sigma_0$ from VW in eqs. (7) and (8) is the essential part of the central potential, whereas the quantity $\Sigma_S - \Sigma_0$ from W, through the term entering $W'\kappa$, determines the spin-orbit (SO) potential in eq. (7).

1.3. Breaking mechanisms for spin and pseudospin symmetries

The solutions of eq. (7) with the same number of nodes (n) of G and the same value of l form a SD, whereas the solutions of eq. (8) with the same number of nodes ($\tilde{n}$) of F and the same value of $\tilde{l}$ form a PSD. As κ depends on j, the κ terms entering equations (7) and (8), to which we shall refer to as the $G-\kappa$ and $F-\kappa$ terms, respectively, are, formally, responsible for the splitting of the SDs and the PSDs, i.e., they break the SS and the PSS. Actually, in this latter case, the splitting of the PSDs can be avoided by an appropriate choice of the self-energies Σ_S and Σ_0 [22, 23].

From eqs. (7) and (8), it is evident that the $G-\kappa$ and $F-\kappa$ terms are also responsible for the breaking of the relations $G_a \propto G_b$ and $F_{a'} \propto F_{b'}$ for the two partner of the SDs (a, b) and PSDs (a', b'), respectively. However, if W is constant $W'/W = 0$ (there is no SO interaction) and the SS is restored with $G_a \propto G_b$ (we shall denote this particular form of SS as SS*). Similarly, if V is constant $V'/V = 0$ and the PSS is restored with $F_{a'} \propto F_{b'}$ (we shall denote this particular form of PSS as PSS*).

Although eqs. (7) and (8) look very similar, there is an essential difference between them. Indeed, $V(r)$ becomes zero at some point (r_0) in the nuclear surface, so that, for $r \to r_0$, $V'/V \propto [(r-r_0)r]^{-1}$ and the $F-\kappa$ term, $\frac{V'}{V}\frac{\kappa}{r}F$, in equation (8) is singular at $r = r_0$.

At present, the PSS is considered slightly broken in nuclei due to some of the following hypotheses:
1) The magnitude of $\Sigma_S + \Sigma_0$ is small [3]-[7],
2) The $F - \kappa$ term is small [8]-[11], or
3) The different contributions to the energy in equation (8) partially compensate each other [14], [18]-[26].

In Sect. 2-5, we analyse the hypotheses 1) and 2), showing that they present very serious problems to justify the approximate PSS observed in heavy nuclei, and we compare the spin and pseudospin symmetries to shed new light for explaining the PSS. In Sect. 6, we give a simple explanation of the PSS, which is valid for both the relativistic and non-relativistic approximations. Finally, the conclusions are summarized in Sect. 7.

2. Pseudospin symmetry and the magnitude of $\Sigma_S + \Sigma_0$

In what follows, we shall designate by "a" and "b" the two states of a SD or PSD with $\kappa < 0$ and $\kappa > 0$, respectively. As shown in ref. [4], in the limit $\Sigma_S + \Sigma_0 = 0$, two PS partners, a and b, have the same energy and F_a and F_b are identical up to a phase (from eq. (8) one can immediately see that $F_a \propto F_b$), i.e., we have exact PSS* (in references [3]-[11] only the PSS* has been considered). But the condition $\Sigma_S + \Sigma_0 = 0$ does not allow bound states (except for models with unrealistic values of $\Sigma_S - \Sigma_0$).

In the inner part of real nuclei, $\Sigma_S + \Sigma_0 \approx -60$ MeV and can be considered small (in relation to the nucleon mass M or $|\Sigma_S - \Sigma_0|$) but it is not zero. Then, in references [3]-[7], it is claimed that approximate PSS*, i.e., $e_a \simeq e_b$ and $F_a \approx F_b$, can be expected on the ground of the smallness of $|\Sigma_S + \Sigma_0|$. However, one can see that this cannot be the main reason for the observed approximate PSS* in nuclei considering, for example, the solution of a Dirac equation for the single-nucleon states with appropriate values of $\Sigma_S + \Sigma_0$ but taking $\Sigma_S - \Sigma_0 = 0$ (i.e., the SO interaction is neglected). Under these conditions, it is clear that the PSDs (mainly those with $\tilde{n} < 3$) do not satisfy the property of quasi-degeneracy imposed by the PSS, in spite that $|\Sigma_S + \Sigma_0|$ is realistic (see, for example, figure 1 or any other shell model scheme obtained in the non-relativistic framework without SO interaction). This argument and the apparent effect of the potential $\Sigma_S - \Sigma_0$ in figure 1 also suggest that the SO interaction plays an essential role in the achievement of the quasi-degeneracy of the PSDs in the nuclear shell model.

To understand better the effect of the SO interaction, we have presented in table 1 the SP energies of the neutron PSDs of the ^{208}Pb nucleus with the DHA (as explained in the table caption) for two intensities of the SO potential: for the realistic one (column ϵ) and for other choice with a stronger strength than the physical one (column ϵ^*). These results show that the condition of quasidegeneracy of the PSDs, $\epsilon_a \simeq \epsilon_b$, for the PSS (or PSS*) is strongly improved as the magnitude of the potential $\Sigma_S - \Sigma_0$ is considerably increased with respect to the realistic values. For the realistic case, it can be seen that, as the $\tilde{n}$ number of a PSD increases, the relation $\epsilon_a \simeq \epsilon_b$ and, consequently, the PSS are satisfied better, but PSDs satisfying $\epsilon_a \simeq \epsilon_b$ with a small value of $\tilde{n}$ can be found as well [18]-[21]. Moreover, in ref. [20], we have also found that the relation $F_a \approx F_b$ (and, consequently, the PSS*) improves, without exception, as $\tilde{n}$ increases. We have mainly attributed this fact to a more important role of the term F'' in eq. (11), though F_a and F_b always differ from each other considerably near the singularity point r_0 (see figures 2 and 3).

However, we have verified that, if we multiply the central potential $\Sigma_S + \Sigma_0$ in the Dirac eq. (5) by a reduction factor ($RF < 1$), neither the PSS nor the PSS* are, necessarily, improved [21]. This conclusion can be extracted from figure 4, where we show the energies of the neutron PSD of the ^{40}Ca nucleus as a function of RF. This result is not an exception and all PSDs that become degenerate ($\epsilon_a = \epsilon_b$) for a given magnitude of the quantity $\Sigma_S + \Sigma_0$ split if this

Table 1. Results for the neutron pseudospin doublets (PSDs) of the ^{208}Pb nucleus [21, 22], obtained with the Dirac Hartree approximation (set NL-SH [31]). $\tilde{n}$, $\tilde{l}$ and κ are given in the text. $\varepsilon(\kappa)$ is the contribution of $\frac{V'}{V}\frac{\kappa}{r} \times F$ to the single-particle energy (SPE) ϵ. ϵ^* is the SPE obtained increasing the spin-orbit interaction by replacement of the self-consistent values of $\Sigma_S - \Sigma_0$ by $2 \times (\Sigma_S - \Sigma_0)$ in W and making the transformation $r \to 1.4 \times r$ to get a single-particle spectrum with a density of levels similar to the self-consistent one (ϵ). Energies are given in MeV.

PSD	$\tilde{n}$	state	$\tilde{l}$	κ	$\varepsilon(\kappa)$	ϵ	ϵ^*	$\epsilon(\exp)$
1	2	$2s_{1/2}$ (a)	1	-1	0.03	-41.42	-38.29	
		$1d_{3/2}$ (b)	1	2	-0.46	-44.67	-40.43	
2	2	$2p_{3/2}$ (a)	2	-2	-0.31	-30.91	-27.61	
		$1f_{5/2}$ (b)	2	3	0.13	-35.16	-29.54	
3	2	$2d_{5/2}$ (a)	3	-3	-1.12	-20.62	-18.18	
		$1g_{7/2}$ (b)	3	4	0.60	-24.75	-18.94	
4	3	$3s_{1/2}$ (a)	1	-1	-0.30	-17.96	-14.68	
		$2d_{3/2}$ (b)	1	2	0.59	-18.87	-14.67	
5	2	$2f_{7/2}$ (a)	4	-4	-2.50	-10.80	-10.52	-10.1
		$1h_{9/2}$ (b)	4	5	1.66	-13.88	-9.90	-10.8
6	3	$3p_{3/2}$ (a)	2	-2	-0.85	-7.85	-7.17	-8.3
		$2f_{5/2}$ (b)	2	3	1.23	-8.68	-7.14	-7.9

magnitude changes, in particular, when it decreases (it is just what happens with the PSDs 4 and 6 in the column e^* of table 1). Thus, the relation $\epsilon_a \simeq \epsilon_b$ can get worse as $|\Sigma_S + \Sigma_0|$ decreases. Furthermore, F_a becomes very different from F_b as the magnitude of $\Sigma_S + \Sigma_0$ decreases when ϵ_a or ϵ_b is close to the continuum. This is due to the fact that both states, except in very particular cases, approach the continuum, as functions of the magnitude of $\Sigma_S + \Sigma_0$, differently. Then, when a state of a PSD becomes unbound, its $F(r)$ component spreads all over the space, whereas its partner remains bound with a localized wave function.

We have also realized that, when $\Sigma_S + \Sigma_0$ is a Coulomb potential and $|\Sigma_S + \Sigma_0| \to 0$, ϵ_a and ϵ_b do not approach each other faster than the energies of other couples of states (not forming a PSD) do, F_a and F_b remaining also different in this limit. The different result found firstly in [3] was due to an error made by the author in the interpretation of the principal quantum number [25]. Thus, the exact PSS obtained for $\Sigma_S + \Sigma_0 = 0$ does not ensure, in general, that for $|\Sigma_S + \Sigma_0|$ small we could expect approximate PSS. These arguments indicate that neither the PSS nor the PSS* can be based on the smallness of $|\Sigma_S + \Sigma_0|$.

3. Spin symmetry and the magnitude of $\Sigma_S - \Sigma_0$

In the limit $\Sigma_S - \Sigma_0 = 0$, two states a and b of a SD have the same energy $\epsilon_a = \epsilon_b$ (i.e., the nucleus exhibits exact SS) and, moreover, it is evident from eq. (7) that $G_a \propto G_b$ (i.e., the nucleus satisfies the condition for exact SS*). In real nuclei, $|\Sigma_S - \Sigma_0|$ is large, therefore, in this case, no good approximate SS or SS* has been expected [6]. In fact, nuclei have large SO splittings. However, as suggested by the almost perturbative character of the $G - \kappa$ term predicted in [20] and confirmed in [21], the G functions of the spin partners are very similar[1]

[1] This result was unexpected by those authors that have been explained the approximate PSS (or PSS*) on the base of the smallness of $|\Sigma_S + \Sigma_0|$, by analogy to the large SO splittings due to the large magnitude of $|\Sigma_S - \Sigma_0|$ (see, for example, ref. [6]).

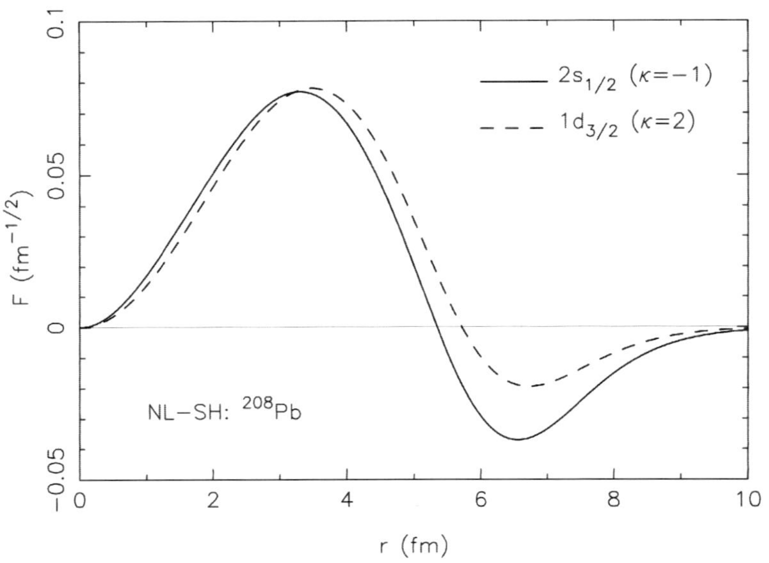

Figure 2. The small components $F(r)$ of the Dirac spinor for the two neutron states of the $(2s_{1/2}, 1d_{3/2})$ PSD of the ^{208}Pb nucleus with the NL-SH set [31] (these results are taken from ref. [21]).

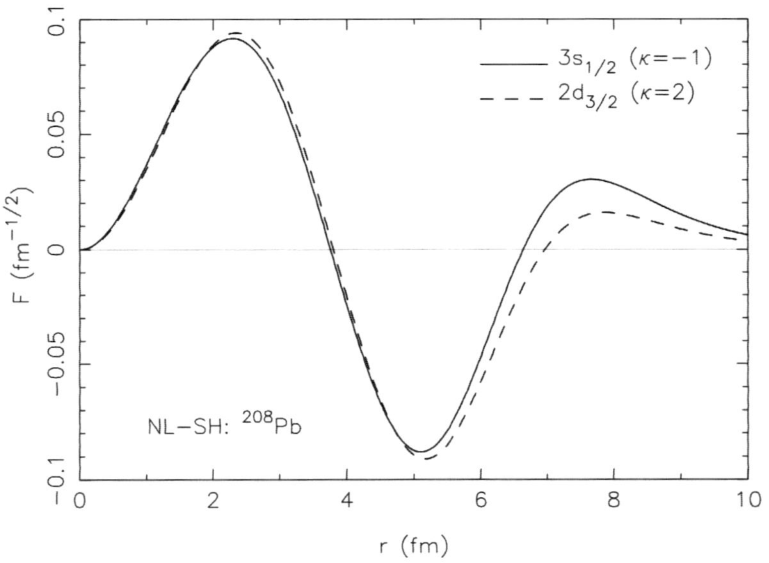

Figure 3. The same as figure 2 but for the $(3s_{1/2}, 2d_{3/2})$ PSD.

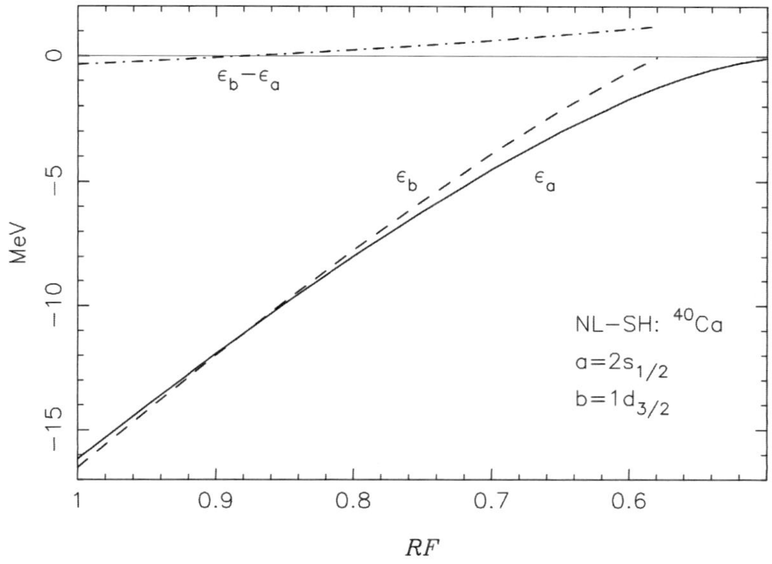

Figure 4. The single-particle energies ϵ_a and ϵ_b of the two states $a = 2s_{1/2}$ and $b = 1d_{3/2}$ of the neutron PSD of the ^{40}Ca nucleus and the difference $\epsilon_a - \epsilon_b$ as a function of the reduction factor RF of $\Sigma_S + \Sigma_0$ in the potential V for the NL-SH set [31] (these results are taken from ref. [21]).

(actually, much more similar than the small components F of the PS partners [21, 6], as can be seen in figures 5 and 6). Moreover, if $|\Sigma_S - \Sigma_0|$ decreases, the SO splittings decrease and the G functions of a SD become more similar, i.e., both the SS and the SS* are progressively improved because, the SS and the SS* are, actually, identical. These results indicate that the relation between the SS or SS* and $|\Sigma_S - \Sigma_0|$ is very different to the relation between the PSS or PSS* and $|\Sigma_S + \Sigma_0|$, due to the different behaviour of the $G - \kappa$ and $F - \kappa$ terms in equations (7) and (8), respectively.

4. Pseudospin symmetry and the $F - \kappa$ term

As we have explained above, the $F - \kappa$ term breaks the PSS* but not, necessarily, the PSS. The PSS has been considered slightly broken in nuclei by several authors [8]-[11] because the $F - \kappa$ term is small. More precisely, the condition required is that the magnitude of the $F - \kappa$ term should be much smaller than that of the pseudocentrifugal barrier (PCB)×F, i.e.:

$$|\frac{V'}{V}\frac{\kappa}{r}| \ll \frac{\tilde{l}(\tilde{l}+1)}{r^2}. \tag{12}$$

In this case, one can expect that the energy difference between two PS partners, which have the same value of $\tilde{l}$, be much smaller than the difference of energies between two states with different values of $\tilde{l}$ that, consequently, belong to different PSDs. The condition given by eq. (12) has been considered less restrictive than the relation $\Sigma_S \approx -\Sigma_0$ (or $\Sigma_S + \Sigma_0 \approx C_{ps} = constant$).

As $V(r_0) = 0$, the $F - \kappa$ term is large around r_0 and, consequently, F_a and F_b differ mostly around r_0 (see figures 2 and 3). It is remarkable that the $F - \kappa$ term together with the one originated from the PCB control the behaviour of the F functions in the nuclear surface, both terms being essential. In fact, the relation between $\tilde{l}$ and κ given by eq. (10) must be exactly fulfilled to get physical $F(r)$ functions (if this relation is slightly violated, the continuous solutions

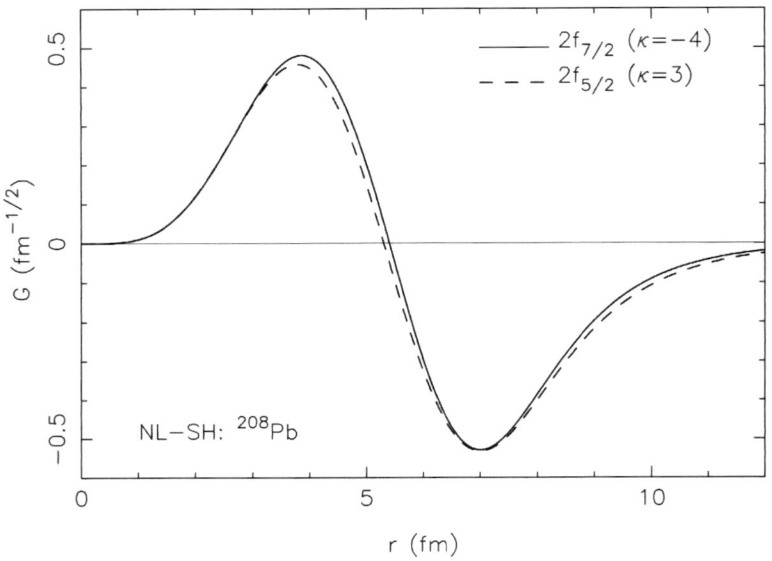

Figure 5. The big components $G(r)$ of the Dirac spinor for the two neutron states of the $(2f_{7/2}, 2f_{5/2})$ SD of the ^{208}Pb nucleus with the NL-SH set [31] (these results are taken from ref. [21]).

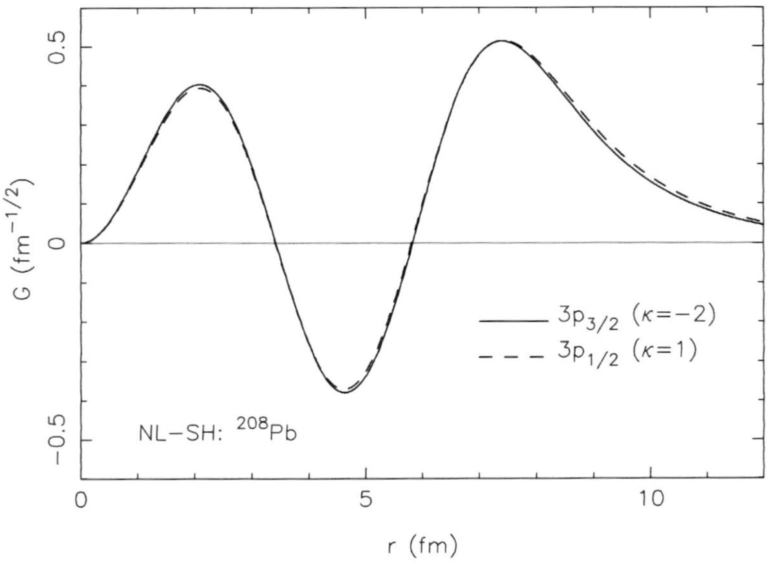

Figure 6. The same as figure 4 but for the $(3p_{3/2}, 3p_{1/2})$ SD.

Journal of Physics: Conference Series **128** (2008) 012034 doi:10.1088/1742-6596/128/1/012034

of eq. (8), $F(r)$, with a continuos derivative, become zero at the singularity point r_0 [20]-[22]). Then, the $F - \kappa$ term should not be considered much smaller than the PCB, in contrast to the conclusions established in references [8]-[11].

However, the $F - \kappa$ term is roughly an odd function around r_0. This allows a small PS splitting being F_a and F_b quite different. In fact, the property $\epsilon_a = \epsilon_b$, within a realistic DHA, needs the condition $F_a \neq F_b$, since, if $F_a(r) = F_b(r)$, the corresponding PS splitting would be due just to the contribution of the $F - \kappa$ term, which is, necessary, different from zero (the exact PSS* is forbidden in this case). Then, we have quasi-degenerate PSDs, not because the $F - \kappa$ term is small, but due to the compensation of its contribution by those coming from the other terms entering eq. (8) different from the $F - \kappa$ term, the precise balance of this compensation depending on the Σ_S and Σ_0 self-energies [19]-[21]. We can say the PSS is an accidental symmetry for these reasons.

Table 1 shows that the contribution of the $F - \kappa$ term to the splittings of the PSDs is not very large, except for the 5th PSD. However, it is surprising that this contribution, except for the 1st PSD, has a sign opposite to that of the splitting itself. This is due to the effect that the $F - \kappa$ term produces on F. It is large enough to compensate the contribution of the $F - \kappa$ term to ϵ and, furthermore, to generate the PSDs splittings [19, 21, 24].

5. Spin symmetry and the $G - \kappa$ term

The $G - \kappa$ term breaks the SS and, of course, the SS*. Since $W'/W \simeq (\Sigma'_S - \Sigma'_0)/2M$, the $G - \kappa$ term can be considered as a "small" relativistic term. In fact, the relation $G_a \approx G_b$ is very well satisfied in all cases, improving also as the number of nodes of $G_{a,b}$ increases (see figures 5 and 6) and, though the SO splittings are large, the $G - \kappa$ term behaves almost in a perturbative way. We can conclude that the $G - \kappa$ term, which breaks the condition $G_a = G_b$ (and, consequently, the SS*) and is considered large, produces a smaller effect in nuclei than the $F - \kappa$ term, which breaks the condition $F_a = F_b$ (and, consequently, the PSS*) and has been considered small by some groups [8]-[11].

6. Pseudospin symmetry as a result of compensation of different effects

6.1. Original concept of the pseudospin symmetry

We have seen that *the conditions* $\epsilon_a \simeq \epsilon_b$ and $F_a \approx F_b$, *for the two states of a* PSD, *are not strictly dependent on each other.* We can use this fact as an argument to recover the original concept of the PSS that grounds on the degeneracy of the PSDs, no matter whether F_a and F_b are proportional each other or not. That is why we have proposed in refs. [22, 23] an explanation of the PSS in real nuclei starting from a situation of exact degeneracy of the PSDs, being F_a and F_b not proportional to each other. In the starting model, $\Sigma_S - \Sigma_0 = 0$ (i.e., there is no SO interaction), whereas $\Sigma_S + \Sigma_0$ is a harmonic oscillator potential. This model, defined by the Hamiltonian H_0, produces degenerate SDs with $G_a \propto G_b$ and degenerate PSDs but with F_a not proportional to F_b [15]. Then, although a more realistic quantity $\Sigma_S + \Sigma_0$ breaks the degeneracy of the PSDs, the choice of an appropriate potential $\Sigma_S - \Sigma_0 \neq 0$ partially restores the quasidegeneracy of the PSDs. Thus, we have a continuous (*almost perturbative*[2]) way that allows to connect the model satisfying exact degeneracy of the PSDs with realistic models of nuclei fulfilling approximate PSS.

[2] This transition is made in two steps: a) the change of the oscillator harmonic potential to a more realistic one and b) the inclusion of the SO potential. Both steps can be realized in an almost perturbative way (we have realized above that this is the case for the SO potential).

Journal of Physics: Conference Series **128** (2008) 012034 doi:10.1088/1742-6596/128/1/012034

6.2. Pseudospin symmetry breaking term

Since the consideration of the $F-\kappa$ term as the PSS breaking term requires the two conditions $\epsilon_a \simeq \epsilon_b$ and $F_a \approx F_b$ to be directly dependent on each other, our present form of understanding the PSS indicates that the $F-\kappa$ term is not the appropriate choice for the PSS breaking term in this model. Let us see how we can define an appropriate breaking term in accordance with the present interpretation of the PSS. Let us write the SP Hamiltonian corresponding to eq. (2) as $H = H_0 + H_1$, where H_0 exhibits exact degeneracy of PSDs and H_1 represents the corresponding breaking term.

If the Dirac equation corresponding to H is given by eq. (4), and we write $W = W_0 + W_1$, $V = V_0 + V_1$ and $\epsilon = \epsilon_0 + \epsilon_1$, where W_0, V_0 and ϵ_0 represent, respectively, the potentials entering eq. (4) and the SP energy corresponding to H_0, we have $W_0 = 2M + \epsilon_0$, $V_0 = \Omega - \epsilon_0$, Ω being an appropriate harmonic oscillator potential for a given nucleus; $W_1 = \Sigma_S - \Sigma_0 + \epsilon_1$ and $V_1 = \Sigma_S + \Sigma_0 - \Omega - \epsilon_1$.

From the Dirac eq. (4) or from eq. (8), we get the following equivalent equation for the small component F:

$$-F'' + \left[\frac{V_0' + V_1'}{V_0 + V_1}\left(\frac{F'}{F} - \frac{\kappa}{r}\right) + \frac{\tilde{l}(\tilde{l}+1)}{r^2} + (V_0 + V_1)(W_0 + W_1)\right] F = 0, \tag{13}$$

where the $F-\kappa$ term breaks, explicitly, the relation $F_a \propto F_b$, even if $V_1 = W_1 = 0$. However, in this case, we have $\epsilon_{a0} = \epsilon_{b0}$ [15] (indicating that the $F-\kappa$ term is not the appropriate PSS breaking term). The fact that the $F-\kappa$ term is not a perturbative term facilitates this equality[3]. The self-consistency effects due to the $F-\kappa$ term, which breaks the proportionality between F_a and F_b, allow to compensate, exactly, the contribution of the $F-\kappa$ term to the pseudospin-orbit splittings. It is important to remind that this situation of exact degeneracy of the PSDs is compatible with bound nuclei, on the contrary to what happens with the exact PSS* considered in refs. [3]-[7]. When the potentials V_1 and W_1 entering eq. (13) are considered, their effects are almost perturbative and the real nuclei with broken PSS are recovered.

6.3. Condition $F_a \approx F_b$ for the two pseudospin partners

The model can explain the relation $\epsilon_a \simeq \epsilon_b$ for the pseudospin partners, but still the question remains, how can we explain the relation $F_a \approx F_b$ found also in relativistic calculations for PSDs with $\tilde{n} \geq 3$? The similarity between F_a and F_b for the two states of a PSD in the inner region of the nuclei can be explained, independently from $\epsilon_a \simeq \epsilon_b$, because the effects of the $F-\kappa$ term, responsible for the lack of proportionality between F_a and F_b, are mainly important in the nuclear surface due to the singularity of the $F-\kappa$ term at r_0. Furthermore, since the role of the terms containing F' and, mainly, F'' in eqs. (8) or (13) increases with the number of nodes $\tilde{n}$ of F [21], F_a and F_b become more similar as $\tilde{n}$ increases and, generally, the same happens with ϵ_a and ϵ_b (although the relation $F_a \approx F_b$ cannot justify the small splitting of PSDs found in calculations [19, 21]).

The improvement of the relation $\epsilon_a \simeq \epsilon_b$ with $\tilde{n}$ can be understood taking into account that: (A) the splitting of the PSDs produced by the deviation of $\Sigma_S + \Sigma_0$ from a harmonic oscillator potential (i.e. by V_1) decreases with $\tilde{n}$ inside a major shell (see figure 1), and (B) the SO interaction in realistic relativistic models, which partially compensates the effect of V_1, is not strong enough to approach the two states of the PSDs with $\tilde{n} \leq 2$ what would be necessary to reach a relation $\epsilon_a \simeq \epsilon_b$ so good as for the PSDs with $\tilde{n} \geq 3$. The results of table 1 show that when the strength of the SO interaction is increased (column ϵ^*), the PSS is improved, mainly for the PSDs with $\tilde{n} = 2$ (which, for $\Sigma_S - \Sigma_0 = 0$, show sizable PS orbit splittings).

[3] Notice that our present form of understanding the PSS is based on the fact that the $F-\kappa$ term is large rather than small. The case considered here supplies also an example of the independence between the two relations $\epsilon_{a0} = \epsilon_{b0}$ and $F_a \propto F_b$.

6.4. Pseudospin symmetry in the non-relativistic framework
It is important to notice that the explanation for the PSS given in this section is not based on the relativistic character of the model used. It can be formulated in similar terms for the non-relativistic models, starting from the non-relativistic SP harmonic oscillator potential without SO interaction. This model produces also degenerate SDs and PSDs. When a more realistic SP potential is used, the degeneracy of the PSDs is broken, but it is partially restored, as in the relativistic case, if an adequate SO interaction is added by hand. Thus, the PSS can be explained in a simple and similar way in the relativistic and non-relativistic approximations, although the SO interaction, which plays an important role in this explanation, is more naturally incorporated in the relativistic models. It is important that the explanation of the PSS given in the relativistic framework be valid also in the the non-relativistic case, since, for realistic models, the quantities V_1 and W_1 cannot be neglected in the non-relativistic limit[4].

7. Conclusions
Our conclusions can be summarized in the following points:
1.- The PSS in nuclei is not due to the smallness (or flatness) of the central potential $\Sigma_S + \Sigma_0$ or to a small $F - \kappa$ term, but, rather, it is related to the partial compensation of the contributions of different terms entering eq. (8). This compensation is possible because the $F - \kappa$ term, which diverges in the nuclear surface, produces effects by any means perturbative, in particular, in the F wave functions of the Dirac spinors. In this compensation, the SO interaction plays an important role, since, partially, restores the PSS, which is broken due to the deviation of the real $\Sigma_S + \Sigma_0$ from a harmonic oscillator potential. It is important to remark, that the mechanisms allowing the compensation, which is crucial to explain the PSS, do not have an exclusively relativistic character, except in what it concerns the SO interaction (if one considers that this interaction has a particular relativistic character).
2.- The $F - \kappa$ term, although "large", is the appropriate breaking term for the PSS* but not for the PSS (as implicitly indicates conclusion 1).
3.- Though the $F - \kappa$ term is not small, the PSS and the PSS* are improved as $\tilde{n}$ increases due, in particular, to an increasing role of the F'' term in eq. (8) with the increase of $\tilde{n}$. Thus, the PSS and the PSS* (with essential differences between F_a and F_b in the nuclear surface) show up themselves more often in the PSDs of heavy nuclei.
4.- The PSS is favoured because the $F - \kappa$ term is roughly an odd function around its singularity point r_0, whereas the relation $F_a \approx F_b$ is favoured by the flatness of $\Sigma_S + \Sigma_0$ inside the nucleus.

Acknowledgments
This work has been supported by the DGESIC grand FIS2005-04033. One of the authors (L.N.S.) is grateful to the University of Cantabria for hospitality.

References
[1] Arima A, Harvey M, Shimizu K 1969 *Phys. Lett.* B **30** 517
[2] Hecht K T, Adler A 1969 *Nucl. Phys.* A **137** 129
[3] Ginocchio J N 1997 *Phys. Rev. Lett.* **78** 436; 1999 *Phys. Rep.* **315** 231; 1999 *J. Phys. G.: Nucl. Part. Phys.* **25** 617
[4] Ginocchio J N, Madland D G 1998 *Phys. Rev.* C **57** 1167
[5] Ginocchio J N, Leviatan A 2001 *Phys. Rev. Lett.* **87** 072502
[6] Borycki P J, Ginocchio J N, Nazarewicz W, Stoitsov M 2003 *Phys. Rev.* C **68** 014304
[7] Ginocchio J N 2005 *Phys. Rep.* **414** 165; 2005 *Nucl. Phys. News* **15** 13
[8] Meng J, Sugawara-Tanabe K, Yamaji S, Ring P, Arima A 1998 *Phys. Rev.* C **58** R628

[4] Notice that the $F - \kappa$ term cannot be considered either as a genuine relativistic term because it cannot be neglected, in eqs. (8) or (13) for the function F, as the nucleon mass M becomes very large.

Journal of Physics: Conference Series **128** (2008) 012034 doi:10.1088/1742-6596/128/1/012034

[9] Sugawara-Tanabe K, Arima A 1998 *Phys. Rev.* C **58** R3065
[10] Meng J, Sugawara-Tanabe K, Yamaji S, Arima A 1999 *Phys. Rev.* C **59** 154
[11] Sugawara-Tanabe K, Meng J, Yamaji S, Arima A 1999 *J. Phys. G: Nucl. Part. Phys.* **25** 811
[12] Chen T S, Lü H F, Meng J, Zhang S O, and Zhou S G 2003 *Chin. Phys. Lett.* **20** 358
[13] Alberto P, Fiolhais M, Malheiro M, Delfino A, Chiapparini M 2001 *Phys. Rev. Lett.* **86** 5015; 2002 *Phys. Rev.* C **65** 034307
[14] Lisboa R, Malheiro M, Alberto P 2003 *Phys. Rev.* C **67** 054305
[15] Lisboa R, Malheiro M, de Castro A S, Alberto P, Fiolhais M 2004 *Phys. Rev.* C **69** 024319
[16] Gambhir Y K, Maharana J P and Warke C S 1998 *Eur. Phys. J. A* **3** 255
[17] Lalazissis G A, Gambhir Y K, Maharana J P, Warke C S, Ring P 1998 *Phy. Phys.* C **58** R45
[18] Marcos S, Savushkin L N, López-Quelle M, Ring P 2000 *Phys. Rev.* C **62**, 054309
[19] Marcos S, López-Quelle M, Niembro R, Savushkin L N, Bernardos P 2001 *Phys. Lett.* B **513** 30
[20] Marcos S, López-Quelle M, Niembro R, Savushkin L N, Bernardos P 2003 *Eur. Phys. J.* A **17** 173
[21] Marcos S, López-Quelle M, Niembro R, Savushkin L N 2004 *Eur. Phys. J.* A **20** 443
[22] Marcos S, López-Quelle M, Niembro R, Savushkin L N 2005 *J. Phys G: Nucl. Part. Phys.* **31** S1551
[23] Marcos S, Fomenko V N, López-Quelle M, Niembro R, Savushkin L N 2005 *Eur. Phys. J.* A **26** 253
[24] López-Quelle M, Savushkin L N, Marcos S, Bernardos P, Niembro R 2003 *Nucl. Phys.* A **727** 269
[25] López-Quelle M, Marcos S, Savushkin L N, Niembro R 2005 *Recent Res. Dev. Physics* **6** 29
[26] López-Quelle M, Savushkin L N, Marcos S and Niembro R 2005 *J. Phys G: Nucl. Part. Phys.* **31** S1911
[27] Bahri C, Draayer J P and Moszkowski S A 1992 *Phys. Rev. Lett.* **68** 2135
[28] Piekarewicz J 2007 *J. of Phys. G: Nucl. and Part.* **34** 467
[29] Bouyssy A, Marcos S and Van Thieu Pham 1984 *Nucl. Phys.* **A422** 541
[30] Serot B S and Walecka J D 1986 *Advances in Nuclear Physics* **16**, ed J W Negele and E Vogt (New York: Plenum)
[31] Sharma M M, Nagarajan M A, Ring P 1993 *Phys. Lett.* B **312** 377

Journal of Physics: Conference Series **128** (2008) 012035 doi:10.1088/1742-6596/128/1/012035

The large nonlinearity scale limit of an information-theoretically motivated nonlinear Schrödinger equation

Le-Huy Nguyen, Hai-Siong Tan and Rajesh R Parwani
Department of Physics, National University of Singapore, Kent Ridge, Singapore.
E-mail: parwani@nus.edu.sg

Abstract. A nonlinear Schrödinger equation, that had been obtained within the context of the maximum uncertainty principle, has the form of a difference-differential equation and exhibits some interesting properties. Here we discuss that equation in the regime where the nonlinearity length scale is large compared to the deBroglie wavelength; just as in the perturbative regime, the equation again displays some universality. We also briefly discuss stationary solutions to a naturally induced discretisation of that equation.

1. Introduction

Physicists study nonlinear Schrödinger equations for two purposes: (i) as effective equations in such fields as condensed matter [1] and (ii) to probe a potential deviation from exact linearity in quantum mechanics [2].

The manner in which the nonlinear Schrödinger equations are arrived at in the literature is varied. Here we will focus on the nonlinear equation that was obtained using information theoretic arguments [3]. Such an approach, often termed the "maximum entropy method" or more generally "maximum uncertainty method", is often used in statistical mechanics to deduce the form of the probability distribution. The philosophy is that one should choose the probability distributions with minimum bias while satisfying relevant constraints [4].

The maximum uncertainty approach had been used to understand the usual linear Schrödinger equation [5]. Of course in such an approach one must motivate the information, or inverse uncertainty, measure that is used to quantify the entropy (uncertainty) associated with the probability distribution. Just as the Gibbs-Shannon measure is the simplest one that satisfies certain axioms in statistical mechanics, the Fisher measure emerges as the simplest one that satifies axioms appropriate for classical ensemble dynamics, and hence for deducing non-relativistic quantum mechanics as a single parameter extension of classical ensemble dynamics [6].

One could use the approach of [6] to obtain generalisations of the Fisher measure, involving higher derivatives, that satisfy the same axioms but which involve more parameters. However the natural constraint that the equation be invariant under a scaling of the wavefunction then leads to nonpolynomial terms $\sim \partial p/p$, with $p(x)$ the probability density; and since the nonlinearity is expected to be phenmenologically small, the nonlinear corrections would lead generally to ill-defined singular terms in perturbation theory.

A different approach to generalising the Fisher measure is to choose the simplest nonsingular measure that contains the Fisher measure in some limit, satisfies the various axioms and which also has some physical motivation. That is how the regularised Kullback-Lieber measure came to be used and the associated nonlinear Schrödinger equation derived [3].

For simplicity we consider here only a single quantum particle in one space dimension, see [3] for the general case. The nonlinear equation is

$$i\hbar\frac{\partial\psi}{\partial t} = -\frac{\hbar^2}{2m}\frac{\partial^2\psi}{\partial x^2} + V(x)\psi + F_1(p)\psi\,, \tag{1}$$

with

$$F_1(p) \equiv Q_{1NL} - Q \tag{2}$$

$$\tag{3}$$

and $p(x) = \psi^\star(x)\psi(x)$; also

$$\mathcal{E}L^2 = \frac{\hbar^2}{4m}\,, \tag{4}$$

and

$$Q = -\frac{\hbar^2}{2m}\frac{1}{\sqrt{p}}\frac{\partial^2\sqrt{p}}{\partial x^2}\,. \tag{5}$$

Let

$$p_\pm(x) \equiv p(x \pm \eta L)\,, \tag{6}$$

where the dimensionless parameter η takes values $0 < \eta \leq 1$, then

$$Q_{1NL} = \frac{\mathcal{E}}{\eta^4}\left[\ln\frac{p}{(1-\eta)p+\eta p_+} + 1 - \frac{(1-\eta)p}{(1-\eta)p+\eta p_+} - \frac{\eta p_-}{(1-\eta)p_- + \eta p}\right]\,. \tag{7}$$

In addition to regulating potential singularities, η may also be viewed as parametrising a family of nonlinear theories, with the usual linear quantum mechanics recovered at $\eta = 0$.

The nonlinear Schrödinger equation (1) resembles a differential-difference equation as the evolution of the wavefunction depends, any fixed time, not just on knowledge at the point x but also at neighbouring points a finite distance away, $x \pm \eta L$. The nonlinearity is also non-polynomial. Nevertheless, the equation shares a number of important properties with the linear Schrödinger equation, such as the conservation of probability and existence of the usual plane wave solutions.

2. Perturbative regime

It was found [7] that for smooth external potentials $V(x)$, treating the nonlinearity as a perturbation results in energy shifts which depend significantly on whether or not the unperturbed states have nodes. Those that do, have their energies shifted by a larger amount

$$\delta E \approx \frac{\hbar^2|L|\pi}{6m}\,\sqrt{\eta(1-\eta)}\,\,(1-4\eta)\sum_{p=1}^{N} C^2_{np} + O(L/a)^2, \tag{8}$$

where a is the characteristic length scale of the linear theory, and L/a is the perturbation parameter. The coefficients C_{np} depend on the slope of the unperturbed wavefunctions near the

Journal of Physics: Conference Series **128** (2008) 012035 doi:10.1088/1742-6596/128/1/012035

nodes, and hence on the external potential. Notice that the dependence of the energy shifts on η is universal, resulting in positive energy shifts for small η and negative shifts for larger values.

One may also try to fix the value of η [8]. If one assumes η to be a parameter that "flows" in the space of theories, then one may see whether there is some value that minimises the energy of the system. Indeed, the energy shifts mentioned above reach a unique global minimum at

$$\eta_m \approx 0.80\,. \tag{9}$$

At this value of η the leading energy shifts are *negative* so that the nonlinearity reduces the energy of the original linear system.

3. Nonperturbative regime

For L large, $p(x \pm L) \sim p(L) \sim 0$, and so the time-independent nonlinear Schrödinger equation for an eigenstate reduces to

$$E\phi(x) = V(x)\phi(x) - \frac{\mathcal{E}}{\eta^4}\log(1-\eta)\phi(x)\,, \tag{10}$$

where $V(x)$ is a given external potential. This equation suggests that $\phi(x)$ can be nonzero only at those points where the potential vanishes. For a potential with a single minimum, this means that the wavefunctions collapse to a delta function and the energy is given in this limit by

$$E = -\frac{\mathcal{E}}{\eta^4}\log(1-\eta) \tag{11}$$

As a function of η this energy attains a minimum at the universal, that is $V(x)$ independent, value

$$\eta \sim 0.9. \tag{12}$$

Notice also that because of (4), the energy goes to zero as $L \to \infty$. A variational calculation gives a quick, though approximate, interpolation between the small and large L regimes. We did this for the SHO potential using trial states which are SHO stationary states but with the usual deBroglie wavelength $a = \sqrt{\frac{\hbar}{m\omega}}$ replaced by a variational parameter b [9]. The dimensionless quantity $c = b/a$ was varied to minimise the action, which depends on the free parameters η and $\epsilon = L/a$. The energy functional was then evaluated at the minimum for each (η, ϵ).

For small ϵ the results are in agreement with the perturbative results as expected. For larger ϵ we found that the wavefunction shrinks and expands suddenly at certain (η dependent) values of ϵ before eventually shrinking ($c \to 0$) to a highly localised state as suggested by (10). The oscillatory behaviour of the wavefunction might be an artifact of the variational calculation. Figures (1-3) summarise some of the variational results. (The dependence of the energy on η is qualitatively similar to that in the perturbative regime).

4. Discretised states

Since the nonlinear equation already has the form of a difference-differential equation, it is most natural to consider the fully discretised equation, with $p(x+L) = p_{n+1}$ etc., as another approach to study nonperturbative aspects of the equation [10]. We remark that if L is related to gravity, as speculated in [3], then this discretisation of space suggested by the structure of the equation is consistent with discretisations implied by other approaches to quantum gravity.

Consider first the the time-independent nonlinear equation for $\eta = 1$. We have

$$E \;=\; \mathcal{E}\left[\ln\frac{p_n}{p_{n+1}} + 1 - \frac{p_{n-1}}{p_n}\right]. \tag{13}$$

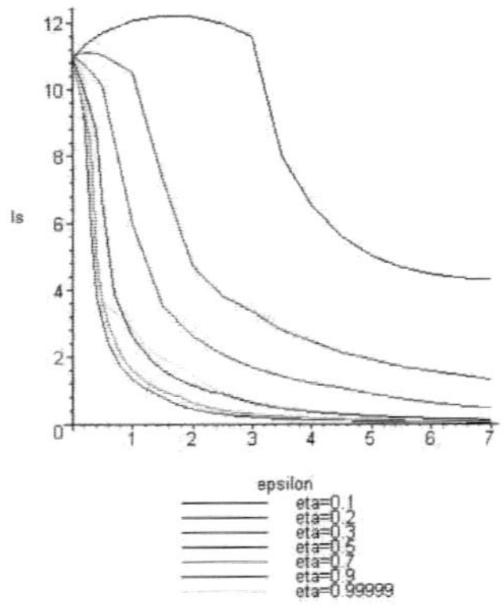

Figure 1. The variationally computed energy versus ϵ. Plots are similar for the ground state and the example of an excited state,$n = 5$, studied here. Generally, for fixed ϵ, the energy decreases as η increases.

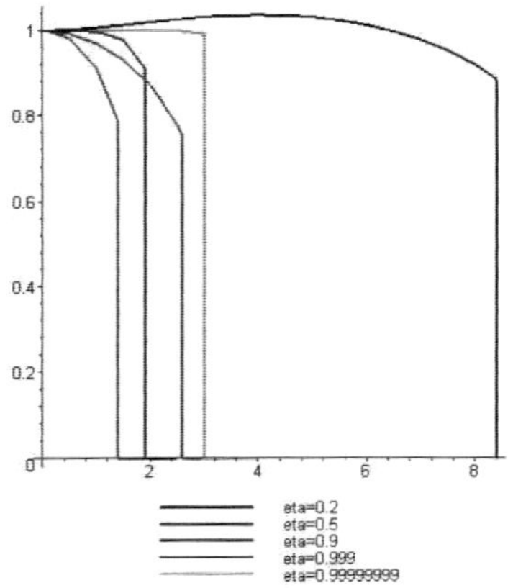

Figure 2. The normalised width c of the variational ground state versus ϵ. Looking at the intercept on the x-axis from the right, the curves correspond to the following values of η : $0.1, 0.2, 0.999999, 0.5, 0.999, 0.9$.

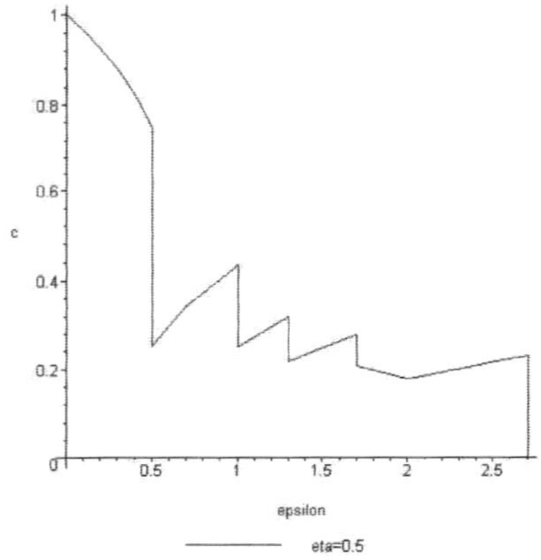

Figure 3. The normalised width c of the variational excited ($n = 5$) state versus ϵ.

If we seek solutions on the full infinite line for which p_n is bounded for all n, then by using the iterative nature of (13) one easily shows that that this is only possible for $E > 0$, and even then there is a lower bound on n (to keep the probability positive). *That is, the discretised space is necessarily truncated.* Similar results hold for the general regularised case, $\eta \neq 1$, and also for the $q-$ deformed equation in Ref.[3].

5. Summary and outlook

The fact that the nonlinear equation (1) was motivated through a philosophically appealing information theoretic approach suggests that the equation might eventually be of physical relevance, for example as an effective equation in one or both of the domains mentioned in the first paragraph of the introduction. As such, it is of interest to investigate the properties of that equation.

We found that for large $\epsilon = L/a$ the nonlinearity acts as a strongly attractive force, causing a localisation of the wavefunctions. The resultant energy of all eigenstates is minimised at a universal value of $\eta \sim 0.9$, which is larger than the value $\eta \sim 0.8$ found (for states with nodes) in the perturbative regime of ϵ. A natural question is whether for any L, there is a $V(x)$-independent η value for which the energy of eigenstates is minimised.

There are several other properties of the nonlinear equation that have yet to be studied. Of particular interest are its time-dependent states: dynamics of wavepackets and possible solitary waves. We invite interested readers to the adventure.

Acknowledgments

R.P. thanks the organisers for their hospitalty and for staging a stimulating event.

References

[1] Sulem C and Sulem P 1999 *The nonlinear Schrödinger equation:self-focusing and wave collapse* (Springer).
[2] For some reviews see Svetlichny G 2004 *Preprint* quant-ph/0410036;
Carroll R 2004 *Preprint* quant-ph/0401082 and references therein.
[3] Parwani R 2005 *Ann. Phys.* **315** 419
[4] Jaynes E T 1957 *Phys. Rev.* **106** 620; **108** 171;
Jaynes E T 2004 *Probability Theory, The Logic of Science* (Cambridge University Press).
[5] Reginatto M 1998 *Phys. Rev.* A **58** 1775 ; Erratum *ibid.* 1999 **60** 1730;
Frieden B R 1988 *J. Mod. Opt* **35** 1297;
Frieden B R 1989 *Am. J. Phys.* **57** 1004
[6] Parwani R 2005 *J. Phys. A:Math. Gen.* **38** 6231;
Parwani R 2006 *Int. J. Theor. Phys.* **45** 1901
[7] Parwani R and Tabia G 2007 *J. Phys. A: Math. Theor.* **40** 5621-5635
[8] Parwani R 2007 *Theoretical and Mathematical Physics* **152(1)** 1012-1016
[9] Hasson R and Richards D 2000 *Preprint* quant-ph/0012068v1
[10] Parwani R and Tan H S 2007 *Phys. Lett.* A **363** 197-201

The preparation time in a scattering experiment

P Bryant

CCQS, Department of Physics, University of Texas at Austin, Austin, TX 78712

E-mail: pbryant@physics.utexas.edu

Abstract. A quantum mechanical theory with time asymmetry intrinsic to states (or observables) features the concept of an initial time of the state and thus a preparation time of the physical system represented by the state. This special time is investigated in the context of scattering theory, where, in standard quantum mechanics, the physical meaning of a preparation time has remained obscure. In an experiment, the preparation time corresponds to an ensemble of times of scattering marking the times in the laboratory when one scattering projectile interacts with one target quantum.

1. Introduction

Standard quantum mechanics has time evolution for states (in the Schrödinger picture) or for observables (in the Heisenberg picture) that extends from $-\infty < t < \infty$. It therefore makes predictions for experimental (Born) probabilities also for $-\infty < t < \infty$. Such predictions are not intuitive because, in an experiment, an observable can be measured in a system represented by a state only after the system has been prepared at some finite time. There exists, however, a causal quantum theory, which makes predictions of Born probabilities only for times $t \geq 0$, where $t = 0$ corresponds to the time at which the quantum system has been prepared [1].

To determine if the concept of a preparation time incorporates naturally into quantum mechanical scattering theory, one must understand what this special time represents phenomenologically. Section 2 is a brief motivation for a theory that addresses the preparation time of physical systems. Section 3 is a discussion of the non-relativistic scattering cross section and its relation to the preparation time. In Section 4, the preparation time of systems represented by scattered states is identified for a scattering experiment.

2. Conjecturing an Arrow of Time

For simplicity, let us discuss systems represented by pure states described by a state vector ϕ (or density operator $\rho = |\phi\rangle\langle\phi|$.) Let the observable be represented by a vector ψ (or the corresponding observable $\Lambda = |\psi\rangle\langle\psi|$.) In the theory, the time evolution of these quantum mechanical entities is given for states by the Schrödinger equation

$$i\hbar\frac{\partial}{\partial t}\phi = H\phi \tag{1}$$

and for observables by the Heisenberg equation

$$i\hbar\frac{\partial}{\partial t}\psi = -H\psi, \quad \text{or} \quad i\hbar\frac{\partial}{\partial t}\Lambda(t) = [\Lambda(t), H]. \tag{2}$$

In standard quantum mechanics, one uses as a boundary condition for these dynamical equations the Hilbert space axiom:

$$\{\phi\} = \{\psi\} = \text{Hilbert space} = \mathcal{H}. \tag{3}$$

Then, by the Stone-von Neumann theorem [2, 3], the solutions of the dynamical equations (1) or (2) are

$$\phi(t) = e^{-\frac{iHt}{\hbar}}\phi_0 , \qquad -\infty < t < \infty, \tag{4}$$

for states, or

$$\psi(t) = e^{\frac{iHt}{\hbar}}\psi_0 , \qquad -\infty < t < \infty, \tag{5}$$

for observables. This is time evolution given for states by the unitary group

$$U^\dagger(t) = e^{-\frac{iHt}{\hbar}} , \qquad -\infty < t < \infty, \tag{6}$$

and for observables by the unitary group

$$U(t) = e^{\frac{iHt}{\hbar}} , \qquad -\infty < t < \infty. \tag{7}$$

The group product is

$$U^\dagger(t_1)U^\dagger(t_2) = U^\dagger(t_1 + t_2). \tag{8}$$

For every evolution $U(t)$ there exists the inverse

$$U^\dagger(t)^{-1} = U^\dagger(-t). \tag{9}$$

The unitary group evolution of states (or of observables) is symmetric in time; one can use any value of the time parameter, $-\infty < t < \infty$, to drive the evolution of a state (or, in the Heisenberg picture, of an observable) forward or backward.

A reversible time evolution for the physical systems *represented* by states is not intuitive or observed, but the reversible time evolution (4) or (5) is generally accepted. In a quantum mechanical experiment, the calculated, measurable quantity is (in the Schrödinger picture) the probability to find the observable $\Lambda = |\psi\rangle\langle\psi|$ in the time-varying state $\phi(t)$. This quantity is the Born probability:

$$\mathcal{P}_{\phi(t)}(\Lambda) = \text{Tr}\left(\Lambda|\phi(t)\rangle\langle\phi(t)|\right) = \text{Tr}\left(|\psi\rangle\langle\psi|\phi(t)\rangle\langle\phi(t)|\right) = |\langle\psi|\phi(t)\rangle|^2. \tag{10}$$

Note that the Born probability to find the observable Λ in the state $\phi(t)$ has time evolution given by the state $\phi(t)$. Comparing (10) with (4) and (6), one finds

$$\mathcal{P}_{\phi(t)}(\Lambda) = |\langle\psi|\phi(t)\rangle|^2 = |\langle\psi|U^\dagger(t)\phi_0\rangle|^2, \qquad -\infty < t < \infty. \tag{11}$$

Again, one can use any value between $-\infty$ and ∞ for the time parameter. One has not only reversible time evolution for states, but also reversible time evolution for the predictions of the experimentally measured quantities.

Phenomenologically, however, one knows that a measurement cannot be made before an experimental system has been prepared. In terms of the theoretical objects, an observable Λ cannot be measured in the state $\phi(t)$ before the system represented by the state $\phi(t)$ has been prepared at some finite time $t_{prep} > -\infty$. This is a statement of the preparation-registration arrow of time, and it emphasizes the notion of a finite preparation time. In time symmetric quantum mechanics, the theoretical predictions of experimental quantities (11) also hold for times $-\infty < t < t_{prep}$.

Journal of Physics: Conference Series **128** (2008) 012036 doi:10.1088/1742-6596/128/1/012036

A quantum theory providing asymmetric time evolution for states, which matches the phenomenologically observed, asymmetric time evolution of physical systems, and having as a feature the preparation time, has been obtained [1, 4]. By changing the boundary conditions (3) for the dynamical equations of quantum mechanics,[1] thus modifying slightly the axiom (3) of standard quantum theory, one finds the time evolution to be [1, 8]

$$\phi(t) = e^{-\frac{iH^\times t}{\hbar}}\phi_0\,, \qquad 0 \le t < \infty, \tag{12}$$

for states or

$$\psi(t) = e^{\frac{iHt}{\hbar}}\psi_0\,, \qquad 0 \le t < \infty, \tag{13}$$

for observables.[2] This is no longer time evolution given by the unitary group $U^\dagger(t)$, but is rather given by a *semigroup*,

$$U^\times(t) = e^{-\frac{iH^\times t}{\hbar}}\,, \qquad 0 \le t < \infty, \tag{14}$$

which is the time evolution operator for states, and by another *semigroup*,

$$U(t) = e^{\frac{iHt}{\hbar}}\,, \qquad 0 \le t < \infty, \tag{15}$$

which is the time evolution operator for observables.[3] While the group product is still defined,

$$U^\times(t_1)U^\times(t_2) = U^\times(t_1 + t_2), \tag{16}$$

being a semigroup means that the inverse, $U^\times(t)^{-1}$, of an element, $U^\times(t)$ with $t > 0$, does not exist. Simply stated, one cannot evolve the state vector representing a physical system backward in time. The Born probability is calculated for $0 \le t < \infty$ only:

$$\mathcal{P}_{\phi(t)}(\Lambda) = |\langle\psi|\phi(t)\rangle|^2 = |\langle\psi|U^\times(t)\phi_0\rangle|^2, \qquad 0 \le t < \infty. \tag{17}$$

As will be discussed in the next section, one identifies the time $t = 0$ in (17) as the preparation time of the physical system represented by the state vector $\phi(t)$. Therefore, one no longer has a Born probability for finding an observable in a state before the system represented by that state is prepared. With the time asymmetric quantum theory, calculations of experimental quantities naturally incorporate the preparation time.

3. Scattering Experiments

The asymmetric time evolution provided by the semigroup operators (14) and (15) has been called *intrinsic* time asymmetry [9, 4], to distinguish it from the extrinsic time asymmetry investigated for open quantum systems [10, 11]. Because the Born probability (17) exists only for time $t \ge 0$, the time $t = 0$ is the time when an observable is just ready to be detected. Because a system described by the state $\phi(t)$ must be prepared before an observable can be detected in it, one identifies the time $t = 0$ as the preparation time of that microphysical system

[1] Specifically, one differentiates mathematically between in-states, ϕ^+, defined by a preparation apparatus (accelerator), and out-observables, ψ^-, defined by a registration apparatus (detector) [5, 6]. The sets of energy wave functions for in-states and out-observables are chosen to be $\{\langle^+E|\phi^+\rangle\} = (H^2_- \cap \mathcal{S})|_{\mathbb{R}_+}$ and $\{\langle^-E|\psi^-\rangle\} = (H^2_+ \cap \mathcal{S})|_{\mathbb{R}_+}$, respectively, where $H^2_\mp$ are the Hardy function spaces, and $\mathcal{S}$ is the Schwartz space [7].

[2] Technically, the time in (12) or (13) is limited from below by some finite time $\tilde{t}$: $t \ge \tilde{t} > -\infty$. Let us choose $\tilde{t} = 0$.

[3] The operator notation $A^\times$ signifies that the operator is an extension of a Hilbert space selfadjoint operator $A = A^\dagger$ onto the space dual to the space of vectors describing states and observables.

described by the state $\phi(t)$. At its moment of preparation, the system is represented by the state vector at $t = 0$: $\phi(t = 0)$.

For a scattering experiment, one has a detector, represented by the observable $\Lambda = |\psi^-\rangle\langle\psi^-|$, which is built to detect that observable of a scattered, microphysical system represented by a scattered state called $\phi^+(t)$. The experimentally observable quantity is the differential cross section, which is calculated as

$$d\sigma(\theta, \vartheta) = \frac{\left(\begin{array}{l} \text{transition probability per unit time} \\ \text{for scattering quanta into } \Delta\Omega \end{array} \right)\Big|_{t=0}}{\left(\begin{array}{l} \text{incident probability per} \\ \text{unit time and unit area} \end{array} \right)}. \tag{18}$$

Here, $\Delta\Omega$ is the solid angle subtended by the detector.

One calculates the transition probability rate in (18) at the time $t = 0$ [12]. The transition probability is given, according to (17), by

$$\mathcal{P}_{\phi^+(t)}\big(|\psi^-\rangle\langle\psi^-|\big) = |\langle\psi^-|\phi^+(t)\rangle|^2, \qquad 0 \leq t < \infty. \tag{19}$$

If one is to identify the time $t = 0$ in (18) as the semigroup time $t = 0$ in (19), as seems natural, then the transition probability rate in (18) is calculated precisely at the *initial* time of a scattered state, which coincides with the preparation time of a scattered, microphysical system. The initial time of a scattered state is therefore the time of scattering of the system it represents.

A consequence of the existence of a preparation time is the necessary, theoretical distinction between the time associated with a state $\phi^+(t)$ and the time associated with any external state or reservoir with which $\phi^+(t)$ might interact. The time $t = 0$ is the semigroup time of (14), which is associated with a particular state, and it is independent of the time parameterizing the evolution of any external state. For the purpose of discussion, take the external state for a scattering experiment to be the laboratory. Even in the non-relativistic case, one distinguishes between the microscopic time t of $\phi^+(t)$ and the macroscopic time marked by the clocks on the wall of the laboratory, t_{lab}. It is only the microscopic time, belonging to the state $\phi^+(t)$ (or, in the Heisenberg picture, to the observable $\psi^-(t)$) that is bounded by $t \geq 0$.

4. Ensemble of Times

Consider the theoretical description of a scattering experiment [13, 14]. The controlled in-state vector, called $\phi^{in}(t)$, represents the incoming, projectile system prepared, up to phase, by an accelerator. The uncontrolled out-state vector, $\phi^+(t)$, represents a microphysical, scattered system, and it is defined by the controlled in-state as well as by the scattering interaction. This microphysical, scattered system is prepared at the time of scattering, which is the time of beam crossing.[4] Typical scattering experiments consist of multiple beam crossings occurring over a span of days or years, as counted by the macroscopic time of the lab.

On the microphysical level, the uncontrolled out-state represents a scattered system with characteristics specified by the vector $\phi^+(t)$. One does not distinguish between two instances of that state (two different microphysical systems), prepared in an experiment, if the only difference between them is the macroscopic laboratory time, t_{lab}, of scattering. In other words, if a scattered, microphysical system is prepared during a beam crossing today, it is considered phenomenologically identical to one prepared during an identical beam crossing yesterday. These

[4] "time of beam crossing" here refers to the moment a single bunch of projectile particles arrives at the target (fixed-target experiment) or to the moment when a bunch of projectiles moving one direction crosses a bunch moving in the opposite direction (collider experiment).

separate yet phenomenologically identical instances of the state (microphysical systems) are all represented by $\phi^+(t)$.

Furthermore, every state $|\phi^+(t_1)\rangle$ evolves over time into a unique state $|\phi^+(t_2)\rangle$:

$$|\phi^+(t_2)\rangle = U^\times(t_2 - t_1)|\phi^+(t_1)\rangle, \quad t_2 > t_1. \tag{20}$$

One can choose to describe two different instances of the state $\phi^+(t)$ by two different vectors ϕ_a^+ and ϕ_b^+. However, if one wishes the two microphysical systems, and thus the two vectors, to be equivalent at some later time t_{meas} when a measurement is to be made, then by the linearity of the unitary time operators (14)

$$\begin{aligned} |\phi_a^+(t_{meas})\rangle - |\phi_b^+(t_{meas})\rangle = \quad 0 &= U^\times(t_{meas})|\phi_a^+\rangle - U^\times(t_{meas})|\phi_b^+\rangle \\ &= U^\times(t_{meas})\big(|\phi_a^+\rangle - |\phi_b^+\rangle\big). \end{aligned} \tag{21}$$

Therefore, $|\phi_a^+\rangle = |\phi_b^+\rangle$, and the two different scattering systems are in fact represented by the same state.[5] It follows that the state $\phi^+(t)$ represents an ensemble of microphysical, scattered systems.

The ensemble of systems contains members prepared during various beam crossings at various macroscopic times in the lab. One is free to consider the macroscopic time in the lab at which a given member of the ensemble was prepared, or scattered. This macroscopic time would correspond to a specific time of beam crossing, and to the ensemble of *microphysical* systems would belong an ensemble of *macrophysical* times of scattering. On the microphysical level, however, every member of the ensemble represented by $\phi^+(t)$ is prepared at the microphysical time $t = 0$.

In the non-relativistic case, one can relate the microscopic and macroscopic times quite easily. Let t^i denote the macroscopic time corresponding to the i-th member of the ensemble. Then t_0^i would be the macroscopic time of the preparation of the i-th member of the ensemble. One can write

$$t = t^i - t_0^i. \tag{22}$$

The microphysical time is t. The ensemble of macroscopic preparation times is $\{t_0\}$.

Physicists preparing data from a scattering experiment are aware of the ensemble, which is an intuitive notion. Experimental results are gathered from an ensemble of events occurring over a period of days or years. Of course, quantum mechanical calculations result in probabilities, and they do not address individual microphysical systems, which are particular instances of a state. The individual times within the ensemble $\{t_0\}$ are not reproducible.

5. Conclusion

The time asymmetric theory of quantum mechanics includes the theoretical notion of a special time, $t = 0$, corresponding to the preparation time of a system represented by a state. This time incorporates naturally into the description of scattering experiments. Its existence emphasizes the theoretical difference between time belonging to a scattered state, $\phi^+(t)$, and the time belonging to the state of an external system such as the laboratory. Phenomenologically, the preparation time corresponds to an ensemble of macroscopic *times of scattering* marking the time in the laboratory when one interaction event between one projectile and one target quantum occurs.

[5] If the microphysical system represented by ϕ_a^+ is prepared, say, at a *microphysical* time Δt later than the system represented by ϕ_b^+ is prepared, the same argument gives $|\phi_a^+\rangle = U^\times(\Delta t)|\phi_b^+\rangle$, and the same conclusion follows. Note that this microphysical (though finite) time Δt is independent of the macroscopic times in the lab at which the systems represented by ϕ_a^+ and ϕ_b^+ were prepared.

Acknowledgments
The author would like to thank the organizers of the Fifth International Symposium on Quantum Theory and Symmetries, and in particular Manuel Gadella and Mariano del Olmo, for their kind hospitality. The author is also grateful for the advice and support from Arno Bohm and Yoshihiro Sato.

References
[1] Bohm A, Loewe M and Van de Ven B 2003 *Fortschritte der Physik* **51** 551
[2] Stone M H 1932 *The Annals of Mathematics* **33** 643
[3] von Neumann J 1932 *The Annals of Mathematics* **33** 567
[4] Bohm A, Antoniou I and Kielanowski P 1995 *Journal of Mathematical Physics* **36** 2593–2604
[5] Bohm A 1981 *Journal of Mathematical Physics* **22** 2813–2823
[6] Bohm A 1978 *Letters in Mathematical Physics* **3** 455
[7] Gadella M 1983 *Journal of Mathematical Physics* **24** 1462–1469
[8] Bohm A, Kielanowski P and Wickramasekara S 2006 *Annals of Physics* **321** 2299–2317 (*Preprint* arXiv:quant-ph/0510060)
[9] Bohm A, Maxson S, Loewe M and Gadella M 1997 *Physica A Statistical Mechanics and its Applications* **236** 485–549
[10] Davis E B 1976 *Quantum Theory of Open Systems* (Academic Press Inc.)
[11] Petruccione F and Breuer H 2002 *The Theory of Open Quantum Systems* (Oxford University Press)
[12] Gell-Mann M and Goldberger M L 1953 *Phys. Rev.* **91** 398–408
[13] Taylor J R 1972 *Scattering Theory: The Quantum Theory on Nonrelativistic Collisions* (John Wiley & Sons, Inc.)
[14] Newton R G 2002 *Scattering Theory of Waves and Particles* 2nd ed (Dover Publications)

Journal of Physics: Conference Series **128** (2008) 012037 doi:10.1088/1742-6596/128/1/012037

Unstable quantum oscillator

María G. Espinosa and Piotr Kielanowski

Departamento de Física, Centro de Investigación
y de Estudios Avanzados del IPN, Mexico

E-mail: `mespinosa@fis.cinvestav.mx` and `kiel@fis.cinvestyav.mx`,

Abstract. We consider a model for an unstable quantum oscillator. The energy levels, the level widths and wave functions of the unstable states have been explicitly found. The obtained results can be applied for the description of the properties of such systems as quantum wells or series of molecular or nuclear resonances.

1. Introduction

Oscillator [1] is one of the most important systems in physics, because it describes the motion in the vicinity of a stable equilibrium. For this reason on can observe in nature many examples of oscillatory motion, both in the classical and quantum domain. The quantum oscillator appears at all levels: molecular, nuclear and in elementary particles and these systems are usually unstable. For this reason it is important to consider an unstable quantum oscillator. In this paper we will discuss a simple one dimensional model with the oscillator potential for $x \leq 0$ and a free motion for $x > 0$. At $x = 0$ we put a semi-transparent potential wall in the form of the attractive $\delta(x)$ potential with regulated strength. Such a system has a set of (almost) equidistant energy levels with complex eigenvalues that means that they are unstable. One can find the explicit values of the real and imaginary part of the energy and determine how they depend on the parameters of the potential.

The organization of the paper is the following. In Section 2 we introduce the Hamiltonian of our system and give equations for the Green's function from which we will determine the spectrum and the eigenfunctions. In Section 3 we find explicit solutions for the Green's function from which in Section 4 we find energy levels and eigenfunctions. We also give a plot of the real and imaginary part of the lowest energy levels. Section 5 contains conclusions.

2. Description of the system

The $-\Omega\,\delta(x)$ potential has interesting properties. When $\Omega < 0$ then it has exactly one bound state and for either sign of Ω it is semi transparent and as $\Omega \to \infty$ it becomes totally opaque.

Let us define the potential $V_{\mathrm{b}}(x)$ of the oscillatory barrier in the following way

$$V_{\mathrm{b}}(x) = \begin{cases} \dfrac{m\omega^2 x^2}{2} & \text{for } x \leq 0 \\ 0 & \text{for } x > 0. \end{cases} \tag{1}$$

Obviously there are no bound states for the potential defined in (1).

We now define the Hamiltonian H of the unstable oscillator

$$H = \frac{p^2}{2m} + V_\mathrm{b}(x) - \Omega\,\delta(x). \tag{2}$$

The potential of this oscillator is drawn in Figure 1 from which one can see that it allows the formation of the quasi bound states to the left of the semi-transparent wall produced by the $-\Omega\,\delta(x)$ part of the potential. In the limit $\Omega \to \infty$ the wall at $x = 0$ becomes totally impenetrable and the system has an infinite number of the oscillator bound states with energies $\hbar\omega(2n+3)/2$, *i.e.* the oscillator states that are antisymmetric under the space reflection $x \to -x$.

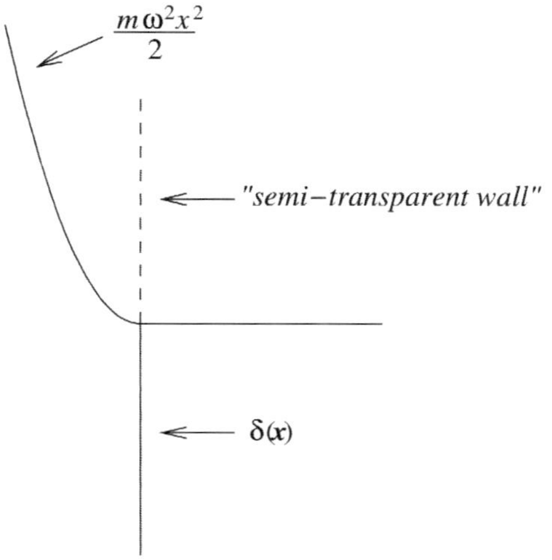

Figure 1. The potential of the unstable oscillator.

One can look for the eigenfunctions and the eigenvalues of the Hamiltonian in many possible ways but we will choose the method that is physically the most appealing *i.e.* by the determination of the Green's function of our system [3] and then by fixing the positions of the poles of the Green's function we find the eigenvalues of energy. The eigenfunctions are determined from the behaviour of the Green's function in the vicinity of the pole.

The boundary conditions for the Green's function that lead to the determination of the eigenvalues of the quasi bound states are the following

$$\mathcal{G}(x,x';E) \to \begin{cases} \text{is bounded} & \text{for } x \to -\infty \\ \sim \mathrm{e}^{\mathrm{i}kx} & \text{for } x \to +\infty. \end{cases} \tag{3}$$

We will look for the Green's function for our system in two steps:

(i) We determine the Green's function $\mathcal{G}_0(x,x';E)$ of the system without the semi-transparent wall of the pure oscillatory barrier.

(ii) The Green's function for the full Hamiltonian is given by the following relation

$$\mathcal{G}(x,x';E) = \mathcal{G}_0(x,x';E) + \frac{\mathcal{G}_0(x,0;E)\mathcal{G}_0(0,x';E)}{\frac{1}{\Omega} - \mathcal{G}_0(0,0;E)}. \tag{4}$$

Now, from (4) we see that the zeros of the denominator as the function of energy

$$\frac{1}{\Omega} - \mathcal{G}_0(0,0;E) = 0 \tag{5}$$

are the positions of the poles of the Green's function $\mathcal{G}(x,x';E)$ and thus correspond to the eigenvalues of the energy of the Hamiltonian.

Journal of Physics: Conference Series **128** (2008) 012037 doi:10.1088/1742-6596/128/1/012037

3. Green's function

We will determine the energy levels of our unstable oscillator defined in (2) from the Green's function which fulfils the equation

$$(H - E)\mathcal{G}(x, x'; E) = \delta(x - x'). \tag{6}$$

As explained earlier we have to find first the Green's function of the Schrödinger equation without the semi-transparent barrier $\mathcal{G}_0(x, x'; E)$ from the equation

$$(H_0 - E)\mathcal{G}_0(x, x'; E) = \left(\frac{p^2}{2m} + V_{\rm b}(x) - E\right)\mathcal{G}_0(x, x'; E) = \delta(x - x') \tag{7}$$

which fulfils the same boundary condition (3) as $\mathcal{G}(x, x'; E)$. In order to determine the function $\mathcal{G}_0(x, x'; E)$ one has to find two solutions of the Schrödinger equation

$$(H_0 - E)\psi^{\pm\infty}(x) = 0 \tag{8}$$

with the following behaviour at $\pm\infty$

$$\begin{array}{ll} \psi^{-\infty}(x) \text{ is bounded} & \text{for } x \to -\infty \\ \psi^{+\infty}(x) \sim {\rm e}^{{\rm i}kx} & \text{for } x \to +\infty \end{array} \tag{9}$$

and then the function $\mathcal{G}_0(x, x'; E)$ is equal

$$\mathcal{G}_0(x, x'; E) = C_0 \begin{cases} \psi^{-\infty}(x)\psi^{+\infty}(x') & \text{for } x \le x' \\ \psi^{+\infty}(x)\psi^{-\infty}(x') & \text{for } x \ge x', \end{cases} \tag{10}$$

where C_0 is a suitable constant which can be determined from the Wronskian of the two solutions $\psi^{\pm\infty}(x)$.

Equation (8) is of the oscillator type for $x \le 0$ and for $x \ge 0$ it is the equation for the free motion. Both types of equations have known solutions and thus the functions $\psi^{\pm\infty}(x)$ are equal

$$\psi^{-\infty}(x) = \begin{cases} {\rm e}^{-\frac{1}{2}\xi^2}\left({}_1F_1\left(\frac{1-2\epsilon}{4}, \frac{1}{2}; \xi^2\right) - 2|\xi|\frac{\Gamma\left(\frac{3-2\epsilon}{4}\right)}{\Gamma\left(\frac{1-2\epsilon}{4}\right)}{}_1F_1\left(\frac{3-2\epsilon}{4}, \frac{3}{2}; \xi^2\right)\right) \\ \text{for } x \le 0 \\ \frac{1}{2}\left(1 - \frac{2{\rm i}}{k}\sqrt{\frac{m\omega}{\hbar}}\frac{\Gamma\left(\frac{3-2\epsilon}{4}\right)}{\Gamma\left(\frac{1-2\epsilon}{4}\right)}\right){\rm e}^{{\rm i}kx} + \text{c.c.} \\ \text{for } x \ge 0. \end{cases} \tag{11}$$

and

$$\psi^{+\infty}(x) = \begin{cases} {\rm e}^{-\frac{1}{2}\xi^2}\left({}_1F_1\left(\frac{1-2\epsilon}{4}, \frac{1}{2}; \xi^2\right) - {\rm i}k|\xi|\sqrt{\frac{\hbar}{m\omega}}{}_1F_1\left(\frac{3-2\epsilon}{4}, \frac{3}{2}; \xi^2\right)\right) \\ \text{for } x \le 0 \\ {\rm e}^{{\rm i}kx} \quad \text{for } x \ge 0. \end{cases} \tag{12}$$

Here ${}_1F_1(\alpha, \beta; z)$ is the confluent hypergeometric function, $\Gamma(z)$ is the Euler's gamma function and ξ, ϵ and k are equal

$$\xi = \sqrt{\frac{m\omega}{\hbar}}x, \quad \epsilon = \frac{E}{\hbar\omega}, \quad k = \frac{\sqrt{2mE}}{\hbar} = \sqrt{\frac{m\omega}{\hbar}}\sqrt{2\epsilon}.$$

Journal of Physics: Conference Series **128** (2008) 012037 doi:10.1088/1742-6596/128/1/012037

Finally the constant C_0 is calculated to be

$$C_0 = \mathrm{i}\frac{2m}{\hbar^2}\frac{1}{k + 2\mathrm{i}\sqrt{\frac{m\omega}{\hbar}}\frac{\Gamma(\frac{3-2\epsilon}{4})}{\Gamma(\frac{1-2\epsilon}{4})}} \tag{13}$$

and this completes the calculation of the Green's function.

4. Energy levels and eigenfunctions

The poles on the real axis of the complex energy plane of the Green's function correspond to the energies of the bound states of the system and the continuum of states produces a discontinuity on the real energy axis. From the Green's function $\mathcal{G}_0(x, x'; E)$ in (10) and the dependence on energy E of the constant C_0 in (13) we conclude that the system without the semi-transparent wall has a continuum of states on the positive real axis of energy E, because $k \sim \sqrt{E}$. The full Green's function is given in (4) and from this it follows that the spectrum of the unstable oscillator will consist of the continuum $E \geq 0$ and the complex or real poles obtained from the zeros of the denominator, equation (5).

Non penetrable wall

Let us first discuss the limiting case $\Omega \to +\infty$, *i.e.* when the wall becomes opaque. In such a case, following (5), the poles of the Green's function are obtained from the equation

$$\mathcal{G}_0(0, 0; E) = C_0\psi^{-\infty}(0)\psi^{+\infty}(0) = i\frac{2m}{\hbar^2}\frac{1}{k + 2\mathrm{i}\sqrt{\frac{m\omega}{\hbar}}\frac{\Gamma(\frac{3-2\epsilon}{4})}{\Gamma(\frac{1-2\epsilon}{4})}} = 0. \tag{14}$$

which can be easily solved, using the fact that the gamma function $\Gamma(z)$ has simple poles for negative integers. From this we get that the energies of the bound states are

$$\frac{3-2\epsilon}{4} = -n \quad \Rightarrow \quad \epsilon = \frac{3}{2} + 2n \quad \Rightarrow \quad E = \hbar\omega\left(\frac{1}{2} + (2n+1)\right). \tag{15}$$

The energies in equation (15) correspond to the odd states of the harmonic oscillator. This result was expected from the elementary analysis of our system and the limiting property of the $\delta(x)$ potential.

Unstable states

For finite values of Ω the $\delta(x)$ potential is semi-transparent and the states within the potential well are not stable. The pole position in energy E of the full Green's function determined from (5) gives the energy and width of unstable state. Explicitly the equation for the pole positions is the following

$$\frac{1}{\Omega} - \mathrm{i}\frac{2m}{\hbar^2}\frac{1}{k + 2\mathrm{i}\sqrt{\frac{m\omega}{\hbar}}\frac{\Gamma(\frac{3-2\epsilon}{4})}{\Gamma(\frac{1-2\epsilon}{4})}} = 0. \tag{16}$$

which written using the dimensionless variables has the form

$$\frac{1}{\Omega\sqrt{\frac{m}{\omega\hbar^3}}} - \frac{2\mathrm{i}}{\sqrt{2\epsilon} + 2\mathrm{i}\frac{\Gamma(\frac{3-2\epsilon}{4})}{\Gamma(\frac{1-2\epsilon}{4})}} = 0. \tag{17}$$

Equation (17) has a series of solutions and for the dimensionless parameter $\Omega\sqrt{\frac{m}{\omega\hbar^3}}$ large, the real part of the solution ϵ is close to the value of ϵ_n for the opaque wall, given in (15). The positions of the lowest complex poles for $\Omega\sqrt{\frac{m}{\omega\hbar^3}} = 10$ are given in Figure 2. ¿From this figure one can see that the width of the oscillator grows with the value of the quantum number n.

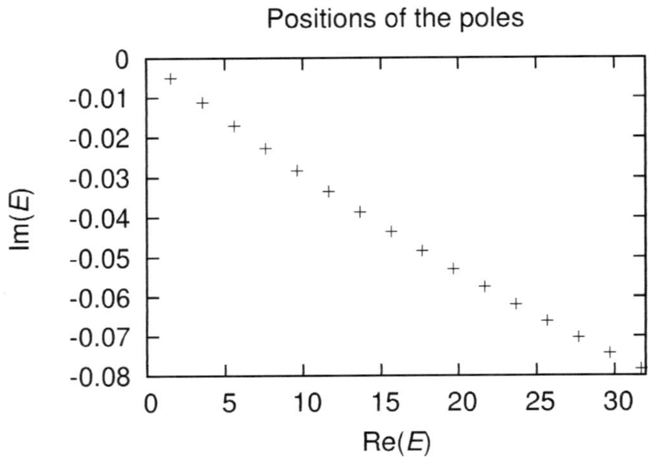

Figure 2. Positions of the poles of an unstable oscillator with the strength of the $\delta(x)$ wall equal to $\Omega\sqrt{\frac{m}{\omega\hbar^3}} = 10$.

Approximate formula for the positions of the poles

One can find approximate roots of equation (17) using the asymptotic behaviour of the Euler's gamma function. We will be looking for a solution in the vicinity of the n-th pole and denote the position of the pole by

$$\epsilon_n = \frac{3}{2} + 2n + \Delta\epsilon_n. \tag{18}$$

We then have

$$\frac{\Gamma\left(\frac{3-2\epsilon_n}{4}\right)}{\Gamma\left(\frac{1-2\epsilon_n}{4}\right)} = \frac{\Gamma\left(-n-\frac{\Delta\epsilon_n}{2}\right)}{\Gamma\left(-n-\frac{1}{2}-\frac{\Delta\epsilon_n}{2}\right)} = \frac{\cos\left(\frac{\pi\Delta\epsilon_n}{2}\right)\Gamma\left(n+\frac{3}{2}+\frac{\Delta\epsilon_n}{2}\right)}{\sin\left(\frac{\pi\Delta\epsilon_n}{2}\right)\Gamma\left(n+1+\frac{\Delta\epsilon_n}{2}\right)}$$
$$\sim \frac{\cos\left(\frac{\pi\Delta\epsilon_n}{2}\right)}{\sin\left(\frac{\pi\Delta\epsilon_n}{2}\right)}\left(n+\frac{3}{4}+\frac{\Delta\epsilon_n}{2}\right)^{\frac{1}{2}} \times \left[1+\mathrm{O}\left(\frac{1}{n^2}\right)\right] \text{ for large } n. \tag{19}$$

After inserting (19) into equation (17) one obtains

$$\frac{1}{\Omega\sqrt{\frac{m}{\omega\hbar^3}}} - \frac{2i}{\sqrt{3+4n+2\Delta\epsilon_n}\left(1+\mathrm{i}\cot\frac{\pi\Delta\epsilon_n}{2}\right)} = 0 \tag{20}$$

from which one obtains an approximate expression for the n-th pole position

$$\epsilon_n = \frac{3}{2} + 2n + \Delta\epsilon_n \approx \frac{3}{2} + 2n + \frac{2}{\pi}\arctan\left(\frac{1}{\mathrm{i}+2\Omega\sqrt{\frac{m}{\omega\hbar^3}}\frac{1}{\sqrt{3+4n}}}\right) \tag{21}$$

Formula (21) is rather accurate even for low values of n, *e.g.* for n=5, the precise value obtained from (17) is $\Delta\epsilon_5 = (0.1431 - i\,0.0335)$, while the one obtained from (21) is $\Delta\epsilon_5 = (0.1423 - i\,0.0329)$. For larger values of n the approximation becomes better.

5. Single bound state pole

This unstable harmonic oscillator has one stable state that has its origin in a single stable state of the $\delta(x)$ well. To find the energy of this state we analyse equation (17) for large negative values of ϵ. In such a case we have

$$k = \sqrt{\frac{m\omega}{\hbar}}\sqrt{2\epsilon} = \mathrm{i}\sqrt{\frac{m\omega}{\hbar}}\sqrt{2|\epsilon|} \quad \text{and} \quad \frac{\Gamma\left(\frac{3-2\epsilon}{4}\right)}{\Gamma\left(\frac{1-2\epsilon}{4}\right)} \sim \sqrt{\frac{-\epsilon}{2}} = \sqrt{\frac{|\epsilon|}{2}}$$

and after inserting these relations into (17) we obtain the position of the stable pole

$$|\epsilon| = \frac{1}{2}\frac{m\Omega^2}{\omega\hbar^3}$$

which gives the following value of the energy of the bound state

$$E = -\frac{1}{2}\frac{m\Omega^2}{\hbar^2}. \tag{22}$$

It is worth to note that the energy of this stable state coincides with the one of the pure $\delta(x)$ well without the oscillator potential.

Wave function of unstable state

The generic form of the Green's function near the pole $E = E_0$ of a bound or unstable state has the form

$$\frac{\psi(x)\psi(x')}{E - E_0}. \tag{23}$$

From (4) one can see that the part of the Green's function of our oscillator that develops the pole has the form

$$\frac{\mathcal{G}_0(x,0;E)\mathcal{G}_0(0,x';E)}{\frac{1}{\Omega} - \mathcal{G}_0(0,0;E)} \tag{24}$$

so it means that the wave function of an unstable state whose pole position is E_n has the form

$$\psi_n(x) \sim \mathcal{G}_0(x,0;E_n) \tag{25}$$

and what remains is the calculation of the normalization of the wave function. The denominator of equation (24) in the vicinity of the pole $E = E_n$ has the form

$$\frac{1}{\Omega} - \mathcal{G}_0(0,0;E) \approx N_n^2(E - E_n), \tag{26}$$

where N_n is the normalization of the wave function. From (26) one immediately obtains

$$N_n^2 = \frac{\partial}{\partial E}\left(\frac{1}{\Omega} - \mathcal{G}_0(0,0;E)\right)\bigg|_{E=E_n} = -\frac{\partial}{\partial E}\left(\mathcal{G}_0(0,0;E)\right)\bigg|_{E=E_n}. \tag{27}$$

Thus the normalization N_n of the wave function $\psi_n(x)$ is equal

$$N_n^2 = \frac{d\epsilon}{dE}\frac{\partial}{\partial\epsilon}\left(2i\sqrt{\frac{m}{\omega\hbar^3}}\frac{1}{\sqrt{2\epsilon} + 2i\frac{\Gamma(\frac{3-2\epsilon}{4})}{\Gamma(\frac{1-2\epsilon}{4})}}\right)\Bigg|_{E_n} \tag{28}$$

and the wave function of the unstable state at E_n is equal

$$\psi_n(x) = \frac{\mathcal{G}_0(x,0;E_n)}{N_n}. \tag{29}$$

Journal of Physics: Conference Series **128** (2008) 012037 doi:10.1088/1742-6596/128/1/012037

6. Conclusions

We have presented a model of an unstable quantum oscillator. The spectrum of such a model consists of a continuum $E \geq 0$ and a series of unstable levels that are in correspondence with the energy levels of an oscillator. Additionally this model has one stable state that has its origin in the stable state of the $\delta(x)$ potential that is the source of the semi-transparent wall. The wave functions of the unstable states were also explicitly obtained with proper normalization. This normalization cannot be obtained from the direct method of the determination of the wave functions as the eigenfunctions of the Schrödinger equation.

An unstable oscillator can have application in nuclear physics in the study of the equidistant series of nuclear resonances and also in molecular physics, where appear series of unstable levels. An interesting application can also be in the study of quantum wells, where appear semi-transparent parabolic walls.

Acknowledgments

MGE would like to thank CONACyT (Mexico) as well as the financial support of Cinvestav.

References

[1] Schiff L I 1968 *Quantum Mechanics* (New York: McGraw-Hill)
[2] Ludviksonn A 1987 *J. Phys. A* **20** 4733
[3] Economou E N 1983 *Green's Functions in Quantum Physics* (Berlin: Springer Verlag)

V International Symposium on Quantum Theory and Symmetries IOP Publishing
Journal of Physics: Conference Series **128** (2008) 012038 doi:10.1088/1742-6596/128/1/012038

Gamow vectors: miscellaneous results

M Gadella

Departamento de Física Teórica, Atómica y Optica. Universidad de Valladolid
Facultad de Ciencias, c. Real de Burgos, s.n.
47011 Valladolid, Spain.

E-mail: manuelgadella@yahoo.com.ar

Abstract. In this note, we discuss some mathematical properties of Gamow vectors. We start with the definition and properties of the approximate Gamow vectors that are Hilbert space vectors with similar properties to Gamow vectors and are defined by integration over the Hamiltonian spectrum and not over the whole real line, as is the case with ordinary Gamow vectors. We also discuss some properties of Gamow dyads, which are density operators constructed with Gamow vectors.

1. Introduction and motivation

The aim of this communication is to discuss some properties of Gamow vectors for quantum resonances. Here, we limit ourselves to nonrelativistic resonances following a model that has been extensively discussed in [1, 2, 3], where the definition, construction and basic properties of Gamow vectors are discussed in detail. This model for resonances requires of some basic assumptions that we can briefly summarize as follows: i.) Resonances are produced in a process of quantum resonance scattering with "free" Hamiltonian H_0 and total Hamiltonian $H = H_0+V$. ii.) We assume asymptotic completeness and the existence of Møller wave operators. iii.) Concerning the S operator, we are assuming that in the energy representation, it is given by a two valued meromorphic function $S(E)$ on a complex variable E. In the language of Riemann surfaces, $S(E)$ is a meromorphic function defined on a two sheeted Riemann surface. iv) Then, resonances are given by pairs of complex poles located in the analytic continuation of $S(E)$ on the second sheet. Each pair of conjugate poles have the form $z_R = E_R - i\Gamma/2$ and $z_R^* = E_R + i\Gamma/2$, with $E_R > 0$ and $\Gamma > 0$.

Further discussions on resonance scattering can be found in [4, 5, 6, 7].

For a resonance with resonance poles at z_R and z_R^*, we have respective Gamow vectors. To the resonance pole z_R corresponds the decaying Gamow vector, which can be defined [8, 9] as a state vector that decays exponentially with decay rate Γ or as an eigenvector of the total Hamiltonian H with eigenvalue z_R. Both notions are equivalent. Similarly, the growing Gamow vector grows exponentially with rate Γ and is an eigenvector of the total Hamiltonian H with eigenvalue z_R^*. One is the time reversal of the other. From this notion, it is clear that neither of the Gamow vectors can be represented by a vector in a Hilbert space.

In the sequel, we shall assume that we have only one resonance with resonance poles at z_R and z_R^* as above. This idea can bring all the properties of the resonance scattering with a maximum of simplicity [10] and can be implemented with the use of Friedrichs model [11] that has the important property of being exactly solvable.

Journal of Physics: Conference Series **128** (2008) 012038 doi:10.1088/1742-6596/128/1/012038

In resonance scattering, a state $\psi^{\rm in}$ is prepared in the remote past and enters in the interaction region, where a resonance is created. Pictorially, we can imagine that the state represents a particle and the idea that a resonance is created by assuming that the particle stays within the interaction region a time large compared with the time it would have spent if no interaction existed. As the particle abandons the interaction region it has a state represented by the vector $\psi^{\rm out}$, where $S\psi^{\rm in} = \psi^{\rm out}$. Both, $\psi^{\rm in}$ and $\psi^{\rm out}$ evolve freely, which means that their time evolution is governed by the free Hamiltonian H_0.

However, we cannot detect in general the whole outgoing wave, but instead its projection into a region where a detector was placed. This projection gives a state that, when properly normalized, has the form $\varphi^{\rm out}$. The amplitude between the output and the state we are really measuring is given by $\langle\varphi^{\rm out}|\psi^{\rm out}\rangle = \langle\varphi^{\rm out}|S\psi^{\rm in}\rangle$. We shall consider this amplitude as the main tool in the construction of the Gamow vectors [1, 10].

As we assume that the Møller operators exists, we can write: $\psi^- := \mathbf{\Omega}_{\rm IN}\psi^{\rm in}$ and $\varphi^+ = \mathbf{\Omega}_{\rm OUT}\varphi^{\rm out}$. Since $S = \mathbf{\Omega}^\dagger_{\rm OUT}\mathbf{\Omega}_{\rm IN}$, where the dagger represents the adjoint, we have the following formula for our amplitude:

$$\langle\varphi^{\rm out}|S\psi^{\rm in}\rangle = \langle\varphi^+|\psi^-\rangle\,. \tag{1}$$

In terms of the energy, (1) reads

$$\int_0^\infty \langle\varphi^{\rm out}|E\rangle\, S(E)\, \langle E|\psi^{\rm in}\rangle\, dE = \langle\varphi^{\rm out}|S\psi^{\rm in}\rangle = \langle\varphi^+|\psi^-\rangle\,, \tag{2}$$

where $|E\rangle$ are the generalized eigenvectors of the free Hamiltonian, $H_0|E\rangle = E|E\rangle$, $E\in[0,\infty)$ [1].

The use of generalized eigenvectors a la Dirac [12] and the notion of Gamow vectors shows the need of having a structure bigger than the Hilbert space. This notion is the rigged Hilbert space, which is, as is well known, a triplet $\mathbf{\Phi}\subset\mathcal{H}\subset\mathbf{\Phi}^\times$, where $\mathbf{\Phi}$ is a locally convex topological vector space, dense in the Hilbert space $\mathcal{H}$ and endowed with a topology strictly finer (with more open sets and therefore with less convergent sequences) that the topology on $\mathcal{H}$. The space $\mathbf{\Phi}^\times$ is the antidual of $\mathbf{\Phi}$ or space of all the continuous antilinear mappings from $\mathbf{\Phi}$ into the set of complex numbers.

Thus, we can construct [2, 3] two rigged Hilbert spaces $\mathbf{\Phi}^\pm\subset\mathcal{H}\subset(\mathbf{\Phi}^\pm)^\times$ so that we can assume that $\varphi^+\in\mathbf{\Phi}^+$ and $\psi^-\in\mathbf{\Phi}^-$, where $\mathcal{H}$ is the Hilbert space of scattering states of the total Hamiltonian H (i.e., the absolutely continuous space of H). The spaces $\mathbf{\Phi}^\pm$ are constructed so that the following conditions are satisfied:

i.) The total Hamiltonian H reduces them, i.e., $H\mathbf{\Phi}^\pm\subset\mathbf{\Phi}^\pm$.

ii.) The total Hamiltonian H is a continuous operator on $\mathbf{\Phi}^\pm$.

iii.) In the energy representation, in which the total Hamiltonian is the multiplication operator (for simplicity we assume no degeneration), the functions in $\mathbf{\Phi}^+$ and in $\mathbf{\Phi}^-$ are represented by restrictions to the real axis of Schwartz functions which are also Hardy functions [2, 3] on the upper and on the lower half planes respectively[1]. These are the so called very well behaved functions [18]. Their values on the positive semiaxis determine both their values on the negative semiaxis and also on the corresponding half plane [19].

iv.) For each real number E, there exists a vector $|E^\pm\rangle\in(\mathbf{\Phi}^\pm)^\times$ such that $H|E^\pm\rangle = E|E^\pm\rangle$, i.e., $|E^\pm\rangle$ is a generalized eigenvector[2] of H. If w^+ and w^- are in the open upper and lower half

[1] The use of Hardy functions has lately received some criticism in the sense that they are incompatible with quantum mechanics [13]. This criticism is however nonsense [14, 15, 16, 17].

[2] The extension of H into the antiduals is done via the duality formula

$$\langle H\varphi^\pm|F^\pm\rangle = \langle\varphi^\pm|HF^\pm\rangle\,, \qquad \forall\,\varphi\in\mathbf{\Phi}^\pm\,, \qquad \forall\,F^\pm\in(\mathbf{\Phi}^\pm)^\times\,.$$

of the complex plane, there are respective vectors $|w^\pm\rangle \in (\mathbf{\Phi}^\pm)^\times$ such that $H|w^\pm\rangle = w^\pm|w^\pm\rangle$. If $E \in [0,\infty)$, i.e., E is in the absolutely continuous spectrum of H, we have the Lippmann-Schwiger equations, $|E^+\rangle = \mathbf{\Omega}_{\rm OUT}|E\rangle$ and $|E^-\rangle = \mathbf{\Omega}_{\rm IN}|E\rangle$. These expressions can be generalized to similar expressions where E runs into the whole real line [2, 3]. This generalization plays no role in our discussion here.

v.) We have the following duality formulas:

$$\langle\varphi^{\rm out}|E\rangle = \langle\mathbf{\Omega}_{\rm OUT}\varphi^{\rm out}|\mathbf{\Omega}_{\rm OUT}|E\rangle = \langle\varphi^+|E^+\rangle$$

$$\langle\psi^{\rm in}|E\rangle = \langle\mathbf{\Omega}_{\rm IN}\psi^{\rm in}|\mathbf{\Omega}_{\rm IN}|E\rangle = \langle\psi^-|E^-\rangle\,. \tag{3}$$

Note that $\langle E|\psi^{\rm in}\rangle = \langle\psi^{\rm in}|E\rangle^* = \langle E^-|\psi^-\rangle = \langle\psi^-|E^-\rangle^*$, where the star denotes complex conjugation. These formulas can be extended to all real values of E [2].

After all these considerations, (2) can be written as [2, 3]

$$\langle\varphi^+|\psi^-\rangle = \int_0^\infty \langle\varphi^{\rm out}|E\rangle\, S(E)\,\langle E|\psi^{\rm in}\rangle\, dE = \int_0^\infty \langle\varphi^+|E^+\rangle\, S(E)\,\langle E^-|\psi^-\rangle\, dE$$

$$= 2\pi i s_R \langle\varphi^+|z_R^+\rangle\langle z_R^{*-}|\psi^-\rangle - \int_{-\infty}^0 \langle\varphi^+|E^+\rangle\, S_{II}(E)\,\langle E^-|\psi^-\rangle\, dE\,, \tag{4}$$

where:

i.) The complex number s_R is the residue of $S(E)$ at the pole z_R.

ii.) The functionals $|z_R^+\rangle$ and $|z_R^{*-}\rangle$ belong to the antiduals $(\mathbf{\Phi}^+)^\times$ and $(\mathbf{\Phi}^-)^\times$ respectively. They have the properties: $H|z_R^+\rangle = z_R|z_R^+\rangle$ and $H|z_R^{*-}\rangle = z_R^*|z_R^{*-}\rangle$. In addition,

$$e^{-itH}|z_R^+\rangle = e^{-itE_R}\, e^{-t\Gamma/2}|z_R^+\rangle\,, \text{if } t>0; \quad e^{-itH}|z_R^{*-}\rangle = e^{-itE_R}\, e^{+t\Gamma/2}|z_R^{*-}\rangle\,, \text{if } t<0\,,$$

so that $|z_R^+\rangle$ and $|z_R^{*-}\rangle$ are the decaying and growing Gamow vector respectively. These vectors can be written in integral form as[3] [1, 10]

$$|z_R^+\rangle = \sqrt{\frac{\Gamma}{2\pi}} \int_{-\infty}^\infty \frac{|E^+\rangle\, dE}{(E-E_R)+i\Gamma/2}\,, \qquad |z_R^{*-}\rangle = \sqrt{\frac{\Gamma}{2\pi}} \int_{-\infty}^\infty \frac{|E^-\rangle\, dE}{(E-E_R)-i\Gamma/2}\,. \tag{5}$$

As Gamow vectors are functionals, equations (5) should be understood as

$$\langle\varphi^+|z_R^+\rangle = \sqrt{\frac{\Gamma}{2\pi}} \int_{-\infty}^\infty \frac{\langle\varphi^+|E^+\rangle\, dE}{(E-E_R)+i\Gamma/2}\,, \qquad \langle\psi^-|z_R^{*-}\rangle = \sqrt{\frac{\Gamma}{2\pi}} \int_{-\infty}^\infty \frac{\langle\psi^-|E^-\rangle\, dE}{(E-E_R)-i\Gamma/2}\,, \tag{6}$$

for all $\varphi^+ \in \mathbf{\Phi}^+$ and $\psi^- \in \mathbf{\Phi}^-$.

Note that the integration goes from $-\infty$ to ∞ and this is possible because $\langle\varphi^+|E^+\rangle$ and $\langle\psi^-|E^-\rangle$ are Hardy functions [2, 3]. Gamow vectors $|z_R^+\rangle$ and $|z_R^{*-}\rangle$ are **not** in the Hilbert space $\mathcal{H}$. The complex conjugate of (4) gives:

[3] However, we are not strictly using the notation in [1, 10, 2] here. The signs for the vectors are changed while the signs for the spaces remain the same. Then, we use the same signs for spaces and their vectors and, with our choice for these signs, we endorse the intuitive idea of using sign plus for the future and sign minus for the past.

$$\langle\psi^-|\varphi^+\rangle = -2\pi i s_R^* \langle\psi^-|z_R^{*-}\rangle\langle z_R^+|\varphi^+\rangle - \int_{-\infty}^0 \langle\psi^-|E^-\rangle\, S_{II}^*(E)\, \langle E^+|\varphi^+\rangle\, dE\,. \tag{7}$$

Finally, if we omit in (4) the arbitrary vectors φ^+ and ψ^-, we obtain:

$$I = 2\pi i s_R |z_R^+\rangle\langle z_R^{*-}| - \int_{-\infty}^0 |E^+\rangle\, S_{II}(E)\, \langle E^-|\, dE\,. \tag{8}$$

This is the identity operator from $\mathbf{\Phi}^-$ into $(\mathbf{\Phi}^+)^\times$. We call $\mathcal{L}(\mathbf{\Phi}^-,(\mathbf{\Phi}^+)^\times)$ the space of linear continuous operators from $\mathbf{\Phi}^-$ into $(\mathbf{\Phi}^+)^\times$. Then, one can show [20, 21] that both terms in the right hand side of (8) belong to $\mathcal{L}(\mathbf{\Phi}^-,(\mathbf{\Phi}^+)^\times)$. In Section 3, we present a proof that the first term in the right hand side of (8) is continuous (it is obviously linear). Then, the second one must be continuous as the identity always is.

Analogously, from (7) one can get

$$I = -2\pi i s_R^* |z_R^{*-}\rangle\langle z_R^+| - \int_{-\infty}^0 |E^-\rangle\, S_{II}^*(E)\, \langle E^+|\, dE\,, \tag{9}$$

which is in $\mathcal{L}(\mathbf{\Phi}^+,(\mathbf{\Phi}^-)^\times)$ as well as both terms in the right hand side of (9). Note that both identities in (8) and (9) are different.

Operators of the form $|z_R^+\rangle\langle z_R^{*-}|$ and $|z_R^{*-}\rangle\langle z_R^+|$ are called **Gamow dyads** and were considered in [21] as a first attempt to construct a Liouvillian formalism for resonances.

In addition to (8) and (9), we have the following expressions for all integer n [20]:

$$H^n = 2\pi i s_R z_R^n\, |z_R^+\rangle\langle z_R^{*-}| - \int_{-\infty}^0 E^n\, |E^+\rangle\, S_{II}(E)\, \langle E^-|\, dE\,, \tag{10}$$

with $H^n \in \mathcal{L}(\mathbf{\Phi}^-,(\mathbf{\Phi}^+)^\times)$ and

$$H^n = -2\pi i s_R^* (z_R^*)^n\, |z_R^{*-}\rangle\langle z_R^+| - \int_{-\infty}^0 E^n\, |E^-\rangle\, S_{II}^*(E)\, \langle E^+|\, dE\,, \tag{11}$$

with $H^n \in \mathcal{L}(\mathbf{\Phi}^+,(\mathbf{\Phi}^-)^\times)$. See Section 3.

So far, the notion of Gamow dyad for simple pole resonances. This notion can be extended to multiple pole resonances [20, 21] and the generalization of (10) and (11) will be given in Section 3. The integral term in the above formulas is called the background term. This term is responsible for the non exponential decay of state vectors in Hilbert space [22] and it is always present.

This communication is organized as follows: In the next section, we shall define approximate Gamow vectors and give their properties and their relation with Gamow vectors. In Section 3, we discuss some properties of Gamow dyads. Finally, we give some concluding remarks.

2. Approximate Gamow vectors

Although the absolutely continuous spectrum of H is the positive semiaxis, the construction of Gamow vector given formula (5) implies the integration on the whole real line. This is possible because the space of test functions here considered is given by a space of Hardy functions. This integration over the whole real axis is necessary to preserve the exponential decay, which is equivalent to the Breit-Wigner energy distribution. However, as the continuous spectrum of H coincides with $[0,\infty)$, we propose to study the effect that a truncated Breit-Wigner energy distribution may have in the definition of the Gamow vector. This motivates the use of the notion of approximate Gamow vector, which can be introduced by means of the following formula:

$$f^D := \sqrt{\frac{\Gamma}{2\pi}} \int_0^\infty \frac{|E^+\rangle}{E - E_R + (i\Gamma)/2}\, dE\,, \tag{12}$$

where $H|E^+\rangle = E|E^+\rangle$. In fact, f^D is a functional on $\mathbf{\Phi}^+$ defined by its action on an arbitrary $\phi^+ \in \mathbf{\Phi}^+$ as

$$\langle \phi^+ | f^D\rangle := \sqrt{\frac{\Gamma}{2\pi}} \int_0^\infty \frac{\langle \phi^+|E^+\rangle}{E - E_R + (i\Gamma)/2}\, dE\,. \tag{13}$$

The integral in (13) is well defined because the numerator is a Schwartz function and $(E - E_R + (i\Gamma)/2)^{-1}$ is a square integrable function in the variable E. The use of the Schwarz inequality shows that f^D is a continuous functional. In fact, it is obviously antilinear and

$$|\langle \phi^+ | f^D\rangle| \le \sqrt{\frac{\Gamma}{2\pi}} \left\{ \int_0^\infty \frac{1}{|E - E_R + (i\Gamma)/2|^2}\, dE \right\}^{1/2} \left\{ \int_0^\infty |\langle \phi^+|E^+\rangle|^2\, dE \right\}^{1/2}$$

$$= C \left\{ \int_{-\infty}^\infty |\langle \phi^+|E^+\rangle|^2\, dE \right\}^{1/2} = C\, ||\langle \phi^+|E^+\rangle||_{L^2(\mathbb{R})}\,. \tag{14}$$

Since the norm in $L^2(\mathbb{R})$ is one of the seminorms that define the topology on $\mathbf{\Phi}^+$ [2, 3], we conclude that f^D is a continuous antilinear functional on $\mathbf{\Phi}^+$ and hence it belongs to the antidual, $f^D \in (\mathbf{\Phi}^+)^\times$.

Gamow vectors are not normalizable and therefore they do not belong to Hilbert space. However, our approximate decaying Gamow vectors are normalizable and henceforth vectors in the Hilbert space as we can readily show:

$$||f^D||^2 = \langle f^D | f^D\rangle = \frac{\Gamma}{2\pi} \int_0^\infty \int_0^\infty \frac{\langle E'|E\rangle\, dE'\, dE}{(E' - E_R - (i\Gamma)/2)(E - E_R + (i\Gamma)/2)}$$

$$= \frac{\Gamma}{2\pi} \int_0^\infty \int_0^\infty \frac{\delta(E - E')\, dE'\, dE}{(E' - E_R - (i\Gamma)/2)(E - E_R + (i\Gamma)/2)}$$

$$= \frac{\Gamma}{2\pi} \int_0^\infty \frac{dE}{(E - E_R)^2 + \Gamma^2/4} = \frac{\Gamma}{2\pi} \int_0^\infty \frac{dE/(\Gamma/2)^2}{\left[\frac{E-E_R}{\Gamma/2}\right]^2 + 1}\,. \tag{15}$$

With the change of variable

$$x = \frac{2(E - E_R)}{\Gamma} \Longrightarrow dx = \frac{2dE}{\Gamma}\,, \tag{16}$$

the last integral in (15) becomes

$$\frac{1}{\pi} \int_{-2E_R/\Gamma}^\infty \frac{dx}{x^2 + 1} = \frac{1}{\pi} \left(\frac{\pi}{2} + \arctan\left(\frac{2E_R}{\Gamma} \right) \right)\,, \tag{17}$$

which is a positive number smaller than one. Note that

$$\lim_{2E_R/\Gamma \mapsto \infty} ||f^D|| = 1\,. \tag{18}$$

In the sequel, we study some interesting properties of the approximate decaying Gamow vector f^D.

Proposition.- In the weak topology on $(\mathbf{\Phi}^+)^\times$, we have that:

$$\lim_{2E_R/\Gamma \mapsto \infty} f^D = i(2\pi\Gamma)^{1/2}\,|z_R^+\rangle\,, \tag{19}$$

where $|z_R^+\rangle$ is the ordinary decaying Gamow vector (5).

Proof.- For brevity, let us call $\psi^D := i(2\pi\Gamma)^{1/2}\,|z_R^+\rangle$. According to the definition of the weak topology, proving (19) is equivalent to show that

$$p_\varphi(f^D - \psi^D) := |\langle\varphi^+|f^D\rangle - \langle\varphi^+|\psi^G\rangle| \longmapsto 0 \qquad \text{as} \qquad \frac{2E_R}{\Gamma} \longmapsto \infty\,, \tag{20}$$

for any $\varphi^+ \in \mathbf{\Phi}^+$. Thus, ($z_R = E_R - i\Gamma/2$ is the position of the resonance pole)

$$p_\varphi(f^D - \psi^D) = \left| \frac{\Gamma}{\sqrt{2\pi}} \int_0^\infty \frac{\langle\varphi^+|E^+\rangle\,dE}{E - z_R} - \frac{i(2\pi)^{1/2}\Gamma}{2\pi i} \int_{-\infty}^\infty \frac{\langle\varphi^+|E^+\rangle\,dE}{E - z_R} \right|$$

$$= \frac{\Gamma}{\sqrt{2\pi}} \left| \int_{-\infty}^0 \frac{\langle\varphi^+|E^+\rangle\,dE}{E - z_R} \right| \le \frac{\Gamma}{\sqrt{2\pi}} \left\{ \int_{-\infty}^0 |\langle\varphi^+|E^+\rangle|^2\,dE \right\}^{1/2} \left\{ \int_{-\infty}^0 \frac{dE}{|E - z_R|^2} \right\}^{1/2}. \tag{21}$$

This last integral can be easily evaluated:

$$\int_{-\infty}^0 \frac{dE}{|E - z_R|^2} = \int_{-\infty}^0 \frac{dE}{(E - E_R)^2 + (\Gamma/2)^2} = \int_{-\infty}^0 \frac{dE/(\Gamma/2)^2}{\left[\frac{E - E_R}{\Gamma/2}\right]^2 + 1}\,. \tag{22}$$

With the change of variable given by (2.5), the latter integral yields to

$$\frac{2}{\Gamma} \int_{-\infty}^{-2E_R/\Gamma} \frac{dx}{x^2 + 1}\,. \tag{23}$$

Thus,

$$p_\varphi(f^D - \psi^D) \le \frac{1}{\pi} \left\{ \int_{-\infty}^0 |\langle\varphi^+|E^+\rangle|^2\,dE \right\}^{1/2} \int_{-\infty}^{-2E_R/\Gamma} \frac{dx}{x^2 + 1}\,. \tag{24}$$

The integral between brackets is constant for fixed φ^+. The second integral goes to zero as $2E_R/\Gamma \longmapsto \infty$. This proves our result.

We conclude that the approximate decaying Gamow vector, that is a regular Hilbert space vector, is a reasonable approximation to the decaying Gamow vector either when the resonance energy is high or when the width is very narrow. In this cases, the decaying Gamow vector can be very well, within a reasonable accuracy, replaced by the approximate Gamow vector. These are the good news.

Now, they come the bad news. One of the problems that Gamow vectors have to represent truly state vectors is the difficulty to define a mean value of the energy. It was Berggren who first tried to give a solution to this problem [23], which is also discussed in [24, 3]. The bad news are that we find the same difficulty when dealing with approximate decaying Gamow vectors, notwithstanding these are Hilbert space vectors. The reason is that f^D does not belong to the domain of the total Hamiltonian H. The proof of this statement is quite simple:

$$Hf^D = \sqrt{\frac{\Gamma}{2\pi}} \int_0^\infty \frac{H|E^+\rangle}{E - E_R + (i\Gamma)/2}\, dE = \sqrt{\frac{\Gamma}{2\pi}} \int_0^\infty \frac{E|E^+\rangle}{E - E_R + (i\Gamma)/2}\, dE\,, \tag{25}$$

Then, taking the square norm, we get:

$$||Hf^D||^2 = \frac{\Gamma}{2\pi} \int_0^\infty \frac{E^2\, dE}{(E - E_R)^2 + \Gamma^2/4} = \infty\,, \tag{26}$$

so that Hf^D is not a vector in Hilbert space. In this circumstances, we cannot have a clear definition of $\langle f^D | Hf^D \rangle$. Even if we pose:

$$\langle f^D | Hf^D \rangle := \frac{\Gamma}{2\pi} \int_0^\infty \int_0^\infty \frac{E\langle E'|E\rangle\, dE\, dE'}{(E' - E_R - i\Gamma/2)(E - E_R + i\Gamma/2)} = \frac{\Gamma}{2\pi} \int_0^\infty \frac{E\, dE}{(E - E_R)^2 + \Gamma^2/4} \tag{27}$$

and perform the change of variable given by (16), we get:

$$\langle f^D | Hf^D \rangle = \frac{E_R}{\pi} \int_{-2E_R/\Gamma}^\infty \frac{dx}{x^2 + 1} + \frac{\Gamma}{2\pi} \int_{-2E_R/\Gamma}^\infty \frac{x\, dx}{x^2 + 1}\,. \tag{28}$$

The second integral in (28) diverges. In the limit $2E_R/\Gamma \longmapsto \infty$, this second integral has a vanishing Cauchy principal value. In this limit, the first integral is equal to π and this gives $\langle f^D | Hf^D \rangle \longmapsto E_R$, a result already obtained in [24].

Needless to say that a similar definition can be given for the approximate growing Gamow vector f^G. Its properties are similar to f^D. In addition, if T is the time reversal operator, one can show that $T|E^\pm\rangle = |E^\mp\rangle$ [25] and therefore, $Tf^D = f^G$ and $Tf^G = f^D$.

Finally, note that f^D cannot decay exponentially as is a vector in Hilbert space. The decaying formula for f^D is

$$\langle f^D | e^{-itH} f^D \rangle = \frac{\Gamma}{2\pi} \int_0^\infty \frac{e^{-itE_R}\, dE}{(E - E_R)^2 + \Gamma^2/4}\,. \tag{29}$$

Again, if we use the change of variables (16) and take the limit $2E_R/\Gamma \longmapsto \infty$ in (29), we obtain $e^{-itE_R}\, e^{-t\Gamma/2}$, by using the properties of the Fourier transform. This decaying behavior coincides with the decaying mode of $|z_R^+\rangle$, an expected result.

3. Gamow dyads: some properties

The use of Gamow dyads for multiple pole resonances suggest us the convenience of changing the notation used so far for Gamow vectors and that has its origin in the work of Arno Bohm [10]. Although this notation can be generalized for multiple pole resonances [27], we believe that a new one is simpler and even more practical. Thus, we write:

$$\sqrt{2\pi i s_R}\, |z_R^+\rangle = |f_0\rangle\,, \qquad \sqrt{-2\pi i s_R^*}\, |z_R^{*+}\rangle = |\widetilde{f}_0\rangle\,. \tag{30}$$

This notation has been already used in a certain number of publications, see for instance [20, 21, 26, 3]. If the resonance is described by a pole of order n, then, there are n decaying Gamow vectors, $|f_0\rangle, |f_1\rangle, \dots |f_{n-1}\rangle$ and n growing Gamow vectors, $|\widetilde{f}_0\rangle, |\widetilde{f}_1\rangle, \dots |\widetilde{f}_{n-1}\rangle$. Only $|f_0\rangle$ decays exponentially and only $|\widetilde{f}_0\rangle$ grows exponentially. The exponential decay of $|\widetilde{f}_1\rangle, \dots |\widetilde{f}_{n-1}\rangle$ as well as the exponential grow of $|\widetilde{f}_1\rangle, \dots |\widetilde{f}_{n-1}\rangle$ is modified by a multiplicative polynomial on time [27, 20, 26, 3].

The study of properties of multiple pole resonances is mostly due to A. Mondragón and his group [28]. They use the name of *degenerate resonances* instead of multiple pole resonances.

In the case of the presence of a multiple pole resonance, with resonance poles at z_R and z_R^* with multiplicity equal to p (both complex conjugate resonance poles must have the same multiplicity), formulas (10) and (11) can be easily generalized

$$H^n = \sum_{k=0}^{p-1} z_R^n |f_k\rangle\langle\widetilde{f}_k| + \sum_{s=1}^{p-2} |f_s\rangle\langle\widetilde{f}_{s+1}| - \int_{-\infty}^{0} E^n S_{II}(E)|E^+\rangle\langle E^-|\,dE \tag{31}$$

and

$$H^n = \sum_{k=0}^{p-1} (z_R^*)^n |\widetilde{f}_k\rangle\langle f_k| + \sum_{s=1}^{p-2} |\widetilde{f}_s\rangle\langle f_{s+1}| - \int_{-\infty}^{0} E^n S_{II}^*(E)|E^-\rangle\langle E^+|\,dE\,, \tag{32}$$

respectively, where n is a natural number. Exactly as in the case of a simple pole resonance, the operator in (31) is in $\mathcal{L}(\mathbf{\Phi}^-, (\mathbf{\Phi}^+)^\times)$ and the operator in (32) is in $\mathcal{L}(\mathbf{\Phi}^+, (\mathbf{\Phi}^-)^\times)$, as we shall show in the last theorem of this section. Note that in both cases the background integral remains the same.

In order to abbreviate the notation, we shall use

$$|w\rangle := \sqrt{-S_{II}(w)}\,|w^+\rangle\,, \qquad |\widetilde{w}\rangle := \sqrt{-S_{II}^*(w)}\,|w^-\rangle\,, \qquad w \equiv E\,, \tag{33}$$

so that the background terms are written as

$$\int_{-\infty}^{0} w\,|w\rangle\langle\widetilde{w}|\,dw\,, \qquad \int_{-\infty}^{0} w\,|\widetilde{w}\rangle\langle w|\,dw\,, \tag{34}$$

for (31) and (32) respectively.

The background term cannot be eliminated and this is a consequence of our next result.

Theorem.- We cannot find $f(w) \in L^p(\mathbb{R}^-)$ with $1 \le p \le \infty$, $\mathbb{R}^- \equiv (-\infty, 0]$, nor $f(w)$ equal to a polynomial such that

$$|f_0\rangle\langle\widetilde{f}_0| = \int_{-\infty}^{0} f(w)\,|w\rangle\langle\widetilde{w}|\,dw\,. \tag{35}$$

Proof.- Pick a $\psi^- \in \mathbf{\Phi}^-$ such that $\langle\psi^-|\widetilde{f}_0\rangle \neq 0$. Let φ^+ arbitrary in $\mathbf{\Phi}^+$. Then, using (33) and omitting the irrelevant sign minus, we have

$$\langle\varphi^+|f_0\rangle\langle\widetilde{f}_0|\psi^-\rangle = \int_{-\infty}^{0} f(w)S_{II}(w)\,\langle\varphi^+|w^+\rangle\langle w^-|\psi^-\rangle\,dw\,. \tag{36}$$

The functions $\langle\varphi^+|w^+\rangle$ and $\langle w^-|\psi^-\rangle$ are in the Schwartz space by hypothesis [2, 3]. Also, by hypothesis, the function $S_{II}(w)$ is polynomially bounded at the infinite. This means that for large $|w|$, there exists a polynomial $P(w)$ such that $|S_{II}(w)| \le |P(w)|$ [2, 3]. Then, if $f(w)$ is either in $L^p(\mathbb{R}^-)$ or is a polynomial, the function $g(w) := f(w)S_{II}(w)\langle w^-|\psi^-\rangle \in L^2(\mathbb{R}^-)$, i.e. it is square integrable in the negative semiaxis. For simplicity, let us write $\varphi^+(w) := \langle w^+|\varphi^+\rangle = \langle\varphi^+|w^+\rangle^*$. Then, (36) and the Schwarz inequality give

$$|\langle\varphi^+|f_0\rangle|\,|\langle\widetilde{f}_0|\psi^-\rangle| \le \left|\int_{-\infty}^{0} [\varphi^+(w)]^*\,g(w)\,dw\right| \le ||g(w)||_-\,||\varphi^+(w)||_-\,, \tag{37}$$

where $||-||_-$ denotes the norm in $L^2(\mathbb{R}^-)$.

Journal of Physics: Conference Series **128** (2008) 012038 doi:10.1088/1742-6596/128/1/012038

We recall that $\langle\varphi^+|f_0\rangle$ is defined as follows: if $\varphi^\#(w) := [\varphi^+(w)]^*$ for real w, then $\langle\varphi^+|f_0\rangle = \varphi^\#(z_R)$. Thus, (37) shows that for any $\varphi^+ \in \mathbf{\Phi}^+$, there exists a positive constant such that

$$|\varphi^\#(z_R)| = |\langle\varphi^+|f_0\rangle| \leq C\,||\varphi^+(w)||_- . \tag{38}$$

According to our hypothesis, $\varphi^+(w) \in \mathcal{H}^2_+ \cap \mathcal{S}$, where $\mathcal{H}^2_+$ is the space of Hardy functions in the lower half plane [29] and $\mathcal{S}$ is the Schwartz space [2, 3]. This means in particular that $\varphi^+(w)$ is defined for all real values of w. We have shown in [2] that the Gamow vector $|f_0\rangle$ cannot be a continuous functional on $L^2(\mathbb{R}^+)$, so that the inequality $|\varphi^\#(z_R)| = |\langle\varphi^+|f_0\rangle| \leq C\,||\varphi^+(w)||_+$ for all $\varphi^+ \in \mathcal{H}^2_+ \cap \mathcal{S}$ is false, where $||-||_+$ denotes the norm on $L^2(\mathbb{R}^+)$. A similar argument shows that (38) is also false and this contradicts the hypothesis (35). Our result is proven. ♠

Corollary 1.- In the case of multiple pole resonances, one can use a similar argument to show that under the hypothesis of the theorem, the identity

$$|f_k\rangle\langle\widetilde{f}_k| = \int_{-\infty}^0 f(w)\,|w\rangle\langle\widetilde{w}|\,dw \tag{39}$$

is also false for natural k.

Corollary 2.- This result can also be extended to any linear combination of dyads of the form $|f_k\rangle\langle\widetilde{f}_k|$.

For simplicity, let us prove this result for the simplest case $|f_0\rangle\langle\widetilde{f}_0| + |f_1\rangle\langle\widetilde{f}_1|$. The proof for any other linear combinations of dyads is similar. We need to show that the relation

$$\langle\varphi^+|f_0\rangle\langle\widetilde{f}_0|\psi^-\rangle + \langle\varphi^+|f_1\rangle\langle\widetilde{f}_1|\psi^-\rangle = \int_{-\infty}^0 f(w)S_{II}(w)\,\langle\varphi^+|w^+\rangle\langle w^-|\psi^-\rangle\,dw \tag{40}$$

is impossible for arbitrary $\varphi^+ \in \mathbf{\Phi}^+$ and $\psi^- \in \mathbf{\Phi}^-$. This case can be obviously reduced to the previous one, provided that we find one $\psi^- \in \mathbf{\Phi}^-$ such that $\langle\widetilde{f}_0|\psi^-\rangle = 0$ and $\langle\widetilde{f}_1|\psi^-\rangle \neq 0$ and another with $\langle\widetilde{f}_0|\psi^-\rangle \neq 0$ and $\langle\widetilde{f}_1|\psi^-\rangle = 0$.

Note that for real w, $\psi^\#(w) := [\psi^-(w)]^*$ with $\psi^\#(w) \in \mathcal{H}^2_+ \cap \mathcal{S}$ and $\langle\varphi^+|f_0\rangle = \psi^\#(z_R^*)$ and $\langle\widetilde{f}_1|\psi^-\rangle = (\psi^\#)'(z_R^*)$, where the prime means derivative. Thus,

i.) we have to find a $\psi^\#(w) \in \mathcal{H}^2_+ \cap \mathcal{S}$ such that $\psi^\#(z_R^*) = 0$ and $(\psi^\#)'(z_R^*) \neq 0$ and

ii.) we have to find a $\psi^\#(w) \in \mathcal{H}^2_+ \cap \mathcal{S}$ such that $\psi^\#(z_R^*) \neq 0$ and $(\psi^\#)'(z_R^*) = 0$.

Note that there are always functions $\varphi^+(w) \in \mathcal{H}^2_+ \cap \mathcal{S}$ such that $\varphi^+(z_R^*) \neq 0$. If this were false and all $\varphi^+(w) \in \mathcal{H}^2_+ \cap \mathcal{S}$ fulfilled $\varphi^+(z_R^*) = 0$, by the Titchmarsh theorem we would have:

$$\varphi^+(z_R^*) = \frac{1}{2\pi i}\int_{-\infty}^{\infty} \frac{\varphi^+(w)\,dw}{E - z_R^*} = 0\,. \tag{41}$$

This means that the arbitrary function $\varphi^\#(w) = [\varphi^+(w)]^* \in \mathcal{H}^2_- \cap \mathcal{S}$ is orthogonal in $L^2(\mathbb{R})$ to $(E - z_R^*)^{-1}$. Since $\mathcal{H}^2_- \cap \mathcal{S}$ is dense in $\mathcal{H}^2_-$ and $L^2(\mathbb{R}) = \mathcal{H}^2_+ \oplus \mathcal{H}^2_-$ [29] this implies that $(E - z_R^*)^{-1} \in \mathcal{H}^2_+$, which is false because $(E - z_R^*)^{-1}$ is not analytic on the open upper half plane. Then,

i.) Take $\eta(w) \in \mathcal{H}^2_+ \cap \mathcal{S}$ with $\eta(z_R^*) \neq 0$ and define $\psi^\#(w) := (w - z_R^*)\,\eta(w)$. Clearly, $\psi^\#(z_R^*) = 0$ and $(\psi^\#)'(z_R^*) = \eta(z_R^*) \neq 0$.

ii.) Take now $\psi^\#(w) := (w - z_R^*)\,\eta'(w) - \eta(w)$, with the same $\eta(w)$ as in i.). Then, $\psi^\#(z_R^*) = -\eta(z_R^*) \neq 0$ and $(\psi^\#)'(z_R^*) = 0$.

Thus, corollary 2 is proven.

Remark.- The proof of the above theorem applies also to those $f(w)$ for which $f(w)\,S_{II}(w)\,\langle w^-|\psi^-\rangle \in L^2(\mathbb{R}^-)$. This include, in particular, an $f(w)$ continuous and

Journal of Physics: Conference Series **128** (2008) 012038 doi:10.1088/1742-6596/128/1/012038

polynomially bounded at the infinite. In addition, if $f(w) = \delta(w-\alpha)$ with α real, our result is also true. A similar result can be obtained for dyads of the form $|\widetilde{f_k}\rangle\langle f_k|$.

The conclusion of the above results is clear, we cannot avoid the background term, which can not be reduced to a Gamow of a linear combination of Gamows. This shows that the integral term cannot be eliminated in the expressions for H^n in (31) and (32). This remark includes also the integral term in the evolution operator. The explicit for for this operator in the decaying case ($t > 0$) and multiple pole resonance is

$$U_t = \int_{-\infty}^{0} e^{-iwt}\,|w\rangle\langle\widetilde{w}|\,dw + e^{-itz_R}\left[\sum_{k=0}^{n-1}|f_k\rangle\langle\widetilde{f_k}|\right.$$

$$\left. -it|f_1\rangle\langle\widetilde{f_2}| + \frac{(-it)^2}{2}\,|f_2\rangle\langle\widetilde{f_3}| + \dots \frac{(-it)^{n-2}}{(n-2)!}\,|f_{n-2}\rangle\langle\widetilde{f}_{n-1}|\right]. \tag{42}$$

This operator belongs to $\mathcal{L}(\mathbf{\Phi}^-,(\mathbf{\Phi}^+)^\times)$ for $t > 0$. In the case of a simple pole resonance, only the term with $k = 0$ survives from (42). For the growing part a similar formula holds, which is time reversal of (42) [30] and is an operator in $\mathcal{L}(\mathbf{\Phi}^+,(\mathbf{\Phi}^-)^\times)$. Note that the function $f(w)$ in the integral term is of the type mentioned in the remark as it has modulus equal to one.

Next, we are giving a result concerning some mathematical properties of the Gamow dyads.

Theorem.- **1**.- The operator $\left\{\begin{array}{c}|f_k\rangle\langle\widetilde{f_k}| \\ |\widetilde{f_k}\rangle\langle f_k|\end{array}\right\}$ belongs to $\left\{\begin{array}{c}\mathcal{L}(\mathbf{\Phi}^-,(\mathbf{\Phi}^+)^\times) \\ \mathcal{L}(\mathbf{\Phi}^+,(\mathbf{\Phi}^-)^\times)\end{array}\right\}$, for $k = 0,1,\dots,n-1$ (assuming that the resonance pole is of order n), although it cannot be extended to an operator in $\left\{\begin{array}{c}\mathcal{L}(\mathbf{\Phi}^+)^\times \\ \mathcal{L}(\mathbf{\Phi}^-)^\times\end{array}\right\}$, which are the spaces of linear operators on $\left\{\begin{array}{c}(\mathbf{\Phi}^+)^\times \\ (\mathbf{\Phi}^-)^\times\end{array}\right\}$ continuous with the weak topology.

2.- The operator $\int_{-\infty}^{0} e^{-itw}\,|w\rangle\langle\widetilde{w}|\,dw$ is in $\mathcal{L}(\mathbf{\Phi}^-,(\mathbf{\Phi}^+)^\times)$ for **all** values of time t. Same is true for $\int_{-\infty}^{0} e^{-itw}\,|\widetilde{w}\rangle\langle\, w|\,dw$ and $\mathcal{L}(\mathbf{\Phi}^+,(\mathbf{\Phi}^-)^\times)$.

3.- The operators H^n in (10) and (31) belong to $\mathcal{L}(\mathbf{\Phi}^-,(\mathbf{\Phi}^+)^\times)$ and the operators H^n in (11) and (32) belong to $\mathcal{L}(\mathbf{\Phi}^+,(\mathbf{\Phi}^-)^\times)$ for all $n = 0,1,2,\dots$.

Proof.- 1.- We present the proof for $|f_0\rangle\langle\widetilde{f_0}|$ only, because the proof for other dyads is similar. We recall that a linear operator $B : \mathbf{\Phi}^- \longrightarrow (\mathbf{\Phi}^+)^\times$ is continuous if and only if for any continuous seminorm p on $(\mathbf{\Phi}^+)^\times$, there exists a constant C and n seminorms on $\mathbf{\Phi}^-$, $p_1, p_2, \dots, p_n$ such that

$$p(B\psi^-) \le C\{p_1(\psi^-) + p_2(\psi^-) + \dots + p_n(\psi^-)\}. \tag{43}$$

Take an arbitrary $\psi^- \in \mathbf{\Phi}^-$ and consider its image by $|f_0\rangle\langle\widetilde{f_0}|$, $|f_0\rangle\langle\widetilde{f_0}|\psi^-\rangle \in (\mathbf{\Phi}^+)^\times$. Then, (43) reads here as

$$p(|f_0\rangle\langle\widetilde{f_0}|\psi^-\rangle) \le C\{p_1(\psi^-) + p_2(\psi^-) + \dots + p_n(\psi^-)\}. \tag{44}$$

We have to show that the inequality (44) holds. Since $(\mathbf{\Phi}^-)^\times$ is endowed with the weak topology with respect to the dual pair $(\mathbf{\Phi}^-,(\mathbf{\Phi}^-)^\times)$, which in our case coincides with the strong topology because $(\mathbf{\Phi}^-)^\times$ is the dual of a nuclear Fréchet space, there is a $\eta^- \in \mathbf{\Phi}^-$ such that $p(\psi^-) = |\langle\eta^-|\psi^-\rangle|$, for all $\psi^- \in \mathbf{\Phi}^-$. Thus,

$$p(|f_0\rangle\langle\widetilde{f_0}|\psi^-\rangle) = |\langle\eta^-|f_0\rangle|\,|\langle\widetilde{f_0}|\psi^-\rangle|. \tag{45}$$

Since $|\langle\eta^-|f_0\rangle| = |\eta^\#(z_R^*)|$ is independent of ψ^-, (44) and (45) implies that $|f_0\rangle\langle\widetilde{f_0}|$ is continuous if and only if

$$|\langle\widetilde{f_0}|\psi^-\rangle| \leq C'\{p_1(\psi^-) + p_2(\psi^-) + \ldots + p_n(\psi^-)\}\,, \tag{46}$$

where C' is a constant. Then, carry (46) into (45) and (44) is proven. Thus, the continuity of $|\widetilde{f_0}\rangle$ on $\mathbf{\Phi}^-$ shows the existence of the constant C' and the seminorms $p_1, p_2, \ldots, p_n$, so that the continuity of $|f_0\rangle\langle\widetilde{f_0}|$ is proven.

To show that $|f_0\rangle\langle\widetilde{f_0}|$ cannot be in $\mathcal{L}(\mathbf{\Phi}^-)^\times$, we assume the contrary. Then, (44) and hence (46) would be true being this time $p_1, p_2, \ldots, p_n$ continuous seminorms on $(\mathbf{\Phi}^-)^\times$ (note that $\mathbf{\Phi}^- \subset (\mathbf{\Phi}^-)^\times$). If this were the case, the mapping $\psi^- \longmapsto \langle\widetilde{f_0}|\psi^-\rangle$ would have been continuous even when $\mathbf{\Phi}^-$ has the topology inherited from $(\mathbf{\Phi}^-)^\times$, which is false, since this mapping is not continuous when $\mathbf{\Phi}^-$ has the stronger Hilbert space topology [2].

Remark.- Note that this means in particular that objects like $\langle f_0|f_0\rangle$ and $\langle\widetilde{f_0}|\widetilde{f_0}\rangle$ cannot be defined in principle. At least, no by means of continuous extensions. These objects have been proposed in the literature (see [31] and references therein), although they have a doubtful sense.

2.- The technique for proving this second part is essentially the same of what we have applied in the former. Only that we have to take into account that the function $S_{II}(w)$ is bounded on the negative semiaxis of the second Riemann sheet [5]. Then, if p is a seminorm on $(\mathbf{\Phi}^+)^\times$ and $\eta^+ \in \mathbf{\Phi}^+$ with $p(\varphi^+) = |\langle\eta^+|\varphi^+\rangle|$, for all $\varphi^+ \in \mathbf{\Phi}^+$, we have

$$p\left(\int_{-\infty}^0 e^{-itw}\,|w\rangle\langle\widetilde{w}|\psi^-\rangle\,dw\right) = \left|\int_{-\infty}^0 e^{-itw}\,\langle\eta^+|w\rangle\langle\widetilde{w}|\psi^-\rangle\,dw\right|$$

$$\leq \int_{-\infty}^0 |\langle\eta^+|w\rangle|\,|\langle\widetilde{w}|\psi^-\rangle|\,dw = \int_{-\infty}^0 |S_{II}(w)|\,|\langle\eta^+|w^+\rangle|\,|\langle w^-|\psi^-\rangle|\,dw$$

$$\leq C\int_{-\infty}^0 |\langle\eta^+|w^+\rangle|\,|\langle w^-|\psi^-\rangle|\,dw \leq C\int_{-\infty}^\infty |\langle\eta^+|w^+\rangle|\,|\langle w^-|\psi^-\rangle|\,dw$$

$$\leq C\,||\eta^+||\,||\psi^-||\,, \tag{47}$$

where $||-||$ is the $L^2(\mathbb{R})$ norm. This proves the second part. We want to stress that this functional is well defined for *all* values of t and not only for $t > 0$.

3.- For $n = 0$, $H^0 = I$ the identity, which is continuous because the topology in the antiduals is weaker than the topology on the Hilbert space, which is weaker than the topology on the test spaces $\mathbf{\Phi}^\pm$ (note that $\mathbf{\Phi}^\pm \subset \mathcal{H} \subset (\mathbf{\Phi}^\mp)^\times$, which are similar to rigged Hilbert spaces). Then, we discuss the nontrivial case $n = 1, 2, \ldots$. If we note that $f(w)S_{II}(w)\langle w^-|\psi^-\rangle \in L^2(\mathbb{R}^-)$, using a similar argument as in the second part of the present theorem, the result follows. The continuity of the operators H^n is proven in all cases.

Concluding remarks

This communication deals with two problems in relation with Gamow vectors. In the first case, we define the notion of approximate Gamow vectors by restricting the integration in the usual definition of Gamow vectors to the positive real axis. This makes sense from the point of view that the positive semiaxis is the continuous spectrum of the Hamiltonian and integration over the whole real line needs analytic continuation of wave functions in the energy representation. Then, when we integrate over the interval $[0, \infty)$, we obtain vectors in the Hilbert space, which in some sense approximate the Gamow vectors and for this reason are called the approximate Gamow vectors. Approximate and exact Gamow vectors coincide in the weak limit when E_R/Γ goes to $+\infty$, where E_R is the resonant energy and Γ the width.

In the second half, we study some mathematical properties of the so called Gamow dyads. In particular, we study the continuity of these objects as operators on a space of test vectors into a space of functionals. We also discuss the nontriviality of the background in the sense that it cannot be described as a Gamow dyad or a linear combination of Gamow dyads (in the case of multiple pole resonances) and consequently, it is something different from a resonance. This fact is well known, but this point of view is new.

Acknowledgements
Partial financial support is acknowledged to the Junta de Castilla y León Project VA013C05, the Ministry of Education and Science of Spain projects MTM2005-09183 and FIS2005-03988.

References

[1] Bohm A 1994 *Quantum Mechanics: Foundations and Applications* (Springer, Berlin)
[2] Bohm A, Gadella M 1989 *Dirac kets, Gamow vectors and Gelfand triplets*, (Springer Lecture Notes in Physics, **348**. Springer Verlag, Berlin)
[3] Civitarese O, Gadella M 2004 *Physics Reports* **396** 43
[4] Newton RG 1982 *Scattering Theory of Waves and Particles* (Springer, Berlin)
[5] Nussenzveig H M 1972 *Causality and Dispersion Relations* (Academic, New York)
[6] Amrein W O, Jauch J, Sinha B 1977 *Scattering Theory in Quantum Mechanics* (Benjamin, Reading MA)
[7] Baumgärtel H, Wollenberg H 1983 *Mathematical Scattering Theory* (Birkhäuser, Bassel, Boston, Sttutgart)
[8] Gamow G 1958 *Zeitschrift für Physik* **51** 204
[9] Nakanishi N 1958 *Progress on Theoretical Physics* **19** 607
[10] Bohm A 1981 *Journal of Mathematical Physics* **21** 2813
[11] Friedrichs K O 1948 *Communications in Pure and Applied Mathematics* **1** 361
[12] Gadella M, Gómez F 2002 *Foundations of Physics* **32** 815
[13] de la Madrid R 2006 *Journal of Physics A: Mathematical and Theoretical* **39** 9255
[14] Gadella M, Wickramasekara S 2007 *Journal of Physics A: Mathematical and Theoretical* **40** 4665
[15] Gadella M, Wickramasekara S arXiv:0707.0778.
[16] Baumgärtel H arXiv:0704.3529v1.
[17] Gadella M, Gómez F, Wickramasekara S, *Hardy class functions for potential scattering and decay, Reports on Mathematical Physics* (in press).
[18] Bohm A, Kaldass H, Wickramasekara S 2003 *Fortschritte der Physik* **51** 569
[19] van Winter C 1974 *Journal of Mathematical Analysis* **47** 633
[20] Gadella M 1997 *International Journal of Theoretical Physics* **36** 2271
[21] Gadella M, Laura R 2001 *International Journal of Quantum Chemistry* **81** 307
[22] Fonda L, Ghirardi G C, Rimini A 1978 *Reports on Progress in Physics* **41** 587
[23] Berggren T 1968 *Nuclear Physics A* **109** 265
[24] Civitarese O, Gadella M, Id Betan R 1999 *Nuclear Physics A* **660** 255
[25] Castagnino M, Gadella M, Id Betán R, Laura R 2001 *Journal of Physics A: Mathematical and General* **34**, 10067
[26] Antoniou I, Gadella M, Pronko G P 1998 *Journal of Mathematical Physics* **39** 2459
[27] Bohm A, et al, 1997 *Journal of Mathematical Physics* **38**, 6072
[28] Hernández E, Jáuregui A, Mondragón A 2000 *Journal of Physics A: Mathematical and General* **33** 4507; Hernández E, Mondragón A, Jáuregui A 2002 *Revista Mexicana de Física* **48** 6; Hernández E, Jáuregui A, Mondragón A 2003 *Physica Review A* **67** 022721; Hernández E, Jáuregui A, Mondragón A 2005 *Physical Review E* **72** 026221.
[29] Koosis P 1980 *Introduction to* $\mathcal{H}^p$ *spaces* (Cambridge, UK)
[30] Gadella M, de la Madrid R 1999 *International Journal of Theoretical Physics* **38** 93
[31] Castagnino M, Gadella M, Gaioli F, Laura R 1999 *International Journal of Theoretical Physics* **38** 2823

Journal of Physics: Conference Series **128** (2008) 012039 doi:10.1088/1742-6596/128/1/012039

Eigenfunction expansions and scattering theory in rigged Hilbert spaces[1]

F Gómez-Cubillo

Dpt. de Análisis Matemático, Universidad de Valladolid. Facultad de Ciencias, 47011 Valladolid, Spain

E-mail: fgcubill@am.uva.es

Abstract. The work reviews some mathematical aspects of spectral properties, eigenfunction expansions and scattering theory in rigged Hilbert spaces, laying emphasis on Lippmann-Schwinger equations and Schrödinger operators.

1. Introduction

The abstract eigenfunction expansion problem for selfadjoint operators has been widely investigated and there is a rich literature on the subject. See, for instance, [1–5]. Generalized eigenfunctions corresponding to values in the continuous spectrum are not in the Hilbert space $\mathcal{H}$ the operator is defined on and auxiliary spaces are necessary. The method depends on the introduction of a so-called *rigged Hilbert space* or *Gelfand triplet*

$$\Phi \subset \mathcal{H} \subset \Phi^{\times} .$$

The subspace Φ of $\mathcal{H}$ is provided with its own linear topology τ, stronger than that one induced by $\mathcal{H}$, so that its adjoint space $\Phi^{\times}$, the set of all continuous antilinear forms on (Φ, τ), includes $\mathcal{H}$. The eigenfunctions will be elements of $\Phi^{\times}$. The method is introduced in Sections 2 and 3 below, from spectral measures to eigenfunction expansions, with certain Radon-Nykodym derivatives as guiding elements.

The main objects in scattering theory [6–8] are the wave operators $\Omega_{\pm}$. They are defined in the time-dependent theory by (strong, weak, Abelian) limits of the form

$$\Omega_{\pm} := \lim_{t\to\pm\infty} e^{itH} e^{-itH_0} ,$$

where H_0, H are selfadjoint operators acting in a Hilbert space $\mathcal{H}$. The stationary scattering theory defines wave operators in terms of the resolvents $(H_0 - z)^{-1}$, $(H - z)^{-1}$. The central questions of existence, completeness and relationship between the various wave operators have been extensively studied. Section 4 is an introductory review of the subject.

Mathematical reasons suggesting the study of wave operators result from the fact that $\Omega_{\pm}$ are intertwining operators for H_0, H. The existence of a nontrivial intertwining operator implies the unitary equivalence of nontrivial parts of H_0 and H. The intertwining property of wave

[1] Dedicated to Professor Arno Bohm on the occasion of his seventieth anniversary

Journal of Physics: Conference Series **128** (2008) 012039 doi:10.1088/1742-6596/128/1/012039

operators is in the basis of the deep connection between scattering theory and spectral analysis. In particular, eigenfunction expansions are main tools in scattering theory and, conversely, wave operators are in the core of abstract Lippmann-Schwinger equations to obtain generalized eigenfunctions. A rigorous description of these deep connections in the framework of rigged Hilbert spaces is given in Section 5 below.

Finally, by way of application, Section 6 deals with Schrödinger pairs $H_0 = -\Delta$, $H = -\Delta + V$ in $L^2(\mathbb{R}^n)$, usual in quantum mechanics. An extensive mathematical work has been devoted to eigenfunction expansions and scattering theory for Schrödinger operators, mainly for short-range multiplicative perturbations $V(x)$. Rigged Hilbert spaces of weighted L^2 spaces

$$L^{2,s}(\mathbb{R}^n) \subset L^2(\mathbb{R}^n) \subset L^{2,-s}(\mathbb{R}^n)$$

for suitable $s > 0$ work dealing with such short-range potentials $V(x)$.

2. Spectral measures and spectral forms

Let $\mathcal{H}$ be a separable Hilbert space with scalar product $(\cdot,\cdot)$. The classical version of the spectral theorem [9] associates to each normal operator A defined on $\mathcal{H}$ a Borel spectral measure $(\sigma(A), \mathcal{B}, P)$, where $\sigma(A)$ is the spectrum of A, $\mathcal{B}$ the Borel σ-algebra of $\sigma(A)$ and $P(E)$ the orthogonal projection corresponding to each set $E \in \mathcal{B}$. Every pair $f, h \in \mathcal{H}$ defines the complex measure

$$\mu_{f,h}(E) := (f, P(E)h) , \quad (E \in \mathcal{B}) .$$

When $f = h$ we shall write $\mu_f := \mu_{f,f}$.

For each element g of $\mathcal{H}$ denote by $\mathcal{H}_g$ the closed subspace of $\mathcal{H}$ defined by

$$\mathcal{H}_g = \mathrm{adh}\, \{f \in \mathcal{H} : f = P(E)g\},$$

where E runs over $\mathcal{B}$ and *adh* denotes adherence. A family of vectors $\{g_j\}_{j=1}^m$ in $\mathcal{H}$, where $m \in \{1, 2, \ldots, \infty\}$, is called a *generating system* of $\mathcal{H}$ with respect to P if $\mathcal{H}$ is the orthogonal sum of the spaces $\mathcal{H}_{g_j}$:

$$\mathcal{H} = \bigoplus_{j=1}^{m} \mathcal{H}_{g_j} .$$

By the *type* $[\mu]$ of a measure μ defined on $(\sigma(A), \mathcal{B})$ we understand the equivalence class of all measures ν on $(\sigma(A), \mathcal{B})$ such that μ and ν are mutually absolutely continuous, that is, they have the same class of null sets. We shall write $[\nu] \succ [\mu]$ to indicate that a measure μ is absolutely continuous with respect to a measure ν and, in this case, $d\mu/d\nu$ shall denote the Radon-Nykodym derivative of μ with respect to ν.

A nonzero vector $g \in \mathcal{H}$ is of *maximal type* with respect to the spectral measure P if for each $f \in \mathcal{H}$, $[\mu_g] \succ [\mu_f]$. In this case, g is called a maximal vector. Such maximal vectors always exist provided that $\mathcal{H}$ is separable. The type $[\mu_g]$ of a maximal vector is called the *spectral type* of P and shall be denoted by $[P]$.

It has been proved in [10] that, given a σ-finite nonnegative scalar measure μ on $(\sigma(A), \mathcal{B})$ such that $[\mu] = [P]$ and a generating system $\{g_j\}_{j=1}^m$ of $\mathcal{H}$ with respect to P, then, for each $f, h \in \mathcal{H}$ and $E \in \mathcal{B}$,

$$(f, P(E)h) = \sum_{j=1}^{m} \int_E \frac{d\mu_{g_j}}{d\mu}(\lambda)\, \frac{d\mu_{f,g_j}}{d\mu_{g_j}}(\lambda)\, \frac{d\mu_{g_j,h}}{d\mu_{g_j}}(\lambda)\, d\mu(\lambda) . \tag{1}$$

The Radon-Nykodym derivatives in Eq. (1) are defined for almost every $\lambda \in \sigma(A)$ with respect to measures μ and μ_{g_j}, the null sets depending on f, h and g_j.

Journal of Physics: Conference Series **128** (2008) 012039 doi:10.1088/1742-6596/128/1/012039

For each $j \in \{1, 2, \ldots, m\}$ let $\{\phi_{j,k}\}_{k=1}^{d_j}$ and orthonormal basis of $\mathcal{H}_{g_j}$. Then

$$\Phi := \text{span}\{\phi_{j,k} : j \in \{1, 2, \ldots, m\},\, k = 1, 2, \ldots, d_j\} \tag{2}$$

is a dense subspace of $\mathcal{H}$ and

$$s(\lambda; \varphi, \phi) := \sum_{j=1}^{m} \frac{d\mu_{g_j}}{d\mu}(\lambda)\, \frac{d\mu_{\varphi,g_j}}{d\mu_{g_j}}(\lambda)\, \frac{d\mu_{g_j,\phi}}{d\mu_{g_j}}(\lambda) \tag{3}$$

is well-defined for every $\varphi, \phi \in \Phi$ and $\lambda \in \hat{\Lambda}$, where

$$\begin{aligned}\hat{\Lambda} := \{ \quad & \lambda \in \sigma(A) : \tfrac{d\mu_{g_j}}{d\mu}(\lambda) \text{ and } \tfrac{d\mu_{g_j,\phi_{j,k}}}{d\mu_{g_j}}(\lambda) \text{ exist and are finite} \\ & \text{for } j \in \{1, 2, \ldots, m\} \text{ and } k = 1, 2, \ldots, d_j\}\,. \end{aligned} \tag{4}$$

Definition 2.1 In general, by a *spectral system* $(\Lambda, \mathcal{B}, \mu, \mathcal{H}, P)$ we mean a spectral measure P on a measurable space $(\Lambda, \mathcal{B})$, with values in the set of all orthogonal projections in a complex Hilbert space $\mathcal{H}$, together with a σ-finite nonnegative scalar measure μ on $(\Lambda, \mathcal{B})$. By a standard process, P is decomposed into the absolutely continuous part P_{ac} and the singular part P_s with respect to μ.

The basic elements of the Kato-Kuroda scattering theory [11] are the *spectral forms* for spectral systems:

Definition 2.2 Let $(\Lambda, \mathcal{B}, \mu, \mathcal{H}, P)$ be an spectral system. A *spectral form* for this system is a complex function on $\hat{\Lambda} \times \Phi \times \Phi$, where Φ is a subspace of $\mathcal{H}$ and $\hat{\Lambda} \subseteq \Lambda$ belongs to $\mathcal{B}$, with the following properties:

(i) For each $\phi, \varphi \in \Phi$, $\lambda \mapsto s(\lambda; \varphi, \phi)$ is μ-integrable in $\hat{\Lambda}$ and its integral on each $E \subseteq \hat{\Lambda}$, $E \in \mathcal{B}$, is equal to $(\varphi, P_{ac}(E)\phi)$, i.e.,

$$(\varphi, P_{ac}(E)\phi) = \int_E s(\lambda; \varphi, \phi)\, d\mu(\lambda).$$

(ii) For each $\lambda \in \hat{\Lambda}$, the function $\varphi, \phi \mapsto s(\lambda; \varphi, \phi)$ is a nonnegative Hermitian form on $\Phi \times \Phi$. We shall write $s(\lambda; \phi)$ for $s(\lambda; \phi, \phi)$.

The subspace Φ is usually called a *spectral subspace* and the subset $\hat{\Lambda}$ a *spectral core* of the spectral system.

Clearly, if Φ, s and $\hat{\Lambda}$ are respectively defined by (2), (3) and (4), then $\mu(\sigma(A)\backslash\hat{\Lambda}) = 0$ and the function $s : \hat{\Lambda} \times \Phi \times \Phi \to \mathbb{C}$ is a *spectral form* of the spectral system $(\sigma(A), \mathcal{B}, \mu, \mathcal{H}, P)$.

3. Eigenfunction expansions

Here the theory of eigenfunction expansions is formulated in an abstract way following ideas by Gelfand-Vilenkin [3], Gelfand-Shilov [2] and Kato-Kuroda [11]. The method depends on the introduction of a rigged Hilbert space $\Phi \subset \mathcal{H} \subset \Phi^{\times}$. The eigenfunctions will be elements of $\Phi^{\times}$.

In what follows the value of $\psi^{\times} \in \Phi^{\times}$ at $\phi \in \Phi$ shall be denoted by $\langle\phi|\psi^{\times}\rangle$, and we shall write $\langle\psi^{\times}|\phi\rangle = \langle\phi|\psi^{\times}\rangle^*$ (the asterisk * denotes complex conjugation). The dual pair $\langle\phi|\psi^{\times}\rangle$ is antilinear in ϕ and linear in $\psi^{\times}$.

Let us begin by considering a selfadjoint operator A in $\mathcal{H}$ and, for simplicity, assume that A has simple spectrum. Let P be the spectral measure of A and let g be any generating vector, so that $\mathcal{H} = \mathcal{H}_g$ and

$$\mu_g(\lambda) = (g, P(\lambda)g)\,.$$

Journal of Physics: Conference Series **128** (2008) 012039 doi:10.1088/1742-6596/128/1/012039

Suppose first that the spectrum of A consist only of certain eigenvalues $\sigma(A) = \{\lambda_1, \lambda_2, \ldots\}$. Then, in the sense of norm in $\mathcal{H}$, the limit

$$\frac{dPg}{d\mu_g}(\lambda) = \lim_{\delta\to 0} \frac{P[\lambda, \lambda+\delta)g}{\mu_g[\lambda, \lambda+\delta)}$$

exists for each $\lambda \in \mathbb{R}$, is null for $\lambda \notin \sigma(A)$, and the elements $\left.\frac{dPg}{d\mu_g}\right|_{\lambda_1}$, $\left.\frac{dPg}{d\mu_g}\right|_{\lambda_2}$, ... form a complete system of eigenvectors of the operator A.

Suppose now that for the operator A there exists a point λ of the spectrum which is not an eigenvalue. Then the derivative $\left.\frac{dPg}{d\mu_g}\right|_{\lambda}$ does not exist in the space $\mathcal{H}$. However, if the rigging $\Phi \subset \mathcal{H} \subset \Phi^\times$ is carried out in a suitable manner, then this derivative could exists in the wider space $\Phi^\times$:

$$\frac{dPg}{d\mu_g}(\lambda) = \psi^\times(\lambda) \in \Phi^\times .$$

Futher, it turns out that $\psi^\times(\lambda)$ is a generalized eigenfunctional of the operator A in the following sense:

$$\langle A\phi|\psi^\times(\lambda)\rangle = \lambda\, \langle\phi|\psi^\times(\lambda)\rangle = \lambda\, \frac{d\mu_{\phi,g}}{d\mu_g}(\lambda)\,, \quad (\phi \in \Phi)\,,$$

(here we assume that $\Phi \subseteq \mathcal{D}(A)$ and $A\Phi \subseteq \Phi$). The system of generalized eigenfunctionals obtained in this way proves to be orthogonal and complete in the sense that a sort of Parseval equation holds:

$$||\phi||^2_{\mathcal{H}} = \int |\langle\phi|\psi^\times(\lambda)\rangle|^2\, d\mu_g(\lambda)\,, \quad (\phi \in \Phi)\,.$$

In general, eigenfunctions expansions are defined to consider more refined representations than the spectral ones constructed above:

Definition 3.1 An *eigenfunction expansion* for an spectral system $(\Lambda, \mathcal{B}, \mu, \mathcal{H}, P)$ with spectral form $(\hat{\Lambda}, \Phi, s)$ is defined in the following terms:

(E1) There exist a σ-finite measure space (Γ, Σ, ρ), a partial isometry $\mathcal{F}$ of $\mathcal{H}$ onto $L^2(\rho)$ with initial set $\mathcal{H}^{ac}$ and a measurable function $\lambda : \Gamma \to \Lambda$ such that

$$[\mathcal{F}\alpha(P)h](\xi) = \alpha(\lambda(\xi))[\mathcal{F}h](\xi)\,, \quad (\rho\text{-a.e. } \xi \in \Gamma)\,,$$

for each $h \in \mathcal{H}$ and $\alpha \in L^\infty(\mu)$, $\alpha(P) := \int_\Lambda \alpha(\lambda)\, P(d\lambda)$.

(E2) There is a complex-valued function ψ on $\Gamma \times \Phi$ such that for each fixed $\xi \in \Gamma$, $\phi \mapsto \psi(\xi;\phi)$ is linear and for each fixed $\phi \in \Phi$,

$$\psi(\xi;\phi) = [\mathcal{F}\phi](\xi)\,, \quad (\rho\text{-a.e. } \xi \in \Gamma)\,.$$

(E3) Φ is a topological vector space (tvs) and $\phi \mapsto \psi(\xi;\phi)$ is continuous on Φ for each $\xi \in \Gamma$. In this case we shall write $\psi(\xi;\phi) = \langle\psi^\times(\xi)|\phi\rangle$, where $\psi^\times(\xi) \in \Phi^\times$. Each $\psi(\xi)$ shall be called an *eigenfunction* of P.

Example 1 Let $\mathcal{H} = L^2(\mathbb{R}^n)$ and $\mathcal{F}$ the Fourier-Plancherel transform from $L^2(\mathbb{R}^n)$ onto itself, so that Γ is another copy of $\mathbb{R}^n$, Σ the Borel σ-algebra and ρ the Lebesgue measure. Let λ be the map from $\Gamma = \mathbb{R}^n$ into $\hat{\Lambda} = \Lambda = \mathbb{R}^+$ given by $\lambda(\xi) = |\xi|^2 = \sum_{j=1}^n \xi_j^2$, $\mathcal{B}$ the Borel σ-algebra of Λ, μ the Lebesgue measure on Λ. Let $P_0(E) = \mathcal{F}^{-1}\hat{P}_0(E)\mathcal{F}$, where $\hat{P}_0(E)$ is the operator of multiplication by the characteristic function $\chi_{w^{-1}(E)}$, and $\Phi = L^1(\mathbb{R}^n) \cap L^2(\mathbb{R}^n)$ with the L^1-topology. Then $\psi_0(\xi)$ is given by the function $(2\pi)^{-n/2}\,\mathrm{e}^{i\xi\cdot x}$ in the sense that

$$\langle \psi_0^\times(\xi)|\phi\rangle = (2\pi)^{-n/2}\int_{\mathbb{R}^n} \phi(x)\,\mathrm{e}^{-i\xi\cdot x}\,dx\,, \quad (\phi\in\Phi)\,,$$

The spectral measure P_0 is the one associated with the selfadjoint realization of $-\Delta = -\sum_{j=1}^n \partial^2/\partial x_j^2$ in $L^2(\mathbb{R}^n)$ and the $\psi_0(\xi)$ are the eigenfunctions of this operator in the usual sense. In particular, for $n=3$, a spectral form is (see Section 15.14 in [1])

$$s_0(\lambda;\varphi,\phi) = (4\pi^2)^{-1}\int_{\mathbb{R}^3\times\mathbb{R}^3} \frac{\sin(\lambda^{1/2}|x-y|)}{|x-y|}\,\phi(x)\,\varphi^*(y)\,dx\,dy\,, \quad (\lambda\in\hat\Lambda,\ \varphi,\phi\in\Phi)\,.$$

In particular, for a spectral system $(\Lambda,\mathcal{B},\mu,\mathcal{H},P)$ with $[\mu]=[P]$ one can consider the spectral form $(\hat\Lambda,\Phi,s)$ given by the formulae (2), (3) and (4), and take the space $L^2(\Gamma,\Sigma,\rho)$ in Definition 3.1 to be $L^2(\hat\Lambda\times\{1,2,\ldots,m\},\mu\times d)$, where d denotes the discrete measure on $\{1,2,\ldots,m\}$. In this situation a complete system of generalized eigenfunctions is given by

$$\langle\psi^\times(\lambda\times k)|\phi\rangle := \sqrt{\frac{d\mu_{g_k}}{d\mu}(\lambda)}\,\frac{d\mu_{g_k,\phi}}{d\mu_{g_k}}(\lambda)\,, \quad (\phi\in\Phi,\ \lambda\in\hat\Lambda,\ k\in\{1,2,\ldots,m\})\,. \tag{5}$$

The minimal topology on Φ for which each $\psi^\times(\lambda\times k)$ is in $\Phi^\times$ is just the one generated by the functionals $\psi^\times(\lambda\times k)$. See [10] for details.

In this context renewed versions of the nuclear and inductive spectral theorems have been proved in [12]. These results assert that dual pairs $\langle\Phi,\Phi^\times\rangle$ with nuclear or countable Hilbert-Schmidt inductive topologies on Φ are *universal riggings* in the sense that the antiduals spaces $\Phi^\times$ contain complete systems of generalized eigenfunctions for any Vitali spectral measure.

There is another version of the spectral theorem in rigged Hilbert spaces due to Yu.M. Berezanskii [5], the so-called *projection spectral theorem*. Given a normal operator A on a Hilbert space $\mathcal{H}$, Berezanskii considers a chain of spaces $D\subseteq\mathcal{H}_+\subseteq\mathcal{H}\subseteq\mathcal{H}_-$ associated with A in a certain sense and proves the existence of a nonnegative operator $S(\lambda):\mathcal{H}_+\to\mathcal{H}_-$ such that $I^\dagger P(\Lambda)I = \int_\Lambda S(\lambda)\,d\rho(\lambda)$, where I is the injection of $\mathcal{H}_+$ into $\mathcal{H}$, $I^\dagger$ its adjoint, $\rho(\Lambda) = \mathrm{Tr}(I^\dagger P(\Lambda)I)$ and P is the spectral measure associated to A. The projection spectral theorem assures that, almost everywhere with respect to a special modification of the measure ρ, the range of $S(\lambda)$ consists of all generalized eigenvectors with eigenvalue λ of A. The connections of the projection spectral theorem with the nuclear spectral theorem are discussed in [13].

The construction of eigenfunction expansions for selfadjoint operators given by Poerschke, Stolz and Weidmann in [14] is closely related to the above approaches. For a selfadjoint operator A defined on $\mathcal{H}$ they start from the spectral representation of A and use triplets of the form $\mathcal{H}_+(T)\subset\mathcal{H}\subset\mathcal{H}_-(T)$, where T is a selfadjoint operator in $\mathcal{H}$ with $T\geq 1$ and $\mathcal{H}_\pm(T)$ are the Hilbert spaces obtained by completion of $\mathcal{D}(T)$ for the scalar products $(f,g)_\pm := (T^{\pm1}f, T^{\pm1}g)_{\mathcal{H}}$. If there is a bounded continuous function $\gamma:\mathbb{R}\to\mathbb{C}$ with $|\gamma|>0$ on $\sigma(A)$ and such that $\gamma(A)T^{-1}$ is a Hilbert-Schmidt operator, then there exist a complete system of generalized eigenfunctions of A in $\mathcal{H}_-(T)$.

4. Wave operators

Given a pair (H_0,H) of selfadjoint operators in a Hilbert space $\mathcal{H}$, the time-dependent scattering theory deals with limits of $e^{itH}e^{-itH_0}$ as $t\to\pm\infty$ in strong, weak or Abelian sense, paying attention to the absolutely continuous subspaces of $\mathcal{H}$ with respect to both operators H_0 and H and Lebesgue measure $d\lambda$. In what follows such subspaces shall be denoted by $\mathcal{H}_0^{ac}$ and $\mathcal{H}^{ac}$, and the orthogonal projections from $\mathcal{H}$ onto them by P_0^{ac} and P^{ac}.

Journal of Physics: Conference Series **128** (2008) 012039 doi:10.1088/1742-6596/128/1/012039

Definition 4.1 The *strong wave operators* for the pair (H_0, H) are defined as

$$\Omega^s_\pm := \operatorname*{s-lim}_{t\to\pm\infty} \mathrm{e}^{itH}\, \mathrm{e}^{-itH_0}\, P_0^{ac}\,.$$

In the same way are defined the *weak* and *Abelian wave operators* $\Omega^w_\pm$ and $\Omega^a_\pm$, substituting in the above expression strong limits (s-lim) by weak (w-lim) and Abelian (a-lim) ones. Among the Abelian limits one can distinguish between strong, weak and absolute Abelian limits.

The stationary theory define the wave operators in terms of the resolvents $R_0(z) = (H_0 - z)^{-1}$ and $R(z) = (H - z)^{-1}$.

Definition 4.2 If there exist dense subspaces Φ_0 and Φ of $\mathcal{H}$ such that for each $\phi_0 \in \Phi_0$ and $\phi \in \Phi$ the limits

$$\lim_{\epsilon\to 0} \pi^{-1}\epsilon\,(R(\lambda \pm i\epsilon)\phi, R_0(\lambda \pm i\epsilon)\phi_0)$$

exist for a.a. $\lambda \in \mathbb{R}$ (with respect to Lebesgue measure, the corresponding null sets depending on ϕ_0 and ϕ), the *stationary wave operators* $\Omega^{est}_\pm$ for the pair (H_0, H) are defined in $\Phi_0 \times \Phi$ by the sesquilinear forms

$$(\phi, \Omega^{est}_\pm \phi_0) := \int_{-\infty}^{\infty} \lim_{\epsilon\to 0} \pi^{-1}\epsilon\,(R(\lambda \pm i\epsilon)\phi, R_0(\lambda \pm i\epsilon)\phi_0)\, d\lambda\,. \tag{6}$$

If they exist, the (strong, weak, abelian, stationary) wave operators $\Omega_\pm$ are bounded, satisfy the *intertwining property*

$$P^{ac}(E)\,\Omega_\pm = \Omega_\pm\, P_0^{ac}(E)\,, \qquad \text{for every Borel set } E \subset \mathbb{R}\,, \tag{7}$$

and $\mathcal{H} \ominus \mathcal{H}_0^{ac} \subset \ker(\Omega_\pm)$ and $\mathrm{Ran}(\Omega_\pm) \subset \mathcal{H}^{ac}$. In particular the strong and stationary wave operators are partial isometries with initial subspace $\mathcal{H}_0^{ac}$. The wave operators $\Omega_\pm$ are said to be *complete* when

$$\ker(\Omega_\pm) = \mathcal{H} \ominus \mathcal{H}_0^{ac} \qquad \text{and} \qquad \mathrm{rang}(\Omega_\pm) = \mathcal{H}^{ac}\,.$$

The books by Baumgärtel-Wollenberg [7], Yafaev [8] and Reed-Simon [6] deal at length with the problem of determining conditions ensuring the existence and completeness of the various wave operators as well as with the relationship between them. Typical criteria have been given for trace class perturbations or in terms of time-falloff conditions or a factorization of the perturbation and the relative regularity of the factors.

The central result of scattering theory for trace class perturbations is the *Kato-Rosenblum theorem* [15, 16]: For selfadjoint operators H_0 and H with trace class difference, $V = H - H_0 \in \mathcal{S}_1$, the strong wave operators $\Omega^s_\pm$ exist and are complete. In particular, the absolutely continuous parts $\mathcal{H}_0^{ac}$ and $\mathcal{H}^{ac}$ are unitarily equivalent.

The condition $V \in \mathcal{S}_1$ is too much restrictive in most applications. There are some less demanding local trace class criteria. For example, if $P(\Lambda)VP_0(\Lambda) \in \mathcal{S}_1$ for every bounded interval Λ and $R(z) - R_0(z)$ is a compact operator for some (and then for all) $z \in \rho(H_0) \cap \rho(H)$, then the strong wave operators $\Omega^s_\pm$ exist and are complete [8, Th.6.4.5].

Additional results with trace class conditions can be found in [17, Section X.4], [6, Section XI.3], [7, Chapter 16] and [8, Chapter 6].

One of the simplest time-falloff criteria is known as *Cook's criterium* [8, Th.2.5.1]: Suppose that $\mathcal{D}(H_0) \subseteq \mathcal{D}(H)$ and that for some set $D_0 \subset \mathcal{D}(H_0) \cap \mathcal{H}_0^{ac}$ dense in $\mathcal{H}_0^{ac}$ one has

$$\int_0^{\pm\infty} ||Ve^{-tH_0} f_0||\, dt < \infty, \qquad (f_0 \in D_0)\,.$$

Then the strong wave operators $\Omega^s_\pm$ exist. More time-falloff criteria can be found in [7, Chapter 15].

In order to introduce factorization criteria, for each selfadjoint operator H in $\mathcal{H}$, with spectral measure P and resolvent $R(z) = (H-z)^{-1}$, consider $\delta = \delta_H$ the operator-valued function $\delta = \delta_H$ defined by

$$\begin{aligned} \delta(\lambda, \epsilon) := &\ (2\pi i)^{-1}\,[R(\lambda + i\epsilon) - R(\lambda - i\epsilon)] \\ = &\ \pi^{-1}\epsilon\, R(\lambda + i\epsilon)\, R(\lambda - i\epsilon)\,, \end{aligned} \qquad (\lambda \in \mathbb{R},\ \epsilon > 0)\,.$$

For $f, g \in \mathcal{H}$, one has

$$\lim_{\epsilon\to 0}(g, \delta(\lambda,\epsilon)f) = \frac{d(g, P(\lambda)f)}{d\lambda} = \frac{d\mu_{g,f}}{d\lambda}\,, \qquad \text{for a.a.}\ \lambda \in \mathbb{R}\,. \tag{8}$$

The null set depending on f and g, relation (8) does not imply the existence of the weak limit of $\delta(\lambda,\epsilon)$ as $\epsilon \to 0$. Actually, $\delta(\lambda,\epsilon)$ has not bounded weak limit for any $\lambda \in \mathbb{R}$ because

$$||\delta(\lambda,\epsilon)|| = \pi^{-1}\epsilon \sup_{u\in\mathbb{R}}[(u-\lambda)^2 + \epsilon^2]^{-1} = (\pi\epsilon)^{-1}.$$

But, for suitable operators G, the weak limit of $G\delta(\lambda,\epsilon)$ can exist. This leads to the following definitions.

Definition 4.3 Let H be a selfadjoint operator in a Hilbert space $\mathcal{H}$. An operator G from $\mathcal{H}$ into a Hilbert space $\mathcal{G}$ is called *H-bounded* if $\mathcal{D}(H) \subseteq \mathcal{D}(G)$ and there are positive constants a and b such that $||Gf|| \le a\,||f|| + b\,||Hf||$ for every $f \in \mathcal{D}(H)$. This condition is equivalent to the boundeness of $GR(z)$ for every $z \in \rho(H)$. An H-bounded operator G is called *H-regular in weak sense* if there exist positive constants $C(\lambda)$ such that

$$||G\delta(\lambda,\epsilon)G^\dagger|| \le C(\lambda)\,, \qquad \text{for a.a.}\ \lambda \in \mathbb{R}\,. \tag{9}$$

An H-bounded operator G is called *H-regular* when

$$\sup_{\lambda\in\mathbb{R},\epsilon>0} ||G\delta(\lambda,\epsilon)G^\dagger|| < \infty\,. \tag{10}$$

Condition (9) is equivalent to the existence of the weak limit $\text{w-lim}_{\epsilon\to0}\, G\delta_H(\lambda,\epsilon)G^\dagger$ for a.a. $\lambda \in \mathbb{R}$, and also implies that $\text{w-lim}_{\epsilon\to0}\, G\delta_H(\lambda,\epsilon)$ exists for a.a. $\lambda \in \mathbb{R}$. The H-regularity of G can also be given in terms of the group e^{-itH} or the resolvent $R(z)$: An H-bounded operator G is H-regular if and only if

$$\sup_{f\in\mathcal{D}(H),||f||=1} \int_{-\infty}^{\infty} ||Ge^{-itH}f||^2\,dt < \infty$$

or

$$\sup_{||f||=1,\epsilon>0} \int_{-\infty}^{\infty} (||GR(\lambda - i\epsilon)f||^2 + ||GR(\lambda + i\epsilon)f||^2)\,d\lambda < \infty\,.$$

The last condition implies that the function $GR(z)f$, vector-valued in $\mathcal{G}$, belongs to the Hardy classes $\mathcal{H}^2_\pm(\mathbb{R}, d\lambda; \mathcal{G})$. Then, for every $f \in \mathcal{H}$, there exists the strong limits $\text{s-lim}_{\epsilon\to0}\, GR(\lambda \pm i\epsilon)f$ for a.a. $\lambda \in \mathbb{R}$. More equivalent conditions to (9) and (10) can be found in Sections 5.1 and 4.3 of Yafaev [8].

Assume the perturbation admits a factorization of the form[2]

$$V := H - H_0 = G^\dagger G_0\,, \tag{11}$$

[2] This relation is understood as equality of the corresponding sesquilinear forms: For $f_0 \in \mathcal{D}(H_0)$ and $f \in \mathcal{D}(H)$, one has $v[f, f_0] := (Hf, f_0) - (f, H_0 f_0) = (Gf, G_0 f_0)$.

where $G_0 : \mathcal{H}_0 \to \mathcal{G}$, $G : \mathcal{H} \to \mathcal{G}$ with $\mathcal{G}$ an auxiliary Hilbert space. If G_0 is H_0-bounded and G is H-bounded, then

$$\pi^{-1}\epsilon\,(R(\lambda \pm i\epsilon)f, R_0(\lambda \pm i\epsilon)f_0) = (f, \delta_H(\lambda,\epsilon)f_0) + (G\delta_H(\lambda,\epsilon)f, G_0R_0(\lambda \pm i\epsilon)f_0)\,.$$

From (8), the first summand in the right converges to $d(f, P(\lambda)Jf_0)/d\lambda$ as $\epsilon \to 0$ for a.a. $\lambda \in \mathbb{R}$. Thus, if the functions $G_0R_0(\lambda \pm i\epsilon)f_0$ and $G\delta(\lambda,\epsilon)f$ converge for f_0 and f in dense sets of $\mathcal{H}_0^{ac}$ and $\mathcal{H}^{ac}$, one in strong sense and the other in weak sense, the conditions in Definition 4.2 are satisfied and the stationary wave operators $\Omega_\pm^{est}$ exist –and in the same way if we consider the functions $G_0\delta_0(\lambda,\epsilon)f_0$ and $GR(\lambda \pm i\epsilon)f$–. In particular, if G_0 is H_0-regular and G is H-regular, one in weak sense, the stationary wave operators $\Omega_\pm^{est}$ exist. In this case, if any other wave operators (strong, weak, Abelian) exist, they coincide with the stationary ones. See Sections 5.2 and 5.3 of [8] for details.

When G_0 is H_0-regular and G is H-regular, the strong wave operators $\Omega_\pm^s$ exist and are complete [8, Th.4.5.1]. In this case the stationary wave operators $\Omega_\pm^{est}$ also exist, coincide with $\Omega_\pm^s$ and relations (6) are satisfied for every $\phi_0, \phi \in \mathcal{H}$.

Standard stuff on relative smoothness can be found in [7, Chapter 17] and [8, Chapters 4, 5].

5. Wave operators vs eigenfunction expansions

There is a close relationship between the wave operators for a pair of selfadjoint operators H_0, H and the eigenfunction expansions of H_0, H. The key property is the intertwining relation (7).

Let P_0, P be the spectral measures of H_0, H and let P_0^{ac}, P^{ac} be their absolute continuous parts with respect to the Lebesgue measure $d\lambda$. Consider a direct integral spectral decomposition of P_0^{ac}

$$\mathcal{H}^{ac}_{d\lambda,H_0} = \int_{\hat\sigma_0}^{\oplus} \mathcal{H}^0_\lambda\, d\lambda$$

and a structural isomorphism $\mathcal{V}_0 : \mathcal{H}_0^{ac} \longrightarrow \mathcal{H}^{ac}_{d\lambda,H_0}$ performing a spectral representation of P_0^{ac}, i.e., a unitary operator $\mathcal{V}_0$ such that

$$\mathcal{V}_0 P_0^{ac}(E) = \chi_E \mathcal{V}_0\,, \qquad \text{for every Borel set } E \subset \mathbb{R}\,, \tag{12}$$

where χ_E denotes the characteristic function of E. For any wave operators $\Omega_\pm$, from (7) and (12) it is obvious that the operators $\mathcal{V}_\pm$ defined by

$$\mathcal{V}_\pm := \mathcal{V}_0 \Omega_\pm^\dagger : \mathcal{H}^{ac} \longrightarrow \mathcal{H}^{ac}_{d\lambda,H_0}$$

satisfy the intertwining property

$$\mathcal{V}_\pm P^{ac}(E) = \chi_E \mathcal{V}_\pm\,.$$

Thus, if the operators $\Omega_\pm : \mathcal{H}_0^{ac} \to \mathcal{H}^{ac}$ are unitary, then $\mathcal{V}_\pm$ are also unitary and perform direct integral representations of H^{ac} in $\mathcal{H}^{ac}_{d\lambda,H_0}$. Moreover, since $\mathcal{V}_0^\dagger\mathcal{V}_0 = P_0^{ac}$, the operators $\Omega_\pm$ can be reconstructed from $\mathcal{V}_\pm$:

$$\Omega_\pm = \mathcal{V}_\pm^\dagger \mathcal{V}_0\,.$$

Following the work of Birman and Yafaev [8, Th.5.5.1], assume that the perturbation $V = H - H_0$ admits a factorization of the form (11). If G_0 is H_0-regular in weak sense, the operators

$$\begin{array}{rccc} Z_0(\lambda; G_0): & \mathcal{G} & \longrightarrow & \mathcal{H}^0_\lambda \\ & \psi & \mapsto & [\mathcal{V}_0 G_0^\dagger \psi](\lambda) \end{array}$$

are well-defined bounded operators for a.a. $\lambda \in \mathbb{R}$, for every $f \in \mathcal{H}$ the derivative $dG_0P_0(\lambda)f/d\lambda$ exists in weak sense and $dG_0P_0(\lambda)f/d\lambda = [Z_0(\lambda, G_0)]^\dagger \mathcal{V}_0 f(\lambda)$ for a.a. $\lambda \in \Lambda$. If also the strong limits

$$\operatorname{s-lim}_{\epsilon \to 0} GR(\lambda \pm i\epsilon)\phi =: GR(\lambda \pm i0)\phi$$

exist for a.a. $\lambda \in \mathbb{R}$ for each ϕ in a dense subspace Φ of $\mathcal{H}^{ac}$, then, as we have seen, the stationary wave operators $\Omega_\pm^{est}$ exist. Let us define at first in Φ the operators $\mathcal{V}_\pm := \mathcal{V}_0[\Omega_\mp^{est}]^\dagger : \mathcal{H} \to \mathcal{H}^{ac}_{d\lambda, H_0}$. Then $\mathcal{V}_\pm$ are bounded operators with a unique extension to $\mathcal{H}$ and, for each $f \in \mathcal{H}$, they verify (in weak sense):

$$[\mathcal{V}_\pm f](\lambda) = [\mathcal{V}_0 f](\lambda) - Z_0(\lambda, G_0)GR(\lambda \pm i0)f\,, \quad \text{for a.a.}\, \lambda \in \mathbb{R}\,,$$

the null-set depending on f.

In general, let μ be a Borel σ-finite nonnegative scalar measure μ on $\mathbb{R}$ and consider respective direct integral decompositions $\mathcal{H}^{ac}_{\mu,H_0} = \int^\oplus_{\hat\sigma_0} \mathcal{H}^0_\lambda \, d\mu$, $\mathcal{H}^{ac}_{\mu,H} = \int^\oplus_{\hat\sigma} \mathcal{H}_\lambda \, d\mu$ over spectral cores $\hat\sigma_0$, $\hat\sigma$ of P_0^{ac}, P^{ac} with corresponding structural isomorphisms $\mathcal{V}_0$, $\mathcal{V}$ (now the spectral cores and absolutely continuous parts with respect to μ). Then any bounded operator $\Omega : \mathcal{H}_0 \to \mathcal{H}$ satisfying the intertwining property (7) diagonalizes in these representations, i.e., the operator

$$\mathcal{V}\Omega\mathcal{V}_0^\dagger : \mathcal{H}^{ac}_{\mu,H_0} \longrightarrow \mathcal{H}^{ac}_{\mu,H} \tag{13}$$

acts as multiplication by an operator-valued function $U(\lambda) : \mathcal{H}^0_\lambda \to \mathcal{H}_\lambda$. One has $||\Omega|| = \operatorname{ess\,sup} ||U(\lambda)||$ and, for μ-a.a. $\lambda \in \hat\sigma_0 \backslash \hat\sigma$, the operators $U(\lambda)$ are the null-operator. Furthermore, if $\Omega : \mathcal{H}_0^{ac} \to \mathcal{H}^{ac}$ is unitary, then the operators $U(\lambda)$ are unitary for μ-a.a. $\lambda \in \hat\sigma_0 = \hat\sigma$. In this case, each measurable orthonormal basis $\{e_j^0(\lambda)\}$ of $\mathcal{H}^{ac}_{\mu,H_0}$ leads to a measurable orthonormal basis $\{e_j(\lambda)\}$ of $\mathcal{H}^{ac}_{\mu,H}$ through the relation

$$e_j(\lambda) = U(\lambda)\, e_j^0(\lambda)\,, \quad \text{for } j \in \{1, 2, \ldots, m\} \text{ and } \mu\text{-a.a.}\, \lambda \in \hat\sigma_0 = \hat\sigma\,. \tag{14}$$

The connection of measurable orthonormal bases $e_j^0(\lambda)$, $e_j(\lambda)$ with generalized eigenfunctions $\psi_0^\times(\lambda, j)$, $\psi^\times(\lambda, j)$ –see Eq. (5)– is established in [10]: There exist respective generating systems $\{g_j^0\}_{j=1}^m$, $\{g_k\}_{j=1}^m$ of $\mathcal{H}_0^{ac}$, $\mathcal{H}^{ac}$ with respect to P_0^{ac}, P^{ac} such that

$$\langle \psi_0^\times(\lambda, j)|\cdot\rangle = (e_j^0(\lambda), [\mathcal{V}_0 \cdot](\lambda))_{\mathcal{H}^0_\lambda} = \sqrt{\frac{d\mu_{g_j^0}}{d\mu}(\lambda)}\, \frac{d\mu_{g_j^0,\cdot}}{d\mu_{g_j^0}}(\lambda)\,,$$

$$\langle \psi^\times(\lambda, j)|\cdot\rangle = (e_j(\lambda), [\mathcal{V} \cdot](\lambda))_{\mathcal{H}_\lambda} = \sqrt{\frac{d\mu_{g_j}}{d\mu}(\lambda)}\, \frac{d\mu_{g_j,\cdot}}{d\mu_{g_j}}(\lambda)\,.$$

One has

$$(e_j^0(\lambda), [\mathcal{V}_0 \cdot](\lambda))_{\mathcal{H}^0_\lambda} = (e_j(\lambda), [\mathcal{V}\Omega \cdot](\lambda))_{\mathcal{H}_\lambda}\,, \qquad (e_j(\lambda), [\mathcal{V} \cdot](\lambda))_{\mathcal{H}_\lambda} = (e_j^0(\lambda), [\mathcal{V}_0\Omega^\dagger \cdot](\lambda))_{\mathcal{H}^0_\lambda}\,,$$

and the relationship between the generating systems $\{g_j^0\}_{j=1}^m$, $\{g_j\}_{j=1}^m$ is of the form

$$\Omega\Big(\int \phi\, dP^0\, g_j^0\Big) = \int \phi\, dP\, g_j\,, \quad \text{for } j \in \{1, 2, \ldots, m\} \text{ and } \phi \in L^\infty(\mu)\,.$$

Kato-Kuroda [11] construct direct integral spectral representations in a simple way from the spectral forms as follows: Let $(\Lambda, \mathcal{B}, \mu, \mathcal{H}, P)$ be a spectral system with spectral form $(\hat\Lambda, \Phi, s)$.

Journal of Physics: Conference Series **128** (2008) 012039 doi:10.1088/1742-6596/128/1/012039

For each $\lambda \in \hat{\Lambda}$, $s(\lambda; \cdot, \cdot)$ defines a semi-inner product in Φ. Let $\mathcal{N}_\lambda$ be the set of all ϕ with $s(\lambda; \phi) = 0$. Then the quotient space $\Phi/\mathcal{N}_\lambda$ is a pre-Hilbert space with the inner product induced by $s(\lambda; \cdot, \cdot)$. Denote by $\tilde{\Phi}_\lambda$ its completion, by $(\cdot, \cdot)_\lambda$ and $|| \cdot ||_\lambda$ the inner product and norm in $\tilde{\Phi}_\lambda$ and by q_λ the quotient map of Φ onto $\Phi/\mathcal{N}_\lambda \subset \tilde{\Phi}_\lambda$. Let $\tilde{\Phi}$ be the vector space of all vector fields $\tilde{\phi} = \{\tilde{\phi}(\lambda)\}_{\lambda \in \hat{\Lambda}}$ with $\tilde{\phi}(\lambda) \in \tilde{\Phi}_\lambda$, identifying elements equal μ-a.e. By a *quasi-simple function* $\tilde{\phi}$ one means a function of the form (finite sum) $\tilde{\phi}(\lambda) = \sum \alpha_k(\lambda)\, \phi_k$ with $\alpha_k \in L^\infty(\mu)$ and $\phi_k \in \Phi$. A function $\tilde{\phi} \in \tilde{\Phi}$ is said to be *s-measurable* if there is a sequence $\{\tilde{\phi}_n\}$ of quasi-simple functions on $\hat{\Lambda}$ such that

$$\lim_{n \to \infty} ||\tilde{\phi}(\lambda) - q_\lambda \phi_n(\lambda)||_\lambda = 0 \quad \text{for } \mu\text{-a.a. } \lambda \in \hat{\Lambda}\,.$$

The direct integral $\mathcal{H}_{\mu,\Phi}$ is the Hilbert space of all s-measurable elements $\tilde{\phi} \in \tilde{\Phi}$ such that $||\tilde{\phi}||^2_{\mathcal{H}_{\mu,\Phi}} = \int_{\hat{\Lambda}} ||\tilde{\phi}(\lambda)||^2_\lambda \, d\mu(\lambda) < \infty$ with inner product $(\tilde{\varphi}, \tilde{\phi})_{\mathcal{H}_{\mu,\Phi}} := \int_{\hat{\Lambda}} (\tilde{\varphi}(\lambda), \tilde{\phi}(\lambda))_\lambda \, d\mu(\lambda)$. Quasi-simple functions are densely embedded in $\mathcal{H}_{\mu,\Phi}$ and the structural isomorphism $\mathcal{V} : \mathcal{H}^{ac} \to \mathcal{H}_{\mu,\Phi}$ is the unitary extension of the isometry $\tilde{\phi} \mapsto \{q_\lambda \tilde{\phi}(\lambda)\}_{\lambda \in \hat{\Lambda}}$ defined for quasi-simple $\tilde{\phi}$.

In what follows $\mathcal{H}_\Phi$ shall denote the smallest closed subspace of $\mathcal{H}$ containing Φ an reducing P. Its absolutely continuous part with respect to μ, $\mathcal{H}^{ac}_\Phi := P^{ac}\mathcal{H}_\Phi$, also reduces P. $\mathcal{H}^{ac}_\Phi = \mathcal{H}^{ac}$ if Φ generates $\mathcal{H}$, i.e, if $\mathcal{H}_\Phi = \mathcal{H}$.

General wave operators $\Omega : \mathcal{H}_1 \to \mathcal{H}_2$ are defined between two spectral systems $(\Lambda, \mathcal{B}, \mu, \mathcal{H}_j, P_j)$ with spectral forms $(\hat{\Lambda}, \Phi_j, s_j)$ and corresponding direct integrals $\mathcal{H}_{\mu,\Phi_j}$, $j = 1, 2$, where the measure space $(\Lambda, \mathcal{B}, \mu)$ and $\hat{\Lambda} \subseteq \Lambda$ are common to both systems. A general wave operator Ω is a partial isometry with initial subspace $\mathcal{H}^{ac}_{\Phi_1}$ and final subspace contained in $\mathcal{H}^{ac}_{\Phi_2}$, which intertwines P_1 and P_2. Kato-Kuroda [11] obtain such general wave operators Ω going in inverse direction from (14) to (13). They derive $U(\lambda)$ from converging sequences of operators defined between the auxiliary spaces Φ_j and give conditions leading to abstract Lippmann-Schwinger equations between the eigenfunctions in the antidual spaces $\Phi_j^\times$. To be precise, assume the following conditions:

(A1) Φ_j is a topological vector space with its own topology and there is a sequence of *approximating spectral forms* s_{jn}, $n = 1, 2, \ldots$, on $\hat{\Lambda} \times \Phi_j \times \Phi_j$, in the following sense: For each $\lambda \in \hat{\Lambda}$, $s_{jn}(\lambda; \cdot, \cdot)$ is an equicontinuous family of nonnegative Hermitian form on $\Phi_j \times \Phi_j$ and $\lim_{n\to\infty} s_{jn}(\lambda; \phi, \varphi) = s_j(\lambda; \phi, \varphi)$ for each $\phi, \varphi \in \Phi_j$.

(A2) For each $\lambda \in \hat{\Lambda}$, there is an equicontinuous sequence $\{K_n(\lambda)\}$ of operators on Φ_1 to Φ_2 such that $s_{2n}(\lambda; K_n(\lambda)\phi) = s_{1n}(\lambda; \phi)$ for $\phi \in \Phi_1$; for each $\phi \in \Phi_1$ and n, the function $K_n(\cdot)\phi$ is s-measurable on $\hat{\Lambda}$; for each $\lambda \in \hat{\Lambda}$ and $\phi \in \Phi_1$, $\{K_n(\lambda)\phi\}_n$ is a Cauchy sequence in Φ_2. Moreover, the operators $K_n(\lambda)$ are sobrejective and $K_n(\lambda)^{-1} : \Phi_2 \to \Phi_1$ satisfy the above conditions with subscripts 1, 2 interchanged.

Under conditions (A1) and (A2) a general wave operator Ω exists and is complete on $\mathcal{H}^{ac}_{\Phi_1}$ to $\mathcal{H}^{ac}_{\Phi_2}$. Furthermore, the operators $K_n(\lambda) : \Phi_1 \to \Phi_2$ being continuous, there exist the operators $K_n(\lambda)^\times : \Phi_2^\times \to \Phi_1^\times$ defined by the relations

$$\langle K_n(\lambda)\phi_1 | \phi_2^\times \rangle = \langle \phi_1 | K_n(\lambda)^\times \phi_2^\times \rangle\,, \qquad (\phi_1 \in \Phi_1,\, \phi_2^\times \in \Phi_2^\times)\,.$$

Since $\{K_n(\lambda)\phi\}_n$ is Cauchy for each $\phi \in \Phi_1$, $\lim \langle \phi | K_n(\lambda)^\times \varphi^\times \rangle = \lim \langle K_n(\lambda)\phi ! \varphi^\times \rangle$ exists for each $\phi \in \Phi_1$ and $\varphi^\times \in \Phi_2^\times$. Since $\{\langle \cdot | K_n(\lambda)^\times \varphi^\times \rangle\}_n$ is equicontinuous, the limit is continuous and can be written as $\langle \cdot | \phi^\times \rangle$ for a unique $\phi^\times \in \Phi_1^\times$. The relation $\phi^\times =: K(\lambda)^\times \varphi^\times$ defines an operator $K(\lambda)^\times : \Phi_2^\times \to \Phi_1^\times$. In other words,

$$\text{w}^*\text{-}\lim_{n \to \infty} K_n(\lambda)^\times = K(\lambda)^\times\,,$$

Journal of Physics: Conference Series **128** (2008) 012039 doi:10.1088/1742-6596/128/1/012039

where w*-lim denotes weak* limit. Similarly, w*-$\lim_{n\to\infty}[K_n(\lambda)^{-1}]^\times = [K(\lambda)^\times]^{-1}$.

If, moreover, the system 1 has an eigenfunction expansion, i.e., satisfies conditions (E1), (E2) and (E3) of Definition 3.1, with partial isometry $\mathcal{F}_1 : \mathcal{H}_1 \to L^2(\rho)$ and eigenfunctions $\psi_1^\times(\xi) \in \Phi_1^\times$, then the system 2 also has an eigenfunction expansion with the same measure space (Γ, Σ, ρ), partial isometry

$$\mathcal{F}_2 := \mathcal{F}_1\Omega^\dagger : \mathcal{H}_2 \to L^2(\rho)$$

and eigenfunctions $\psi_2^\times(\xi) \in \Phi_2^\times$ given by

$$\psi_2^\times(\xi) := [K(w(\xi))^\times]^{-1}\psi_1^\times(\xi) \qquad (K(w(\xi))^\times \psi_2^\times(\xi) = \psi_1^\times(\xi)),$$

the *Lippmann-Schwinger equations in abstract form*.

Particular methods using an auxiliary Banach space Φ were proposed by Howland [18, 19] and Retjo [20]. In these works complete wave operators of stationary character are defined in a direct way and systems of generalized eigenfunctions for H_0 and H exist in $\Phi^\times$ satisfying the original Lippmann-Schwinger equations [21]. These methods are applied to Φ-regular pairs and gentle perturbations.

6. Schrödinger operators

In this Section we shall be concerned with eigenfunction expansions and scattering theory for Schrödinger pairs $[-\Delta, -\Delta + V(x)]$ in $L^2(\mathbb{R}^n)$, $n \in \mathbb{N}$. Here $L^2 = L^2(\mathbb{R}^n)$ is the space of square integrable complex-valued functions on $\mathbb{R}^n$ and $\Delta = \sum_j \partial^2/\partial x_j^2$ is the Laplacian operator, the derivatives taken in distributional sense.

For any real s we shall consider the *weighted* L^2 *space*

$$L^{2,s} = L^{2,s}(\mathbb{R}^n) := \{u(x) : (1 + |x|^2)^{s/2}\, u(x) \in L^2(\mathbb{R}^n)\},$$

where $x = (x_1, \ldots, x_n) \in \mathbb{R}^n$, $|x|^2 = \sum_j x_j^2$. In $L^{2,s}$ we introduce the norm $||u||_{0,s} := ||(1 + |x|^2)^{s/2}\, u(x)||_{L^2}$. Obviously, $(L^{2,s})^\times = L^{2,-s}$ and the dual pair is given by $\langle u, v\rangle = \int u(x)^* v(x)\, dx$, $u \in L^{2,s}$ and $v \in L^{2,-s}$. The *weighted Sobolev* L^2 *spaces* are given by

$$\mathcal{H}_{m,s} = \mathcal{H}_{m,s}(\mathbb{R}^n) = \{u(x) : D^\alpha u \in L^{2,s},\ 0 \le |\alpha| \le m\}.$$

Here for a multi-index $\alpha = (\alpha_1, \ldots, \alpha_n)$ of order $|\alpha| = \sum_j \alpha_j$ the derivatives $D^\alpha = D_1^{\alpha_1} \cdots D_n^{\alpha_n}$, $D_j = -i\partial/\partial x_j$, are taken in distributional sense. $\mathcal{H}_{m,s}$ is a Hilbert space under the norm $||u||_{m,s}^2 := \sum_{|\alpha|\le m} ||D^\alpha u||_{0,s}^2$. The usual Sobolev spaces $\mathcal{H}_{m,0}$ will also be denoted by $\mathcal{H}_m$.

Definition 6.1 A real function $V(x) \in L^2_{loc}(\mathbb{R}^n)$ is said to belong to the class SR (*short-range*) if for some $\epsilon > 0$ the multiplication mapping

$$u(x) \mapsto (1 + |x|)^{1+\epsilon}\, V(x)\, u(x)$$

defines a compact operator from $\mathcal{H}_2(\mathbb{R}^n)$ into $L_2(\mathbb{R}^n)$.

A real function V verifying

$$\sup_{x\in\mathbb{R}^n}\left[(1 + |x|)^{2+2\epsilon} \int_{|y-x|\le 1} |V(y)|^2\, |y - x|^{-n+\mu}\, dy\right] < \infty \tag{15}$$

for some $\epsilon > 0$ and $0 < \mu < 4$ is of class SR. Note that (15) holds in particular for $V \in L^p_{loc}(\mathbb{R}^n)$, with $p = 2$ for $n \le 3$, $p > n/2$ for $n \ge 4$, and such that

$$V(x) = O(|x|^{-1-\epsilon}) \quad \text{as } |x| \to \infty.$$

Denote by H_0 the selfadjoint realization of the Laplacian $-\Delta$ in $L^2(\mathbb{R}^n)$, with domain $\mathcal{D}(H_0) = \mathcal{H}_2(\mathbb{R}^n)$. For $V(x)$ of class SR the multiplication operator $u \mapsto V(x)\,u(x)$ is H_0-compact. Then $-\Delta + V(x)$ admits a unique selfadjoint realization in $L^2(\mathbb{R}^n)$, which we shall denote by H. Both operators H_0 and H possess the same domain of definition $\mathcal{H}_2(\mathbb{R}^n)$ and also the same essential spectrum which consists of the interval $[0,\infty)$. The spectrum of H is $\sigma(H) = [0,\infty) \cup \{\lambda_j\}$ with $\{\lambda_j\}$ a discrete set of negative eigenvalues with finite multiplicity and having zero as its only possible limit point. Furthermore, H lacks continuous singular spectrum and the set $e_+ = e_+(H)$ of all positive eigenvalues of H is a discrete set of eigenvalues of finite multiplicity with 0 and ∞ as the only possible limit points. See [17, Section IV.5.6] and [22] for details. Let $e_+^{1/2} = e_+(H)^{1/2} := \{k \in (0,\infty) : k^2 \in e_+(H)\}$ and $\mathcal{N} = \mathcal{N}(H) := \{\xi \in \mathbb{R}^n : |\xi|^2 \in e_+(H)\} \cup \{0\}$.

The eigenfunction expansion for H_0 has been given in Example 1. For potentials $V(x)$ satisfying condition (15) Agmon [22] proves the existence of two (essentially unique) families $\psi_\pm(x,\xi)$ of generalized eigenfunctions of $H = -\Delta + V(x)$ defined for every $\xi \in \mathbb{R}^n \backslash \mathcal{N}$ and of class $L^2_{loc}(\mathbb{R}^n \times \mathbb{R}^n \backslash \mathcal{N})$. For every fixed ξ the functions $\psi_\pm(x,\xi)$ belong to $C(\mathbb{R}^n_x) \cap \mathcal{H}_2^{loc}(\mathbb{R}^n_x)$ and satisfies the differential equation

$$[-\Delta_x + V(x) - |\xi|^2]\,\psi_\pm(x,\xi) = 0\,, \tag{16}$$

that is, $\psi_\pm(x,\xi)$ are eigenfunctions of H corresponding to $|\xi|^2$.

The *eigenfunction expansion theorem* is given in the following terms: There exist two bounded linear maps $\mathcal{F}_\pm : L^2(\mathbb{R}^n) \to L^2(\mathbb{R}^n)$ such that:

(a) Let P_{ac} and P_p denote the respective projections onto the absolutely continuous (with respect to Lebesgue measure) and point (eigenvalue) parts of $L^2(\mathbb{R}^n)$ with respect to H, so that $L^2(\mathbb{R}^n) = P_{ac}L^2(\mathbb{R}^n) \oplus P_pL^2(\mathbb{R}^n)$. Then $\operatorname{Ker}\mathcal{F}_\pm = P_pL^2(\mathbb{R}^n)$ and the restriction of $F_\pm$ to $P_{ac}L^2(\mathbb{R}^n)$ is a unitary operator from $P_{ac}L^2(\mathbb{R}^n)$ onto $L^2(\mathbb{R}^n)$.

(b) For any $f \in L^2(\mathbb{R}^n)$,

$$[\mathcal{F}_\pm f](\xi) = (2\pi)^{-n/2} \lim_{N\to\infty} \int_{|x|<N} f(x)\,\psi_\pm(x,\xi)^*\,dx\,,$$

$$[\mathcal{F}_\pm^\dagger f](x) = (2\pi)^{-n/2} \lim_{j\to\infty} \int_{K_j} f(\xi)\,\psi_\pm(x,\xi)\,d\xi\,,$$

where K_j is an increasing sequence of compacts sets such that $\cup_j K_j = \mathbb{R}^n \backslash \mathcal{N}$.

(c) Let $M_{|\cdot|^2}$ denote the multiplication operator by $|\xi|^2$. Then for every $f \in \mathcal{D}(H)$

$$P_{ac}Hf = \mathcal{F}_\pm^\dagger M_{|\cdot|^2}\mathcal{F}_\pm f\,.$$

Then the *wave operators* $\Omega_\pm$ for the pair (H_0, H) exist and are complete ($\operatorname{Rang}\Omega_\pm = P_{ac}L^2(\mathbb{R}^n)$). They are given by

$$\Omega_\pm := \operatorname*{s-lim}_{t\to\pm\infty} e^{itH} e^{-itH_0} = \mathcal{F}_\pm^\dagger \mathcal{F}\,, \tag{17}$$

with $\mathcal{F}$ the ordinary Fourier transform.

The methods in [22] are of stationary nature. Consider the resolvents $R_0(z) = (H_0 - z)^{-1}$ and $R(z) = (H - z)^{-1}$ as analytic operator valued functions on $\mathbb{C}\backslash\overline{\mathbb{R}_+}$ and $\mathbb{C}\backslash\sigma(H)$, respectively, with values in $B(L^{2,s}, \mathcal{H}_{2,-s})$ for any $s > 1/2$. For potentials V of class SR the following limits exist in the uniform operator topology of $B(L^{2,s}, \mathcal{H}_{2,-s})$:

$$R_0(\lambda \pm i0) := \lim_{\epsilon\downarrow 0} R_0(\lambda \pm i\epsilon)\,, \qquad (\lambda \in \mathbb{R}_+)\,,$$

$$R(\lambda \pm i0) := \lim_{\epsilon\downarrow 0} R(\lambda \pm i\epsilon)\,, \qquad (\lambda \in \mathbb{R}_+ \backslash e_+(H))\,.$$

This result is usually known as the *limiting absorption principle.*

Introduce polar coordinates $\xi = k\omega$ with $k = |\xi|$ and $\omega = \xi/k$, and write: $\psi_\pm(x,k,\omega) = \psi_\pm(x,k\omega)$. Then for fixed $(x,k) \in \mathbb{R}^n \times \mathbb{R}_+ \backslash e_+^{1/2}$ the function $\psi_\pm(x,k,\omega)$ belongs to $L^2(S^{n-1})$. For any function g in $L^2(S^{n-1})$ define

$$\psi_0^g(x,k) := \int_{S^{n-1}} \exp(ik\omega\cdot x)\, g(\omega)\, dS_\omega\,, \quad \psi_\pm^g(x,k) := \int_{S^{n-1}} \psi_\pm(x,k\omega)\, g(\omega)\, dS_\omega\,.$$

Then for any fixed $k \in \mathbb{R}_+\backslash e_+^{1/2}$ the function $\psi_\pm^g(x,k)$ lies in $\mathcal{H}_{2,-s}(\mathbb{R}^n) \cap C^{0,-s}(\mathbb{R}^n)$ for any $s > 1/2$, satisfies the differential equation $[-\Delta_x + V(x) - k^2]\,\psi_\pm^g(x,k) = 0$ and the Lippmann-Schwinger equation

$$\psi_\pm^g(x,k) = \psi_0^g(x,k) - R(k^2 \mp i0)[V(\cdot)\,\psi_0^g(\cdot,k)](x)\,.$$

Eigenfunction expansions for particular short-range potential Schrödinger operators were already obtained by Povzner [23], who considered potentials $V(x)$ bounded from below and vanishing outside a compact set in $\mathbb{R}^3$. Povzner's eigenfunctions for $H = -\Delta + V(x)$ are of the form $\psi = (2\pi)^{-3/2}\exp(-i\xi x) + \psi_1$, where $\psi_1 = O(|x|^{-1})$ and $|x|(d\psi_1/d|x| - i|\xi|\psi_1) = O(|x|^{-1})$ as $|x| \to \infty$. Faddeev [24] generalized Povzner's results for potentials $V(x)$ such that $V(x) \in \mathrm{Lip}^1$, $V(x) = O(|x|^{-3-\varepsilon})$ and $\operatorname{grad} V(x) = O(1/|x|^{-3-\varepsilon})$ as $|x| \to \infty$, $(\varepsilon > 0)$. Later Ikebe [25] considered potentials $V(x)$ in $L^2(\mathbb{R}^3)$ satisfying $V(x) = O(|x|^{-2-\varepsilon})$ as $|x| \to \infty$, which, except for a finite number of singular points, are locally Hölder-continuous. Alsholm and Schmidt [26] extended the above works in $\mathbb{R}^n$, $n \geq 3$, requiring integral, rather than pointwise, smoothness and decay conditions on $V(x)$, and obtained simplified proofs under more general hypotheses. A detailed study of potentials $V : \mathbb{R}^3 \to \mathbb{R}$ such that $V(x) = (1+|x|)^{-s}[B_1(x) + B_2(x)]$ with $B_1 \in L^2$, $|B_2(x)| \leq \beta(1+|x|)^{-\frac{1}{2}-\delta}$ for some $s > 3/2$, $\delta > 0$ and $0 \leq \beta < \infty$ (roughly speaking, absolutely square-integrable V decreasing to zero more rapidly than $|x|^{-2-\varepsilon}$ for some $\varepsilon > 0$ as $|x| \to \infty$) can be found in [27].

Kuroda [28] studied real valued potentials $V(x)$ on $\mathbb{R}^n$, $n \geq 3$, such that there exists $s > n$ with

$$\begin{aligned}(1+|x|)^{s/2}V(x) \in L^p(E_1)\,, \quad E_1 &:= \{x : |(1+|x|)^{s/2}V(x)| \leq 1\}\,,\\ (1+|x|)^{s/2}V(x) \in L^q(E_2)\,, \quad E_2 &:= \{x : |(1+|x|)^{s/2}V(x)| > 1\}\,,\end{aligned}$$

for some $p < 2n$ and $q > n/2$ (if $n \geq 4$) or $q = 2$ (if $n = 3$). Under the above conditions H defines a selfadjoint operator on $\mathcal{D}(H) = \mathcal{D}(H_0) = \mathcal{H}_2(\mathbb{R}^n)$ and the spectrum of H consists of the absolutely continuous part (with respect to Lebesgue measure) extending from 0 to ∞, a finite number of nonpositive eigenvalues and, possibly, a positive singular spectrum confined in a closed null set e_s. Moreover, for any $\xi \in \mathbb{R}^n$ such that $|\xi|^2 \notin e_s$ the Lippmann-Schwinger equation

$$\psi_\pm(x,\xi) = (2\pi)^{-n/2}\exp(ix\cdot\xi) - \int_{\mathbb{R}^n} R_n^\pm(x,y;|\xi|)\, q(y)\, \psi_\pm(y,\xi)\, dy$$

has a unique solution in $L^{2,-s}(\mathbb{R}^n)$. Here the kernels $R_n^\pm(x,y;k)$ are the outgoing and incoming fundamental solutions of $-\Delta - k^2$, which can be expressed in terms of the Hankel functions $H_\cdot^{(+)} = H_\cdot^{(1)}$ and $H_\cdot^{(-)} = H_\cdot^{(2)}$ [1, Formula 13.7.2]:

$$R_n^\pm(x,y;k) = \pm\frac{i}{4}\left(\frac{k}{2\pi\,|x-y|}\right)^{(n-2)/2} H_{(n-2)/2}^{(\pm)}(k\,|x-y|)\,.$$

The functions $\psi_\pm(x,\xi)$ satisfy the differential equation (16) in the distribution sense and serve as kernels of the generalized Fourier transforms $\mathcal{F}_\pm$ associated to the absolutely continuous part

Journal of Physics: Conference Series **128** (2008) 012039 doi:10.1088/1742-6596/128/1/012039

H^{ac} of H. Also in this case relations (17) lead to complete wave operators, so that H^{ac} is unitarily equivalent to H_0. For each fixed ξ, $\psi_\pm(x,\xi)$ is a bounded continuous function of x which differs from $(2\pi)^{-n/2}\exp(ix\cdot\xi)$ by a function tending to 0 uniformly as $|x|\to\infty$.

Similar results were obtained by Shenk and Thoe [29] in exterior domains for real valued potentials $V(x)$ on $\mathbb{R}^n$, $n\geq 2$, such that $(1+|x|)^{(n+1+\epsilon)/2}V(x)$ ($\epsilon>0$ fixed but arbitrary) is uniformly α-Hölder continuous on $\mathbb{R}^n$ $(0<\alpha\leq 1)$ and tends to zero as $|x|\to\infty$.

The abstract results in Poerschke-Stolz-Weidmann [14] (see Section 3) are applied to Schrödinger operators in $L^2(\mathbb{R}^n)$, $n\leq 3$. For potentials $V:\mathbb{R}^n\to\mathbb{R}$ which are uniformly locally in $L^2(\mathbb{R}^n)$, generalized eigenfunctions for the Schrödinger operator $H=-\Delta+V(x)$ in $L^2(\mathbb{R}^n)$ are found belonging to $(1+|x|^2)^{s/2}\mathcal{H}_1\subset L^{2,-s}\cap\mathcal{H}_1^{loc}$ with $s>n/2$.

Later Poerschke and Stolz [30] reversed the original procedure and established eigenfunction expansions as a result of the existence and intertwining property of (modified (channel)) wave operators. The existence of wave operators $\Omega_\pm$ is taken as granted and used to construct expansions on the ranges of $\Omega_\pm$. They are able to include singular potentials and even perturbations by first and second order differential operators, the eigenfunctions remaining in weighted Sobolev spaces. This reversed procedure was first adopted by Jensen and Kitada [31] mainly for smooth potentials.

Notably, the spectrum $\sigma(H)$ of H on $L^2(\mathbb{R}^n)$, $n\in\mathbb{N}$, may be characterized (up to a set of spectral measure zero) as the set of values for which a polynomially bounded generalized eigenfunction exists. This fact was proved in [32, 33] for potentials $V=V_+-V_-$ with $V_\pm\geq 0$, $V_+\in K_n^{loc}$ and $V_-\in K_n$, where, for $n\in\mathbb{N}$, K_n is the class defined as follows: If $n\geq 3$, $f\in K_n$ iff

$$\limsup_{\alpha\downarrow 0}\ \sup_x \int_{|x-y|\leq\alpha} |x-y|^{-(n-2)}\,|f(y)|\,dy=0\,;$$

if $n=2$, replace $|x-y|^{-(n-2)}$ by $\log|x-y|^{-1}$ and take $\alpha\leq 1$; if $n=1$, K_1 is the set of f with $\sup_x\int|x-y|\leq 1|f(y)|\,dy<\infty$. We say that $f\in K_n^{loc}$ iff $fg\in K_n$ for all $g\in C_0^\infty(\mathbb{R}^n)$, the space of smooth functions with compact support.

For such type of potentials $V=V_+-V_-$ Simon [34, Th.C.5.4] showed that there exists an eigenfunction expansion for H with eigenfunctions $\psi(x,\lambda)$ satisfying $|\psi(x,\lambda)|\leq C\,(1+|x|^2)^{s/2}$ for $s>n/2$ and some constant C. Furthermore, the semigroup $\{e^{-tH}: t\geq 0\}$ is defined on $L^{2,s}$. Denote its generator by H_s. One has $\sigma(H_s)=\sigma(H)$ for every $s\in\mathbb{R}$ and the adjoint H_s^* of H_s with respect to the dual pair $\langle L^{2,s},L^{2,-s}\rangle$ coincides with H_{-s}. Moreover, assume also that $V\in L^2_{loc}$ and let $\psi\in L^{2,-s}$ for some $s>0$ such that $(\psi,Hg)=\lambda\,(\psi,g)$ for all $g\in C_0^\infty$. Then $\psi\in\mathcal{D}(H_{-s})$, $H_{-s}\psi=\lambda\psi$ and $\lambda\in\sigma(H_{-s})=\sigma(H)$. See [33].

A detailed study of the class K_n, its connections with L^p properties, functional integrals, and regularity properties of eigenfunctions can be found in [34].

Acknowledgments

The author wishes to acknowledge kind invitation from QTS-5 organizers and financial support from JCyL-project VA013C05 (Castilla y León) and MEC-project FIS2005-03989 (Spain).

References

[1] Titchmarsh E C 1958 *Eigenfunction expansions associated with second-order differential equations. Vol. 2* (Oxford: Clarendon Press)
[2] Gelfand I M and Shilov G E 1967 *Generalized functions. Vol. 3: Theory of differential equations* (New York-London: Academic Press)
[3] Gelfand I M and Vilenkin N Y 1964 *Generalized functions. Vol. 4: Applications of harmonic analysis* (New York-London: Academic Press)
[4] Berezanskii Y M 1968 *Expansions in eigenfunctions of selfadjoint operators* (*Translations of Mathematical Monographs* no 17) (Providence, R.I.: American Mathematical Society)

[5] Berezanskii Y M, Sheftel Z G and Us G F 1996 *Functional analysis* (*Operator Theory: Advances and Applications* no 86) (Basel: Birkhäuser-Verlag)
[6] Reed M and Simon B 1979 *Methods of modern mathematical physics. Vol. III: Scattering theory* (New York-London: Academic Press)
[7] Baumgärtel H and Wollenberg M 1983 *Mathematical scattering theory* (*Operator Theory: Advances and Applications* no 9) (Basel: Birkhäuser-Verlag)
[8] Yafaev D R 1992 *Mathematical scattering theory. General theory* (*Translations of Mathematical Monographs* no 105) (Providence: American Mathematical Society)
[9] von Neumann J 1955 *Mathematical foundations of quantum mechanics* (Princeton: Princeton University Press)
[10] Gadella M and Gómez F 2002 *Found. Phys.* **32** 815–869
[11] Kato T and Kuroda S 1970 *Functional analysis and related fields* (New York: Springer) pp 99–131
[12] Gadella M and Gómez F 2005 *Bull. Sci. Math.* **129** 567–590
[13] Berezanskii Y M 1985 *Ukrain. Mat. Zh.* **37** 146–154
[14] Poerschke T, Stolz G and Weidmann J 1989 *Math. Z.* **202** 397–408
[15] Kato T 1957 *Proc. Japan. Acad. Ser. A Math. Sci.* **33** 260–264
[16] Rosenblum M 1957 *Pacific J. Math.* **7** 997–1010
[17] Kato T 1976 *Perturbation theory for linear operators* 2nd ed (*Grundlehren der Mathematischen Wissenschaften* no 132) (Berlin-New York: Springer-Verlag)
[18] Howland J S 1967 *J. Math. Anal. and Appl.* **20** 22–47
[19] Howland J S 1968 *J. Funct. Anal.* **2** 1–23
[20] Retjo P A 1967 *Math. Anal. Appl.* **20** 145–187
[21] Lippmann B A and Schwinger J 1950 *Phys. Rev.* **79** 469–480
[22] Agmon S 1975 *Ann. Scuola Norm. Sup. Pisa Cl. Sci. (4)* **2** 151–218
[23] Povzner A Y 1955 *Dokl. Akad. Nauk SSSR (N.S.)* **104** 360–363
[24] Faddeev L D 1957 *Vestnik Leningrad. Univ. Ser. Mat. Meh. Astr.* **12** 164–172
[25] Ikebe T 1960 *Arch. Rational Mech. Anal.* **5** 1–34
[26] Alsholm P and Schmidt G 1971 *Arch. Rational Mech. Anal.* **40** 281–311
[27] Amrein W, Sinha K and Jauch J 1977 *Scattering theory in quantum mechanics* (*Lecture Notes and Supplements in Physics* no 16) (Reading, Mass.-London-Amsterdam: W. A. Benjamin, Inc.)
[28] Kuroda S T 1967 *Proc. Nat. Acad. Sci. U.S.A.* **57** 1213–1217
[29] Shenk N and Thoe D 1971 *J. Math. Anal. Appl.* **36** 313–351
[30] Poerschke T and Stolz G 1993 *Math. Z.* **212** 337–357
[31] Jensen A and Kitada H 1988 *Math. Z.* **199** 1–13
[32] Kovalenko V F and Semenov J A 1978 *Russian Math. Surveys* **33** 119–157
[33] Simon B 1981 *J. Funct. Anal.* **42** 347–355
[34] Simon B 1982 *Bull. Amer. Math. Soc. (N.S.)* **7** 447–526

V International Symposium on Quantum Theory and Symmetries IOP Publishing
Journal of Physics: Conference Series **128** (2008) 012040 doi:10.1088/1742-6596/128/1/012040

Dirac's causality leads to time asymmetry

Yoshihiro Sato[1] **and Hitoshi Kan**[2]

[1]Department of physics, University of Texas at Austin, Austin, TX78712, USA
[2]Hakodate National College of Technology, Tokura-cho 14-1, Hakodate, 042-8501, Japan

E-mail: [1]satoyosh@physics.utexas.edu, [2]kan@physics.utexas.edu

Abstract. On the basis of Dirac's causality, we will show that the time evolution is limited to a semigroup. The abstract vector space for states and (yes-or-no) observables are then not the entire Hilbert space but its particular dense subspaces, called Hardy spaces. The Hardy spaces and their functional spaces together make the Hardy rigged Hilbert spaces, which is also called the time-asymmetric boundary condition (TABC). We will illustrate the usage of the TABC with the neutral kaon decay experiment.

1. Quantum mechanical causality

1.1. Preparation-registration arrow of time

For our foundation, we adopt an objective formulation of quantum mechanics by G. Ludwig and his school [1] where preparation and registration of microsystems are discussed in detail. Microsystem is a small system such as an electron, positron, ion, atom, or molecule that obeys the principles of the quantum mechanics. In practice they are first prepared by an preparation apparatus such as a particle accelerator, then measured by a registration apparatus such as particle counters. A typical arrangement is given in Fig. 1. In such an experiment,

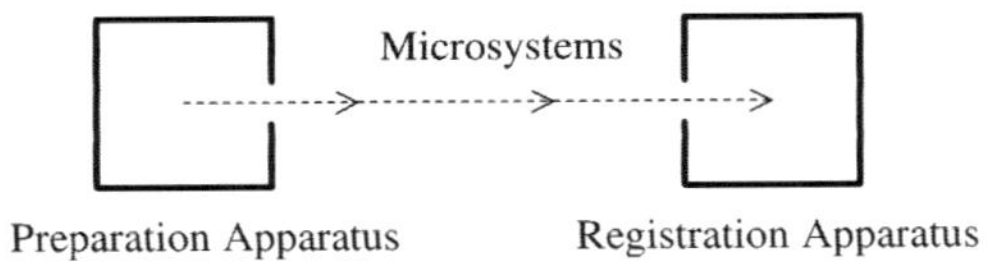

Figure 1. Preparation and registration of a microsystem.

every microsystem must first be prepared before it can be registered. This statement is called *preparation-registration arrow of time* [2].

In terms of states and observables, the preparation-registration arrow of time is described as follows: In a laboratory one prepares a microsystem in a state $|\phi\rangle$ at an instant of time. This process can be repeated for preparation of N identical microsystems in the same state $|\phi\rangle$. Let us denote the time of the experimenter's clock, the laboratory time continuum, by T. The laboratory times at which these N microsystems are prepared form an ensemble, that we call *the ensemble of beginnings of time* [3],

$$\{T_1, T_2, T_3, \cdots, T_N\}, \tag{1}$$

Journal of Physics: Conference Series **128** (2008) 012040 doi:10.1088/1742-6596/128/1/012040

where the subscripts refer to each of the N microsystems. One registers these individual microsystems by an observable $\mathcal{O}$ at the other instants of time,

$$\{T_1', T_2', T_3', \cdots, T_N'\}. \tag{2}$$

By the ensembles (1) and (2), the preparation-registration arrow of time is expressed as

$$T_i \leq T_i' \quad \text{for } i = 1, 2, 3, \cdots, N. \tag{3}$$

1.2. Dirac's causality and semigroup time evolution

The preparation-registration arrow of time (3) defines causally allowed order of events. Here one may ask a question: Can this causality be extended to measurements made beyond the preparation or the registration?

In order to answer this question we quote a causality argument given by Dirac [4], that hereafter we refer to *Dirac's causality*:

"Causality applies only to a system which is left undisturbed. If a system is small, we cannot observe it without producing serious disturbance and hence we cannot expect to find any causal connexion between the results of our observation".

This causality justifies calling Eq. (1) the beginning of time: an observation isolates a microsystem from any other observations performed previously. Hence causality applies only to a time interval between the preparation and the registration during which the microsystem is left undisturbed.

In terms of states and observables, this is the statement that dynamical equations in quantum mechanics [4] are applicable only between the preparation and the registration. The time evolution of the state $|\phi(t)\rangle$ in the Schrödinger picture is described by the Schrödinger equation,

$$i\frac{d}{dt}|\phi(t)\rangle = H\,|\phi(t)\rangle, \tag{4a}$$

on the other hand, the time evolution of the observable $\mathcal{O}$ is described by the Heisenberg equation

$$i\frac{d}{dt}\mathcal{O}(t) = [\mathcal{O}(t), H], \tag{4b}$$

where t is the time parameter. Because of Dirac's causality, this time parameter is defined only between the preparation time T_i and the registration time T_i' for each of the N microsystems. This means that t describes the time intervals,

$$t_i \equiv T_i' - T_i \quad \text{for } i = 1, 2, 3, \cdots, N, \tag{5}$$

which eventually form an ensemble,

$$t : \{t_1, t_2, t_3, \cdots, t_N\}. \tag{6}$$

According to the preparation-registration arrow of time (3), each of the time intervals satisfy $t_i \geq 0$. This leads to the *time asymmetry* expressed as

$$0 \leq t < \infty. \tag{7}$$

This time asymmetry requires that the dynamical equations (4) have to be integrated under some suitable boundary condition to yield the *semigroup* time evolution [5],

$$|\phi(t)\rangle = e^{-iHt}|\phi\rangle \text{ for } 0 \leq t < \infty \text{ in the Schrödinger picture}, \tag{8a}$$

$$\mathcal{O}(t) = e^{iHt}\mathcal{O}e^{-iHt} \text{ for } 0 \leq t < \infty \text{ in the Heisenberg picture}, \tag{8b}$$

with the initial condition $|\phi(t=0)\rangle = |\phi\rangle$ and $\mathcal{O}(t=0) = \mathcal{O}$. This is the general result obtained from the preparation-registration arrow of time and Dirac's causality.

Journal of Physics: Conference Series **128** (2008) 012040 doi:10.1088/1742-6596/128/1/012040

2. Transition probability

The transition probability, or also called *Born probability* [5], is described by an special observable which satisfy $\mathcal{O}_\psi^2 = \mathcal{O}_\psi$, whose eigenvalues are only 1 and 0 [4, 6]. This is expressed by a normalized vector $|\psi\rangle$ (which is the eigenvector of $\mathcal{O}_\psi$) as

$$\mathcal{O}_\psi = |\psi\rangle\langle\psi|. \tag{9}$$

Let us call such an observable $\mathcal{O}_\psi$, and also the vector $|\psi\rangle$, *yes-or-no observable.* In the Heisenberg picture the time evolution of $\mathcal{O}_\psi$ with the initial condition $\mathcal{O}_\psi(t = 0) = \mathcal{O}_\psi$ is given by substituting Eq. (9) into Eq. (8b) as

$$\mathcal{O}_\psi(t) = e^{iHt}|\psi\rangle\langle\psi|e^{-iHt} \equiv |\psi(t)\rangle\langle\psi(t)|, \tag{10a}$$

where

$$|\psi(t)\rangle = e^{iHt}|\psi\rangle \text{ for } 0 \leq t < \infty. \tag{10b}$$

By this observable, the transition probability $\mathcal{P}(t)$ that $\mathcal{O}_\psi$ is found in the state $|\phi\rangle$ is then given by the expectation value,

$$\mathcal{P}(t) = \langle\phi(t)|\mathcal{O}_\psi|\phi(t)\rangle = |\langle\psi|\phi(t)\rangle| \text{ in the Schrödinger picture,} \tag{11a}$$

and

$$\mathcal{P}(t) = \langle\phi|\mathcal{O}_\psi(t)|\phi\rangle = |\langle\psi(t)|\phi\rangle| \text{ in the Heisenberg picture.} \tag{11b}$$

This probability is bound as $0 \leq \mathcal{P}(t) \leq 1$ by Schwartz inequality, for example, in the Schrödinger picture,

$$0 \leq |\langle\psi|\phi(t)\rangle|^2 \leq \langle\psi|\psi\rangle\langle\phi(t)|\phi(t)\rangle = 1 \tag{12}$$

for the normalized vectors $\langle\phi(t)|\phi(t)\rangle = \langle\psi|\psi\rangle = 1$.

3. Boundary conditions

In this section we discuss a boundary condition to integrate the dynamical equations (4). They are the abstract vector spaces for states and that for the yes-or-no observable.

3.1. Hilbert space

In mathematical physics the Hilbert space H has been taken for the abstract vector space for states [7],

$$|\phi\rangle \in \mathsf{H}. \tag{13}$$

In the Hilbert space, the Stone-von Neumann theorem [8] states that for a Hamiltonian H selfadjoint on H the Schrödinger equation is integrated to an unitary *group* time evolution,

$$|\phi(t)\rangle = e^{-iHt}|\phi\rangle \text{ for } -\infty < t < \infty. \tag{14}$$

This time evolution disagrees with the semigroup time evolution (8a).

One may think that this discrepancy can be resolved by choosing a suitable observable so that the expectation value $\langle\mathcal{O}\rangle_t \equiv \langle\phi(t)|\mathcal{O}|\phi(t)\rangle$ vanishes for $-\infty < t \leq 0$. This possibility, however, is denied by a theorem proved by Hegerfeldt [9], which states that for a positive operator $\mathcal{O}$ on H either of the following cases are possible:

(i) $\langle \mathcal{O} \rangle_t \neq 0$ for almost all t, and the set of such t's is dense and open, or
(ii) $\langle \mathcal{O} \rangle_t = 0$ for all t.

From this follows is that the transition probability given by the operator (9) is such that

$$\mathcal{P}(t) \text{ does not vanish for } -\infty < t < 0 \text{ unless } \mathcal{P}(t) \text{ is identically zero for } -\infty < t < \infty. \tag{15}$$

Thus there seems no way to obtain the semigroup time evolution in the Hilbert space.

3.2. Time asymmetric boundary condition (TABC) for states

There is a boundary condition that gives the semigroup time evolution (8); it is called the time asymmetric boundary condition [5], which uses not the entire Hilbert space H but its particular dense subspace, the Hardy space Φ_-, for states

$$|\phi^+\rangle \in \Phi_- \subset \mathsf{H}, \tag{16}$$

where the superscript "+" on the state vector designates the element of Φ_-. This restriction leads to a rigged Hilbert space (RHS) of the Hardy space [10],

$$\left.\begin{matrix} |\phi^+\rangle \\ \text{or} \\ \langle\phi^+| \end{matrix}\right\} \in \Phi_- \subset \mathsf{H} \subset \Phi_-^\times \ni \left\{\begin{matrix} \langle^+E\,a| \\ \text{or} \\ |E\,a^+\rangle \end{matrix}\right. \tag{17}$$

Here $|E\,a^+\rangle$ ($\langle^+E\,a|$) is the energy eigenket (eigenbra), the functional over the Hardy space Φ_-, which satisfies the generalized eigenvalue equation,

$$H_-^\times |E\,a^+\rangle = E\,|E\,a^+\rangle \tag{18}$$

$$\mathcal{A}^\times\,|E\,a^+\rangle = a\,|E\,a^+\rangle \tag{19}$$

where $H_-^\times$ is the unique extension of the Hamiltonian H_- defined on Φ_- to $\Phi_-^\times$, and $\mathcal{A}$ is the set of operators that makes together with the Hamiltonian a complete set of commuting observables (CSCO), e.g., angular momentum operator $\boldsymbol{J}^2$ and J_3. By this Hardy RHS the state vector is expanded as

$$|\phi^+\rangle = \sum_a \int_0^\infty dE\,|E\,a^+\rangle\,\phi_a^+(E), \tag{20}$$

where $\phi_a^+(E) \equiv \langle^+E\,a|\phi^+\rangle = \overline{\langle\phi^+|E\,a^+\rangle}$ is the state energy wave function [5].

This boundary condition is characterized by the analyticity of the energy wave function, that is, $\phi_a^+(E)$ is the *Hardy function from below* $\mathcal{S}\cap\mathcal{H}_-|_{\mathbb{R}_+}$, analytic in the lower-half complex energy plane satisfying [5]

$$\int_{-\infty}^{\infty} dE\,\left|\phi_a^+(E+iy)\right|^2 < \infty \text{ for any } y < 0. \tag{21}$$

This is a necessary and sufficient condition that the wave function satisfies the Hilbert transform [11], also called *dispersion relation* [11, 12] in physics,

$$\mathrm{Re}\,\phi_a^+(E) = -\frac{1}{\pi}\mathrm{P}\int_{-\infty}^{\infty} d\omega\,\frac{\mathrm{Im}\,\phi_a^+(\omega)}{\omega - E} \tag{22a}$$

$$\mathrm{Im}\,\phi_a^+(E) = \frac{1}{\pi}\mathrm{P}\int_{-\infty}^{\infty} d\omega\,\frac{\mathrm{Re}\,\phi_a^+(\omega)}{\omega - E} \tag{22b}$$

The semigroup time evolution is the immediate consequence of the analyticity (21). Integration of the Schrödinger equation in Φ_- is given by

$$
\begin{aligned}
|\phi^+(t)\rangle &= e^{-iH_-t}\,|\phi^+\rangle \\
&= \sum_a \int_0^\infty dE\,|E\,a^+\rangle\langle^+E\,a|e^{-iH_-t}\phi^+\rangle \\
&= \sum_a \int_0^\infty dE\,|E\,a^+\rangle\left[e^{-iEt}\phi_a^+(E)\right]. \qquad (23)
\end{aligned}
$$

For this time dependent state vector to stay in the Hardy space Φ_-, the time dependent wave function must satisfy Eq. (21),

$$
\int_{-\infty}^{\infty} dE\,\left|e^{-i(E+iy)t}\phi_a^+(E+iy)\right|^2 = \int_{-\infty}^{\infty} dE\,e^{2yt}\,\left|\phi_a^+(E+iy)\right|^2 < \infty \text{ for any } y < 0. \qquad (24)
$$

This integral converges only for $0 \le t < \infty$, so that we obtain the semigroup time evolution

$$
|\phi^+(t)\rangle = e^{-iH_-t}|\phi^+\rangle \text{ for } 0 \le t < \infty. \qquad (25)
$$

This is exactly the same as Eq. (8a).

3.3. *TABC for yes-or-no observables and transition probability*

We need yet another boundary condition that leads to the semigroup time evolution (10b) for the yes-or-no observable; this is the other Hardy RHS [5],

$$
\left.\begin{array}{c} |\psi^-\rangle \\ \text{or} \\ \langle\psi^-| \end{array}\right\} \in \Phi_+ \subset \mathsf{H} \subset \Phi_+^\times \ni \left\{\begin{array}{c} \langle^-E\,a| \\ \text{or} \\ |E\,a^-\rangle \end{array}\right. \qquad (26)
$$

where the superscript "-" designates the element of this Hardy RHS. By this RHS the yes-or-no observable is expanded as

$$
|\psi^-\rangle = \sum_a \int_0^\infty dE\,|E\,a^-\rangle\,\psi_a^-(E), \qquad (27)
$$

where $\psi_a^-(E) \equiv \langle^-E\,a|\psi^-\rangle$ is the energy wave function of the yes-or-no observable. This wave function is the *Hardy function from above* $\mathcal{S}\cap\mathcal{H}_+|_{\mathbb{R}_+}$, analytic in the upper-half complex energy plane satisfying

$$
\int_{-\infty}^{\infty} dE\,\left|\psi_a^-(E+iy)\right|^2 < \infty \text{ for any } y > 0. \qquad (28)
$$

This is a necessary and sufficient condition that the wave function satisfies another (complex conjugated) dispersion relation,

$$
\mathrm{Re}\,\psi_a^-(E) = \frac{1}{\pi}\mathrm{P}\int_{-\infty}^{\infty} d\omega\,\frac{\mathrm{Im}\,\psi_a^-(\omega)}{\omega - E} \qquad (29a)
$$

$$
\mathrm{Im}\,\psi_a^-(E) = -\frac{1}{\pi}\mathrm{P}\int_{-\infty}^{\infty} d\omega\,\frac{\mathrm{Re}\,\psi_a^-(\omega)}{\omega - E} \qquad (29b)
$$

In the same manner as from Eq. (23) to Eq. (24), the time evolution of the yes-or-no observable in the Heisenberg picture is given by

$$|\psi^-(t)\rangle = e^{iH_+t}|\psi^-\rangle \text{ for } 0 \le t < \infty, \tag{30}$$

where H_+ is the selfadjoint Hamiltonian restricted to Φ_+. This time evolution is exactly the same as Eq. (10b).

By the TABC for states (17) and the TABC for yes-or-no observable (26), the transition probability is given by

$$\mathcal{P}(t) = \left|\langle\psi^-|\phi^+(t)\rangle\right|^2 \text{ in the Schrödinger picture} \tag{31a}$$

$$= \left|\langle\psi^-(t)|\phi^+\rangle\right|^2 \text{ in the Heisenberg picture} \tag{31b}$$

for $0 \le t < \infty$.

3.4. Decaying state in the TABC

In addition to that the TABC satisfies the quantum mechanical causality condition, the time asymmetry (7), there exist in $\Phi_+^\times$ Gamow kets [10, 5] that describe decaying, or quasi-stable, states. Gamow ket is the eigenket of the Hamiltonian $H_+^\times$ satisfying the generalized eigenvalue equation

$$H_+^\times|z_D\, a_D\,^-\rangle = (E_D - i\Gamma/2)\,|z_D\, a_D\,^-\rangle, \tag{32}$$

where $E_D > 0$ and $\Gamma > 0$, and a_D is the particular set of eigenvalues of $\mathcal{A}^\times$ that characterizes the decaying state. The time evolution of the Gamow ket is given by

$$e^{-iH_+^\times t}|z_D\, a_D\,^-\rangle = e^{-\Gamma t/2}e^{-iE_Dt}|z_D\, a_D\,^-\rangle \text{ for } 0 \le t < \infty \text{ only.} \tag{33}$$

With the Gamow ket $|z_D\, a_D\,^-\rangle$, the state vector is expanded as

$$|\phi^+(t)\rangle = \sum_{i=1}^{N_D} e^{-\Gamma_i t/2}e^{-iE_{D_i}t}|z_{D_i}\, a_{D_i}\,^-\rangle + |B(t)^-\rangle, \tag{34}$$

where N_D is the number of decaying states and

$$|B(t)^-\rangle \equiv \sum_a \int_0^{-\infty} dE\,|E\,a^-\rangle\langle^-E\,a|\phi^+(t)\rangle \tag{35}$$

is called the background ket [5]. With the expansion (34) the transition probability (31) is given in the Schrödinger picture by

$$\mathcal{P}(t) = \left|\sum_{i=1}^{N_D} e^{-\Gamma_i t/2}e^{-iE_{D_i}t}\langle\psi^-|z_{D_i}\, a_{D_i}\,^-\rangle + \langle\psi^-|B(t)^-\rangle\right|^2. \tag{36}$$

4. Neutral kaon decay experiment

In order to illustrate the usage of the TABC we discuss the neutral kaon decay experiment.

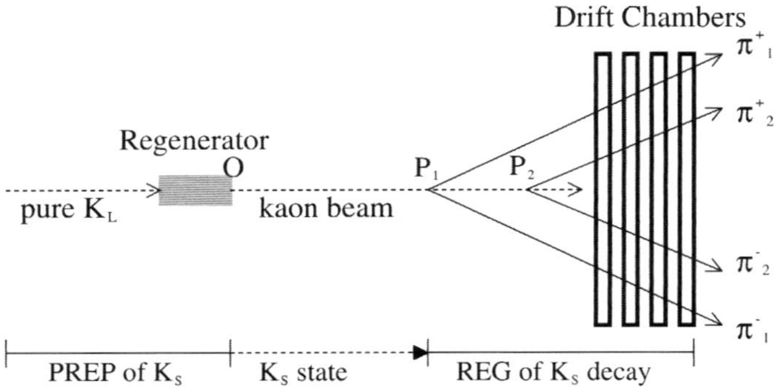

Figure 2. The kaon regeneration experiment. The kaon beam after the regenerator is a coherent mixture of K_S and K_L. O is the downstream edge of the regenerator, and P_i is the decay vertex of i-th kaon reconstructed from the trajectories of the $\pi_i^+\pi_i^-$ pair.

4.1. K_S lifetime measurement

A precise measurements of the lifetime of the short-lived kaon K_S has been made by Fermilab experiment 731 [13]. The experiment uses the technique of the coherent regeneration of K_S. The preparation of the K_S beam is performed by impinging pure K_L beam on a slab of material, such as boron carbide, called the regenerator (Fig. 2). In this process the regenerator actually produces coherent mixture of K_S and K_L but the regenerated K_S flux dominates over the K_L flux. Therefore, for simplicity, we neglict the contribution from K_L in the following discussion.

As shown in Fig. 2, the regenerated K_S beam immediately proceeds to the vacuum decay volume, and its decay products are observed by a detector located downstream. To be specific, let us focus on the charged 2π decay mode, $K_S \to \pi^+\pi^-$. The pairs of π^+ and π^- decayed from the K_S beam are detected by a series of drift chambers that register trajectories and momenta of the individual pions. The decay vertexes of kaon, $\{\mathrm{P}_1, \mathrm{P}_2, \cdots, \mathrm{P}_N\}$, are reconstructed from the obtained data set by extrapolating the trajectories of the corresponding pion pairs, and the momenta of kaon $\{\boldsymbol{p}_1, \boldsymbol{p}_2, \cdots, \boldsymbol{p}_N\}$ are determined by the conservation rule,

$$\boldsymbol{p}_i = \boldsymbol{p}_i^+ + \boldsymbol{p}_i^- , \tag{37}$$

where $\boldsymbol{p}_i^+$ and $\boldsymbol{p}_i^-$ are the momenta of i-th π^+ and π^-, respectively. From these decay vertexes and the momenta the proper times τ_i of individual i-th kaon is calculated by the relativistic kinematics,

$$\tau_i = \frac{m\, d_i}{|\boldsymbol{p}_i|} , \tag{38}$$

where m is the mass of K_S and d_i is the distance from the edge of the regenerator to the i-th reconstructed decay vertex, $d_i = \overline{\mathrm{OP}_i}$. Finally, the set of proper times for N events of individual kaon decay is obtained,

$$\{\tau_1, \tau_2, \cdots, \tau_N\} , \tag{39}$$

with $N \sim \mathcal{O}(10^6)$, the order of magnitude of the number of kaons in the beam.

The lifetime τ_S for the $\pi^+\pi^-$ mode is extracted from the observed proper times (39) by data fitting with the phenomenological formula of the instantaneous rate $R(\tau)$ for kaon decays [13],

$$R(\tau) \propto |\rho|^2 \, e^{-\tau/\tau_S} , \tag{40}$$

Journal of Physics: Conference Series **128** (2008) 012040 doi:10.1088/1742-6596/128/1/012040

where ρ is called the K_S amplitude determined by the regeneration process.

In the following section, we will reproduce the same formula by quantum mechanics with the TABC.

4.2. The preparation and the registration of K_S

The analysis of the experiment just briefly described is based on the assumption that kaons and pions behave as (classical) particles. Here we shall give a quantum mechanical account to the experiment, namely in terms of the states and observables.

Kaons are prepared by the regenerator[1] in the same quantum state $|\phi_K^+\rangle$. This preparation is completed at the edge of the regenerator (O in Fig. 2) from which the kaons emerge. This means that the ensemble of beginnings of time $t = 0$ of this quantum state is the ensemble of laboratory times at which the kaon start to emerge[2],

$$t = 0 \, : \, \{T_1, T_2, \cdots, T_N\} \text{ at the edge of the regenerator O.} \tag{41}$$

After this point O of preparation, the kaons propagate in vacuum so that they are left free of any influence besides the weak interaction in themselves.

The kaon state $|\phi_K^+\rangle$ describes a decaying K_S state. This is written by the complex basis vector expansion with a *relativistic* Gamow ket $|K_S{}^-\rangle$ satisfying the (generalized) eigenvalue equation [14],

$$H^\times |K_S{}^-\rangle = \gamma \left(m - \frac{i}{2\tau_S} \right) |K_S{}^-\rangle, \tag{42}$$

where γ is the Lorentz factor, and m_S is the mass of K_S. In terms of this Gamow ket the kaon state vector is expanded as[3]

$$|\phi_K^+\rangle = |K_S{}^-\rangle + |B_K^-\rangle, \tag{43}$$

where $|B_K^-\rangle$ is the background term [14]. In the Schrödinger picture, this vector evolves in time. If we are on the rest-frame of the kaon then $\gamma = 1$ holds and the time parameter t is taken to be the proper time τ of the kaon. Then from Eqs. (42) and (43), we obtain the time evolution of state in the rest-frame,

$$|\phi_K^+(\tau)\rangle = e^{-i(m-i/2\tau_S)\tau} \, |K_S{}^-\rangle + |B_K^-(\tau)\rangle \text{ for } \tau \geq 0. \tag{44}$$

The registration of the kaon decay into $\pi^+\pi^-$ is described by the yes-or-no observable $|\psi_{\pi^+\pi^-}^-\rangle \in \Phi_+$ as whether the kaon has decayed (yes) or not (no). This observable, however, does not represent actual measurements by the series of drift chambers, because the kaon decays in vacuum without being "looked at" [2]; the decay is "observed" only afterword as a *reconstructed* vertex extracted from the data of the pions. Thus the disturbance accompanying this "observation" made on the microsystem (kaon) is not due to interaction with a measuring apparatus but due to the weak interaction responsible for the spontaneous decay of the

[1] To a good approximation, we have neglected the recoil of the nuclei of the regenerator in the forward scattering with K_L.

[2] This $t = 0$ applies not only to a single run, but also to a whole runs of the experiment; the number of elements in the ensemble is the number of the total events that recorded in the period of the experiment performed under the same condition. (This is no longer the case, however, if any settings of beam, regenerator, or any other preparation related variables are significantly changed.)

[3] By the approximation that the momentum spread of the kaon is very small [14].

microsystem. On this basis, the decay of the kaon state $|\phi_K^+\rangle$ is described by the registration of the observable $|\psi^-_{\pi^+\pi^-}\rangle$ at the various laboratory times

$$\{T_1', T_2', \cdots, T_N'\} \text{ at the decay vertexes } \{\mathrm{P}_1, \mathrm{P}_2, \cdots, \mathrm{P}_N\} \text{ respectively,} \tag{45}$$

which defines the ends of time evolution of the kaon state. Thus we have the ensemble of (proper) time intervals during which each of the kaons is left undisturbed, i.e., undecayed, as

$$\tau \; : \; \{\tau_1, \tau_2, \cdots, \tau_N\}, \tag{46}$$

each obtained by

$$\tau_i = \frac{T_i' - T_i}{\gamma_i} \quad \text{with} \quad \gamma_i = \frac{\sqrt{\boldsymbol{p}_i^2 + m^2}}{m}. \tag{47}$$

This set of time intervals is exactly the same as Eq. (39).

The transition probability that the observable $|\psi^-_{\pi^+\pi^-}\rangle$ is found in the state $|\phi_K^+\rangle$ is given by

$$\mathcal{P}(\tau) = \left|\langle\psi^-_{\pi^+\pi^-}|\phi_K^+(\tau)\rangle\right|^2. \tag{48}$$

With Eq. (44), this reproduces the formula for the decay rate (40) as follows: If the effect of the background term is too small to be observed, namely,

$$\left|\langle\psi^-_{\pi^+\pi^-}|B^-\rangle\right|^2 \ll \left|\langle\psi^-_{\pi^+\pi^-}|K_S\rangle\right|^2, \tag{49}$$

we obtain the exponential decay (the Weisskopf-Wigner approximation),

$$\mathcal{P}(\tau) \cong e^{-\tau/\tau_S} \left|\langle\psi^-_{\pi^+\pi^-}|K_S\rangle\right|^2. \tag{50}$$

From this, the decay rate is given by

$$R(\tau) = -\frac{d\mathcal{P}(\tau)}{d\tau} = \frac{\left|\langle\psi^-_{\pi^+\pi^-}|K_S\rangle\right|^2}{\tau_S} e^{-\tau/\tau_S}. \tag{51}$$

Because the K_S flux depends on the regenerator, $\left|\langle\psi^-_{\pi^+\pi^-}|K_S\rangle\right|^2 \propto |\rho|^2$ holds and we reproduce Eq. (40).

5. Summary

Semigroup time evolution is a general requirement from the preparation-registration arrow of time and Dirac's causality. The semigroup is attained by the time-asymmetric boundary condition (TABC) which is characterized by the energy wave functions satisfying dispersion relations. Since TABC includes quasi-stable (decaying) states as Gamow kets, stable particles and quasi-stable particles are described in the same footing by the TABC. This leads to a physical interpretation that decay phenomenon of a quasi-stable particle is one of the "observations" that disturbs a quantum system.

Acknowledgments

This work was supported by Fundaċion Universidades de Castilla y Leòn.

Journal of Physics: Conference Series **128** (2008) 012040 doi:10.1088/1742-6596/128/1/012040

References

[1] Ludwig G 1985 *Foundations of Quantum Mechanics*, Vol. I and II (Berlin: Springer-Verlag)
[2] Bohm A and Harshman N L 1998 *Irreversibility and causality* ed Bohm A, Doebner H and and Kielanowski P (Berlin Heidelberg: Springer)
[3] Bohm A, Kielanowski P and Wickramasekara S 2006 *Ann. Phys.* **321** 2299
[4] Dirac P A M 1958 *The Principles of Quantum Mechanics* 4th ed (Oxford: Oxford University Press)
[5] Bohm A, Maxson M, Loewe M and Gadella M 1997 *Physica* A **236** 485
[6] Haag R 1996 *Local Quantum Physics: Fields, Particles, Algebras* 2nd ed (Berlin: Springer)
[7] von Neumann J 1955 *Mathematica Foundations of Quantum Mechanics* (Princeton: Princeton University Press)
[8] Stone M H 1932 *Ann. Math.* **33** 643
von Neumann J 1932 *Ann. Math.* **33** 567
[9] Hegerfeldt G C 1994 *Phys. Rev. Lett.* **72** 596
Hegerfeldt G C 1998 *Irreversibility and causality* Bohm A, Doebner H and Kielanowski P ed (Berlin Heidelberg: Springer)
[10] Bohm A and Gadella M 1989 *Dirac Kets, Gamow Vectors, and Gel'fand Triplets* Lecture Notes in Physics Vol.348 (Berlin: Springer)
[11] Goldberger M L and Watson K M 1964 *Collision Theory* (New York: Wiley)
[12] Jackson J D 1999 *Classical Electrodynamics* 3rd ed. (New York: John Wiley & sons)
[13] Gibbons K L et al 1997 *Phys. Rev.* D **55** 6625
[14] Bohm A, Kaldass H and Wickramasekara S 2003 *Fortschr. Phys.* **51** 569
Bohm A, Kaldass H and Wickramasekara S 2003 *Fortschr. Phys.* **51** 604

Journal of Physics: Conference Series **128** (2008) 012041 doi:10.1088/1742-6596/128/1/012041

General structures of Hamiltonian hierarchies

R C dos Anjos [1], E Drigo Filho[1] and R M Ricotta[2]

[1] UNESP, Universidade Estadual Paulista - IBILCE

[2] FATEC/SP, Faculdade de Tecnologia de São Paulo - CEETEPS-UNESP

E-mail: elso@ibilce.unesp.br, regina@fatecsp.br

Abstract. Three dimensional exactly solvable quantum potentials for which an extra term of form $\frac{1}{r^2}$ has been added are shown to maintain their functional form which allows the construction of the Hamiltonian hierarchy and the determination of the spectra of eigenvalues and eigenfunctions within the Supersymmetric Quantum Mechanics formalism. For the specific cases of the harmonic oscillator and the Coulomb potentials, known as pseudo-harmonic oscillator and pseudo-Coulomb potentials, it is shown here that the inclusion of the new term corresponds to rescaling the angular momentum and it is responsible for maintaining their exact solvability.

1. Introduction

In the context of supersymmetric quantum mechanics, [1], the Hamiltonian hierarchy, introduced by Sukumar, [2], is an important algebraic structure that introduces a formalism which allows us to solve the Schrödinger equation, not only for the exactly solvable potentials but for the partially solvable ones, [3]-[7].

Few potentials in three dimensions are known to permit the construction of their complete hierarchy. The harmonic oscillator and the Coulomb potentials are examples of this property. It has been observed that the inclusion of an extra term of form $\frac{1}{r^2}$ in these potentials is such that their exact solvability is preserved, [8]. They are so called pseudo-harmonic oscillator and pseudo-Coulomb potentials, respectivelly, [9]-[15]. Variations of the pseudo-Coulomb potentials are the Kratzer potential, [16], or Mie-type potential, [17]. The study of these potentials is motivated by the fact that the new term is followed by an extra parameter, useful for applications in quantum chemistry and atomic and molecular physics, [17] -[25]. The reason why the factorization is possible is because the term added to the potential acts as a rescaling of the angular momentum barrier term. In this way, the exact solvability and shape invariance of both potentials are preserved.

In this work the Hamiltonian hierarchy of the pseudo-harmonic and pseudo-Coulomb potentials in three dimensions are shown and their spectra of eigenvalues and eigenfunctions are indicated.

2. The Hamiltonian hierarchy

Through the super-algebra, for a given Hamiltonian H_1, factorized in terms of the bosonic operators, it is possible to construct its hierarchy of Hamiltonians. For the general spontaneously

Journal of Physics: Conference Series **128** (2008) 012041 doi:10.1088/1742-6596/128/1/012041

broken supersymmetric case we have

$$H_1 = -\frac{d^2}{dx^2} + V_1(x) = A_1^+ A_1^- + E_0^{(1)} \tag{1}$$

where $E_0^{(1)}$ is the lowest eigenvalue, written in $\frac{\hbar^2}{2m}$ units. The bosonic operators are defined by

$$A^{\pm} = \mp\frac{d}{dx} + W_1(x) \tag{2}$$

where the superpotential $W_1(r)$ satisfies the Riccati equation

$$W_1^2 - W_1' = V_1(x) - E_0^{(1)}. \tag{3}$$

The eigenfunction for the lowest state is related to the superpotential W_1 by

$$\Psi_0^{(1)}(x) = Nexp(-\int_0^x W_1(\bar{x})d\bar{x}). \tag{4}$$

The supersymmetric partner Hamiltonian is given by

$$H_2 = A_1^- A_1^+ + E_0^{(1)} = -\frac{d^2}{dr^2} + (W_1^2 + W_1') + E_0^{(1)}. \tag{5}$$

Thus, factorizing H_2 in terms of a new pair of bosonic operators, $A_2^{\pm}$ we get,

$$H_2 = A_2^+ A_2^- + E_0^{(2)} = -\frac{d^2}{dx^2} + (W_2^2 - W_2') + E_0^{(2)} \tag{6}$$

where $E_0^{(2)}$ is the lowest eigenvalue of H_2 and W_2 satisfy the Riccati equation,

$$W_2^2 - W_2' = V_2(x) - E_0^{(2)}. \tag{7}$$

Thus a whole hierarchy of Hamiltonians can be constructed, with simple relations connecting the eigenvalues and eigenfunctions of the n-members, [2],

$$H_n = A_n^+ A_n^- + E_0^{(n)} \tag{8}$$

$$A_n^{\pm} = \mp\frac{d}{dx} + W_n(x) \tag{9}$$

$$\Psi_n^{(1)} = A_1^+ A_2^+ ... A_n^+ \psi_0^{(n+1)}\ , \quad E_n^{(1)} = E_0^{(n+1)} \tag{10}$$

where $\Psi_0^{(1)}(x)$ is given by (4).

3. The pseudo-harmonic oscillator

The potential for this case is given by

$$V_{PH} = (br + \frac{a}{r})^2 = \frac{a^2}{r^2} + 2ab + b^2 r^2. \tag{11}$$

The factorized radial Hamiltonian H_1, first member of the hierarchy, is given by

$$H_1 = A_1^+ A_1^- + E_0^{(1)} = -\frac{d^2}{dr^2} + V_1(r) \tag{12}$$

where $V_1(r)$ incorporates the potential barrier term, $V_1(r) = V_{PH} + l(l+1)/r^2$.

In order to simplify the calculation, we rescale the angular momentum l to a new constant γ,

$$a^2 + l(l+1) = \gamma(\gamma+1) \tag{13}$$

whose solution for γ is

$$\gamma = -\frac{1}{2} + \frac{1}{2}\sqrt{1 + 4(a^2 + l(l+1))}. \tag{14}$$

Writing the superpotential as

$$W_1(r) = \beta r - \frac{\alpha}{r} \tag{15}$$

and substituting it into equation (3) we arrive at

$$\alpha = \gamma + 1\ ,\ \beta = b. \tag{16}$$

As a result, the groundstate energy is given by

$$E_0^{(1)} = 2ab + 2\gamma b + 3b. \tag{17}$$

In terms of the parameters, the superpotential (15) is rewritten as

$$W_1(r) = br - \frac{\gamma+1}{r}. \tag{18}$$

With the above superpotential through the equation (4), a convergent ground state wavefunction. It is given by

$$\Psi_0^{(1)}(r) = Ne^{-\frac{ab}{\gamma+1}r}r^{\gamma+1}. \tag{19}$$

Repeating the same procedure we can evaluate the $(n+1)$-th member of the hierarchy. The results are the following, [8],

$$W_{n+1} = br - \frac{\gamma+n+1}{r}, \tag{20}$$

the ground state wavefunction is

$$\Psi_0^{(n+1)}(r) = Ne^{-\frac{b}{2}r^2}r^{\gamma+n+1}. \tag{21}$$

and the ground state energy is

$$E_0^{(n+1)} = 2ab + 2\gamma b + (4n+3)b, \tag{22}$$

where γ is defined by (14). Thus the potentials of the hierarchy are given by

$$V_{n+1} = b^2r^2 + \frac{(\gamma+n)(\gamma+n+1)}{r^2} + 2nb + 2ab. \tag{23}$$

From the above results we notice that this is a shape invariant potential, as defined by [26].

Journal of Physics: Conference Series **128** (2008) 012041 doi:10.1088/1742-6596/128/1/012041

4. Pseudo-Coulomb potential

The pseudo-Coulomb potential is given, in atomic units, as

$$V_{PC} = c - \frac{b}{r} + \frac{a^2}{r^2} \tag{24}$$

and the factorized Hamiltonian associated is given by

$$H_1 = A_1^+ A_1^- + E_0^{(1)} = -\frac{d^2}{dr^2} + V_1(r). \tag{25}$$

$V_1(r)$ is the potential of the first Hamiltonian of the hierarchy, $V_1(r) = V_{PC} + l(l+1)/r^2$. Similarly as in the previous case, we rescale the angular momentum l to a new constant γ, as in (13), whose solution for γ is exactly like (14). Writing the superpotential as

$$W_1(r) = \beta - \frac{\alpha}{r} \tag{26}$$

and substituting it into equation (3) we arrive at

$$\alpha = \gamma + 1\ , \quad \beta = \frac{b}{2(\gamma+1)}. \tag{27}$$

As a result, the energy is given by

$$E_0^{(1)} = c - \beta^2 = c - \frac{b^2}{4(\gamma+1)^2}. \tag{28}$$

In terms of the parameters, the superpotential (26) is rewritten as

$$W_1(r) = \frac{b}{2(\gamma+1)} - \frac{\gamma+1}{r} \tag{29}$$

The convergent ground state wavefunction is

$$\Psi_0^{(1)}(r) = N e^{-\frac{b}{2(\gamma+1)}r} r^{\gamma+1}. \tag{30}$$

Repeating the same procedure we can evaluate the $(n+1)$-th member of the hierarchy. The results are the following, [8],

$$W_{n+1} = \frac{b}{2(\gamma+n+1)} - \frac{\gamma+n+1}{r} \tag{31}$$

or given in terms of the potentials

$$V_{n+1} = c - \frac{b}{r} + \frac{(\gamma+n)(\gamma+n+1)}{r^2} \tag{32}$$

with groundstate wavefunctions

$$\Psi_0^{(n+1)}(r) = N e^{-\frac{b}{2(\gamma+n+1)}r} r^{\gamma+n+1}. \tag{33}$$

The corresponding ground state energy

$$E_0^{(n+1)} = c - \frac{b^2}{4(\gamma+n+1)^2} \tag{34}$$

For the particular case where

$$c^2 \to b^2\ , \quad b \to 2ab \tag{35}$$

the pseudo-Coulomb potential takes the usual form as shown in the literature, [15], [24]. Once again we notice that this pseudo-Coulomb potential is shape invariant in the sense of [26].

5. Conclusions

The Hamiltonian hierarchy of pseudo-harmonic and pseudo-Coulomb potentials were evaluated. These potentials are constructed through the introduction of a term proportional to $\frac{1}{r^2}$ into the original harmonic oscillator and Coulomb potentials, respectively. However, the introduction of this additional term, important for studies of molecular systems, is such that it acts as a centrifugal term. This is the reason why the exact solvability and the shape invariance of both potentials are preserved. It is important to remark the fact that the method allows us to evaluate the spectra of energy and eigenfunctions of the original potential.

Acknowledgments

EDF acklowledges partial financial support from CNPq.

6. References

[1] Witten E 1981 *Nucl. Phys.* **B 188** 513
[2] Sukumar C V 1985 *J. Phys. A: Math. Gen.* **18** L57
[3] Drigo Filho E and Ricotta R M 1995 *Mod. Phys. Lett.* **A10** 1613
[4] Fernandez D J, Negro J and Del Olmo M A 1996 *Annals of Phys.* **252** 386
[5] Gonul B, Ozer O, Cancelik Y and M. Kocak 2000 *Phys. Lett.* **A 275** 238
[6] Drigo Filho E and Ricotta R M 2001 *Braz. J. Phys.* **31** 334
[7] Fakhri H and Chenaghlou A 2004 *J. Phys. A: Math. Gen.* **37** 8545
[8] Drigo Filho E and Ricotta R M 2007 to appear in Physica Scripta.
[9] Ballhausen C J 1988 *Chem. Phys. Lett.* **146** 449
[10] Büyükkiliç F, Demirhan D and Özeren S F 1992 *Chem. Phys. Lett.* **194** 9
[11] Wang L Y, Gu X Y, Ma Z Q and Dong S H 2002 *Found. Phys. Lett.* **15** 569
[12] Chen C Y and Dong S H 2005 *Phys. Lett.* **A 335** 374
[13] Ikhdair S M and Sever R, quant-ph/0611065, quant-ph/0611183, quant-ph/0702052, quant-ph/0703042, quant-ph/0703131
[14] Setare M R and Karimi E 2007 *Phys. Scr.* **75** 90
[15] Cheng Y F and Dai T Q 2007 *Phys. Scr.* **75** 274
[16] Kratzer A, Z. Physics **3** (1920) 289
[17] Erkoç S and Sever R 1986 *Phys. Rev.* **D 33** 588
[18] Popov P 2001 *J. Phys. A: Math. Gen.* **34** 5283 and 2002 *J. Phys. A: Math. Gen.* **35** 7205
[19] Schmiedmayer J and Scrinzi A 1996 *Phys. Rev.* **A 54** R2525
[20] Hartmann H and Schuch D 1980 *Int. J. Quantum Chem.* **18** 125
[21] Kibler M and Negadi T 1984 *Int. J. Quantum Chem.* **26** 405
[22] Kibler M and Winternitz P 1987 *J. Phys. A: Math. Gen.* **20** 4097
[23] Carpido-Bernido M V 1991 *J. Phys. A: Math. Gen.* **24** 3013
[24] Berkdemir C, Berkdemir A and Han J G 2006 *Chem. Phys. Lett.* **417** 326
[25] Berkdemir C, 2007 *Am. J. Phys.* **75** 81
[26] Gedenshtein L and Krive I V 1985 *So. Phys. Usp.* **28** 645

Beyond conventional factorization: Non-Hermitian Hamiltonians with radial oscillator spectrum

Ivan Cabrera-Munguia and Oscar Rosas-Ortiz
Departamento de Física, Cinvestav, AP 14-740, 07000 México DF, Mexico
E-mail: rcabrera@fis.cinvestav.mx and orosas@fis.cinvestav.mx

Abstract. The eigenvalue problem of the spherically symmetric oscillator Hamiltonian is revisited in the context of canonical raising and lowering operators. The Hamiltonian is then factorized in terms of two not mutually adjoint factorizing operators which, in turn, give rise to a non-Hermitian radial Hamiltonian. The set of eigenvalues of this new Hamiltonian is exactly the same as the energy spectrum of the radial oscillator and the new square-integrable eigenfunctions are complex Darboux-deformations of the associated Laguerre polynomials.

1. Introduction

Factorization is a powerful method to compute eigenvalues and recurrence relations for solutions of differential equations in mathematical physics (see e.g. [1–5]). Introduced in Quantum Mechanics (QM) since Fock, Schrödinger and Dirac times, some of the antecedents of this algebraic method can be found in the geometrical formalism of Darboux and Bäcklund. It is also notable that the (canonical) Fock-Dirac ladder operators have been of big influence in the development of contemporary Physics over the years [6]. The method, as formalized by Infeld and Hull [7], was extended by Mielnik to embrace factorizing operators of the harmonic oscillator Hamiltoninan which do not create nor annihilate occupation numbers in the Fock states [8]. Applications of the Mielnik's method to radial problems were reported first by Fernández [9] and connections with the Supersymmetric formalism of QM were found by Nieto [10] and Sukumar [11]. The relationship between factorizing and Darboux-deforming potentials in QM was extensively studied by Andrianov and coworkers [12]. Nowadays, the factorization method is of the highest utility in constructing new exactly solvable potentials in QM (see e.g. [13, 14] and references quoted therein). In particular, recent results show that the method can be used in solving the problem of complex potentials with real spectrum [15–20] as well as in the study of resonances and Gamow-Siegert states [21–23].

In the one-dimensional case there is a unique expression for the canonical ladder operators since the energy levels are not degenerated. For instance, the Fock-Dirac operators transform an eigenfunction of the harmonic oscillator into another one and factorize, at the same time, the corresponding Hamiltonian. However, in a general situation, the pair of operators which factorize a given Hamiltonian do not necessarily correspond to the shift operators. Distorted versions of the harmonic oscillator, for example, require higher order raising and lowering operators although the Hamiltonian is factorized by first order differential operators [8, 24]. On the other hand, it is usual to find degenerate energy levels in higher dimensions and several kinds of shift operators are necessary to intertwine the wave-functions [2]. A particular case is of special

Journal of Physics: Conference Series **128** (2008) 012042 doi:10.1088/1742-6596/128/1/012042

interest: spherical symmetry in central potentials induces energy degeneracies under rotations (different orientations of the angular momentum vector lead to the same energy). Hence, the ladder operators are labelled by the azimuthal quantum number ℓ and their action mainly affects the ℓ–dependence of the solutions (see for instance [7, 9, 25–27]). Thus, they intertwine solutions of potentials which differ in one unit of the angular momentum.

In this work we are interested in the three dimensional isotropic oscillator. The Hamiltonian of this system can be factorized in four different forms by means of a basic set of canonical ladder operators [28]. Each of these products preserves the adjointness of the Hamiltonian since the factors are mutually adjoint. In contrast, we are going to express the radial oscillator Hamiltonian as the product of two not mutually adjoint factorizing operators. The advantage of these new operators is that they give rise to non-Hermitian radial Hamiltonians for which the point spectrum is exactly the same as the spherically symmetric oscillator one.

In Section 2 conventional factorizations of the isotropic oscillator are revisited. Explicitly derived, the four canonical ladder operators are used to construct the basis of physical wave-functions (orthonormal, associated Laguerre polynomials) and the energy spectrum. For a given ℓ, the creation or annihilation of nodes in the wave-functions require second order differential operators which do not factorize the Hamiltonian. In Section 3 a pair of not mutually adjoint factorizing operators is introduced to extend the number of factorizations in which the radial oscillator Hamiltonian can be expressed. These same operators are used to construct a new Hamiltonian which is not self adjoint under the inner product of the Laguerre–Hilbert Space $\mathcal{H}$. The relevant aspect of our results is that the eigenfunctions of the new Hamiltonian are square-integrable in $\mathcal{H}$. It is also introduced a pair of ladder operators which increase or decrease the energy of this new system. As we shall see, these shift operators are necessarily fourth order differential operators. Finally, the paper is closed with some concluding remarks in Section 4.

2. Radial Oscillator Revisited

The time-independent Schrödinger equation for the isotropic oscillator $V(r) = r^2$, in appropriate units and after separation of variables, reduces to

$$H_\ell \, \phi(r,\ell) = E(\ell) \, \phi(r,\ell) \tag{1}$$

where the azimuthal quantum number ℓ is a non-negative integer, $E(\ell)$ is twice the dimensionless energy eigenvalue and the radial Hamiltonian reads:

$$H_\ell \equiv -\frac{d^2}{dr^2} + V_\ell(r) = -\frac{d^2}{dr^2} + \frac{\ell(\ell+1)}{r^2} + r^2. \tag{2}$$

The effective potential $V_\ell(r)$ has the domain $D_V = [0, \infty)$ and the functions $\phi(r,\ell)$ are defined in terms of the usual radial wave-function $\phi(r,\ell) = rR(r,\ell)$. To be physically interpretable, these last functions have to satisfy $\int_0^{+\infty} |R(r,\ell)|^2 r^2 dr < \infty$. In the sequel, whenever there is no confusion we shall use the shortcut notation $f \equiv f(r,\ell,\varepsilon_\ell,\ldots)$, keeping implicit the dependence of f on r, ℓ, ε_ℓ, and other possible variables and parameters.

Conventional factorization is useful in solving equation (1) by expressing (2) as the product of two mutually adjoint first order differential operators

$$H_\ell = a_\ell^\dagger a_\ell + \varepsilon_\ell; \qquad a_\ell^\dagger := -\frac{d}{dr} + \alpha(r,\ell); \qquad (a_\ell^\dagger)^\dagger = a_\ell \tag{3}$$

with ε_ℓ a real (*factorization*) constant to be fixed and $\alpha(r,\ell)$ a function satisfying the Riccati equation

$$-\alpha' + \alpha^2 = V_\ell - \varepsilon_\ell. \tag{4}$$

A particular solution of equation (4) is immediate:

$$\alpha = r - \frac{\ell+1}{r}, \qquad \varepsilon_\ell = 2\ell + 3. \tag{5}$$

This *superpotential* α leads to one of the canonical forms of the *factorizing* operators:

$$a_\ell = \frac{d}{dr} - \frac{\ell+1}{r} + r, \qquad a_\ell^\dagger = -\frac{d}{dr} - \frac{\ell+1}{r} + r. \tag{6}$$

The first advantage of the method is clear by noticing that a solution of $a_\ell\, \phi_0(r,\ell) = 0$ gives a solution of (1) belonging to the eigenvalue $E_0(\ell) := E_0^{(\ell)} = \varepsilon_\ell$. The calculation leads to

$$\phi_0(r,\ell) := \phi_0^{(\ell)}(r) = C_0^{(\ell)}\, r^{\ell+1} e^{-r^2/2} \tag{7}$$

which is free of nodes and becomes zero at the edges of D_V. Hence, $C_0^{(\ell)}$ can be taken as the normalization constant and, by applying the Sturm oscillation theorem [29], we know that $E_0^{(\ell)} = 2\ell + 3$ is the ground state energy of H_ℓ. Now, it is useful to reverse the product (3) to get

$$H_{\ell+1} = a_\ell\, a_\ell^\dagger + \varepsilon_\ell - 2 := a_\ell\, a_\ell^\dagger + \theta_\ell. \tag{8}$$

Notice that the azimuthal quantum number ℓ plays the role of a parameter in all the previous expressions. Hence, by changing $\ell \to \ell - 1$ and $\ell \to \ell + 1$ in (8) and (3) respectively we also obtain

$$H_\ell = a_{\ell-1} a_{\ell-1}^\dagger + \theta_{\ell-1}, \qquad H_{\ell+1} = a_{\ell+1}^\dagger\, a_{\ell+1} + \varepsilon_{\ell+1}. \tag{9}$$

As a consequence, the following intertwining relationships are true

$$a_\ell\,(H_\ell - 2) = H_{\ell+1}\, a_\ell, \qquad a_{\ell-1}^\dagger\,(H_\ell + 2) = H_{\ell-1}\, a_{\ell-1}^\dagger. \tag{10}$$

Thereby, if $\phi(r,\ell)$ is eigenfunction of H_ℓ with eigenvalue $E(\ell)$, then $\phi(r,\ell+1) \propto a_\ell\, \phi(r,\ell)$ and $\phi(r,\ell-1) \propto a_{\ell-1}^\dagger\, \phi(r,\ell)$ respectively satisfy the eigenvalue equation of $H_{\ell+1}$ and $H_{\ell-1}$ with eigenvalue $E(\ell+1) = E(\ell) - 2$ and $E(\ell-1) = E(\ell) + 2$.

Let us take full advantage of these last results. After applying the R.H.S. relationship (10) on the ground state $\phi_0^{(\ell)}$, we arrive at the function

$$\phi_1^{(\ell-1)} \propto a_{\ell-1}^\dagger\, \phi_0^{(\ell)} = (-2r^2 + 2\ell + 1)\, \phi_0^{(\ell-1)} \tag{11}$$

which has a single node at $r = \sqrt{(2\ell+1)/2} \in D_V$ and satisfies the Schrödinger equation

$$H_{\ell-1}\, \phi_1^{(\ell-1)} = (E_0^{(\ell)} + 2)\, \phi_1^{(\ell-1)}. \tag{12}$$

Function (11) can be rewritten as

$$\phi_1^{(\ell-1)} \propto \phi_0^{(\ell-1)} \left\{ r^{-2(\ell-1/2)}\, e^{r^2} \left(\frac{1}{2r}\frac{d}{dr} \right) r^{2(\ell+1/2)}\, e^{-r^2} \right\}. \tag{13}$$

After an intermediary change of independent variable $r^2 = x$, one arrives at

$$\phi_1^{(\ell-1)}(r) = C_1^{(\ell-1)}\, \phi_0^{(\ell-1)}(r)\, L_1^{(\ell-1)+1/2}(r^2) \tag{14}$$

where $C_1^{(\ell-1)}$ is a normalization constant and

$$L_n^\nu(x) = \frac{1}{n!} x^{-\nu} e^x \frac{d^n}{dx^n}(e^{-x} x^{n+\nu}), \qquad \nu > -1 \tag{15}$$

is the associated Laguerre polynomial of degree n. A similar procedure leads to

$$\phi_2^{(\ell-2)}(r) = C_2^{(\ell-2)} \phi_0^{(\ell-2)}(r) L_2^{(\ell-2)+1/2}(r^2) \propto a_{\ell-2}^\dagger \phi_1^{(\ell-1)}(r) \tag{16}$$

which is the eigenfunction of $H_{\ell-2}$ with eigenvalue $E_2^{(\ell-2)} = E_1^{(\ell-1)} + 2 = E_0^{(\ell)} + 4$. The procedure can be applied s times to get the eigenfunction of $H_{\ell-s}$ with eigenvalue $E_s^{(\ell-s)} = E_0^{(\ell)} + 2s$:

$$\phi_s^{(\ell-s)} = C_s^{(\ell-s)} \left(\prod_{k=0}^{s-1} a_{\ell-s+k}^\dagger \right) \phi_0^{(\ell)} := C_s^{(\ell-s)} \left(a_{\ell-s}^\dagger \cdots a_{\ell-2}^\dagger a_{\ell-1}^\dagger \right) \phi_0^{(\ell)}, \quad s = 1, 2, \ldots \tag{17}$$

Changing $\ell \to \ell + s$ and using $L_0^\nu \equiv 1$, this last expression leads to

$$\phi_s^{(\ell)}(r) = C_s^{(\ell)} \phi_0^{(\ell)}(r) L_s^{\ell+1/2}(r^2), \qquad s = 0, 1, \ldots \tag{18}$$

Function (18) is the physical solution of (1) with eigenvalue $E_s^{(\ell)} = E_0^{(\ell+s)} + 2s = \varepsilon_{\ell+s} + 2s = 2(2s+\ell) + 3$. The presence of nodes in $\phi_s^{(\ell)}$ is due to the zeros of the polynomial $L_s^{\ell+1/2}$, the number of which is equal to s. By taking $n = 2s + \ell = 1, 0, \ldots$, we can use the conventional notation $\phi_s^{(\ell)}(r) \equiv \phi_n(r)$ and $E_s^{(\ell)} \equiv E_n = 2n + 3$.

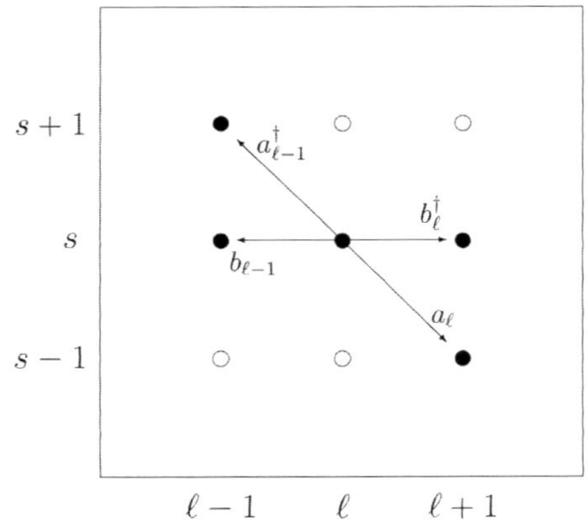

Figure 1. Lattice representation in the $(s \times \ell)$–plane for the action of the four canonical factorization operators a_ℓ, $a_\ell^\dagger$, b_ℓ, $b_\ell^\dagger$, of the radial oscillator.

Remark that the action of a_ℓ on an arbitrary function (18), different from $\phi_0^{(\ell)}$, produces $a_\ell \phi_s^{(\ell)} = \phi_{s-1}^{(\ell+1)}$. Thus a_ℓ annihilates one node of the wave-function at the cost of increasing in one unit the azimuthal quantum number ℓ. This can be represented in the Cartesian $(s \times \ell)$–plane as the diagonal displacement $(s, \ell) \to (s-1, \ell+1)$; a schematic picture of which is shown in Figure 1. The action of $a_{\ell-1}^\dagger$ reverses the previous result; this is represented by $(s, \ell) \to (s+1, \ell-1)$. Notice that a particular situation occurs for $\ell = 0$: The R.H.S. relationship (10) prohibits the

Journal of Physics: Conference Series **128** (2008) 012042 doi:10.1088/1742-6596/128/1/012042

action of $a_0^\dagger$ on any eigenfunction $\phi_s^{(0)}$ of H_0 because it produces the forbidden value $\ell = -1$. However, function (18) can be also derived by applying the appropriate product of annihilation operators $\{a_\ell\}$ on a convenient excited state of H_0. The straightforward calculation leads to

$$\phi_s^{(\ell)} = C_s^{(\ell)} \left(\prod_{k=0}^{\ell-1} a_{\ell-1-k} \right) \phi_{s+\ell}^{(0)}, \qquad \ell \neq 0, \quad s = 0, 1, \ldots \tag{19}$$

Another canonical factorization, different from (3) and (9), is still possible for H_ℓ:

$$H_\ell = b_\ell \, b_\ell^\dagger + \kappa_\ell, \qquad b_\ell^\dagger := -\frac{d}{dr} + \gamma(r, \ell), \qquad (b_\ell^\dagger)^\dagger = b_\ell \tag{20}$$

where the factorization constant κ_ℓ is to be fixed and the superpotential $\gamma(r, \ell)$ fulfills

$$\gamma' + \gamma^2 = V_\ell - \kappa_\ell. \tag{21}$$

The immediate solution of this last equation reads

$$\gamma = r + \frac{\ell + 1}{r}, \qquad \kappa_\ell = -\varepsilon_\ell = -(2\ell + 3). \tag{22}$$

Hence, the reverse product (20) is given by

$$H_{\ell+1} = b_\ell^\dagger \, b_\ell - \theta_\ell \tag{23}$$

and the new intertwining relationships are

$$b_{\ell-1} \, (H_\ell - 2) = H_{\ell-1} \, b_{\ell-1}, \qquad b_\ell^\dagger \, (H_\ell + 2) = H_{\ell+1} \, b_\ell^\dagger. \tag{24}$$

The action of $b_\ell^\dagger$ on the eigenfunction $\phi_s^{(\ell)}$ of H_ℓ with eigenvalue $E_s^{(\ell)}$ produces a square-integrable eigen-solution of $H_{\ell+1}$ belonging to the eigenvalue $E_s^{(\ell)} + 2 = 2(2s + \ell + 1) + 3 \equiv E_s^{(\ell+1)}$. Notice that the number of nodes is preserved; we write $\phi_s^{(\ell+1)} = C_s^{(\ell+1)} \, b_\ell^\dagger \, \phi_s^{(\ell)}$. Such an operation is represented in Figure 1 as the horizontal displacement $(s, \ell) \to (s, \ell + 1)$. On the other hand, the function $\phi_s^{(\ell-1)} = C_s^{(\ell-1)} \, b_{\ell-1} \, \phi_s^{(\ell)}$ is eigenfunction of $H_{\ell-1}$ with eigenvalue $E_s^{(\ell-1)} = E_s^{(\ell)} - 2$. Thus, the action of $b_{\ell-1}$ is represented by the displacement $(s, \ell) \to (s, \ell - 1)$ in the $(s \times \ell)$–plane, as it is also shown in Figure 1.

Expressions (3), (9), (20) and (23) recover the canonical factorizations of H_ℓ reported by Fernández, Negro and del Olmo in 1996 [28] (see also [30, 31] and references quoted therein). As we have shown, the four canonical ladder operators a_ℓ, $a_\ell^\dagger$, b_ℓ, $b_\ell^\dagger$, induce displacements in the $(s \times \ell)$–plane by either creating (annihilating) a node in the wave-function at the cost of decreasing (increasing) the azimuthal quantum number in one unit or by preserving the number of nodes but changing the value of ℓ in one unit. Although all these transformations involve a horizontal component in the displacements of the point (s, ℓ), purely vertical displacements are also possible. Indeed, we have a pair of mutually adjoint second order differential operators defined as follows:

$$\begin{aligned} (S_\ell^\dagger)^\dagger = S_\ell &:= a_{\ell-1} \, b_{\ell-1} = b_\ell \, a_\ell \\ &= -H_\ell + 2r\tfrac{d}{dr} + 2r^2 + 1. \end{aligned} \tag{25}$$

These operators increase or decrease in one unit the number of nodes in the wave-functions

$$S_\ell^\dagger \, \phi_s^{(\ell)} \propto \phi_{s+1}^{(\ell)}, \qquad S_\ell \, \phi_s^{(\ell)} \propto \phi_{s-1}^{(\ell)}, \tag{26}$$

Journal of Physics: Conference Series **128** (2008) 012042 doi:10.1088/1742-6596/128/1/012042

and intertwine eigenfunctions of H_ℓ for which the energy differs in four units

$$S_\ell H_\ell = (H_\ell + 4)S_\ell, \qquad S_\ell^\dagger H_\ell = (H_\ell - 4)S_\ell^\dagger. \tag{27}$$

Hence, we get the following commutation rules

$$[S_\ell, H_\ell] = 4S_\ell, \qquad [S_\ell^\dagger, H_\ell] = -4S_\ell^\dagger. \tag{28}$$

It is clear that the Hamiltonian H_ℓ is not factorized by S_ℓ and $S_\ell^\dagger$:

$$S_\ell S_\ell^\dagger = (H_\ell - \varepsilon_\ell + 4)(H_\ell + \varepsilon_\ell), \qquad S_\ell^\dagger S_\ell = (H_\ell + \varepsilon_\ell - 4)(H_\ell - \varepsilon_\ell). \tag{29}$$

However, there is a simple commutation rule

$$[S_\ell, S_\ell^\dagger] = 8H_\ell. \tag{30}$$

Figure 2 shows the schematic representation of two arbitrary but contiguous energy levels of H_ℓ; the corresponding points of the $(s \times \ell)$–plane are also depicted. As indicated, the operator $S_\ell^\dagger$ scales up the energy to the closer excited one and $S_\ell^\dagger$ operates in the opposite direction.

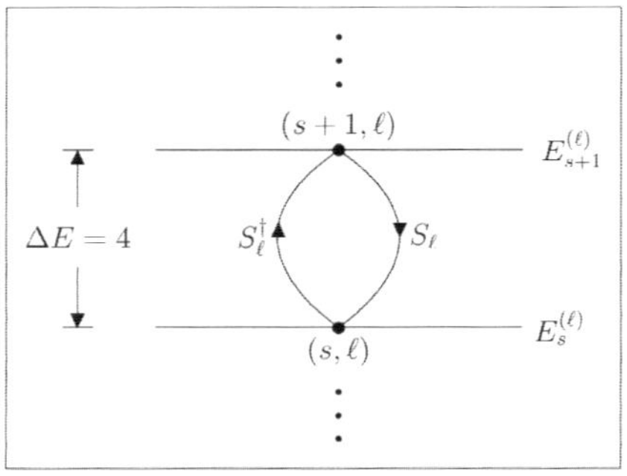

Figure 2. The ladder operators $S_\ell^\dagger$ and S_ℓ respectively create and annihilate nodes in the eigenfunctions of H_ℓ. Notice that displacements in the $(s \times \ell)$–plane are purely vertical, thus the value of the azimuthal quantum number ℓ is fixed.

3. Complex Factorization

In this section we are going to go a step further in the factorization of Hamiltonian (2). Let us analyze the product

$$H_\ell = A_\ell B_\ell + \epsilon_\ell \tag{31}$$

where ϵ_ℓ is now a complex factorization constant with non-trivial imaginary part (from now on $\Re(z)$ and $\Im(z)$ respectively stand for the real and imaginary parts of $z \in \mathbb{C}$ while $\overline{z}$ corresponds to its complex conjugate). The new factorizing operators are established in the context of a 'refined factorization' [32] to read

$$A_\ell = -\frac{d}{dr} + \beta(r, \ell), \qquad B_\ell = \frac{d}{dr} + \beta(r, \ell) \tag{32}$$

with β a complex function satisfying the Riccati equation

$$-\beta' + \beta^2 = V_\ell - \epsilon_\ell. \tag{33}$$

The first aspect which distinguishes (31) from (3) is that A_ℓ and B_ℓ are not mutually adjoint in the Hilbert space $\mathcal{H} = L^2(D_V)$ spanned by the vectors (14). However, as H_ℓ is selfadjoint, $H_\ell = B_\ell^\dagger A_\ell^\dagger + \overline{\epsilon}_\ell$ must be true. The straightforward calculation shows that this last expression leads to the complex conjugation of equation (33), the solution of which is immediate since $V_\ell(r)$ is real.

As in the previous section, the factors in (31) lead to solutions of the involved eigenvalue problem. In particular, the function $u(r,\ell)$, annihilated by B_ℓ, is also solution of the Schrödinger equation

$$H_\ell\, u = \epsilon_\ell\, u. \tag{34}$$

Since ϵ_ℓ is complex and H_ℓ is an Hermitian operator, it is clear that u is not square-integrable in D_V. However, this function is very useful because the non-linear Riccati equation (33) can be mapped into (34) by means of the logarithmic derivative $\beta = -\frac{d}{dr}\ln u$. The appropriate transformations allow to express the solutions of (34) in terms of confluent hypergeometric functions ${}_1F_1(a, c; z)$ (see e.g. [3, 4]). In particular, we shall use

$$u(r,\ell) = c_\ell\, r^{\ell+1} e^{-r^2/2} {}_1F_1\left(\frac{\ell}{2} + \frac{3}{4} - \frac{\epsilon_\ell}{4}, \ell + \frac{3}{2}; r^2\right) \tag{35}$$

with c_ℓ an arbitrary integration constant which will be fixed as 1. Hence, we have

$$\beta(r,\ell) = r - \frac{\ell+1}{r} - 2r\left(\frac{2\ell + 3 - \epsilon_\ell}{4\ell + 6}\right) \frac{{}_1F_1\left(\frac{\ell}{2} + \frac{7}{4} - \frac{\epsilon_\ell}{4}, \ell + \frac{5}{2}; r^2\right)}{{}_1F_1\left(\frac{\ell}{2} + \frac{3}{4} - \frac{\epsilon_\ell}{4}, \ell + \frac{3}{2}; r^2\right)}. \tag{36}$$

The well known analytical properties of the ${}_1F_1$ functions lead to

$$\beta \approx \begin{cases} -\frac{\ell+1}{r} & r \to 0 \\ -r & r \to +\infty \end{cases} \tag{37}$$

The reverse product (31) produces a new second order differential operator

$$B_\ell\, A_\ell + \epsilon_\ell = H_\ell + 2\beta' \tag{38}$$

which is a non-Hermitian Hamiltonian in $\mathcal{H}$ since β is complex. The new potential

$$v(r,\ell,\epsilon_\ell) := V_\ell(r) + 2\beta'(r,\ell) \tag{39}$$

is mainly a real function at the edges of D_V:

$$\begin{cases} v(r \to 0, \ell, \epsilon_\ell) = V_{\ell+1}(r \to 0) + i\,\frac{4\Im(\epsilon_\ell)}{4\ell+6} \\ v(r \to \infty, \ell, \epsilon_\ell) = V_\ell(r \to \infty) - 2 \end{cases} \tag{40}$$

Figure 3 shows the global aspect of the new potentials $v(r,\ell,\epsilon_\ell)$ for $\ell = 0$ and several values of ϵ_0. Notice that the Darboux-deformation (39) of the initial potential V_ℓ is characterized by a zone $M \subset D_V$ surrounding the origin. Namely, $M = [0, r_0)$, with r_0 a finite number depending on ℓ and ϵ_ℓ. In this *deformation zone* M, the function $\Re(v)$ is merely a distortion of $V_{\ell+1}$ while

Journal of Physics: Conference Series **128** (2008) 012042 doi:10.1088/1742-6596/128/1/012042

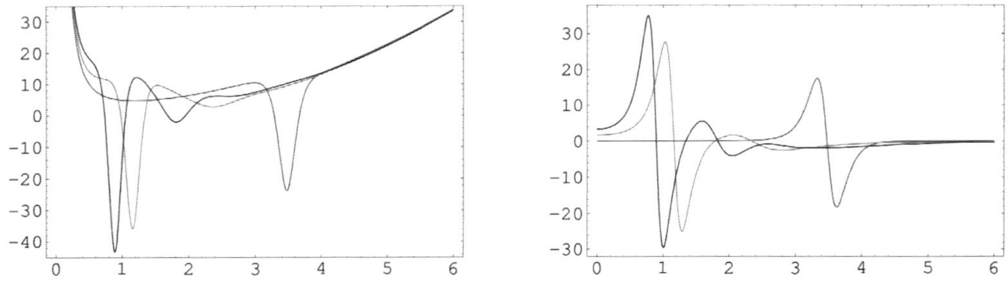

Figure 3. Real (left) and imaginary (right) parts of the complex potential (39) obtained from $V_0(r)$ with a factorization constant $\epsilon_0 = 3 + i\,10^{-3}$ (blue curve), $\epsilon_0 = 7 + i\,\frac{5}{2}$ (red curve) and $\epsilon_0 = 11 + i5$ (black curve). Observe that greater values of $\Re(\epsilon_0)$ induce larger displacements to the left in the deformation

$\Im(v)$ is a nontrivial position-dependent function. For the sake of notation, we shall write $v_{\ell+1}$ for the complex potential defined in (39). Thereby, the non-Hermitian Hamiltonian (38) reads

$$h_{\ell+1} = B_\ell\, A_\ell + \epsilon_\ell = -\frac{d^2}{dr^2} + v_{\ell+1}. \tag{41}$$

Once we have obtained the factorizations (31) and (41), it is immediate to arrive at the following intertwining relationships:

$$B_\ell\, H_\ell = h_{\ell+1}\, B_\ell, \qquad A_\ell\, h_{\ell+1} = H_\ell\, A_\ell. \tag{42}$$

Thus, the action of B_ℓ on $\phi_s^{(\ell)}$ gives a function $\psi_s^{(\ell+1)}(r, \epsilon_\ell) \propto B_\ell\, \phi_s^{(\ell)}(r)$, which satisfies the Schrödinger equation

$$h_{\ell+1}\, \psi_s^{(\ell+1)} = E_s^{(\ell)}\, \psi_s^{(\ell+1)}. \tag{43}$$

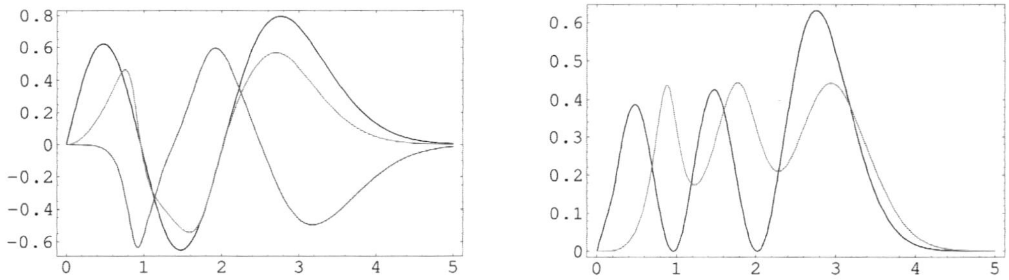

Figure 4. Left: The second excited eigenstate $\phi_2^{(0)}(r)$ of H_0 (black curve) and the real (blue curve) and imaginary (red curve) parts of the corresponding Darboux-deformation $\psi_2^{(1)}(r, \epsilon_0)$ for $\epsilon_0 = 11 + i\,5$. Right: The Born probability densities of $\phi_2^{(0)}(r)$ (black curve) and $\psi_2^{(1)}(r, \epsilon_0)$.

The straightforward calculation shows that this new function $\psi_s^{(\ell+1)}$ is square-integrable in D_V. Figure 4 depicts the initial eigenfunction $\phi_s^{(\ell)}$ corresponding to $\ell = 0$, $s = 2$, as well as its Darboux-deformation $\psi_2^{(1)}(r, \epsilon_0 = 11 + i\,5)$. The function $\phi_2^{(0)}$ corresponds to the second excited state of H_0, accordingly it shows two nodes in D_V. However, although $\psi_2^{(1)}$ corresponds to the second excited energy $E_2^{(0)} = 11$ of h_1, it is notable that this function is free of nodes. Thereby, the usual correspondence between the number of nodes and the level of excitation of a given energy eigenstate is missing in the new functions. Despite this fact, the normalized function $|\psi_s^{(\ell+1)}|^2$ can be put in correspondence with the Born's probability density.

In the previous section we realize that the action of the canonical factorizing operators a_ℓ and $a_\ell^\dagger$ on a given physical state $\phi_s^{(\ell)}$ respectively reduces and increases in one unit the number of nodes (b_ℓ and $b_\ell^\dagger$, on the other hand, do not change the parameter s). In contrast, nodes disappear in the complex Darboux-deformations of physical states, as established by the set of equations (32), (36) and (42). In this way, the parameter s in $\psi_s^{(\ell+1)}$ is nothing but the heritage of the number of nodes of the initial wave-function $\phi_s^{(\ell)}$. Another important consequence of the absence of this kind of zeros is that the new set of square-integrable functions $\{\psi_s^{(\ell+1)}\}$ is not orthogonal. However, it is also possible to construct ladder operators which increase or decrease the energy of this system.

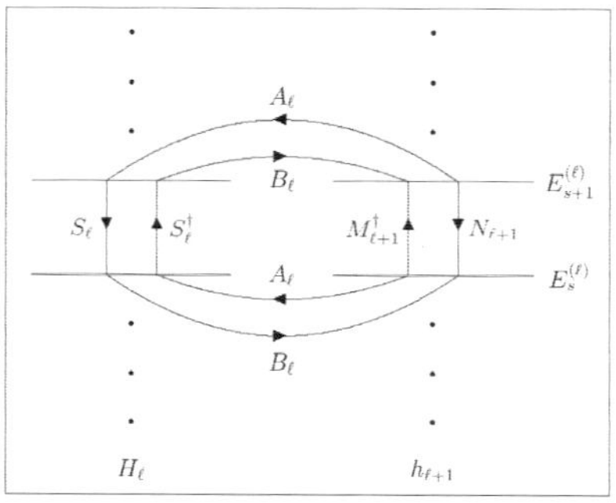

Figure 5. Schematic representation of the intertwining operators A_ℓ and B_ℓ, as they are applied on the energy levels of H_ℓ and $h_{\ell+1}$. The construction of the creation and annihilation operators of $h_{\ell+1}$ is respectively depicted as the composites $M_{\ell+1}^\dagger = B_\ell\, S_\ell^\dagger\, A_\ell$ and $N_{\ell+1} = B_\ell\, S_\ell\, A_\ell$.

In analogy with (27), let us analyze the intertwining relationships

$$N_{\ell+1}\,(h_{\ell+1} - 4) = h_{\ell+1} N_{\ell+1}, \qquad M_{\ell+1}^\dagger\,(h_{\ell+1} + 4) = h_{\ell+1} M_{\ell+1}^\dagger. \tag{44}$$

The above equations give place to the commutation rules

$$[N_{\ell+1}, h_{\ell+1}] = 4N_{\ell+1}, \qquad [M_{\ell+1}^\dagger, h_{\ell+1}] = -4M_{\ell+1}^\dagger \tag{45}$$

which are the same as those fulfilled by H_ℓ, $S_\ell^\dagger$ and S_ℓ in (28). Thus, $M_{\ell+1}^\dagger$ and $N_{\ell+1}$ should respectively play the role of the raising and lowering operators for the eigenvalues of $h_{\ell+1}$. From (44), it is clear that $N_{\ell+1}\,\psi_s^{(\ell+1)}$ is eigenfunction of $h_{\ell+1}$ with eigenvalue $E_s^{(\ell)} - 4 = E_{s-1}^{(\ell)}$. Thus, $N_{\ell+1}$ lowers the energy of the system in four units by annihilating one unit in the parameter s. Hence, we write $\psi_{s-1}^{(\ell+1)} \propto N_{\ell+1}\,\psi_s^{(\ell+1)}$. In a similar way we get $\psi_{s+1}^{(\ell+1)} \propto M_{\ell+1}^\dagger\,\psi_s^{(\ell+1)}$. Let us remember that S_ℓ and $S_\ell^\dagger$ were constructed in terms of vertical displacements of the points in the $(s \times \ell)$–plane. In the present case, we have to solve the algebraic equations (44–45) and Figure 5 is of special utility. The simplest solution is given by the following products

$$M_{\ell+1}^\dagger = B_\ell\, S_\ell^\dagger\, A_\ell, \qquad N_{\ell+1} = B_\ell\, S_\ell\, A_\ell, \qquad (M_{\ell+1}^\dagger)^\dagger = N_{\ell+1}, \tag{46}$$

which can be verified by algebraic operations. Since $M^{\dagger}_{\ell+1}$ and $N_{\ell+1}$ are fourth order differential operators, they do not factorize the Hamiltonian h_ℓ. The corresponding products are fourth degree polynomials factorized as follows

$$\begin{aligned} M^{\dagger}_{\ell+1} N_{\ell+1} &= (h_{\ell+1} - \epsilon_\ell - 4)(h_{\ell+1} + 2\ell - 1)(h_{\ell+1} - 2\ell - 3)(h_{\ell+1} - \epsilon_\ell) \\ N_{\ell+1} M^{\dagger}_{\ell+1} &= (h_{\ell+1} - \epsilon_\ell + 4)(h_{\ell+1} - 2\ell + 1)(h_{\ell+1} + 2\ell + 3)(h_{\ell+1} - \epsilon_\ell) \end{aligned} \tag{47}$$

Finally, it is remarkable that, in contrast with $S^{\dagger}_\ell$ and S_ℓ, the ladder operators $M^{\dagger}_{\ell+1}$ and $N_{\ell+1}$ are not mutually adjoint (it is a heritage of the factorizing operators A_ℓ and B_ℓ). A similar situation gave place to the *Distorted Heisenberg-Weyl Algebra* reported in [24]. Research in this direction is in progress.

4. Concluding Remarks

Conventional factorization is useful to solve the eigenvalue problem of the spherical symmetric harmonic oscillator. The corresponding Hilbert space $\mathcal{H} = L^2(D_V)$ is spanned by the physical solutions which, in turn, are written in terms of orthonormal associated Laguerre polynomials. A new kind of factorizing operators has been introduced to express the radial oscillator Hamiltonian H_ℓ as the product of two not mutually adjoint operators plus a complex constant. The reversed factorization produces a new Hamiltonian which is not Hermitian under the inner product of the associated Laguerre polynomials. However, the corresponding eigenvalue equation has been solved for the radial oscillator energy spectrum. Although the new solutions are not orthogonal, they are square-integrable in $\mathcal{H}$. Hence, the wave-functions so obtained can be put in correspondence with the Born probability density.

The complex potentials of the non-Hermitian Hamiltonians reported here are such that their imaginary part is not a simple constant but a nontrivial function of the position. There is a well defined subset of $D_V = [0, \infty)$ in which this 'complex deformation' of the radial oscillator potential is relevant. A similar situation was reported for complex Coulomb-like [20, 21] and complex linear oscillator-like [19] interactions. Recent results show that, in a 'free particle background', this kind of potentials behaves as an optical device which both refracts and absorbs light waves [23]. Then, it is interesting to analyze the situation in a different background, namely in the radial oscillator one. On the other hand, in Section 3, it has been shown that the wave-functions of these complex potentials do not have a clear correspondence between the number of nodes and the level of energy excitation. However, our current studies show that this correspondence should be established by means of the real and imaginary parts of the new functions [33] (see also [34]). The improvements of these last results are going to be reported elsewhere.

Acknowledgments.
The support of CONACyT project 24233-50766-F is acknowledged.

References

[1] Miller W 1968 *Lie Theory and Special Functions* (New York: Academic Press)
[2] de Lange O L and Raab R E 1991 *Operator Methods in Quantum Mechanics* (Oxford: Oxford University Press)
[3] Negro J, Nieto L M and Rosas-Ortiz O 2000 *J Math Phys* **41** 7964
[4] Rosas-Ortiz O, Negro J and Nieto L M 2003 *Rev Mex Fis* **49 Suppl 1** 88
[5] Dong S H 2007 *Factorization Method in Quantum Mechanics*, Fundamental Theories of Physics 150 (The Netherlands: Springer)
[6] Mielnik B and Rosas-Ortiz O 2004 *J Phys A: Math Gen* **37** 10007
[7] Infeld L and Hull T E 1951 *Rev Mod Phys* **23** 21

Journal of Physics: Conference Series **128** (2008) 012042 doi:10.1088/1742-6596/128/1/012042

[8] Mielnik B 1984 *J Math Phys* **25** 3387
[9] Fernández D J 1984 *Lett Math Phys* **8** 337
[10] Nieto M M 1984 *Phys Lett B* **145** 208
[11] Sukumar C V 1985 *J Phys* A: *Math Gen* **18** L57;
Sukumar C V 1985 *J Phys* A: *Math Gen* **18** 2917;
Sukumar C V 1985 *J Phys* A: *Math Gen* **18** 2937
[12] Andrianov A A, Borisov N V and Ioffe M V 1984 *JETP Lett* **39** 93;
Andrianov A A, Borisov N V and Ioffe M V 1984 *Phys Lett* A **105** 19;
Andrianov A A, Borisov N V, Ioffe M V and Eides M I 1984 *Theor Math Phys* **61** 965;
Andrianov A A, Borisov N V and Ioffe M V 1984 *Theor Math Phys* **61** 1078
[13] de Lima Rodriguez R 2002 The quantum mechanics SUSY algebra: an introduction review *Preprint* hep-th/0205017
[14] Special issue dedicated to the subject of the International Conference on Progress in Supersymmetric Quantum Mechanics (PSQM'03) ed I Aref'eva, D J Fernández, V Hussin, J Negro, L M Nieto and B Samsonov 2004 *J Phys A*: *Math Gen* **37**, Number 43
[15] Bagrov V G, Ochanov I N and Samsonov B F 1995 *J Moscow Phys Soc* **5** 191
[16] Baye D, Lévai G and Sparenberg J M 1996 *Nucl Phys* A **599** 435
[17] Andrianov A A, Ioffe M V, Cannata F and Dedonder J P 1999 *Int J Mod Phys* A **14** 2675
[18] Bagchi B, Mallik S and Quesne C 2001 *Int J Mod Phys* A **16** 2859
[19] Fernández D J, Muñoz R and Ramos A 2003 *Phys Lett* A **308** 11
[20] Rosas-Ortiz O and Muñoz R 2003 *J Phys* A: *Math Gen* **36** 8497
[21] Rosas-Ortiz O 2007 *Rev Mex Fis* **53 Suppl 2** 103
[22] Fernández-García N 2007 *Rev Mex Fis* **53 Suppl 4** 42
[23] Fernández-García N and Rosas-Ortiz O 2008 Gamow-Siegert functions and Darboux-deformed short range potentials *Ann Phys* doi:10.1016/j.aop.2007.11.002
[24] Fernández D J, Nieto L M and Hussin V 1994 *J Phys* A: *Math Gen* **27** 3547;
Fernández D J, Nieto L M and Rosas-Ortiz O 1995 *J Phys* A: *Math Gen* **28** 2693;
Rosas-Ortiz J O 1996 *J Phys* A: *Math Gen* **29** 3281
[25] Newmarch J D and Golding R M 1978 *Am J Phys* **46** 658
[26] Dongpei Z 1987 *J Phys* A: *Math Gen* **20** 4331
[27] Rosas-Ortiz J O 1998 *J Phys* A: *Math Gen* **31** L507;
Rosas-Ortiz J O 1998 *J Phys* A: *Math Gen* **31** 10163
[28] Fernández D J, Negro J and del Olmo M A 1996 *Ann Phys* **252** 386
[29] Berezin F A and Shubin M A 1991 *The Schrödinger equation* (Dordrecht: Kluwer Academic Publishers)
[30] Stahlhofen A A 1998 *Phys Lett* A **241** 298
[31] Liu I F, Lei Y A and Zeng J Y 1998 *Phys Lett* A **241** 300
[32] Negro J, Nieto L M and Rosas-Ortiz O 2000 *J Phys* A: *Math Gen* **33** 7207
[33] Cabrera-Munguia I 2007 M.Sc. Thesis (Mexico: Cinvestav monographs)
[34] Cabrera-Munguia I 2007 *AIP Conf Proc* **960** 50

Journal of Physics: Conference Series **128** (2008) 012043 doi:10.1088/1742-6596/128/1/012043

Second-order SUSY partners of the trigonometric Pöschl-Teller potentials

Alonso Contreras-Astorga[1] **and David J. Fernández C.**[2]
Departamento de Física, Cinvestav, A.P. 14-740, 07000 México D.F., Mexico
E-mail: [1]`acontreras@fis.cinvestav.mx`, [2]`david@fis.cinvestav.mx`

Abstract. The second-order supersymmetry transformations are used to generate Hamiltonians with known spectra departing from the trigonometric Poschl-Teller potentials.

1. Introduction

The supersymmetric quantum mechanics (SUSY QM) is the simplest technique to design systems whose Hamiltonians have prescribed energy spectra [1–13]. In this procedure, departing from an initial solvable Hamiltonian H, it can be constructed a new solvable one $\widetilde{H}$ whose levels differ slightly from those of H. By iterating appropriately this method, one can construct Hamiltonians whose spectra are arbitrarily close to any desired one.

Contrasting with the standard first-order technique, in which it is possible to manipulate just the ground state energy, the second-order SUSY QM offers additional possibilities of spectral manipulation [2, 12]: i) two new levels can be 'created' between a pair of neighbor physical ones E_{i-1}, E_i of H; ii) two neighbor physical energies can be deleted; iii) one level can be created at an arbitrary position; iv) one physical energy can be deleted; v) there is not modification of the initial spectrum.

The SUSY methods have been extensively applied to potentials defined on the full real line or the positive semi-axis. In order to complete the scheme, here we will work in detail the second-order technique for the trigonometric Pöschl-Teller potentials, which are defined in a finite interval [13]. One of our aims is to find additional information about the behavior of the SUSY partner potentials at the boundaries of the x-domain (see e.g. [14]).

2. Second-order supersymmetric quantum mechanics

In the second-order SUSY QM a differential operator of second order intertwines two hermitian Schrödinger Hamiltonians in the way:

$$\widetilde{H}B^{\dagger} = B^{\dagger}H, \tag{1}$$

where

$$H = -\frac{1}{2}\frac{d^2}{dx^2} + V(x), \qquad \widetilde{H} = -\frac{1}{2}\frac{d^2}{dx^2} + \widetilde{V}(x), \tag{2}$$

$$B^{\dagger} = \frac{1}{2}\left(\frac{d^2}{dx^2} - \eta(x)\frac{d}{dx} + \gamma(x)\right). \tag{3}$$

After some manipulations, the intertwining relationship reduces to:

$$\widetilde{V} = V - \eta', \tag{4}$$

$$\gamma = \frac{\eta'}{2} + \frac{\eta^2}{2} - 2V + d, \tag{5}$$

$$\frac{\eta\eta''}{2} - \frac{\eta'^2}{4} + \eta^2\eta' + \frac{\eta^4}{4} - 2V\eta^2 + d\eta^2 + c = 0, \tag{6}$$

with $c, d \in \mathbb{R}$. Given $V(x)$, the new potential $\widetilde{V}(x)$ and $\gamma(x)$ are obtained from (4,5) once we find a solution $\eta(x)$ of (6), which is obtained from the Ansätz

$$\eta' = -\eta^2 + 2\beta\eta + 2\xi, \tag{7}$$

where $\beta(x)$ and $\xi(x)$ are to be determined. After some calculations we get $\xi^2 = c$ and:

$$\beta'(x) + \beta^2(x) = 2[V(x) - \epsilon], \qquad \epsilon = \frac{d+\xi}{2}. \tag{8}$$

We can work as well with a Schrödinger equation, obtained from (8) through $\beta = u'/u$:

$$-\frac{u''}{2} + Vu = \epsilon u. \tag{9}$$

Note that if $c \neq 0$, ξ takes two different values, $\pm\sqrt{c}$, and thus we need to solve the Riccati equation (8) for two factorization energies $\epsilon_{1,2} = (d \pm \sqrt{c})/2$. Then, one constructs algebraically a common solution $\eta(x)$ of the corresponding pair of equations (7). On the other hand, if $c = 0$ one has to solve (8) for $\epsilon = d/2$ and to find then the general solution of the Bernoulli equation resulting for $\eta(x)$ (see (7)). There is an obvious difference between the case with real factorization constant ($c > 0$) and the complex case ($c < 0$), suggesting therefore a classification for the solutions $\eta(x)$ based on the sign of c [15].

2.1. *The real case* ($c > 0$)

Here we have $\epsilon_1, \epsilon_2 \in \mathbb{R}$, $\epsilon_1 \neq \epsilon_2$, the corresponding Riccati solutions of (8) being denoted by $\beta_1(x)$, $\beta_2(x)$ respectively. The resulting expression for $\eta(x)$ becomes:

$$\eta(x) = -\frac{2(\epsilon_1 - \epsilon_2)}{\beta_1(x) - \beta_2(x)} = \frac{2(\epsilon_1 - \epsilon_2)u_1u_2}{W(u_1, u_2)} = \frac{W'(u_1, u_2)}{W(u_1, u_2)}, \tag{10}$$

where $W(f, g) = fg' - f'g$. It is clear from Eqs.(4,10) that the potentials $\widetilde{V}(x)$ have no new singularities if the Wronskian $W(u_1, u_2)$ is nodeless in (x_l, x_r).

Note that the intertwining relationship (1) ensures that, if $\psi_n(x)$ are normalized eigenfunctions of H with eigenvalues E_n, then the normalized eigenfunctions ϕ_n of $\widetilde{H}$ with the same eigenvalues read:

$$\phi_n(x) = \frac{B^\dagger\psi_n(x)}{\sqrt{(E_n - \epsilon_1)(E_n - \epsilon_2)}}. \tag{11}$$

However, in general the set $\{\phi_n(x), n = 0, 1, \ldots\}$ is not complete [12], since there are two mathematical eigenstates $\phi_{\epsilon_i}(x)$ of $\widetilde{H}$ with eigenvalues ϵ_i, $i = 1, 2$, belonging as well to the kernel of $B = (B^\dagger)^\dagger$ which could be normalized and therefore should be added, i.e.,

$$B\phi_{\epsilon_j} = 0, \qquad \widetilde{H}\phi_{\epsilon_j} = \epsilon_j\phi_{\epsilon_j}, \qquad j = 1, 2, \tag{12}$$

$$\Rightarrow \quad \phi_{\epsilon_1} \propto \frac{\eta}{u_1} \propto \frac{u_2}{W(u_1, u_2)}, \qquad \phi_{\epsilon_2} \propto \frac{\eta}{u_2} \propto \frac{u_1}{W(u_1, u_2)}. \tag{13}$$

As a consequence, several interesting spectral changes can be done through the second-order SUSY QM, and some of them are worth to be mentioned [2,12].

a) For $\epsilon_2 < \epsilon_1 < E_0$, by taking a nodeless u_1 and u_2 with one node it turns out that $W(u_1, u_2)$ is nodeless, ϕ_{ϵ_1}, ϕ_{ϵ_2} are normalizable and thus $\mathrm{Sp}(\widetilde{H}) = \{\epsilon_2, \epsilon_1, E_0, E_1, \dots\}$.

b) For $E_{i-1} < \epsilon_2 < \epsilon_1 < E_i$, $i = 1, 2, \dots$ with u_2, u_1 having $i+1$, i nodes respectively, $W(u_1, u_2)$ becomes nodeless, ϕ_{ϵ_1}, ϕ_{ϵ_2} are normalizable and $\mathrm{Sp}(\widetilde{H}) = \{E_0, \dots, E_{i-1}, \epsilon_2, \epsilon_1, E_i, \dots\}$.

c) For $\epsilon_2 = E_{i-1}$, $\epsilon_1 = E_i$, $u_2 = \psi_{i-1}$, $u_1 = \psi_i$, the Wronskian is nodeless but ϕ_{ϵ_1}, ϕ_{ϵ_2} are non-normalizable. Thus, $\mathrm{Sp}(\widetilde{H}) = \{E_0, \dots, E_{i-2}, E_{i+1}, \dots\}$.

2.2. *The complex case* ($c < 0$) [16, 17]

Now $\epsilon \equiv \epsilon_1 \in \mathbb{C}$, $\epsilon_2 = \bar{\epsilon}$, and the real $\eta(x)$ resulting of (7) becomes:

$$\eta(x) = -\frac{2\mathrm{Im}(\epsilon)}{\mathrm{Im}[\beta(x)]} = \frac{w'(x)}{w(x)}, \qquad w(x) = \frac{W(u, \bar{u})}{2(\epsilon - \bar{\epsilon})}, \tag{14}$$

where $\beta(x)$ is the Riccati solution of (8) associated to ϵ. Once again, $w(x)$ must be nodeless in (x_l, x_r) to avoid extra singularities in $\widetilde{V}(x)$. Since $w(x)$ is non-decreasing monotonic ($w'(x) = |u(x)|^2$), a sufficient condition ensuring the lack of zeros is

$$\lim_{x\to x_l} u(x) = 0 \qquad \text{or} \qquad \lim_{x\to x_r} u(x) = 0. \tag{15}$$

Transformation functions obeying (15), associated to complex values of ϵ, generate real potentials $\widetilde{V}(x)$ isospectral to $V(x)$. The corresponding normalized eigenfunctions $\phi_n(x)$ can be obtained from (11) by the substitutions $\epsilon_1 = \epsilon$, $\epsilon_2 = \bar{\epsilon}$.

2.3. *The confluent case* ($c = 0$) [18–20]

Here we have $\xi = 0$, $\epsilon \equiv \epsilon_1 = \epsilon_2 \in \mathbb{R}$. The general solution of the Bernoulli equation resulting from (7) reads:

$$\eta(x) = \frac{w'(x)}{w(x)}, \tag{16}$$

$$w(x) = \widetilde{w}_0 + \int e^{2\int \beta(x)dx} dx = w_0 + \int_{x_0}^{x} [u(y)]^2 \, dy, \tag{17}$$

where x_0 is a fixed point in $[x_l, x_r]$. Once again, $w(x)$ must be nodeless in order that $\widetilde{V}(x)$ has no singularities in (x_l, x_r). Since $w(x)$ is non-decreasing monotonic ($w'(x) = u^2(x)$), the simplest choice ensuring a nodeless $w(x)$ is to take $u(x)$ satisfying either

$$\lim_{x\to x_l} u(x) = 0, \qquad I_- = \int_{x_l}^{x_0} [u(y)]^2 \, dy < \infty, \tag{18}$$

or

$$\lim_{x\to x_r} u(x) = 0, \qquad I_+ = \int_{x_0}^{x_r} [u(y)]^2 \, dy < \infty. \tag{19}$$

In both cases one can determine a w_0-domain for which $w(x)$ is nodeless.

The expressions for the eigenfunctions $\phi_n(x)$ of $\widetilde{H}$ associated to the initial eigenvalues E_n can be obtained from equation (11) by taking $\epsilon = \epsilon_1 = \epsilon_2$. However, the spectrum of $\widetilde{H}$ depends

Journal of Physics: Conference Series **128** (2008) 012043 doi:10.1088/1742-6596/128/1/012043

once again on the normalizability of the eigenfunction ϕ_ϵ of $\widetilde{H}$ associated to ϵ which is also annihilated by B, with explicit expression given by:

$$\phi_\epsilon(x) \propto \frac{\eta(x)}{u(x)} = \frac{u(x)}{w(x)}. \tag{20}$$

Note that, for $\epsilon > E_0$, one can find solutions satisfying either (18) or (19) and such that the $\phi_\epsilon(x)$ given in (20) is normalized, i.e., the confluent algorithm allows to embed single energy levels *above* the ground state energy of H.

3. Trigonometric Pöschl-Teller potentials

Let us apply the previous techniques to the trigonometric Pöschl-Teller potentials [13, 21]:

$$V(x) = \frac{(\lambda-1)\lambda}{2\sin^2(x)} + \frac{(\nu-1)\nu}{2\cos^2(x)}, \qquad \lambda, \nu > 1, \tag{21}$$

defined on $(0, \pi/2)$, which are illustrated in gray in Figure 1. The eigenfunctions $\psi_n(x)$ of H, satisfying the boundary conditions $\psi_n(0) = \psi_n(\pi/2) = 0$, and the corresponding eigenvalues are given by:

$$\psi_n(x) = c_n \sin^\lambda(x) \cos^\nu(x)\ {}_2F_1(-n, n+\mu; \lambda + 1/2; \sin^2(x)), \tag{22}$$

$$E_n = \frac{(\mu+2n)^2}{2}, \qquad n = 0, 1, 2, \ldots, \qquad \mu = \lambda + \nu, \tag{23}$$

where c_n are the normalization factors.

3.1. *Second-order SUSY transformations*

Let us explore briefly the three cases of the classification discussed at section 2 (for illustrations of the new potentials see the dashed, dotted and continuous black curves in Figure 1).

3.1.1. Real case. For $\epsilon_1, \epsilon_2 \in \mathbb{R}$, several possibilities of modifying the initial spectrum of H are available.

a) Deletion of two neighbor physical levels. Let us take first $\epsilon_1 = E_i$, $\epsilon_2 = E_{i-1}$, $u_1(x) = \psi_i(x)$, $u_2(x) = \psi_{i-1}(x)$ (see equation (22)). The second-order SUSY partner of $V(x)$ thus takes the form:

$$\widetilde{V}(x) = \frac{\lambda(\lambda+3)}{2\sin^2(x)} + \frac{\nu(\nu+3)}{2\cos^2(x)} - [\ln(W_{i,i-1})]'', \tag{24}$$

where

$$W_{i,i-1} = W({}_2F_1(-i, i+\mu; \lambda + \tfrac{1}{2}; \sin^2(x)), {}_2F_1(-i+1, i-1+\mu; \lambda+\tfrac{1}{2}; \sin^2(x))) \tag{25}$$

is nodeless in $(0, \pi/2)$.

b) Creation of two new levels. First we find the general solution of the Schrödinger equation (9) with the potential (21) for an arbitrary factorization energy ϵ:

$$\begin{aligned} u(x) = \sin^\lambda(x)\cos^\nu(x)[A\, {}_2F_1\left(\tfrac{\mu}{2} + \sqrt{\tfrac{\epsilon}{2}}, \tfrac{\mu}{2} - \sqrt{\tfrac{\epsilon}{2}}; \lambda + \tfrac{1}{2}; \sin^2(x)\right) \\ + C \sin^{1-2\lambda}(x)\ {}_2F_1\left(\tfrac{1+\nu-\lambda}{2} + \sqrt{\tfrac{\epsilon}{2}}, \tfrac{1+\nu-\lambda}{2} - \sqrt{\tfrac{\epsilon}{2}}; \tfrac{3}{2} - \lambda; \sin^2(x)\right)], \end{aligned} \tag{26}$$

with asymptotic behavior

$$u(x) \underset{x\to 0}{\sim} C\sin^{1-\lambda}(x), \qquad u(x) \underset{x\to\frac{\pi}{2}}{\sim} (Aa+Cb)\cos^{1-\nu}(x), \tag{27}$$

where

$$a = \frac{\Gamma(\lambda+\frac{1}{2})\Gamma(\nu-\frac{1}{2})}{\Gamma(\frac{\mu}{2}+\sqrt{\frac{\epsilon}{2}})\Gamma(\frac{\mu}{2}-\sqrt{\frac{\epsilon}{2}})}, \qquad b = \frac{\Gamma(\frac{3}{2}-\lambda)\Gamma(\nu-\frac{1}{2})}{\Gamma(\frac{1+\nu-\lambda}{2}+\sqrt{\frac{\epsilon}{2}})\Gamma(\frac{1+\nu-\lambda}{2}-\sqrt{\frac{\epsilon}{2}})}.$$

Without loosing generality, let us take $C = 1$ and $A = -b/a + D$ so that, for $E_{i-1} < \epsilon < E_i$, the solution $u(x)$ will have $i + 1$ nodes for $D < 0$ while it will have i nodes for $D > 0$.

Let us implement now the second-order SUSY transformation for $E_{i-1} < \epsilon_2 < \epsilon_1 < E_i$, the corresponding seed solutions are given by (26) with $D_2 < 0$, $D_1 > 0$, which ensures that the Wronskian is nodeless. It is convenient to express the u solutions in the way:

$$u_1(x) = \sin^{1-\lambda}(x)\cos^{1-\nu}(x)v_1(x), \qquad u_2(x) = \sin^{1-\lambda}(x)\cos^{1-\nu}(x)v_2(x), \tag{28}$$

where $v_i(x),\ i = 1,2$ are non-divergent for $x \in [0,\pi/2]$. Thus

$$\widetilde{V}(x) = \frac{(\lambda-1)(\lambda-4)}{2\sin^2(x)} + \frac{(\nu-1)(\nu-4)}{2\cos^2(x)} - \ln[W(v_1,v_2)]''. \tag{29}$$

For an illustration of these potentials, see the dotted curve in Figure 1.

3.1.2. Complex case. For $\epsilon \in \mathbb{C}$ the expression given in (26) is still valid. The condition (15), required to avoid the zeros in the Wronskian, can be satisfied in two different ways. In the first place we select $A = 1,\ C = 0$. The two singularities induced in the new potential can be managed by expressing

$$u(x) = \sin^{\lambda}(x)\cos^{1-\nu}(x)v(x). \tag{30}$$

Therefore:

$$\widetilde{V}(x) = \frac{\lambda(\lambda+3)}{2\sin^2(x)} + \frac{(\nu-1)(\nu-4)}{2\cos^2(x)} - \ln[w(x)]'', \tag{31}$$

$$w(x) = \frac{W(v,\bar{v})}{2(\epsilon-\bar{\epsilon})}. \tag{32}$$

The second solution satisfying (15) arises by taking $C = 1,\ A = -b/a$, and now it is convenient to express

$$u(x) = \sin^{1-\lambda}(x)\cos^{\nu}(x)v(x). \tag{33}$$

The second-order SUSY partner of the trigonometric Pöschl-Teller potential becomes now:

$$\widetilde{V}(x) = \frac{(\lambda-1)(\lambda-4)}{2\sin^2(x)} + \frac{\nu(\nu+3)}{2\cos^2(x)} - \ln[w(x)]''. \tag{34}$$

An example of these potentials is drawn in dashed in Figure 1.

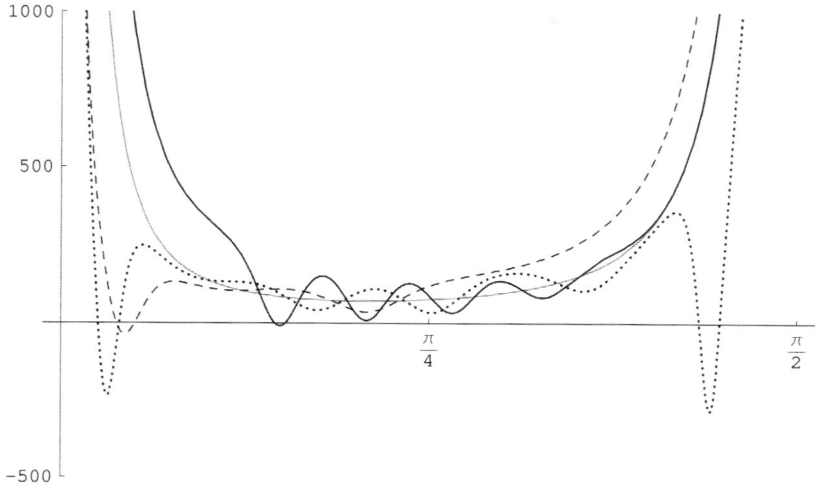

Figure 1. Trigonometric Pöschl-Teller potential for $\lambda = 5$, $\nu = 8$ (gray curve) and its second order SUSY partners arising from the deletion of $E_4 = 220.5$ using the confluent algorithm (black continuous curve), the creation of two levels at $\epsilon_1 = 160.205$ and $\epsilon_2 = 146.205$ (dotted curve), and an isoespectral transformation with $\epsilon = 120.125 + 1.5i$ (dashed curve).

3.1.3. Confluent case. For $\epsilon = \epsilon_1 = \epsilon_2$ let us use once again the expression given in (26). There are two adequate solutions, the first one arises by taking $A = 1,\ C = 0$:

$$u(x) = \sin^{\lambda}(x)\cos^{\nu}(x)\ {}_2F_1\left(\tfrac{\mu}{2} + \sqrt{\tfrac{\epsilon}{2}}, \tfrac{\mu}{2} - \sqrt{\tfrac{\epsilon}{2}}; \lambda + \tfrac{1}{2}; \sin^2(x)\right). \tag{35}$$

A straightforward calculation leads to:

$$w(x) = w_0 + \sum_{m=0}^{\infty} \frac{(\frac{\mu}{2}+\sqrt{\frac{\epsilon}{2}})_m(\frac{\mu}{2}-\sqrt{\frac{\epsilon}{2}})_m}{(\lambda+\frac{1}{2})_m\, m!\,(2\lambda+2m+1)} \sin^{2\lambda+2m+1}(x)$$
$$\times\, {}_3F_2\left(\tfrac{1+\lambda-\nu}{2} - \sqrt{\tfrac{\epsilon}{2}}, \tfrac{1+\lambda-\nu}{2} + \sqrt{\tfrac{\epsilon}{2}}, \lambda + m + \tfrac{1}{2}; \lambda + \tfrac{1}{2}, \lambda + m + \tfrac{3}{2}; \sin^2(x)\right). \tag{36}$$

Note that $w(x)$ is nodeless in $[0, \pi/2]$ for $w_0 > 0$. The SUSY partners of the Pöschl-Teller potentials are thus given by (4) with $\eta(x)$ given by (16,36).

The second solution, appropriate to implement the confluent algorithm, is obtained from (26) with $C = 1,\ A = -b/a$ which, up to a constant factor, reduces to:

$$u(x) = \sin^{\lambda}(x)\cos^{\nu}(x)\ {}_2F_1\left(\tfrac{\mu}{2} + \sqrt{\tfrac{\epsilon}{2}}, \tfrac{\mu}{2} - \sqrt{\tfrac{\epsilon}{2}}; \nu + \tfrac{1}{2}; \cos^2(x)\right). \tag{37}$$

This expression leads straightforwardly to:

$$w(x) = w_0 - \sum_{m=0}^{\infty} \frac{(\frac{\mu}{2}+\sqrt{\frac{\epsilon}{2}})_m(\frac{\mu}{2}-\sqrt{\frac{\epsilon}{2}})_m}{(\nu+\frac{1}{2})_m\, m!\,(2\nu+2m+1)} \cos^{2\nu+2m+1}(x)$$
$$\times\, {}_3F_2\left(\tfrac{1+\nu-\lambda}{2} - \sqrt{\tfrac{\epsilon}{2}}, \tfrac{1+\nu-\lambda}{2} + \sqrt{\tfrac{\epsilon}{2}}, \nu + m + \tfrac{1}{2}; \nu + \tfrac{1}{2}, \nu + m + \tfrac{3}{2}; \cos^2(x)\right). \tag{38}$$

Now $w(x)$ is nodeless in $[0, \pi/2]$ for $w_0 < 0$, and by choosing such a nodeless $w(x)$ one can built up straightforwardly the confluent second-order SUSY partner potentials of $V(x)$.

It is important to observe that, when $\epsilon = E_n$ for some given n, the physical solution (22) and the one given in equation (35) just differ by the factor c_n. Moreover, the infinite summation in (36) truncates in this case at $m = n + 1$; by combining appropriately all this information, it turns out that the w-expression when $u(x) = \psi_n(x)$ becomes:

$$w(x) = w_0 + c_n^2 \sum_{m=0}^{n} \frac{(\mu+n)_m(-n)_m}{(\lambda+\frac{1}{2})_m\, m!\,(2\lambda+2m+1)} \sin^{2\lambda+2m+1}(x) \\ \times {}_3F_2\left(\tfrac{1}{2}-\nu-n, \tfrac{1}{2}+\lambda+n, \lambda+m+\tfrac{1}{2}; \lambda+\tfrac{1}{2}, \lambda+m+\tfrac{3}{2}; \sin^2(x)\right). \quad (39)$$

This function is nodeless in $[0,\pi/2]$ for $w_0 > 0$ or $w_0 < -1$, and in this domain $\widetilde{V}(x)$ becomes isospectral to the initial Pöschl-Teller potential. However, if we choose either $w_0 = 0$ or $w_0 = -1$, the level $\epsilon = E_n$ will not be longer an eigenvalue of $\widetilde{H}$, since the corresponding eigenfunction is non-normalizable. Thus $\mathrm{Sp}(\widetilde{H}) = \{E_0, \ldots, E_{n-1}, E_{n+1}, \ldots\}$, i.e. we have suppressed the initial n-th level (for an illustration, see the black continuous curve in Figure 1).

4. Conclusions

In this paper we have applied successfully the second-order SUSY QM to generate exactly solvable Hamiltonians departing from the trigonometric Pöschl-Teller potentials. Through this technique, it is possible to produce interesting deformations of the quadratic spectrum of H, namely, we can embed pairs of (or single) energy levels between two neighbor initial energies. It has been possible as well to delete two neighbor (or a single) energy levels, as well as to obtain Hamiltonians isospectral to H. Thus, we have shown that a simple mathematical tool is available for designing potentials with an arbitrarily prescribed spectrum.

Acknowledgments

The authors acknowledge the support of Conacyt, project No. 49253-F. ACA acknowledges a Conacyt M. Sc. grant as well as the support of Cinvestav.

References

[1] Andrianov A A, Ioffe M V and Spiridonov V 1993 *Phys Lett* A **174** 273
Andrianov A A, Ioffe M V, Cannata F and Dedonder J P 1995 *Int J Mod Phys* A **10** 2683
[2] Bagrov V G and Samsonov B F 1997 *Phys Part Nucl* **28** 374
Samsonov B F 1999 *Phys Lett* A **263** 274
[3] Fernández D J 1997 *Int J Mod Phys* A **12** 171
Fernández D J, Hussin V and Mielnik B 1998 *Phys Lett* A **244** 309
Cariñena J F, Ramos A and Fernández D J 2001 *Ann Phys* **292** 42
[4] Bagchi B, Ganguly A, Bhaumik D and Mitra A 1999 *Mod Phys Lett* A **14** 27
Bagchi B K 2001 *Supersymmetry in Quantum and Classical Mechanics* (Boca Raton: Chapman & Hall/CRC)
[5] Aoyama H, Sato M and Tanaka T 2001 *Phys Lett* B **503** 423
Aoyama H, Sato M and Tanaka T 2001 *Nucl Phys* B **619** 105
[6] Andrianov A A and Sokolov A V 2003 *Nucl Phys* B **660** 25
Andrianov A A and Cannata F 2004 *J Phys* A **37** 10297
[7] Leiva C and Plyushchay M 2003 JHEP10(2003)069
Plyushchay M 2004 *J Phys* A **37** 10375
[8] Mielnik B and Rosas-Ortiz O 2004 *J Phys* A **37** 10007
[9] Ioffe M V and Nishnianidze D N 2004 *Phys Lett* A **327** 425
[10] González-López A and Tanaka T 2004 *Phys Lett* B **586** 117
[11] Sukumar C V 2005 *AIP Conf Proc* **744** 166
[12] Fernández D J and Fernández-García N 2005 *AIP Conf Proc* **744** 236
[13] Fernández D J, Hussin V and Rosas-Ortiz O 2007 *J Phys* A **40** 6491
[14] Contreras-Astorga A and Fernández D J 2007 *AIP Conf Proc* **960** 55
[15] Fernández D J and Ramos A 2006 *Topics in Mathematical Physics, General Relativity and Cosmology in Honor of Jerzy Plebański* ed H García-Compeán et al (Singapore: World Scientific) p 167
[16] Fernández D J, Muñoz R and Ramos A 2003 *Phys Lett* A **308** 11

[17] Rosas-Ortiz O and Muñoz R 2003 *J Phys* A **36** 8497
[18] Mielnik B, Nieto L M and Rosas-Ortiz O 2000 *J Phys* A **33** 7207
[19] Fernández D J and Salinas-Hernández E 2003 *J Phys* A **36** 2537
[20] Fernández D J and Salinas-Hernández E 2005 *Phys Lett* A **338** 13
[21] Antoine J P, Gazeau J P, Monceau P, Klauder J R and Penson K A 2001 *J Math Phys* **42** 2349

V International Symposium on Quantum Theory and Symmetries IOP Publishing
Journal of Physics: Conference Series **128** (2008) 012044 doi:10.1088/1742-6596/128/1/012044

Optical potentials using resonance states in Supersymmetric Quantum Mechanics

Nicolás Fernández-García and Oscar Rosas-Ortiz
Departamento de Física, Cinvestav, AP 14-740, 07000 México DF, Mexico
E-mail: jnicolas@fis.cinvestav.mx and orosas@fis.cinvestav.mx

Abstract. Complex potentials are constructed as Darboux-deformations of short range, radial nonsingular potentials. They behave as optical devices which both refracts and absorbs light waves. The deformation preserves the initial spectrum of energies and it is implemented by means of a Gamow-Siegert function (resonance state). As straightforward example, the method is applied to the radial square well. Analytical derivations of the involved resonances show that they are 'quantized' while the corresponding wave-functions are shown to behave as bounded states under the broken of parity symmetry of the related one-dimensional problem.

1. Introduction

Solutions of the Schrödinger equation associated to complex eigenvalues $\epsilon = E_\alpha$ and satisfying purely outgoing conditions are known as Gamow-Siegert functions [1, 2]. These solutions represent a special case of scattering states for which the 'capture' of the incident wave produces delays in the scattered wave. The 'time of capture' can be connected with the lifetime of a decaying system (resonance state) which is composed by the scatterer and the incident wave. Then, it is usual to take $\text{Re}(\epsilon)$ as the binding energy of the composite while $\text{Im}(\epsilon)$ corresponds to the inverse of its lifetime. The Gamow-Siegert functions are not admissible as physical solutions into the mathematical structure of Quantum Mechanics since, in contrast with conventional scattering wave-functions, they are not finite at $r \to \infty$. Thus, such a kind of functions is acceptable in Quantum Mechanics only as a convenient model to solve scattering equations. However, because of the resonance states relevance, some approaches extend the formalism of quantum theory so that they can be defined in a precise form [3–9].

In this paper the Gamow-Siegert functions are analyzed not to precisely represent decaying systems but to be used as the cornerstone of complex Darboux-transformations in Supersymmetric Quantum Mechanics (Susy-QM). The 'unphysical' behaviour of these solutions plays a relevant role in the transformation: While the Gamow-Siegert function u diverges at $r \to \infty$, the limit of its logarithmic derivative $\beta = -u'/u$ at $r \to \infty$ is a complex constant. This last function is used to deform the initial potential into a complex function. It is worth noticing that complex eigenvalues ϵ have been used as factorization constants in Susy-QM (see for instance [10–18] and the discussion on 'atypical models' in [19]). However, as far as we know, until the recent results reported in [20–22] the connection between Susy-QM and resonance states has been missing. Our interest in the present work is two-fold. First, we want to show that appropriate approximations lead to analytical expressions for the real and imaginary parts

Journal of Physics: Conference Series **128** (2008) 012044 doi:10.1088/1742-6596/128/1/012044
of the resonance energies in the case of radial short range potentials. Second, we want to enlarge the number of applications of the Darboux transformations in constructing new exactly solvable models in Quantum Mechanics.

In Section 2 the main aspects of solving the Schrödinger equation for radial, nonsingular short range potentials are reviewed. The relevance of the scattering amplitude $S(k)$ in the analysis of both the bounded and scattering wave-functions is clearly stated. In Section 3 the analytical properties of $S(k)$ as a function of the complex kinetic parameter k are studied. It is shown that poles of the scattering amplitude which live in the lower k–plane are connected with the Gamow-Siegert functions while the poles on the positive imaginary axis lead to bounded physical energies. The general aspects of the complex Darboux-transformations are included in Section 4. All the previous developments are then applied to the square radial well and new analytical expressions for the involved resonance energies are reported in Section 5. We conclude the paper with some concluding remarks in Section 6.

2. Short range radial potentials revisited

2.1. *General considerations*

Let us consider the Hamiltonian of one particle in the external, spherically symmetric field $U(\vec{r}) = U(r)$. The Schrödinger equation reduces to the eigenvalue problem:

$$H\psi(\vec{r}) \equiv \left(-\frac{\hbar^2}{2m}\Delta + U(r)\right)\psi(\vec{r}) = \mathcal{E}\psi(\vec{r}). \tag{1}$$

We are interested in potentials which decrease more rapidly than $1/r$ at large r. Indeed, $U(r)$ is a short-range potential (i.e., there exists $a \in \text{Dom}(U) := D_U$ such that $U(r) = 0$ at $r > a$) which is nonsingular (i.e., $U(r)$ satisfies $r^2U(r) \to 0$ at $r \to 0$). The wave function $\psi(\vec{r})$ must be single-valued and continuous everywhere in D_U. Since the operators H, L^2, L_3 and P commute with each other, in spherical coordinates the Schrödinger equation (1) has solutions of the form

$$\psi_{\ell m}(\vec{r}) = R_\ell(r)Y_{\ell m}(\theta,\varphi) \tag{2}$$

where θ and φ are respectively the polar and the azimuthal angles, $R_\ell(r)$ is a function depending on r and $Y_{\ell m}(\theta,\varphi)$ stands for the spherical harmonics. For simplicity in calculations we shall use the function $u_\ell(r) = rR_\ell(r)$. The introduction of (2) into equation (1) yields:

$$\left[-\frac{d^2}{dr^2} + \frac{\ell(\ell+1)}{r^2} + v(r)\right]u_\ell(r) := \left[-\frac{d^2}{dr^2} + V_\ell(r)\right]u_\ell(r) = k^2u_\ell(r) \tag{3}$$

with $v = 2mU/\hbar^2$ and $k^2 = E \equiv 2m\mathcal{E}/\hbar^2$ the dimensionless expressions for the potential and the kinetic parameter respectively. As usual, the effective potential V_ℓ is defined as the potential v plus the centrifugal barrier $\ell(\ell+1)/r^2$. Physical solutions $\psi(\vec{r})$ of (1) should be constructed with functions u_ℓ fulfilling the following boundary conditions:

$$\begin{cases} u_\ell(r) = 0 & r \to 0 \\ \frac{u_\ell(r)}{r} < \infty & r \to \infty \end{cases} \tag{4}$$

Furthermore, the function u_ℓ and its derivative u'_ℓ have to be continuous for $r > 0$ since (3) includes second order derivatives. As we shall see, the study of the involved matching conditions is simplified by introducing a function β_ℓ as follows

$$\beta_\ell(r) := -\frac{u'_\ell(r)}{u_\ell(r)} = -\frac{d}{dr}\ln u_\ell(r). \tag{5}$$

Journal of Physics: Conference Series **128** (2008) 012044 doi:10.1088/1742-6596/128/1/012044

In general we will assume that the energy spectrum is composed by negative (bounded) and positive (scattering) energies. For the sake of notation we shall use the symbol $\phi_\ell(r)$ for the solutions of (3) which satisfy the boundary conditions (4); the symbol $u_\ell(r)$ will stand for general mathematical solutions of equation (3).

2.2. *Bases of solutions*

For arbitrary ℓ and $r > a$, the appropriate basis of solutions can be written in the form

$$u_\ell^{(+)}(r) = ikr\, h_\ell^{(1)}(kr), \qquad u_\ell^{(-)}(r) = -ikr\, h_\ell^{(2)}(kr) \tag{6}$$

where $h_\nu^{(1)}$ and $h_\nu^{(2)}$ are the spherical Hänkel functions of order $\nu = \ell$ (see e.g. [23]). For large values of kr in the region $r > a$, the functions (6) behave as follows

$$u_\ell^{(\pm)}(r) \approx \exp\left[\pm i\left(kr - \frac{\ell\pi}{2}\right)\right]. \tag{7}$$

Thereby, $u_\ell^{(+)}$ represents a diverging wave and describes particles moving with speed $\vartheta = k$ in all directions from the origin. In turn, the function $u_\ell^{(-)}$ represents a converging wave and describes particles moving with speed ϑ towards the origin.

It is worth noticing that the Schrödinger equation (3) is invariant under the change $k \to -k$. Hence, if $u_\ell(r,k)$ is solution of (3) for $E = k^2$, then $u_\ell(r,-k)$ is also a solution for the same energy. However, these last functions must differ at most in a constant factor because the solutions of (3) have to be single-valued. In contrast, up to a global phase, they interchange their roles at $r \to \infty$, as indicated by the relationship

$$u_\ell^{(\pm)}(r,k) = e^{i\pi\ell}\, u_\ell^{(\mp)}(r,-k), \qquad r \to \infty. \tag{8}$$

On the other hand, the basis of solutions at $r < a$ may be written

$$u_\ell^{(1)}(r) = \frac{2\ell+1}{\Lambda_\ell}\, kr\, j_\ell(kr), \qquad u_\ell^{(2)}(r) = -\Lambda_\ell\, kr\, n_\ell(kr), \qquad \Lambda_\ell := \frac{2^{-\ell}\sqrt{\pi}}{\Gamma(\ell+1/2)} \tag{9}$$

where j_ℓ and n_ℓ are respectively the spherical Bessel and Neumann functions of order ℓ. For very small values of r we have

$$u_\ell^{(1)}(r \to 0) \approx (kr)^{\ell+1}, \qquad u_\ell^{(2)}(r \to 0) \approx \frac{1}{(kr)^\ell}. \tag{10}$$

2.2.1. Bounded states. Imaginary values of the kinetic parameter $k = \pm i\sqrt{|E|} \equiv \pm i\kappa$ correspond to negative values of the energy $E = k^2 = -\kappa^2$. As usual, we shall consider k to be in the upper complex k–plane I_+. In this way $u_\ell^{(-)}(r,\kappa)$ does not satisfy the boundary condition at $r \to \infty$ and $u_\ell^{(2)}(r,\kappa)$ does not fulfill the boundary condition at $r = 0$. Hence, the physical solutions should be constructed with $u_\ell^{(1)}(r,\kappa)$ and $u_\ell^{(+)}(r,\kappa)$. We write:

$$\phi_\ell(r,\kappa) = \begin{cases} \zeta_\ell(\kappa)\, u_\ell^{(1)}(r,\kappa)\, g_\ell(r,\kappa) & r < R \\[2ex] \xi_\ell(\kappa)\, u_\ell^{(+)}(r,\kappa) & r \geq R \end{cases} \tag{11}$$

where the intermediary function $g_\ell(r,\kappa)$ is equal to 1 at $r = 0$ and depends on the potential. The coefficients $\zeta_\ell(\kappa)$ and $\xi_\ell(\kappa)$ are complex numbers while the matching condition $\beta_\ell(a,\kappa) = \kappa$

Journal of Physics: Conference Series **128** (2008) 012044 doi:10.1088/1742-6596/128/1/012044

corresponds to a transcendental equation which is fulfilled for a set of discrete values of the kinetic parameter κ_n, $n = 0, 1, 2, \ldots$ Under these conditions the functions $\phi_\ell(r, \kappa_n)$ are finite and the integral

$$\int_0^{+\infty} |\phi_\ell(r, \kappa_n)|^2 dr \tag{12}$$

converges. Thereby, $|\phi_\ell(r, \kappa_n)|^2$ can be identified as the involved Born's probability density.

2.2.2. Scattering states. Positive energies $E = k^2$ correspond to real values of k and both solutions (7) remain finite in $r \geq a$. As a consequence, both of them are physically acceptable in this region. To analyze these solutions first let us consider the case $\ell = 0$ and, for the sake of simplicity, let us drop this subindex from the functions $u_\ell(r)$. The general solution in the free of interaction zone ($r > a$) may be written as

$$u_{\text{out}}(r) = \gamma(k) \left[u^{(-)}(r) - S(k) u^{(+)}(r) \right]. \tag{13}$$

The constant $\gamma(k)$ is the amplitude of the converging wave and does not depend on the potential $v(r)$. The scattering amplitude $S(k)$, in contrast, strongly depends on the potential and encodes the information of the scattering phenomenon. The matching condition between $u_{\text{out}}(r)$ and the solution $u_{\text{in}}(r)$ in the interaction zone reads

$$\beta_{\text{out}}(a) = \beta_{\text{in}}(a) \tag{14}$$

and can be satisfied by appropriate values of γ and S. Indeed, equation (14) is equivalent to the following matrix array

$$\left. \begin{pmatrix} u_{\text{in}} & u^{(+)} \\ u'_{\text{in}} & u'^{(+)} \end{pmatrix} \right|_{r=a} \begin{pmatrix} \gamma^{-1} \\ S \end{pmatrix} = \left. \begin{pmatrix} u^{(-)} \\ u'^{(-)} \end{pmatrix} \right|_{r=a} \tag{15}$$

the solution of which is given by the system of equations

$$\gamma = \left. \frac{W\left(u_{\text{in}}, u^{(+)}\right)}{2ik} \right|_{r=a}, \qquad S = \left. \frac{W\left(u_{\text{in}}, u^{(-)}\right)}{W\left(u_{\text{in}}, u^{(+)}\right)} \right|_{r=a}, \tag{16}$$

where $W(\cdot, \cdot)$ stands for the Wronskian of the involved functions.

To get a better idea of the roles played by each one of the terms in equation (13) let us consider the free motion ($v(r) = 0$) in the whole of D_v. From equations (6) and (9) we realize that a regular solution at the origin may be written as follows

$$u_{\text{free}}(r) \equiv u^{(-)}(r) - u^{(+)}(r) = -2ikr\, j_0(kr) = -2i \sin(kr). \tag{17}$$

This term can be introduced into equation (13) to get

$$u_{\text{out}}(r) = \gamma(k) \left[u_{\text{free}}(r) - (S(k) - 1) u^{(+)}(r) \right]. \tag{18}$$

Thus, the total wave function after scattering u_{out} is given by the wavepacket one would have if there were no scattering γu_{free} plus a scattered term $-\gamma(S-1)u^{(+)}$. This last example shows the important role played by the scattering amplitude in the analysis of positive energies.

On the other hand, the potential we are dealing with is neither a sink nor a source of particles, then the condition $|S(k)| = 1$ (elastic scattering) is true and we have

$$S(k) = e^{2i\delta(k)}. \tag{19}$$

Journal of Physics: Conference Series **128** (2008) 012044 doi:10.1088/1742-6596/128/1/012044

The scattering phase $\delta(k)$ represents the difference in phase of the outgoing parts. The situation is clear if one introduces (17) and (19) into equation (18) to get

$$u_{\text{out}}(r \to \infty) = -2i\gamma(k)\, e^{i\delta(k)}\, \sin(kr + \delta). \tag{20}$$

As we have indicated, the scattering states include $k \in \mathbb{R}$ since $E > 0$. However, the function (13) is also valid if k is one of the imaginary values of the kinetic parameter leading to the bounded energies $E_n = -\kappa_n^2$. Indeed, this case provides further information about the scattering amplitude. For $k_n = i\kappa_n$, the function (13) at $r \to \infty$ reads

$$u_{\text{out}}(r \to \infty, \kappa_n) \sim -2i\gamma(i\kappa_n)\left[e^{\kappa_n r} - S(i\kappa_n)e^{-\kappa_n r}\right]. \tag{21}$$

The first term in square brackets never vanishes ($\kappa_n > 0$), then the coefficient S must be singular at the complex point $k_n = i\kappa_n$. In other words, the points k_n are nothing but poles of the scattering amplitude. Since these points are in the upper complex k–plane I_+, we realize that $S(k)$ is a regular function in I_+ except at k_n, $n = 0, 1, 2, \ldots$ We shall use these results in the next sections.

Finally, similar expressions can be found for arbitrary azimuthal quantum numbers ℓ. A straightforward calculation gives

$$u_\ell(r) = \gamma_\ell(k)\left(u_\ell^{(-)}(r) - S_\ell(k)u_\ell^{(+)}(r)\right), \qquad r > a \tag{22}$$

the asymptotic behaviour of which reads

$$u_\ell(r \to \infty) \approx -2i\gamma_\ell(k)e^{i\delta_\ell(k)}\, \sin\left(kr + \delta_\ell(k) - \frac{\ell\pi}{2}\right). \tag{23}$$

The above discussed properties of the scattering amplitude $S(k)$ are easily generalized to the case of arbitrary angular momentum $S_\ell(k)$.

3. Gamow-Siegert functions

3.1. Analytic continuation of the scattering amplitude $S_\ell(k)$

In the previous section we realized that S_ℓ is a regular function in I_+ except at the points $k = i\kappa$ leading to bound states of the energy. In order to get a well behaved scattering amplitude in the whole of the complex k–plane, this function must be extended to be analytic in I_-. With this aim, first let us construct an arbitrary linear combination of the functions (7), it reads

$$u_\ell(r) = \zeta_\ell(k)\, u_\ell^{(-)}(r) - \xi_\ell(k)\, u_\ell^{(+)}(r) \equiv \zeta_\ell(k)\left[u_\ell^{(-)}(r) - S_\ell(k)u_\ell^{(+)}(r)\right]. \tag{24}$$

This function is regular at the origin if $u_\ell(r = 0) = 0$, then we arrive at the following relationship

$$S_\ell(k) = \frac{\xi_\ell(k)}{\zeta_\ell(k)} = \left.\frac{u_\ell^{(-)}(r)}{u_\ell^{(+)}(r)}\right|_{r=0}. \tag{25}$$

Now, after a change of sign in k, the coefficients of the single-valued function (24) should satisfy

$$-e^{i\pi\ell}\xi_\ell(-k) = \alpha\zeta_\ell(k), \qquad e^{i\pi\ell}\zeta_\ell(-k) = -\alpha\xi_\ell(k) \tag{26}$$

where α is a proportionality factor and we have used equation (8). In this way, the expression (25) leads to

$$S_\ell(k)\, S_\ell(-k) = 1. \tag{27}$$

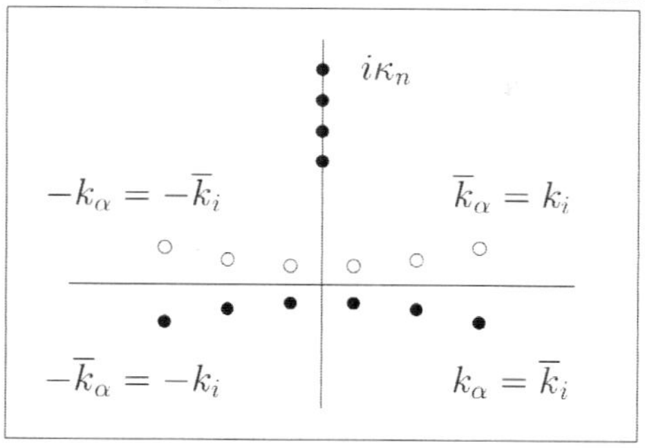

Figure 1. Schematic representation of the poles (disks) and the zeros (circles) of the scattering amplitude $S_\ell(k)$ in the complex k–plane. Bounded energies $E_n = -\kappa_n^2$ correspond to the poles located on the positive imaginary axis.

Let $k_i \in I_+$ be a zero of S_ℓ. From equation (27) we notice that the scattering amplitude is well behaved in I_- except at the point $-k_i$, for which $S_\ell(-k_i) \to \infty$. Thus, $-k_i \in I_-$ is a pole of S_ℓ.

On the other hand, if the kinetic parameter k is real, then the complex conjugate of any solution of equation (3) is also a solution for the same energy. However, the solution is unique so these last functions differ at most in a global phase. Thereby, in a similar form as in the previous case, we get

$$S_\ell(k)\,\overline{S}_\ell(k) = 1 \tag{28}$$

where the bar stands for complex conjugation. This last equation is a consequence of the elastic scattering we are dealing with since $|S_\ell(k)| = 1$. However, as k is real, also equation (28) has to be extended to the whole of the complex k–plane. The natural condition reads

$$S_\ell(k)\,\overline{S}_\ell(\overline{k}) = 1. \tag{29}$$

It is clear that $\overline{S}_\ell(\overline{k}_i)$ diverges because $k_i \in I_+$ is a zero of S_ℓ. In other words, $\overline{k}_i \in I_-$ is a pole of S_ℓ since $|S_\ell(\overline{k}_i)| \to \infty$. Moreover, as $-k_i$ is a pole of S_ℓ, from equation (29) we also notice that $\overline{S}_\ell(-\overline{k}_i) = 0$. Thus, $-\overline{k}_i$ is another zero of the scattering amplitude. In summary, if k_α (k_i) is a pole (zero) of S_ℓ, then $-\overline{k}_\alpha$ ($-\overline{k}_i$) is another pole (zero) while $\overline{k}_\alpha$ and $-k_\alpha$ ($\overline{k}_i$ and $-k_i$) are zeros (poles) of the scattering amplitude. In this way $S_\ell(k)$ is a meromorphic function of the complex kinetic parameter k, with poles restricted to the positive imaginary axis (bound states) and to the lower half-plane I_- (see Figure 1).

3.2. Resonance states

Let us express one of the poles of the scattering amplitude as $k_\alpha = \alpha_1 - i\alpha_2$, with α_1 an arbitrary real number and $\alpha_2 > 0$. The function (22), evaluated for k_α at $r \to \infty$ reads

$$u_\ell(r \to \infty, k_\alpha) \approx \gamma_\ell(k_\alpha) e^{-i(\alpha_1 r - \ell\pi/2)} \left[e^{-\alpha_2 r} - S_\ell(k_\alpha) e^{\alpha_2 r} \right]. \tag{30}$$

Observe that the first term in brackets vanishes for large values of $-\alpha_2 r$. Hence, there remains only the scattering, purely outgoing wave $e^{\alpha_2 r}$. As a consequence, the particle is maximally scattered by the potential field $v(r)$ and the wave function is not "well behaved" since it does not fulfill the boundary condition at $r \to \infty$. However, as we have seen, this last "unphysical behaviour" is not only natural but necessary to study the elastic scattering process of a particle

Journal of Physics: Conference Series **128** (2008) 012044 doi:10.1088/1742-6596/128/1/012044

by a short range, nonsingular radial field interaction. Wave functions behaving like (30) are known as Gamow-Siegert solutions of the Schrödinger equation (3). We write

$$u_\ell^{(GS)}(r) = \begin{cases} \theta_\ell \, g_\ell(r, k_\alpha) \, u_\ell^{(1)}(k_\alpha r) & r < R \\ \\ -\gamma_\ell(k_\alpha) S_\ell(k_\alpha) u_\ell^{(+)}(r, k_\alpha) & R \leq r \end{cases} \tag{31}$$

where θ_ℓ is a constant and the intermediary function $g_\ell(r, k_\alpha)$ is equal to 1 at $r = 0$ and depends on the potential. Besides the matching condition (16), the Gamow-Siegert functions fulfill the purely outgoing boundary condition:

$$\lim_{r \to +\infty} \left\{ \beta_\ell^{(GS)}(r) + i k_\alpha \right\} = 0. \tag{32}$$

Our interest is now addressed to the Gamow-Siegert functions not precisely as describing a decaying system but as appropriate mathematical tools to study the construction of complex Darboux-deformed potentials.

4. Complex-Darboux deformations of radial potentials

4.1. General considerations

In order to throw further light on the function $\beta_\ell(r)$ we may note that (5) transforms the Schrödinger equation (3) into a Riccati one

$$-\beta_\ell'(r) + \beta_\ell^2(r) + \epsilon_\ell = V_\ell(r) \tag{33}$$

where the energy k^2 has been changed for the arbitrary number ϵ_ℓ. Remark that (33) is not invariant under a change in the sign of the function beta:

$$\beta_\ell'(r) + \beta_\ell^2(r) + \epsilon_\ell = V_\ell(r) + 2\beta_\ell'(r). \tag{34}$$

These last equations define a Darboux transformation $\widetilde{V}_\ell(r) \equiv \widetilde{V}_\ell(r, \epsilon_\ell)$ of the initial potential $V_\ell(r)$. If ϵ_ℓ is a nontrivial complex number, then the Darboux-deformation is necessarily a complex function

$$\widetilde{V}_\ell(r) = V_\ell(r) + 2\beta_\ell'(r) \equiv V_\ell(r) - 2\frac{d^2}{dr^2} \ln \varphi_\epsilon(r) \tag{35}$$

where the transformation function $u_\ell(r, \epsilon) \equiv \varphi_\epsilon(r)$ is the general solution of (3) for the complex eigenvalue ϵ_ℓ. The solutions $y_\ell(r) = y_\ell(r, \epsilon_\ell, \mathcal{E})$ of the non-Hermitian Schrödinger equation

$$-y_\ell''(r) + \widetilde{V}_\ell(r) \, y_\ell(r) = \mathcal{E} y_\ell(r) \tag{36}$$

are easily obtained

$$y_\ell(r) \propto \frac{\mathrm{W}(\varphi_\epsilon(r), u_\ell(r))}{\varphi_\epsilon(r)}, \tag{37}$$

where $u_\ell(r)$ is eigensolution of equation (3) with eigenvalue $\mathcal{E}$.

Journal of Physics: Conference Series **128** (2008) 012044 doi:10.1088/1742-6596/128/1/012044

4.2. New exactly solvable optical potentials

Let the transformation function $\varphi_\epsilon(r)$ be a Gamow-Siegert function (31). The *superpotential* $\beta_\ell^{(GS)}(r, k_\alpha)$ and the new potential $\widetilde{V}_\ell(r)$ behave at the edges of D_v as follows

$$\beta_\ell^{(GS)}(r,k_\alpha) = \begin{cases} -\frac{\ell+1}{r} & r\to 0 \\ -ik_\alpha & r\to +\infty \end{cases} \qquad \Rightarrow \qquad \widetilde{V}_\ell(r) = \begin{cases} V_{\ell+1}(r) & r\to 0 \\ V_\ell(r) & r\to +\infty \end{cases} \tag{38}$$

It is remarkable that even for an initial potential with $\ell = 0$ the corresponding Darboux-deformation contains a nontrivial centrifugal term in the interaction zone $r < a$. Notice also that the new potential is mainly a real function at the edges of D_v. Thus, the function $\mathrm{Im}(\widetilde{V}_\ell)$ is zero at the origin and vanishes at $r \to +\infty$. In general $\mathrm{Im}(\widetilde{V}_\ell)$ oscillates along the whole of $(0, +\infty)$ according with the level of excitation of the Gamow-Siegert solution $u_\ell^{(GS)}(r, k_\alpha)$: The higher the level of excitation the greater the number of oscillations. Such a behaviour implies a series of maxima and minima in $\mathrm{Im}(\widetilde{V}_\ell)$ which can be analyzed in terms of the optical bench [19].

As regards the physical solutions of the new potential it is immediate to verify that, in all cases (bounded and scattering states of the initial potential), the corresponding transformations fulfill the boundary condition at $r = 0$. The analysis of the behaviour at $r \to \infty$ is as follows.

4.2.1. Scattering states. Let $u_\ell(r)$ be a scattering state of the initial potential in $r > a$. At $r \to +\infty$, the corresponding Darboux-Deformation (37) behaves as follows

$$\lim_{r\to+\infty} y_\ell(r) = -i\gamma_\ell(k)\left[(k_\alpha + k)u^{(-)}(r) - (k_\alpha - k)S_\ell(k)u^{(+)}(r)\right]_{r\to+\infty} \tag{39}$$

where we have used equation (22). As we can see, the Darboux-deformations of u_ℓ behave as scattering states at $r > a$.

If the scattering state $u_\ell(r)$ is a Gamow-Siegert function and $k \neq k_\alpha, \overline{k}_\alpha, -k_\alpha, -\overline{k}_\alpha$, the Darboux-deformation preserves the purely outgoing condition (32). Special cases are:

(i) $k = k_\alpha \Rightarrow$ the Wronskian in (37) is zero and there is no transformation.
(ii) $k = \overline{k}_\alpha \Rightarrow S_\ell(\overline{k}_\alpha) = 0$, then $y_\ell(r)$ is an exponentially decreasing function.
(iii) $k = -k_\alpha \Rightarrow S_\ell(-k_\alpha) = 0$ and the coefficient of $u^{(-)}(r)$ vanishes, then $y_\ell(r) = 0$.
(iv) $k = -\overline{k}_\alpha \Rightarrow S_\ell(-\overline{k}_\alpha) \to \infty$, then $y_\ell(r)$ fulfills the purely outgoing condition (32).

4.2.2. Bounded states. If $I_+ \ni k = i\kappa = i\sqrt{|E|}$ the function $u_\ell(r, \sqrt{|E|})$ corresponds to one of the bounded states (11). Thereby, from equation (37) we get

$$\lim_{r\to+\infty} y_\ell(r, \sqrt{|E|}) \approx (ik_\alpha - \sqrt{|E|})\, \xi_\ell(\sqrt{|E|})\, e^{-i\frac{\ell\pi}{2}}\, e^{-\sqrt{|E|}\,r}. \tag{40}$$

Thus, the Darboux-deformations of bounded wave-functions vanishes at $r \to \infty$. However, although this new functions are square-integrable, they do not form an orthogonal set in the Hilbert space spanned by the initial square-integrable wave-functions [14] (see also [24] and the 'puzzles' with self orthogonal states [25]).

In the next section we are going to solve the Schrödinger equation for the radial square well. We shall get the solutions corresponding to complex energy eigenvalues and purely outgoing boundary conditions. Thus, we shall construct the involved Gamow-Siegert functions.

5. Gamow-Siegert states of the radial square well

Let us consider the potential

$$v(r) = \begin{cases} -v_0 & r < a \\ 0 & a \le r \end{cases} \tag{41}$$

with a and v_0 positive real numbers. The regular solution of the Schrödinger equation (3) for this potential may be written

$$u_\ell(r) = \begin{cases} \theta qr\, j_\ell(qr) & r < a \\ -i\gamma_\ell(k)\, kr\left[h_\ell^{(2)}(kr) + S_\ell(k) h_\ell^{(1)}(kr)\right] & a \le r \end{cases} \tag{42}$$

where the *interaction parameter* q is defined by $q^2 = v_0 + k^2$. For simplicity, we shall analyze the s-wave solutions ($\ell = 0$) for which the matching conditions (16) lead to

$$S(q) = -\left[\frac{ik\sin(qa) + q\cos(qa)}{ik\sin(qa) - q\cos(qa)}\right] e^{-2ika}, \qquad \gamma(k) = -\frac{\theta e^{ika}}{2ik}[ik\sin(qa) - q\cos(qa)]. \tag{43}$$

Let us take $\theta = 2ik$. The scattering amplitude $S(k)$ has zeros and poles which respectively correspond to the roots of the transcendental equations

$$\frac{iq}{k} = \tan(qa), \qquad -\frac{iq}{k} = \tan(qa). \tag{44}$$

Indeed, equations (44) correspond to the even and odd quantization conditions for negative energies if $k = i\kappa \in I_+$, just as it has been discussed in the previous sections. However, we are looking for points k_α in the complex k–plane such that $\mathrm{Re}(k_\alpha) \neq 0$ and $\mathrm{Im}(k_\alpha) \neq 0$. Thereby, if k_α is a pole of S, the converging wave in (42) vanishes. Then the Gamow-Siegert function reads

$$u_{\ell=0}^{(\mathrm{GS})}(r) = \begin{cases} 2ik_\alpha \sin[q(k_\alpha) r] & r < a \\ 2ik_\alpha \sin[q(k_\alpha)a]\, e^{ik_\alpha(r-a)} & a \le r \end{cases} \tag{45}$$

In order to solve the second equation in (44) let us rewrite the scattering amplitude as

$$S(k) = -\left[\frac{e^{2iqa}(k+q) + (q-k)}{e^{2iqa}(k-q) - (q+k)}\right] e^{-2ika}. \tag{46}$$

In this way, the poles of S are also the solutions of the transcendental equation

$$e^{2iqa} = \frac{|k|^2 - |q|^2 + 2\mathrm{Re}(k\overline{q})}{|k-q|^2}. \tag{47}$$

If the interaction parameter q is real, then $\sin(2qa) = 0$ and we obtain the following quantization rule for the energy

$$q_n = \frac{\pi n}{2a} \quad \Rightarrow \quad E_n = \left(\frac{n\pi}{2a}\right)^2 - v_0, \quad n = 0, 1, \ldots \tag{48}$$

On the other hand, let k and q be the complex numbers $k = K_1 + iK_2$ and $q = Q_1 + iQ_2$. Then $q^2 = v_0 + \epsilon = (v_0 + \epsilon_1) + i\epsilon_2$, with $\epsilon = k^2$. Hence we have

$$Q_1^2 - Q_2^2 = v_0 + \epsilon_1, \qquad 2Q_1Q_2 = \epsilon_2 \tag{49}$$

Journal of Physics: Conference Series **128** (2008) 012044 doi:10.1088/1742-6596/128/1/012044

and

$$K_1^2 - K_2^2 = \epsilon_1, \qquad 2K_1K_2 = \epsilon_2. \tag{50}$$

The imaginary part of equation (47) is $\sin(2Q_1a) = 0$ and we obtain the following quantization rule

$$Q_{1,n} = \frac{\pi n}{2a}, \quad n = 0, 1, \ldots \tag{51}$$

Therefore, if $|Q_2| << 1$, the first equation in (49) shows that equation (51) leads to $\epsilon_1 \approx E_n$. The second equation in (49), on the other hand, gives a value of ϵ_2 which is not necessarily equal to zero. In this way, we shall study the poles of S for which the complex points ϵ are such that $\mathrm{Re}(\epsilon) \approx E_n$ and $\mathrm{Im}(\epsilon) \approx -\frac{\Gamma}{2}$, with $0 < \Gamma << 1$. The main problem is then to approximate the adequate value of Γ as connected with the discreteness of Q_1 in (51). With this aim, for a given finite value of a we take $|Q_2|a << 1$. A straightforward calculation shows that the second equation in (44) uncouples into the system

$$Q_1 = \sqrt{v_0}\sin(Q_1a)\cosh(Q_2a), \qquad Q_2 = \sqrt{v_0}\cos(Q_1a)\sinh(Q_2a). \tag{52}$$

In turn, the second one of these last equations reduces to

$$\cos(Q_1a) \approx \frac{1}{a\sqrt{v_0}} = \frac{1}{\eta}. \tag{53}$$

Now, let $|\delta| << 1$ be a correction of Q_1 around the quantized values (51), that is $Q_1a \approx Q_{1,n}a + \delta$. Then, equation (53) leads to

$$\delta_n = -\frac{\sin(\pi n/2)}{\eta}, \quad n \text{ odd}. \tag{54}$$

Thus, only odd values of n are allowed into the approximation we are dealing with. Now, to ensure small values of Q_2 we propose $Q_2a = \sum_{m=1}^{+\infty} \lambda_m/\eta^m$. From (49) we have

$$\frac{\epsilon_1}{v_0} \approx -1 + \left(\frac{aQ_{1,n}}{\eta}\right)^2 - \frac{2aQ_{1,n}\sin(\pi n/2)}{\eta^3} + \frac{1-\lambda_1^2}{\eta^4} - \frac{2\lambda_1\lambda_2}{\eta^5} - \cdots \tag{55}$$

In order to cancel the term including η^{-4} we take $\lambda_1 = \pm 1$ (the appropriate sign will be fixed below). Higher exponents in the power of $1/\eta$ are dropped by taking $\lambda_{m>1} = 0$. In order to get $\epsilon_1 \to E_n$, after introducing (51) into (55), we realize that the condition $\eta >> 1$ allows us to drop the term including η^{-3}. Thus, at the first order in the approximation of Q_2a we get $\epsilon_1 \approx E_n$.

It is also necessary to have regard to the positiveness of energy ϵ_1, which is ensured whenever n exceeds a minimum value. Let us take $n := n_{\mathrm{inf}} + m$, $m \in \mathbb{Z}^+$, where n_{inf} is the ceiling function of $2\eta/\pi$, i.e. $n_{\mathrm{inf}} = \left\lceil \frac{2\eta}{\pi} \right\rceil = \left\lceil \frac{2a\sqrt{v_0}}{\pi} \right\rceil$. Since n is odd, the integer m is even (odd) if n_{inf} is odd (even). We finally arrive at

$$\mathrm{Re}(\epsilon) \approx \left(\left[\frac{(n_{\mathrm{inf}}+m)\pi}{2a\sqrt{v_0}}\right]^2 - 1\right) v_0, \quad m = \begin{cases} 0, 2, 4, \ldots & n_{\mathrm{inf}} \text{ odd} \\ 1, 3, 5, \ldots & n_{\mathrm{inf}} \text{ even} \end{cases} \tag{56}$$

On the other hand, it is worth noticing that the introduction of $Q_2a = \pm 1/\eta$ and (54) into the second equation in (49) leads to the conclusion that $Q_2a = +1/\eta$ has to be dropped. The expression for the imaginary part of the complex energy ϵ is calculated as follows. First, let us rewrite equations (52) in the form

$$K_1 = \pm\sqrt{v_0}\sin(Q_1a)\sinh(Q_2a), \qquad K_2 = \pm\sqrt{v_0}\cos(Q_1a)\cosh(Q_2a). \tag{57}$$

Journal of Physics: Conference Series **128** (2008) 012044 doi:10.1088/1742-6596/128/1/012044

Since $\epsilon_2 = -\Gamma/2$, from equations (50) we realize that the sign of K_1 must be the opposite one of K_2. The introduction of (53) into the second equation in (57) leads to

$$K_2 \approx \pm\frac{1}{a}\left(1+\frac{1}{2\eta^2}\right) \tag{58}$$

where we have used $Q_2 a \approx -1/\eta$. If $\eta >> 1$ we get $aK_2 \approx \pm 1$. Then, this last result together with the real part of the factorization constant ϵ gives

$$K_1^2 = \epsilon_1 + K_2^2 \approx \epsilon_1 + \frac{1}{a^2}. \tag{59}$$

Thus, for $a^2\epsilon_1 > 1$ we obtain $K_1 \approx \pm\sqrt{\epsilon_1}$. Finally, the introduction of (58) and (59) into the second equation of (50) produces

$$\mathrm{Im}(\epsilon) = -\frac{\Gamma}{2} \equiv -2|K_1|\,|K_2| \approx -\frac{2}{a}\sqrt{\mathrm{Re}(\epsilon)}. \tag{60}$$

In Figure 2 we show the absolute value of the scattering amplitude close to one of its poles k_α and the corresponding zero $\bar{k}_\alpha$. Some of the first resonances are reported in Table 1 for different values of the potential strength v_0 and the cutoff a.

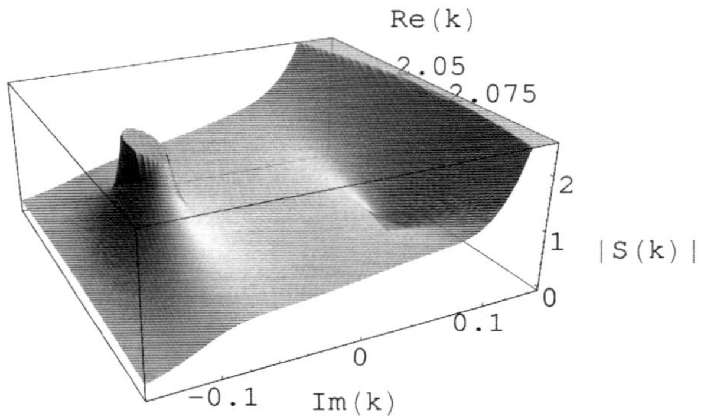

Figure 2. The absolute value of the scattering amplitude $S(k)$ close to the pole k_α (and the zero $\bar{k}_\alpha$) which corresponds to the first resonance $E_\alpha = k_\alpha^2$ of the square well reported in Table 1 for $n_{\mathrm{inf}} = 64$. The limits of our approach give the number $k_\alpha = 2.063412 - i\,0.099882$.

5.1. Complex Darboux-deformations of the radial square well

Figure 3 shows the global behaviour of a typical Gamow-Siegert function associated with the radial square well. Notice the exponential growing of the amplitude for $r > a$. This kind of solutions are used in (35) to complex Darboux-deform the square well as it is shown in Figure 4.

The cardiod-like behaviour seems to be a profile of the complex potentials derived by means of Darboux-deformations (compare with [20] and [22]). In this case, the complex potential shows concentric cardiod curves for values of r inside the interaction zone, one of them is shown at the left of Figure 4. The real and imaginary parts of $\widetilde{V}_\ell(r)$ are then characterized by oscillations and changes of sign depending on the position in $r < a$. As a consequence, these complex potentials behave as an optical device which both refracts and absorbs light waves (see details in [22] and

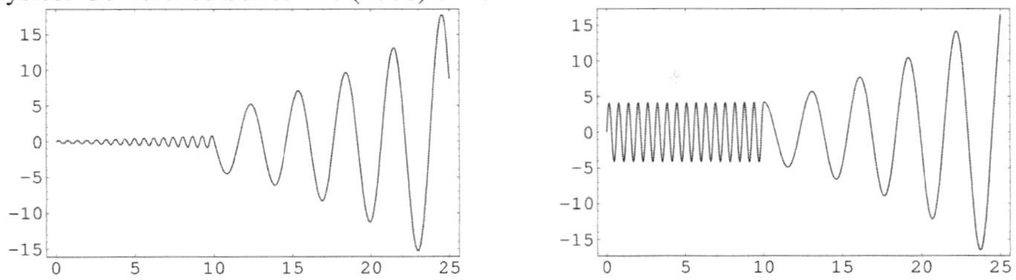

Figure 3. The real (left) and imaginary (right) parts of the Gamow-Siegert function (45) associated with the first resonance state reported in Table 1 for $n_{\rm inf} = 64$.

discussions on the optical bench in [19]). The presence of this kind of oscillations is also noted at distances slightly greater than the cutoff $r = a$. Hence, the complex Darboux-deformations (35) are short range potentials which enlarge the initial 'interaction zone' as it is shown in the right part of Figure 4.

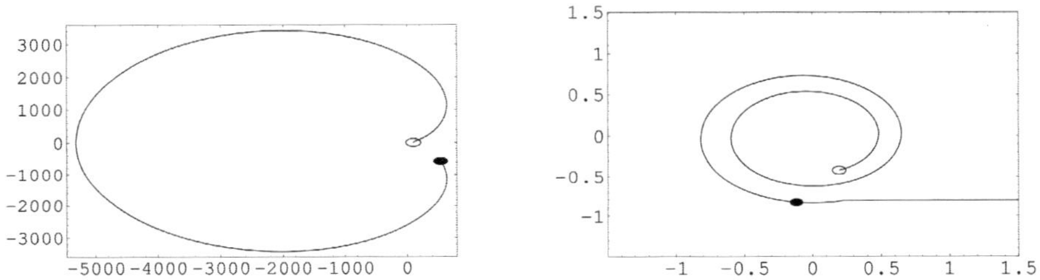

Figure 4. The Argand-Wessel diagram of the complex Darboux-deformed square well with $v_0 = 100$ and $a = 10$. Left: Detail of the cardiod-like behaviour of the new potential between $r = 9.8$ (disk) and $r = 10$ (circle). Right: The disk is evaluated at $r = 10.1$ and the circle at $r = 13$. As complementary information: $\widetilde{V}_0(r=10) = 100.2255 - i\,0.8076$ and $\widetilde{V}_0(r=10.1) = -0.1157 - i\,0.8306$.

Table 1. The first four resonance energies for the radial square well of intensity $v_0 = 100$ and cutoff $a = 10$ ($n_{\rm inf} = 64$) and for the radial square well of intensity $v_0 = 1000$ and cutoff $a = 10$ ($n_{\rm inf} = 202$). Notice that, in each case, the even values of m obtained for the related one-dimensional problem are missing (see e.g. Table A.1 of reference [22]).

	$n_{\rm inf} = 64$	$n_{\rm inf} = 202$
$m = 1$	$04.247696 - i\,0.412198$	$16.791319 - i\,0.819544$
$m = 3$	$10.761635 - i\,0.656098$	$36.925312 - i\,1.215324$
$m = 5$	$17.472966 - i\,0.836013$	$57.256697 - i\,1.513363$
$m = 7$	$24.381689 - i\,0.987556$	$77.785474 - i\,1.763921$

6. Concluding remarks

We have studied the elastic scattering process of a particle by short range, nonsingular radial field interactions. The poles of the involved scattering amplitude $S(k)$ play a relevant role in the construction of physical solutions connected with either bounded or scattering states of the energy. It has been shown that poles k_α in the lower half complex k–plane lead to the "unphysical" Gamow-Siegert functions which are necessary to depict the behaviour of the scattering phenomenon.

The Gamow-Siegert functions were used to transform the radial short range potentials into complex ones which preserve the initial energy spectrum. These new potentials are 'opaque' in the sense that they simultaneously emit and absorb flux, just as an optical device which both refracts and absorbs light waves. The transformation preserves the square-integrability of the solutions at the cost of producing non-orthogonal sets of wave-functions. It is also notable that scattering states are transformed into scattering states while deformed Gamow-Siegert functions can be either a new Gamow-Siegert function or an exponentially decreasing function depending on the involved kinetic parameter k and the pole k_α.

The method has been applied to the radial square well in the context of s-waves ($\ell = 0$). Analytical expressions were derived for the corresponding complex energies in the long lifetime limit (i.e., for small values of $|\mathrm{Im}(E_\alpha)| = \Gamma/2$) and, as a consequence, such energies fulfill a 'quantization rule'. In contrast with the 'even' and 'odd' resonances of short range potentials defined in the whole of the straight-line (see e.g. [22]), the resonances of radial short range potentials are labelled by only odd (positive) integers. This result enforces the interpretation of Gamow-Siegert functions as representing quasi-bounded states: The bounded spectrum of a potential which is invariant under the action of the parity operator $v(x) = v(-x)$ includes odd and even functions. The symmetry is broken by adding an impenetrable wall at the negative part of the straight line and only the odd solutions are preserved. As we have shown, the same is true for resonance states.

Finally, the results reported in this paper are complementary to our previous work [22]. It is remarkable that explicit derivations of Gamow-Siegert functions are barely reported in the literature, not even for simple models like those studied here (however see [5, 20, 21, 26–29]). We hope our approach has shed some light onto the solving of the Schrödinger equation for complex energies and functions fulfilling the purely outgoing boundary condition.

Acknowledgments.
The support of CONACyT project 24233-50766-F is acknowledged. The authors are grateful to M. Gadella and D. Julio for interesting discussions.

References
[1] Gamov G 1928 *Z Phys* **51** 204
[2] Siegert AJF 1939 *Phys Rev* **56** 750
[3] Bohm A, Gadella M and Mainland GB 1989 *Am J Phys* **57** 1103
[4] Bohm A and Gadella M 1989 *Lecture Notes in Physics* vol 348 (New York: Springer)
[5] de la Madrid R and Gadella M 2002 *Am J Phys* **70** 626
[6] Civitarese O and Gadella M 2004 *Phys Rep* **396** 41
[7] Aguilar J and Combes JM 1971 *Commun Math Phys* **22** 269
[8] Balslev E and Combes JM 1971 *Commun Math Phys* **22** 280
[9] Simon B 1972 *Commun Math Phys* **27** 1
[10] Cannata F, Junker G and Trost J 1988 *Phys Lett A* **246** 219
[11] Andrianov AA, Ioffe MV, Cannata F and Dedonder JP 1999 *Int J Mod Phys A* **14** 2675
[12] Bagchi B, Mallik S and Quesne C 2001 *Int J Mod Phys A* **16** 2859
[13] Fernández DJ, Muñoz R and Ramos A 2003 *Phys Lett A* **308** 11
[14] Rosas-Ortiz O and Muñoz R 2003 *J Phys A: Math Gen* **36** 8497
[15] Muñoz R 2005 *Phys Lett A* **345** 287
[16] Samsonov BF and Pupasov AM 2006 *Phys Lett A* **356** 210
[17] Samsonov BF 2006 *Phys Lett A* **358** 105
[18] Cabrera-Munguia I and Rosas-Ortiz O 2007 *this issue*
[19] Mielnik B and Rosas-Ortiz O 2004 *J Phys A: Math Gen* **37** 10007

[20] Rosas-Ortiz O 2007 *Rev Mex Fis* **53 Suppl 2** 103
[21] Fernández-García N 2007 *Rev Mex Fis* **53 Suppl 4** 42
[22] Fernández-García N and Rosas-Ortiz O 2008 *Ann Phys* doi:10.1016/j.aop.2007.11.002
[23] Abramowitz M and Stegun I 1972 *Handbook of Mathematical Functions* (New York: Dover Pub Inc)
[24] Ramírez A and Mielnik B 2003 *Rev Mex Fis* **49 Suppl 2** 130
[25] Sokolov AV, Andrianov AA and Cannata F 2006 *J Phys A: Math Gen* **9** 10207
[26] Webber TA, Hammer CL and Zidell VS 1982 *Am J Phys* **50** 839
[27] Sprung DWL and Wu H 1996 *Am J Phys* **64** 136
[28] Antoniou IE, Gadella M, Hernandez E, Jauregui A, Melnikov Y, Mondragón A and Pronko GP 2001 *Chaos, Sol. and Fractals* **12** 2719
[29] de la Madrid R 2004 *J Phys A: Math Gen* **37** 8129

Journal of Physics: Conference Series **128** (2008) 012045 doi:10.1088/1742-6596/128/1/012045

Solvable $\mathcal{PT}$-symmetric potentials in 2 and 3 dimensions

Géza Lévai

Institute of Nuclear Research of the Hungarian Academy of Sciences (ATOMKI), Debrecen, Pf. 51, Hungary 4001

E-mail: levai@atomki.hu

Abstract. The solution of non-central $\mathcal{PT}$-symmetric potentials is discussed by the separation of the variables in polar and angular coordinates. Conditions are formulated to guarantee the separation of the variables and the $\mathcal{PT}$ symmetry of the potential. The original eigenvalue equation is separated into one-dimensional Schrödinger-type differential equations. The importance of the boundary conditions, especially that of the periodic boundary condition of the azimuthal equation is pointed out. Further conditions leading to exact solutions of the whole problem are also formulated. An example combining the harmonic oscillator and the Scarf I potential in the radial and polar equation is discussed in detail, and the bound-state wave functions and the energy eigenvalues are derived. The spectrum exhibits partial degeneracies similar to those observed in the spectrum of the isotropic harmonic oscillator.

1. Introduction

Since the introduction of $\mathcal{PT}$-symmetric quantum mechanics [1] much effort has been devoted to exploring the physical significance of this theory and its connection with ordinary Hermitian quantum mechanics. It was found that $\mathcal{PT}$-symmetric quantum mechanics is a special case of pseudo-Hermiticity [2] and this can explain some of the unusual features of $\mathcal{PT}$-symmetric systems, such as the partly or fully real energy spectrum and the conservation of norm. The probabilistic interpretation of $\mathcal{PT}$-symmetric quantum mechanics, and in general pseudo-Hermitian models also received much attention, and positive definit metric operators have been cosntructed [3]. Although non-Hermitian theories that make use of modified metric operators have been known previously under various names, *e.g.* pseudo- [4], quasi- [5] and crypto-Hermiticity [6], the role of $\mathcal{PT}$-symmetric quantum mechanics is indisputable in reviving interest in these theories.

Exactly solvable quantum mechanical models are indispensable in gaining insight in the physical background of any model, and this is also the case with $\mathcal{PT}$-symmetric quantum mechanics. The $\mathcal{PT}$-symmetric versions of real potentials have been constructed, and have been used to explore various aspects of $\mathcal{PT}$ symmetry, such as identifing potentials with real and complex eigenvalues [7, 8, 9, 10, 11, 12, 13], describing the mechanism of the spontaneous breakdown of $\mathcal{PT}$ symmetry [14, 15], combining $\mathcal{PT}$-symmetry with supersymmetric [10, 16] and algebraic [17, 18, 19] techniques, determining the normalization constants and pseudo-norm of $\mathcal{PT}$-symmetric potentials [15, 20, 21], *etc.*

The scope of investigations has been extended from one-dimensional non-relativistic problems, in various directions, such as scattering solutions [18, 22], periodic structures [23], coupled channels [24], more particles [25] and relativistic wave equations [26]. After some first efforts [27, 28, 29] the systematic extension of $\mathcal{PT}$-symmetric potentials in higher dimensions has also started [30, 31]. In these studies the separation of the variables has been applied and the multi-dimensional problem has been reduced to one-dimensional eigenvalue problems resembling the Schrödinger equation. The importance of the angular sector in introducing the spontaneous breakdown of $\mathcal{PT}$ symmetry has been pointed out in the case of both two- [30] and three-dimensional [31] models. Similar studies have been performed earlier on real non-central potentials [32], nevertheless, the $\mathcal{PT}$ symmetry requirement makes these studies rather more complicated. The previous works focused more on the formal solution of $\mathcal{PT}$-symmetric potentials in two and three dimensions, and actual solvable examples with complete analytical solutions have not been given. The aim of the present work is to fill this gap by presenting explicit examples for solvable models.

In section 2 the general formalism is presented for both two- and three-dimensional potentials and conditions are formulated for the implementation of $\mathcal{PT}$ symmetry. Section 3 deals with further conditions allowing for the exact solution of the separated eigenvalue equations in terms of known solvable potentials, and it also contains the explicit wave functions and energy eigenvalues of a concrete example. Finally in section 4 a summary of the results is given.

2. Separation of the variables and conditions for $\mathcal{PT}$ symmetry

Let us consider the Schrödinger equation with a general non-central potential and constant mass

$$\left(\frac{\mathbf{p}^2}{2m}+V(\mathbf{r})\right)\psi(\mathbf{r}) = -\frac{\hbar^2}{2m}\Delta\psi(\mathbf{r})+V(\mathbf{r})\psi(\mathbf{r}) = E\psi(\mathbf{r}) \ . \tag{1}$$

In what follows we specify (1) for two and three spatial dimensions and investigate the conditions under which its formal solution is possible by the separation of the angular and radial variables. We also formulate the conditions for the $\mathcal{PT}$ symmetry of the potential, which also implies the $\mathcal{PT}$ symmetry of the corresponding Hamiltonian, as the kinetic term is always $\mathcal{PT}$-symmetric. Without the loss of generality we can use the units $2m = \hbar = 1$.

2.1. $\mathcal{PT}$-symmetric Hamiltonians in 2 dimensions

The Schrödinger equation (1) expressed in terms of polar coordinates in two dimensions is

$$\frac{1}{\rho}\frac{\partial}{\partial\rho}\left(\rho\frac{\partial\psi}{\partial\rho}\right)+\frac{1}{\rho^2}\frac{\partial^2\psi}{\partial\varphi^2}-V(\rho,\varphi)\psi+E\psi=0 \ . \tag{2}$$

With the substitution

$$\psi(\rho,\varphi)=\rho^{-1/2}\phi(\rho)\tau(\varphi) \tag{3}$$

the equation

$$\phi''\tau+\frac{1}{\rho^2}\phi\tau''-\left(V(\rho,\varphi)-\frac{1}{4\rho^2}-E\right)\phi\tau=0 \tag{4}$$

is obtained, where prime denotes derivation with respect to the corresponding single variable and $\rho\in[0,\infty)$ and $\varphi\in[0,2\pi]$. The separation of the radial and angular variables is possible, if $\tau(\varphi)$ satisfies the equation

$$\tau''=(K(\varphi)-k)\tau \ . \tag{5}$$

In this case the potential can be written as the sum of a central and a non-central term as

$$V(\rho,\varphi)=V_0(\rho)+\frac{1}{\rho^2}K(\varphi) \ . \tag{6}$$

Journal of Physics: Conference Series **128** (2008) 012045 doi:10.1088/1742-6596/128/1/012045

The condition for $\mathcal{PT}$ symmetry is then

$$V(\rho,\varphi) = V^*(\rho,\varphi+\pi) \ , \tag{7}$$

which means that the central potential component $V_0(\rho)$ has to be real and $K(\varphi)$ has to satisfy

$$K^*(\varphi+\pi) = K(\varphi) \ . \tag{8}$$

Equations (5) and (8) indicate that the $\mathcal{PT}$ symmetry of the full two-dimensional potential (6) generates a $\mathcal{PT}$-symmetric Hamiltonian in the angular variable too, although the space reflection operator assigned to this problem has to be defined as

$$\mathcal{P}_\varphi \varphi = \varphi + \pi \ . \tag{9}$$

Furthermore, the problem has to be defined on the $\varphi \in [0, 2\pi]$ domain with periodic boundary conditions $\tau(\varphi) = \tau(\varphi+2\pi)$ and $\tau'(\varphi) = \tau'(\varphi+2\pi)$. Note that $\tau(\varphi)$ need not vanish at the boundaries, so the solutions of (5) are less restricted than those of a proper Schrödinger equation defined on a finite domain.

The eigenvalue k of equation (5) can be real or complex, depending on whether the $\mathcal{PT}$ symmetry of this angular problem is unbroken or spontaneously broken. It is also important to note that (8) implies *different* constraints on the real and imaginary component of $K(\varphi)$: $K_R(\varphi+\pi) = K_R(\varphi)$ and $K_I(\varphi+\pi) = -K_I(\varphi)$. In the case of a periodic $K(\varphi)$ this means that the periodicity of the two components have to be different, with even and odd number of periods for $K_R(\varphi)$ and $K_I(\varphi)$. This is analogous to the usual $V^*(-x) = V(x)$ $\mathcal{PT}$ symmetry requirement in one dimension, where the real and imaginary potential components have to be even and odd functions of the coordinate, respectively.

Separating the angular parts, we are left with the equation

$$-\phi'' + \left[V_0(\rho) + \left(k - \frac{1}{4}\right)\frac{1}{\rho^2}\right]\phi - E\phi = 0 \ , \tag{10}$$

in which the eigenvalue k of the angular equation (5) appears explicitly. Equation (10) is similar to the radial Schrödinger equation obtained in the case of real central potentials, and the boundary conditions also have to be chosen similarly. It is notable though, that the "angular momentum" can now take on complex values too if the $\mathcal{PT}$ symmetry of equation (5) is spontaneously broken. In this case the energy eigenvalues E will also be complex, which means that the $\mathcal{PT}$ symmetry of the *whole* two-dimensional system will also be spontaneously broken. Taking the complex conjugate of (10) it is straightforward to see that there will be two complex conjugate energy eigenvalues originating from the complex conjugate values k and k^*. This is similar to the situation in the case of the spontaneous breakdown of the $\mathcal{PT}$ symmetry in one-dimensional problems.

2.2. $\mathcal{PT}$-symmetric Hamiltonians in 3 dimensions

Using polar coordinates equation (1) takes the form

$$\frac{1}{r^2}\frac{\partial}{\partial r}\left(r^2\frac{\partial\psi}{\partial r}\right) + \frac{1}{r^2}\frac{\partial^2\psi}{\partial\theta^2} + \frac{1}{r^2}\cot\theta\frac{\partial\psi}{\partial\theta} + \frac{1}{r^2\sin^2\theta}\frac{\partial^2\psi}{\partial\varphi^2} - V(r,\theta,\varphi)\psi + E\psi = 0 \ . \tag{11}$$

Factorizing the solutions as

$$\psi(r,\theta,\varphi) = r^{-1}\phi(r)\sin^{-1/2}\omega(\theta)\tau(\varphi) \ , \tag{12}$$

where $r \in [0, \infty)$, $\theta \in [0, \pi]$ and $\varphi \in [0, 2\pi]$ we obtain the three-dimensional analogue of equation (4):

$$\phi''\omega\tau + \frac{1}{r^2}\phi\omega''\tau + \frac{1}{r^2\sin^2\theta}\phi\omega\tau'' - \left(V(r,\theta,\varphi) - \frac{1}{4r^2} - \frac{1}{4r^2\sin^2\theta} - E\right)\phi\omega\tau = 0 \,. \tag{13}$$

The separation of the variables is possible, if $\tau(\varphi)$ satisfies (5) as in the two-dimensional case, while for $\omega(\theta)$ a similar second-order differential equation holds:

$$\omega'' = (P(\theta) - p)\omega \,, \tag{14}$$

with the difference that the boundary conditions need not be periodic. Similarly to $\tau(\varphi)$, $\omega(\theta)$ need not vanish at the boundaries either, so, again solutions beyond the physical wave functions of one-dimensional confined potentials are allowed.

Making use of the angular functions $K(\varphi)$ and $P(\theta)$, the non-central potential can be written as

$$V(r,\theta,\varphi) = V_0(r) + \frac{K(\varphi)}{r^2\sin^2\theta} + \frac{P(\theta)}{r^2} + \frac{1}{r^2\sin^2\theta}\left(\frac{1}{4} - k\right) \,, \tag{15}$$

Here $V_0(r)$ is a spherical potential, which appears in a radial equation

$$-\phi'' + \left[V_0(r) + \frac{1}{r^2}\left(p - \frac{1}{4}\right)\right]\phi - E\phi = 0 \,. \tag{16}$$

Similarly to the two-dimensional case, equation (16) is formally identical with a radial Schrödinger equation in which the $l(l+1)$ angular momentum term is replaced by $p - 1/4$, where p can be real or imaginary, depending on the solution of the angular equation (14). It is notable that p is related to the "angular momentum" via $p = (l + 1/2)^2$.

The conditions under which the non-central potential (15) is $\mathcal{PT}$-symmetric are determined by the $\mathcal{PT}$ symmetry of the angular equations (5) and (14). Considering that the $\mathcal{P}$ operator acts like $\mathcal{P} : \mathbf{r} \to -\mathbf{r}$, the $\mathcal{PT}$ operation can be factorized into angular terms as

$$\mathcal{P} = \mathcal{P}_\theta \mathcal{P}_\varphi \,, \tag{17}$$

$$\mathcal{P}_\theta \theta = \pi - \theta \,, \tag{18}$$

where $\mathcal{P}_\varphi$ is the same as in the two-dimensional case: (9). From equations (17), (18) and (9) it is seen that the $\mathcal{PT}$ transform of (15) is

$$V(r,\theta,\varphi) = V^*(r, \pi - \theta, \varphi + \pi) \tag{19}$$

$$= V_0^*(r) + \frac{K^*(\varphi + \pi)}{r^2\sin^2\theta} + \frac{P^*(\pi - \theta)}{r^2} + \frac{1}{r^2\sin^2\theta}\left(\frac{1}{4} - k^*\right) \,. \tag{20}$$

This implies the following restrictions of the quantities appearing in the expression (15):

$$V_0(r) = V_0^*(r) \,, \tag{21}$$

$$P^*(\pi - \theta) = P(\theta) \,, \tag{22}$$

$$K^*(\varphi + \pi) = K(\varphi) \,, \tag{23}$$

$$k^* = k \,. \tag{24}$$

This means that the $\mathcal{PT}$ symmetry of the non-central potential (15) hinges on the $\mathcal{PT}$ symmetry of the angular equations, while further conditions require the reality of the central potential component and that of k, *i.e.* the eigenvalue of the equation (5). It is important to note that the eigenvalue p of the polar equation (14) need not be real. Note that if p is real, then (16) becomes an ordinary radial Schrödinger equation with real angular momenta and real energy eigenvalue E. On the contrary, if p is complex, then the energy eigenvalue of (16) and thus that of the *whole* Hamiltonian (1) E will be complex, *i.e.* the $\mathcal{PT}$ symmetry of (1) will be spontaneously broken. Furthermore, in this case E^* will also be an eigenvalue belonging to p^*, as can be seen from the complex conjugation of (16). This was also the case in two dimensions, where k played the role of p.

3. Conditions for exact solvability

The results of section 2 indicate that the general solution of the Schrödinger equation is possible by solving the angular and radial equations such that the energy eigenvalues E of the Schrödinger equation (1) will be supplied by the energy eigenvalues of the radial equation, in which the role of the "angular momentum" is played by the eigenvalue of a radial equation (that of (5) for $d = 2$ and that of (14) for $d = 3$). The energy eigenvalues E will be real or complex depending on whether the "angular momentum" is real or complex. Furthermore in the three-dimensional case the eigenvalue of the azimuthal equation (5) appear explicitly in the non-central component of the potential function (15), so further considerations have to be made if we expect the potential to be state-independent. In this section we discuss the conditions under which the exact discrete solutions of the non-central $\mathcal{PT}$-symmetric potentials can be given, and illustrate the results with some examples. It turns out that the solutions of equation (5) play a crucial role in both the two- and the three-dimensional cases.

3.1. Solving the radial equations

Due to the analogous structure of equations (10) and (16) it is reasonable to discuss their solution jointly. Exact solution of this type of equation with arbitrary value of the "angular momentum" and the principal quantum number is possible for the harmonic oscillator, Coulomb and square well potentials. Usually these potentials are discussed with integer number of the angular momentum l, although the results are easily generalized to non-integer real values too, as is the case with the Kratzer potential, for example [33]. Table 1 contains the quantities relevant to the harmonic oscillator and Coulomb potentials, together with the appropriate substitutions that lead to the exact solution of equations (10) and (16). Note that the formulae allow complex values of the "angular momentum" too, which occur as complex conjugate pairs in the case of the spontaneous breakdown of $\mathcal{PT}$ symmetry. The wave functions are expressed in terms of generalized Laguerre polynomials [34] in both cases. Note that in the substitution for l the indefinite sign comes from the square root taken from $k = (l + \frac{1}{2})^2$ and $p = (l + \frac{1}{2})^2$.

When the radial potential is the infinite square well, the solutions can be expressed in terms of Bessel functions the order of which is related to l [33]. However, the energy eigenvalues cannot be expressed in closed form in this case, because they are determined by the zeros of the Bessel functions. Nevertheless, the square well potential can be used to construct non-central $\mathcal{PT}$-symmetric potentials with exact solutions.

Some solutions can also be obtained for arbitrary k or p for quasi-exactly solvable (QES) potentials [37] in the sense that the first few solutions (up to a given principal quantum number) can be determined exactly then. There are many more solvable potentials for the special value of $k = 1/4$ or $p = 1/4$ (these are the ones solvable in the usual framework for s waves only), but these are unimportant from the point of view of constructing $\mathcal{PT}$-symmetric potentials.

3.2. Solving the angular equation in $d = 2$

The most important prescriptions here are the unusual $\mathcal{PT}$ symmetry requirement (23) of the "potential" $K(\varphi)$ and the periodic boundary conditions of the solutions.

The simplest choice is applying the *real* infinite square well as a special $\mathcal{PT}$-symmetric potential. Then $K(\varphi) = 0$ so (6) reduces to the real central potential $V_0(\rho)$. The general

Table 1. The energy eigenvalues and wave functions of the harmonic oscillator and Coulomb potentials with the substitutions necessary to obtain the exact solutions of the radial Schrödinger equations (10) and (16) in two and three dimensions. The notations of [35] and [36] are followed.

	Harmonic oscillator	Coulomb potential
Potential	$\frac{\omega^2}{4}x^2 + \frac{l(l+1)}{x^2}$	$-\frac{e^2}{x} + \frac{l(l+1)}{x^2}$
E_n	$(2n + l + \frac{3}{2})\omega$	$-\frac{e^4}{4(n+l+1)^2}$
$\phi_n(x)$	$C^{HO}x^{l+1}\exp(-\omega x^2/4)L_n^{(l+\frac{1}{2})}(\frac{\omega}{2}x^2)$	$C^{C}x^{l+1}\exp(-\frac{e^2 x}{2(n+l+1)})L_n^{(2l+1)}(\frac{e^2 x}{n+l+1})$
x for $d = 2$	ρ	ρ
l for $d = 2$	$\pm k^{1/2} + \frac{1}{2}$	$\pm k^{1/2} + \frac{1}{2}$
x for $d = 3$	r	r
l for $d = 2$	$\pm p^{1/2} + \frac{1}{2}$	$\pm p^{1/2} + \frac{1}{2}$

solutions can be written in terms of the exponential functions $e^{\pm im\varphi}$, but in order to construct $\tau(\varphi)$ functions that are the eigenfunctions of the $\mathcal{P}_\varphi\mathcal{T}$ operator (see (9)) with unit eigenvalue, a special combination of them has to be taken:

$$\tau_m(\varphi) = \frac{i^m}{(2\pi)^{1/2}}\cos(m\varphi) \ . \tag{25}$$

These wave functions are $\mathcal{PT}$-normalized as $\langle\tau_n|\mathcal{P}_\varphi|\tau_m\rangle = \delta_{nm}(-1)^m$ and the corresponding energy eigenvalues are $k = k_m = m^2$. Nothing changes essentially if $K(\varphi)$ is chosen to be a real constant: this case corresponds to a redefinition of the "angular momentum" as the relation of k and m is changed.

Further solutions can be obtained by appling exactly solvable $\mathcal{PT}$-symmetric potentials defined on a finite domain, such as the $\mathcal{PT}$-symmetric Scarf I [20] or Rosen–Morse I [21] potentials. Due to the requirement (23), however, these potentials have to be defined in two (or even number of) separate domains $\varphi \in [0, 2\pi]$, with periodic boundary conditions. One problem with these potentials is that they possess inverse-square-type singularity at the boundaries, which will thus separate the segments from each other by an impenetrable wall. This can be avoided by using parameters that result in a weakly attractive singularity at the boundaries, which would, in principle allow communication between the individual segments. It is notable that considering these potentials under the usual $\mathcal{PT}$-symmetry requirements and with solutions vanishing at the boundaries, only real energy eigenvalues can be obtained [20, 21].

Let us now turn to semi-analytically solvable potentials. Examples for this are the combination of imaginary step potentials on a ring [28, 29]. Due to the periodic boundary conditions the energy eigenvalues of these systems asymptotically go to those of the infinite square well, but the number and arrangement of the steps are also reflected in the local relative position of the levels. The solutions up to 20 or so can be easily determined for these potentials by graphical and analytical methods. It was also found that by increasing non-Hermiticity (*i.e.* the height of the steps), complex-energy solutions appear at a certain point [29]. With the use of these potentials in (5) therefore at least a finite set of the solutions of non-central 2-dimensional $\mathcal{PT}$-symmetric potentials can be handled. It has to be noted that the potential in [28] does not obey (23), because it has even number of imaginary steps on the ring, nevertheless, the results are instructive to the case considered here too.

Further possibility could be applying $\mathcal{PT}$-symmetric arrangements of Dirac delta potentials [23] or attempting to construct the Lamé-type potentials [38] with $\mathcal{PT}$ symmetry.

Journal of Physics: Conference Series **128** (2008) 012045 doi:10.1088/1742-6596/128/1/012045

3.3. Solving the angular equations in $d = 3$

Equation (5) appears in the three-dimensional case as the azimuthal equation, and since the boundary conditions are the same as in the two-dimensional case, the solutions outlined in subsection 3.2 can be used here too. An important difference is, however, that the eigenvalue k of equation (5) appears now in the potential function (15), rather than in the radial equation (16), where its role is taken over by p, the eigenvalue of equation (14).

In what follows, therefore we discuss the exact solutions of the polar equation (14). When discussing the exact solution of this equation, it is worthwhile to modify the formalism in order to handle the θ-dependent terms in (15) more consistently. Let us modify (14) such that we introduce in it a term that depends on the eigenvalue k of equation (5):

$$\omega'' = (P(\theta) - p)\omega \equiv \left(\widetilde{P}(\theta) + \frac{k}{\sin^2\theta} - p\right)\omega \ . \tag{26}$$

Then (15) is formally simplified to a form which does not contain the constant k explicitly:

$$V(r,\theta,\varphi) = V_0(r) + \frac{K(\varphi)}{r^2\sin^2\theta} + \frac{\widetilde{P}(\theta)}{r^2} + \frac{1}{4r^2\sin^2\theta} \ . \tag{27}$$

The state-independence of $V(r,\theta,\varphi)$ is thus achieved at the price of transferring k-dependence into (26). However, when our objective is finding solvable potentials in 3 dimensions, the closed solutions of (26) also have to be determined. Unfortunately, the range of exactly solvable potentials containing a $\sin^{-2}\theta$ term is rather limited. In fact, considering potentials that are non-singular within the $\theta \in (0,\pi)$ domain, it is resticted to the Scarf I and Rosen–Morse I potentials (see *e.g.* [7]).

As a further aspect of this modification we note that prescribing the $\mathcal{PT}$-invariance of $V(r,\theta,\varphi)$ leads to the (21) and (23) conditions and the $\widetilde{P}^*(\pi-\theta) = \widetilde{P}(\theta)$ requirement. This is equivalent with the already established conditions, *i.e.* (22) and (24), of which the latter one can be dropped if $P(\theta)$ is a potential that contains a $\sin^{-2}\theta$ term. In summary, the modification of the formalism by (26) has no effect on the results as long as we are interested in exactly solvable $\mathcal{PT}$-symmetric potentials in 3 dimensions.

Considering the importance of the $\mathcal{PT}$-symmetric Scarf I [20] and Rosen–Morse I [21] potentials from the point of view of constructing exactly solvable $\mathcal{PT}$-symmetric potentials in three dimensions, it seems worthwhile to present the most important results adapted to the present situation.

In order to use the results of [20], first a coordinate shift $\theta = x + \frac{\pi}{2}$ is necessary to define the Scarf I potential in the right domain $\theta \in [0,\pi]$. Obviously, this does not influence the results in any ways. Applying the $\mathcal{PT}$-symmetric Scarf I potential in (14) one then gets

$$P(\theta) = \left(\frac{\alpha^2+\beta^2}{2} - \frac{1}{4}\right)\frac{1}{\sin^2\theta} - \frac{\alpha^2-\beta^2}{2}\frac{\cos\theta}{\sin^2\theta} \tag{28}$$

and

$$p_j = \left(j + \frac{\alpha+\beta+1}{2}\right)^2 \ , \tag{29}$$

while the bound-state solutions are expressed in terms of Jacobi polynomials as

$$\omega_j(\theta) = C_j^{SI}(1+\cos\theta)^{\alpha/2+1/4}(1-\cos\theta)^{\beta/2+1/4}P_j^{(\beta,\alpha)}(\cos\theta) \ . \tag{30}$$

The normalization constants have been determined in [7] such that $\mathcal{P}_\theta\omega_j(\theta) = \omega_j(\theta)$. The corresponding expressions for the $\mathcal{PT}$-symmetric Rosen–Morse I potential [21] are

$$P(\theta) = \left(j + \frac{\alpha+\beta}{2}\right)\left(j + \frac{\alpha+\beta}{2} + 1\right)\frac{1}{\sin^2\theta} + \mathrm{i}\frac{\alpha^2-\beta^2}{2}\cot\theta \ , \tag{31}$$

Table 2. Expressions for the potential (15), the α and β parameters and the relation of the "angular momentum" l of the radial equations with p, when the $\mathcal{PT}$-symmetric Scarf I and Rosen–Morse I potentials are used in the polar equation (14).

	Scarf I	Rosen–Morse I
$V(r,\theta,\varphi)$	$V_0(r)+\frac{K(\varphi)}{r^2\sin^2\theta}+\frac{A}{r^2\sin^2\theta}-\frac{B\cos\theta}{r^2\sin^2\theta}$	$V_0(r)+\frac{K(\varphi)}{r^2\sin^2\theta}+\frac{A}{r^2\sin^2\theta}+\frac{\mathrm{i}B}{r^2}\cot\theta$
A	$\frac{\alpha_m^2+\beta_m^2}{2}-k_m$, real	$\left(j+\frac{\alpha_{jm}+\beta_{jm}+1}{2}\right)^2-k_m$, real
B	$\frac{\alpha_m^2-\beta_m^2}{2}$, imaginary	$\frac{\alpha_{jm}^2-\beta_{jm}^2}{2}$, real
α	$\alpha_m=\pm(A+B+k_m)^{1/2}$	$\alpha_{jm}=\pm\left[(A+k_m)^{1/2}-j-\frac{1}{2}\right]$ $\pm\frac{B}{2}\left[(A+k_m)^{1/2}-j-\frac{1}{2}\right]^{-1}$
β	$\beta_m=\pm(A-B+k_m)^{1/2}$	$\beta_{jm}=\pm\left[(A+k_m)^{1/2}-j-\frac{1}{2}\right]$ $\mp\frac{B}{2}\left[(A+k_m)^{1/2}-j-\frac{1}{2}\right]^{-1}$
$(l+\frac{1}{2})^2$	$p_{jm}=\left(j+\frac{\alpha_m+\beta_m+1}{2}\right)^2$	$p_{jm}=\left(\frac{\alpha_{jm}+\beta_{jm}}{2}\right)^2+\left(\frac{\alpha_{jm}-\beta_{jm}}{2}\right)^2$

$$p_j=\left(\frac{\alpha_j+\beta_j}{2}\right)^2+\left(\frac{\alpha_j-\beta_j}{2}\right)^2 \tag{32}$$

and

$$\omega_j(\theta)=C_j^{RMI}(1+\mathrm{i}\cot\theta)^{\alpha_j/2}(1-\mathrm{i}\cot\theta)^{\beta_j/2+1/4}P_j^{(\alpha_j,\beta_j)}(\mathrm{i}\cot\theta)\ . \tag{33}$$

Note that in (31) we used the original form of the potential, which contains the principal quantum number j. Obviously, the j-dependence of the potential has to be cancelled by the appropriate choice of the parameters (hence the indices in α_j and β_j), however, the above expression is more suited to the application of the Rosen–Morse I potential in the present context.

Finally, we note that in principle the Pöschl–Teller I potential could also be applied in the polar equation as another trigonometric potential containing the $\sin^{-2}\theta$ term [35, 7], however, in this case the other term is $\cos^{-2}\theta$, which would introduce a singularity at $\theta=\pi/2$. Defining it only in one of the domains and then extending the length scale as $\theta\to 2\theta$ would only simply recover the Scarf I potential.

3.4. Illustration

As an example, let us consider the potential

$$V(r,\theta,\varphi)=\frac{\omega^2}{4}r^2+\frac{A}{r^2\sin^2\theta}-\frac{B\cos\theta}{r^2\sin^2\theta} \tag{34}$$

which is nothing but (15) with the harmonic oscillator, the Scarf I potential and the free motion in the r, θ and φ variables, respectively. Since $K(\varphi)=0$, the solution of the azimuthal equation (5) is given by (25) with $k_m=m^2$, so this latter quantity has to be substituted into α_m and β_m in Table 2.

Making use of the results from Table 1 and equations (30) and (25), the solution of the potential (34) are given by

$$\begin{aligned}\psi(r,\theta,\varphi) &= \frac{\mathrm{i}^m}{(2\pi)^{1/2}}C_n^{HO}C_j^{SI}(1+\cos\theta)^{\alpha/2+1/4}(1-\cos\theta)^{\beta/2+1/4}P^{(\beta,\alpha)}(\cos\theta)_j \\ &\quad \times r^{l_m+1}\exp(-\omega r^2/4)L_n^{(l_m+\frac{1}{2})}(\frac{\omega}{2}r^2)\cos(m\varphi)\ ,\end{aligned} \tag{35}$$

where

$$\alpha_m = \pm(A+B+k_m)^{1/2} \tag{36}$$
$$\beta_m = \pm(A-B+k_m)^{1/2} \tag{37}$$
$$l_m = \pm\left(j+\frac{\alpha_m+\beta_m+1}{2}\right)-\frac{1}{2} \tag{38}$$

and B is imaginary. It is seen that there are several possible choices for choosing the signs in the formulae. It can be shown that the solutions of the ordinary harmonic oscillator (or any central potential) can be obtained as a special case if opposite signs are chosen in equations (36) and (37), while in (38) the positive sign has to taken.

The energy eigenvalues of this potential will then be

$$E_{njm} = \left(2n+j+\frac{\alpha_m+\beta_m}{2}+\frac{3}{2}\right)\omega \ . \tag{39}$$

The spectrum will be real or complex depending on whether the $\alpha_m^* = \beta_m$ or the $\alpha_m^* = -\beta_m$ relation holds.

Due to the somewhat complicated dependence of α_m and β_m on m, the spectrum is considerably less simple than in the case of the ordinary isotropic harmonic oscillator. Nevertheless, there are degeneracy patterns in it, as when m is fixed, states with the same value of $2n+j$ have degenerate eigenvalues. This is similar to the degeneracy observed in the energy spectrum of the isotropic harmonic oscillator. Operators increasing/decreasing n with one unit and decreasing/increasing j by two units can be constructed, so in any fixed-m subspace similar operators can act as in the case of the isotropic harmonic oscillator (see *e.g.* [39]).

Similar results could be obtained applying the Coulomb potential as $V_0(r)$, and both radial potentials could be combined with the Rosen–Morse I potential in the polar angle variable. The degeneracy pattern of the energy spectrum would be less rich in that case, because the "angular momentum" depends on j in a more complicated form (see Table 2), so finding combinations of n and j that lead to the same energy are less probable.

Finally, the free motion in the azimuthal variable can also be replaced with more general potentials, as has been discussed in subsection (3.2). The dependence on the results on m becomes more complicated then (including the case when even there is no closed formula for k_m in terms of m), so the complexity of the situation certainly increases.

4. Summary and outlook

We discussed non-central $\mathcal{PT}$-symmetric quantum potentials in 2 and 3 dimensions by separating the variables in radial and angular variables. The original problem can then be reduced to two or three ordinary Schrödinger equation eigenvalue problem with various boundary conditions. The $\mathcal{PT}$ symmetry of the whole potential also introduces conditions for the "potentials" and eigenvalues appearing in the separated eigenvalue equations. The role of the boundary conditions, and in particular, that of the periodic boundary condition characterizing the azimuthal equation has been discussed. It was shown that although usual one-dimensional potentials can be used to formulate the angular equations, some care has to be taken in adapting these systems to the actual situation.

We also analyzed the possible exact solutions of the separate eigenvalue equations. The radial equation turned out to be similar to the ordinary radial Schrödinger equation, with the difference that the "angular momentum" appearing in it can take on complex, as well as real values. In the former case the energy eigenvalues appear in complex conjugate pairs, as is the case also in one-dimensional $\mathcal{PT}$-symmetric potentials. Exact solutions of the radial equation can be obtained using the harmonic oscillator and the Coulomb potentials. It is notable that

Journal of Physics: Conference Series **128** (2008) 012045 doi:10.1088/1742-6596/128/1/012045

in one-dimensional $\mathcal{PT}$-symmetric quantum mechanics the Coulomb potential can be discussed only on trajectories outside the real x axis, because otherwise the boundary conditions cannot be enforced [40, 7].

The analysis of the polar equation showed that its exact solutions are possible with potentials containing the $\sin^{-2}\theta$ term, and this indicated the important role of the Scarf I and Rosen–Morse I potentials.

Finally, we constructed an exactly solvable $\mathcal{PT}$-symmetric potential using the harmonic oscillator, the Scarf I potential and the the free particle in the radial, polar and azimuthal coordinates and determined its bound-state wave functions and energy eigenvalues. The energy spectrum exhibited degeneracies similar to those observed in the case of the (real) isotropic harmonic oscillator, although the spectrum is separated to subspaces defined by the quantum number m.

4.1. Acknowledgments
This work was supported by the OTKA grant No. T49646 (Hungary).

References
[1] Bender C M and Boettcher S 1998 *Phys. Rev. Lett.* **80** 5243
[2] Mostafazadeh A 2002 *J. Math. Phys.* **43** 205
Mostafazadeh A 2002 *J. Math. Phys.* **43** 2814
Mostafazadeh A 2002 *J. Math. Phys.* **43** 3944
[3] Scholtz F G, Geyer H B and Hahne F J W 1992 *Ann. Phys. (N.Y.)* **213** 74
Bender C M, Brody D C and Jones H F 2002 *Phys. Rev. Lett.* **89** 270401
Bender C M 2006 *J. Phys. A: Math. Gen.* **39** 9993
Mostafazadeh A 2006 *J. Phys. A: Math. Gen.* **39** 10171
Scholtz F G and Geyer H B 2006 *J. Phys. A: Math. Gen.* **39** 10189
Figueira de Morrison Faria C and Fring A 2006 *Czech. J. Phys.* **56** 899
Mostafazadeh A 2006 *J. Math. Phys.* **47** 072103
[4] Dirac P A M 1942 *Proc. Roy. Soc. London* **180** 1
Pauli W 1943 *Rev. Mod. Phys.* **15** 175
Gupta S N 1950 *Proc. Roy. Soc. London* **63** 681
Bleuer K 1950 *Helv. Phys. Acta* **23** 567
Sudarshan E C G 1961 *Phys. Rev.* **123** 2183
Lee T D and Wick G C 1969 *Nucl. Phys. B* **9** 209
[5] Scholtz F G, Geyer H B and Hahne F J W 1992 *Ann. Phys. (N.Y.)* **213** 74
[6] Smilga A V 2007 arXiv:0706.4064
[7] Lévai G and Znojil M 2000 *J. Phys. A: Math. Gen.* **33** 7165
[8] Lévai G and Znojil M 2001 *Mod. Phys. Lett. A* **30** 1973
[9] Znojil M 1999 *Phys. Lett. A* **259** 220
[10] B. Bagchi and R. Roychoudhury: *J. Phys. A: Math. Gen.* **33** (2000) L1
[11] Znojil M 2000 *J. Phys. A: Math. Gen.* **33** L61
[12] Lévai G, Sinha A and Roy P 2003 *J. Phys. A: Math. Gen.* **36** 7611
[13] Sinha A, Lévai G and Roy P 2004 *Phys. Lett. A* **322** 78
[14] Ahmed Z 2001 *Phys. Lett. A* **282** 343
[15] Lévai G, Cannata F and Ventura A 2002 *Phys. Lett. A* **300** 271
[16] Lévai G and Znojil M 2002 *J. Phys. A: Math. Gen.* **35** 8793
[17] Bagchi B and Quesne C 2000 *Phys. Lett. A* **273** 285
[18] Lévai G, Cannata F and Ventura A 2001 *J. Phys. A: Math. Gen.* **34** 839
[19] Lévai G, Cannata F and Ventura A 2002 *J. Phys. A: Math. Gen.* **35** 5041
[20] Lévai G 2006 *J. Phys. A: Math. Gen.* **39** 10161
[21] Lévai G 2007 submitted to *Phys. Lett. A*
[22] Cannata F and Ventura A 2006 *Czech. J. Phys.* **56** 943
[23] Ahmed Z 2001 *Phys. Lett. A* **286** 231
Cerveró J M and Rodriguez A 2004 *J. Phys. A: Math. Gen.* **37** 10167
[24] Znojil M 2006 *J. Phys. A: Math. Gen.* **39** 441
[25] Znojil M and Tater M 2001 *J. Phys. A: Math. Gen.* **34** 1793

[26] Şimşek M and Egrifes H 2004 *J. Phys. A: Math. Gen.* **37** 4379
Egrifes H and Sever R 2005 *Phys. Lett. A* **344** 117
[27] Bender C M, Dunne G V, Meisinger P N and Şimşek M 2001 *Phys. Lett A* **281** 311
[28] Znojil M 2003 *J. Phys. A: Math. Gen.* **36** 7825
[29] Jakubský V and Znojil M 2004 *Czech. J. Phys.* **54** 1101
[30] Lévai G 2007 *J. Phys. A: Math. Theor.* **40** F273
[31] Lévai G 2008 *J. Phys. A: Math. Theor.* **41** 4015
[32] Khare A and Bhaduri R K 1994 *Am. J. Phys.* **62** 1008
Bülent G and Zorba I 2000 *Phys. Lett. A* **269** 83
Kerimov G A 2006 *J. Phys. A: Math. Gen.* **39** 1183
[33] Flügge S 1971 *Practical Quantum Mechanics* (Berlin: Springer)
[34] Abramowitz M and Stegun I A 1970 *Handbook of Mathematical Functions* (New York: Dover)
[35] Lévai G 1989 *J. Phys. A: Math. Gen.* **22** 689
[36] Cooper F, Khare A and Sukhatme U 1995 *Phys. Rep.* **251** 267
[37] Ushveridze A G 1994 *Quasi-Exactly Solvable Models in Quantum Mechanics* (Bristol and Philadelphia: IOP Publishing)
[38] Khare A and Sukhatme U 2004 *J. Phys. A: Math. Gen.* **37** 10037
Fernandez D J and Ganguly A 2007 *Ann. Phys.* **322** 1143
[39] Wybourne B G 1974 *Classical Groups for Physicists* (New York: Wiley)
[40] Znojil M and Lévai G 2000 *Phys. Lett. A* **271** 327

Journal of Physics: Conference Series **128** (2008) 012046 doi:10.1088/1742-6596/128/1/012046

Quantum toboggans: models exhibiting a multisheeted $\mathcal{PT}-$symmetry

Miloslav Znojil

Nuclear Physics Institute AS CR, 250 68 Řež, Czech Republic

E-mail: znojil@ujf.cas.cz

Abstract. A generalization of the concept of $\mathcal{PT}-$symmetric Hamiltonians $H = p^2 + V(x)$ is presented. For the usual analytic potentials $V(x)$ (with singularities) and for the recently widely accepted "$\mathcal{PT}-$symmetric" asymptotic boundary conditions for wave functions $\psi(x)$ (selected inside a pair of complex wedges generalizing the usual $x \to \pm\infty$ asymptotics), non-equivalent quantum toboggans are defined as integrated along topologically different paths $\mathcal{C}$ of coordinates $x \in \mathbb{C}$.

1. Introduction

Among non-Hermitian Hamiltonians $H \neq H^\dagger$ defined, say, in a suitable "auxiliary" Hilbert space $\mathcal{H}^{(aux)}$, a privileged subfamily is formed by their $\eta-$pseudohermitian special cases which are such that $H^\dagger = \eta\, H\, \eta^{-1}$ in terms of a suitable operator $\eta \neq I$. Whenever $\eta = \mathcal{P}$ happens to coincide with the operator of parity, we arrive at the increasingly popular [1] $\mathcal{PT}-$symmetric quantum models H.

In general, *all* of the above-mentioned Hamiltonians H can happen to possess a real and discrete spectrum of energies $E_0 \leq E_1 \leq \ldots$. In such a case we may always imagine that while our $H \neq H^\dagger$ (acting in $\mathcal{H}^{(aux)}$) is just one of many possible operator representations of such a spectrum of bound-state energies, there may exist a "true" physical Hilbert space $\mathcal{H}^{(phys)}$ and, in it, an orthonormalized basis of kets $\{\, |n\rangle \,\}_{n=0,1,\ldots}$ such that a certain operator $H^{(phys)}$ defined by its spectral representation,

$$H^{(phys)} = \sum_{n=0}^{\infty} |\, n\rangle\, E_n\, \langle n| \qquad (1)$$

can be declared the "true", mathematically correct representant of the Hamiltonian of our original quantum system. Indeed, the latter operator is manifestly *self-adjoint* in $\mathcal{H}^{(phys)}$ and all the postulates of Quantum Mechanics are satisfied.

There exist multiple practical applications of such an idea of the use of two different Hilbert spaces in parallel (cf., e.g., the most recent and up-to-date review [1]). One of their most characteristic shared features is that while $H^{(phys)}$ is a technically very complicated operator in $\mathcal{H}^{(phys)}$, its "equivalent" representation H in $\mathcal{H}^{(aux)}$ is, by assumption, "very simple", in comparison at least [2].

In the latter sentence, we used the quotation marks on purpose: "very simple" need not mean trivial. Even worse, the "equivalence" of the mutual mapping between H and $H^{(phys)}$ is particularly conventional a concept. Indeed, it is obvious that in the very second paragraph of our present text the readers may have noticed that our choice of the basis $\{\,|n\rangle\,\}_{n=0,1,\ldots}$ was *entirely* arbitrary.

Of course, a responsible attitude towards the latter ambiguity problem unifies various applications of the formalism which may be sampled as ranging from nuclear physics [2] and field theory [1] far beyond the territory of quantum theories, involving even the terrains as distant as random matrices [3], cosmology [4] or classical optics [5], electrodynamics [6] and magnetohydrodynamics [7]. In each of these contexts, the ambiguity problem finds its specific resolution. In particular, in the narrower domain of quantum theories themselves, various additional conditions are usually imposed, relying on various persuasive phenomenological arguments as summarized, in compact form, in refs. [2, 8] (for Quantum Mechanics) and [1] (mainly in the context of field theories).

Once a suitable map between $\mathcal{H}^{(aux)}$ and $\mathcal{H}^{(phys)}$ has been established (this is not to be discussed here), the most persuasive distinction between these spaces can be seen in the prohibitively complicated form of the operator $H^{(phys)}$ when compared with H. Typically, in the specific, $\mathcal{PT}-$symmetric quantum models $H = \mathcal{P}\,H^\dagger\,\mathcal{P}$, the manifestly self-adjoint $H^{(phys)}$ is even fairly difficult to define. In contrast, H is often selected as an ordinary differential operator, easily and efficiently tractable by many standard mathematical techniques. *Pars pro toto*, we recommend the readers to have a look at the "first nontrivial", exactly solvable $\mathcal{PT}-$symmetric harmonic oscillator [9] for illustration.

On the latter, $\mathcal{PT}-$symmetric background there is still a lot of space for the study of some "slightly" more complicated though still mathematically tractable Hamiltonians H. In this spirit we intend to pay attention here to the newly introduced [10] and developed [11, 12] family of the models called "quantum toboggans" (QT). Our brief review and introduction to this subject will be divided in section 2 (on the concept of quantum toboggans), section 3 (on QT constructions), section 4 (on the QT models with more branch points) and summary.

2. The concept of quantum toboggans

During a "prehistory" of our theme, several complex potentials $V(x)$ have been shown to generate real bound-state spectra, be it an imaginary cubic $V(x) \sim ix^3$ at $|x| \gg 1$ [13] or a negative quartic $V(x) \sim -x^4$ at $|x| \gg 1$ [14], both exhibiting $\mathcal{PT}-$symmetry (note that $\mathcal{P}$ changes parity in this context, $x \to -x$, while the complex conjugation $\mathcal{T}$ mimics time reversal).

In some sense, the "history" commenced in 1993 when Bender and Turbiner published a letter where certain standard ordinary differential Schrödinger equations

$$\left(-\frac{d^2}{dx^2} + V(x)\right)\,\psi(x) = E\,\psi(x) \tag{2}$$

were declared physical *even when* complemented by certain anomalous, complexified Dirichlet asymptotic boundary conditions [15]. Indeed, once you assume a suitable form of analyticity of $V(x)$ you may require, in a mathematically consistent manner, that

$$\psi\left(-\varrho\cdot e^{i\,\theta_{(left)}}\right) = \psi\left(+\varrho\cdot e^{i\,\theta_{(right)}}\right) = 0 \qquad \varrho \to +\infty \tag{3}$$

not only for the usual $\theta_{(left)} = \theta_{(left)} = 0$ but also at *any* non-vanishing angles $\theta_{(left)} \neq 0 \neq \theta_{(left)}$.

The next and decisive step towards possible explicit and nontrivial applications of such a more or less trivial mathematics in physics has been made by Bender and Boettcher [16] who presented a persuasive numerical and semiclassical support for their conjecture that in many similar cases the spectrum *can* remain real and, hence, observable. More specifically, they employed and recommended a special choice of the conditions (3) setting, in our present notation,

$$\theta_{(left)} = -\theta_{(right)}\,. \tag{4}$$

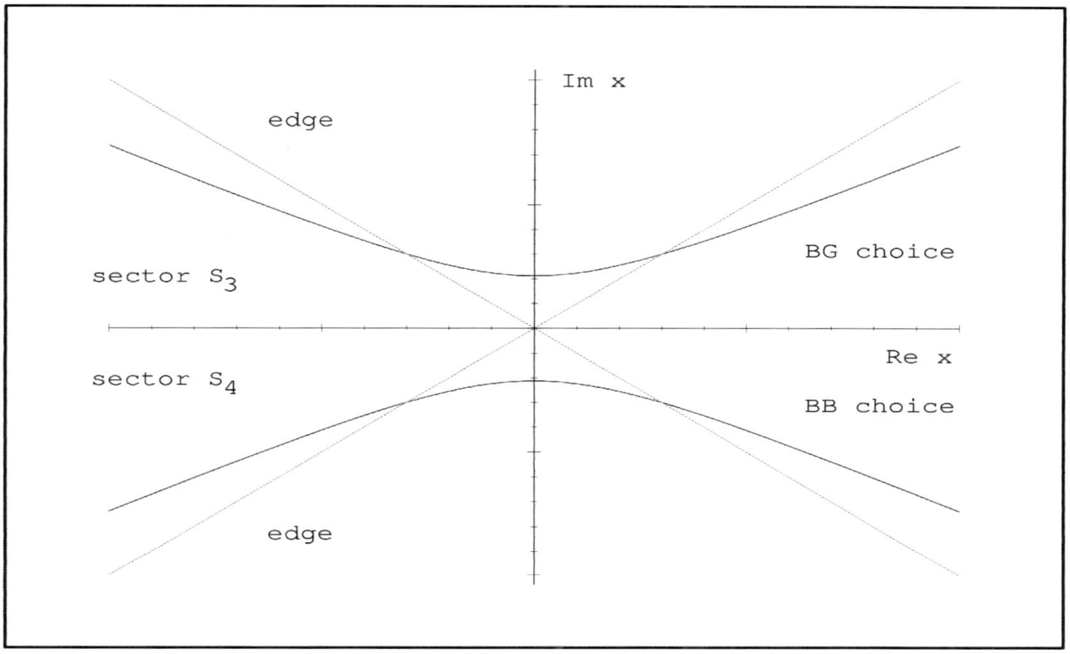

Figure 1. Complex contours of coordinates (BG = choice made in ref. [14], BB = choice made in ref. [19]).

They found out, purely empirically, that such a postulate seems to offer very good chances that the spectrum remains real. In the other words, in a way transferring certain numerical experience [17] and known mathematical tricks (cf., e.g., the textbooks [18] or the Buslaev's and Grecchi's paper [14]) into a much more ambitious physical project. They recommended to work with the "curved", complex paths $\mathcal{C}$ of coordinates x which remain left-right symmetric in the complex plane of x, i.e., which are, in an obvious sense, "$\mathcal{PT}-$symmetric" (cf. their sample given here in Figure 1).

In a way fortunate for the subsequent quick development of the subject [20], the reality of spectra in many $\mathcal{PT}-$symmetric models has been fairly soon proved in an entirely rigorous manner [21]. The way has been opened for the birth of quantum toboggans [10].

The mathematical essence of quantum toboggans (QT, [11]) is easy to explain briefly. On an intuitive level one immediately sees that in the case of the presence of a branch point in $\psi(x)$ (say, at $x^{(BP)} = 0$), the "standard" integration paths as sampled in Figure 1 can be replaced, e.g., by the "tobogganic spiral" of Figure 2. On a more abstract level one simply has to recognize that the current (and mathematically very strong) assumptions concerning the analytic behaviour of $V(x)$ may be perceivably weakened without any real changes in our understanding of the $\mathcal{PT}-$symmetry-related problems, conjectures and constructions. *In nuce*, we may admit that in the complex plane of x our potentials may be allowed to have singularities which are distributed in a left-right symmetric manner in $\mathbb{C}$.

In all the cases with *more than one* branch point (cf. [11, 12] or section 4 below), the discussion of the non-equivalent QT paths would be technically complicated though still very similar to

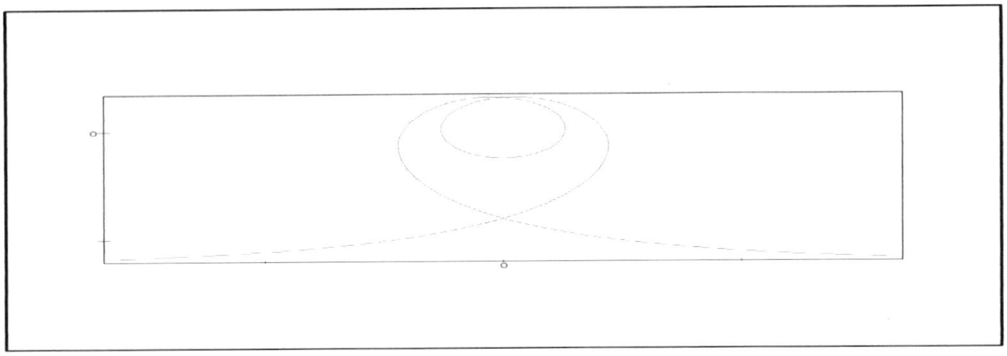

Figure 2. Tobogganic contour $\mathcal{C}^{(N)}$ with $N = 2$.

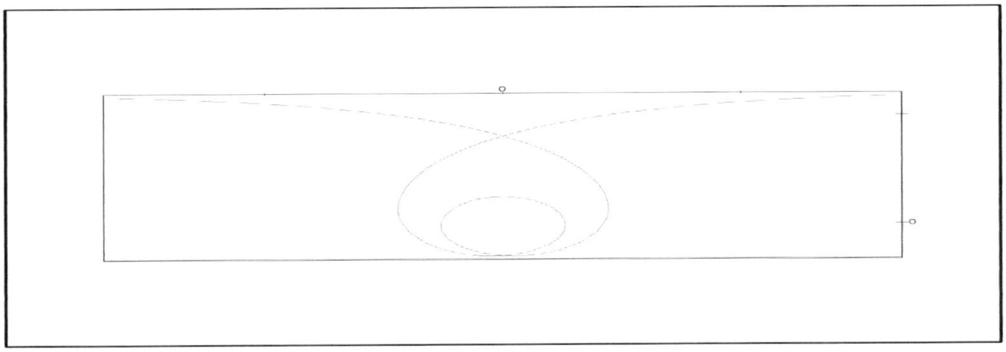

Figure 3. The complex conjugate version of the contour of Figure 2.

the single-branch-point simplest case. For this reason, let's now stay just in the simplest QT = QT1 case with holomorphic $V(x)$ possessing the centrifugal-like pole $\sim \ell(\ell+1)/x^2$ in the origin. Obviously, this model generates also a branch point $x^{(BP)} = 0$ in $\psi(x)$ generated by this centrifugal-type term in the potential. In order to classify the related non-equivalent integration paths $\mathcal{C}$ in eq. (2), one must check how many times they turn around the centre before they start approaching their $\varrho \gg 1$ asymptotics.

This check is also sufficient. Indeed, as long as the coordinates $x \in \mathcal{C}$ are complex and, by assumption, $0 = x^{(BP)} \notin \mathcal{C}$, we may just let the angles $\theta_{(left)}$ and $\theta_{(right)}$ vary *beyond* the interval $(-\pi/2, \pi/2)$. In particular, in the specific $\mathcal{PT}-$symmetric cases as defined by eq. (4), we shall accept a convenient convention concerning the (say, counterclockwise) orientation of the winding of the curves $\mathcal{C}$. Moreover, on the zeroth Riemann sheet of the total Riemann surface pertaining to our QT bound states $\psi(x)$ we shall fix our QT curves $\mathcal{C}$ as crossing the imaginary axis below zero and in the left-right direction so that, for example, the curves running in the opposite direction (like the one displayed in Figure 3) will not be considered here as independent new models.

3. The simplest quantum-toboggan models with the single branch point

The overall winding number N of $\mathcal{C}$ is highly relevant for the specification of the boundary conditions (3). In the presence of just single branch point $x^{(BP)}$, the knowledge of the winding number N in $\mathcal{C} = \mathcal{C}^{(N)}$ specifies the QT1 bound-state problem completely. We may visualize our QT1 spirals $\mathcal{C}^{(N)}$ as curves which are parametrized by an angle $\phi \in (-\Phi, \Phi)$ with a positive radius $\varrho = \varrho(\phi)$. In such a setting, the simplest definition of the $\mathcal{PT}-$symmetry of the curve $\mathcal{C}^{(N)}$ may be based on the symmetry requirement $\varrho(\phi) = \varrho(-\phi)$ [10].

For the most elementary $\mathcal{PT}-$symmetric harmonic-oscillator Schrödinger QT1 equation

$$\left(-\frac{d^2}{dx^2} + \frac{\ell(\ell+1)}{x^2} + x^2\right)\psi(x) = E\,\psi(x)$$

$\mathcal{PT}-$symmetric path $\mathcal{C}^{(N)}$ N-times encircles $x = 0$. At the trivial $N = 0$ this model may be defined along the straight contour

$$\mathcal{C}^{(0)} = \{x \mid x = t - i\,\varepsilon,\, t \in I\!\!R\}$$

and one finds that at any $\alpha(\ell) = \ell + 1/2$ it generates "twice as many" bound-state levels than its half-line predecessor (cf. [9]),

$$E = E_{n,\ell,\pm} = 4n + 2 \pm 2\alpha(\ell)\,, \qquad n = 0, 1, \ldots\,.$$

Various anharmonic non-tobogganic generalizations of this solvable model with various potentials such that $\mathrm{Re}\,V(x) = +\mathrm{Re}\,V(-x)$ and $\mathrm{Im}\,V(x) = -\mathrm{Im}\,V(-x)$ cease to be unique since even several non-tobogganic asymptotic boundary conditions

$$\psi(\pm\mathrm{Re}\ L + i\,\mathrm{Im}\ L) = 0\,,$$

$$|L| \gg 1 \quad \text{or} \quad |L| \to \infty\,.$$

can prove non-equivalent.

In the tobogganic cases characterized by a nonvanishing winding-number integer $N \neq 0$, the curves $\mathcal{C} = \mathcal{C}^{(N)}$ lie on a multisheeted Riemann surface. They can be parametrized there by an angle $\varphi \in (-(N+1)\pi,\, N\pi)$ as, e.g.,

$$\mathcal{C}^{(N)} = \left\{x = \varepsilon\,\varrho(\varphi, N)\,e^{i\,\varphi}\,,\, \varepsilon > 0\right\}\,,$$

$$\varrho(\varphi, N) = \sqrt{1 + \tan^2\frac{\varphi + \pi/2}{2N+1}}\,.$$

A deeper discussion of some features of quantum toboggans can be facilitated by the quasi-exact solvability of the underlying potential chosen, say, in the asymptotically decadic form

$$V(x) = x^{10} + \text{asymptotically smaller terms}$$

where [22]

$$\psi(x) = e^{-x^6/6 + \text{asymptotically smaller terms}}\,.$$

We may reparametrize

$$\psi(x) = \exp\left[-\frac{1}{6}\varrho^6\cos 6\varphi + \ldots\right]$$

and see that there exist as many as five non-tobogganic versions of this model, with angles in the eligible complex wedges

$$\Omega_{(first\ right)} = \left(-\frac{\pi}{2}+\frac{\pi}{12}, -\frac{\pi}{2}+\frac{3\pi}{12}\right),$$

$$\Omega_{(first\ left)} = \left(-\frac{\pi}{2}-\frac{\pi}{12}, -\frac{\pi}{2}-\frac{3\pi}{12}\right),$$

$$\Omega_{(third\ right)} = \left(-\frac{\pi}{2}+\frac{5\pi}{12}, -\frac{\pi}{2}+\frac{7\pi}{12}\right), \quad \ldots$$

$$\ldots \quad \Omega_{(fifth\ left)} = \left(-\frac{\pi}{2}-\frac{9\pi}{12}, -\frac{\pi}{2}-\frac{11\pi}{12}\right).$$

It is worth noting that *all* of the models of this type can be interpreted as equivalent to their non-tobogganic partners confined by a *different* "effective" potential. For this purpose one can simply perform a $\mathcal{PT}$−symmetric change of variables in the "initial" $\mathcal{PT}$−symmetric model

$$\left[-\frac{d^2}{dx^2} - (ix)^2 + \lambda\, W(ix)\right] \psi(x) = E(\lambda)\, \psi(x)$$

where

$$W(ix) = \mathbf{\Sigma}\, g_\beta (ix)^\beta$$

and where one sets

$$ix = (iy)^\alpha\,, \qquad \psi(x) = y^\varrho\, \varphi(y).$$

Once we use the freedom in the choice of $\alpha > 0$ we have

$$i\, dx = i^\alpha \alpha y^{\alpha-1}\, dy, \qquad \frac{(iy)^{1-\alpha}}{\alpha}\frac{d}{dy} = \frac{d}{dx}.$$

This gives an equivalent, "Sturmian" problem in an intermediate differential equation form

$$y^{1-\alpha}\frac{d}{dy} y^{1-\alpha}\frac{d}{dy}\, y^\varrho\, \varphi(y) +$$

$$+ i^{2\alpha}\alpha^2 \left[-(iy)^{2\alpha} + \lambda\, W[(iy)^\alpha] -\right.$$

$$\left. -E(\lambda)\right]\, y^\varrho\, \varphi(y) = 0\,.$$

Here, the first term

$$y^{1-\alpha}\frac{d}{dy} y^{1-\alpha}\frac{d}{dy}\, y^{[(\alpha-1)/2]}\, \varphi(y) =$$

$$= y^{2+\varrho-2\alpha}\frac{d^2}{dy^2}\, \varphi(y) + \varrho(\varrho-\alpha) y^{\varrho-2\alpha}\, \varphi(y)\,,$$

"behaves" at the specific

$$\varrho = \frac{\alpha-1}{2}$$

so that the new equation preserves the same Schrödinger form:

$$-\frac{d^2}{dy^2}\, \varphi(y) + \frac{\alpha^2-1}{4y^2}\, \varphi(y) +$$

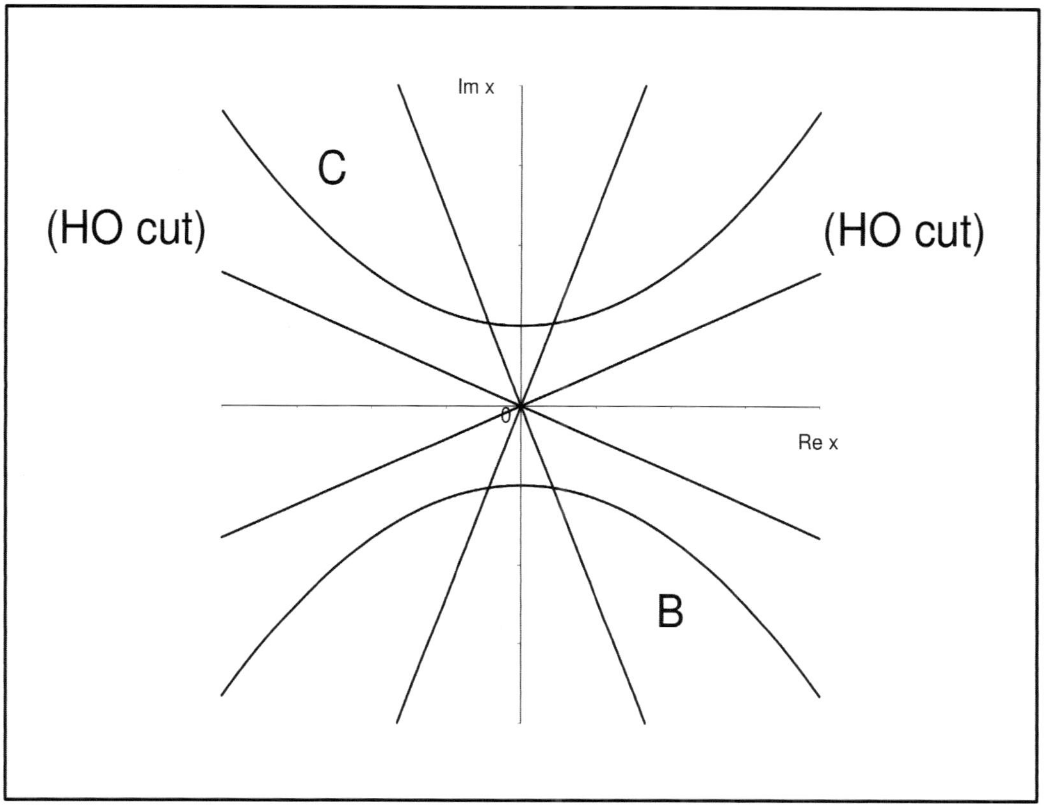

Figure 4. Sextic oscillator as a map of a $\mathcal{PT}$−symmetric harmonic-oscillator contour $\mathcal{C}^{(0)}$ (curve B) and of a tobogganic harmonic oscillator contour $\mathcal{C}^{(1)}$ (curve C).

$$+(iy)^{2\alpha-2}\alpha^2\left[-(iy)^{2\alpha}+\lambda\,W[(iy)^{\alpha}]\,\varphi(y)=\right.$$
$$=(iy)^{2\alpha-2}\alpha^2\,E(\lambda)\,\varphi(y)\,.$$

What was important is that the change of variables changed also the range of the angle in $\mathcal{C}^{(N)}$ so that by the choice of α one can diminish the winding number N. In this manner, many polynomial potentials prove interrelated. E.g., with $\alpha = 1/2$ we get the quadratic oscillator $V_g(y) = -(iy)^2 + i\,g_1\,y + g_{-1}\,(iy)^{-1} + g_{-2}\,(iy)^{-2}$ from the sextic oscillator

$$-\frac{d^2}{dx^2}\,\varphi(x)+\frac{\ell(\ell+1)}{x^2}\,\varphi(x)+V_f(x)\,\varphi(x)=E\,\varphi(x)\,,$$
$$V_f(x)=x^6+f_4\,x^4+f_2\,x^2+f_{-2}\,x^{-2},$$

etc (cf. Figure 4).

In the conclusion of this section let us emphasize that the $\mathcal{PT}$−symmetry in the presence of the single branch point can be based on the introduction of the two different parity-like operators $\mathcal{P}^{(\pm)} : x \to x \cdot \exp(\pm i\pi)$ as well as of the two eligible rotation-type innovations $\mathcal{T}^{(\pm)}$ of the time reversal.

4. General quantum toboggans with more branch points

Once you study the bound-state problem

$$-\frac{\hbar^2}{2m}\frac{d^2}{dx^2}\,\psi_n(x) + V(x)\,\psi_n(x) = E_n\,\psi_n(x)\,, \tag{5}$$

you may set $\hbar = 2m = 1$ and choose, say, the potential with two second-order poles,

$$V(x) = V_{regular}(x) + \frac{G}{(x-1)^2} + \frac{G^*}{(x+1)^2}\,.$$

In this way you arrive at the wave functions

$$\psi^{(general)}(x) = c_+\psi^{(special)}_{(+)}(x) + c_-\psi^{(special)}_{(-)}(x) \tag{6}$$

with the two (in general, complex) branch points at, say, $x = x^{(BP)} = \pm 1$ [23]. Thus, the Riemann surface $\mathcal{R}$ of $\psi(x)$ becomes composed of *many* sheets $\mathcal{R}_k$.

4.1. The harmonic-oscillator example of a bitoboggan

Without boundary conditions our differential Schrödinger equation comprises many eigenvalue problems at once [15]. In the QT = QT2 models with two branch points they are all in a one-to-one correspondence with our selection of the QT paths $x^{(\varrho)}(s)$. In a way introduced in ref. [12], they may be classified by certain "winding" or "knotting" descriptors ϱ. These $x^{(\varrho)}(s)$ connect their asymptotes while passing through a compact domain of x which contains all the singularities of $V(x)$.

In the most transparent non-tobogganic case one can stay on the single Riemann sheet and, for illustrative purposes, pick up the harmonic oscillator with $\psi_n^{(\pm|\alpha|)}(x) = x^{1/2\pm|\alpha|}\,\exp(+x^2/2)\times$ a polynomial, considered as integrated along U-shaped paths

$$y^{(U)}(s) = x^{(U)}(s) + \mathrm{i}\,\varepsilon = \left\{ \begin{array}{ll} |\,s|\,e^{-11\,\mathrm{i}\,\pi/8}\exp[\mathrm{i}\,\xi(s)], & s \ll -1\,, \\ |\,s|\,e^{3\mathrm{i}\,\pi/8}\exp[\mathrm{i}\,\xi(s)], & s \gg 1 \end{array} \right. \tag{7}$$

on which $x(s) = y(s) - \mathrm{i}\,\varepsilon$, $\xi(s) \in (-\pi/8, \pi/8)$ and

$$\lim_{s\to\pm,\infty} \psi\left[x^{(U)}(s)\right] = 0\,. \tag{8}$$

In the alternative, tobogganic cases with $N = 1$, $\xi \in (-\pi/8, \pi/8)$ and $\delta > 0$ in

$$x^{(N=1)}(s) = y^{(N=1)}(s) - \mathrm{i}\,\varepsilon = \left\{ \begin{array}{ll} |\,s-\eta|\,e^{-13\,\mathrm{i}\,\pi/8}\exp[\mathrm{i}\,\xi(s)] + \mathrm{i}\,\delta, & s \ll -1\,, \\ |\,s-\eta|\,e^{5\mathrm{i}\,\pi/8}\exp[\mathrm{i}\,\xi(s)] + \mathrm{i}\,\delta, & s \gg 1 \end{array} \right. \tag{9}$$

we may consider the paths encircling two branch points by winding

- counterclockwise around $x^{(BP)}_{(-)}$ (to be marked by a letter L),
- counterclockwise around $x^{(BP)}_{(+)}$ (letter R),
- clockwise around $x^{(BP)}_{(-)}$ ($Q = L^{-1}$),
- clockwise around $x^{(BP)}_{(+)}$ ($P = R^{-1}$).

In this way, a four-letter alphabet can be used to label all paths $x = x^{(\varrho)}(s)$ by words ϱ of length $2N$ and of a concatenated form $\varrho = \Omega \bigcup \Omega^T$ which is due to the underlying $\mathcal{PT}$−symmetry $L \leftrightarrow R$.

Thus, besides the symbol $\varrho = \emptyset$ for the non-tobogganic case one has four possibilities at $N = 1$, viz.,

$$\Omega \in \left\{L\,,L^{-1}\,,R\,,R^{-1}\right\}\,, \qquad N=1\,,$$

$$\varrho \in \left\{LR\,,L^{-1}R^{-1}\,,RL\,,R^{-1}L^{-1}\right\}\,, \qquad N=1\,,$$

or the following dozen cases at $N = 2$,

$$\Omega \in \left\{LL, LR, RL, RR, L^{-1}R, R^{-1}L, LR^{-1}\,,\right.$$

$$\left. RL^{-1}\,, L^{-1}L^{-1}, L^{-1}R^{-1}, R^{-1}L^{-1}, R^{-1}R^{-1}\right\}$$

(with four items $LL^{-1}\,,L^{-1}L\,,RR^{-1}$, $R^{-1}R$ not allowed among the $4^2 = 16$ eligible ones), etc [12].

4.2. Rectifiable bitobogganic contours with subscripted descriptors $\varrho = \varrho_0$

Let us remind the readers that in the presence of a mere single branch point we did set

$$\mathrm{i}\,x = (\mathrm{i}\,z)^2\,,\ \psi_n(x) = \sqrt{z}\,\varphi_n(z)\,.$$

implying the *strict equivalence* of the *harmonic* QT1 oscillator to its manifestly $\mathcal{PT}$−symmetric *sextic-oscillator* partner

$$\left(-\frac{d^2}{dz^2} + 4z^6 + 4E_n z^2 + \frac{4\alpha^2 - 1/4}{z^2}\right)\varphi(z) = 0$$

defined along the following, *manifestly non-tobogganic* path

$$\mathcal{C} = \left\{ \begin{array}{ll} \sqrt{|\,s-\eta|}\,e^{-9\,\mathrm{i}\,\pi/16}\exp[\mathrm{i}\,\xi(s)/2] + \mathcal{O}(\delta/\eta), & s \ll -1\,, \\ \sqrt{|\,s-\eta|}\,e^{\mathrm{i}\,\pi/16}\exp[\mathrm{i}\,\xi(s)/2] + \mathcal{O}(\delta/\eta), & s \gg 1 \end{array} \right. . \qquad (10)$$

In the presence of the *pair* of the branch points, say, $x^{(BP)} = \pm 1$, the simplest changes of variables can be employed again. To the Schrödinger equation

$$\left[-\frac{d^2}{dx^2} + \frac{\ell(\ell+1)}{(x-1)^2} + \frac{\ell(\ell+1)}{(x+1)^2} + V(ix)\right]\psi(x) =$$

$$= E\,\psi(x)\,.$$

this often enables us to assign the rectified partner

$$\left[-\frac{d^2}{dz^2} + U_{eff}(\mathrm{i}\,z)\right]\varphi(z) = 0$$

where

$$U_{eff}(\mathrm{i}\,z) = U(\mathrm{i}\,z) + \frac{\mu(\mu+1)}{(z-1)^2} + \frac{\mu(\mu+1)}{(z+1)^2} \equiv$$

$$\equiv U(\mathrm{i}\,z) + 2\,\frac{\mu(\mu+1)[1-(\mathrm{i}\,z)^2]}{[1+(\mathrm{i}\,z)^2]^2}\,.$$

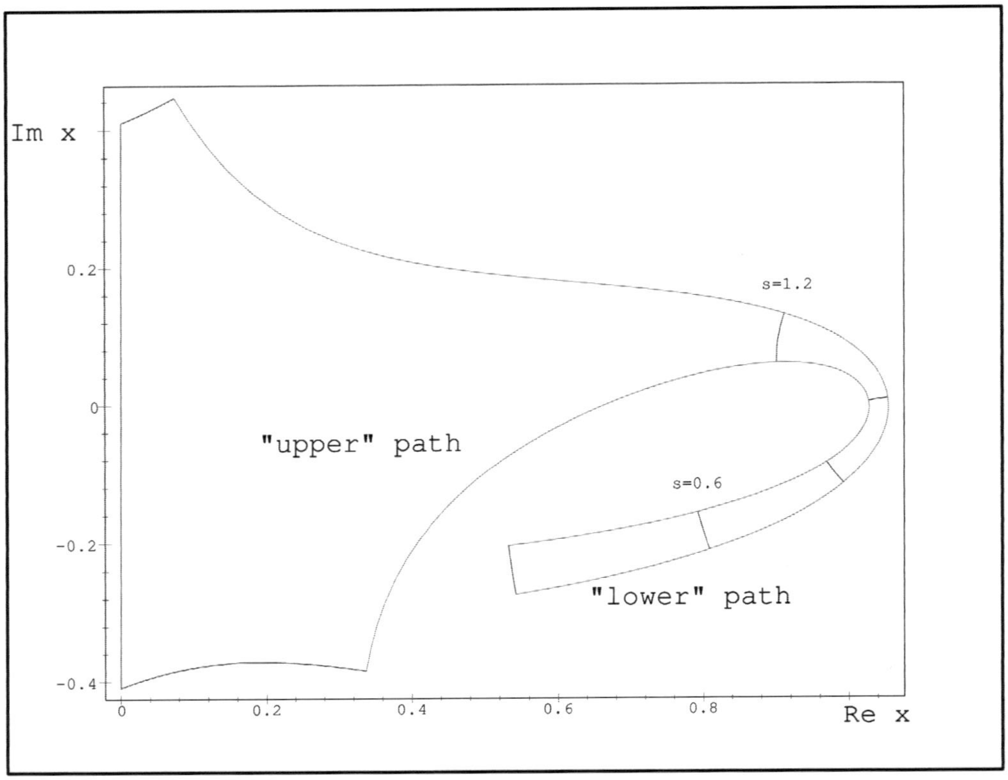

Figure 5. Two bitoboggans ($\kappa = 2.4$, $s \in (0.4, 1.4)$).

Proof
Using an implicit rectification formula

$$1 + (ix)^2 = \left[1 + (iz)^2\right]^\kappa , \qquad \kappa > 1$$

we reveal that $z = -\mathrm{i}\,\varrho$ gets mapped upon itself. Hence, one can recommend the use of the explicit rectification formula $x = -\mathrm{i}\sqrt{(1-z^2)^\kappa - 1}$. As a result, certain effective non-tobogganic potentials are obtained. Their construction is routine since

$$\frac{d}{dx} = \beta(z)\frac{d}{dz}, \qquad \beta(z) = -\mathrm{i}\,\frac{\sqrt{(1-z^2)^\kappa - 1}}{\kappa\, z\,(1-z^2)^{\kappa-1}}\,.$$

Thus, we may set $\psi(x) = \chi(z)\,\varphi(z)$ with $\chi(z) = const/\sqrt{\beta(z)}$ [24] and get $V_{eff}(\mathrm{i}x) = V(\mathrm{i}x) + 2\,\ell(\ell+1)[1-(\mathrm{i}x)^2]/[1+(\mathrm{i}x)^2]^2$ in

$$\left(-\beta(z)\frac{d}{dz}\,\beta(z)\frac{d}{dz} + V_{eff}[ix(z)] - E\right)\chi(z)\,\varphi(z) = 0\,.$$

or

$$U_{eff}(\mathrm{i}\,z) = \frac{V_{eff}[\mathrm{i}\,x(z)] - E_n}{\beta^2(z)} + \frac{\beta''(z)}{2\,\beta(z)} - \frac{[\beta'(z)]^2}{4\,\beta^2(z)}$$

Journal of Physics: Conference Series **128** (2008) 012046 doi:10.1088/1742-6596/128/1/012046

in the standard Schrödinger equation. QED.

Graphically, it is interesting to reconstruct the shapes of the tobogganic pull-backs (cf. Figure 5 for ilustration). For this purpose, in the vicinity of the negative imaginary axis we may consider the mapping

$$z = -\mathrm{i}\, r\, e^{\mathrm{i}\,\theta} \quad \longrightarrow \quad x = -\mathrm{i}\, \left[\left(1 + r^2\, e^{2\,\mathrm{i}\,\theta}\right)^{\kappa} - 1\right]^{1/2} .$$

At the small radii r it degenerates to the mere multiplication by a constant $\sqrt{\kappa}$. At the larger radii we arrive at a more complicated knot-like shapes of $x^{\varrho_0}(s)$, tractable easily by computer graphics, via their definition as a pullback of the straight-line $z(s) = s - \mathrm{i}\,\varepsilon$. In this approach the winding number N proves fairly sensitive to the value of shift ε. In Figure 5 we see the clear acceleration of the winding after transition from $\varepsilon = \varepsilon_{upper} = 0.15$ to $\varepsilon = \varepsilon_{lower} = 0.20$.

5. Summary

In the context of mathematics and in a way paralleling the birth of interest in $\mathcal{PT}$ symmetry, the concept of quantum toboggans could in fact be also understood as almost trivial in mathematics. In physics, the first hints for its introduction resulted from several independent mathematical sources. The first one has been the above-cited Buslaev's and Grecchi's paper [14] where a *formal necessity* of a branch point involved *all* their analytic bound-state wave functions $\psi_n(x)$. A clearer understanding of this hint (reflecting the necessity of an effective kinematical centrifugal force in more dimensions) came a few years later when attention has been turned to an "unperturbed" harmonic-oscillator special case of their model [9]. In parallel, virtually the same branch points in wave functions resurfaced during the proofs of the reality of the energies [21], during the studies of the quasi-exact solvability of certain $\mathcal{PT}-$symmetric models [22] and after a supersymmetrization of certain $\mathcal{PT}-$symmetric Hamiltonians [23].

Let us summarize: what should be remembered in the context of mathematics is the new use of the changes of variables in Schrödinger equations. This can rectify the QT paths of coordinates and may also lead to some new and nonstandard *feasible* calculations.

In the parallel phenomenological model-building context, the perspective of new physics may be expected to be derived from the prospective use of the tobogganic paths $\mathcal{C}^{(N)}$. This could throw new light not only on "innovated" bound states (of a "topological" origin) but also on the very unusual scattering-type states [11].

Acknowledgments

Work supported by the GAČR grant Nr. 202/07/1307, by the MŠMT "Doppler Institute" project Nr. LC06002 and by the Institutional Research Plan AV0Z10480505.

References

[1] Bender C M 2007 *Reports on Progress in Physics* **70** 947
[2] Scholtz F G, Geyer H B and Hahne F J W 1992 *Ann. Phys. (NY)* **213** 74
[3] Jarosz A and Nowak M A 2006 *J. Phys. A: Math. Gen.* **39** 10087
[4] Andrianov A A, Cannata F and Kamenshchik A Y 2006 *J. Phys. A: Math. Gen.* **39** 9975
[5] Berry M V 2004 *Czech. J. Phys.* **54** 1039
[6] Mostafazadeh A and Loran F 2008 *Europhys. Lett.* **81** 10007
[7] Kirillov O and Günther U 2006 *J. Phys. A: Math. Gen.* **39** 10057
[8] Mostafazadeh A 2002 *J. Math. Phys.* **43** 205 and 2814 and 3944
[9] Znojil M 1999 *Phys. Lett.* **A 259** 220
[10] Znojil M 2005 *Phys. Lett.* **A 342** 36
[11] Znojil M2006 *J. Phys. A: Math. Gen.* **39** 13325
[12] Znojil M 2008 *Phys. Lett.* **A 372** 584
[13] Caliceti E, Graffi S and Maioli M 1980 *Commun. Math. Phys.* **75** 51
[14] Buslaev V and Grecchi V 1993 *J. Phys. A: Math. Gen.* **26** 5541
[15] Bender C M and Turbiner A 1993 *Phys. Lett.* **A 173** 442

[16] Bender C M and Boettcher S 1998 *Phys. Rev. Lett.* **80** 5243
[17] Bessis D 1992 private communication
[18] Sibuya Y 1975 *Global Theory of Second Order Linear Differential Equation with Polynomial Coefficient* (Amsterdam: North Holland)
[19] Bender C M and Boettcher S 1998 *J. Phys. A: Math. Gen.* **31** L273
[20] Znojil M 2006 *Czech. J. Phys.* **56** 977
[21] Dorey P, Dunning C and Tateo R 2001 *J. Phys. A: Math. Gen.* **34** 5679
Shin K C 2002 *Commun. Math. Phys.* **229** 543
[22] Znojil M 2000 *J. Phys. A: Math. Gen.* **33** 6825
[23] Sinha A and Roy P 2004 *Czech. J. Phys.* **54** 129
[24] Liouville L 1837 *J. Math. Pures Appl.* **1** 16

Journal of Physics: Conference Series **128** (2008) 012047 doi:10.1088/1742-6596/128/1/012047

Representations of fuzzy torus

N Aizawa[1] **and R Chakrabarti**[2]

[1] Department of Mathematics and Information Sciences, Graduate School of Science, Osaka Prefecture University, Nakamozu Campus, Sakai, Osaka 599-8531, Japan

[2]Department of Theoretical Physics, University of Madras, Guindy Campus, Chennai 600 025, India

E-mail: aizawa@mi.s.osakafu-u.ac.jp

Abstract. A classification of Hermitian representations for the recently introduced fuzzy torus algebra is presented. This is carried out by regarding the fuzzy torus algebra as a q-deformation of parafermion. In addition to the known representations, new representations of both finite and infinite dimension are found. Using the infinite dimensional representation, coherent state for the fuzzy torus is constructed. Dirac operator on commutative torus is also discussed.

1. Introduction

Fuzzy spaces provide a method of regularizing the quantum filed theory allowing an alternate route to the naive lattice regularization. It has been observed that the use of such spaces has some advantages (see for example [1]). Fuzzy spaces also provide one of the models of noncommutative manifold on which one can develop noncommutative geometry [2]. The basic idea of fuzzy spaces is that the algebras of functions on a smooth manifold, as well as their local coordinates, are replaced with associative algebras of finite dimensional matrices. The size of the matrices is arbitrary. In the limit of infinite dimensional matrices, the geometry on commutative manifold is recovered.

The typical example of the fuzzy spaces and the most extensively studied one may be the fuzzy 2-sphere [3]. Many other examples of fuzzy spaces have been constructed so far. We shall mention some of them below. Regarding the ordinary 2-sphere S^2 as a one-dimensional complex projective space $\mathbb{C}P^1$, construction of the fuzzy S^2 has been extended to higher dimensional fuzzy projective spaces [4]. The fuzzy $\mathbb{C}P^n$ opens a way to other fuzzy spaces. Fuzzy version of S^3 are extracted from fuzzy $\mathbb{C}P^3$[5]. Exploiting the observation that $\mathbb{C}P^3$ is an S^2 bundle over S^4, the fuzzy S^4 has been constructed [6]. Fuzzy version of S^4 is also discussed by using tensor product representation of $Spin(5)$ [7] and the Hopf fibration $S^7 \to S^4$ [8]. Fuzzy S^5 has been established by utilizing a $U(1)$ fibration over a fuzzy $\mathbb{C}P^2$[9]. Fuzzy orbifolds [10] and fuzzy toric varieties [11] are obtained as a subset of fuzzy complex projective spaces. It has been shown that fuzzy S^2 and fuzzy $\mathbb{C}P^n$ have their supersymmetric extensions [12, 13]. Further examples are found in the investigations of fuzzy unitary Grassmannian spaces [14, 15] and fuzzy orthogonal Grassmannian spaces [16]. Fuzzy orthogonal Grassmanians are used to extract fuzzy spheres of higher dimensions in [16]. Furthermore, fuzzy version of disc [17] and flag manifolds including supersymmetric extension [18] have been discussed.

It should be noted that except some specific cases the matrices representing the coordinates of fuzzy space do not form a closed algebra. In this sense, some of the aforementioned fuzzy spaces are unsatisfactory. Nevertheless, one can discuss filed theory on such spaces by suppressing unwanted terms in functional integral. Recently, it was proposed a general recipe for constructing fuzzy version of compact connected Riemann surfaces of arbitrary genus as closed algebras [19]. The genus one case (torus) was explicitly carried out and the authors of [19] defined the fuzzy torus as a nonlinear C-algebra with three generators. The fuzzy torus algebra T_F^2 has been recast as a q-deformed Lie algebra in [20]. This may be understood as a linearization of the nonlinear T_F^2 algebra by using the q-deformed commutators.

In the present work, we are studying the fuzzy torus algebra from a different viewpoint. We regard T_F^2 as a q-deformed parafermion algebra which is different from the q-deformations of parafermion or paraboson discussed in the literature [21, 22, 23, 24, 25, 26]. Using this approach, we classify the Hermitian representations of T_F^2. In addition to recovering the known finite dimensional representations [19, 20], families of finite and infinite dimensional representations are newly found. Using the newly found infinite dimensional representations, coherent state for the fuzzy torus is constructed. Coherent states play an important role in noncommutative geometry, since one can define noncommutative star product using coherent states [27, 28]. We also give Dirac operator on the commutative torus in our setting. This is the first step to construct Dirac operator on fuzzy torus. Dirac operator is one of the fundamental building blocks of noncommutative geometry [2] and there is no need to say its physical importance. In many literature discussing Dirac operator on torus, the torus is defined by giving cyclic condition on flat plane. It means that such torus has vanishing curvatures. On the contrary, our torus does have nonvanishing curvatures.

This paper is organized as follows. Next section is a brief review of the construction of fuzzy Riemann surfaces proposed in [19]. The definition of the fuzzy torus is also given there. In §3, the fuzzy torus algebra is reformed into a kind of q-deformation of parafermion algebra. Using this form, possible Hermitian representations are derived. The results of §3 are also reported in [29]. In §4, coherent state, which is an eigenstate of the annihilation operator of the q-deformed parafermion, is given. New deformation of the Bessel functions is introduced to formulate the coherent state. We discuss Dirac operator on commutative torus in §5. §6 is devoted to concluding remarks.

2. Fuzzy version of Riemann surfaces

In this section, we briefly review the work of [19] which is a construction of fuzzy version of compact connected Riemann surfaces as a nonlinear algebra. The construction consists of two steps.

The first step is an embedding of Riemann surfaces in $\mathbb{R}^3$. Using the Cartesian coordinates (x, y, z) in $\mathbb{R}^3$, the genus zero Riemann surface S^2 is described by a constrained polynomial $C(x, y, z) = 0$, where $C(x, y, z) = x^2 + y^2 + z^2 - 1$. It is shown, with the aid of the regular value theorem and Morse theory, that similar polynomial description is possible for any Riemann surfaces of higher genus. Let us consider the following polynomial with a positive constant c :

$$C(x, y, z) = (P(x) + y^2)^2 + z^2 - c, \tag{2.1}$$

$$P(x) = \sum_{j=0}^{2k} a_j x^j, \qquad a_{2k} > 0.$$

The constrained polynomial $C(x, y, z) = 0$ describes a genus g Riemann surface if the equations $P(x) \pm \sqrt{c} = 0$ have the certain number of nondegenerate solutions. The required number of solutions for $P(x) + \sqrt{c} = 0$ is $2g$ and for $P(x) - \sqrt{c} = 0$ is two.

The second step is a Poisson bracket and its quantization. It is known that for arbitrary function $\ell : \mathbb{R}^3 \to \mathbb{R}$

$$\{f, g\}_{\mathbb{R}^3} = \nabla \ell \cdot (\nabla f \times \nabla g), \tag{2.2}$$

Journal of Physics: Conference Series **128** (2008) 012047 doi:10.1088/1742-6596/128/1/012047

defines a Poisson bracket for functions on $\mathbb{R}^3$. In the present case, we take $\ell = C(x,y,z)/2$. Then the Poisson brackets between the coordinates read

$$\{x,y\}_{\mathbb{R}^3} = z, \qquad \{y,z\}_{\mathbb{R}^3} = P'(x)(P(x)+y^2), \qquad \{z,x\}_{\mathbb{R}^3} = 2y(P(x)+y^2), \tag{2.3}$$

where $P'(x)$ denotes the derivative of $P(x)$. Next we apply Dirac quantization to these Poisson brackets, namely, the Poisson brackets are replaced with commutators. Ordering problem arise in the right hand sides of (2.3). The following is the proposed ordering in [19]:

$$\begin{aligned}
&[X,Y] = i\hbar Z, \\
&[Y,Z] = i\hbar \sum_{n=1}^{2g} a_n \sum_{j=0}^{n-1} X^j \left(P(X)+Y^2\right) X^{n-1-j}, \\
&[Z,X] = i\hbar \{Y, P(X)+Y^2\},
\end{aligned} \tag{2.4}$$

where the quantized coordinates are denoted by capital letters and the parameter $\hbar$ imparts noncommutativity. Its commutative limit is given by $\hbar \to 0$. One can verify that the polynomial algebra (2.4) satisfy the Jacobi identity.

For the case of torus ($g=1$), the polynomial (2.1) reads:

$$C(x,y,z) = (x^2+y^2-\mu)^2 + z^2 - c, \tag{2.5}$$

where $\mu = -a_0$ is a real parameter. The relation $C(x,y,z) = 0$ describes a torus for $\sqrt{c} < \mu$ and a squashed sphere for the domain $-\sqrt{c} < \mu < \sqrt{c}$. The fuzzy torus algebra is obtained from (2.4):

$$\begin{aligned}
[X,Y] = i\hbar Z, \qquad [Y,Z] &= i\hbar\{X,\varphi\}, \qquad [Z,X] = i\hbar\{Y,\varphi\}, \\
\varphi &\equiv X^2+Y^2-\mu,
\end{aligned} \tag{2.6}$$

and the constraint

$$C_F \equiv \varphi^2 + Z^2 = c. \tag{2.7}$$

It is easily observed that the element C_F is the center of the algebra. This is consistent with the constraint (2.7). In [19], the parameter c is set equal to unity. Following [20], we, however, keep its value arbitrary throughout this paper.

3. Hermitian representations of fuzzy torus algebra

3.1. Fuzzy torus and q-Deformed parafermion

The fuzzy torus has been obtained as a nonlinear algebra of three generators in the previous section. Our next question is what kinds of representations are available for the fuzzy torus algebra. Two classes of representations have already been derived in [19]. In this section, we consider the problem more systematic way and try to classify Hermitian representations. To this end, we recast the relations in (2.6) as a form which is interpreted as a q-deformed parafermion.

Introducing the complex variable $W = X + iY$ and its Hermitian adjoint $W^\dagger$, the defining relations (2.6) are reexpressed as

$$[W,W^\dagger] = 2\hbar Z, \qquad [Z,W] = \hbar\{W,\varphi\}, \qquad [Z,W^\dagger] = -\hbar\{W^\dagger,\varphi\}, \tag{3.1}$$

$$\varphi \equiv \frac{1}{2}\{W,W^\dagger\} - \mu. \tag{3.2}$$

Eliminating Z via the first equation in (3.1), it is observed [19] that the operators $W, W^\dagger$ obey the following trilinear relation and its Hermitian adjoint:

$$W(W^\dagger)^2 - 2\frac{1-\hbar^2}{1+\hbar^2}W^\dagger W W^\dagger + (W^\dagger)^2 W = \mu\frac{4\hbar^2}{1+\hbar^2}W^\dagger. \tag{3.3}$$

The deformation parameter q is introduced [20] by the relation

$$2\frac{1-\hbar^2}{1+\hbar^2} = q + q^{-1}. \tag{3.4}$$

It follows that the complex parameter q is of unit magnitude: $|q| = 1$. The commutative $\hbar \to 0$ limit now corresponds to $q \to 1$. We now scale the conjugate variables W and $W^\dagger$ as follows

$$a = \left(\frac{2}{|\mu|(2-q-q^{-1})}\right)^{1/2} W, \qquad a^\dagger = \left(\frac{2}{|\mu|(2-q-q^{-1})}\right)^{1/2} W^\dagger. \tag{3.5}$$

For a fixed μ above scaling is well-defined in the *noncommutative* $q \neq 1$ *i.e.* $\hbar \neq 0$ regime. The trilinear relation (3.3) now assumes two distinct forms depending on the sign of the parameter μ. For the choice $\mu < 0$, it reads

$$a(a^\dagger)^2 - [2]_q\, a^\dagger a a^\dagger + (a^\dagger)^2 a = -2a^\dagger, \tag{3.6}$$

where the q-number is defined as usual

$$[x]_q = \frac{q^x - q^{-x}}{q - q^{-1}}. \tag{3.7}$$

For the alternate $\mu > 0$ case, the relation (3.3) may be recast as

$$a(a^\dagger)^2 - [2]_q\, a^\dagger a a^\dagger + (a^\dagger)^2 a = 2a^\dagger. \tag{3.8}$$

The trilinear relation (3.6) may be regarded as q-deformation of the parafermion [30]. The relation (3.8) has the same form albeit with an opposite sign on the right hand side. One may regard this as a variant of the q-deformed parafermion. The transformation (3.5) is singular for a fixed μ in the limit $q \to 1$. In this scenario the classical $q = 1$ parafermions do not arise. There is, however, an interesting possibility

$$|\mu| \sim (2 - q - q^{-1})^{-1}, \tag{3.9}$$

which allows taking the $q \to 1$ limit in (3.5). This limit is pertinent in understanding some of the representations discussed below.

3.2. Representations

Using the q-paraferimionic form of the fuzzy torus algebra given in the previous subsection, we study possible Hermitian representations. Our strategy is as follows: we start with the smallest number of assumptions and construct the most general representation. Then, further assumptions are imposed to obtain possible subclasses of the representations. As observed in [19], the operators $aa^\dagger$ and $a^\dagger a$ commute, and, therefore, they are simultaneously diagonalizable. We thus assume the existence of a normalized state $|0\rangle$ such that

$$aa^\dagger\,|0\rangle = p\,|0\rangle\,, \qquad a^\dagger a\,|0\rangle = r\,|0\rangle\,, \qquad \langle 0|0\rangle = 1. \tag{3.10}$$

Other states may be obtained by repeated actions of a or $a^\dagger$ on $|0\rangle$. The states are required to be normalizable:

$$|n\rangle = \mathcal{N}_n (a^\dagger)^n |0\rangle, \qquad |-n\rangle = \mathcal{N}_{-n} a^n |0\rangle \quad \forall\ n > 0, \tag{3.11}$$

where $\mathcal{N}_{\pm n}$ are the normalization constants. Normalization of the $|\pm 1\rangle$ states puts constraints on the values of p and r:

$$\|\, |1\rangle \,\|^2 = |\mathcal{N}_1|^2 \langle 0|\, a a^\dagger \,|0\rangle = |\mathcal{N}_1|^2 p, \qquad \|\, |-1\rangle \,\|^2 = |\mathcal{N}_{-1}|^2 \langle 0|\, a^\dagger a \,|0\rangle = |\mathcal{N}_{-1}|^2 r.$$

It follows that $p, r > 0$, and we choose $\mathcal{N}_1 = p^{-1/2}$, $\mathcal{N}_{-1} = r^{-1/2}$. For the states of $|\pm n\rangle\, \forall n \geq 2$, we use the relations (3.6) or (3.8) appropriately for evaluating the norm. For the regime $\mu < 0$, the following relations are verified inductively:

$$a\,|n\rangle = \left([n]_q\, p - [n-1]_q\, r - 2 \left[\frac{n-1}{2} \right]_q [n]_{\sqrt{q}} \right)^{1/2} |n-1\rangle \equiv A_n\, |n-1\rangle, \tag{3.12}$$

$$a^\dagger\,|n\rangle = A_{n+1}\, |n+1\rangle, \tag{3.13}$$

and

$$a\,|-n\rangle = A_{-n}\, |-n-1\rangle, \qquad a^\dagger\,|-n\rangle = A_{-n+1}\, |-n+1\rangle. \tag{3.14}$$

The diagonal operators read

$$a^\dagger a\, |\pm n\rangle = A^2_{\pm n}\, |\pm n\rangle, \qquad a a^\dagger\, |\pm n\rangle = A^2_{\pm n+1}\, |\pm n\rangle. \tag{3.15}$$

For the choice $\mu > 0$ the construction of states proceeds as before:

$$a\,|n\rangle = \left([n]_q\, p - [n-1]_q\, r + 2 \left[\frac{n-1}{2} \right]_q [n]_{\sqrt{q}} \right)^{1/2} |n-1\rangle \equiv \mathsf{A}_n\, |n-1\rangle, \tag{3.16}$$

$$a^\dagger\,|n\rangle = \mathsf{A}_{n+1}\, |n+1\rangle, \tag{3.17}$$

while the relations (3.14) and (3.15) hold with the replacement of the normalization constant A_n by A_n. The Casimir operator is quartic in the variables $W,\ W^\dagger$. Its eigenvalues may be computed via the results obtained above:

$$\frac{4}{\mu^2\left(2 - [2]_q\right)}\, C_F\, |\pm n\rangle = \begin{cases} \left(p^2 + r^2 - [2]_q\, pr + 2(p+r) + \dfrac{4}{2 - [2]_q} \right) |\pm n\rangle, & \mu < 0, \\[2ex] \left(p^2 + r^2 - [2]_q\, pr - 2(p+r) + \dfrac{4}{2 - [2]_q} \right) |\pm n\rangle, & \mu > 0. \end{cases} \tag{3.18}$$

We have *formally* obtained a double-sided infinite dimensional representation with two parameters: p, r. Our representation has close kinship with the symplecton realization [31] of the boson calculus. Reflecting the symmetry of (3.6) (or (3.8)) and its adjoint under the exchange of a and $a^\dagger$, the representation has the symmetry structure $A_{n+1}(p,r) = A_{-n}(r,p)$. The usefulness of the above representations becomes evident below where we impose restrictions for obtaining the subclasses of the above infinite dimensional representation. These restrictions terminate the infinite series of state vectors.

Assuming that the state $|0\rangle$ is annihilated by the operator a, we have, $r = 0$. All states $|-n\rangle$ labeled by negative integers are eliminated while the semi-infinite series of states $\{|n\rangle\, | n = 0, 1, 2, \ldots \infty\}$ are retained. This is a Fock-type representation constructed on the vacuum $|0\rangle$. The number of parameters is reduced to one.

Journal of Physics: Conference Series **128** (2008) 012047 doi:10.1088/1742-6596/128/1/012047

A key requirement for the existence for the classes of representations discussed below is that their *Hermiticity needs to be maintained.* This, in turn, requires the constants $A^2_{\pm n}, \mathsf{A}^2_{\pm n} \forall n > 0$ to be *nonnegative*. Keeping this in mind, we study all the possibilities in order.

(1) Truncated Fock-type representation

To obtain finite dimensional Fock-type representations, we further assume that there exist a positive integer N such that

$$A_N = 0 \quad \text{for} \quad \mu < 0, \qquad \mathsf{A}_N = 0 \quad \text{for} \quad \mu > 0. \tag{3.19}$$

The representation space is spanned by N independent states: $|0\rangle, |1\rangle, \ldots, |N-1\rangle$ with the highest state satisfying $a^\dagger |N-1\rangle = 0$. For instance, the N dimensional representation for $\mu < 0$ reads

$$a^\dagger = \begin{pmatrix} 0 & A_{N-1} & 0 & \cdots & 0 \\ & 0 & A_{N-2} & \cdots & 0 \\ & & \ddots & \ddots & \vdots \\ & & & 0 & A_1 \\ & & & & 0 \end{pmatrix}. \tag{3.20}$$

The null condition (3.19) requires

$$[N]_q \, p \mp 2 \left[\frac{N-1}{2}\right]_q [N]_{\sqrt{q}} = 0, \tag{3.21}$$

where the upper (lower) sign refers to $\mu < 0$ ($\mu > 0$). The relation (3.21) holds for two cases:

(i) A generic value of

$$q = \exp(i\theta) \tag{3.22}$$

leads to a N-dependent order parameter p:

$$p = \pm 2 \left[\frac{N-1}{2}\right]_q \frac{[N]_{\sqrt{q}}}{[N]_q} = \pm \left(\frac{\tan(N\theta/2)}{\tan(\theta/2)} - 1\right). \tag{3.23}$$

This case corresponds to the 'string solutions' of [19, 20]. The angle θ is restricted by the requirement of nonnegativity of the elements A^2_n (A^2_n) $\forall n \in (1, \ldots, N-1)$:

$$A^2_n = -\mathsf{A}^2_n = \frac{\sin^2(n\theta/2)}{\sin^2(\theta/2)} \left(\frac{\tan(N\theta/2)}{\tan(n\theta/2)} - 1\right). \tag{3.24}$$

For the regime $\mu < 0$ the positivity $A^2_n > 0$ holds in the domain $-\pi/N < \theta < \pi/N$, whereas for the choice $\mu > 0$ the positivity $\mathsf{A}^2_n > 0$ is satisfied, for instance, in the restricted region $\pi/N < \theta < \pi/(N-1)$, $-\pi/(N-1) < \theta < -\pi/N$. This class of representation matrices are symmetric with respect to the minor diagonal: $A_{N-k} = A_k$, $\mathsf{A}_{N-k} = \mathsf{A}_k$.

(ii) The root of unity values of $q = \exp(i2\pi/N)$ restricts $p\,(> 0)$ over a range. The solutions corresponding to this class is novel. We list some low dimensional Hermitian representations for $\mu > 0$ as examples:

$$a^\dagger = \begin{pmatrix} 0 & \sqrt{2-p} & 0 \\ & 0 & \sqrt{p} \\ & & 0 \end{pmatrix}, \quad a^\dagger = \begin{pmatrix} 0 & \sqrt{2-p} & 0 & 0 \\ & 0 & \sqrt{2} & 0 \\ & & 0 & \sqrt{p} \\ & & & 0 \end{pmatrix},$$

$$a^\dagger = \begin{pmatrix} 0 & \sqrt{2-p} & 0 & 0 & 0 \\ & 0 & \sqrt{2(1+(2-p)\sin\frac{\pi}{10})} & 0 & 0 \\ & & 0 & \sqrt{2(1+p\sin\frac{\pi}{10})} & 0 \\ & & & 0 & \sqrt{p} \\ & & & & 0 \end{pmatrix}. \tag{3.25}$$

The elements $\mathsf{A}_n^2\, \forall n \in (1, \ldots, N-1)$ for the representations (3.25) read

$$\mathsf{A}_n^2 = \frac{\sin^2(n\,\pi/N)}{\sin^2(\pi/N)}\,(1+(p-1)\,\mathsf{f}_n), \qquad \mathsf{f}_n = \frac{\tan(\pi/N)}{\tan(n\,\pi/N)}. \tag{3.26}$$

The sequence $\{\mathsf{f}_n | n = 1, \ldots, N-1\}$ is bounded as $-1 \leq \mathsf{f}_n \leq 1$. The Hermiticity requirement now restricts the order parameter as $0 < p < 2$. The value of p being positive definite, the Hermitian representations of this class do not exist for $\mu < 0$.

(2) Cyclic representations

The cyclic finite dimensional representations may be ensured by identifying the states $|N\rangle$ and $|0\rangle$. In particular, this makes the eigenvalues of the operators $aa^\dagger$ and $a^\dagger a$ on the states $|N\rangle$ and $|0\rangle$ same. For the $\mu > 0$ case, we obtain

$$\mathsf{A}_{N+1}^2 = p, \qquad \mathsf{A}_N^2 = r. \tag{3.27}$$

Eliminating r from the relations in (3.27), we obtain

$$[N]^2_{\sqrt{q}}\, p = \frac{2}{2-[2]_q}\,[N]^2_{\sqrt{q}}.$$

For the choice $[N]_{\sqrt{q}} \neq 0$, it follows $p = 2(2-[2]_q)^{-1}$. It then turns out that $r = p$, and $\mathsf{A}_n = \sqrt{p}\ \ \forall n$. This leads to $Z\,|n\rangle = 0$ [19]. This case is uninteresting. Alternate choice $[N]_{\sqrt{q}} = 0$, *i.e.* $q^N = 1$ yields a class of two-parametric (p, r) N-dimensional representations:

$$a^\dagger = \begin{pmatrix} 0 & \sqrt{2-p-r} & 0 \\ & 0 & \sqrt{p} \\ \sqrt{r} & & 0 \end{pmatrix}, \quad a^\dagger = \begin{pmatrix} 0 & \sqrt{2-p} & 0 & 0 \\ & 0 & \sqrt{2-r} & 0 \\ & & 0 & \sqrt{p} \\ \sqrt{r} & & & 0 \end{pmatrix}, \tag{3.28}$$

$$a^\dagger = \begin{pmatrix} 0 & \sqrt{2(1+r\sin\frac{\pi}{10})-p} & 0 & 0 & 0 \\ & 0 & \sqrt{2(1+(2-p-r)\sin\frac{\pi}{10})} & 0 & 0 \\ & & 0 & \sqrt{2(1+p\sin\frac{\pi}{10})-r} & 0 \\ & & & 0 & \sqrt{p} \\ \sqrt{r} & & & & 0 \end{pmatrix}.$$

These representations corresponds to the 'loop solutions' in [19, 20], and possess the symmetry

$$\mathsf{A}_n(p, r) = \mathsf{A}_{N+1-n}(r, p). \tag{3.29}$$

To examine their nonnegativity the elements $\{\mathsf{A}_n^2 | n \in (1, \ldots, N)\}$ may be recast by isolating their symmetric part $\mathcal{S}_n(p+r)$ as follows

$$\mathsf{A}_n^2 = \mathcal{P}_n\, p + \mathcal{S}_n(p+r), \quad \mathcal{P}_n = [n]_q + [n-1]_q, \quad \mathcal{S}_n(\mathcal{X}) = -[n-1]_q\, \mathcal{X} + 2\left[\frac{n-1}{2}\right]_q [n]_{\sqrt{q}}. \tag{3.30}$$

Due to the symmetry (3.29) we only need to check the nonnegativity of the elements A_n^2 for $n = 1, \ldots, N/2$ $((N+1)/2)$ for an even (odd) N. In this domain we have $\mathcal{P}_n \geq 0$, and

$$\mathcal{S}_n(p+r) = \frac{\sin^2((n-1)\pi/N)}{\sin^2(\pi/N)}\,(1 + (1-p-r)\,\mathsf{f}_{n-1}), \tag{3.31}$$

where the sequence $\{\mathsf{f}_n\}$ has been defined in (3.26). In the present context the entries of the sequence $\{\mathsf{f}_{n-1}\}$, where $n = 2, \ldots, N/2$ $((N+1)/2)$ for an even (odd) N, is bounded as $0 < \mathsf{f}_{n-1} \leq 1$. This yields the parametric values $0 < p, 0 < r, p+r < 2$ for the required nonnegativity. For the odd-N case the symmetric element reads

$$\mathsf{A}^2_{\frac{N+1}{2}} = \frac{1}{2\,\cos(\pi/N)}\left(\frac{1}{2\,\sin^2(\pi/2N)} - p - r\right). \tag{3.32}$$

It may be noticed that the solutions obtained in (3.25) may be obtained as the $r \to 0$ limiting case of the present cyclic representations. The corresponding matrix structures are, however, of different ranks, and in that sense inequivalent. Moreover the cyclic property of the solutions (3.28) is *lost* when $r = 0$ is substituted in them. Hermitian cyclic representations do not exist for the $\mu < 0$ case as the elements $\{A_n^2 | n = 1, \ldots, N\}$ are not nonnegative. For instance, we have

$$A^2_{\frac{N}{2}}\Big|_{\text{even } N} = -r - \frac{1}{\sin^2(\pi/N)}, \qquad A^2_{\frac{N+1}{2}}\Big|_{\text{odd } N} = -\frac{1}{2\cos(\pi/N)}\left(\frac{1}{2\,\sin^2(\pi/2N)} + p + r\right). \tag{3.33}$$

(3) Infinite dimensional representations

Turning towards the infinite dimensional representations we discuss the special cases where we notice the existence of such Hermitian representations. The identity

$$2\left[\frac{n-1}{2}\right]_q [n]_{\sqrt{q}} = [n]^2_{\sqrt{q}} - [n]_q$$

allows us to rewrite the elements A_n^2 for the Fock states ($r = 0$) in the $\mu > 0$ case as

$$\mathsf{A}_n^2 = [n]^2_{\sqrt{q}} + (p-1)\,[n]_q. \tag{3.34}$$

For the choice $p = 1$, therefore, the Hermiticity of the infinite set of basis states with $n = 1, 2, \ldots \infty$ is guaranteed. To establish a range of values of the order parameter p around $p = 1$ that preserves the positivity $\mathsf{A}_n^2 > 0$, we use (3.22) to obtain

$$\mathsf{A}_n^2 = \frac{\sin^2(n\theta/2)}{\sin^2(\theta/2)}\,(1 + (p-1)\,\mathfrak{f}_n), \qquad \mathfrak{f}_n = \frac{\tan(\theta/2)}{\tan(n\theta/2)}. \tag{3.35}$$

The entries of the sequence $\{\mathfrak{f}_n | n = 1, 2, \ldots \infty\}$ for any positive *integral* value of n remain *finite* for the choice $\theta = 2\pi/\mathcal{N}$, where $\mathcal{N}$ is an *irrational* number. Employing the finite maximum and minimum entries of the said sequence a range of values of p may now be easily established for which Fock-state representations are Hermitian and well-defined.

Another possible scenario is to turn to the quasi-classical states mentioned in the context of (3.9). Using the classical limit $q \to 1$ in the context of the (p, r) two-parametric states for the regime $\mu > 0$ we obtain

$$\mathsf{A}_n^2 = n^2 + (p-r-1)\,n + r \quad \forall\, n = 1, 2, \ldots \infty. \tag{3.36}$$

Journal of Physics: Conference Series **128** (2008) 012047 doi:10.1088/1742-6596/128/1/012047

For the choice $p \geq r+1$, these two-parametric states maintain $\mathsf{A}_n^2 > 0$ ensuring the Hermiticity of the representations. These infinite dimensional quasi-classical states reflect, in some sense, noncommutative spaces bordering on the commutative regime. Hermitian infinite dimensional representations for the $\mu < 0$ case do not exist.

We have studied so far the case of $\mu \neq 0$. To complete the study of Hermitian representations, we next investigate the case of $\mu = 0$. Although the algebra of this case does not correspond to a deformed parafermion, one can repeat the analysis used earlier. We use the trilinear relation

$$W(W^\dagger)^2 - [2]_q W^\dagger W W^\dagger + (W^\dagger)^2 W = 0, \tag{3.37}$$

and assume a state $|0\rangle$ subject to the conditions

$$WW^\dagger |0\rangle = p\,|0\rangle\,, \qquad W^\dagger W\,|0\rangle = r\,|0\rangle\,, \qquad \langle 0|0\rangle = 1. \tag{3.38}$$

Using the same arguments applied earlier the positivity $p, r > 0$ follows, and the states are given by

$$W\,|n\rangle = \Big([n]_q p - [n-1]_q r\Big)^{1/2} |n-1\rangle \equiv w_n\,|n-1\rangle\,, \tag{3.39}$$

$$W^\dagger\,|n\rangle = w_{n+1}\,|n+1\rangle\,. \tag{3.40}$$

This leads to

$$W^\dagger W\,|n\rangle = w_n^2\,|n\rangle\,, \qquad WW^\dagger\,|n\rangle = w_{n+1}^2\,|n\rangle\,. \tag{3.41}$$

The eigenvalue of the Casimir operator reads

$$C_F\,|n\rangle = \frac{p^2 + r^2 - [2]_q pr}{2 - [2]_q}\,|n\rangle\,. \tag{3.42}$$

Relations parallel to (3.39) - (3.42) also hold for the state $|-n\rangle$.

Fock-type representation is obtained by requiring $W\,|0\rangle = 0$ *i.e.* $r = 0$. Implementing $w_N = 0$ we may truncate the Fock-type representation and obtain a finite N dimensional one. There is a slight difference from the case of $\mu \neq 0$. The equation

$$w_N = \sqrt{[N]_q p} = 0, \quad p > 0$$

may be solved only for the choice $[N]_q = 0$. Namely, it is possible to obtain finite dimensional Hermitian representations from the Fock-type one for $q^{2N} = 1$ *i.e.* $q = \exp(i\pi/N)$. The expression for w_n^2 is found to be nonnegative:

$$w_n^2 = \frac{\sin(n\pi/N)}{\sin(\pi/N)}\,p > 0 \quad \forall n \in (1, \ldots, N-1). \tag{3.43}$$

Such a truncation for a generic q does not exist.

Now we investigate the possible realizations of the cyclic representations for the $\mu = 0$ case. Conditions for cyclic representation are given by

$$w_N^2 = r, \qquad w_{N+1}^2 = p. \tag{3.44}$$

Above equations may be recast as

$$[N]_q p = ([N-1]_q + 1)r, \qquad ([N+1]_q - 1)p = [N]_q r. \tag{3.45}$$

Table 1. Hermitian representations of T_F^2

		$\mu<0$	$\mu=0$	$\mu>0$
infinite dimensional		no	no	yes
truncated	$q^N\neq 1$	yes	no	yes
	$q^N=1$	no	yes	yes
cyclic		no	no	yes

Eliminating r from (3.45) we obtain a relation for p:

$$q^{-N}(1-q^N)^2p=0.$$

Therefore the cyclic representations may exist only if $q^N=1$. By computation, however, it may be seen that negative values of w_n^2 exist:

$$w^2_{\frac{N}{2}}\Big|_{\text{even }N}=-r,\qquad w^2_{\frac{N+1}{2}}\Big|_{\text{odd }N}=-\frac{p+r}{2\cos(\pi/N)}. \tag{3.46}$$

This eliminates the possibility of having Hermitian cyclic representations for the $\mu=0$ case.

Lastly, paralleling the $\mu>0$ case the infinite dimensional representations for the choice $\mu=0$ exist for the said semi-classical states in the $q\to 1$ limit as the elements

$$w_n^2\Big|_{q\to 1}=n(p-r)+r\quad \forall n=1,2,\dots\infty \tag{3.47}$$

are nonnegative for $p\geq r$.

The Hermitian representations of the fuzzy torus algebra T_F^2 are summarized in Table 1.

4. Coherent state for infinite dimensional representation

The existence of the infinite dimensional representations for $\mu>0$ encourages us to define a coherent state for fuzzy torus as an eigenstate of the annihilation operator a. We consider the $r=0$ case here. The coherent state $|\zeta\rangle$, $\zeta\in\mathbb{C}$ is defined as follows:

$$a|\zeta\rangle=\zeta|\zeta\rangle,\qquad |\zeta\rangle=\sum_{n=0}^{\infty}c_n(\zeta)|n\rangle. \tag{4.1}$$

The coefficients c_n satisfy the recurrence relation:

$$c_{n+1}=\frac{\zeta}{\mathsf{A}_n}c_n, \tag{4.2}$$

which is easily solved to give the expression:

$$c_n=\zeta^n\left(\prod_{k=1}^{n}\mathsf{A}_k\right)^{-1}c_0. \tag{4.3}$$

The coefficient c_0 is determined by the normalization condition $\langle\zeta|\zeta\rangle = 1$:

$$\frac{1}{c_0^2} = \sum_{n=0}^{\infty} \frac{|\zeta|^{2n}}{\prod_{k=1}^{n} \mathsf{A}_k^2} = \sum_{n=0}^{\infty} \frac{|\zeta|^{2n}}{[n]_q! \prod_{k=1}^{n} \left(\frac{[k]_q}{\{k\}_{\sqrt{q}}^2} + p - 1 \right)}, \qquad \{n\}_q = \frac{q^n + q^{-n}}{q + q^{-1}} \xrightarrow[q \to 1]{} 1. \tag{4.4}$$

Comparing the expression (4.4) with the definition of modified Bessel function of the first kind:

$$I_m(2x) = \sum_{n=0}^{\infty} \frac{x^{m+2n}}{n!(n+m)!}, \tag{4.5}$$

the following deformation of $I_m(2x)$ is inspired:

$$\mathscr{I}_m^{(q)}(2x) = \sum_{n=0}^{\infty} \frac{x^{m+2n}}{[n]_q! \prod_{k=1}^{n} \left(\frac{[k]_q}{\{k\}_{\sqrt{q}}^2} + m \right)}. \tag{4.6}$$

In terms of (4.6), the coherent state reads:

$$|\zeta\rangle = \left(\frac{|\zeta|^{p-1}}{\mathscr{I}_{p-1}^{(q)}(2|\zeta|)} \right)^{1/2} \sum_{n=0}^{\infty} \frac{\zeta^n}{\prod_{k=1}^{n} \mathsf{A}_k} |n\rangle . \tag{4.7}$$

The overlap of the coherent states is also given in terms of (4.6):

$$\langle\xi|\eta\rangle = \frac{\mathscr{I}_{p-1}^{(q)}(2\sqrt{\bar{\xi}\zeta})}{\sqrt{\mathscr{I}_{p-1}^{(q)}(2|\xi|)\, \mathscr{I}_{p-1}^{(q)}(2|\zeta|)}} e^{-i\phi(p-1)/2}, \tag{4.8}$$

where $\phi = \arg(\zeta) - \arg(\xi)$. If we set $\xi = -\zeta$, then the overlap (4.8) is a real quantity having a simple expression:

$$\langle -\zeta|\zeta\rangle = \frac{\mathcal{J}_{p-1}^{(q)}(2|\zeta|)}{\mathscr{I}_{p-1}^{(q)}(2|\zeta|)}, \tag{4.9}$$

where $\mathcal{J}_{p-1}^{(q)}(2x)$ is a deformation of Bessel function of the first kind defined by

$$\mathcal{J}_m^{(q)}(2x) = \sum_{n=0}^{\infty} \frac{(-1)^n x^{m+2n}}{[n]_q! \prod_{k=1}^{n} \left(\frac{[k]_q}{\{k\}_{\sqrt{q}}^2} + m \right)}. \tag{4.10}$$

Since the overlap (4.9) is real, one can define orthogonal cat-type states for the fuzzy torus:

$$|\pm\rangle = \frac{1}{\sqrt{2(1 \pm \langle -\zeta|\zeta\rangle)}} (\, |\zeta\rangle \pm |-\zeta\rangle \,). \tag{4.11}$$

We defined the coherent states (4.7) as an eigenstate of the annihilation operator a. If the states have the property of decomposition of unity, they may be used to define a star product on the fuzzy torus along the line developed in [28]. Detailed analysis of the coherent states will be published elsewhere.

Journal of Physics: Conference Series **128** (2008) 012047 doi:10.1088/1742-6596/128/1/012047

5. Dirac operator on commutative torus

We have obtained algebraic relations and representations for fuzzy torus. What we need now is Dirac operator on fuzzy torus. Physical motivation of this is, of course, introduction of fermions into noncommutative space. Mathematically, Dirac operator together with algebra and Hilbert space on which the algebra acts consists the so-called spectral triple, that is a basic ingredient of noncommutative geometry [2]. Dirac operator on fuzzy space may have some different approaches. This is the case of fuzzy 2-sphere [32, 33, 34, 35]. In any approaches one will take, the fuzzy Dirac operator has to be reduced to the ordinary one in the commutative limit.

As a preliminary to Dirac operator on fuzzy torus, we derive Dirac operator on our commutative torus (2.5). We use the following parameterization of the torus:

$$x = (\mu + r\sin\theta)^{1/2}\cos\varphi, \qquad y = (\mu + r\sin\theta)^{1/2}\sin\varphi, \qquad z = r\cos\theta. \tag{5.1}$$

It is easy to read off the metric tensor from the parameterization (5.1):

$$g_{\mu\nu} = \begin{pmatrix} \frac{r^2\cos^2\theta}{4g_{22}} + r^2\sin^2\theta & 0 \\ 0 & \mu + r\sin\theta \end{pmatrix}. \tag{5.2}$$

The suffices 1 and 2 stand for θ and φ, respectively. The zweibein which connects curved torus and flat tangent plane on it is given by

$$\sigma_\mu^a = \begin{pmatrix} \sqrt{g_{11}} & 0 \\ 0 & \sqrt{g_{22}} \end{pmatrix}. \tag{5.3}$$

The zweibein satisfies the required relation $e_\mu^a\, e_\nu^b\, \delta_{ab} = g_{\mu\nu}$, where and in the sequel summation over repeated indices is understood. In two dimensional space, Clifford algebra which provides the Dirac gamma matrices in flat space is realized by the Pauli matrices. Then the gamma matrices on torus are given by

$$\gamma^1 = \frac{1}{\sqrt{g_{11}}}\sigma_x, \qquad \gamma^2 = \frac{1}{\sqrt{g_{22}}}\sigma_y, \tag{5.4}$$

where σ_x, σ_y are Pauli matrices. These curved gamma matrices satisfy the relation : $\{\gamma^\mu, \gamma^\nu\} = g^{\mu\nu}$. We now have enough data to write down the covariant derivative for spinors on the torus. After some algebra, we obtain the followings:

$$D_1 = \partial_\theta, \qquad D_2 = \partial_\varphi + \Gamma_2, \qquad \Gamma_2 = -\frac{i\partial_\theta g_{22}}{4\sqrt{g_{11}g_{22}}}\sigma_z. \tag{5.5}$$

Thus the Dirac operator on the torus (2.5) reads:

$$i\gamma^\mu D_\mu = \frac{i}{\sqrt{g_{11}}}\sigma_x\left(\partial_\theta + \frac{\partial_\theta g_{22}}{4g_{22}}\right) + \frac{i}{\sqrt{g_{22}}}\sigma_y\partial_\varphi. \tag{5.6}$$

If we solve spectral problem of the operator (5.6), it may be useful in the investigation of Dirac operator on fuzzy torus. However, eigenvalues and eigenfunctions of (5.6) may have its own interest. This work is under progress.

Journal of Physics: Conference Series **128** (2008) 012047 doi:10.1088/1742-6596/128/1/012047

6. Concluding remarks

We introduced a q-parafermionic viewpoint to fuzzy tours. This helps us to do systematic study of Hermitian representations of the fuzzy torus algebra. It was observed that the fuzzy torus algebra have both finite and infinite dimensional representations. Remarkably, depending on the region of μ, there is a substantial difference in possible representations. The infinite dimensional representations existing for $\mu > 0$ were used to define coherent states. The coherent sates lead to the introduction of new deformation of the Bessel functions. Investigation of the deformed Bessel functions may have its own interest in the context of possible generalization of the known special functions and their relations to algebraic structures. We want to define Dirac operator on the fuzzy torus. The task, however, seems to have some difficulty, since Dirac operator in commutative limit has enough complicated form as seen in §5. This reflects the structure of the torus polynomial (2.5). One can take another polynomial description for torus which admits Dirac operator of simpler form. For instance, one may take as follows:

$$(x^2 + y^2 + z^2 + \mu^2 - c)^2 = 4\mu^2(x^2 + y^2). \tag{6.1}$$

Applying the quantization procedure of §2 to (6.1), we obtain an another fuzzy torus algebra. However, the algebra is very messy in form and its representations may not be easily obtained.

Similar difficulty arise when we go up to higher genus algebras. Even the genus two case, the fuzzy analogue is defined by the algebra of high complexity. What we learnt from the present study is that in order to study properties and representations of fuzzy algebras for higher genus, it may be helpful to map the algebras to something else such as higher order parafermions.

Acknowledgments

The work of N.A. is partially supported by the grants-in-aid from JSPS, Japan (Contract No. 18540380). Other author (R.C.) is partially supported by DAE (BRNS), Government of India.

References

[1] Balachandran A P, Kürkçüoğlu S and Vaidya S 2005 *Lectures on Fuzzy and Fuzzy SUSY Physics, Preprint* hep-th/0511114
[2] Connes A 1994 *Noncommutative Geometry* (Academic Press)
[3] Madore J 1992 *Class. Quant. Grav.* **9** 69
[4] Balachandran A P, Dolan B P , Lee J, Martin X and O' Connor D 2002 *J. Geom. Phys.* **43** 184
[5] Dolan B P and O'Connor D 2003 *JHEP* **0310** 060
[6] Abe Y 2004 *Phys. Rev.* **D70** 126004
[7] Medina J and O'Connor 2003 *JHEP* **0311** 051
[8] Gross H, Klimčík C and Prešnajder P 1996 *Comm. Math. Phys.* **180** 429
[9] Lozano Y and Rodoríguez-Gómez D 2005 *JHEP* **0508** 044
[10] Martin X 2004 *Fuzzy orbifolds, Preprint* hep-th/0405060]
[11] Sämann C 2006 *Fuzzy toric geometries, Preprint* hep-th/0612173
[12] Grosse H and Reiter G 1998 *J. Geom. Phys.* **28** 349
[13] Ivanov E, Mezincescu L and Townsend P K 2003 *Fuzzy* $\mathbb{C}P^{(n|m)}$ *as a quantum superspace, Preprint* hep-th/0311159
[14] Bordemann M, Brischle M, Emmrich C and Waldmann S 1996 *Lett. Math. Phys.* **36** 357
[15] Dolan B P and Jahn O 2003 *Int. J. Mod. Phys.* **A18** 1935
[16] Dolan B P, O'Connor D and Prešnajder P 2004 *JHEP* **0402** 055
[17] Lizzi F, Vitale P and Zampini A 2003 *JHEP* **0308** 057
[18] Murray S and Sämann C 2006 *Quantization of flag manifolds and their supersymmetric extensions, Preprint* hep-th/0611328
[19] Arnlind J, Bordemann M, Hofer L, Hoppe J and Shimada H 2006 *Fuzzy Riemann surfaces, Preprint* hep-th/0602290
[20] Nakayama 2006 *Phys. Lett.* **B 638** 283
[21] Palev T D 1994 *Lett. Math. Phys.* **31** 151
[22] Floreanini R and Vinet L 1990 *J. Phys.* **A**:*Math. Gen.* **23** L1019

Journal of Physics: Conference Series **128** (2008) 012047 doi:10.1088/1742-6596/128/1/012047

[23] Celeghini E, Palev T D and Tarlini M 1991 *Mod. Phys. Lett.* **B5** 187
[24] Palev T D 1993 *J. Phys.* **A**:*Math. Gen.* **26** L1111
[25] Hadjiivanov L K 1993 *J. Math. Phys.* **34** 5476
[26] Beckers J and Debergh N 1991 *J. Phys.* **A**:*Math. Gen.* **24** L1277
[27] Gross H and Prešnajder P 1993 *Lett. Math. Phys.* **28** 239
[28] Alexanian G, Pinzul A and Stern A 2001 *Nucl. Phys.* **B600** 531
[29] Aizawa N and Chakrabarti R 2007 *J. Phys.* **A**:*Math. Theor.* **40** 10021
[30] Ohnuki Y and Kamefuchi S 1982 *Quantum Field Theory and Parastatistics* (University of Tokyo Press)
[31] Biedenharn L C and Louck J D 1971 *Anal. Phys.* **63** 459
[32] Gross H and Prešnajder P 1995 *Lett. Math. Phys.* **33** 171
[33] Carow-Watamura U and Watamura S 1997 *Comm. Math. Phys.* **183** 365
[34] Carow-Watamura U and Watamura S 2000 *Comm. Math. Phys.* **212** 395
[35] Balachandran A P and Immirzi G 2003 *Phys. Rev.* **D68** 065023

Journal of Physics: Conference Series **128** (2008) 012048 doi:10.1088/1742-6596/128/1/012048

Accelerated path integral calculations for many-body systems

A Balaž, I Vidanović and A Bogojević

Scientific Computing Laboratory, Institute of Physics, Belgrade, Serbia

E-mail: antun@phy.bg.ac.yu

Abstract. We present a generalization of a recently developed method for accelerated Monte Carlo calculation of path integrals of generic many particle systems. Using this method and the derived hierarchy of effective actions we calculate the energy spectra of a two particle model with quartic interaction for several values of coupling and demonstrate agreement with analytical results governing the increase in efficiency of the new method.

Keywords: Path integral, Effective action, Many-body systems

1. Introduction

Path integrals present a compact formalism for dealing with quantum theories [1, 2, 3]. In analytical approaches, path integrals allow for a generalization and extension of quantization methods to more complex physical systems, including systems with no classical counterparts [4]. Path integrals give an ideal mathematical framework for dealing with symmetries (e.g. gauge symmetry, symmetry breaking, anomalies), and for deriving nonperturbative results [5, 6]. Originally introduced in quantum mechanics and later most widely used in high energy theory, path integrals are today used in almost all areas of physics, ranging from the condensed matter, atomic, molecular and nuclear physics, to the physics of polymers, biophysics, and chemistry. Moreover, path integrals are starting to play important roles in several areas of mathematics and in modern finance [7], especially in option pricing applications [8, 9]. The downside of the formalism is that the mathematical properties of path integrals are not sufficiently well understood, and that an extremely small number of path integrals can be solved exactly [10]. An extensive overview of the path integral formalism and its various applications can be found in [11]. Although functional formalism has been used to derive many general approximation techniques (e.g. perturbation theory, semiclassical expansion, variational methods) along with a host of model-specific approximation techniques, many interesting models cannot be analyzed analytically in sufficient detail, with required precision, or in a given sector, and need to be treated numerically.

The definition of path integrals as a limit of multiple integrals over a discretized theory makes their numerical evaluation quite natural. The most applicable and efficient numerical method for such calculations is based on various Monte Carlo (MC) techniques [12]. Path Integral Monte Carlo (PIMC) simulations represent a general tool for numerical evaluation of properties of complex quantum and statistical systems. PIMC techniques dominate over other numerical methods for calculating integrals of high dimensionality. Reviews of some key PIMC variants (Metropolis, Green's function MC, Diffusion MC, Fourier PIMC) can be found in refs. [12, 13]. Notwithstanding the successes of these modern numerical integration methods, path integrals remain notoriously demanding of computing. This presents the main obstacle for tackling many important physical problems. The slow convergence of numerical methods is the direct result of our limited analytical knowledge of the properties of path integrals. Substantial new speedup can only follow from key new analytical input.

Considerable research effort has been devoted to the development of approaches that will enable faster numerical convergence to the continuum. Efficient implementation of PIMC algorithms, coupled with various physical system approximations, has resulted in important advances in the application of path integrals to real-world problems [14, 15, 13]. Another line of research has been the construction of discretized effective actions that speed up convergence of discretized physical quantities to the continuum limit. This has been done through the application of generalized Trotter formulas [16, 17, 18, 19], improvements in short-time propagation [20, 21, 22], and the application of various model-specific improvements to the action [23, 24]. A comparison of several of these approaches is given in ref. [25].

A recently developed method for analytical construction of improved discretized effective actions [26, 27, 28], based on the study of relationship between discretizations of different coarseness [29, 30], has led to substantial speedup of numerical path-integral calculations of several orders of magnitude for transition amplitudes, partition functions and expectation values. The discretized effective actions within this approach are labeled by a level number p, leading to $1/N^p$ convergence to the continuum limit, where N is the discretization coarseness. Up to now, the method has been limited to one-particle one-dimensional systems. The key new step that we present in this paper is the generalization of this formalism to generic non-relativistic many-particle systems in arbitrary dimensions.

The organization of the paper is as follows: Section 2 introduces the hierarchy of effective discretized actions and gives a brief overview of previously obtained results. The new and extended (many particles, higher dimensions) analytical approach to the construction of these effective actions is presented in Section 3. Section 4 presents numerical results that demonstrate the validity of the analytically derived speedup in convergence. The efficiency of the new approach is further demonstrated in Section 5 through the calculation of energy spectra of several models, which extend the results previously obtained in refs. [31, 32]. We conclude the paper with a brief summary of obtained results, indicating what we see to be the next steps in our line of research and its future applications. The Appendix gives the level $p = 5$ discretized effective action, leading to $1/N^5$ convergence to the continuum limit for path integral calculations of general quantum many-body systems in arbitrary number of dimensions. Higher level effective actions have been derived and are available on our web site [33].

2. Effective actions

We present a method for systematically increasing the efficiency of PIMC calculations of a general quantum theory. We illustrate the details of the derivation on the case of a non-relativistic quantum system consisting of M distinguishable particles in d spatial dimensions, with Hamiltonian of the form $\hat{H} = \hat{K} + \hat{V}$, where $\hat{K}$ is the usual kinetic energy operator, and $\hat{V}$ is the potential describing interparticle interactions and interactions with external fields. The above set of theories encompasses a large number of physically relevant models. However, as we shall see from the derivations, the method is in principle applicable to all quantum theories.

In the path integral formalism, the imaginary time amplitude $A(a, b; T)$ for a transition from initial state[1] $|a\rangle$ to final state $|b\rangle$ in time T is given as the $N \to \infty$ limit of the discretized

[1] The set of positions of all particles in coordinate representation is described by an Md dimensional vector.

Journal of Physics: Conference Series **128** (2008) 012048 doi:10.1088/1742-6596/128/1/012048

amplitude[2]

$$A_N(a,b;T) = \frac{1}{(2\pi\varepsilon)^{\frac{MNd}{2}}} \int \mathrm{d}q_1 \cdots \mathrm{d}q_{N-1}\; e^{-S_N} \,. \qquad (1)$$

In this expression, N is the discretization coarseness (number of time slices), while S_N is the naively discretized action,

$$S_N = \sum_{n=0}^{N-1} \varepsilon \left(\sum_{i=1}^{M} \frac{1}{2} \left(\frac{\delta_{n,i}}{\varepsilon} \right)^2 + V(\bar{q}_n) \right) . \qquad (2)$$

Vectors with one index (e.g. q_n) represent the set of positions of all particles after n time steps $\varepsilon = T/N$, while vectors with two indices (e.g. $q_{n,i}$) represent d-dimensional positions of particle i at time step $n\varepsilon$. We have also introduced discretized velocities $\delta_{n,i} = q_{n+1,i} - q_{n,i}$ and mid-point coordinates $\bar{q}_n = (q_n + q_{n+1})/2$. The $N \to \infty$ limit of the above discretized amplitude gives the continuum amplitude $A(a,b;T)$ in the mid-point prescription, symbolically written as

$$A(a,b;T) = \int_{q(0)=a}^{q(T)=b} [\mathrm{d}q]\; e^{-S[q(t)]} \,. \qquad (3)$$

The above definition of path integral requires the transition from continuum to discretized theory, i.e. the introduction of coarseness N. Although well suited for numerical calculations, expressions given by equation (1) converge to the continuum very slowly, typically as $O(1/N)$, and it is quite natural to look for a discretization scheme leading to better convergence. It was shown in [29] that there is an ideal discretization (determined by the ideal discretized action S^*), giving for a general theory the exact (continuum limit) result for *any* discretization coarseness N

$$A_N^*(a,b;T) = A(a,b;T) \,.$$

Note that for free particles S^* is in fact just the naively discretized action.

The defining relation for path integrals as the continuum limit of discretized amplitudes given by equation (1) follows from the completeness relation (decomposition of unity)

$$A(a,b;T) = \int \mathrm{d}q_1 \cdots \mathrm{d}q_{N-1}\; A(a,q_1;\varepsilon) \cdots A(q_{N-1},b;\varepsilon) \,, \qquad (4)$$

through the substitution of short-time amplitudes $A(q_n, q_{n+1}; \varepsilon)$ calculated to first order in time step ε. A faster converging result may be obtained by evaluating the amplitudes under the integral to higher orders in ε. From the above completeness relation, it follows that the ideal discretized action S^* leads to exact propagation in time, and is given in terms of the exact amplitude, according to

$$A(q_n, q_{n+1}; \varepsilon) = (2\pi\varepsilon)^{-\frac{Md}{2}}\; e^{-S_n^*} \,. \qquad (5)$$

The ideal discretized action S^* is simply the sum of expressions S_n^*.

The knowledge of the ideal action S^* is equivalent to the knowledge of the exact expression for the amplitude A. At first, this would seem to indicate that nothing is to be gained by equation (5). This is, however, not the case. We will use equation (5) to input new analytical information into our numerical procedure by calculating amplitudes within some available analytical approximation scheme. For example, short time propagation can be calculated as a power series expansion in ε, which starts from the naive action (2). The details of this derivation have been inspired by [11] and are presented in Section 3. One can also attack the

[2] Throughout the paper, we use system of units in which $\hbar$ and masses of all particles are set to unity. If needed, those quantities can be easily recovered in all expressions using dimensional analysis.

problem of solving equation (5) using various other approximative schemes, especially in the case when short time approximation is not appropriate. The application of the Feynman and Kleinert variational approach [34, 35], further developed in [36, 37, 38, 39, 40], could prove very useful is such cases.

We can look at the problem of speeding up the numerical convergence of discretized path integral expressions from another point of view. Purely considering the discretization procedure, we can immediately conclude that (2) is not unique, and that the freedom we have in the discretization procedure can be effectively used to speed up numerical convergence of discretized path integral calculations.

The first type of freedom is associated with the prescription used. For the class of theories considered, i.e. theories without ordering ambiguities, we have the freedom to choose the points in which to evaluate the potential V. For example, we could choose the point $\alpha q_n+(1-\alpha)q_{n+1}$ for any $\alpha\in[0,1]$. As is well known, different choices correspond to different ordering prescriptions in the operator formalism. The choice of the middle point $\bar{q}_n$ ($\alpha=1/2$) is the most common one. It corresponds to the symmetric or Weyl ordering of operators $\hat{p}$ and $\hat{q}$, and so always leads to a hermitian expression for the hamiltonian $\hat{H}$. Two other prescriptions are also often used. The left ordering prescription evaluates the potential at q_n ($\alpha=1$), the left boundary of the interval $[q_n,q_{n+1}]$ (in the operator formalism this corresponds to taking the $\hat{p}$'s to the left of the $\hat{q}$'s in all the products that appear in the hamiltonian). Similarly, one defines the right ordering prescription ($\alpha=0$). Although they lead to somewhat simpler looking expressions, the left and right prescriptions do not in general give hermitian hamiltonians. However, for the considered class of theories, all prescriptions lead to the same continuum amplitude – the discretized amplitudes do differ, but they tend to the same continuum limit. However, as we shall see at the end of this section, the choice of prescription does somewhat affect the speed of convergence.

The second, and more important freedom related to our choice of discretized action has to do with the introduction of additional terms that explicitly vanish in the continuum limit. In contrast to the first type of freedom in the discretization procedure which is present only if there are no ordering ambiguities, this second type of freedom exist for all theories. This freedom is much wider than the first one, allowing for a systematic improvement of numerical convergence of path integral calculations, and the construction of a hierarchy of discretized actions $S_N^{(p)}$, denoted by level number p, giving improved convergence

$$A_N^{(p)}(a,b;T)=A(a,b;T)+O(1/N^p). \qquad (6)$$

We designate such improved discretized actions as effective actions. For example, the term

$$\sum_{n=0}^{N-1}\sum_{i=1}^{M}\varepsilon\,\delta_{n,i}^2\,g(\bar{q}_{n,i})\,,$$

where g is regular function when $\varepsilon\to 0$, does not change the continuum physics since it goes over into

$$\varepsilon^2\sum_{i=1}^{M}\int_0^T \mathrm{d}t\,\dot{q}_i^2\,g(q_i)\,,$$

i.e. it vanishes as ε^2. Although such additional terms do not change the continuum physics they do affect the speed of convergence to that continuum limit.

In a previous series of publications [26, 27, 28] we had constructed the effective actions $S^{(p)}$ through a study of the relation between discretizations of different coarseness. This investigation was limited to one-dimensional one-particle models with Hamiltonians of the form

$\hat{H} = \frac{1}{2}\hat{p}^2 + V(\hat{q})$. In the next section we present a new, more general approach which can be applied to more complex systems (many particles, higher dimensions, general Hamiltonian). Where applicable, the two methods are equivalent: $S^{(p)}$ is just the solution of equation (5) truncated to ε^p.

Before going into further details about the new method, we give a brief overview of earlier results related to the speeding up of the convergence of path integral calculations. Key results regarding numerical evaluation of path integrals are presented in the reviews of Barker and Henderson [14], Kalos and Whitlock [12], and Ceperley [13], as well as in the papers by Pollock and Ceperley [15]. More recent results regarding the efficient implementation of PIMC algorithms are available in refs. [41, 42, 43, 44, 45]

As already stated, the mid-point prescription using the naively discretized action gives

$$A_N(a,b;T)^{mid} = A(a,b;T) + O(1/N)\,,$$

for all a and b. On the other hand, in the naive left prescription the diagonal amplitude for $a \to a$ transitions converges much faster

$$A_N(a,a;T)^{left} = A(a,a;T) + O(1/N^2)\,.$$

This behavior can be easily shown both analytically and numerically. We note in passing that this is strongly related to the well known result for the partition function evaluated using naive discretization in the left prescription [25], which follows directly from the above amplitude by integrating over a (to get the trace) and writing the time of propagation T as the inverse temperature β,

$$Z_N(\beta)^{left} = Z(\beta) + O(1/N^2)\,.$$

However, going back to the language of amplitudes, it is also easy to show that the amplitudes for different initial and final states converge slower, i.e. for $a \neq b$ we have again

$$A_N(a,b;T)^{left} = A(a,b;T) + O(1/N)\,.$$

Note that the $1/N^2$ speed of convergence of off-diagonal amplitudes in the left prescription can be recovered very easily. It was shown in ref. [27] that for all a and b we have

$$\frac{1 + e^{\varepsilon(V(a)-V(b))}}{2}\, A_N(a,b;T)^{left} = A(a,b;T) + O(1/N^2)\,.$$

For a long time the state of the art result was $1/N^4$ convergence of discretized partition functions obtained by Takahashi and Imada [17], and Li and Broughton [18]. In these papers the authors used a generalized form [16] of the Trotter formula to increase the speed of convergence of the discretized partition function. Their final result is a derivation of a formula for the effective potential

$$V^{eff} = V + \frac{1}{24}\varepsilon^2 (V')^2\,.$$

The authors showed that by using this effective potential (in the left prescription) one gets

$$Z_N(\beta)^{eff} = Z(\beta) + O(1/N^4)\,.$$

Let us note that the crucial step in the derivation of the above effective potential from the generalized Trotter formula uses the cyclic property of the trace, i.e. the above increase in the speed of convergence only holds for the partition function and not the amplitudes. A direct numerical simulation shows that amplitudes calculated using this effective potential converge just as fast as the amplitudes in the naive left prescription, i.e. as $1/N$. Said another way, it is only the integral over all the diagonal amplitudes that has the $O(1/N^4)$ behavior and not any individual amplitude.

Journal of Physics: Conference Series **128** (2008) 012048 doi:10.1088/1742-6596/128/1/012048

3. New approach

As already outlined, the standard definition of path integrals as limits of time-sliced discretized expressions is based on the completeness relation (4). As we have seen, the next step is the approximate calculation of short-time amplitudes $A(q_n, q_{n+1}; \varepsilon)$ as a power series in ε. The use of the naive action in equation (1) corresponds to keeping only the dominant term for the short-time amplitude. This is what gives the $1/N$ convergence to standard path integral expressions[3]. Calculating the short time amplitudes exactly we find

$$A(q_n, q_{n+1}; \varepsilon) = (2\pi\varepsilon)^{-\frac{Md}{2}}\, e^{-S_n^*(q_n, q_{n+1}; \varepsilon)}, \tag{7}$$

where the ideal discretized effective action equals

$$S_N^* = \sum_{n=0}^{N-1} S_n^*(q_n, q_{n+1}; \varepsilon)\,.$$

$1/N^p$ convergence of path integrals is obtained by evaluating amplitudes up to order ε^p,

$$A(q_n, q_{n+1}; \varepsilon) = (2\pi\varepsilon)^{-\frac{Md}{2}} \left(e^{-S_n^{(p)}(q_n, q_{n+1}; \varepsilon)} + O(\varepsilon^{(p+1)}) \right). \tag{8}$$

Inserting the above result into the formula for S_N^*, we obtain the level p effective action $S_N^{(p)}$ as

$$S_N^{(p)} = \sum_n S_n^{(p)}(q_n, q_{n+1}; \varepsilon)\,. \tag{9}$$

A simple shift of integration variable $q = \xi + x$ about a fixed referent trajectory ξ gives

$$A(q_n, q_{n+1}; \varepsilon) = e^{-S_n[\xi]} \int_{x(-\varepsilon/2)=0}^{x(\varepsilon/2)=0} [\mathrm{d}x]\; e^{-\int_{-\varepsilon/2}^{\varepsilon/2} \mathrm{d}s \left(\frac{1}{2}\dot{x}^2 + U(x;\xi)\right)}\,. \tag{10}$$

Note that we have also shifted the time from $t \in [n\varepsilon, (n+1)\varepsilon]$ to $s \in [-\varepsilon/2, \varepsilon/2]$. The referent trajectory ξ satisfies the same boundary conditions as q. As a result, the new integration variable x vanishes at the boundaries. The action $S_n[\xi]$ is defined as

$$S_n[\xi] = \int_{-\varepsilon/2}^{\varepsilon/2} \mathrm{d}s \left(\frac{1}{2}\dot{\xi}^2 + V(\xi) \right),$$

and $U(x;\xi) = V(\xi + x) - V(\xi) - x\ddot{\xi}$, with dots representing derivatives over time s. The amplitude may now be written as

$$A(q_n, q_{n+1}; \varepsilon) = \frac{e^{-S_n[\xi]}}{(2\pi\varepsilon)^{\frac{Md}{2}}} \left\langle e^{-\int_{-\varepsilon/2}^{\varepsilon/2} \mathrm{d}s\; U(x;\xi)} \right\rangle, \tag{11}$$

where $\langle ... \rangle$ denotes the expectation value with respect to the free massless particle action. The above expression holds for any choice of referent trajectory ξ.

The class of theories considered is free of ordering ambiguities, i.e. different ordering prescriptions yield the same continuum results. We will present a derivation of the discretized effective actions in the midpoint prescription. We stress that the same calculation can be done just as easily in any other ordering prescription. Since we are using series expansion in small

[3] When calculating partition functions in the left prescription one has an additional symmetry which gets rid of odd powers in ε. As a result, in this case naive discretization converges as $1/N^2$.

Journal of Physics: Conference Series **128** (2008) 012048 doi:10.1088/1742-6596/128/1/012048

time step ε, to retain all terms of the desired order it is necessary to take into account that the short time propagation of the considered class of theories satisfies the diffusion relation $\delta^2_{n,i} \propto \varepsilon$. Therefore, keeping all terms proportional to $\varepsilon^k \delta^{2l}_{n,i}$ with $k + l \leq p$ will be sufficient for obtaining the desired precision.

The free particle expectation value in equation (11) is calculated using the series expansion

$$\left\langle e^{-\int \mathrm{d}s\, U(x;\xi)} \right\rangle = 1 - \int \mathrm{d}s\, \langle U(x;\xi) \rangle + \frac{1}{2} \int\int \mathrm{d}s\mathrm{d}s' \langle U(x;\xi) U(x';\xi') \rangle + \ldots ,$$

with shorthand notation $x' = x(s'), \xi' = \xi(s')$. By expanding $U(x;\xi)$ around the referent trajectory ξ, we get

$$U(x;\xi) = x_i(\partial_i V(\xi) - \ddot{\xi}_i) + \frac{1}{2} x_i x_j \partial_i \partial_j V(\xi) + \ldots . \tag{12}$$

From now on, indices i and j go over all Md components of positions of particles, and we assume summation over repeated indices. Expectation values of products $\langle x_i(s) \ldots x_j(s') \rangle$ can be calculated in the usual way, by introducing a generating functional for the free-particle theory whose propagator is given by:

$$\begin{aligned} \Delta(s,s')_{i,j} = & \frac{\delta_{ij}}{\varepsilon} \theta(s - s') \left(\frac{\varepsilon}{2} - s\right) \left(\frac{\varepsilon}{2} + s'\right) \\ + & \frac{\delta_{ij}}{\varepsilon} \theta(s' - s) \left(\frac{\varepsilon}{2} + s\right) \left(\frac{\varepsilon}{2} - s'\right) . \end{aligned} \tag{13}$$

Using Wick's theorem, one easily finds $\langle x_i(s) \rangle = 0$, $\langle x_i(s) x_j(s') \rangle = \Delta(s,s')_{i,j}$, etc. Note that the calculation of the generating functional (and also of the expectation values) is the same, irrespective of the choice of ξ. In all cases the action and the boundary conditions for the field x are the same, and so the propagator is always given by (13). However, different choices of ξ are related to different approximation techniques: the choice of classical trajectory for ξ corresponds to the semiclassical expansion, while the choice of a linear referent trajectory leads to short-time expansion.

In order to perform the remaining integrations over s in (11), due to the explicit dependence of the referent trajectory on s, we first expand the potential and all its derivatives in (12) around some reference point. We choose $\bar{q}_n$ as that reference point, corresponding to the mid-point prescription. Once one chooses the referent trajectory $\xi(s)$, all expectation values in (11) are given in terms of quadratures.

By choosing linear referent trajectories $\xi(s) = \bar{q}_n + \frac{\delta_n}{\varepsilon} s$ and calculating up to terms of order ε^2 (level $p = 2$), we obtain:

$$S_n[\xi] = \varepsilon \left(\frac{1}{2} \frac{\delta_n^2}{\varepsilon^2} + V(\bar{q}_n) + \frac{\delta_{n,i}\delta_{n,j}}{24} \partial^2_{i,j} V(\bar{q}_n) \right) + O(\varepsilon^3) , \tag{14}$$

$$\left\langle e^{-\int \mathrm{d}s\, U(x;\xi)} \right\rangle = 1 - \frac{\varepsilon^2}{12} \Delta V(\bar{q}_n) + O(\varepsilon^3) = e^{-\frac{\varepsilon^2}{12}\Delta V(\bar{q}_n)} + O(\varepsilon^3) . \tag{15}$$

From these expressions we see that the level $p = 2$ discretized effective action is

$$S_N^{(p=2)} = \sum_{n=0}^{N-1} \varepsilon \left(\frac{1}{2} \left(\frac{\delta_n}{\varepsilon} \right)^2 + V(\bar{q}_n) + \frac{\varepsilon}{12} \Delta V(\bar{q}_n) + \frac{\delta_{n,i}\delta_{n,j}}{24} \partial^2_{i,j} V(\bar{q}_n) \right) . \tag{16}$$

In the above equation, the Laplacian operator consists of the sum of second partial derivatives over all Md components of vectors of particle positions. The explicit analytical expression for

Journal of Physics: Conference Series **128** (2008) 012048 doi:10.1088/1742-6596/128/1/012048

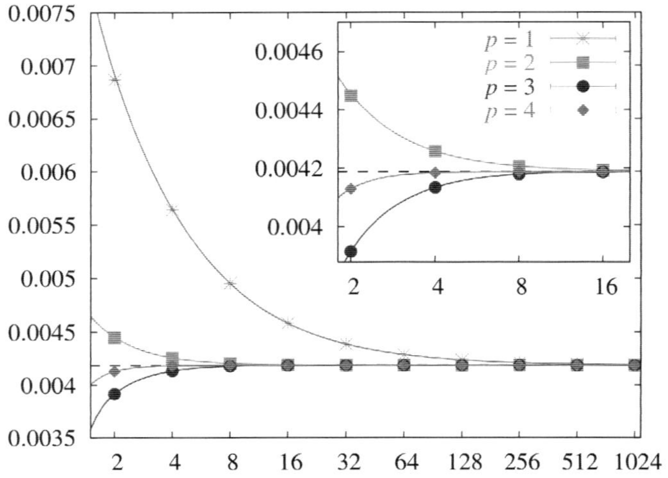

Figure 1. Convergence of the discretized amplitudes $A_N^{(p)}$ to the continuum as functions of N for $p = 1, 2, 3, 4$ for the system of two particles in two dimensions in a quartic potential (17), $g_1 = 10$, $g_2 = 0$, with time of propagation $T = 1$ and initial and final states $a = (0, 0; 0.2, 0.5)$, $b = (1, 1; 0.3, 0.6)$. The number of MC samples was 10^6. Horizontal black line represents the continuum limit, while solid lines correspond to the fitted functions given by (18).

the many-particle discretized effective action at level $p = 5$ is given in the Appendix. There are no obstacles in going to higher values of p. The derived expressions become algebraically more complex, and the calculations are best conducted using some of the available packages for symbolic calculus (e.g. Mathematica). Higher level effective actions can be found on our web site [33].

4. Numerical results

In this section we present results of numerical PIMC simulations that confirm the analytically derived speedup in convergence of discretized path integrals. To do this, we have conducted a series of PIMC simulations of transition amplitudes for a two-dimensional system of two particles interacting through potential

$$V(\vec{r}_1, \vec{r}_2) = \frac{1}{2}(\vec{r}_1 - \vec{r}_2)^2 + \frac{g_1}{24}(\vec{r}_1 - \vec{r}_2)^4 + \frac{g_2}{2}(\vec{r}_1 + \vec{r}_2)^2 . \tag{17}$$

Numerical simulations, based on our SPEEDUP [33] PIMC code, have been performed for different values of couplings g_1 and g_2 and for variety of initial and final states. The associated continuum limit amplitudes $A^{(p)}$ have been estimated by fitting polynomials in $1/N$ to the discretized values $A_N^{(p)}$, according to the analytically derived relation (6):

$$A_N^{(p)} = A^{(p)} + \frac{B^{(p)}}{N^p} + \frac{C^{(p)}}{N^{p+1}} + \ldots . \tag{18}$$

For all values of p the fitted continuum values $A^{(p)}$ agree within the error bars. The obtained $1/N$ dependence gives explicit verification of the analytically derived increase in convergence. The typical case is illustrated in Figure 1. The expected $1/N^p$ convergence may be most easily discerned from Figure 2, the plot of the deviations of discretized amplitudes from the continuum limit. As can be seen, the increase of level p leads to an ever faster approach to the continuum.

As a result of the newly presented method, the usual simulations (in which one calculates specific physical quantities such as the one in Figure 1) proceed much faster than by using standard methods. On the other hand, Figure 2 is itself time consuming since it illustrates subdominant behavior. For this reason the figures contain only results obtained by effective action to level $p = 4$. Note, however, that the $p = 4$ curve corresponds to a precision of four decimal places even for an extremely coarse discretization such as $N = 2$.

In usual PIMC calculations one always chooses the number of MC samples so that the stochastic error of numerical results is of the order of deviations from the continuum limit. In Figure 2 the number of MC samples had to be much larger, in order for deviations from the continuum limit to be clearly visible.

5. Energy spectra

The improved convergence of path integral expressions for amplitudes leads directly to the same kind of improvement in the convergence of discretized partition functions, owing to the relation

$$Z_N(\beta) = \int \mathrm{d}q A_N(q, q; \beta)\,, \tag{19}$$

where the inverse temperature β plays the role of the time of propagation T. The partition function is the central object for obtaining information about statistical systems. In addition, the partition function offers a straightforward way for extracting information about low lying energy levels of a system. This follows from evaluating the trace in the definition of the partition function in the energy eigen-basis

$$Z(\beta) = \sum_{n=0}^{\infty} d_n e^{-\beta E_n}\,, \tag{20}$$

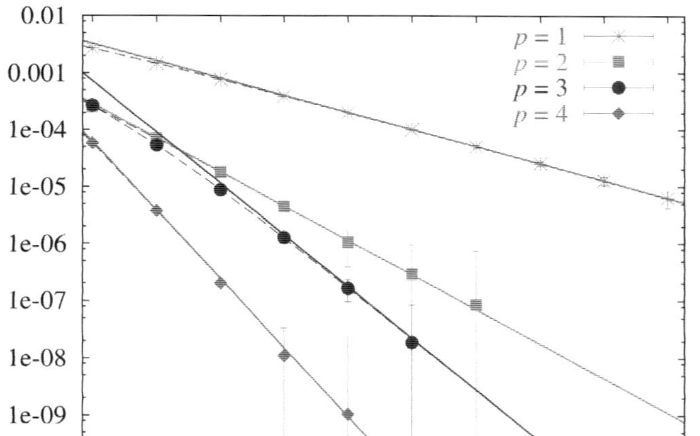

Figure 2. Deviations from the continuum limit $|A_N^{(p)} - A|$ as functions of N for $p = 1, 2, 3, 4$ for the system of two particles in two dimensions in quartic potential $g_1 = 10$, $g_2 = 0$, with time of propagation $T = 1$ and initial and final states $a = (0, 0; 0.2, 0.5)$, $b = (1, 1; 0.3, 0.6)$. The number of MC samples was 10^6 ($p = 1$), 10^7 ($p = 2$), 10^9 ($p = 3$), and 10^{10} ($p = 4$). Solid lines give the leading $1/N^p$ behavior, while dashed lines correspond to the fitted functions given by (18).

Journal of Physics: Conference Series **128** (2008) 012048 doi:10.1088/1742-6596/128/1/012048

where E_n and d_n denote corresponding energy levels and degeneracies. A detailed description of the procedure for extracting energy levels from the partition function is given in [31, 32]. The free energy of the system, given by

$$F(\beta) = -\frac{1}{\beta} \ln Z(\beta)\,, \tag{21}$$

tends to the ground state energy E_0 in the large β limit. Similarly, we introduce auxiliary functions

$$F^{(n)}(\beta) = -\frac{1}{\beta} \ln \frac{Z(\beta) - \sum_{i=0}^{n-1} d_i\, e^{-\beta E_i}}{d_n}\,, \tag{22}$$

which can be fitted for large β to

$$f^{(n)}(\beta) = E_n - \frac{1}{\beta} \ln(1 + ae^{-\beta b})\,. \tag{23}$$

and which tend to the corresponding energy level E_n.

In this way, by studying the large β behavior of the free energy and the corresponding auxiliary functions, one obtains the low lying energy spectrum of the model. Note that the procedure for the construction of the auxiliary function $f^{(n)}$ is recursive, i.e. in order to construct $f^{(n)}$ we need to now all the energy levels below E_n. This leads to the accumulation of errors as n increases, and limits the number of levels that can be calculated. The practical limit on how high in the spectrum we can go depends on the precision with which we calculate path integrals. The orders of magnitude increase in precision of the presented method for calculating path integrals can therefore be used to extract numerical information about a larger number of energy levels.

Figures 3 and 4 illustrate the calculation of low lying energy levels using the more efficient PIMC formalism presented in this paper on the case of a two-dimensional system of two distinguishable particles interacting through a quartic potential (17).

Figure 3 demonstrates that the improved convergence of free energies is the same as in the case of amplitudes. Figure 4 illustrates the calculation of the low lying energy levels of the

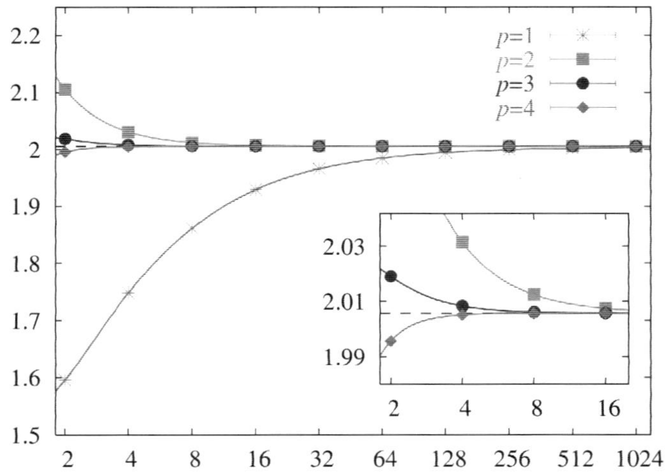

Figure 3. Convergence of discretized free energies $F_N(\beta)$ to the continuum as functions of N for $p = 1, 2, 3, 4$ for the system of two particles in two dimensions in a potential (17), $g_1 = 1$, $g_2 = 1$, $\beta = 1$. The number of MC samples was 10^7.

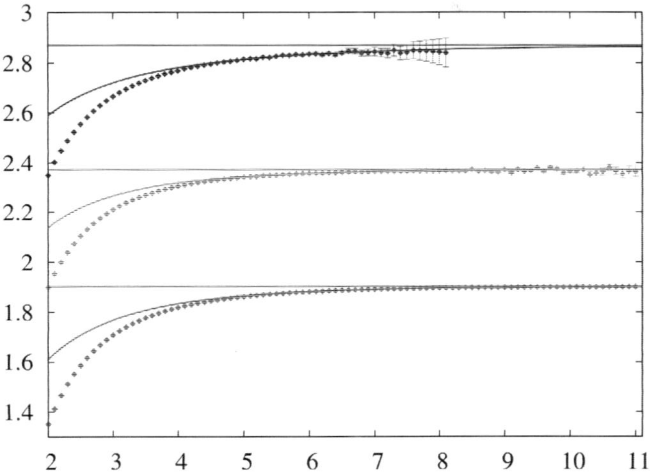

Figure 4. Dependence of the free-energy and auxiliary functions $f^{(1)}$ and $f^{(2)}$ on β for the system in the potential (17), $g_1 = 0$, $g_2 = 1/9$. Horizontal lines correspond to energy levels. Numerical simulations were performed with 10^9 MC samples, level $p = 5$ effective action and $N = 64$.

model. Table 1 gives the calculated energy levels for quartic coupling from $g_1 = 0$ (free theory) to $g_1 = 10$ (strongly interacting theory).

Table 1. Low lying energy levels of the system in the potential given by (17), $g_2 = 1/9$, calculated using $= 10^9$ MC samples, level $p = 5$ effective action and $N = 64$. The degeneracies of the calculated energy levels are found to be $d_0 = 1$, $d_1 = 2$, $d_2 = 3$, $d_3 = 6$.

g_1	E_0	E_1	E_2	E_3
0	1.8857(1)	2.3571(6)	2.83(1)	3.3(2)
0.1	1.9019(2)	2.374(2)	2.82(1)	—
1	2.0228(2)	2.497(3)	2.94(3)	—
10	2.6327(6)	3.098(4)	3.57(3)	—

6. Conclusions

We have presented a derivation of discretized effective actions that lead to substantial, systematical speedup of numerical procedures for the calculation of path integrals of a generic many-particle non-relativistic theory. The derived speedup holds for all path integrals - for transition amplitudes, partition functions, expectation values [46], as well as for calculations of energy levels. The obtained analytical results have been numerically verified through simulations of path integrals for several multi-particle models. The developed calculation scheme has been completed and is ready for applications to relevant problems in condensed matter physics. Further analytical work will focus on the generalization of the outlined scheme to more complex quantum systems: bosonic and fermionic quantum field theory, gauge theories, topologically non-trivial spaces.

Acknowledgments
This work was supported in part by the Ministry of Science of the Republic of Serbia, under project No. OI141035. I. V. also acknowledges support from EU Centre of Excellence grant CX-CMCS. The presented numerical results were obtained on the AEGIS e-Infrastructure whose operation is supported in part by FP6 projects EGEE-II and SEE-GRID-2.

Appendix
The effective discretized action for a general non-relativistic multi-particle system in the mid-point prescription is, to level $p=5$, given by

$$\begin{aligned}
S_N^{p=5} &= \sum\left\{\frac{\delta_i\delta_i}{2\varepsilon^2}+V\right\}_{p=1} \\
&+ \left\{\frac{\varepsilon}{12}\Delta V+\frac{\delta_i\delta_j}{24}\partial^2_{i,j}V\right\}_{p=2} \\
&+ \left\{-\frac{\varepsilon^2}{24}\partial_iV\partial_iV+\frac{\varepsilon^2}{240}\Delta\Delta V+\frac{\varepsilon\delta_i\delta_j}{480}\partial^2_{i,j}\Delta V+\frac{\delta_i\delta_j\delta_k\delta_l}{1920}\partial^4_{i,j,k,l}V\right\}_{p=3} \\
&+ \left\{\frac{\varepsilon^3}{6720}\Delta\Delta\Delta V-\frac{\varepsilon^3}{120}\partial_iV\partial_i\Delta V-\frac{\varepsilon^3}{360}\partial^2_{i,j}V\partial^2_{i,j}V\right. \\
&- \frac{\varepsilon^2\delta_i\delta_j}{480}\partial_kV\partial^3_{k,i,j}V+\frac{\varepsilon^2\delta_i\delta_j}{13440}\partial^2_{i,j}\Delta\Delta V-\frac{\varepsilon^2\delta_i\delta_j}{1440}\partial^2_{i,k}V\partial^2_{k,j}V \\
&+ \left.\frac{\varepsilon\delta_i\delta_j\delta_k\delta_l}{53760}\partial^4_{i,j,k,l}\Delta V+\frac{\delta_i\delta_j\delta_k\delta_l\delta_m\delta_n}{322560}\partial^6_{i,j,k,l,m,n}V\right\}_{p=4} \\
&+ \left\{\frac{\varepsilon^4}{241920}\Delta\Delta\Delta\Delta V-\frac{\varepsilon^4}{1680}\partial^2_{i,j}V\partial^2_{i,j}\Delta V-\frac{17\varepsilon^4}{40320}\partial_i\Delta V\partial_i\Delta V\right. \\
&- \frac{\varepsilon^4}{2240}\partial_iV\partial_i\Delta\Delta V-\frac{\varepsilon^4}{6720}\partial^3_{i,j,k}V\partial^3_{i,j,k}V+\frac{\varepsilon^4}{240}\partial_iV\partial_jV\partial^2_{i,j}V \\
&+ \frac{\varepsilon^3\delta_i\delta_j}{483840}\partial^2_{i,j}\Delta\Delta\Delta V-\frac{\varepsilon^3\delta_i\delta_j}{6720}\partial_kV\partial^3_{i,j,k}\Delta V-\frac{\varepsilon^3\delta_i\delta_j}{10080}\partial^2_{i,k}V\partial^2_{k,j}\Delta V \\
&- \frac{\varepsilon^3\delta_i\delta_j}{10080}\partial^2_{k,l}V\partial^4_{i,j,k,l}V-\frac{\varepsilon^3\delta_i\delta_j}{5040}\partial_k\Delta V\partial^3_{i,j,k}V-\frac{\varepsilon^3\delta_i\delta_j}{20160}\partial^3_{i,k,l}V\partial^3_{k,l,j}V \\
&+ \frac{\varepsilon^2\delta_i\delta_j\delta_k\delta_l}{1935360}\partial^4_{i,j,k,l}\Delta\Delta V-\frac{\varepsilon^2\delta_i\delta_j\delta_k\delta_l}{53760}\partial_mV\partial^5_{m,i,j,k,l}V \\
&- \frac{\varepsilon^2\delta_i\delta_j\delta_k\delta_l}{40320}\partial^2_{i,m}V\partial^4_{m,j,k,l}V-\frac{\varepsilon^2\delta_i\delta_j\delta_k\delta_l}{32256}\partial^3_{i,j,m}V\partial^3_{m,k,l} \\
&+ \left.\frac{\varepsilon\delta_i\delta_j\delta_k\delta_l\delta_m\delta_n}{11612260}\partial^6_{i,j,k,l,m,n}\Delta V+\frac{\delta_i\delta_j\delta_k\delta_l\delta_m\delta_n\delta_p\delta_q}{92897280}\partial^8_{i,j,k,l,m,n,p,q}V\right\}_{p=5}
\end{aligned}$$

References

[1] Feynman R P 1948 *Rev. Mod. Phys.* **20** 367
[2] Feynman R P and Hibbs A R 1965 *Quantum Mechanics and Path Integrals* (New York: McGraw-Hill)
[3] Feynman R P 1972 *Statistical Mechanics* (New York: W. A. Benjamin)
[4] Masujima M 2000 *Path Integral Quantization and Stochastic Quantization*, (New York: Springer)
[5] Itzykson C and Zuber J.-B. 1980 *Quantum Field Theory* (New York: McGraw-Hill)
[6] Coleman S 1985 *Aspects of Symmetry* (Cambridge University Press)
[7] Baaquie B 2004 *Quantum Finance* (Cambridge University Press)
[8] Kleinert H 2002 *Physica A* **311** 536
[9] Kleinert H 2002 *Physica A* **312** 217
[10] Steiner F and Grosche C 1997 *A Table of Feynman Path Integrals* (New York: Springer)
[11] Kleinert H 2004 *Path Integrals in Quantum Mechanics, Statistics, Polymer Physics, and Financial Markets*, 3rd edition (Singapore: World Scientific)
[12] Kalos M H and Whitlock P A 1986 *Monte Carlo Methods* (New York: John Wiley and Sons)

[13] Ceperley D M 1995 *Rev. Mod. Phys.* **67** 279
[14] Barker J A and Henderson D 1976 *Rev. Mod. Phys.* **48** 587
[15] Pollock E L and Ceperley D M 1984 *Phys. Rev. B* **30** 2555
[16] De Raedt H and De Raedt B 1983 *Phys. Rev. A* **28** 3575
[17] Takahashi M and Imada M 1984 *J. Phys. Soc. Jpn.* **53** 3765
[18] Li X P and Broughton J Q 1987 *J. Chem. Phys.* **86** 5094
[19] Suzuki M 1990 *Phys. Lett. A* **146** 319
[20] Makri N and Miller W H 1988 *Chem. Phys. Lett.* **151** 1
[21] Makri N and Miller W H 1989 *J. Chem. Phys.* **90** 904
[22] Makri N 1992 *Chem. Phys. Lett.* **193** 435
[23] Alford M, Klassen T R and Lepage G P 1998 *Phys. Rev. D* **58** 034503
[24] Bond S D, Laird B B and Leimkuhler B J 2003 *J. Comp. Phys.* **185** 472
[25] Krajewski F R and Müser M H 2002 *Phys. Rev. B* **65** 174304
[26] Bogojević A, Balaž A and Belić A 2005 *Phys. Rev. Lett.* **94** 180403
[27] Bogojević A, Balaž A and Belić A 2005 *Phys. Rev. B* **72** 064302
[28] Bogojević A, Balaž A and Belić A 2005 *Phys. Lett. A* **344** 84
[29] Bogojević A, Balaž A and Belić A 2005 *Phys. Rev. E* **72** 036128
[30] Bogojević A, Balaž A and Belić A 2005 *Phys. Lett. A* **345** 258
[31] Stojiljković D, Bogojević A and Balaž A 2006 *Phys. Lett A* **360** 217
[32] Stojiljković D, Bogojević A and Balaž A 2007 Energy levels and expectation values via accelerated path integral Monte Carlo *Proceedings of the QTS-5 Conference* (Valladolid, Spain)
[33] `http://scl.phy.bg.ac.yu/speedup/`
[34] Feynman R P and Kleinert H 1986 *Phys. Rev. A* **34** 5080
[35] Kleinert H 1992 *Phys. Lett. B* **280** 251
[36] Kleinert H 1993 *Phys. Lett. A* **173** 332
[37] Kleinert H, Kürzinger W and Pelster A 1998 *J. Phys. A* **31** 8307
[38] Weissbach F, Pelster A and Hamprecht B 2002 *Phys. Rev. E* **66** 036129
[39] Pelster A, Kleinert H and Schanz M 2003 *Phys. Rev. E* **67** 016604
[40] Brandt S F and Pelster A 2005 *J. Math. Phys.* **46** 112105
[41] Predescu C 2004 *Phys. Rev. E* **69** 056701
[42] Predescu C 2005 *Phys. Rev. E* **71** 045701
[43] Predescu C 2006 *J. Phys. Chem B* **110** 667
[44] Yamamoto T M 2005 *J. Chem. Phys.* **123** 104101
[45] Paulin S, Alastuey A and Dauxois T 2007 *J. Stat. Phys.* **128** 1391
[46] Grujić J, Bogojević A and Balaž A 2006 *Phys. Lett. A* **360** 205

Journal of Physics: Conference Series **128** (2008) 012049 doi:10.1088/1742-6596/128/1/012049

Bases in Lie and Quantum Algebras

A. Ballesteros[1], **E. Celeghini**[2] **and M.A. del Olmo**[3]

[1] Departamento de Física, Universidad de Burgos, E-09006, Burgos, Spain
[2] Departimento di Fisica, Università di Firenze and INFN–Sezione di Firenze, I50019 Sesto Fiorentino, Firenze, Italy
[3] Departamento de Física Teórica, Universidad de Valladolid, E-47005, Valladolid, Spain.

E-mail: angelb@ubu.es, celeghini@fi.infn.it, olmo@fta.uva.es

Abstract.
Applications of algebras in physics are related to the connection of measurable observables to relevant elements of the algebras, usually the generators. However, in the determination of the generators in Lie algebras there is place for some arbitrary conventions. The situation is much more involved in the context of quantum algebras, where inside the quantum universal enveloping algebra, we have not enough primitive elements that allow for a privileged set of generators and all basic sets are equivalent. In this paper we discuss how the Drinfeld double structure underlying every simple Lie bialgebra characterizes uniquely a particular basis without any freedom, completing the Cartan program on simple algebras. By means of a perturbative construction, a distinguished deformed basis (we call it the analytical basis) is obtained for every quantum group as the analytical prolongation of the above defined Lie basis of the corresponding Lie bialgebra. It turns out that the whole construction is unique, so to each quantum universal enveloping algebra is associated one and only one bialgebra. In this way the problem of the classification of quantum algebras is moved to the classification of bialgebras. In order to make this procedure more clear, we discuss in detail the simple cases of $su(2)$ and $su_q(2)$.

1. Introduction

Algebras are, of course, the same object in mathematics and physics. However, from the mathematical viewpoint the interest is focused on the general description of the structure as abstract as possible, while for physical applications the interest is focused on the basic elements of the algebra to which some precise physical meaning is attributed.

In this "physically" motivated work we summarize a research line devoted to individuate, for both Lie and quantum algebras, these basic objects –the generators– in an intrinsic way eliminating every arbitrariness. In this review we do not describe all the technical details of this approach and the interested reader is invited to look at the specific papers quoted in the bibliography [1]–[5].

In the Lie case the situation is quite simple and almost solved more than half century ago. Inside the universal enveloping algebra of a given Lie algebra the distinguished subspace of the algebra is completely determined by the requirement to be closed under commutation relations or, equivalently, to have a primitive coproduct. Also the decomposition of the space of every simple Lie algebra in the form $n_- \oplus h \oplus n_+$ (where h is the abelian Cartan subalgebra and $n_\pm$ are two nihilpotent equivalent subalgebras) is an intrinsic property of interest for both theory and applications. Then the Cartan analysis of n_+ allows to individuate, always without any arbitrariness, the simple roots and to draw with them the Dynkin diagram. However, some freedom remains in the length of the roots and different conventions have been adopted so far: Weyl and Chevalley canonical bases, etc. (see, for instance, [6]). In this context, a way to "solve" the problem of non-simple roots is just to eliminate them from the game: the Chevalley reformulation is indeed made in terms of the Cartan subalgebra, the simple roots and the Serre relations. However, in Physics this is not the usual way to proceed since, in general, all roots (either simple or not) are considered on the some footing and are related to physical objects. In conclusion, few details remained until now unsolved in the construction of the bases within the Cartan program on simple algebras.

As a first result, we discuss here how these problems can be solved by introducing objects more general than algebras: Lie bialgebras and Drinfeld doubles. It is indeed well known that every simple Lie algebra has associated a Lie bialgebra which is "almost" a Drinfeld double [7]. This "approximation" can be in fact removed because every simple algebra of rank r, $\mathcal{A}_r$, is related to the Drinfeld double $\mathcal{A}_r \oplus t_r$, where t_r is an abelian r-dimensional algebra [2, 3]. It turns out that, up to an overall multiplicative factor implicit in the algebra, this Drinfeld double fixes uniquely the normalization factor for all the generators.

The situation is quite different when we attempt to introduce a quantum deformation. Indeed, a general investigation concerning the uniqueness of this quantization process has not been given yet and only some restrictive results for certain deformations of simple Lie algebras have been obtained. As a consequence, although the existence of the quantization is guaranteed for each Lie bialgebra (see [7]), a classification of quantum groups in the spirit of Cartan cannot be found in the literature [1, 2].

Moreover, the infinitesimal counterpart of a quantum group is not well defined. So, we are not dealing with a deformation of a Lie algebra but with a quantum universal enveloping algebra $U_q(g)$, which is a Hopf algebra deformation of the universal enveloping algebra $U(g)$ of the Lie algebra g, i.e. a deformation of an infinite-dimensional object. A possible basis of $U(g)$ is the Poincaré-Birkhoff-Witt (PBW) basis on g (i.e. the set of all ordered monomials of powers of the generators of g). Of course, there are infinitely many basic sets different from g (all related among them by invertible transformations) that originate suitable PBW bases. However g is a special subspace of $U(g)$ and, for this reason the only one considered. But in quantum algebras it is not easy to find a prescription that permits the choice of a privileged bases. Indeed, contrarily to the non deformed case, where inside all the sets of basic elements the vector space of generators is univocally defined, in $U_q(g)$ there is an infinitude of basic sets coexisting on the same footing. This implies that quantum deformations lead us locally (therefore, geometrically) to structures which are quite different from Lie algebras. In particular, when we consider a deformation of an n-dimensional Lie algebra, only the infinite dimensional $U_q(g)$ makes sense in despite that its PBW basis is constructed in terms of a basic set of n elements. This problem of the basis underlies many difficulties encountered when a precise physical/geometrical meaning has to be assigned to $U_q(g)$ as, for instance, in the context of quantum deformations of space-time symmetries. In that case, it is well known that the models so obtained depend on the choice of different bases [8], where different possibilities are related through nonlinear transformations.

The second result that we present here is the solution of the above mentioned problem, providing a universal prescription for the characterization and the explicit construction of a n-dimensional basis, that can be properly called, the quantum analogue of the Lie basis.

To begin with, we analyze the role and properties of the Lie algebra generators within $U(g)$. Among the infinite possible PBW bases, all related by nonlinear invertible transformations, the generators determine the only one that is closed under commutation rules and whose tensor product representations are constructed additively. The latter property can be stated in Hopf-algebraic terms as the Friedrichs theorem [9], which states that the only primitive elements in $U(g)$ (*i.e.* the elements such that $\Delta(X) = \Delta_{(0)}(X) := 1 \otimes X + X \otimes 1$) are just the generators of g as a Lie algebra. In this way, the generators of g become distinguished elements of $U(g)$. However, the situation changes drastically in $U_q(g)$, where the law for the construction of tensor product representations (coproduct) uses nonlinear functions and no primitive bases exist. We show that, among the infinitely many possible bases, there is just one where the coproducts are "as primitive as possible", since all its inessential terms, related to nonlinear transformations in the enveloping algebra, have been removed. Thus, the only changes from the Lie primitive coproducts are those imposed by the consistency between the bialgebra cocommutator δ and the Hopf algebra postulates. This basis (that has, of course, the same dimension than the corresponding Lie algebra) is proposed as the true deformation of the Lie algebra and called "quantum algebra", to be distinguished from the (much larger) quantum universal enveloping algebra.

To develop our approach we recall that Lie group theory is based on analyticity with respect to group parameters. In the same way, analyticity in the deformation parameter(s) will give us the keystone for the identification of the proper quantum algebra, that will be defined as the n-dimensional vector space $(g_q, \Delta) \subset U_q(g)$ obtained as analytical prolongation of the Lie bialgebra $(g, \Delta_{(0)})$. Note that analyticity in the deformation has already played a useful role in quantum algebras, for instance in their contractions [10].

In this analytical prolongation, the cocommutator map of the Lie bialgebra plays a fundamental role since it describes the first order deformation and can be considered as the derivative at the origin of the quantum coproduct. Thus, for a given Lie-Hopf algebra $(g, \Delta_{(0)})$, there are as many different quantizations as inequivalent cocommutators δ. The Hopf algebra postulates together with the assumption that all the results are analytical functions of both the deformation parameter(s), introduced by the bialgebra, and of generators determine uniquely the "analytical" generators.

As a third result we will discuss how the Friedrichs theorem can be extended to quantum algebras. Like in the Lie case, where to a $U(g)$ corresponds one and only one Lie algebra $(g, \Delta_{(0)})$, to a $U_q(g)$ corresponds one and only one quantum algebra (g_q, Δ). Since we prove that (g_q, Δ) is in one-to-one correspondence with (g, δ), then to each bialgebra corresponds one and only one quantization: the classification of quantum groups is reduced to the classification of Lie bialgebras.

Summarizing, we close the commutative diagram by adding the vertex (g_q, Δ)

$$
\begin{array}{ccc}
(g, \Delta_{(0)}) & \xleftarrow{\text{Friedrichs theorem}} & U(g) \\
\big\downarrow q & & \big\downarrow q \\
(g_q, \Delta) & \xleftarrow{\text{Generalized Friedrichs th.}} & U_q(g)
\end{array}
$$

The vertical lines represent the quantization procedure, and the horizontal ones are related to the definition of the basic set of generators of the universal enveloping algebra and of its quantum analogue.

The paper is organized as follows. Section 2 is devoted to Lie bialgebras and Drinfeld doubles, and after a short description of these objects we explicitly discuss the simplest case of $su(2)$, where the length of the root vector turns out to be fixed by the self-dual Drinfeld double structure. The generalization to any simple Lie algebra is briefly discussed. In Section 3 we describe the analytical approach to quantization, which allows to relate, in a perturbative way, the Lie bialgebra (g, δ) and the quantum algebra (g_q, Δ). In this way, we show that quantization not only relates $U_q(g)$ to $U(g)$ but also quite simpler n-dimensional objects like (g_q, Δ) to (g, δ). Again the discussion will be illustrated through the standard deformation of $su(2)$. Starting from the standard bialgebra of $su(2)$ we derive a well defined form of the quantum algebra $su_q(2)$. Section 4 is devoted to revisit the first horizontal line of the diagram. Its generalization, that corresponds to the lower line, is discussed in Section 5. Finally, some conclusions close the paper.

2. Lie bialgebras and Drinfeld doubles

A Lie bialgebra (g, η) is a Lie algebra $g := \{Z_p\}$ endowed with a skew-symmetric linear map (cocommutator) $\eta : g \to g \otimes g$ such that

$$\eta([Z_p, Z_q]) = (\mathrm{ad}_{Z_p} \otimes 1 + 1 \otimes \mathrm{ad}_{Z_p})\eta(Z_q) - (\mathrm{ad}_{Z_q} \otimes 1 + 1 \otimes \mathrm{ad}_{Z_q})\eta(Z_p), \qquad \forall Z_p, Z_q \in g.$$

Thus, besides the commutation relations

$$[Z_p, Z_q] = f_{p,q}^r \, Z_r \,, \tag{1}$$

we have another set of relations

$$\eta(Z_p) = c_p^{q,r} \, Z_q \otimes Z_r. \tag{2}$$

The two sets of structure constants $f_{p,q}^r$ and $c_p^{r,q}$ look quite symmetric and, indeed, on the dual space $g^* := \{z^p\}$ they play an inverted role

$$[z^p, z^q] = c_r^{p,q} \, z^r, \qquad \eta(z^p) = f_{q,r}^p \, z^q \otimes z^r \,. \tag{3}$$

The compatibility relations between the structure constants

$$c_r^{p,q} f_{s,t}^r = c_s^{p,r} f_{r,t}^q + c_s^{r,q} f_{r,t}^p + c_t^{p,r} f_{s,r}^q + c_t^{r,q} f_{s,r}^p \tag{4}$$

allow to combine g and g^* in an unique algebra called a Drinfeld double [7]. This Drinfeld double is, indeed, defined as a Lie algebra $\bar{g}$ such that, $\bar{g} = g \oplus g^*$ as vector space, where g and g^* are disjoint sub-bialgebras, and the crossed commutation rules between g and g^* are defined in terms of the structure tensors $f_{q,r}^p$ and $c_p^{q,r}$ by eq. (5). The Jacobi identities for the algebra $\bar{g}$ are just the compatibility relations (4).

The crossed commutation relations

$$[z^p, Z_q] = f_{q,r}^p z^r - c_q^{p,r} Z_r. \tag{5}$$

are related to the invariance of the pairing

$$\langle Z_p, Z_q \rangle = 0, \qquad \langle Z_p, z^q \rangle = \delta_p^q, \qquad \langle z^p, z^q \rangle = 0. \tag{6}$$

Moreover, the double Lie algebra $\bar{g}$ has a (quasitriangular) double Lie bialgebra structure $(\bar{g}, \delta)$, where its cocommutators are determined by its sub-bialgebras (g, η) and (g^*, η) as

$$\delta(Z_p) = -\eta(Z_p) = -c_p^{q,r} Z_q \otimes Z_r, \qquad \delta(z^p) = \eta(z^p) = f_{q,r}^p z^q \otimes z^r. \tag{7}$$

In the particular case that $c_p^{q,r} = -f_{q,r}^p$ [11] the Drinfeld double gets an additional symmetry, and we call it a self-dual Drinfeld double. It is worthy noticing that the positive and negative Borel subalgebras $b_\pm$ of any simple Lie algebra g have this last property, but, as they have the Cartan subalgebra in common, they cannot be identified **as** g and g^*. Hence, the classical Lie algebras are "almost" Drinfeld doubles [12, 7].

In fact, this problem has been solved with a slight extension of $\bar{g}$ considering the direct sum of $\bar{g}$ with an abelian algebra t_n, where n is the rank of $\bar{g}$ [2]. The most elementary example of self-dual Drinfeld double structure is that of the algebra $su(2) \oplus t_1$. We write $su(2)$ as

$$[J_3, J_\pm] = \pm J_\pm, \qquad [J_+, J_-] = \alpha J_3, \tag{8}$$

where the most common physical convention is $\alpha = 2$, while in mathematical contexts different conventions and a different overall factor are used to write $su(2) := \{h, f, g\}$ relations as $[h, f] = 2f$, $[h, g] = -2g$, $[f, g] = h$. In disagreement with both of them, from $su(2) \oplus t_1$, we will obtain $\alpha = 1$.

The two $su(2)$ Borel subalgebras $s^\pm := \{J_3, J_\pm\}$ are not disjoint (as it should be required by (6)) but, if we introduce a new abelian algebra $t_1 \equiv \mathbb{R}$ with generator I, that commutes with $su(2)$ (i.e. $[I, \cdot] = 0$), we can define two disjoint algebras $b_+ := \{Z_1, Z_2\}$ and $b_- := \{z^1, z^2\}$ that allow us to define a true self-dual Drinfeld double. Let us indeed define

$$Z_1 = \frac{1}{\sqrt{2}}(J_3 + iI), \qquad Z_2 = J_+, \qquad z^1 = \frac{1}{\sqrt{2}}(J_3 - iI), \qquad z^2 = J_-.$$

The commutation relations imposed by $su(2) \oplus t_1$ inside b_+ and b_- are, respectively

$$[Z_1, Z_2] = \frac{1}{\sqrt{2}} Z_2, \qquad [z^1, z^2] = -\frac{1}{\sqrt{2}} z^2,$$

such that, the structure constants are

$$f_{1,2}^2 = -f_{2,1}^2 = -c_2^{1,2} = c_2^{2,1} = \frac{1}{\sqrt{2}}.$$

Compatibility conditions (4) are easily checked. Hence, because $c_k^{i,j} = -f_{i,j}^k$, we have a self-dual Drinfel double. From (5) we get the crossed commutation relations as

$$[Z_1, z^1] = 0, \qquad [Z_1, z^2] = -\frac{1}{\sqrt{2}} z^2, \qquad [Z_2, z^1] = -\frac{1}{\sqrt{2}} Z_2, \qquad [Z_2, z^2] = \frac{1}{\sqrt{2}}(Z_1 + z^1).$$

The cocommutators are

$$\delta(Z_1) = 0, \qquad \delta(Z_2) = \frac{1}{\sqrt{2}} Z_2 \wedge Z_1, \qquad \delta(z^1) = 0, \qquad \delta(z^2) = \frac{1}{\sqrt{2}} z^2 \wedge z^1.$$

Now, we can come back to our original basis. The commutation relations inside $b_\pm$ have been, of course, assumed but the value of α is determined by the Drinfeld double to be exactly $\alpha = 1$. This is equivalent to impose the following relation on the $so(3)$ generators

$$J_\pm \equiv \frac{1}{\sqrt{2}}(J_1 \pm iJ_2)$$

as, from a different point of view, is required by the Killing form.

In this way the full double Lie bialgebra is obtained. Since we are interested in the subalgebra $su(2)$, we consider the trivial representation of t_1 (where $I = 0$)

$$\delta(I) = 0, \qquad \delta(J_3) = 0, \qquad \delta(J_+) = \frac{1}{2} J_+ \wedge J_3, \qquad \delta(J_-) = \frac{1}{2} J_- \wedge J_3. \tag{9}$$

The generalization to larger algebras is straightforward. We display here the results for the A_n series [3]. Let us consider the Lie algebra $gl(n+1) = A_n \oplus h := \{H_i,\ F_{ij}\}$, where h is the Lie algebra generated by $\sum H_i$, we define:

$$\begin{aligned} &[H_i, H_j] = 0, \\ &[H_i, F_{jk}] = (\delta_{ij} - \delta_{ik}) F_{jk}, \\ &[F_{ij}, F_{kl}] = (\delta_{jk} F_{il} - \delta_{il} F_{kj}) + \delta_{jk}\delta_{il}(H_i - H_j). \end{aligned} \tag{10}$$

As before, we introduce an abelian algebra $t_{n+1} := \{I_i\}$ with $n+1$ central generators I_i and we construct the algebra $gl(n+1) \oplus t_{n+1}$ that we show to be a Drinfeld double. With a change of basis similar to the $su(2)$ case, the new generators are

$$Z_i := \frac{1}{\sqrt{2}}(H_i + \mathbf{i} I_i), \quad Z_{ij} := F_{ij}, \qquad z^i := \frac{1}{\sqrt{2}}(H_i - \mathbf{i} I_i), \quad z^{ij} := F_{ji} \quad (i < j). \tag{11}$$

We obtain thus two equivalent soluble Lie algebras $b_\pm$ each of dimension $(n+1)(n+2)/2$

$$b_+ := \{Z_i, Z_{ij}\}, \qquad b_- := \{z^i, z^{ij}\}, \qquad i, j = 1, \ldots, n+1, \quad i < j,$$

such that $b_+ + b_- = gl(n+1) \oplus t_{n+1}$ as vector spaces.

The commutation rules imposed by (10) inside b_+ and b_- are

$$\begin{aligned} &[Z_i, Z_j] = 0, \qquad [Z_i, Z_{jk}] = \tfrac{1}{\sqrt{2}}(\delta_{ij} - \delta_{ik})\, Z_{jk}, \qquad [Z_{ij}, Z_{kl}] = (\delta_{jk} Z_{il} - \delta_{il} Z_{kj}), \\ &[z^i, z^j] = 0, \qquad [z^i, z^{jk}] = -\tfrac{1}{\sqrt{2}}(\delta_{ij} - \delta_{ik})\, z^{jk}, \qquad [z^{ij}, z^{kl}] = -(\delta_{jk} z^{il} - \delta_{il} z^{kj}). \end{aligned} \tag{12}$$

The two algebras b_+ and b_- can be paired by

$$\langle z^i, Z_j \rangle = \delta^i_j, \qquad \langle z^{ij}, F_{kl} \rangle = \delta^i_k \delta^j_l, \tag{13}$$

defining a bilinear form such that both $b_\pm$ are isotropic.

The compatibility relations (4) can be easily checked and we are now able to came back from (12) and (13) to all the formulas (10) but this is strictly related to our definition of the root vectors. A change of scale like $F_{ij} \to k F_{ij}$ will introduce into (10) changes that are is incompatible with the Drinfeld double structure.

This procedure can be easily extended to any arbitrary semi-simple Lie algebra g_n by considering $g_n \oplus t_n$, which can be equipped with a self-dual Drinfeld double structure. In Ref. [3, 4] the explicit form for the classical series A_n, B_n, C_n and D_n can be found.

In conclusion: there is a Drinfeld double behind every simple Lie algebra but only if it is written in the correct basis.

The Cartan program on semi-simple Lie algebras has been thus concluded by fixing, without any freedom, the length of the vectors associated to all the roots. This is the main result coming from the Drinfeld double perspective at a "classical" level.

Journal of Physics: Conference Series **128** (2008) 012049 doi:10.1088/1742-6596/128/1/012049

It is worthy to note that the Drinfeld double fixes everything not only at the Lie algebraic level but also at the Lie bialgebra one: the canonical Lie bialgebra structure for $gl(n+1)\oplus t_{n+1}$ is determined by the cocommutator δ and reads

$$\begin{aligned}
&\delta(I_i) = 0,\\
&\delta(H_i) = 0,\\
&\delta(F_{ij}) = -\tfrac{1}{2}F_{ij}\wedge(H_i-H_j) - \tfrac{\mathrm{i}}{2}F_{ij}\wedge(I_i-I_j) + \textstyle\sum_{k=i+1}^{j-1} F_{ik}\wedge F_{kj}, \qquad i<j,\\
&\delta(F_{ij}) = \tfrac{1}{2}F_{ij}\wedge(H_i-H_j) - \tfrac{\mathrm{i}}{2}F_{ij}\wedge(I_i-I_j) - \textstyle\sum_{k=j+1}^{i-1} F_{ik}\wedge F_{kj}, \qquad i>j.
\end{aligned} \tag{14}$$

We have to observe that the chain of Drinfeld doubles $g_n\oplus t_n \subset g_{n+1}\oplus t_{n+1}$ is preserved at the level of Lie bialgebras. However, although g_n is a subalgebra of $g_n\oplus t_n$, the cocommutator $\delta(g_n)$ does not define a sub-bialgebra since it depends on the extra t_n sector.

3. Analytical deformation

The problem of the generalization of the previous results to quantum algebras is quite more difficult. Strictly speaking, quantum algebras are not deformed Lie algebras in the sense that the initial object to be deformed is $U(g_n)$, the (infinite-dimensional) universal enveloping algebra built on the Lie algebra g_n, and the final object, with similar properties, is the quantum universal enveloping algebra $U_q(g_n)$ (also infinite-dimensional).

The universal enveloping algebra $U(g_n)$ has, as a possible basis, the Poincaré-Birkhoff-Witt (PBW) basis, i.e. the set of all ordered monomials built up on the powers of the generators of g_n. Moreover, the PBW basis is only a particular one among the possible bases all related between them by invertible transformations. The crucial point is that, when considering $U_q(g_n)$, the algebra subset looses all its privileges in a structure where linearity does not make sense. On the contrary, when considering $U(g_n)$, we have the chance that the Lie subset is the only subset that has two relevant properties: first, it is the only n-dimensional set closed under commutation and, second, only the elements of the Lie algebra are primitive. This last property is –at least in physical terms– the most relevant as it implies the additivity of physical observables related to the algebra.

Indeed, $U(g_n)$ and $U_q(g_n)$ are equipped with Hopf algebra structures. The main point is the existence of an isomorphism, called coproduct,

$$\Delta : U(g_n) \to U(g_n)\otimes U(g_n), \qquad \Delta_q : U_q(g_n) \to U_q(g_n)\otimes U_q(g_n).$$

Those elements of $U(g_n)$ such that $\Delta(X) = 1\otimes X + X\otimes 1$ are said to be primitive and the Friedrichs theorem establishes that only the elements of g_n are primitive. Moreover, as the Lie basis elements are cocommutative (i.e. $\Delta(X) = \sigma\circ\Delta(X)$ with $\sigma(A\otimes B) = B\otimes A$), this property is also valid for all $U(g_n)$.

Unfortunately, deformation breaks both properties, primitivity and cocommutativity. So, we looses the canonical procedure to individuate among all the possible basic sets a privileged one. At this point we are compelled to considerer at the same time the whole $U_q(g_n)$ where all the bases are equivalent. Let us recall that three basic sets are usually considered in the literature: Lie basis, crystal basis [13, 14] and the most common one, which for $su_q(2) := \{J_3, J_\pm\}$ [7] is

$$[J_3, J_\pm] = \pm J_\pm, \qquad [J_+, J_-] = \frac{\sinh(zJ_3)}{\sinh(z/2)}; \tag{15}$$

Journal of Physics: Conference Series **128** (2008) 012049 doi:10.1088/1742-6596/128/1/012049

$$\Delta(J_3) = 1 \otimes J_3 + J_3 \otimes 1, \qquad \Delta(J_\pm) = e^{\frac{z}{2}J_3} \otimes J_\pm + J_\pm \otimes e^{-\frac{z}{2}J_3}. \tag{16}$$

The Lie basis of $U_q(g_n)$ describes a situation where both the basic set and its coproduct are closed under commutation relations but the relation between them is different (and quite more complex) from the Lie one. For a physicist this means that we have the same symmetry of the Lie situation but we loose the usual definition of composed system. It could, perhaps, be the appropriate scheme to describe systems of particles with an interaction among them. The crystal basis (often called canonical basis) is, in some sense, the opposite of the Lie basis: while in the Lie basis the algebra is written for $z \to 0$ and all deformation is transferred to the coproduct, in the crystal basis the algebra is obtained for $z \to \infty$ while, as in the Lie basis, the problem of consistency is left to the coproduct. Also in the crystal basis the representations look quite simple and have been applied in statistical mechanics [15] and in genetics [16].

However, it is not an accident that everybody knows $U_q(su(2))$ in terms of the basic set displayed in (15), (16), since this basis is, up to few details, the analytical prolongation to the $su(2)$ generators defined in Sect. 2. More information can be found in Ref. [5], where the discussion is more general and technical.

Analyticity in Lie theory is analyticity in the parameter (dual space of the generators), but here we have to extend the analyticity to the deformation parameters z in such a way that in a well precise limit (usually $z \to 0$) the quantum algebra reduces to the corresponding Lie algebra (without loss of generality, we shall restrict here to the one-parameter case). We also recall that analyticity in quantum algebras is not new since, for instance, contraction theory relies upon this concept [10].

The point is that from the theory of analytical functions we know that there exist infinitely many analytical functions that have the same value in a point and the same derivative in the same point. Consistently, from a Lie bialgebra we can obtain, by analytic prolongation, infinitely many basic sets all of them describing the same $U_q(g)$ and expressed by means of infinitely many free parameters, independent from z. This is coherent with the fact that $U_q(g)$ (as well as $U(g)$) can be described by infinitely many basic sets. But, as discussed in the following, all these free parameters remain present in the limit $z \to 0$: each one of the basic sets of $U_q(g)$ is related by analyticity to a basic set of $U(g)$ and, as $U(g)$ has a privileged basic set in the Lie generators, the basic set related by analyticity to Lie generators is a privileged basic set for $U_q(g)$.

For instance $U(su(2))$ can be defined, as usual, from the $su(2)$ generators, but – equivalently– starting from another basic set like, for instance,

$$Y_\pm = J_\pm + J_3^2 J_\pm, \qquad Y_3 = J_3 + J_3^2. \tag{17}$$

As these relations are (formally) invertible $U(su(2))$ –as well as its deformation– does not change, but $\{Y_\pm, Y_3\}$ is clearly not the best choice for applications. This is exactly the freedom we remove in the definition of the analytical basis by imposing that the deformation starts from the Lie generators and that only the contributions necessary to save the consistency between the bialgebra (g, δ) and the Hopf postulates are introduced. All contribution related to nonlinear transformations in $U(g)$, like (17), are in this way eliminated.

As it is well-known Δ is related to δ by

$$\delta = \lim_{z \to 0} \frac{\Delta - \sigma \circ \Delta}{2z} : \tag{18}$$

the cocommutator δ can be seen as the derivative at the origin of the quantization and $U_q(g)$ is sometimes called a "quantization of $U(g)$ in the direction of δ".

The analytical deformation is introduced in three steps:

(i) We find order-by-order the changes imposed by δ in $\Delta_{(0)}$ to be consistent with coassociativity (21) and we determine in this way the full coproduct Δ.

(ii) By using analyticity and the homomorphism property of Δ (23), we obtain the commutation rules for g_q starting from the known ones of g. Thus, the n-dimensional (g_q, Δ) is constructed.

(iii) From g_q a PBW basis in $U_q(g)$ is built.

In this way we construct an unique connection $(g, \delta) \to (g_q, \Delta) \to U_q(g)$. Since (g, δ) is the limit of (g_q, Δ) and, as it will be show in Sect. 5, every $U_q(g)$ admits one and only one basis g_q, the arrows can be inverted and a one-to-one correspondence is found between (g, δ) and $U_q(g)$. So, equivalences between q-deformations of $U(g)$ imply equivalences among their associated bialgebras and the classification of the $U_q(g)$ is carried to the quite simpler classification of Lie bialgebras.

Analyticity means that the commutation relations of any basic set $\{Y_j\}$ $(j = 1, 2, \ldots, n)$ of $U_q(g)$ (as well as of $U(g)$) are analytical functions of the Y_j and that the quantum coproduct Δ of the Y_j can be written as a formal series

$$\Delta(Y_i) = \sum_{k=0}^{\infty} \Delta_{(k)}(Y_i) = \Delta_{(0)}(Y_i) + \Delta_{(1)}(Y_i) + \ldots \tag{19}$$

where $\Delta_{(k)}(Y_i)$ is a homogeneous polynomial of degree $k+1$ in $1 \otimes Y_j$ and $Y_j \otimes 1$.

There are two properties of Hopf algebras which are relevant in this analytical approach: the coassociativity condition

$$(\Delta \otimes 1 - 1 \otimes \Delta) \circ \Delta(Y_i) = 0, \tag{20}$$

that in a perturbative form will be rewritten as

$$\sum_{j=0}^{k} \left(\Delta_{(j)} \otimes 1 - 1 \otimes \Delta_{(j)}\right) \circ \Delta_{(k-j)}(Y_i) = 0, \qquad \forall k; \tag{21}$$

and the homomorphism property

$$\Delta([Y_i, Y_j]) = [\Delta(Y_i), \Delta(Y_j)]\,, \tag{22}$$

which can be rewritten as

$$\Delta_{(k)}([Y_i, Y_j]) = \sum_{l=0}^{k} \left[\Delta_{(l)}(Y_i), \Delta_{(k-l)}(Y_j)\right], \qquad \forall k\,. \tag{23}$$

To enlighten the construction we discuss explicitly the quantization of the Lie bialgebra $(su(2), \delta)$ described by (8) with $\alpha = 1$ and (9). More details can be found in [5]. The results of this Section show that analyticity chooses the coproduct given by formula (16) among all possible ones [1] and the usual expressions (15) must be replaced by other slightly different ones.

As stated before, let us start by getting the quantum coproduct. The zero-approximation of the quantum coproduct is $\Delta_{(0)}$, while $\Delta_{(1)}$ is essentially δ and the $\Delta_{(k)}$ are obtained adding order by order the contributions imposed by consistency with eq. (21).

Journal of Physics: Conference Series **128** (2008) 012049 doi:10.1088/1742-6596/128/1/012049

The case of J_3 is simple: $\Delta_{(0)}(J_3) = J_3 \otimes 1 + 1 \otimes J_3$ and $\delta(J_3) = 0$, that implies that the anti-cocommutative part of $\Delta_{(1)}(J_3)$ is zero. Thus $\Delta_{(1)}(J_3)$ must be cocommutative. But cocommutative contributions are related to non linear bases like the one of eq. (17) in the Lie algebra and can be removed by coming back, with an appropriate change of basis, to the Lie generators. So (see eq.(31) and Ref. [5] for an exhaustive discussion) we write $\Delta_{(1)}(J_3) = 0$. Then, we have

$$\Delta(J_3) = \Delta_{(0)}(J_3) + \mathcal{O}_{(2)}(J_3),$$

where we have defined $\mathcal{O}_{(m)}(Y_i)$ as a polynomial or degree $> m$ in $1 \otimes Y_j$ and $Y_j \otimes 1$. From eq.(19), we can write

$$\Delta(J_3) = \Delta_{(0)}(J_3) + \Delta_{(2)}(J_3) + \mathcal{O}_{(3)}(J_3).$$

Again, as eq. (21) for $k = 2$ is consistent with $\Delta_{(2)}(J_3) = 0$, we perform a (cubic) change of basis and we write

$$\Delta(J_3) = \Delta_{(0)}(J_3) + \Delta_{(3)}(J_3) + \mathcal{O}_{(4)}(J_3).$$

As this procedure can be iterated for all k, the analytical prescription for J_3 imposes $\Delta_{(k)}(J_3) = 0, \ \ \forall k > 0$. Hence

$$\Delta(J_3) = \Delta_{(0)}(J_3) = J_3 \otimes 1 + 1 \otimes J_3, \tag{24}$$

i.e., the analytical procedure associates a primitive coproduct to any null δ. This could be considered trivial, as $\{J_3\}$ closes an $u(1)$ algebra, but this is not the case since it has been obtained from a precise prescription and it is not, like in [7], an arbitrary choice inside $U(g)$.

For $\Delta(J_+)$ we have

$$\Delta(J_+) = \Delta_{(0)}(J_+) + \Delta_{(1)}(J_+) + \mathcal{O}_{(2)}(J_+).$$

From (18) and the coassociativity condition (21) for $k = 1$, we have that $\Delta_{(1)}(J_+) = \frac{z}{2}\,\delta(J_+)$ as, like in the J_3 case, the possible cocommutative contribution –independent from δ– can be put to zero by a change of basis. We have thus

$$\Delta(J_+) = \Delta_{(0)}(J_+) + \frac{z}{2}\,\delta(J_+) + \Delta_{(2)}(J_+) + \mathcal{O}_{(3)}(J_+).$$

The coassociativity condition (21) for $k = 2$ solved in the unknown $\Delta_{(2)}(J_+)$ gives (again disregarding arbitrary contributions that can be removed by a nonlinear change of basis)

$$\Delta_{(2)}(J_+) = \frac{z^2}{8}(J_3^2 \otimes J_+ + J_+ \otimes J_3^2). \tag{25}$$

Considering next order, we can now write

$$\Delta(J_+) = \Delta_{(0)}(J_+) + \frac{z}{2}\,\delta(J_+) + \Delta_{(2)}(J_+) + \Delta_{(3)}(J_+) + \mathcal{O}_{(4)}(J_+),$$

where $\Delta_{(2)}(J_+)$ is given by eq. (25) and $\Delta_{(3)}$ is the new unknown. Solving the coassociativity condition (21) for $k = 3$ we have

$$\Delta_{(3)}(J_+) = \frac{z^3}{48}(J_3^3 \otimes J_+ - J_+ \otimes J_3^3) \tag{26}$$

and the general formula obtained by iteration, always putting to zero the cocommutative contributions related to nonlinearity in $U(su(2))$ and independent from δ, is

$$\Delta_{(k)}(J_+) = \frac{1}{k!}\left(\frac{z}{2}\right)^k (J_3^k \otimes J_+ + (-1)^k J_+ \otimes J_3^k), \qquad \forall k.$$

The $\Delta_{(k)}$ are easily summed to

$$\Delta(J_+) = e^{\frac{z}{2}J_3} \otimes J_+ + J_+ \otimes e^{-\frac{z}{2}J_3}.$$

The approach is exactly the same for J_- and gives a similar result. Thus, we obtain the analytical quantum coproduct associated to $(su(2), \delta)$

$$\begin{aligned} \Delta(J_3) &= J_3 \otimes 1 + 1 \otimes J_3, \\ \Delta(J_+) &= e^{\frac{z}{2}J_3} \otimes J_+ + J_+ \otimes e^{-\frac{z}{2}J_3}, \\ \Delta(J_-) &= e^{\frac{z}{2}J_3} \otimes J_- + J_- \otimes e^{-\frac{z}{2}J_3}. \end{aligned} \tag{27}$$

that it is exactly the coproduct displayed in expression (16).

Now we have simply to start from the commutators (8) with $\alpha = 1$ and to impose order by order the homomorphism condition for the deformed commutation rules. We thus find

$$[J_3, J_\pm] = \pm J_\pm, \qquad [J_+, J_-] = \frac{1}{z}\sinh(z\, J_3). \tag{28}$$

Note that $\Delta_{(0)}([J_+, J_-]) = \Delta_{(0)}(J_3)$ forbids the introduction of z-dependent normalizations, like $\sinh(z/2)$ of expressions (15), and a factor 2 has been removed at the classical level from the Drinfeld double structure. Expressions (27) and (28) define uniquely the standard analytical deformation of the Cartan basis of $su(2)$ such that the q-generators J_3, J_+, J_- could be called the q-Cartan basis of $su_q(2)$.

By inspection, all the symmetries (for example, $\{J_3, J_+, J_-\} \leftrightarrow \{J_3, -J_+, -J_-\}$) and the embedding conditions (for instance, $su(2) \supset borel(J_3, J_+) \supset u(1) := \{J_3\}$) of the bialgebra (g, δ) are automatically preserved in the quantization (g_q, Δ). Like for the bialgebra only the $u(1)$ generated by J_3 is a sub-quantum-algebra of $su_q(2)$.

4. Perturbative Friedrichs theorem

We give now a constructive proof of the Friedrichs theorem, building explicitly the primitive generators $\{X_j\}$ in terms of an arbitrary set $\{Y_j\}$ on which a PWB basis for the whole $U(g)$ can be built. The elements Y_j are cocommutative but, in principle, non primitive. The machinery consists in repeated changes of bases that allow to obtain each time a better approximation to primitivity where –this is the essential point– the problem is reformulated at each step in terms of the preceding basis. One infinite iteration of the procedure allows to find, among the infinitely many possible bases of $U(g)$, the Lie generators.

In more detail, let us consider $X_i \equiv \lim_{k\to\infty} X_i^k$ were $\{X_i^k\}$ is a basic set that approximates the Lie-Hopf coproducts up to order k. The terms $\mathcal{O}_{(m)}(X_i^k)$ are, in this Section, cocommutative since we are working in $U(g)$. Now any original basic set $\{Y_i\}$ is a zero approximation to $\{X_i\}$, i.e. $X_i^0 := Y_i$. Indeed

$$\Delta(X_i^0) = \Delta_{(0)}(X_i^0) + \mathcal{O}_{(1)}(X_i^0) = X_i^0 \otimes 1 + 1 \otimes X_i^0 + \mathcal{O}_{(1)}(X_i^0). \tag{29}$$

The explicit form consistent with formula (21) of $\mathcal{O}_{(1)}(X_i^0)$ in expression (29) is

$$\mathcal{O}_{(1)}(X_i^0) = \sum d_i^{jl}\ (X_j^0 \otimes X_l^0 + X_l^0 \otimes X_j^0)\ +\ \mathcal{O}_{(2)}(X_i^0),$$

where d_i^{lj} are arbitrary constants. From expression (19) we get

$$\mathcal{O}_{(1)}(X_i^0) = \Delta_{(1)}(X_i^0) + \mathcal{O}_{(2)}(X_i^0), \tag{30}$$

but if we define (see [5] for details) the next approximation

$$X_i^1 := X_i^0 - \sum d_i^{jl}\,\{X_j^0, X_l^0\} \tag{31}$$

we obtain a coproduct for X_i^1 with vanishing first order contributions, i.e.

$$\Delta(X_i^1) = \Delta_{(0)}(X_i^1) + \mathcal{O}_{(2)}(X_i^0).$$

Eq. (31) allows to rewrite $\mathcal{O}_{(2)}(X_i^0)$ in terms of X_i^1 as $\mathcal{O}_{(2)}(X_i^1)$

$$\Delta(X_i^1) = \Delta_{(0)}(X_i^1) + \mathcal{O}_{(2)}(X_i^1).$$

Now, defining a new change of basis X_i^2 we obtain, in terms of the same X_i^2, the coproduct of the second approximation X_i^2 to the generators as

$$\Delta(X_i^2) = \Delta_{(0)}(X_i^2) + \mathcal{O}_{(3)}(X_i^2),$$

which is free form both first and second order contributions. The procedure can now be iterated and the $\Delta_m(X_i^{m-1})$ contribution eliminated through a new change of basis that affects the higher orders only. The residual term becomes $\mathcal{O}_{(m+1)}(X_i^m)$ and we get the m-order approximation to the Lie generators

$$\Delta(X_i^m) = \Delta_{(0)}(X_i^m) + \mathcal{O}_{(m+1)}(X_i^m).$$

The true generators of the Lie algebra g are (formally) recovered in the limit

$$X_i := \lim_{m\to\infty} X_i^m$$

and, in agreement with the Friedrichs theorem, their coproduct is the primitive one

$$\lim_{m\to\infty} \Delta(X_i^m) = \lim_{m\to\infty} \Delta_{(0)}(X_i^m) = \Delta_{(0)}(X_i) = \Delta(X_i) = X_i \otimes 1 + 1 \otimes X_i.$$

This coproduct is an algebra homomorphism with respect to the Lie commutation rules

$$\Delta[X_i, X_j] = [X_i, X_j] \otimes 1 + 1 \otimes [X_i, X_j],$$

and the n generators of the Lie algebra are univocally identified in a constructive manner within $U(g)$, pushing away order by order the corrections to a primitive coproduct.

The central point of this perturbative approach to Friedrichs theorem (as well as to its following extension to $U_q(g)$) is that at each order all the relations can be rewritten in terms of the corresponding approximations of the generators.

Journal of Physics: Conference Series **128** (2008) 012049 doi:10.1088/1742-6596/128/1/012049

5. Unicity of deformation

This novel proof of Friedrich's Theorem has been described here because the procedure that allows to individuate the generators of the quantum algebra g_q, among the infinitely many possible bases within $U_q(g)$, is exactly the same that allows to individuate the generators of the Lie algebra g among the infinitely many possible bases of $U(g)$. The perturbative approach allows thus a generalization of the Friedrichs theorem to quantum algebras.

Indeed, the construction introduced in Sect. 4 works also when $\delta \neq 0$. Hence, it provides a prescription for the construction of the almost primitive generators starting from an arbitrary set of basic elements of any $U_q(g)$. In this way we obtain (g_q, Δ) from $U_q(g)$ and we close the diagram displayed in the Introduction.

Let us sketch the proof of such result. As in the undeformed case, let $\{Y_j\}$ be an arbitrary set of basic elements that determine $U_q(g)$, whose classical limit is a Lie bialgebra with $\delta \neq 0$. Eqs. (29) and (30) are still valid, but now we have to add to the cocommutative part of $\Delta_{(1)}$ the anti-cocommutative contribution given by δ

$$\Delta(X_i^0) = \Delta_{(0)}(X_i^0) + z\,\delta(X_i^0) + \sum d_i^{jl}\ \left(X_j^0 \otimes X_l^0 + X_l^0 \otimes X_j^0\right)\ + \mathcal{O}_{(2)}(X_i^0).$$

Like in Sect. 4 we define $X_i^1 := X_i^0 - \sum d_i^{jl}\,\{X_j^0, X_l^0\}$. As this change of variables does affect the δ contribution only to higher orders, the differences between $\delta(X_i^0)$ and $\delta(X_i^1)$ can be included in $\mathcal{O}_{(2)}(X_i^0)$ (or, equivalently, $\mathcal{O}_{(2)}(X_i^1)$). So,

$$\Delta(X_i^1) = \Delta_{(0)}(X_i^1) + z\,\delta(X_i^1) + \mathcal{O}_{(2)}(X_i^1).$$

From eq. (19), we introduce $\Delta_{(2)}(X_i^1)$

$$\mathcal{O}_{(2)}(X_i^1) = \Delta_{(2)}(X_i^1) + \mathcal{O}_{(3)}(X_i^1).$$

Similarly to $\Delta_{(1)}$, the second order $\Delta_{(2)}$ contains two contributions: the first one – characteristic of the deformations– is proportional to z^2, and is constrained by the consistency between δ and the coassociativity condition. The second one is only related to the definition of the basis in $U(g)$ and can be removed by another change of basis that does not modify the form of the z-dependent terms since the modifications can be included in $\mathcal{O}_{(3)}(X_i^2)$.

This procedure can be iterated. Thus, once the problem is solved for $\Delta_{(m-1)}$, a contribution to $\Delta_{(m)}$ proportional to z^m (cocommutative for m even and anti-cocommutative for m odd) is found. Indeed the unessential z-independent terms are removed, exactly as in the case $\delta = 0$, with a change of basis that does not affect the form of the previous z-depending terms because the introduced changes are always of orders higher of z^m and thus pushed out in $\mathcal{O}_{(m+1)}$. For $m \to \infty$ the same coproducts derived from $(g, \Delta_{(0)})$ in Sect. 3 are found. The deformed commutation rules are then imposed by the homomorphism (23). Hence, we have closed the lower row of the diagram, finding that $U_q(g)$ has as one of its basic sets the analytical basis (g_q, Δ) obtained as the analytical continuation in z of the Lie generators.

6. Concluding remarks

The main result of this paper is the definition and the operative prescription for the complete construction of the basis in semi-simple algebras, both Lie and quantum. In the Lie case our result end the Cartan program by fixing the few normalizations adjusted before by convention. The situation is completely different in quantum algebras where also the basic sub-space of $U_q(g)$, corresponding to the Lie algebra, has to be found. We solve the problem by means of analyticity: by starting from the cocommutator δ given by the Lie bialgebra, a perturbative method based on cocommutativity and followed by the appropriate sum of the series allows us to determine the full expressions for the deformed coproducts. In this way, each primitive coproduct is deformed to its quantum counterpart. A perturbative application of the homomorphims property of the coproduct map allows to start from the Lie form of the commutators and to arrive to their deformations. In this way the full analytical quantization –including non simple roots– is obtained.

As an important consequence, this quantization procedure imposes that the deformation of the commutators remain antisymmetric, i.e. in the analytical basis there is no room for q-commutators, objects without a precise symmetry. This is relevant for physical applications since the commutators are essential elements in Quantum Mechanics as well as in Poisson-Lie structures (whose q-counterparts are not well-defined).

The realization of one and only one quantum analog of the generators has also a relevant byproduct. As $(g, \delta) \leftrightarrows (g_q, \Delta) \leftrightarrows U_q(g)$, a one-to-one correspondence between $U_q(g)$ and (g, δ) is established. In other words, each Lie bialgebra admits one (as it was known) and only one (as stated here) quantization and the problem of classification of semisimple quantum groups can be fully solved by restoring to the classification of the underlying Lie bialgebras.

Finally let us note that the proposed method is constructive and it could be implemented to build by computer the quantum algebra deformation of every Lie bialgebra (g, δ).

Acknowledgments

This work was partially supported by the Ministerio de Educación y Ciencia of Spain (Projects FIS2005-03959 and MTM2007-67389), by the Junta de Castilla y León (Project VA013C05) and by INFN-CICyT (Italy-Spain).

References

[1] Ballesteros A, Celeghini E and del Olmo M A 2004 *J. Phys. A: Math. Gen.* **37** 4231
[2] Ballesteros A, Celeghini E and del Olmo M A 2005 *J. Phys. A: Math. Gen.* **38** 3909
[3] Ballesteros A, Celeghini E and del Olmo M A 2006 *J. Phys. A: Math. Gen.* **39** 9161
[4] Ballesteros A, Celeghini E and del Olmo M A 2007 *J. Phys. A: Math. Theor.* **40** 2013
[5] Celeghini E, Ballesteros A and del Olmo M A 2008 From quantum universal enveloping algebras to quantum algebras *J. Phys. A: Math. Gen.* **41** *in press*; arXiv:0712.0520
[6] Cornwell J F 1984 *Group Theory in Physics*, (London: Academic Press)
[7] Chari V and Pressley A 1994 *A Guide to Quantum Groups* (Cambridge: Cambridge Univ. Press)
[8] Majid S 1995 *Foundations of Quantum Group Theory* (Cambridge: Cambridge Univ. Press)
[9] Jacobson N 1979 *Lie algebras* (New York: Dover)
[10] Celeghini E, Giachetti R, Sorace E and Tarlini M 1992 *Lecture Notes in Math.* vol. 1510, ed P P Kulish (Berlin: Springer) pg.221
[11] Gomez X 2000 *J. Math. Phys.* **41** 4939
[12] Drinfel'd V G 1987 Quantum Groups *Proc. Int. Congress of Mathematicians* (Berkeley 1986) ed A M Gleason (Providence: AMS)
[13] Lusztig G 1990 *J. Amer. Math. Soc.* **3** 447; 1990 *Progr. Theor. Phys. Suppl.* **102** 175
[14] Kashiwara M 1990 *Comm. Math. Phys.* **133** 249; 1993 *Duke Math. J.* **69** 455
[15] Kang S J, Kashiwara M, Misra K C, Miwa T, Nakashima T and Nakayashiki A 1992 *Duke Math. J.* **68** 499
[16] Minichini C and Sciarrino A 2006 *Biosystems* **84** 191.

V International Symposium on Quantum Theory and Symmetries IOP Publishing
Journal of Physics: Conference Series **128** (2008) 012050 doi:10.1088/1742-6596/128/1/012050

Feynman formula for a diffusion of particles with a variable mass in a domain

Yana A Butko[1], **Martin Grothaus**[2] **and Oleg G Smolyanov**[3]

[1] Bauman Moscow State Technichal University, 105005, Moscow, Russia
[2] Technische Universität Kaiserslautern, 67653, Kaiserslautern, Deutschland
[3] Lomonosov Moscow State University, 119992, Moscow, Russia

E-mail: yanabutko@yandex.ru, grothaus@mathematik.uni-kl.de, smolyanov@yandex.ru

Abstract. In the present note we consider a class of second order parabolic equations with position dependent coefficients; such equations describe a diffusion of (quasi) particles with a variable mass. We represent a solution of Cauchy–Dirichlet problem for such class of equations in a bounded domain in the form of a limit of finite dimensional integrals of elementary functions. Such kind of a representation is usually called Feynman formula and can be used for calculations. Finite dimensional integrals in our Feynman formula give approximations for a functional integral over a probability measure on a set of trajectories in the domain where the solution of the considered problem is investigated; this measure is generated by a diffusion process with variable diffusion coefficient and absorption on the boundary, hence, to get Feynman formula also means to get a representation of the solution of the considered problem with the help of a functional integral (such kind of a representation is usually called Feynman–Kac formula).

1. Introduction

In the present note we consider a class of second order parabolic equations with position dependent coefficients; such equations describe a diffusion of (quasi) particles with a variable mass. Quantum analogues of these (quasi) particles appear in models, describing semiconductors, liquid crystals etc.

We represent a solution of Cauchy–Dirichlet problem for such class of equations in a bounded domain in the form of a limit of finite dimensional integrals. Such kind of representations is usually called Feynman formulas. Following Feynman approach, limits in Feynman formulas can be interpreted as functional (path) integrals over suitable measures or pseudomeasures on the set of trajectories in configuration space. Functional integrals, representing solutions of evolutionary equations, are usually called Feynman–Kac formulas. It is Feynman–Kac formulas that allow to investigate properties of evolutionary equations by methods of stochastic analysis.

Hence, to get Feynman formulas is just another way to get Feynman–Kac formulas. On the other hand, Feynman formulas have their own advantages, since in most cases it is possible to represent solutions of evolutionary equations by Feynman formulas, containing elementary functions only, whereas corresponding Feynman–Kac formulas are actually limits of finite dimensional integrals, containing some transitional probabilities which often can not be expressed by elementary functions. Just this situation takes place in case of boundary value problems.

In the present work we obtain a Feynman formula for a diffusion of particles with variable mass in a bounded domain, applying a method developed in works [3] — [5] for investigation of diffusion and quantum evolution of particles with constant mass on Riemannian manifolds. This method is based on Chernoff theorem (see [9], [7]) and allows to investigate a wide class of initial and boundary value problems for evolutionary equations with linear and non-linear configuration space (see [2] — [8]).

2. Preliminaries

Let $f(\cdot,\cdot):[0,\infty)\times\mathbb{R}\to\mathbb{C}$. In papers [1], [10] the evolution of particles with position dependent mass is described by the Hamiltonian $\mathcal{H}$, such that

$$\mathcal{H}f(x)=\mathcal{H}_0 f(x)+V(x)f(x)$$

where

$$\mathcal{H}_0=-\frac{1}{4}(m^{\alpha}(x)\frac{\partial}{\partial x}m^{\beta}(x)\frac{\partial}{\partial x}m^{\gamma}(x)+m^{\gamma}(x)\frac{\partial}{\partial x}m^{\beta}(x)\frac{\partial}{\partial x}m^{\alpha}(x))$$

and $\alpha+\beta+\gamma=-1$. We assume that function $V:\mathbb{R}\to\mathbb{C}$ is continuous and function $m(\cdot):\mathbb{R}\to(0,\infty)$ is twice continuously differentiable on $\mathbb{R}$. If $m(x)\equiv\text{const}$ then operator $\mathcal{H}_0$ coincides with a standart quantum mechanics Hamiltonian of a one-dimensional free particle $\mathcal{H}_0=-\frac{1}{2m}\frac{\partial^2}{\partial x^2}$. Under assumptions above operator $\mathcal{H}_0$ is symmetric and there exist continuous functions $A(\cdot),B(\cdot),C(\cdot):\mathbb{R}\to\mathbb{R}$ such that $A(x)>0$ for any $x\in\mathbb{R}$ and

$$\mathcal{H}_0+V=-\frac{1}{2}A(\cdot)\frac{\partial^2}{\partial x^2}+B(\cdot)\frac{\partial}{\partial x}+C(\cdot).$$

Let $m\in\mathbb{N}$ and A a continuous mapping from $\mathbb{R}^m$ into the space $L(\mathbb{R}^m)$ of linear operators on $\mathbb{R}^m$ such that for any $x\in\mathbb{R}^m$ the operator $A(x)$ is symmetric and positive. Let $\Delta_A(x)$ be the differential operator acting on φ being twice differentiable in $x\in\mathbb{R}^m$ as follows: $(\Delta_A\varphi)(x):=\operatorname{tr}(A(x)\varphi^{(2)}(x))$, where $\varphi^{(2)}(x):=(\operatorname{Hess}\varphi)(x)$ is the Hessian matrix of φ in $x\in\mathbb{R}^m$. Let B be a continuous vector valued function on $\mathbb{R}^m$ and C be a continuous scalar function on $\mathbb{R}^m$. In the sequel we consider the Hamiltonian H defined for φ being twice differentiable in $x\in\mathbb{R}^m$ in the following way:

$$H\varphi(x):=\frac{1}{2}(\Delta_A\varphi)(x)+\langle B(x),\nabla\varphi(x)\rangle+C(x)\varphi(x),$$

where $\langle\cdot,\cdot\rangle$ denotes the euclidean scalar product in $\mathbb{R}^m$.

3. Cauchy–Dirichlet problem for a diffusion of particles with a variable mass in a bounded domain

Let $G\subset\mathbb{R}^m$ be a bounded domain. Its closure and boundary we denote by $\overline{G}$ and ∂G, respectively. For $f:[0,\infty)\times\overline{G}\to\mathbb{R}$ such that $f(t,\cdot)$ is twice differentiable on G for all $t\geq 0$ and $f(\cdot,x)$ is differentiable on $(0,\infty)$ for all $x\in G$ we consider the following Cauchy–Dirichlet problem:

$$\begin{aligned}\frac{\partial f}{\partial t}(t,x)&=Hf(t,x), && t>0,\ x\in G,\\ f(0,x)&=f_0(x), && x\in\overline{G},\\ f(t,x)&=0, && t\geq 0,\ x\in\partial G.\end{aligned}\tag{3.1}$$

Here we assume that $f(t,\cdot)\in C_0(\overline{G})$ for any $t\geq 0$, where $C_0(\overline{G})$ is Banach space of continuous functions on $\overline{G}$ vanishing on the boundary. The norm in $C_0(\overline{G})$ is defined by $\|f\|_{C_0(\overline{G})}=\sup_{x\in\overline{G}}|f(x)|$.

We assume that the domain G and the coefficients A, B and C satisfy sufficient conditions for the existence of a strongly continuous semigroup $(T_t)_{t\geq 0}$ on $C_0(\overline{G})$ providing a solution to (3.1). Then we construct a family $(F(t))_{t\geq 0}$ of integral operators, such that (by Chernoff theorem)

$$T(t) = \lim_{n\to\infty} (F(t/n))^n, \tag{3.2}$$

where the limit is in the strong operator topology in the space $L(C_0(\overline{G}))$ of all continuous linear operators in $C_0(\overline{G})$. Hence, a right hand side of (3.2) gives us a Feynman formula for the solution of the considered Cauchy–Dirichlet problem (3.1).

4. A construction of the family $(F(t))_{t\geq 0}$

Let X be a Banach space, $L(X)$ be the space of all continuous linear operators in X with strong operator topology, $\|\cdot\|$ denotes the operator norm on $L(X)$ and Id the identity operator in X. If $D(T) \subset X$ is a linear subspace and $T : D(T) \to X$ is linear (operator), then $D(T)$ denotes the domain of T.

Definition 4.1 *The derivative at the origin of a function $F : [0, \varepsilon) \to L(X)$, $\varepsilon > 0$, is a linear mapping $F'(0) : D(F'(0)) \to X$ such that*

$$F'(0)g := \lim_{t\searrow 0} t^{-1}(F(t)g - F(0)g),$$

where $D(F'(0))$ is the vector space of all elements $g \in X$ for which the above limit exists.

Theorem 4.2 (Chernoff theorem, see [7], [9]) *Let X be a Banach space, $F : [0,\infty) \to L(X)$ be a (strongly) continuous mapping such that $F(0) = Id$ and $\|F(t)\| \leq e^{at}$ for some $a \in [0,\infty)$ and all $t \geq 0$. Let D be a linear subspace of $D(F'(0))$ such that the restriction of the operator $F'(0)$ to this subspace is closable. Let $(L, D(L))$ be this closure. If $(L, D(L))$ is the generator of a strongly continuous semigroup $(T_t)_{t\geq 0}$, then for any $t_0 > 0$ the sequence $((F(t/n))^n)_{n\in\mathbb{N}}$ converges to $(T_t)_{t\geq 0}$ as $n \to \infty$ in the strong operator topology, uniformly with respect to $t \in [0, t_0]$, i.e., $T_t = \lim_{n\to\infty}(F(t/n))^n$ locally uniformly in $L(X)$.*

A family of operators $(F(t))_{t\geq 0}$ is called *Chernoff equivalent* to the semigroup $(T_t)_{t\geq 0}$ if this family satisfies the assertions of the Chernoff theorem with respect to this semigroup.

Our aim is to construct a family of integral operators $(F(t))_{t\geq 0}$, which is Chernoff equivalent to the semigroup $T(t)$, resolving the Cauchy–Dirichlet problem (3.1).

Let fix some $\varepsilon > 0$. Any function $\varphi \in C_0(\overline{G})$ can be extended to a continuous in $\mathbb{R}^m$ function $\widetilde{\varphi}$ with compact support in G^ε, where G^ε is ε-neighborhood of G in $\mathbb{R}^m$, and $\|\varphi(x)\|_{C_0(\overline{G})} \geq \sup\limits_{x\in\mathbb{R}^m} |\widetilde{\varphi}(x)|$. For each φ a function $\widetilde{\varphi}$ can be chosen in such way that for any $a, b \in \mathbb{R}$ if $\varphi = a\varphi_1 + b\varphi_2$ then $\widetilde{\varphi} = a\widetilde{\varphi_1} + b\widetilde{\varphi_2}$. Such function $\widetilde{\varphi}$ with $\varphi = \widetilde{\varphi}|_{\overline{G}}$ in sequel we call related to φ. Let's consider a set $D := \Big\{\varphi \in C_0(\overline{G}) \,|\, \varphi$ can be extended to some $\widetilde{\varphi} \in C_0^4(G^\varepsilon)$ with $\sup_{x\in\mathbb{R}^m} |\widetilde{\varphi}(x)| \leq \|\varphi\|_{C_0(\overline{G})}$ and $H\varphi \in C_0(\overline{G})\Big\}$, here $C_0^4(G^\varepsilon)$ denotes the space of four times continuously differentiable functions with compact support in G^ε. A set of all $\widetilde{\varphi} \in C_0^4(G^\varepsilon)$, such that $\widetilde{\varphi}|_{\overline{G}} = \varphi \in D$, we denote as $\widetilde{D}$. Hence, $D \subset C_0(\overline{G})$, $\widetilde{D} \subset C_0^4(G^\varepsilon) \subset C(\mathbb{R}^m)$. We denote the generator of $(T_t)_{t\geq 0}$ by (H, D(H)) and assume that the set D is a core of $(H, D(H))$.

Let $a(x) := \det A(x)$. We introduce a family of operators $(F_1(t))_{t\geq 0}$ in $C(\mathbb{R}^m)$ such that $F_1(0) = Id$ and for $t > 0$ the operator $F_1(t)$ is given by the formula:

$$F_1(t)\widetilde{\varphi}(x) = \frac{1}{\sqrt{a(x)(2\pi t)^m}} \int_{\mathbb{R}^m} \exp\{-\frac{\langle A^{-1}(x)(x-y), x-y\rangle}{2t}\}\widetilde{\varphi}(y)dy. \tag{4.1}$$

Journal of Physics: Conference Series **128** (2008) 012050 doi:10.1088/1742-6596/128/1/012050

If A is a constant matrix this family coincides with the semigroup $(e^{\frac{t}{2}\Delta_A})_{t\geq 0}$. If A is not constant it is not so anymore. Nevertheless, the following holds:

Lemma 4.3 *For $\widetilde{\varphi} \in \widetilde{D}$ the derivative of $F_1(t)\widetilde{\varphi}(x)$ at zero coincides with $\frac{1}{2}\Delta_A\widetilde{\varphi}(x)$ uniformly in $x \in \overline{G^\varepsilon}$, i.e., for any $\widetilde{\varphi} \in \widetilde{D}$ the following is valid uniformly with respect to $x \in \overline{G^\varepsilon}$ as $t \searrow 0$:*

$$F_1(t)\widetilde{\varphi}(x) = \widetilde{\varphi}(x) + \frac{1}{2}\Delta_A\widetilde{\varphi}(x) + o(t).$$

Proof: Let $c(t,x) := (a(x)(2\pi t)^m)^{-1/2}$, $t \geq 0$, $x \in \overline{G^\varepsilon}$. Then

$$\frac{1}{t}(F_1(t)\widetilde{\varphi} - \widetilde{\varphi})(x) = \frac{c(t,x)}{t}\cdot$$
$$\cdot\left(\int\limits_{\mathbb{R}^m} \exp\{-\frac{\langle A^{-1}(x)(x-y), x-y\rangle}{2t}\}\widetilde{\varphi}(y)dy - \int\limits_{\mathbb{R}^m} \exp\{-\frac{\langle A^{-1}(x)(x-y), x-y\rangle}{2t}\}\widetilde{\varphi}(x)dy\right) =$$
$$= \frac{c(t,x)}{t}\int_{\mathbb{R}^m} \exp\{-\frac{\langle A^{-1}(x)(x-y), x-y\rangle}{2t}\}(\widetilde{\varphi}(y) - \widetilde{\varphi}(x))dy.$$

Since $\widetilde{\varphi} \in C_0^4(G^\varepsilon) \subset C^4(\mathbb{R}^m)$ a Taylor expansion at the point x yields

$$\widetilde{\varphi}(y) - \widetilde{\varphi}(x) = \widetilde{\varphi}^{(1)}(x)(y-x) + \frac{1}{2}\widetilde{\varphi}^{(2)}(x)(y-x)^2$$
$$+ \frac{1}{3!}\widetilde{\varphi}^{(3)}(x)(y-x)^3 + \frac{1}{4!}\widetilde{\varphi}^{(4)}(\theta x + (1-\theta)y)(y-x)^4,$$

where $\widetilde{\varphi}^{(k)}(x)$ is understood as a k-linear functional on $(\mathbb{R}^m)^k$ and a symbol $\widetilde{\varphi}^{(k)}(x)(y-x)^k$ stands for the result of application of $\widetilde{\varphi}^{(k)}(x)$ to the vector $(y-x) \in \mathbb{R}^m$ taken k times, $\theta \in [0,1]$. Hence,

$$\frac{1}{t}(F_1(t)\widetilde{\varphi} - \widetilde{\varphi})(x)$$
$$= \frac{1}{t}c(t,x)\int_{\mathbb{R}^m} e^{\frac{-\langle A^{-1}(x)(x-y), x-y\rangle}{2t}}(\widetilde{\varphi}^{(1)}(x)(y-x) + \ldots + \frac{1}{4!}\widetilde{\varphi}^{(4)}(\theta x + (1-\theta)y)(y-x)^4)dy =$$
$$= \frac{1}{2t}c(t,x)\int_{\mathbb{R}^m} \exp\{\frac{-\langle A^{-1}(x)(x-y), x-y\rangle}{2t}\}\widetilde{\varphi}^{(2)}(x)(y-x)^2dy +$$
$$+ t\frac{1}{24t^2}c(t,x)\int_{\mathbb{R}^m} \exp\{\frac{-\langle A^{-1}(x)(x-y), x-y\rangle}{2t}\}\widetilde{\varphi}^{(4)}(\theta x + (1-\theta)y)(y-x)^4dy.$$

Since $\widetilde{\varphi}^{(4)}$ is bounded, the last term converges to zero as $t \searrow 0$ uniformly in $x \in \overline{G^\varepsilon}$. As in the case of a constant A (now with a parameter), the remaining term converges to $\frac{1}{2}\Delta_A\widetilde{\varphi}(x)$ as $t \searrow 0$ uniformly in $x \in \overline{G^\varepsilon}$. ■

Let us now define a family of operators $(F_2(t))_{t\geq 0}$ in $C(\mathbb{R}^m)$ such that $F_2(0) = Id$ and for $t > 0$ the operator $F_2(t)$ is given by the formula:

$$F_2(t)\widetilde{\varphi}(x) =$$
$$= \frac{1}{\sqrt{a(x)(2\pi t)^m}}\int_{\mathbb{R}^m} \exp\{\frac{-\langle A^{-1}(x)(x-y), x-y\rangle}{2t}\}\exp\{-\langle A^{-1}B(x), x-y\rangle\}\widetilde{\varphi}(y)dy,$$

where $A^{-1}B(x) := A^{-1}(x)B(x)$.

Journal of Physics: Conference Series **128** (2008) 012050 doi:10.1088/1742-6596/128/1/012050

Lemma 4.4 *For any $\widetilde{\varphi} \in \widetilde{D}$ as $t \searrow 0$ we have uniformly with respect to $x \in \overline{G^{\varepsilon}}$:*

$$F_2(t)\widetilde{\varphi}(x) = \widetilde{\varphi}(x) + \frac{t}{2}\Delta_A\widetilde{\varphi}(x) + t\langle B(x), \nabla\widetilde{\varphi}(x)\rangle + \frac{t}{2}\langle A^{-1}B(x), B(x)\rangle\widetilde{\varphi}(x) + o(t).$$

Proof: Let us notice that if $\widetilde{\varphi} \in \widetilde{D}$, then also $\exp\{-\langle A^{-1}B(x), x - \cdot\rangle\}\widetilde{\varphi} \in \widetilde{D}$ for all $x \in \overline{G^{\varepsilon}}$ and

$$F_2(t)\widetilde{\varphi}(x) = F_1(t)(\exp\{-\langle A^{-1}B(x), x - \cdot\rangle\}\widetilde{\varphi})(x).$$

By Lemma 4.3 we get

$$\begin{aligned} F_2(t)\widetilde{\varphi}(x) &= \\ &= (\exp\{-\langle A^{-1}B(x), x - \cdot\rangle\}\widetilde{\varphi})(x) + \frac{t}{2}\Delta_A(\exp\{-\langle A^{-1}B(x), x - \cdot\rangle\}\widetilde{\varphi})(x) + o(t) = \\ &= \widetilde{\varphi}(x) + \frac{t}{2}\mathrm{tr}\left(A(\widetilde{\varphi}^{(2)} + 2A^{-1}B \otimes \nabla\widetilde{\varphi} + (A^{-1}B \otimes A^{-1}B)\widetilde{\varphi})\right)(x) + o(t) = \\ &= \widetilde{\varphi}(x) + \frac{t}{2}\Delta_A\widetilde{\varphi}(x) + t\langle B(x), \nabla\widetilde{\varphi}(x)\rangle + \frac{t}{2}\langle A^{-1}B(x), B(x)\rangle\widetilde{\varphi}(x) + o(t). \end{aligned}$$

■

Let $s : (0, \infty) \to (0, \infty)$ be a smooth function which monotonically decreases to 0 as $t \searrow 0$ such that $s(t) = o(t)$. For example, $s = c\,\mathrm{arctg}$ for some $0 < c < \frac{1}{\pi}\mathrm{diam}(G)$. Let $G_{s(t)} \subset G$ be defined as $G_{s(t)} = \{x \in G \,|\, \mathrm{dist}(x, \partial G) > s(t)\}$. Let $\psi_{s(t)} : \mathbb{R}^m \to [0, 1]$, $t > 0$ be a family of smooth functions such that $\psi_{s(t)}(x) = 1$ for $x \in G_{s(t)}$, $\psi_{s(t)}(x) = 0$ for $x \in \mathbb{R}^m \setminus G$. Hence the family $\psi_{s(t)}$ approximates the indicator function of the domain G pointwisely as $t \searrow 0$.

Next let us introduce a family of operators $(F_3(t))_{t>0}$ acting on $\widetilde{\varphi} \in C(\mathbb{R}^m)$ as follows:

$$F_3(t)\widetilde{\varphi}(x) = \psi_{s(t)}(x)e^{tV(x)}\widetilde{\varphi}(x),$$

where $V(x) = C(x) - \frac{1}{2}\langle A^{-1}B(x), B(x)\rangle$, $x \in \mathbb{R}^m$. Finally, we define the family of operators $(F(t))_{t\geq 0}$ on $C_0(\overline{G})$ in the following way: $F(0) = Id$ and, for $t > 0$ and for any $\varphi \in C_0(\overline{G})$, $F(t)\varphi = F_3(t)F_2(t)\widetilde{\varphi}$, where $\widetilde{\varphi}$ is related to φ as above.

Lemma 4.5 *For any $\varphi \in D$ as $t \searrow 0$ we have uniformly with respect to $x \in \overline{G}$:*

$$F(t)\varphi(x) = \varphi(x) + \frac{t}{2}\Delta_A\varphi(x) + t\langle B(x), \nabla\varphi(x)\rangle + tC(x)\varphi(x) + o(t).$$

Proof: Since V is a continuous function, by a Taylor expansion we get $e^{tV(x)} = 1 + tV(x) + o(t)$ as $t \searrow 0$ uniformly in $x \in \overline{G}$. Thus, by Lemma 4.4 we get

$$\begin{aligned} F(t)\varphi(x) &= \psi_{s(t)}(x)\big(1 + tV(x) + o(t)\big) \\ &\quad \times \left(\widetilde{\varphi}(x) + \frac{t}{2}\Delta_A\widetilde{\varphi}(x) + t\langle B, \nabla\widetilde{\varphi}\rangle(x) + \frac{t}{2}\langle A^{-1}B, B\rangle\widetilde{\varphi}(x) + o(t)\right) \\ &= \psi_{s(t)}\left(\widetilde{\varphi} + \frac{t}{2}\Delta_A\widetilde{\varphi} + t\langle B, \nabla\widetilde{\varphi}\rangle + tC\widetilde{\varphi}\right)(x) + o(t) = \psi_{s(t)}\left(\widetilde{\varphi} + tH\widetilde{\varphi}\right)(x) + o(t) \end{aligned}$$

for any $\widetilde{\varphi} \in \widetilde{D}$, uniformly in $x \in \overline{G^{\varepsilon}}$.

Let us show that $F(t)\varphi(x) = \varphi(x) + tH\varphi(x) + o(t)$ as $t \searrow 0$, uniformly with respect to $x \in \overline{G}$, for all $\varphi \in D$. Let firstly $x \in G_{s(t)}$. Then $\psi_{s(t)}(x) = 1$ and $F(t)\varphi(x) = \varphi(x) + tH\varphi(x) + o(t)$ as $t \searrow 0$, uniformly with respect to $x \in G_{s(t)}$. If $x \in \partial G$, then

$\psi_{s(t)}(x) = 0$ and $F(t)\varphi(x) = 0 = \varphi(x) + tH\varphi(x)$. Finally let $x \in G \setminus G_{s(t)}$. Then $|F(t)\varphi(x) - \varphi(x) - tH\varphi(x)| = \big|(1 - \psi_{s(t)}(x))(\varphi(x) + tH\varphi(x)) + o(t)\big| \leq |\varphi(x) + tH\varphi(x)| + o(t)$ as $t \searrow 0$. For any $x \in G \setminus G_{s(t)}$ there exist at least one $x_{\text{bd}} \in \partial G$ such that $\text{dist}(x, x_{\text{bd}}) = \text{dist}(x, \partial G) \leq s(t) = o(t)$ as $t \searrow 0$. Since $\varphi \in D$, we have $\varphi(x) = \varphi(x_{\text{bd}}) + \widetilde{\varphi}^{(1)}(c_x)(x - x_{\text{bd}})$, where $c_x = \theta x + (1-\theta)x_{\text{bd}}$ for some $\theta \in [0,1]$ and $\widetilde{\varphi}$ is related to φ as above. Therefore $|\varphi(x)| \leq s(t) \sup_{y\in\mathbb{R}^m} |\widetilde{\varphi}^{(1)}(y)| = o(t)$ as $t \searrow 0$, uniformly with respect to $x \in G \setminus G_{s(t)}$. Since $H\varphi \in C_0(\overline{G})$, also $H\varphi(x) \to 0$ uniformly with respect to $x \in G \setminus G_{s(t)}$ for $t \searrow 0$. Thus $tH\varphi = o(t)$. Therefore $|F(t)\varphi(x) - \varphi(x) - tH\varphi(x)| \leq o(t)$ as $t \searrow 0$, uniformly with respect to $x \in G \setminus G_{s(t)}$. Summarizing, we get for any $\varphi \in D$ and uniformly with respect to $x \in \overline{G}$: $F(t)\varphi(x) = \varphi(x) + tH\varphi(x) + o(t)$ as $t \searrow 0$. ■

Theorem 4.6 *Let $(T_t)_{t\geq 0}$ be the strongly continuous semigroup on $C_0(\overline{G})$ providing a solution to the Cauchy-Dirichlet problem (3.1). Then the family of operators $(F(t))_{t\geq 0}$ is Chernoff equivalent to this semigroup and therefore*

$$T_t = \lim_{n\to\infty} [F(t/n)]^n$$

in $L(C_0(\overline{G}))$, locally uniformly with respect to $t \geq 0$.

Proof: We have to prove that $(F(t))_{t\geq 0}$ is Chernoff equivalent to $(T_t)_{t\geq 0}$. Since by Lemma 4.5 the derivative of $F(t)$ on $C_0(\overline{G})$ at zero coincides with the generator of the semigroup $(T_t)_{t\geq 0}$ on its core D, we only need to estimate the norm of the operators $F(t)$ and to show strong continuity of the family $(F(t))_{t\geq 0}$. The norm is estimated as follows (here $X = C_0(\overline{G})$ and $\widetilde{\varphi}$ is related to φ as above): $||F(t)|| =$

$$= \sup_{||\varphi||_X \leq 1} \sup_{x\in G} \left| \frac{\psi_{s(t)}(x)e^{tV(x)}}{\sqrt{a(x)(2\pi t)^m}} \int_{\mathbb{R}^m} \exp\{\tfrac{-\langle A^{-1}(x)(x-y), x-y\rangle}{2t}\} \exp\{-\langle A^{-1}B(x), x-y\rangle\} \widetilde{\varphi}(y)dy \right| =$$

$$= \sup_{||\varphi||_X \leq 1} \sup_{x\in G} \left| \frac{\psi_{s(t)}(x)e^{tV(x)}}{\sqrt{a(x)(2\pi t)^m}} \int_{\mathbb{R}^m} \exp\{\tfrac{-\langle A^{-1}(x)(x-y+tB(x)), x-y+tB(x)\rangle}{2t}\} \exp\{\tfrac{t}{2}\langle A^{-1}B, B\rangle(x)\} \widetilde{\varphi}(y)dy \right| \leq$$

$$\leq e^{t\sup_{x\in G}|C(x)|} \sup_{x\in G} \left| \frac{1}{\sqrt{a(x)(2\pi t)^m}} \int_{\mathbb{R}^m} \exp\{\tfrac{-\langle A^{-1}(x)(x-y+tB(x)), x-y+tB(x)\rangle}{2t}\} d(y - tB(x)) \right| =$$

$$= e^{t\sup_{x\in G}|C(x)|}.$$

Next let us show strong continuity of the family $(F(t))_{t\geq 0}$. For this it is sufficient to show that $\lim_{t\to t_0} \|F(t)\varphi - F(t_0)\varphi\|_{C_0(\overline{G})} = 0$ for all $\varphi \in D$ and any $t_0 \geq 0$. For $t_0 > 0$ this is true, because all functions in the formula of $F(t)$ are continuous with respect to $t \in (0, +\infty)$, $\widetilde{\varphi}$ is continuous with compact support, and $\exp\{\frac{-\langle A^{-1}(x)(x-y), x-y\rangle}{2t}\}$ is a continuous function of the three arguments $(t, x, y) \in [t_1, t_2] \times \overline{G^\varepsilon} \times \overline{G^\varepsilon}$, where $t_0 \in (t_1, t_2)$, $[t_1, t_2] \subset (0, \infty)$.

If $\varphi \in D$, then

$$\lim_{t\searrow 0} \frac{1}{\sqrt{a(x)(2\pi t)^m}} \int_{\mathbb{R}^m} \exp\{\frac{-\langle A^{-1}(x)(x-y), x-y\rangle}{2t}\} \widetilde{\varphi}(y)d(y) = \varphi(x),$$

uniformly in $x \in \overline{G}$, since for any fixed $x \in \overline{G}$

$$f_t^x(y) = \frac{1}{\sqrt{a(x)(2\pi t)^m}} \exp\{\frac{-\langle A^{-1}(x)(x-y), x-y\rangle}{2t}\}, \quad y \in \mathbb{R}^m,$$

tends to the Dirac function δ_x as $t \searrow 0$ and $\widetilde{\varphi}$ is bounded and uniformly continuous. Hence, $\lim_{t\searrow 0} F(t)\varphi = \varphi$ in $C_0(\overline{G})$ for all $\varphi \in D$. Thus, by Chernoff's theorem the statement follows. ■

5. Feynman and Feynman–Kac formulae

By theorem (4.6) the solution $f(t,x)$ of Cauchy–Dirichlet problem (3.1) with initial condition $f(0,x) = f_0(x)$ can be obtained by the formula $f(t,x) = \lim\limits_{n\to\infty}[F(t/n)]^n f_0(x)$, where for any $\varphi \in C_0(\overline{G})$

$F(t)\varphi(x) = \frac{\psi_{s(t)}(x)e^{tV(x)}}{\sqrt{a(x)(2\pi t)^m}} \int\limits_{\mathbb{R}^m} \exp\{-\frac{\langle A^{-1}(x)(x-y),x-y\rangle}{2t}\}\exp\{-\langle A^{-1}B(x), x-y\rangle\}\widetilde{\varphi}(y)dy.$

Thus,

$$f(t,x) = \lim_{n\to\infty} \int\limits_{\mathbb{R}^m} \dots \int\limits_{\mathbb{R}^m} \left(\prod_{j=1}^{n} \psi_{s(t/n)}(x_j)\right) \exp\{\frac{t}{n}\sum_{j=1}^{n} V(x_{j-1})\}\cdot$$

$$\cdot \exp\{\sum_{j=1}^{n}\langle A^{-1}B(x_{j-1}), x_j - x_{j-1}\rangle\} p_A(\frac{t}{n}, x_0, x_1)...p_A(\frac{t}{n}, x_{n-1}, x_n)\widetilde{f}_0(x_n)dx_1...dx_n, \quad (5.1)$$

where $p_A(t,x,y) = \frac{1}{\sqrt{a(x)(2\pi t)^m}}\exp\{-\frac{\langle A^{-1}(x)(x-y),x-y\rangle}{2t}\}$, $x_0 = x$ and $\widetilde{f}_0$ is related to f_0.

Let a symbol γ_G denote the indicator of the set G, i.e. $\gamma_G(x) = 1$ if $x \in G$ and $\gamma_G(x) = 0$ if $x \notin G$. Neither the speed of convergence of $s(t) \to 0$ (if it is $o(t)$), nor the choice of a family $\{\psi_{s(t)}\}$ approximating γ_G when $t \to 0$ can change the limit in the formula (5.1). Since $\psi_{s(t)}$ is a smooth function with a compact support in G and $\widetilde{f}_0$ is a continuous function with a compact support in G^ε, we actually integrate a continuous function over a compact $\overline{G}^{n-1} \times \overline{G^\varepsilon}$ in the formula (5.1) and, hence, as $\psi_{s(t)} \to \gamma_G$ when $t \to 0$ we can choose $s(t/n)$ such that

$$\Bigg| \int\limits_{\overline{G^\varepsilon}} \int\limits_{\overline{G}^{n-1}} \exp\{\frac{t}{n}\sum_{j=1}^{n} V(x_{j-1})\}\exp\{\sum_{j=1}^{n}\langle A^{-1}B(x_{j-1}), x_j - x_{j-1}\rangle\}p_A(t/n, x_0, x_1)\cdot ...$$

$$... \cdot p_A(t/n, x_{n-1}, x_n)\widetilde{f}_0(x_n)\left(\prod_{k=1}^{n}(\psi_{s(t/n)}(x_{k-1}) - \gamma_G(x_{k-1}))\right) dx_1...dx_n \Bigg| < \frac{1}{n},$$

Then the limit in the formula (5.1) coincides with the following limit:

$$f(t,x) = \lim_{n\to\infty} \int\limits_{\overline{G^\varepsilon}} \int\limits_{\overline{G}^{n-1}} \exp\{\frac{t}{n}\sum_{j=1}^{n} V(x_{j-1})\}\exp\{\sum_{j=1}^{n}\langle A^{-1}B(x_{j-1}), x_j - x_{j-1}\rangle\}\cdot$$

$$\cdot p_A(t/n, x_0, x_1)...p_A(t/n, x_{n-1}, x_n)\widetilde{f}_0(x_n)dx_1...dx_n. \quad (5.2)$$

Let $F_n^x(x_n) = \int\limits_{\overline{G}^{n-1}} \exp\{\frac{t}{n}\sum\limits_{j=1}^{n} V(x_{j-1})\}\exp\{\sum\limits_{j=1}^{n}\langle A^{-1}B(x_{j-1}), x_j - x_{j-1}\rangle\}p_A(t/n, x_0, x_1)...$ $...p_A(t/n, x_{n-1}, x_n)dx_1...dx_{n-1}$. Then $f(t,x) = \lim\limits_{n\to\infty}\int\limits_{\overline{G^\varepsilon}} \widetilde{f}_0(y)F_n^x(y)dy$. Since for any $\varepsilon > 0$ the function $f(t,x)$ does not depend on an extension $\widetilde{f}_0$ of the initial condition f_0 onto $G^\varepsilon \setminus \overline{G}$, we get $f(t,x) = \lim\limits_{\varepsilon\to 0}\lim\limits_{n\to\infty}\int\limits_{\overline{G^\varepsilon}} \widetilde{f}_0(y)F_n^x(y)dy$ and by Lebesgue dominated convergence theorem $f(t,x) = \lim\limits_{\varepsilon\to 0}\int\limits_{\overline{G^\varepsilon}} \widetilde{f}_0(y)\lim\limits_{n\to\infty} F_n^x(y)dy = \int\limits_{\overline{G}} f_0(y)\lim\limits_{n\to\infty} F_n^x(y)dy = \lim\limits_{n\to\infty}\int\limits_{\overline{G}} f_0(y)F_n^x(y)dy$. Hence, the following theorem is proved:

Journal of Physics: Conference Series **128** (2008) 012050 doi:10.1088/1742-6596/128/1/012050

Theorem 5.1 *Let $f(t,x)$ be a solution of Cauchy–Dirichlet problem (3.1). Then the following Feynman formula is true:*

$$f(t,x) = \lim_{n\to\infty} \int_G \dots \int_G \exp\{\frac{t}{n}\sum_{j=1}^{n} V(x_{j-1})\} \exp\{\sum_{j=1}^{n} \langle A^{-1}B(x_{j-1}), x_j - x_{j-1}\rangle\}\cdot$$

$$\cdot\, p_A(t/n, x_0, x_1)\dots p_A(t/n, x_{n-1}, x_n) f_0(x_n) dx_1 \dots dx_n. \quad (5.3)$$

Remark 5.2. One can show in a similar way that the solution of Cauchy Problem

$$\frac{\partial f}{\partial t}(t,x) = Hf(t,x),$$
$$f(0,x) = f_0(x)$$

in the whole space $\mathbb{R}^m$ can be represented by the same Feynman formula (5.3) where integrals over G are substituted by integrals over $\mathbb{R}^m$.

Remark 5.3. Let now

$$\Phi(\xi) = \exp\{-\frac{1}{2}\int_0^t \langle A^{-1}B(\xi(\tau)), B(\xi(\tau))\rangle d\tau\}\cdot$$

$$\cdot \exp\{\int_0^t \langle A^{-1}B(\xi(\tau)), d\xi(\tau)\rangle\} \cdot \exp\{\int_0^t C(\xi(\tau))d\tau\} f_0(\xi(t)).$$

Using a fact that the equation $\frac{\partial f}{\partial t}(t,x) = \frac{1}{2}\Delta_A f(t,x)$ is a Kolmogorov equation for the Ito stochastic differential equation $d\xi(t) = \sqrt{A(\xi(t))}dw(t)$ (where $\sqrt{A(x)}$ means the positive square root), one can show that finite dimensional integrals at right hand side of the Feynman formula (5.3) are finite dimensional approximations of an integral of the function Φ over a measure μ, generated by a diffusion process, in the domain G with absorption on the boundary, governed by this stochastic differential equation. This means that the following Feynman–Kac formula is true:

$$f(t,x) = \int_{C([0,t],G)} \Phi(\xi)\mu(d\xi).$$

It is worth noticing, that just the Feynman formula, containing elementary functions only, gives an effective method to calculate the integral in the Feynman–Kac formula.

Acknowledgments

This work has been supported by RFBR, grant N 06-01-00761.

References

[1] Gadella M, Kuru S, Negro J 2007 Self adjoint Hamiltonians with a mass jump: General matching conditions *Phys. Lett.* A **362** 265-268

[2] Gadella M, Smolyanov O G 2008 Feynman Formulas for Particles with Position-Dependent Mass, *Doklady Math.* **77** 120-124

[3] Weizsacker H v, Smolyanov O G, Wittich O 2000 Diffusion on Compact Riemannian Manifolds and Surface Measures, *Doklady Math.* **61** 230-235

[4] Smolyanov O G, Weizsäcker H v, Wittich O 2000 Brownian Motion on a Manifold as Limit of Stepwise Conditioned Standard Brownian Motions *Stochastic Proceses, Physics and Geometry: New Interplays. II: A Volume in Honor of Sergio Albeverio. Ser. Conference Proceedings. Canadian Math. Society. Providence: AMS,* **29** 589-602

[5] Smolyanov O G, Weizsaecker H v, Wittich O 2007 Surface Measures and Initial Boundary Value Problems Generated by Diffusions with Drift *Doklady Math.* **76** 606-610

[6] Smolyanov O G, Weizsaecker H v, Wittich O 2007 Chernoff's Theorem and Discrete Time Approximations of Brownian Motion on Manifolds *Potent. Anal.* **26** 1-29

[7] Smolyanov O G, Tokarev A G, Truman A 2002 Hamiltonian Feynman path integrals via the Chernoff formula, *J. Math. Phys.* **43** 5161-5171

[8] Butko Ya A 2008 Feynman formulas and functional integrals for diffusion with drift in a domain on a manifold, *Math. Notes* **83** 301-316

[9] Engel K-J, Nagel R 2000 One Parameter Semigroups for Linear Evolution Equations. B.: Springer

[10] Thomsen J, Einevoll G T, Hemmer P C 1989 Operator ordering in effective-mass theory *Phys. Rev.* B **39** 783-788

V International Symposium on Quantum Theory and Symmetries IOP Publishing
Journal of Physics: Conference Series **128** (2008) 012051 doi:10.1088/1742-6596/128/1/012051

On a Curvature-Statistics Theorem

M. Calixto[1,2] and V. Aldaya[2]

[1]Departamento de Matemática Aplicada y Estadística, Universidad Politécnica de Cartagena, Paseo Alfonso XIII 56, 30203 Cartagena, Spain
[2]Instituto de Astrofísica de Andalucía, Apartado Postal 3004, 18080 Granada, Spain

E-mail: Manuel.Calixto@upct.es

Abstract. The spin-statistics theorem in quantum field theory relates the spin of a particle to the statistics obeyed by that particle. Here we investigate an interesting correspondence or connection between curvature ($\kappa = \pm 1$) and quantum statistics (Fermi-Dirac and Bose-Einstein, respectively). The interrelation between both concepts is established through vacuum coherent configurations of *zero modes* in quantum field theory on the compact $O(3)$ and noncompact $O(2,1)$ (spatial) isometry subgroups of de Sitter and Anti de Sitter spaces, respectively. The high frequency limit, is retrieved as a (zero curvature) group contraction to the Newton-Hooke (harmonic oscillator) group. We also make some comments on the physical significance of the vacuum energy density and the cosmological constant problem.

1. Introduction

Quantum vacua are not really empty. We know that zero-point energy, like other non-zero vacuum expectation values, leads to observable consequences like, for instance, the Casimir effect [1], and influences the behaviour of the Universe at cosmological scales, where the vacuum (dark) energy is expected to contribute to the cosmological constant, which affects the expansion of the universe.

In Quantum Field Theory, one expects the vacuum state to be stable under some underlying symmetry group G (viz, the Poincaré group). Then the action of some spontaneously broken symmetry transformations can destabilize the vacuum and make it to *radiate*. Such is the case of the Planckian radiation of the Poincaré invariant vacuum under uniform accelerations, that is, the Unruh effect [2]. Here, the Poincaré invariant vacuum looks the same to any inertial observer but converts into a thermal bath of radiation with temperature

$$T = \hbar a / 2\pi c k_B$$

in passing to a uniformly accelerated frame (a denotes the acceleration, c the speed of light and k_B the Boltzmann constant). These radiation phenomena are usually linked to some kind of global mutilation of the spacetime (namely, existence of horizons). In the reference [3], it was shown that the reason for this radiation is more profound and related to the spontaneous breakdown of the conformal symmetry in *quantum* field theory. From this point of view, a Poincaré invariant vacuum can be regarded as a coherent state of conformal zero modes, which are undetectable ("dark") by inertial observers but unstable under relativistic uniform accelerations (special conformal transformations). There we used the conformal group

Journal of Physics: Conference Series **128** (2008) 012051 doi:10.1088/1742-6596/128/1/012051

in (1+1)-dimensions, $SO(2,2) \simeq SO(2,1) \times SO(2,1)$, which consists of two copies of the pseudo-orthogonal group $SO(2,1)$ (left- and right-moving modes, respectively). Here we study in general the breakdown of a symmetry G in quantum field theory and apply to the particular cases of $O(3), O(2,1)$ and Newton-Hooke groups (previously developed in [4]). We could think of $O(3)$ and $O(2,1)$ as isometry subgroups of the spatial part of de Sitter and Anti de Sitter spaces, with positive and negative curvature κ, respectively.

The organization of the paper is as follows. In Sec. 3, we construct $G = O(3), O(2,1)$ and Newton-Hooke invariant quantum field theories, in a unified manner. To be more precise, we shall work with their double covers $U(2)$ and $U(1,1)$ instead, for convenience. Then we shall break the original symmetry G by choosing a pseudo-vacuum $|\theta\rangle$, which turns out to be a coherent state of zero modes, invariant under a subgroup $G_0 \subset G$. We could think of zero modes as "virtual" particles without "bright" energy (see later on Sec. 3) and undetectable by G_0 observers. However, we shall show that a general symmetry transformation, which incorporates the broken symmetry generators in G/G_0, produces a "rearrangement" of the the G_0-invariant vacuum $|\theta\rangle$ and makes it to *radiate*. In other words, we shall associate a *thermal bath* with the excited (or, let us say, "polarized") G_0-vacuum and we shall show in Sections 4 and 5 that the *mean energy per mode* matches the Bose-Einstein (BE) and Fermi-Dirac (FD) expressions, for the non-compact $U(1,1)$ and compact $U(2)$ isometry groups ($\kappa = \mp 1$), respectively. The limit of high frequencies (or large interlevel energy spacing $\hbar\omega \gg k_B T$) is regained from a contraction to the Newton-Hooke group (zero curvature, $\kappa = 0$), as explained in Sect. 6. The last Section is devoted to some comments on the vacuum energy and its implications in cosmology.

Before, we describe in Sec. 2 the general construction of a G-invariant quantum field theory as a second quantization on a group G and the spontaneous breakdown of G to G_0, the stability subgroup of the degenerated vacuum. We shall provide the essential ingredients to calculate probability distributions and average energies of the ground state excitations.

2. The general context

Let $\mathcal{G} = \{X_\alpha, \alpha = 0, 1, 2, \ldots, l\}$ be the (Lie) algebra of observables of a given quantum system, among which we highlight X_0 as the Hamiltonian operator. Let $\mathcal{H}$ be the (Hilbert) carrier space of a unitary irreducible representation U of the Lie group G. Let us assume that the energy spectrum is discrete and bounded from bellow, that is, there is a vacuum vector $|0\rangle$ whose energy E_0 we set to zero, i.e. $X_0|0\rangle = 0$. Let $B(\mathcal{H}) = \{|n\rangle, n = 0, 1, 2, \ldots\}$ be a (finite or infinite) orthonormal basis of $\mathcal{H}$ made up of energy eigenvectors, $X_0|n\rangle = E_n|n\rangle$.

An important ingredient to construct a G-invariant quantum field theory, as a second-quantized (many-particle) theory, will be the irreducible matrix coefficients

$$U_{mn}(x) \equiv \langle m|U(x)|n\rangle, \tag{1}$$

of the representation $U(x) = e^{i\sum_\alpha x^\alpha X_\alpha}$ of G in the orthonormal basis $B(\mathcal{H})$, where $\{x^\alpha \in \mathbb{R}, \alpha = 0, \ldots, \dim(G) - 1\}$ stands for a coordinate system in G.

Given the Fourier expansion, in energy modes, of a field

$$|\psi\rangle = \sum_{n=0} a_n |n\rangle\,, \tag{2}$$

the Fourier coefficients a_n (resp. $a_n^\dagger$) are promoted to annihilation (resp. creation) operators of energy modes E_n in the second quantized theory, with commutation relations $[a_n, a_m^\dagger] = \delta_{n,m}$. The (finite) action of G on annihilation operators is:

$$a_m \to a'_m = \sum_{n=0} U_{mn}(x) a_n, \tag{3}$$

Journal of Physics: Conference Series **128** (2008) 012051 doi:10.1088/1742-6596/128/1/012051

together with the conjugated expression for creation operators. The infinitesimal generators of this finite action are the second quantized version, $\hat{X}_\alpha$, of the basic observables X_α. They have the following explicit expression in terms of creation and annihilation operators:

$$\hat{X}_\alpha = -i \sum_{m,n=0} a_m^\dagger \left. \frac{\partial U_{mn}(x)}{\partial x^\alpha} \right|_{x=0} a_n. \tag{4}$$

For example, since $|n\rangle$ are energy E_n eigenstates and we have set $E_0 = 0$, the Hamiltonian operator is:

$$\hat{X}_0 = \sum_{n=1} E_n a_n^\dagger a_n. \tag{5}$$

The vacuum $|\hat{0}\rangle$ of the second quantized theory is characterized by being stable under the symmetry group G, i.e. it is annihilated by the symmetry generators

$$\hat{X}_\alpha |\hat{0}\rangle = 0, \ \alpha = 0, \dots, \dim(G) - 1 \tag{6}$$

and also by $a_n|0\rangle = 0, \forall n \geq 0$. Then an orthonormal basis for the Hilbert space of the second quantized theory is constructed by taking the orbit through the vacuum $|\hat{0}\rangle$ of the creation operators $a_n^\dagger$:

$$|q(n_1), \dots, q(n_p)\rangle \equiv \frac{(a_{n_1}^\dagger)^{q(n_1)} \dots (a_{n_p}^\dagger)^{q(n_p)}}{(q(n_1)! \dots q(n_p)!)^{1/2}} |\hat{0}\rangle, \tag{7}$$

where $q(n) \in \mathbb{N}$ denotes the occupation number of the energy level n.

Note that any multi-particle state (7) made up of an arbitrary content of *zero modes*, like

$$|\theta\rangle \equiv \sum_{q=0}^{\infty} \theta_q (a_0^\dagger)^q |\hat{0}\rangle, \ \ \theta_q \in \mathbb{C}, \tag{8}$$

has zero total energy, i.e. $\hat{X}_0|\theta\rangle = 0$, since $[\hat{X}_0, a_0] = 0$. It also verifies $a_n|\theta\rangle = 0, \forall n > 0$. Let us denote by $G_0 \subset G$ the maximal stability (isotropy) subgroup of this "degenerated vacuum" $|\theta\rangle$, of which the Hamiltonian $\hat{X}_0$ is one of its generators.

Actually, the operator a_0 conmutes with the unbroken symmetry generators $\hat{X}_\alpha^{(0)} \in \mathcal{G}_0$ and the creation operators:

$$[a_0, \hat{X}_\alpha^{(0)}] = 0 = [a_0, a_n^\dagger], \ \forall n > 0, \tag{9}$$

so that, by Shur lemma, a_0 behaves as (a multiple of) the identity operator in the broken theory. That is, it is natural to demand a_0 to leave the G_0-invariant vacuum (8) stable, which implies that:

$$a_0|\theta\rangle = \theta|\theta\rangle \Rightarrow |\theta\rangle = e^{\theta a_0^\dagger - \bar{\theta} a_0} |\hat{0}\rangle. \tag{10}$$

Thus, the vacuum of our (spontaneously) broken theory will be a *coherent state of zero modes* (see [5] and [6] for a thorough exposition on coherent states).

Now we show that a general unitary symmetry transformation (3), which incorporates broken symmetry generators in G/G_0, produces a "rearrangement" of this pseudo-vacuum $|\theta\rangle$ and causes it to radiate. In other words, we can associate a *quantum statistical ensemble* with the excited (or, let us say, "polarized") vacuum

$$|\theta'\rangle \equiv e^{\theta a_0'^\dagger - \bar{\theta} a_0'} |\hat{0}\rangle, \tag{11}$$

where a_0' is given by (3).

Indeed, one can compute the average number of particles with energy E_n inside $|\theta'\rangle$ as the expectation value:

$$N_n(x) = \langle\theta'|a_n^\dagger a_n|\theta'\rangle = |\theta|^2|U_{0n}(x)|^2, \tag{12}$$

hence, $|\theta|^2$ is the total average of particles of this quantum statistical ensemble. In the same way, the probability $P_n(p,x)$ of observing p particles with energy E_n in $|\theta'\rangle$ can be calculated as:

$$P_n(q,x) = |\langle q(n)|\theta'\rangle|^2 = \frac{e^{-|\theta|^2}}{q!}|\theta|^{2q}\,|U_{0n}(x)|^{2q} = \frac{e^{-|\theta|^2}}{q!}N_n^q(x). \tag{13}$$

Hence, the relative probability of observing a state with total energy E in the excited vacuum $|\theta'\rangle$ is:

$$P(E) = \sum_{\substack{q_0,\ldots,q_k:\\ \sum_{n=0}^k E_n q_n = E}} \prod_{n=0}^k P_n(q_n,x)\,. \tag{14}$$

For the cases studied in this paper, this distribution function can be factorized as $P(E) = \Omega(E)e^{-\tau E}$, where $\Omega(E)$ is a relative weight proportional to the number of states with energy E and the factor $e^{-\tau E}$ fits this weight properly to a temperature $T = k_B/\tau$.

We shall be primarily interested in the mean values of the basic observables (4). They can be calculated as:

$$\begin{aligned}\mathcal{X}_\alpha &= \langle\theta'|\hat{X}_\alpha|\theta'\rangle = -i|\theta|^2\sum_{m,n=0} U_{0m}(x)\frac{\partial U_{mn}(0)}{\partial x^\alpha}\bar{U}_{0n}(x)\\ &= -i|\theta|^2\left(U(x)\frac{\partial U(0)}{\partial x^\alpha}U^\dagger(x)\right)_{00}.\end{aligned} \tag{15}$$

In particular, the mean energy is simply:

$$\mathcal{X}_0 = |\theta|^2\sum_{n=1}|U_{0n}(x)|^2E_n. \tag{16}$$

We shall see that $\mathcal{X}_0$ matches the usual FD and BE expressions for the compact $G = U(2)$ and non-compact $G = U(1,1)$ cases, respectively.

3. The case of $G_\kappa = O(3), O(2,1)$ and Newton-Hooke groups

The Lie algebra commutators of our basic symmetry group G_κ are:

$$[A_+,A_-] = 2\kappa H - \Xi,\quad [H,A_\pm] = \pm A_\pm,\quad [\Xi,\text{all}] = 0, \tag{17}$$

where H represents the *Hamiltonian*, Ξ plays the role of the *zero-point energy* (or the total number of particles operator in second quantization), $A_\pm$ are *ladder* creation and annihilation operators and $\kappa = \pm1, 0$ is the *curvature* parameter for $U(2)$, $U(1,1)$ and the Newton-Hooke (harmonic oscillator) groups, respectively.

The group $U(1,1)$ is noncompact so, unlike the case of $U(2)$, all its unitary irreducible representations (unirreps) are infinite-dimensional. This group has a number of series of unirreps: principal, discrete and supplementary. We shall consider here only representations of the discrete series where each carrier space $\mathcal{H}_s$ is labelled by the *(conformal) spin* $s = 1/2, 1, 3/2, 2, \ldots$ and is spanned by the orthonormal basis $B(\mathcal{H}_s) = \{|s,n\rangle, n = 0,1,2,\ldots\}$. The action of the operators (17) on this basis vectors is:

$$\begin{aligned}H|s,n\rangle &= n|s,n\rangle, & A_+|s,n\rangle &= \sqrt{(n+1)(2s-\kappa n)}|s,n+1\rangle,\\ \Xi|s,n\rangle &= 2s|s,n\rangle, & A_-|s,n\rangle &= \sqrt{n(2s-\kappa(n-1))}|s,n-1\rangle.\end{aligned} \tag{18}$$

Journal of Physics: Conference Series **128** (2008) 012051 doi:10.1088/1742-6596/128/1/012051

Looking at this representation, we can think of an "exotic" harmonic oscillator with an equispaced energy spectrum $E_n = \epsilon n$, where we have introduced the interlevel energy spacing $\epsilon = \hbar\omega$ to give dimensions. For negative, $\kappa = -1$, and zero, $\kappa = 0$, curvature (i.e., for $U(1,1)$ and the Newton-Hooke groups, respectively), this energy spectrum is unbounded from above, whereas for positive curvature, $\kappa = 1$ (i.e., for $U(2)$), we have a bounded spectrum with $2s+1$ states.

Using the standard Iwasawa decomposition (see e.g. [5]), any group element $U \in G_\kappa$ can be represented as:

$$U(\zeta, \bar{\zeta}, \tau, \varphi) = e^{\zeta A_+ - \bar{\zeta} A_-} e^{i\tau H} e^{i\varphi \Xi}, \tag{19}$$

where $\varphi, \tau \in [0, 2\pi]$ and $\zeta \in \mathbb{C}$. An important ingredient to construct a G_κ invariant quantum field theory, as a second quantization on G_κ, will be the irreducible matrix coefficients of the representation (18) in the orthonormal basis $B(\mathcal{H}_s)$:

$$U^{(s)}_{mn}(\zeta, \bar{\zeta}, \tau, \varphi) \equiv \langle s, m | U(\zeta, \bar{\zeta}, \tau, \varphi) | s, n \rangle. \tag{20}$$

Given the Fourier expansion in modes of a field with (conformal) spin s,

$$|\psi\rangle = \sum_{n=0} a_n |s, n\rangle\,, \tag{21}$$

the Fourier coefficients a_n (resp. $a_n^\dagger$) are promoted to annihilation (resp. creation) operators of energy modes $E_n = \hbar\omega n$ in the second quantized theory, with commutation relations $[a_n, a_m^\dagger] = \delta_{n,m}$. The second quantized version, $\hat{H}, \hat{\Xi}, \hat{A}_\pm$, of the basic operators (17), given by the general expression (4), now reads:

$$\begin{array}{ll} \hat{H} = \sum_n n a_n^\dagger a_n, & \hat{A}_+ = \sum_n \sqrt{(n+1)(2s-\kappa n)}\, a_n^\dagger a_{n+1}, \\ \hat{\Xi} = 2s \sum_n a_n^\dagger a_n, & \hat{A}_- = \sum_n \sqrt{n(2s-\kappa(n-1))}\, a_n^\dagger a_{n-1}, \end{array} \tag{22}$$

where summations start at $n=0$ and are finite or infinite, depending on the curvature κ. Here we highlight the *total energy* operator, $\hat{E} \equiv \hbar\omega\hat{H}$, and the *total number of particles* operator, $\hat{N} \equiv \frac{1}{2s}\hat{\Xi}$.

The vacuum $|\hat{0}\rangle$ of the second quantized theory is characterized by being stable under the basic isometry group G_κ, i.e.

$$U|0\rangle = |0\rangle, \forall U \in G_\kappa \Rightarrow \{\hat{H}, \hat{\Xi}, \hat{A}_+, \hat{A}_-\}|0\rangle = 0, \tag{23}$$

and annihilated by $a_n|0\rangle = 0, \forall n \geq 0$. Then an orthonormal basis for the Hilbert space of the second quantized theory is constructed as in (7). Note that any second quantized state (7) made up of an arbitrary content of *zero modes*, like (8), has zero total energy, i.e. $\hat{H}|\theta\rangle = 0$. It also verifies $\hat{A}_-|\theta\rangle = 0$ and $a_n|\theta\rangle = 0, \forall n > 0$, so that the state (8) behaves as a (degenerated) vacuum under the (unbroken) subgroup $B \subset G_\kappa$ of affine or similitude transformations, generated by $\mathcal{B} = \{\hat{H}, \hat{A}_-\}$. Moreover, given that a_0 acts as (a multiple of) the identity operator in the broken theory, that is, it commutes with:

$$[a_0, \hat{H}] = [a_0, \hat{A}_-] = [a_0, a_n^\dagger] = 0, \forall n > 0, \tag{24}$$

it is natural to demand a_0 to leave the affine vacuum (8) stable, which implies again (10). Thus, the vacuum of our (spontaneously) broken theory will be a *coherent state of zero modes*. The squared modulus of the complex parameter θ has the physical significance of the vacuum expectation value of the number operator in the new vacuum; indeed, one can verify that:

$$\langle\theta|\hat{N}|\theta\rangle = |\theta|^2. \tag{25}$$

Journal of Physics: Conference Series **128** (2008) 012051 doi:10.1088/1742-6596/128/1/012051

As stated in the Introduction, we can think of zero modes as *virtual* particles without ("bright") energy E, undetectable ("dark") by affine observers. However, we shall show that a general unitary symmetry transformation (20), which incorporates the broken symmetry generator $\hat{A}_+$, produces a "rearrangement" of the the affine vacuum $|\theta\rangle$ and makes it to radiate. In other words, we shall associate a *thermal bath* with the excited (or, let us say, "polarized") vacuum (11) and we shall show in Sections 4 and 5 that the *mean energy per mode*, $\langle\theta'|\hat{E}|\theta'\rangle$, matches the usual BE and FD expressions, for the non-compact $U(1,1)$ and compact $U(2)$ isometry groups ($\kappa = \mp 1$), respectively. The limit of high frequencies (or large interlevel energy spacing $\hbar\omega \gg k_B T$) is regained from a contraction to the Newton-Hooke group (zero curvature, $\kappa = 0$), as explained in Sect. 6. Note however that the average number of particles is conserved under this (unitary) transformation, i.e. $\langle\theta'|\hat{N}|\theta'\rangle = |\theta|^2$, because $U^\dagger U = 1$. We shall relate this quantity to a vacuum energy density, the sign of which depends on the curvature κ, and we shall make some comments on the cosmological constant problem in Sect. 7.

4. Broken $U(1,1)$ symmetry: hyperbolic geometry and BE statistics

The irreducible matrix coefficients of the representation (18) in the orthonormal basis $B(\mathcal{H}_s)$ are given in this case ($\kappa = -1$) by:

$$\begin{aligned} U^{(s)}_{mn}(\zeta,\bar\zeta,\tau,\varphi) &= e^{2si\varphi}(1-z\bar z)^s\sqrt{\frac{C^{(s)}_m}{C^{(s)}_n}}\sum_{q=\max(0,m-n)}^{m}\binom{n}{m-q}\binom{2s+n+q-1}{q} \\ &\quad\times(-1)^{n-m+q}e^{in\tau}z^q\bar z^{n-m+q}, \end{aligned} \tag{26}$$

where $C^{(s)}_n = n!/(2s+n-1)!$, $\zeta = |\zeta|e^{i\phi}$ and $z \equiv e^{i\phi}\tanh|\zeta|$ is restricted to the unit disk $D = \{z \in \mathbb{C} : |z| < 1\}$, which is the stereographic projection of the upper sheet of the hyperboloid $\mathbb{H}^2 = \{\vec v \in \mathbb{R}^3 : \vec v^2 = v_0^2 - v_1^2 - v_2^2 = 1, v_0 > 0\}$ onto the complex plane. The correspondence is established through:

$$\vec v = (\cosh(2|\zeta|), \sinh(2|\zeta|)\cos\phi, \sinh(2|\zeta|)\sin\phi).$$

The finite action (3) of $U(1,1)$ on annihilation (resp. creation) operators of zero modes is:

$$a_0 \to a_0' = e^{2si\varphi}(1+z\bar z)^s\sum_{n=0}^{\infty}(-1)^n\binom{2s+n-1}{n}^{1/2}e^{in\tau}\bar z^n a_n, \tag{27}$$

which leads to the excited vacuum:

$$|\theta'\rangle \equiv e^{\theta a_0'^\dagger - \bar\theta a_0'}|0\rangle = e^{-|\theta|^2/2}\sum_{q=0}^{\infty} z^q \sum_{\substack{m_1,\ldots,m_q:\\ \sum_{n=1}^q n m_n = q}} \prod_{n=0}^{q}\frac{R_n^{m_n}}{m_n!}\prod_{n=0}^{q}(a_n^\dagger)^{m_n}|0\rangle\,, \tag{28}$$

where $R_n \equiv \theta(-1)^n e^{-2si\varphi - in\tau}(1-z\bar z)^s\binom{2s+n-1}{n}^{1/2}$ and we set $m_0 \equiv 0$. We have used the general identity:

$$\left(\sum_{n=0}^{\infty}\gamma_n z^n\right)^l = \sum_{q=0}^{\infty}\delta_q z^q,\ \ \delta_0 = \gamma_0^l,\ \ \delta_q = \frac{1}{q\gamma_0}\sum_{s=1}^{q}(sm-q+s)\gamma_s\delta_{q-s}. \tag{29}$$

Journal of Physics: Conference Series **128** (2008) 012051 doi:10.1088/1742-6596/128/1/012051

From (28), we see that the relative probability of observing a state with total energy $E_q = \hbar\omega q$ in the excited vacuum $|\theta'\rangle$ is:

$$\begin{aligned} P_q &= \Omega(E_q)(|z|^2)^q\,, \\ \Omega(E_q) &\equiv \sum_{\substack{m_1,\dots,m_q:\\ \sum_{n=1}^q nm_n = q}} \prod_{n=0}^{q} \frac{|R_n|^{2m_n}}{m_n!}\,. \end{aligned} \tag{30}$$

We can associate a *thermal bath* with this distribution function by noticing that $\Omega(E_q)$ behaves as a relative weight proportional to the number of states with energy $E_q = \hbar\omega q$; the factor $(|z|^2)^q$ fits this weight properly to a temperature T as:

$$(|z|^2)^q = e^{q\log|z|^2} = e^{-\frac{E_q}{k_BT}},\ \ T \equiv -\frac{\hbar\omega}{k_B\log|z|^2}. \tag{31}$$

We could think of (conformal) $U(1,1) \simeq O(1,2)$ transformations as transitions to a uniformly relativistic accelerated frame (see e.g. [3]), so that $T = \frac{\hbar a}{2\pi c k_B}$ is the temperature associated with a given acceleration $a \equiv -2\pi\omega c/\log|z|^2$.

After some intermediate calculations, the expected value of the total energy $\hat{E}$ in the accelerated vacuum $|\theta'\rangle$ proves to be:

$$\begin{aligned} \langle\theta'|\hat{E}|\theta'\rangle &= |\theta|^2(1-|z|^2)^{2s}\sum_{n=0}^{\infty}\hbar\omega n\binom{2s+n-1}{n}|z|^{2n} \\ &= 2s|\theta|^2\hbar\omega\frac{|z|^2}{1-|z|^2} = 2s|\theta|^2\frac{\hbar\omega e^{-\hbar\omega/k_BT}}{1-e^{-\hbar\omega/k_BT}}\,, \end{aligned} \tag{32}$$

which is proportional to the *mean energy per mode* of the B-E statistics. In D spatial dimensions, the number of states with frequency between ω and $\omega + d\omega$ is proportional to ω^{D-1}. Thus, for $D = 3$, the spectral distribution of the radiation of the accelerated vacuum $|\theta'\rangle$ is *Planckian*, i.e. $|\theta'\rangle$ radiates as a *black body*.

Still, we could have introduced a *chemical potential* μ by setting $|z|^2 = e^{(\mu-\hbar\omega)/k_BT}$ with the restriction $\mu < \hbar\omega$, in order to preserve the condition $|z| < 1$.

5. Broken $U(2)$ symmetry: spherical geometry and FD statistics

The irreducible matrix coefficients of the representation (18) in the orthonormal basis $B(\mathcal{H}_s)$ are given in this case ($\kappa = 1$) by:

$$\begin{aligned} U^{(s)}_{mn}(\zeta,\bar\zeta,\tau,\varphi) &= e^{2si\varphi}(1+z\bar z)^{-s}\sqrt{\frac{C^{(s)}_m}{C^{(s)}_n}}\sum_{q=\max(0,m-n)}^{\min(m,2s-n)}\binom{n}{m-q}\binom{2s-n}{q} \\ &\times(-1)^{n-m+q}e^{in\tau}z^q\bar z^{n-m+q}, \end{aligned} \tag{33}$$

where $C^{(s)}_n = \binom{2s}{n}^{-1}$ and $z \equiv e^{i\phi}\tan|\zeta|$ is now related to the stereographic projection of the two-dimensional sphere $\mathbb{S}^2 = \{\vec v \in \mathbb{R}^3 : \vec v^2 = v_0^2 + v_1^2 + v_2^2 = 1\}$ onto the complex plane, given by:

$$\vec v = (\cos(2|\zeta|), \sin(2|\zeta|)\cos\phi, \sin(2|\zeta|)\sin\phi).$$

The finite action of $U(2)$ on annihilation (resp. creation) operators of zero modes is now:

$$a_0 \to a'_0 = e^{2si\varphi}(1+z\bar z)^{-s}\sum_{n=0}^{2s}(-1)^n\binom{2s}{n}^{1/2}e^{in\tau}\bar z^n a_n, \tag{34}$$

which leads to the polarized vacuum $|\theta'\rangle$ according to the formula (11). As for the hyperbolic case, we can associate a thermal bath with the state $|\theta'\rangle$. The difference now is that the factor $|z|^2$ is unbounded from above, i.e. $|z|^2 < \infty$, which means that we have to introduce a non-zero *chemical potential* μ, such that $|z|^2 = e^{(\mu-\hbar\omega)/k_BT}$, or/and allow for *negative temperatures.*

The expected value of the total energy $\hat{E}$ in the polarized vacuum $|\theta'\rangle$ proves to be:

$$\begin{aligned}\langle\theta'|\hat{E}|\theta'\rangle &= |\theta|^2(1+|z|^2)^{-2s}\sum_{n=0}^{2s}\hbar\omega n\binom{2s}{n}|z|^{2n} \\ &= 2s|\theta|^2\hbar\omega\frac{|z|^2}{1+|z|^2} = 2s|\theta|^2\frac{\hbar\omega}{1+e^{(\hbar\omega-\mu)/k_BT}}\,, \end{aligned} \tag{35}$$

which is proportional to the *mean energy per mode* of the FD statistics.

6. Flat geometry and the high frequency limit

Like in the previous two cases, one can compute the finite action of the Newton-Hooke group G_0 on annihilation (resp. creation) operators of zero modes:

$$a_0 \to a_0' = e^{2si\varphi}e^{-2s|z|^2/2}\sum_{n=0}^{\infty}(-1)^n e^{i\tau}\sqrt{\frac{(2s)^n}{n!}}\bar{z}^n\, a_n, \tag{36}$$

where $z \equiv \zeta$ is unbounded, $|z| < \infty$, like for the spherical geometry. The expected value of the total energy $\hat{E}$ in the excited vacuum $|\theta'\rangle$ is now:

$$\begin{aligned}\langle\theta'|\hat{E}|\theta'\rangle &= |\theta|^2 e^{-2s|z|^2}\sum_{n=0}^{\infty}\hbar\omega n\frac{(2s)^n}{n!}|z|^{2n} \\ &= 2s|\theta|^2\hbar\omega|z|^2 = 2s|\theta|^2\hbar\omega e^{(\mu-\hbar\omega)/k_BT}\,, \end{aligned} \tag{37}$$

which is the $\hbar\omega \gg k_BT$ limit of B-E and F-D statistics. Note that, unlike for the B-E and F-D statistics, here we could "reabsorb" the chemical potential μ into the vacuum expectation value of the total number of particles $|\theta|^2$ by setting $|\theta|^2 = e^{-\mu/k_BT}$. Therefore, the introduction of μ in our scheme is only indispensable for the F-D statistics.

7. Cosmological implications

We have separated standard ("bright") energy $\hat{H}$ from vacuum ("dark") energy $\hat{\Xi}$ in our model. However, we could combine both contributions to define a total Hamiltonian,

$$\hat{H}_{\rm tot} = \hat{H} + \frac{\kappa}{2}\hat{\Xi}, \tag{38}$$

such that the map $\hat{H} \mapsto \hat{H}_{\rm tot}$ renders the original commutation relations (17) into: $[\hat{A}_+, \hat{A}_-] = 2\kappa\hat{H}$, for $\kappa = \pm 1$, and $[\hat{A}_+, \hat{A}_-] = -\hat{\Xi}$, for $\kappa = 0$. Hence, the vacuum energy is now given by:

$$\langle\theta|\hat{H}_{\rm tot}|\theta\rangle = \frac{\kappa}{2}\langle\theta|\hat{\Xi}|\theta\rangle = \kappa s|\theta|^2 \tag{39}$$

Although we are dealing with a simplified (toy) model, we feel tempted to link this vacuum energy to a cosmological constant, as is done in modern cosmology. From this point of view, hyperbolic spatial geometries ($\kappa = -1$), like Anti de Sitter space-time, have positive pressure, which causes the expansion of empty space to slow down. On the contrary, for spherical spatial geometries ($\kappa = 1$), like de Sitter space, the expansion of empty space will tend to speed up. For flat space ($\kappa = 0$) we have zero cosmological constant. Note that we can make the parameter $|\theta|^2$ as small as we like, thus eluding huge vacuum energies for either spherical or hyperbolic geometries. Therefore, our derivation of the cosmological constant from the vacuum energy in quantum field theory is free from the traditional drawbacks of "fine-tuning". Moreover, the three spatial geometries, $\kappa = \pm 1, 0$, considered in this article have different qualitative behaviour under vacuum radiation at low frequencies, something that could be used in addition to experimentally discern between these topologies.

We have presented here a quite simplified model, but we think that it grasps the essentials of more involved instances. A proper discussion of all these vacuum phenomena inside the entire conformal group $SO(4, 2)$ will be developed elsewhere.

Acknowledgements

Work partially supported by the MCYT and Fundación Séneca under projects FIS2005-05736-C03-01 and PB/9/FS/02

References

[1] Casimir H B G 1948 *Proc. Kon. Nederland. Akad. Wetensch.* **B51** 793
[2] Unruh W G 1976 *Phys. Rev.* **D14** 870
[3] Aldaya V, Calixto M and Cerveró J M 1999 *Commun. Math. Phys.* **200** 325-354
[4] Calixto M and Aldaya V 2006 *J. Phys.* **A39** L539
[5] Perelomov A 1986 *Generalized Coherent States and Their Apllications* (Berlin: Springer)
[6] Klauder J R and Bo-Sture Skagerstam 1985 *Coherent States: Applications in Physics and Mathematical Physics* (Singapore: World Scientific)

Journal of Physics: Conference Series 128 (2008) 012052 doi:10.1088/1742-6596/128/1/012052

Obtainment of internal labelling operators as broken Casimir operators by means of contractions related to reduction chains in semisimple Lie algebras

R Campoamor-Stursberg
Dpto. Geometría y Topología, Fac. CC. Matemáticas
Universidad Complutense de Madrid, E-28040 Madrid, Spain

E-mail: rutwig@pdi.ucm.es

Abstract. We show that the Inönü-Wigner contraction naturally associated to a reduction chain $\mathfrak{s} \supset \mathfrak{s}'$ of semisimple Lie algebras induces a decomposition of the Casimir operators into homogeneous polynomials, the terms of which can be used to obtain additional mutually commuting missing label operators for this reduction. The adjunction of these scalars that are no more invariants of the contraction allow to solve the missing label problem for those reductions where the contraction provides an insufficient number of labelling operators.

1. Introduction

One of the main applications of group theoretical methods to physical problems is related to classification schemes, where irreducible representations of a Lie group have to be decomposed into irreducible representations of a certain subgroup appearing in some relevant reduction chain

$$\left| \begin{array}{ccccccc} \mathfrak{s} & \supset & \mathfrak{s}' & \supset & \mathfrak{s}'' & \dots \\ \downarrow & & \downarrow & & \downarrow & \\ [\lambda] & & [\lambda'] & & [\lambda''] & \dots \end{array} \right\rangle . \qquad (1)$$

This is the case for dynamical symmetries used for example in nuclear physics, where one objective of the algebraic model is to describe the Hamiltonian (or mass operator in the relativistic frame) as a function of the Casimir operators of the chain elements. The corresponding energy formulae can the easily deduced from the expectation values in the reduced representations. As example, the Gell-Mann-Okubo mass formula can be derived using this ansatz [1]. In many situations, the labels obtained from the reduction (1) are sufficient to solve the problem, e.g., of we require multiplicity free reductions, as used in $SU(N)$ tumbling gauge models [2] or the interacting boson model [3]. However, often the subgroup does not provide a sufficient number of labels to specify the basis states unambigously, due to multiplicities greater than one for induced representations. This turns out to be the rule for non-canonical embeddings and generic representations of $\mathfrak{g}$. For some special types, like totally symmetric or anti-symmetric representations, additional labels are not necessary to solve the problem, and the degeneracies can be solved directly with the available operators.

Many procedures methods have been developed to solve the so-called missing label problem (short MLP), from specific construction of states for the reduction chain to the formal construction of all possible labelling operators using enveloping algebras [4]. The latter procedure allows in theory to find the most general labelling operator, although the effective computation of admissible operators is rather cumbersome. In addition, there is no general criterion to compute the number of operators necessary to generate integrity bases in enveloping algebras.

In the mid seventies, Peccia and Sharp [5] considered an analytic approach to the MLP based on the method used to compute the generalized Casimir invariants of Lie algebras in the commutative frame. This method is very close to the interpretation of Casimir operators as functions that are constant on co-adjoint orbits, and consists essentially on a restriction of this problem. The labelling operators are interpreted as solutions of a certain subsystem of partial differential equations corresponding to the embedded subalgebra. Subgroup scalars are therefore the differential invariants of the subalgebra in a specific realization, and their symmetrization serves to separate multiplicities of reduced representations. Although this approach is, in principle, computationally easier than the pure algebraic approach using operators in the enveloping algebra, the integration of such differential equations is far from being trivial. Moreover, the orthogonality conditions required to the labelling operators, in order to avoid undesirable interactions, must be solved by pure algebraic methods in most cases.

Most of the MLP considered in the literature have been solved or studied from an algebraic or analytic point of view, generally looking for solutions of lowest degrees. In appearance, no attention has been paid to the properties that the embedding imply.[1] The embedding $\mathfrak{s} \supset \mathfrak{s}'$ is conditioned by physical reasons, that is, the choice of the embedding class corresponds to some specific coupling scheme or some relevant internal property that must be emphasized (like angular momentum). Since the embedding determines the branching rules for irreducible representations, it should be expected that labelling operators needed to solve multiplicities are, in some manner, codified by the properties of the symmetry breaking determined by the reduction chain.
When dealing with the MLP algebraically or analytically, it is not immediately clear to which extent we are using the properties of this embedding or the branching rules. Moreover, we can ask whether the obtained labelling operators have some intrinsic or physical meaning at all. Formulated in another way: are the labelling operators of the MLP completely determined by the reduction $\mathfrak{s} \supset \mathfrak{s}'$ (and therefore, by the underlying physics), or are they the result of a formal algebraic/analytic manipulation?

In [7], the missing label problem was approached from a quite general point of view, but having in mind the important fact observed in [8] that symmetry breaking and contractions of Lie algebras have many points in common. This actually is deeply related to the characterization of inhomogeneous algebras obtained from contractions of semisimple algebras [9]. Therefore, any reduction chain $\mathfrak{s} \supset \mathfrak{s}'$ naturally induces a contraction of Lie algebras. It is therefore natural to ask if the operators provided by the contraction $\mathfrak{g}$ (the so-called contracted Casimir operators) can be used to solve the MLP completely. The idea, in a different form, had been used previously for angular momentum subalgebra, and can be seen clearly in the so-called rotor expansion method [4].

The main goal of the general contraction ansatz in labelling problems can be resumed in the following points:

(i) Find a procedure to solve the MLP using explicitly the properties of the embedding $\mathfrak{s} \supset \mathfrak{s}'$ [breaking symmetry $\leftrightarrow$ contractions], that is, without invoking formal operators.

[1] As known, non-equivalent embeddings lead to different branching rules, and therefore to different classification schemes. This is the difference between the Wigner supermultiplet (nuclear LS coupling) and the strange-spin multiplet structure of $\mathfrak{su}(4)$ [6].

(ii) Justify a natural choice of labelling operators as "broken Casimir operators".

(iii) Find a phenomenological explanation for the non-integer expectation values of labelling operators.

In this first development, only the contracted invariants were used to generate labelling operators. This approach was however sufficient to solve many physically relevant missing label problems, and the result were in perfect harmony with those obtained using other techniques. It was also observed that the method can fail to find a sufficient number of missing operators whenever the identity $\mathcal{N}(\mathfrak{g}) = \mathcal{N}(\mathfrak{s}) = n$ is satisfied. The failure is essentially a consequence of an insufficient number of invariants in the contraction.

The aim of this work is to further develop the contraction procedure, but using not only the contracted invariants, but a certain decomposition induced in the Casimir operators of the contracted algebra, which turn out to be subgroup scalars but no more invariants of the contraction. With this decomposition, we are able to surmount the difficulty for cases with insufficient number of contracted operators. A more interesting consequence of this fact is the possibility of explaining the existence of missing label operators of the same degree, as they have already been constructed in the algebraic frame.

2. Classical Casimir operators

Given a presentation $\mathfrak{s} = \left\{X_1, .., X_n \mid [X_i, X_j] = C_{ij}^k X_k\right\}$ of a Lie algebra $\mathfrak{s}$ in terms of generators and commutation relations, we are interested in (polynomial) operators $C_p = \alpha^{i_1..i_p} X_{i_1}..X_{i_p}$ in the generators of $\mathfrak{s}$ such that the constraint $[X_i, C_p] = 0$, $(i = 1..n)$ is satisfied. Such an operator necessarily lies in the centre of the enveloping algebra of $\mathfrak{s}$, and is traditionally referred to as Casimir operator. However, in many dynamical problems, the relevant invariant functions are not polynomials, but rational or even trascendental functions (e.g. the inhomogeneous Weyl group). Therefore the approach with the universal enveloping algebra has to be generalized in order to cover arbitrary Lie groups. The most widely used method is the analytical realization. The generators of the Lie algebra $\mathfrak{s}$ are realized by the differential operators $\widehat{X}_i = C_{ij}^k x_k \frac{\partial}{\partial x_j}$, where the x_i are commuting variables associated to each generator X_i. In this approach, a function $F \in C^\infty(\mathfrak{s}')$ is an invariant of $\mathfrak{s}$ if and only if it satisfies the system of PDEs

$$\widehat{X}_i(F) = 0, \quad i = 1..n. \tag{2}$$

Using the symmetrization map $Sym(x_1^{a_1}..x_p^{a_p}) = \frac{1}{k!}\sum_{\sigma \in S_p} x_{\sigma(1)}^{a_1}...x_{\sigma(p)}^{a_p}$, we recover the Casimir operator for polynomial solutions. With this analytical ansatz, it is easily seen that the number of independent solutions is

$$\mathcal{N}(\mathfrak{s}) = \dim \mathfrak{s} - \text{ rank } \left[C_{ij}^k x_k\right]. \tag{3}$$

As already commented, one of the main applications of Casimir operators of Lie algebras is the labelling of irreducible representations. In a more general approach, irreducible representations of a Lie algebra $\mathfrak{g}$ can be labelled using the eigenvalues of its generalized Casimir invariants [5]. For each representation, the number of internal labels needed is given by

$$i = \frac{1}{2}(\dim \mathfrak{g} - \mathcal{N}(\mathfrak{g})). \tag{4}$$

For the special case of semisimple Lie algebras, this number is deeply related to the number of positive roots of its complexification. If we use some subalgebra $\mathfrak{g} \supset \mathfrak{h}$ to label the basis states of $\mathfrak{g}$, the embedding provides $\frac{1}{2}(\dim \mathfrak{h} + \mathcal{N}(\mathfrak{h})) + l'$ labels, where l' is the number of invariants of

Journal of Physics: Conference Series **128** (2008) 012052 doi:10.1088/1742-6596/128/1/012052

$\mathfrak{g}$ that depend only on variables of the subalgebra $\mathfrak{h}$ [5]. In general, this is still not sufficient to separate multiplicities of induced representations, so that we need to find

$$n = \frac{1}{2}\left(\dim \mathfrak{g} - \mathcal{N}(\mathfrak{g}) - \dim \mathfrak{h} - \mathcal{N}(\mathfrak{h})\right) + l' \tag{5}$$

additional operators, which are commonly called missing label operators. The total number of available operators of this kind is twice the number of needed labels, $m = 2n$. For the case $n > 1$, it remains the problem of determining a set of n mutually commuting operators, as commented before. These operators can be seen to be subgroup scalars, so that the analytical approach is a practical method to find the labelling operators.[2] Considering the realization of $\mathfrak{s} \supset \mathfrak{s}'$ by differential operators indicated in (2), we restrict to the PDEs corresponding to subgroup generators

$$\widehat{X}_i = C_{ij}^k x_k \frac{\partial}{\partial x_j}, \quad 1 \leq i \leq \dim \mathfrak{s}'. \tag{6}$$

Solutions to this system $f(\mathfrak{s}')$ reproduce the differential invariants in $\dim \mathfrak{s}$ dimensions. The total number of solutions of the latter system is given by:

$$\mathcal{N}(f(\mathfrak{s}')) = m + \mathcal{N}(\mathfrak{s}) + \mathcal{N}(\mathfrak{s}') - l'. \tag{7}$$

We observe that (7) refers to the number of functionally independent solutions. Integrity bases, constrained by the coarser algebraic independence, will generally have much more elements than (7), and no general procedure is known to compute its dimension. In addition, to be useful as labelling operators, two (symmetrized) solutions F_1, F_2 of system (6) must satisfy the orthogonality condition $[F_1, F_2] = 0$ and commute with the Casimir operators C_i and D_j of $\mathfrak{s}$ and the subalgebra $\mathfrak{s}'$. Therefore the set of commuting operators $\{C_i, D_j, F_k\}$ serves to label the states unambiguously.

It was proved in [7] that any reduction chain $\mathfrak{s} \supset \mathfrak{s}'$ is naturally associated to an Inönü-Wigner contraction $\mathfrak{s} \rightsquigarrow \mathfrak{g} = \mathfrak{s}' \overrightarrow{\oplus}_R (\dim \mathfrak{s} - \dim \mathfrak{s}') L_1$, where the representation R of $\mathfrak{s}'$ satisfies the constraint[3]

$$\mathrm{ad}\mathfrak{s} = \mathrm{ad}\mathfrak{s}' \oplus R. \tag{8}$$

The contraction is easily seen to be defined by the non-singular transformations

$$\Phi_t(X_i) = \begin{cases} X_i, & 1 \leq i \leq s \\ \frac{1}{t} X_i, & s+1 \leq i \leq n \end{cases}, \tag{9}$$

where $\{X_1, .., X_s, X_{s+1}, .., X_n\}$ is a basis of $\mathfrak{s}$ such that $\{X_1, .., X_s\}$ generates the subalgebra $\mathfrak{s}'$ and $\{X_{s+1}, .., X_n\}$ spans the representation space R. If both $\mathfrak{s}$ and $\mathfrak{s}'$ are semisimple, the contraction is isomorphic to an inhomogeneous Lie algebra with Levi decomposition $\mathfrak{g} = \mathfrak{s}' \overrightarrow{\oplus}_R (\dim \mathfrak{s} - \dim \mathfrak{s}') L_1$. Since it satisfies the condition $[\mathfrak{g}, \mathfrak{g}] = \mathfrak{g}$, this algebra admits a fundamental basis of invariants consisting of Casimir operators. Moreover, by the contraction we have the inequality $\mathcal{N}(\mathfrak{s}) \leq \mathcal{N}(\mathfrak{g})$. The system of PDEs corresponding to this contraction can be divided into two parts:

$$\widehat{X}_i F = C_{ij}^k x_k \frac{\partial F}{\partial x_j} = 0, \qquad 1 \leq i \leq s, \tag{10}$$

$$\widehat{X}_{s+i} F = C_{s+i,j}^{s+k} x_{s+k} \frac{\partial F}{\partial x_j} = 0, \quad 1 \leq i, k \leq n-s, 1 \leq j \leq s. \tag{11}$$

[2] In many cases, the degeneracies of the reduction chain can be solved without using the subgroup scalars, or are determined by a specifical ansatz adapted to the involved groups.

[3] Actually, the branching rule depends on a numerical index j_f that characterizes the embedding class.

Journal of Physics: Conference Series **128** (2008) 012052 doi:10.1088/1742-6596/128/1/012052

The subsystem (10) corresponds to the generators of $\mathfrak{s}'$ realized as subalgebra of $\mathfrak{s}$, while the remaining equations (11) describe the representation. Written in matrix form, the system is given by

$$\begin{pmatrix} 0 & \cdots & C_{1s}^{k}x_k & C_{1,s+1}^{k}x_k & \cdots & C_{1,n}^{k}x_k \\ \vdots & & \vdots & \vdots & & \vdots \\ -C_{1s}^{k}x_k & \cdots & 0 & C_{s,s+1}^{k}x_k & \cdots & C_{s,n}^{k}x_k \\ -C_{s,s+1}^{k}x_k & \cdots & -C_{s,s+1}^{k}x_k & 0 & \cdots & 0 \\ \vdots & & \vdots & \vdots & & \vdots \\ -C_{1n}^{k}x_k & \cdots & -C_{s,n}^{k}x_k & 0 & \cdots & 0 \end{pmatrix} \begin{pmatrix} \partial_{x_1}F \\ \vdots \\ \partial_{x_s}F \\ \partial_{x_{s+1}}F \\ \vdots \\ \partial_{x_n}F \end{pmatrix} = 0.$$

Since the first s first rows reproduce exactly the system of PDEs needed to compute the missing label operators, we conclude that any invariant of $\mathfrak{g}$ is a candidate for missing label operator. In view of this situation, the following questions arise naturally:

(i) Do polynomial functions of the invariants of these algebras suffice to determine n mutually orthogonal missing label operators?

(ii) Can all available operators found by this procedure?

The answer to the first equation is in the affirmative for those cases where the contraction provides a number of independent invariants exceeding the number of needed labelling operators. It fails when these two quantities coincide, which suggests that the procedure has to be refined. The answer to the second question is generally in the negative (see e.g. the algebras with one missing label), although it cannot be excluded that with the refinement proposed in this work we are able to recover a complete set of independent labelling operators for some special reductions. In most cases, only half of the available operators should be expected, since all operators obtained are the result, in some sense, of "breaking" the original Casimir operators. Whether a further refinement allows to obtain additional labelling operators that are independent remains for the moment an unanswered question.

3. Decomposition of Casimir operators

In this section we prove that the contraction induced by the reduction chain induces a decomposition of the corresponding Casimir operators of $\mathfrak{s}$, which allow, among other properties, to determine the invariants of the contraction $\mathfrak{g}$. However, other terms are also relevant for the missing label problem, and constitute the solution to the problem pointed out in [7] when the number of invariants of the contraction is not sufficient. These additional terms do not constitute invariants of the contraction, and where therefore not considered in [7].

We briefly recall the definition of contracted invariants. Since classical Inönü-Wigner contractions are the only type of contractions needed for the labelling problem, we can restrict ourselves to this case [10]. Let $C_p(X_1, ..., X_n) = \alpha^{i_1...i_p}X_{i_1}...X_{i_p}$ be a pth-order Casimir operator of $\mathfrak{s}$. Then the transformed invariant takes the form

$$F(\Phi_t(X_1), .., \Phi_t(X_n)) = t^{n_{i_1}+...+n_{i_p}}\alpha^{i_1...i_p}X_{i_1}...X_{i_p}, \tag{12}$$

where $n_{i_j} = 0, 1$. Taking the maximal power in t,

$$M = \max\left\{n_{i_1} + ... + n_{i_p} \mid \alpha^{i_1..i_p} \neq 0\right\}, \tag{13}$$

the limit

$$\begin{aligned} F'(X_1, .., X_n) &= \lim_{t\to\infty} t^{-M}F(\Phi_t(X_1), ..., \Phi_t(X_n)) \\ &= \sum_{n_{i_1}+...+n_{i_p}=M} \alpha^{i_1...i_p}X_{i_1}...X_{i_p} \end{aligned}$$

provides a Casimir operator of degree p of the contraction $\mathfrak{g}'$. Now, instead of extracting only the term with the highest power of t, we consider the whole decomposition

$$C_p = t^M C_p' + \sum_{\alpha} t^{\alpha}\Phi_{\alpha} + \Phi_0, \tag{14}$$

where $\alpha < M \leq p$ and Φ_0 is a function of the Casimir operators of the subalgebra $\mathfrak{s}'$ (these generators have not been re-scaled). It is straightforward to verify that C_p' is not only an invariant of the contraction $\mathfrak{g}$, but also a solution to the MLP.[4] This first term was central to the argumentation in [7], and allowed to obtain commuting sets of labelling operators. However, the remaining terms can also be individually considered as candidates for labelling operators, as states the following

Proposition 1 *The functions Φ_α are solutions of the missing label problem, that is, they satisfy the system*

$$\widehat{X}_i\Phi_\alpha = C_{ij}^k x_k \frac{\partial \Phi_\alpha}{\partial x_j} = 0, \quad 1 \leq i \leq s. \tag{15}$$

Proof. First of all, the decomposition (14) tells that the Casimir operator C_p can be rewritten as a sum of homogeneous polynomials C_p', Φ_α with the property that C_p' is of homogeneity degree $p - M$ in the variables $\{x_1, .., x_s\}$ associated to subalgebra generators and degree M in the remaining variables $\{x_{s+1}, .., x_n\}$. Accordingly, any Φ_α is of degree $p - \alpha$ in the variables $\{x_1, .., x_s\}$ and α in the $\{x_{s+1}, .., x_n\}$. We denote this by saying that these functions are of bi-degree $(p - \alpha, \alpha)$.

Now the equations (10) corresponding to subalgebra generators remain unaltered by the contraction procedure, since the re-scaling of generators does not affect them.[5] Thus for any $1 \leq i \leq s$ and any homogeneous polynomial Ψ of bi-degree $(p - q, q)$ we obtain

$$\widehat{X}_i\Psi = C_{ij}^k x_k \frac{\partial \Psi}{\partial x_j} + C_{ij+s}^{k+s} x_{k+s} \frac{\partial \Psi}{\partial x_j}, \tag{16}$$

and the result is easily seen to be again a polynomial with the same bi-degree. This means that evaluating $C_p = t^M C_p' + \sum_\alpha t^\alpha \Phi_\alpha + \Phi_0$ is a sum of polynomials of different bi-degree, and since C_p is a Casimir operators, the only possibility is that each term is a solution of the system. We thus conclude that the Φ_α are solutions of (10). ■

The first question that arises from decomposition (14) is how many independent additional solutions we obtain. Since all Φ_α together sum the Casimir operator, some dependence relations must exist.

Lemma 1 *Let C_p be a Casimir operator of $\mathfrak{s}$ of order p. Suppose that*

$$C_p = \Phi_{(p-\alpha_1,\alpha_1)} + ... + \Phi_{(p-\alpha_q,\alpha_q)}, \quad 0 \leq \alpha_i < \alpha_{i+1} \leq p \tag{17}$$

is the decomposition of C_p into homogeneous polynomials of bi-degree (p, q).

(i) *If $\Phi_{(0,p)} \neq 0$, then at most $q - 2$ polynomials $\Phi_{(p-\alpha_j,\alpha_j)}$ are functionally independent on the Casimir operators of $\mathfrak{s}$ and $\mathfrak{s}'$.*

[4] For $t = 1$, equation (14) shows how the Casimir operator decomposes into homogeneous polynomials in the variables of the subalgebra and the complementary space over the original basis.

[5] On the contrary, for the remaining equations the differential operators of the generators corresponding to the representation have been modified, and the equations are dependent on t.

Journal of Physics: Conference Series **128** (2008) 012052 doi:10.1088/1742-6596/128/1/012052

(ii) If $\Phi_{(0,p)} = 0$, then at most $q-1$ polynomials $\Phi_{(p-\alpha_j,\alpha_j)}$ are functionally independent on the Casimir operators of $\mathfrak{s}$ and $\mathfrak{s}'$.

The proof follows at once observing that $\Phi_{(0,p)}$ is a function of the Casimir operators of the subalgebra $\mathfrak{s}'$. The independence on the Casimir operators of $\mathfrak{s}'$ does not imply in general that the $\Phi_{(p-\alpha,\alpha)}$ obtained are all functionally independent between themselves. The number of independent terms depends also on the representation R induced by the reduction [11]. In any case, however, at least one independent term is obtained for any Casimir operator of degree at least three. For the special case of $n=1$ labelling operator, two terms independent on the Casimir operators were found, which allowed to select one as the labelling operator [7]. Once a set of functionally independent solutions to system (10) has been chosen (including the Casimir operators), the first part of the labelling problem is solved. Now, if we want to obtain a set of commuting operators, we have to look for all commutators among the symmetrized operators $\Phi_{(p-\alpha_j,\alpha_j)}$. We denote by $\Phi^{symm}_{(p-\alpha_j,\alpha_j)}$ the symmetrized polynomial. Then $\left[\Phi^{symm}_{(p-\alpha_j,\alpha_j)}, \Phi^{symm}_{(q-\alpha_k,\alpha_k)}\right]$ is a homogeneous polynomial of degree $p+q-1$, and also constitutes a missing label operator. Actually this brackets is expressible as sum of polynomials of different bi-degree, and these terms constitute themselves labelling operators [12]. A procedure to solve the missing label problem can thus be resumed in the following steps:

- Decompose the Casimir operators of $\mathfrak{s}$ of degree $p \geq 3$ with respect to the associated contraction.
- Determine the commutator of all symmetrized polynomials $\Phi^{symm}_{(p-\alpha_j,\alpha_j)}$ with $\alpha_j \neq 0$.
- From those commuting operators, extract n operators that are functionally independent from the Casimir operators of $\mathfrak{s}$ and the subalgebra $\mathfrak{s}'$.

In general, the second step is reduced to pure computation. There is no simple procedure to determine whether two missing label operators are mutually orthogonal, although various symbolic routines have been developed to compute these brackets (see e.g. [13]). In some special circumstances, however, the decomposition (14) can provide orthogonality without being forced to compute the brackets. If for a specific MLP it is known that no solutions of bi-degree (r,s) exists for some fixed $r+s=p+q$, and if we have two labelling operators such that $\left[\Phi^{symm}_{(p-\alpha_j,\alpha_j)}, \Phi^{symm}_{(q-\alpha_k,\alpha_k)}\right]$ is a sum of polynomials of bi-degree (s,r), then the commutation follows at once. This idea was first explored systematically in [12]. We remark that in the commutative frame, it would suffice to show that no polynomial function of bi-degree (r,s) is a solution to subsystem (10).[6]

4. Examples

In this section we show how the decomposition of Casimir operators of higher order provide solutions to missing label problem that could not be solved completely by only using the contraction, or for which no proposed set of labelling operators has been computed yet. We insist on the fact that the main difficulty in the formal approach to the MLP resides in obtaining a sufficient number of (functionally) independent labelling operators, from which a commuting set can be extracted.

4.1. $G_2 \supset \mathfrak{su}(2) \times \mathfrak{su}(2)$

This chain was indicated in [7] to give an insufficient number of labelling operators when only the contraction invariants are considered. Actually, in this case we have $n = \frac{1}{2}(14-2-6-2) = 2$ labelling operators, and the inhomogeneous contraction $G_2 \rightsquigarrow (\mathfrak{su}(2) \times \mathfrak{su}(2)) \overrightarrow{\oplus}_R 8L_1$ preserves

[6] The labelling operator in the enveloping algebra of $\mathfrak{s}$ follows as the symmetrized form of such functions.

the number of invariants. This means that we would only obtain one additional operator, since the (contracted) operator of order two is of no use. Now, the method failed because it did not take into account the decomposition of the sixth order operator into homogeneous polynomials of bi-degree (p,q) in the variables. We show that, with this decomposition, we obtain a complete solution to the MLP related to the chain $G_2 \supset \mathfrak{su}(2) \times \mathfrak{su}(2)$. To this extent, we choose the same tensor basis used in [14] consisting of the generators $\{j_0, j_\pm, k_0, k_\pm, R_{\mu,\nu}\}$ with $\mu = \pm\frac{3}{2}, \pm\frac{1}{2}$, $\nu = \pm\frac{1}{2}$. The generators $R_{\mu,\nu}$ are related to an irreducible tensor representation R of $\mathfrak{su}(2)\times\mathfrak{su}(2)$ of order eight. In this case, the contraction $G_2 \rightsquigarrow (\mathfrak{su}(2) \times \mathfrak{su}(2)) \overrightarrow{\oplus}_R 8L_1$ is obtained considering the transformations:

$$j_0' = j_0,\ j_\pm' = j_\pm,\ k_0' = k_0, k_\pm' = k_\pm,\ R_{\mu,\nu}' = \frac{1}{t} R_{\mu,\nu}.$$

If we decompose now the Casimir operators C_2 and C_6 over the transformed basis, we obtain the following decomposition

$$\begin{aligned} C_2 &= t^2 C_{(2,0)} + C_{(0,2)}, \\ C_6 &= t^6 C_{(6,0)} + t^4 C_{(4,2)} + t^2 C_{(2,4)} + C_{(0,6)}, \end{aligned} \tag{18}$$

where $C_{(0,2)}, C_{(0,6)}$ are functions of the Casimir operators of $\mathfrak{su}(2) \times \mathfrak{su}(2)$. Since $C_{(2,0)}$ is functionally dependent on the invariants of G_2 and $\mathfrak{su}(2) \times \mathfrak{su}(2)$, it is not further useful. Now it can be verified that

$$\frac{\partial(C_2, C_6, C_{21}, C_{22}, C_{(2,4)}, C_{(4,2)})}{\partial\left(k_0, k_-, j_0, j_+, R_{\frac{3}{2},\frac{1}{2}}, R_{-\frac{3}{2},\frac{1}{2}}\right)} \neq 0, \tag{19}$$

where C_{21} and C_{22} are the quadratic Casimir operators of $\mathfrak{su}(2) \times \mathfrak{su}(2)$. This provides us with six independent operators. A long and tedious computation, due to the quite high number of terms before and after symmetrization, shows moreover that the chosen operators commute:

$$\begin{aligned} &\left[C_i, C_{(2,4)}\right] = \left[C_i, C_{(4,2)}\right] = 0, \quad i = 2, 6 \\ &\left[C_{(4,2)}, C_{(2,4)}\right] = 0. \end{aligned} \tag{20}$$

Therefore the set $\left\{C_2, C_6, C_{21}, C_{22}, C_{(2,4)}, C_{(4,2)}\right\}$ can be taken to solve the labelling problem.

It should be remarked that a direct comparison with the operators obtained in [14] is quite difficult, for various reasons. At first, there the scalars in the enveloping algebra were considered, not symmetrizations of functions, which implies that lower order terms where considered when explicitly indicating the labelling operators. On the other hand, we have only distinguished the bi-degree, that is, the degree of the polynomials in the variables of the $\mathfrak{su}(2) \times \mathfrak{su}(2)$ subalgebra and the tensor representation R, while in [14] the order with respect to any of the copies of $\mathfrak{su}(2)$ was considered, resulting in operators labelled with three indices. Therefore the operators $C_{(p,q)}$ considered here correspond to the sum of several scalars there. In addition, our solution contains the term $C^{(114)}$ excluded in [14],[7] confirming that the pair of commuting operators obtained above is different from that found previously. We also remark that a further distinction of the degrees of the polynomials $\Phi_{(a,b)}^{symm}$ in the variables of the $\mathfrak{su}(2)$ copies is not possible due to the contraction.

4.2. The chain $\mathfrak{sp}(6) > \mathfrak{su}(3) \times \mathfrak{u}(1)$

The unitary reduction of the symplectic Lie algebra of rank three has found ample applications in the nuclear collective model [15]. In this case, nuclear states are classified by means of

[7] This is a scalar having degree one in each of the copies of $\mathfrak{su}(2)$ and four in the $R_{\mu,\nu}$ generators.

Journal of Physics: Conference Series **128** (2008) 012052 doi:10.1088/1742-6596/128/1/012052

irreducible representations of $\mathfrak{sp}(6)$ reduced with respect to the unitary subalgebra $\mathfrak{su}(3) \times \mathfrak{u}(1)$. Since the induced representations are not multiplicity free, we have to add $n = 3$ labelling operators to distinguish the states. Generating functions for this chain were studied in [16], but without obtaining explicitly the three required operators. In this section, we will determine a commuting set of labelling operators that solves the MLP for this reduction. As we shall see, this case cannot be solved using only the invariants of the associated contraction.

We will use the Racah realization for the symplectic Lie algebra $\mathfrak{sp}(6, \mathbb{R})$. We consider the generators $X_{i,j}$ with $-3 \leq i, j \leq 3$ satisfying the condition

$$X_{i,j} + \varepsilon_i \varepsilon_j X_{-j,-i} = 0, \tag{21}$$

where $\varepsilon_i = \operatorname{sgn}(i)$. Over this basis, the brackets are given by

$$[X_{i,j}, X_{k,l}] = \delta_{jk} X_{il} - \delta_{il} X_{kj} + \varepsilon_i \varepsilon_j \delta_{j,-l} X_{k,-i} - \varepsilon_i \varepsilon_j \delta_{i,-k} X_{-j,l}, \tag{22}$$

where $-3 \leq i, j, k, l \leq 3$. The three Casimir operators C_2, C_4, C_6 of $\mathfrak{sp}(6, \mathbb{R})$ are easily obtained as the coefficients of the characteristic polynomial

$$|A - T\mathrm{Id}_6| = T^6 + C_2 T^4 + C_4 T^2 + C_6, \tag{23}$$

where

$$A = \begin{pmatrix} x_{1,1} & x_{2,1} & x_{3,1} & -Ix_{-1,1} & -Ix_{-1,2} & -Ix_{-1,3} \\ x_{1,2} & x_{2,2} & x_{3,2} & -Ix_{-1,2} & -Ix_{-2,2} & -Ix_{-2,3} \\ x_{1,3} & x_{2,3} & x_{3,3} & -Ix_{-1,3} & -Ix_{-2,3} & -Ix_{-3,3} \\ Ix_{1,-1} & Ix_{1,-2} & Ix_{1,-3} & -x_{1,1} & -x_{1,2} & -x_{1,3} \\ Ix_{1,-2} & Ix_{2,-2} & Ix_{2,-3} & -x_{2,1} & -x_{2,2} & -x_{2,3} \\ Ix_{1,-3} & Ix_{2,-3} & Ix_{3,-3} & -x_{3,1} & -x_{2,3} & -x_{3,3} \end{pmatrix}. \tag{24}$$

The symmetrized operators give the usual polynomials in the enveloping algebra. Since the unitary algebra $\mathfrak{u}(3)$ is generated by $\{X_{i,j} | 1 \leq i, j \leq 3\}$, in order to write $\mathfrak{sp}(6, \mathbb{R})$ in a $\mathfrak{su}(3) \times \mathfrak{u}(1)$ basis, it suffices to replace the diagonal operators $X_{i,i}$ by suitable linear combinations. Taking $H_1 = X_{1,1} - X_{2,2}$, $H_2 = X_{2,2} - X_{3,3}$ and $H_3 = X_{1,1} + X_{2,2} + X_{3,3}$ we obtain the Cartan subalgebra of $\mathfrak{su}(3)$, while H_3 commutes with all $X_{i,j}$ with positive indices i, j. The invariants over this new basis are simply obtained replacing the $x_{i,i}$ by the corresponding linear combinations of h_i. The contraction $\mathfrak{sp}(6) \rightsquigarrow (\mathfrak{su}(3) \times \mathfrak{u}(1)) \overrightarrow{\oplus}_R 12L_1$, where R is the complementary to $(\mathrm{ad}(\mathfrak{su}(3) \otimes (1))$ in the adjoint representation of $\mathfrak{sp}(6)$:[8]

$$\mathrm{ad}\mathfrak{sp}(6) = (\mathrm{ad}\mathfrak{su}(3) \otimes (1)) \oplus R.$$

The contraction is determined by the transformations

$$H'_i = H_i, \; X'_{i,j} = X_{i,j}, \; X'_{-i,j} = \frac{1}{t} X_{-i,j}, \; X'_{i,-j} = \frac{1}{t} X_{i,-j}, \quad 1 \leq i, j \leq 3. \tag{25}$$

The contraction $(\mathfrak{su}(3) \times \mathfrak{u}(1)) \overrightarrow{\oplus}_R 12L_1$ satisfies $\mathcal{N} = 3$, thus has 3 Casimir operators that can be obtained as contraction of C_2, C_4, C_6. Note however that $n = 3$, thus the invariants of the contraction will provide at most two independent missing label operators. This means that using only the contraction, we cannot solve the MLP for this chain. In order to find a third labelling

[8] More precisely, R decomposes into a sextet and antisextet with $\mathfrak{u}(1)$ weight ± 1 and a singlet with $\mathfrak{u}(1)$ weight 1.

Journal of Physics: Conference Series **128** (2008) 012052 doi:10.1088/1742-6596/128/1/012052

operator, we have to consider the decomposition of the fourth and sixth order Casimir operators of $\mathfrak{sp}(6)$. Over the preceding transformed basis we obtain:

$$\begin{array}{l} C_4 = t^4 C_{(4,0)} + t^2 C_{(2,2)} + C_{(0,4)}, \\ C_6 = t^6 C_{(6,0)} + t^4 C_{(4,2)} + t^2 C_{(2,4)} + C_{(0,6)}, \end{array} \tag{26}$$

where $C_{(k,l)}$ denotes a homogeneous polynomial of k in the variables of R and degree l in the variables of the unitary subalgebra. The $C_{(0,k)}$ are functions of the Casimir operators of $\mathfrak{su}(3) \times \mathfrak{u}(1)$, and therefore provide no labelling operators. We remark that, before symmetrization, $C_{(2,2)}$ has 126 terms, $C_{(2,4)}$ 686 terms, and $C_{(4,2)}$ 444 terms. The symmetrized operators $C_{(2,2)}, C_{(4,2)}$ and $C_{(2,4)}$ can be added to the Casimir operators of $\mathfrak{sp}(6)$ and the subalgebra $\mathfrak{su}(3) \times \mathfrak{u}(1)$, and the 9 operators can be seen to be (functionally) independent.

$$\begin{array}{ll} \left[C_i, C_{(2,2)}\right] = \left[C_i, C_{(4,2)}\right] = \left[C_i, C_{(2,4)}\right] = 0, & i = 2, 4, 6. \\ \left[C_{(2,2)}, C_{(4,2)}\right] = \left[C_{(2,2)}, C_{(2,4)}\right] = \left[C_{(2,4)}, C_{(4,2)}\right] = 0. & \end{array} \tag{27}$$

4.3. Applications to the conformal algebra

Among the many important problems in Physics where the conformal group $SO(2,4)$ plays an impotant role, like the dynamical non-invariance group of hydrogen-like atoms, the application to the periodic charts of neutral atoms in ions was first considered in [17]. This direction was followed to classify chemical elements by various authors [18]. More recently, the conformal group and its invariants are in the centre of the more ambitious program KGR, in order to obtain quantitative predictions of the periodic table of elements [19, 20]. To this extent, the set formed by the three Cartan generators and the Casimir operators (of degrees 2, 3 and 4), which commute between themselves, can be used to label certain physical properties. However, as noted by Racah [21], this set is still not sufficient for classification purposes. We have to add three additional operators[9] to obtain a complete set of commuting operators that solve labelling problems. This follows at once if we consider the missing label problem for the Cartan subalgebra. In this case

$$n = \frac{1}{2}(15 - 3 - 3 - 3) = 3.$$

Therefore the Racah operators can be identified with labelling operators for the reduction chain determined by the Cartan subalgebra. To exemplify the procedure, we compute the Racah operators for the conformal algebra. We use the fact that it is isomorphic to the Lie algebra $\mathfrak{su}(2,2)$. We start from the the $\mathfrak{u}(2,2)$-basis formed by the operators $\{E_{\mu\nu}, F_{\mu\nu}\}_{1\leq\mu,\nu\leq p+q=n}$ with the constraints

$$\begin{array}{r} E_{\mu\nu} + E_{\nu\mu} = 0,\ F_{\mu\nu} - F_{\nu\mu} = 0, \\ g_{\mu\mu} = ((1, 1, -1, -1)\,. \end{array}$$

The brackets are then given by

$$[E_{\mu\nu}, E_{\lambda\sigma}] = g_{\mu\lambda} E_{\nu\sigma} + g_{\mu\sigma} E_{\lambda\nu} - g_{\nu\lambda} E_{\mu\sigma} - g_{\nu\sigma} E_{\lambda\mu} \tag{28}$$

$$[E_{\mu\nu}, F_{\lambda\sigma}] = g_{\mu\lambda} F_{\nu\sigma} + g_{\mu\sigma} F_{\lambda\nu} - g_{\nu\lambda} F_{\mu\sigma} - g_{\nu\sigma} F_{\lambda\mu} \tag{29}$$

$$[F_{\mu\nu}, F_{\lambda\sigma}] = g_{\mu\lambda} E_{\nu\sigma} + g_{\nu\lambda} E_{\mu\sigma} - g_{\nu\sigma} E_{\lambda\mu} - g_{\mu\sigma} E_{\lambda\nu} \tag{30}$$

To recover the conformal algebra, we take the Cartan subalgebra spanned by the vectors $H_\mu = g_{\mu+1,\mu+1} F_{\mu\mu} - g_{\mu\mu} F_{\mu+1,\mu+1}$ for $\mu = 1..3$. The centre of $\mathfrak{u}(p,q)$ is obviously generated by $g^{\mu\mu} F_{\mu\mu}$.

[9] If r denotes the dimension of a semisimple Lie algebra $\mathfrak{s}$ and l its rank, the number $f = \frac{1}{2}(r - 3l) = 3$ is usually referred to as the Racah number.

Proposition 2 *A maximal set of independent Casimir invariants of* $\mathfrak{su}(2,2)$ *is given by the coefficients* C_k *of the characteristic polynomial* $|IA - \lambda \mathrm{Id}_N| = \lambda^4 + \sum_{k=2}^{4} D_k \lambda^{4-k}$*, where*

$$A = \begin{pmatrix} -I(\frac{3}{4}h_1 - \frac{1}{2}h_2 + \frac{1}{4}h_3) & -e_{12} - If_{12} & e_{13} + If_{13} & e_{14} + If_{14} \\ e_{12} - If_{12} & I(\frac{1}{4}h_1 + \frac{1}{2}h_2 - \frac{1}{4}h_3) & e_{23} + If_{23} & e_{24} + If_{24} \\ e_{13} - If_{13} & e_{23} - If_{23} & I(\frac{1}{4}h_1 - \frac{1}{2}h_2 - \frac{1}{4}h_3) & e_{34} + If_{34} \\ e_{14} - If_{14} & e_{24} - If_{24} & -e_{34} + If_{34} & I(\frac{1}{4}h_1 - \frac{1}{2}h_2 + \frac{3}{4}h_3) \end{pmatrix}. \tag{31}$$

The classical Casimir operators are obtained symmetrizing the functions C_k. In order to compute the Racah operators, we consider the MLP for the chain $\mathfrak{h} \subset \mathfrak{su}(2,2)$, where $\mathfrak{h}$ denotes the Cartan subalgebra. The corresponding contraction[10] is defined by the non-singular transformations

$$H_i' = \frac{1}{t} H_i, \quad i = 1, 2, 3.$$

According to this contraction, the Casimir operators decompose as follows:

$$\begin{array}{l} C_2 = t^2 C_{(2,0)} + C_{(0,2)}, \\ C_3 = t^3 C_{(3,0)} + t^2 C_{(2,1)} + C_{(0,3)}, \\ C_4 = t^4 C_{(4,0)} + t^3 C_{(3,1)} + t^2 C_{(2,2)} + C_{(0,4)}, \end{array} \tag{32}$$

where the $C_{(0,i)}$ are functions of h_1, h_2, h_3. The functions I_{ij} are all solutions to the MLP. In order to complete the set of orthogonal operators $\{H_1, H_2, H_3, C_2, C_3, C_4\}$ with three mutually commuting labelling operators, we first extract those triples that are functionally independent from the Casimir operators of $\mathfrak{su}(2,2)$ and the h_i. We can take for example $C_{(3,0)}, C_{(4,0)}, C_{(3,1)}$. Since

$$\frac{\partial(H_1, H_2, H_3, C_2, C_3, C_4, C_{(3,0)}, C_{(4,0)}, C_{(3,1)})}{\partial(h_1, h_2, h_3, e_{12}, e_{13}, e_{14}, f_{23}, f_{24}, f_{34})} \neq 0, \tag{33}$$

these operators are independent. A somewhat more laborious computation shows that the symmetrized forms of $C_{(2,1)}, C_{(4,0)}, C_{(3,1)}$ satisfy the commutators

$$\begin{array}{ll} [C_i, C_{(3,0)}] = [C_i, C_{(4,0)}] = [C_i, C_{(3,1)}] = 0, & i = 1, 2, 3. \\ [C_{(3,0)}, C_{(4,0)}] = [C_{(3,0)}, C_{(3,1)}] = [C_{(4,0)}, C_{(3,1)}] = 0. & \end{array} \tag{34}$$

Since the $C_{(i,j)}$ are solutions to the MLP, they commute with any function of the generators H_i. In contrast to the previous cases, the length of the labelling operators never exceeds 70 terms before symmetrization. In conclusion, the set $\{H_1, H_2, H_3, C_2, C_3, C_4, C_{(3,0)}, C_{(4,0)}, C_{(3,1)}\}$ is complete formed by commuting operators. We observe that linear combinations of the three Racah operators are of potential use to describe chemical and physical properties like ionization energy, atomic volume or magnetic properties. Since this identification relies heavily on experimental data [18], it remains to compute the corresponding eigenvalues for irreducible representations (IRREPs) of $\mathfrak{su}(2,2)$, which constitutes a quite hard numerical problem. This task is in progress.

5. Conclusions

The method of contraction is useful to solve the MLP when the number of invariants of the contraction associated to the reduction chain $\mathfrak{s} \supset \mathfrak{s}'$ exceeds the number of needed labelling operators. In the case where the invariants of the inhomogeneous contraction do not suffice to

[10] In this case, the contraction is no more an inhomogeneous Lie algebra. The procedure remains however valid, which suggests that it could also be valid for non-semisimple algebras.

find a complete solution of the missing label problem, it is expectable that labelling operators of the same degree appear. This suggests that further terms of the Casimir operators of $\mathfrak{s}$ that disappear during the contraction can be useful to complete the set of missing label operators. We have shown that the contraction induces a decomposition of the Casimir operators, the terms of which are all solutions to the MLP. From these terms a set of n independent labelling operators can be extracted, reducing the problem to determine which combinations are mutually orthogonal. In this sense, the method proposed in [7] is a first approximation to solve the MLP using the properties of reduction chains, which however turns out to be useful in most practical cases. The bi-degree of the Casimir operators of a Lie algebra with respect to the variables associated to the generators of a subalgebra are therefore a relevant tool to obtain and classify these labelling operators, although further distinction of terms, for example when the subalgebra consists of various copies, is also convenient to deduce additional operators.[11] This subdivision cannot however be deduced from the contraction, since all generators of the subalgebra play the same role.

Some important aspects of the decomposition method of Casimir operators based on the contractions and its use in labelling problems are specially emphasized:

- The solutions provide a "natural" choice for the labelling operators. Their interpretation as "broken" Casimir operators confers them a certain physical meaning, in contrast to operators obtained by pure algebraic means, where the physical interpretation of the operator is often not entirely clear.
- The decomposition provides also a consistent explanation to the question why a number of reduction chains give labelling operators of the same degree. This fact is directly related to an insufficient number of invariants in the contraction associated to the chain.
- This could probably explain why the eigenvalues of such labelling operators are not integers, as already indicated by Racah [21]. It follows from the decomposition that the eigenvalues of the labelling operators contribute to the eigenvalues of the Casimir operators. In this context, the interpretation of a labelling operator as "broken" Casimir operator leads to the idea of "broken" integer eigenvalues.

Some questions still remain open, namely, whether there exist reductions $\mathfrak{s} \supset \mathfrak{s}'$ for which the method followed here provides all available labelling operators. An answer in this direction implies to find the general solution to the MLP for each considered chain. Nowadays, only for a few number of algebras these computations have been carried out completely [4, 22]. A complete study of all physically relevant reduction chains involving simple Lie algebras up to some fixed rank would certainly provide new insights to this problem. On the other hand, in can also not be excluded that for reduction chains with a great number of labelling operators, the terms of the decomposition are not sufficient to construct a set of independent labelling operators. To which extent the invariants of the contraction not appearing as contracted operators play a role must still be analyzed.[12]

Another problem, still in progress, is to obtain complete sets of commuting operators for all simple Lie algebras, using the MLP determined by the Cartan subalgebra. The commented application to the periodic charts of atoms in only one of the problems where this special type of reductions have been shown to be of interest in developing algebraic models in molecular physics or nuclear spectroscopy [23].

[11] This turns out to be the case for the chain $\mathfrak{su}(4) \supset \mathfrak{su}(2) \times \mathfrak{su}(2)$ [4].

[12] Such situations appear, e.g., considering a simple algebra of high rank and regular subalgebras of low rank. If the induced representation (8) contains copies of the trivial representation, then the generators associated to these will play the role of labelling operators. It happens moreover that these generators cannot be obtained contracting the Casimir operators of $\mathfrak{s}$. This situation is however unlike to appear in some physically interesting case.

Journal of Physics: Conference Series **128** (2008) 012052 doi:10.1088/1742-6596/128/1/012052

Acknowledgments
This work was partially supported by the research project MTM2006-09152 of the Ministerio de Educación y Ciencia. The author expresses his gratitude to M. R. Kibler and Ch. Quesne for useful discussions and remarks, as well as to G. Vitiello and J. Van der Jeugt for additional references.

References
[1] Okubo S 1962 *Prog. Theor. Phys.* **27** 949
Goldberg H and Lehrer-Ilamed Y 1962 *J. Math. Phys.* **4** 501
Burakovsky L and Horwitz L P 1997 *Found. Phys. Lett.* **10** 131
Burakovsky L and Goldman T 1997 (hep-ph/9708498)
[2] Süalp G and Kaptanoglu S 1983 *Ann. Phys.* **147** 460
Kitazawa N 1994 *Tumbling and Technicolor Theory* DPNU-94-03 (Nagoya: Japan).
[3] Iachello F and Arima A 1987 *The interacting boson model*, (Cambridge Univ. Press, Cambridge).
[4] Elliott J P 1958 *Proc. Roy. Soc. Lond. A* **245** 128, 562
Bargmann V and Moshinsky M 1965 *Nucl. Phys.* **23** 177
Green H S and Bracken A J 1974 *Int. J. Theor. Phys.* **11** 157
Judd B R, Miller W, Patera J and Winternitz P 1974 *J. Math. Phys.* **15** 1787
Sharp R T 1975 *J. Math. Phys.* **16**, 2050
Quesne Ch 1976 *J. Math. Phys.* **17**, 1452; 1977 **18**, 1210
De Meyer H, Vanden Berghe G, Van der Jeugt J and De Wilde P 1985 *J. Math. Phys.* **26** 2124
Rowe D J, Le Blanc R and Repka J 1989 *J. Phys. A: Math. Gen.* **22** L309
Hecht K T 1994 *J. Phys. A: Math. Gen.* **27** 3445
Turner P S, Rowe D J and Repka J 2006 *J. Math. Phys.* **47** 023507
[5] Peccia A and Sharp R T 1976 *J. Math. Phys.* **17** 1313
[6] Wigner E P 1937 *Phys. Rev.* **51** 106
Lipkin H J 1964 *Phys. Rev. Lett.* **13** 590
Khanna F C and Umezawa H 1994 *Chinese J. Phys.* **32** 1317
von Isacker P and Juillet O 1999 *Nuclear Phys.* **A564** 739
Valencia J P and Wu H C 2004 *Braz. J. Phys.* **34** 837
[7] Campoamor-Stursberg 2007 *J. Phys. A: Math. Theor.*, to appear (hep-th/arXiv:0706.2581)
[8] Vitiello G and de Concini C 1976 Nuclear Phys. B**116** 141
Anderson J T 1981 *Phys. Rev.* **D23** 1856
Celeghini E, Tarlini M and Vitiello G 1984 *Nuovo Cimento A* **84** 19
[9] Campoamor-Stursberg R 2006 *J. Phys. A: Math. Gen.* **39** 2325
——2007 *J. Phys. A: Math. Theor.* **40** 5355
[10] Segal I E 1951 *Duke Math. J.* **18** 221
Inönü E and Wigner E P 1953 *Proc. Nat. Acad. Sci U.S.A.* **39** 510
Domokos G and Tindle G L 1968 *Commun. Math. Phys.* **7** 160
[11] Campoamor-Stursberg R 2003 *J. Phys. A: Math. Gen.* **36** 1357
[12] Van der Jeugt 1984 *J. Math. Phys.* **25** 1221
[13] De Meyer H, Vanden Berghe G and De Wilde P 1987 *Comput. Phys. Comm.* **44** 197
[14] Hughes J W B and Van der Jeugt J 1985 *J. Math. Phys.* **26** 894
[15] Rosensteel G 1977 *Int. J. Theor. Phys.* **16** 63
Graber J L and Rosensteel G 2003 *Phys. Rev. C* **68** 014301
[16] Gaskell R, Rosensteel G and Sharp R T 1981 *J. Math. Phys.* **22** 2732
[17] Barut A O 1972 in *The Structure of Matter* (Proc. Rutherford Centennial Symp., 1971) (Christchurch, New Zealand)
[18] Rumer Yu B and Fet A I 1971 *Teor. Mat. Fiz.* **9** 203
Konopel'chenko B G and Rumer Yu B 1979 *Uspekhi Fiz. Nauk* **129** 339
Byakov V M, Kulakov Yu I, Rumer Yu B and Fet A I 1976 Preprint ITEP-26, ITEP-90
Kibler M R 1989 *J. Mol. Struct.* **187** 83
Carlson C M, Hefferlin R A and Zhuvikin G V 1995 *Analysis of Group Theoretical Periodic Systems of Molecules using Tabulated Data* Joint Report SC/SPBU-2
Kibler M R 2004 *The Mathematics of the Periodic Table*, ed D H Rouvray and R B King (Baldock: Research Studies Press)
[19] Kibler M R 2004 *Mol. Phys.* **102** 1221
[20] Kibler M R, private communication

[21] Racah G 1949 *Phys. Rev.* **76** 1352
——1951 *Group Theory and Spectroscopy,* (Princeton Univ. Press, New Jersey)
Fano U and Racah G 1959 *Irreducible tensorial Sets* (Academic Press: New York)
[22] Partensky A and Maguin C 1978 *J. Math. Phys.* **19** 511
[23] Frank A and van Isacker P 1994 *Algebraic Methods in Molecular and Nuclear Structure Physics* (Wiley: New York)
Iachello F and Levine R D 1995 *Algebraic Theory of Molecules* (Oxford Univ. Press: Oxford)

Journal of Physics: Conference Series 128 (2008) 012053 doi:10.1088/1742-6596/128/1/012053

On position-dependent mass harmonic oscillators

S Cruz y Cruz[1,2], **J Negro**[3] **and L M Nieto**[3]

[1] Departamento de Física, Cinvestav, AP 14-740, México 07000 DF, Mexico
[2] Sección de Estudios de Posgrado e Investigación, UPIITA-IPN, Av. IPN 2580, CP 07340 México DF, Mexico
[3] Departamento de Física Teórica, Atómica y Óptica, Universidad de Valladolid, 47071 Valladolid, Spain

E-mail: sara@fis.cinvestav.mx, jnegro@fta.uva.es, luismi@metodos.fam.cie.uva.es

Abstract. The classical and quantum position-dependent mass harmonic oscillators are constructed by means of the supersymmetric approach. The correspondence between the classical and the quantum Hamiltonians is used to fix the ordering of the kinetic term in the quantum framework. Some examples are given for different types of mass function.

1. Introduction

In the study of the electronic properties of semiconductors, it is well known that the envelope wave function of a single electron fulfills the Schrödinger equation (the effective mass wave equation) with a Hamiltonian

$$H = -\frac{\hbar^2}{2m}\frac{d^2}{dx^2} + E + V(x)\,, \tag{1}$$

where E is the edge of the conduction band and $V(x)$ a potential term which includes the interaction with an impurity (if any) inside the semiconductor, an external potential, etc [1, 2]. In this Hamiltonian, m stands for the effective mass of the electron in the lattice. In the case of nonuniform semiconductors, both, the effective mass m, as well as the edge of the conduction band E, depend on the position. Therefore, it is necessary to establish the effective mass wave equation in an appropriate form [3–8], as it is clear that the Hamiltonian (1) is not a Hermitian operator when $m = m(x)$. The position-dependence of the edge of the conduction band can be absorbed into the potential term, but since the effective mass $m(x)$ and the momentum operator p do not commute any longer, a problem arises in stating the correct form of the kinetic term. A general expression for a Hermitian kinetic energy operator is

$$T = \frac{1}{2}m^a p\, m^{2b} p\, m^a, \tag{2}$$

where $a + b = -1/2$. For solving a particular problem we require, therefore, to know the value of a or b. Nevertheless, it is not clear how to fix both parameters.

In general, the Hamiltonians with position-dependent mass (PDM) have been widely studied due to its multiple applications, not only in the theory of semiconductors. Indeed, they also appear in the study of quantum dots [9] and quantum liquids [10], as well as in the description

Journal of Physics: Conference Series **128** (2008) 012053 doi:10.1088/1742-6596/128/1/012053

of non linear oscillators [11, 12], to mention just a few fields. A very important dilemma in this context is how to deal with the ordering ambiguity of the kinetic term (2) in the quantum case. This question has been addressed using different approaches (see for instance [13, 14]). The aim of this work is to establish a kind of criterion to chose a proper ordering in the particular case of the PDM harmonic oscillator. Our treatment will be based on a comparison between the classical and the quantum cases. We will also show that this choice enables to extend, to this system, many of the well known properties of the constant mass (CM) case in a very simple way, reproducing some results already obtained from other points of view.

The paper is organized as follows. In Section 2, we present a short review of the classical CM harmonic oscillator. Section 3 is devoted to the study of the classical PDM harmonic oscillator by means of the canonical transformation connecting this problem with the CM case. In Section 4 we illustrate the formalism by considering some particular mass functions. In Section 5 it is shown that there is a particular ordering in the kinetic term of the quantum PDM Hamiltonian giving an expression for the potential which coincides with the classical one. This allows us to solve the problem in a very simple form by means of the correspondence between CM and PDM systems. In Section 6 these results are illustrated using the same mass functions introduced before in the classical analysis. Finally, we present some concluding remarks.

2. The classical constant mass harmonic oscillator

Since the comparison between the CM and the PDM harmonic oscillators will be the keystone of our approach, let us start by presenting some basic features of the CM harmonic oscillator. Indeed, let us consider the Hamiltonian (choosing, without loss of generality, $m = \omega = 1$)

$$\mathcal{H} = \frac{P^2}{2} + \frac{X^2}{2}, \tag{3}$$

where X, P stand for the classical position and momentum variables. The phase space trajectories can be constructed in a purely algebraic way: observe that $\mathcal{H}$ can be factorized in the form

$$\mathcal{H} = a^+ a^- = a^- a^+, \qquad a^\pm = \frac{1}{\sqrt{2}}(\mp iP + X), \tag{4}$$

with $a^\pm$ fulfilling $(a^-)^* = a^+$ (* denoting complex conjugation). It turns out that these functions close the Heisenberg algebra with Poisson brackets [15]

$$i\{a^-, a^+\} = 1, \qquad i\{\mathcal{H}, a^\pm\} = \pm a^\pm. \tag{5}$$

This algebraic structure allows us to construct non-autonomous integrals of motion of the form

$$Q^\pm = a^\pm e^{\mp it} = \frac{1}{\sqrt{2}}(\mp iP + X)\, e^{\mp it} \tag{6}$$

satisfying $(Q^-)^* = Q^+$, and $Q^+Q^- = \mathcal{H}$. On the other hand, it is well known that the Hamiltonian is a conserved quantity, indeed the total energy E of the system. Therefore, letting $Q^+ = \sqrt{E}e^{i\phi}$, with ϕ a constant phase fixed by the initial conditions, it is possible to obtain the well known phase trajectories for the CM harmonic oscillator, namely

$$X(t) = \sqrt{2E}\cos(t + \phi), \qquad P(t) = -\sqrt{2E}\sin(t + \phi), \tag{7}$$

which are concentric circumferences centered at the origin and having radius $\sqrt{2E}$. The position and momentum of a particle in this potential can take arbitrarily large values just by choosing the energy E in the proper way.

3. The classical position dependent mass harmonic oscillator

A generic PDM classical Hamiltonian has the standard form

$$\mathcal{H} = \frac{p^2}{2m(x)} + \mathcal{V}(x), \tag{8}$$

where $m(x)$ is an arbitrary position-dependent function and the potential $\mathcal{V}(x)$, which depends on the form of $m(x)$, has to be determined. With this aim, let us suppose that the Hamiltonian $\mathcal{H}$ can be factorized in terms of two functions $\mathcal{A}^{\pm}(x,p)$ of the form

$$\mathcal{A}^{\pm}(x,p) = \mp i \frac{p}{\sqrt{2m(x)}} + \mathcal{W}(x). \tag{9}$$

In this way

$$\mathcal{H} = \mathcal{A}^{+}\mathcal{A}^{-} = \mathcal{A}^{-}\mathcal{A}^{+} = \frac{p^2}{2m(x)} + \mathcal{W}^2(x), \tag{10}$$

and then, the potential $\mathcal{V}(x)$ and the position-dependent function $\mathcal{W}(x)$ are related by

$$\mathcal{V}(x) = \mathcal{W}^2(x). \tag{11}$$

Next, we demand that the functions $\mathcal{H}$, $\mathcal{A}^{\pm}$ close the Heisenberg algebra with Poisson brackets as in the CM case (5). It is not difficult to show that

$$i\left\{\mathcal{A}^{-},\mathcal{A}^{+}\right\} = \frac{2\mathcal{W}(x)'}{\sqrt{2m(x)}}, \qquad i\left\{\mathcal{H},\mathcal{A}^{\pm}\right\} = \pm\frac{2\mathcal{W}(x)'}{\sqrt{2m(x)}}\mathcal{A}^{\pm}, \tag{12}$$

so the former condition fixes $\mathcal{W}(x)$ as

$$\mathcal{W}(x) = \frac{1}{\sqrt{2}}\left(\int^x \sqrt{m(t)}dt + X_0\right), \tag{13}$$

where X_0 is an integration constant. The potential $\mathcal{V}(x)$ has then the form

$$\mathcal{V}(x) = \frac{1}{2}\left(\int^x \sqrt{m(t)}dt + X_0\right)^2, \tag{14}$$

and hence, the classical PDM harmonic oscillator Hamiltonian (8) is

$$\mathcal{H} = \frac{p^2}{2m(x)} + \frac{1}{2}\left(\int^x \sqrt{m(t)}dt + X_0\right)^2. \tag{15}$$

Notice that it has a very similar form to the CM Hamiltonian (3). Indeed, if we perform the canonical transformation

$$X(x) = \int^x \sqrt{m(t)}dt + X_0, \qquad P(x,p) = \dot{X}(x) = \frac{p}{\sqrt{m(x)}}, \tag{16}$$

the PDM Hamiltonian (15) transforms into the CM one (3). Observe that the constant X_0 in (13) determines the position of the origin of the potential in the variable X. We fix this constant in such a way that

$$\mathcal{V}(x=0) = 0. \tag{17}$$

Journal of Physics: Conference Series **128** (2008) 012053 doi:10.1088/1742-6596/128/1/012053

The classical PDM problem, for the harmonic oscillator, can then be reduced to the CM problem, where the new "position" and "momentum" variables are related to the mass function by means of (16). For instance, the phase space trajectories $x(t)$, $p(t)$ are determined as

$$x(t) = X^{-1}\left(\sqrt{2E}\cos(t+\phi)\right), \qquad p(t) = -\sqrt{2E}\sqrt{m(x(t))}\sin(t+\phi), \tag{18}$$

and can be either harmonic or not, depending on the form of $m(x)$. It is worthwhile to mention that they could also be obtained by means of the integrals of motion

$$\mathcal{Q}^{\pm} = \frac{1}{\sqrt{2}}\left(\mp iP(x,p) + X(x)\right)e^{\mp it} = \frac{1}{\sqrt{2}}\left(\mp i\frac{p}{\sqrt{m(x)}} + \int^{x}\sqrt{m(t)}dt + X_0\right)e^{\mp it} = \mathcal{A}^{\pm}e^{\mp it}. \tag{19}$$

In order to construct the phase space trajectories for some particular cases, it is necessary to specify the form of the mass function. We can choose it basically in an arbitrary way; however, since our treatment is based on the correlation between the CM and the PDM systems, it is more natural to choose $m(x)$ in such a way that we can make a complete correspondence between both harmonic oscillators. We refer particularly to the canonical transformation (16) connecting them. Observe that for some choices of $m(x)$, the first equation in (16) may not map the mass natural domain $\mathcal{D}(m)$ onto the whole real line, as it is required if $X(x)$ represent the position of the CM harmonic oscillator. With this in mind, we should take into account the following facts [16]:

(A) Suppose the mass function has $\mathbb{R}$ as its domain. Then, $m(x)$ should be such that $X(x)$ becomes divergent as $|x| \to \infty$; in this way, we assure that the range $\mathcal{R}(X) = \mathbb{R}$. For instance, if $m(x) \sim 1/x^q$ (with $q > 0$) as $|x|$ grows, then

$$X(x) \sim \frac{1}{x^{(q-2)/2}} \tag{20}$$

is divergent for $q < 2$. If $q = 2$, then $X(x) \sim \ln|x|$.

(B) Suppose, on the contrary, that $m(x)$ is only defined on a finite interval (a, b). Then, $\mathcal{R}(X) = \mathbb{R}$ if $X(x)$ is divergent in both limits of this interval. For instance, if $m(x)$ presents a singularity, *i.e.*, $m(x) \sim 1/(x-x_0)^q$ (with $q > 0$) as $x \to x_0$, then

$$X(x) \sim \frac{1}{(x-x_0)^{(q-2)/2}}, \tag{21}$$

is now divergent for $q > 2$. Again, if $q = 2$, then $X(x) \sim \ln|x|$.

If $m(x)$ does not satisfy these conditions, this approach is still valid but the solutions to the problem will present some differences with respect to the CM harmonic oscillator solutions. We will see that this fact also has consequences on the wave functions of the corresponding quantum PDM Hamiltonian.

4. Examples of classical position dependent mass harmonic oscillator

As illustrative examples, we present now the cases of some masses considered before, for some authors, in the study of PDM systems in different contexts. These include some forms of $m(x)$ depending on two real parameters, m_0 and λ, which do not satisfy the conditions mentioned at the end of the previous section.

In the first instance, let us consider the mass

$$m_1(x) = m_0\left(\frac{1+\lambda+x^2}{1+x^2}\right)^2, \tag{22}$$

which is a function free of singularities in the real line, *i.e.*, $\mathcal{D}(m_1) = \mathbb{R}$. It takes its maximum value $m_0(1+\lambda)^2$ at $x=0$ and tends to a constant value m_0 as $|x| \to \infty$. The function (22) is represented in Figure 1(a). The "position" is obtained from (16)–(17). It is not difficult to show that $X_0 = 0$ and that

$$X_1(x) = \sqrt{m_0}\,[x + \lambda \arctan x] \tag{23}$$

takes arbitrary values in $\mathbb{R}$. The potential has the form

$$\mathcal{V}_1(x) = \frac{m_0}{2}\,[x + \lambda \arctan x]^2 , \tag{24}$$

and it can be seen from Figure 1(b), that it is just a slight deformation of the CM harmonic oscillator potential. Also, some plots of the phase trajectories are represented in Figure 1(c). They are given by the expression

$$\frac{m_0}{2}\,[x + \lambda \arctan x]^2 + \frac{1}{2m_0}\left(\frac{1+x^2}{1+\lambda - x^2}\right)^2 p^2 = E, \tag{25}$$

and they seem quite similar to the usual harmonic oscillator. Notice that, in the limit $\lambda \to 0$, the CM harmonic oscillator is recovered. The mass (22) has been used in many works due to its physical behavior [17–20], since it represent a system which has almost a constant mass all over the space except for a small region near the origin, where its mass softly increases.

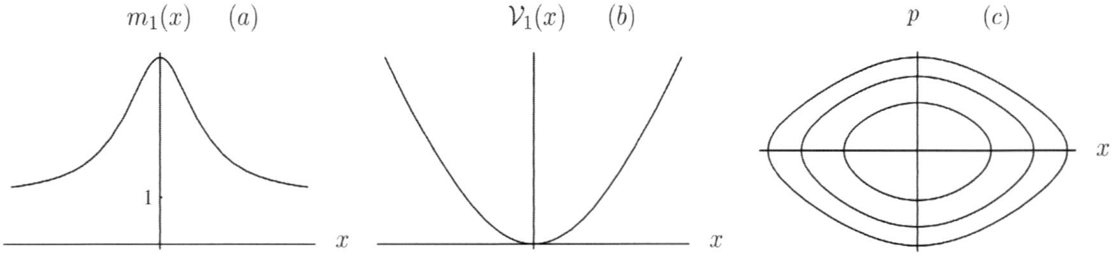

Figure 1. The mass $m_1(x)$ (a), the potential $\mathcal{V}_1(x)$ (b) and some phase space trajectories (c) for the following values of parameters: $m_0 = \lambda = 1$, $\phi = 0$ and $E = 0.3, 0.7, 1.1$.

In the second place, we consider the mass

$$m_2(x) = m_0 \tanh^2(\lambda x) . \tag{26}$$

This function, represented in Figure 2(a), is also free of singularities in the real line ($\mathcal{D}(m_2) = \mathbb{R}$). It is constant in almost all the straight real line, but contrary to the previous case, it rapidly vanishes in a small region near the origin.

In this case

$$X_2(x) = \frac{\sqrt{m_0}}{\lambda}\,\mathrm{sign}(x)\ \ln(\cosh \lambda x) , \tag{27}$$

takes also arbitrary values in $\mathbb{R}$. The potential

$$\mathcal{V}_2(x) = \frac{m_0}{2\lambda^2}\ \ln^2(\cosh \lambda x) , \tag{28}$$

is shown in Figure 2(b) and the phase trajectories

$$\frac{m_0}{2\lambda^2}\ \ln^2(\cosh \lambda x) + \frac{1}{2m_0}\frac{p^2}{\tan^2 \lambda x} = E \tag{29}$$

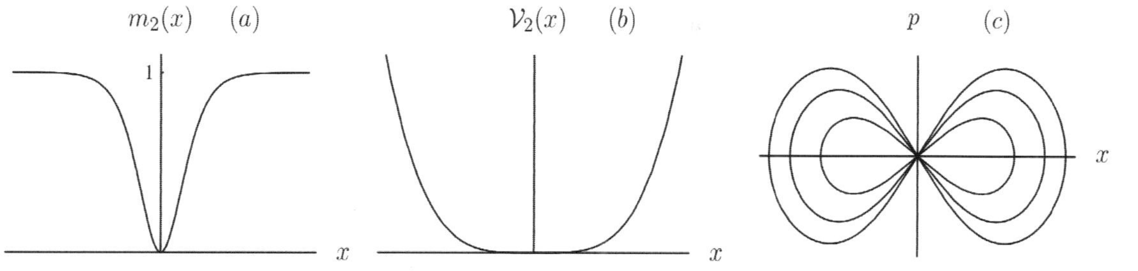

Figure 2. The mass $m_2(x)$ (a), the potential $\mathcal{V}_2(x)$ (b) and some phase space trajectories (c) for the following values of parameters: $m_0 = \lambda = 1$, $\phi = 0$ and $E = 0.3, 0.7, 1.1$.

are radically deformed near the origin, showing the fact that the mass vanishes at $x = 0$ (see Figure 2(c)). The CM case is now recovered in the limit $\lambda \to \infty$. The mass (26) has been used to illustrate the mechanism of evaluating the Green functions for PDM systems [21].

Finally, we consider the case

$$m_3(x) = \frac{m_0}{1 - (\lambda x)^2}\,. \tag{30}$$

This mass function has two singularities, its domain is now $\mathcal{D}(m_3) = (-1/\lambda, 1/\lambda)$, and it rapidly grows as x tend to the limits of the domain, while it reaches its minimum value at $x = 0$ (see Figure 3(a)).

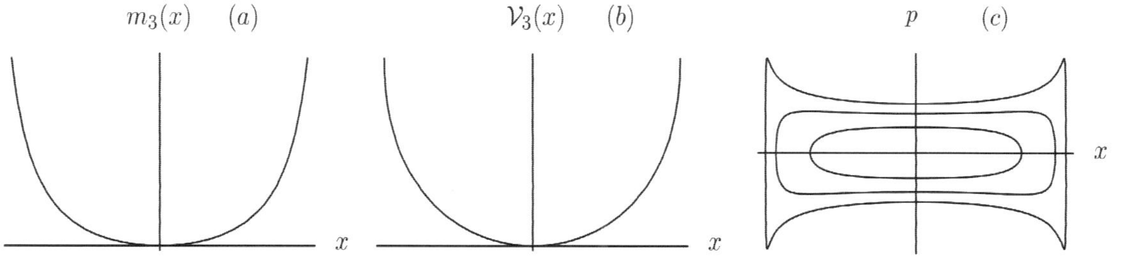

Figure 3. The mass $m_3(x)$ (a), the potential $\mathcal{V}_3(x)$ (b) and some phase space trajectories (c) for the following values of parameters: $m_0 = \lambda = 1$, $\phi = 0$ and $E = 0.3, 0.7, 1.1$.

Observe that, even when the mass presents singularities at $|x| = 1/\lambda$, the variable

$$X_3(x) = \frac{\sqrt{m_0}}{\lambda} \arcsin \lambda x \tag{31}$$

is not divergent at this points and takes values only in the interval $\left[-\frac{\sqrt{m_0}\pi}{2\lambda}, \frac{\sqrt{m_0}\pi}{2\lambda}\right]$. The finite range potential

$$\mathcal{V}_3(x) = \frac{m_0}{2\lambda^2} \arcsin^2 \lambda x, \tag{32}$$

takes its maximum value $\mathcal{V}_{max} = \dfrac{m_0\pi^2}{8\lambda^2}$ at $x = \pm 1$, as it is shown in Figure 3(b). Nevertheless, due to the fact that the mass becomes infinite at this points, the particle is still confined in the region determined by $\mathcal{D}(m_3)$, even when its energy $E > \mathcal{V}_{max}$. The plots of the phase space trajectories

$$\frac{m_0}{2\lambda^2} \arcsin^2 \lambda x + \frac{1}{2m_0}\left(1 - (\lambda x)^2\right) p^2 = E \tag{33}$$

Journal of Physics: Conference Series **128** (2008) 012053 doi:10.1088/1742-6596/128/1/012053

illustrate well this situation in Figure 3(c): the motion is confined to a zone from which the system can not escape since at the boundaries its mass tends to infinity. This fact becomes more evident as the energy of the system takes larger values, in such a way that, for $E > \mathcal{V}_{max}$, the momentum becomes infinity. The CM harmonic oscillator is recovered, in this case, in the limit $\lambda \to 0$. Masses of this kind appears in the study of singular oscillators (see, *e.g.* [11, 12]).

Observe that, in the previous examples, the "position" $X(x)$ and the potential $\mathcal{V}(x)$ are, respectively, odd and even functions due to the fact that $m(x)$ is an even function of x in all the cases.

5. The quantum position-dependent mass harmonic oscillator

It is worthwhile, as in the classical framework, to begin this section with some words about the CM quantum harmonic oscillator. The Hamiltonian has the form ($\hbar = m = \omega = 1$)

$$h = -\frac{1}{2}\frac{d^2}{dy^2} + \frac{1}{2}y^2 = b^+b^- + \frac{1}{2}, \tag{34}$$

where y and $-id/dy$ are the position and momentum operators in the coordinate representation. The creation and annihilation operators

$$b^\pm = \frac{1}{\sqrt{2}}\left(\mp\frac{d}{dy} + y\right), \tag{35}$$

together with the Hamiltonian h, close the Heisenberg algebra

$$\left[b^-, b^+\right] = 1, \qquad \left[h, b^\pm\right] = \pm b^\pm. \tag{36}$$

These commutation relations allow us to determine the spectral values and wave functions:

$$E_n = n + \frac{1}{2}, \qquad \phi_n(y) = \frac{1}{\sqrt{n!}}\left(b^+\right)^n \phi_0(y) = \frac{1}{\sqrt{\pi^{1/2}2^n n!}} e^{-y^2/2} H_n(y), \tag{37}$$

with $\phi_0(y)$ the ground state (the eigenstate state annihilated by b^-), and $H_n(y)$ the Hermite polynomials.

In the PDM case, as stated before, there exist an ordering ambiguity of p and $m(x)$ in the kinetic term of the Hamiltonian due to the fact that, these variables do not commute in this framework. A general, Hermitian, kinetic term has the form (2)

$$T_a = -\frac{1}{2}m^a\frac{d}{dx}m^{2b}\frac{d}{dx}m^a, \tag{38}$$

with $a + b = -1/2$. Then, the generic PDM Hamiltonian is

$$H_a = -\frac{1}{2}m^a\frac{d}{dx}m^{2b}\frac{d}{dx}m^a + V_a(x), \tag{39}$$

where the potential $V_a(x)$, as in the classical case, is a position-dependent function which has to be determined.

It turns out that the ambiguity parameters can be transferred from the kinetic to the potential term just by choosing a particular ordering and redefining the potential in the proper form. In many works, this particular ordering is fixed *a priori* without taking into account the particular problem which is to be solved. Here, we will first determine the potential, and its form will suggest the proper ordering for solving this problem.

Using the supersymmetric approach to construct the potential $V_a(x)$ [16–18], we suppose that H_a can be factorized in terms of two linear operators $A_a^{\pm}$:

$$H_a = A_a^+ A_a^- + \frac{1}{2}. \tag{40}$$

These operators have the form

$$A_a^- = \frac{1}{\sqrt{2}} m^b \frac{d}{dx} m^a + W_a(x), \tag{41}$$

$$A_a^+ = -\frac{1}{\sqrt{2}} m^a \frac{d}{dx} m^b + W_a(x), \tag{42}$$

where $W_a(x)$ is a position-dependent function to be determined. Observe $A_a^{\pm}$ can also be written as

$$A_a^- = \frac{1}{\sqrt{2}} \left[\frac{1}{\sqrt{m}} \frac{d}{dx} + \frac{a}{\sqrt{m}} (\ln m)' \right] + W_a\,, \tag{43}$$

$$A_a^+ = -\frac{1}{\sqrt{2}} \left[\frac{1}{\sqrt{m}} \frac{d}{dx} + \frac{b}{\sqrt{m}} (\ln m)' \right] + W_a\,, \tag{44}$$

so defining a new differential operator

$$\mathbf{D} = \frac{1}{\sqrt{m}} \frac{d}{dx}, \tag{45}$$

they take very simple forms:

$$A_a^- = \frac{1}{\sqrt{2}} \left[\mathbf{D} + a \left(\mathbf{D} \ln m \right) \right] + W_a\,, \tag{46}$$

$$A_a^+ = -\frac{1}{\sqrt{2}} \left[\mathbf{D} + b \left(\mathbf{D} \ln m \right) \right] + W_a\,. \tag{47}$$

Equation (40) implies that $V_a(x)$ and $W_a(x)$ are related by the expression

$$V_a(x) = \frac{1}{\sqrt{2}} \left[\frac{4a+1}{2} W_a \left(\mathbf{D} \ln m \right) - \mathbf{D} W_a \right] + W_a^2 + \frac{1}{2}. \tag{48}$$

It is not difficult to show that

$$\left[A_a^-, A_a^+ \right] = \sqrt{2} \mathbf{D} W_a + \frac{4a+1}{4} \left(\mathbf{D}^2 \ln m \right); \tag{49}$$

then, demanding that $A_a^{\pm}$ fulfill the Heisenberg algebra, we find the following expression for $W_a(x)$

$$W_a(x) = \frac{1}{\sqrt{2}} \left[\int^x \sqrt{m} dt + y_0 - \frac{4a+1}{4} \left(\mathbf{D} \ln m \right) \right], \tag{50}$$

where y_0 is an integration constant which has to be chosen (as in the classical case) in such a way that the origin of the potential occurs at $x = 0$. The potential has then the form:

$$V_a(x) = \frac{1}{2} \left[\left(\int^x \sqrt{m} dt + y_0 \right)^2 + \frac{4a+1}{4} \left(\mathbf{D}^2 \ln m \right) - \left(\frac{4a+1}{4} \mathbf{D} \ln m \right)^2 \right]. \tag{51}$$

Journal of Physics: Conference Series **128** (2008) 012053 doi:10.1088/1742-6596/128/1/012053

At this point it is important to mention that $V_a(x)$ depends explicitly on the parameter a, and the subindex labels different potentials for different choices of the ordering in (38). However, the Hamiltonian H_a does not explicitly depends on this parameter, and the subindex only labels a particular ordering of T_a. This means that the eigenvalues and eigenvectors of H_a will not depend on a, so there is some freedom in choosing this parameter. Remember, on the other hand that this ambiguity is not present in the classical case, where the dynamical variables are commuting functions of x, p. This fact suggest that we can fix the ordering in the quantum case by making a correspondence of the potential $V_a(x)$ with its classical counterpart $\mathcal{V}(x)$. Observe that, for $a = -1/4$, both, classical and quantum potentials has the same form, as it happens in the CM harmonic oscillator. In this sense we state that the PDM harmonic oscillator Hamiltonian is given by

$$H = -\frac{1}{2}\sqrt[4]{m}\,\mathbf{D}^2\frac{1}{\sqrt[4]{m}} + \frac{1}{2}\left(\int^x \sqrt{m}dt + y_0\right)^2 = A^+A^- + \frac{1}{2}, \tag{52}$$

where

$$A^\pm = \frac{1}{\sqrt{2}}\left(\mp\sqrt[4]{m}\,\mathbf{D}\frac{1}{\sqrt[4]{m}} + \int^x \sqrt{m}dt + y_0\right). \tag{53}$$

The algebraic properties of $A^\pm$ have the following consequences: (a) the spectrum E_n of H is the same as for the CM harmonic oscillator, (b) the wave functions $\psi_n(x)$ are given by:

$$\psi_n(x) = \frac{1}{\sqrt{n!}}\left(A^+\right)^n \psi_0(x), \tag{54}$$

where $\psi_0(x)$ is the ground state defined as the wave function annihilated by A^-:

$$A^-\psi_0(x) = \frac{1}{\sqrt{2}}\left(\sqrt[4]{m}\,\mathbf{D}\frac{1}{\sqrt[4]{m}} + \int^x \sqrt{m}dt + y_0\right)\psi_0(x) = 0. \tag{55}$$

In order to integrate this equation we can make the substitution

$$\psi_0(x) = \sqrt[4]{m}\,\phi_0\left(\int^x \sqrt{m}dt + y_0\right),$$

then (55) transforms into

$$\frac{1}{\sqrt{2}}\left(\mathbf{D} + \int^x \sqrt{m}dt + y_0\right)\phi_0\left(\int^x \sqrt{m}dt + y_0\right) = 0\,. \tag{56}$$

We can easily observe that this equation is the same defining the ground state ϕ_0 for the CM harmonic oscillator, where y is now replaced by the integral of the square root of the mass function (actually, if one makes the change of variable $y = \int\sqrt{m}dx + y_0$, then $\mathbf{D} \equiv d/dy$). The rest of the wave functions can be constructed by applying consecutively the operator A^+ to the ground state:

$$\psi_n(x) = \sqrt[4]{m}\frac{1}{\sqrt{n!}}\left(\mathbf{D} + \int^x \sqrt{m}dt + y_0\right)^n \phi_0\left(\int^x \sqrt{m}dt + y_0\right). \tag{57}$$

By comparing (57) with (37) we can conclude that the wave functions $\psi_n(x)$ of the PDM harmonic oscillator are given in terms of the CM harmonic oscillator wave functions $\phi_n(x)$ in the form

$$\psi_n(x) = \sqrt[4]{m}\,\phi_n\left(\int^x \sqrt{m}dt + y_0\right). \tag{58}$$

This last expression is a well known result called the *point canonical transformation* [22, 23], we should emphasize that in this formalism it comes directly from the algebraic approach used to construct the wave functions.

It is not difficult to show that the eigenstates (58) are square integrable, though, not necessarily normalized functions. Let

$$y(x) = \int^{x} \sqrt{m}\, dt + y_0 , \tag{59}$$

where $\mathcal{D}(y) = \mathcal{D}(m)$ and the range of $y(x)$ is $\mathcal{R}(y)$. Then

$$\int_{\mathcal{D}(m)} |\psi_n(x)|^2\, dx = \int_{\mathcal{R}(y)} |\phi_n(y)|^2\, dy \leq 1 \tag{60}$$

since the CM harmonic oscillator wave functions $\phi_n(y)$ are normalized. Therefore, in the case $y(x)$ (which corresponds to $X(x)$ in the classical framework) ranges all over $\mathbb{R}$, the functions (58) are square integrable as well as normalized eigenstates of the PDM harmonic oscillator; in any other case, even when they are square integrable functions, they will not be normalized.

6. Examples of quantum position dependent mass harmonic oscillator

In this section we will present the cases of the masses chosen before, in the classical approach, in order to illustrate our results in the quantum case. Figures 4, 5 and 6 show the potential $V(x)$ and the first three wave functions $\psi_n(x)$ and probability densities $\rho_n(x) = |\psi_n(x)|^2$ for each case, with $\lambda = 1$.

In the first place we consider the mass $m_1(x)$ given in (22). The wave functions

$$\psi_n(x) = \frac{1}{\sqrt{\pi^{1/2} 2^n n!}} \sqrt{\frac{1+\lambda+x^2}{1+x^2}}\; e^{-\frac{1}{2}[x+\lambda \arctan x]^2} H_n(x + \lambda \arctan x) , \tag{61}$$

look quite similar to the usual harmonic oscillator eigenfunctions $\phi_n(y)$ given in (37) (see Figure 4). In this case $\mathcal{R}(y) = \mathbb{R}$, so the eigenstates are bounded, square integrable and normalized functions.

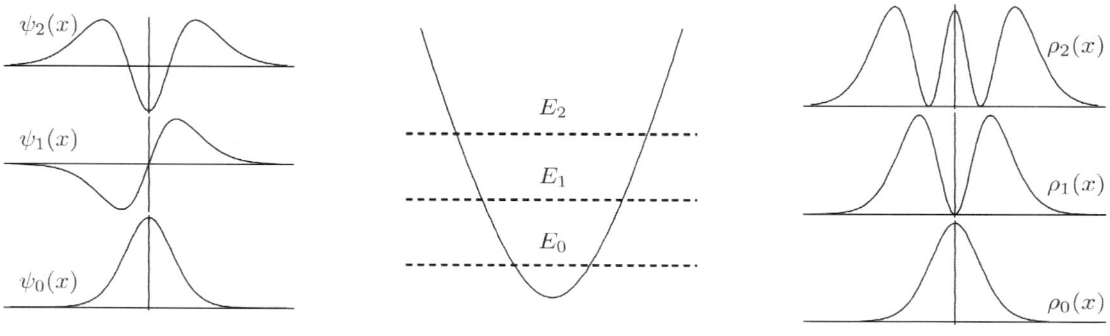

Figure 4. The potential $V(x)$ and its first three wave functions $\psi_n(x)$ and probability densities $\rho_n(x) = |\psi_n(x)|^2$ for the mass $m_1(x)$ with parameters $m_0 = \lambda = 1$.

In the second place, for $m_2(x)$ in (26), the wave functions

$$\psi_n(x) = \frac{1}{\sqrt{\pi^{1/2} 2^n n!}} \sqrt{|\tanh \lambda x|}\; e^{-\frac{1}{2}[\frac{1}{\lambda} \ln \cosh \lambda x]^2} H_n\left(\frac{1}{\lambda} \operatorname{sign}(x)\, \ln(\cosh \lambda x)\right) , \tag{62}$$

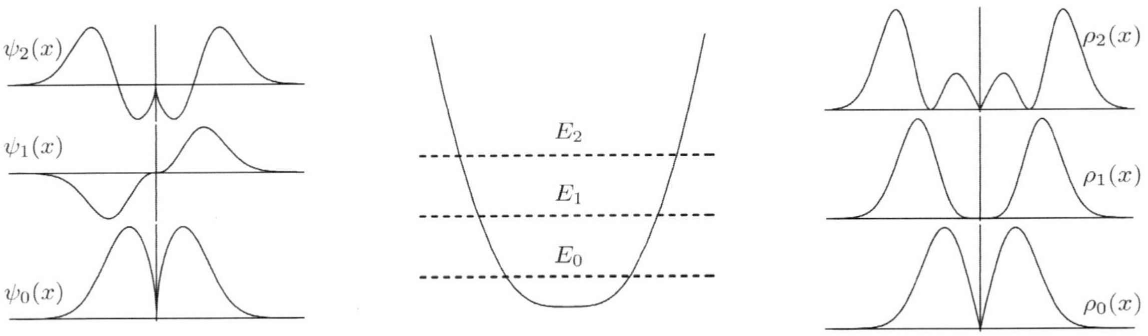

Figure 5. The potential $V(x)$ and its first three wave functions $\psi_n(x)$ and probability densities $\rho_n(x) = |\psi_n(x)|^2$ for the mass $m_2(x)$ with parameters $m_0 = \lambda = 1$.

are bounded, square integrable and also normalized functions since the range $\mathcal{R}(y) = \mathbb{R}$ (see Figure 5). Observe, additionally, that the probability of finding the system near the origin tends to zero due to the factor $\sqrt[4]{m}$ in the wave functions.

Next, for $m_3(x)$ in (30), it is evident from Figure 6 that the corresponding wave functions given by

$$\psi_n(x) = \frac{1}{\sqrt{\pi^{1/2}2^n n!}} \frac{1}{\sqrt[4]{1-(\lambda x)^2}} \, e^{-\frac{1}{2}\left[\frac{1}{\lambda}\arcsin\lambda x\right]^2} H_n\left(\frac{1}{\lambda}\arcsin\lambda x\right), \tag{63}$$

are not bounded functions in $\mathcal{D}(m_3)$. This is because $\mathcal{R}(y) \neq \mathbb{R}$, and the exponential factor can not compensate the divergencies of the factor $\sqrt[4]{m}$ near the boundaries. Even so, it can be easily checked that these wave functions are square integrable, though, not normalized.

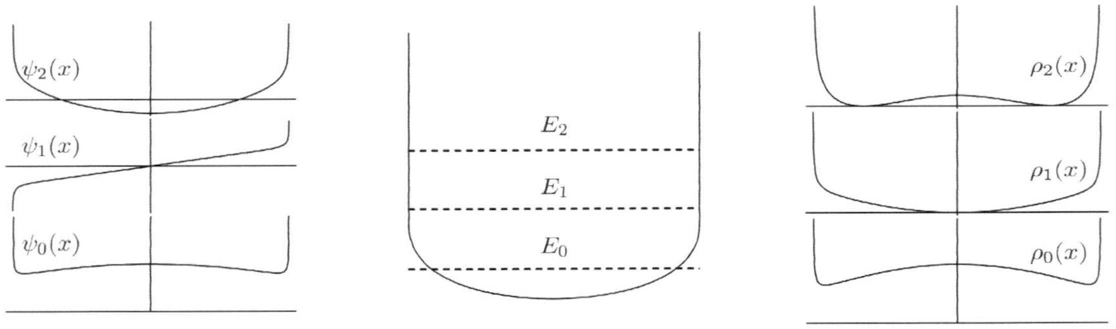

Figure 6. The potential $V(x)$ and its first three wave functions $\psi_n(x)$ and probability densities $\rho_n(x) = |\psi_n(x)|^2$ for the mass $m_3(x)$ with parameters $m_0 = \lambda = 1$.

Finally, it is worthwhile to mention that, correspondingly to the classical case, the parity of the mass function $m(x)$ in these examples imply that the wave functions $\psi_n(x)$ and the probability densities $\rho_n(x)$ are, respectively, odd and even functions of x in all cases.

7. Concluding remarks

He have studied the PDM harmonic oscillator from the classical as well as the quantum points of view by means of the supersymmetric approach. The correspondence to the CM counterpart is given by a canonical transformation in the classical case. The quantum Hamiltonian can be decomposed in many different forms, corresponding to different orderings of the kinetic term.

Journal of Physics: Conference Series **128** (2008) 012053 doi:10.1088/1742-6596/128/1/012053

However, only a very particular decomposition gives the same potential term as for the classical case. Since the quantum Hamiltonian is constructed by making this correlation to the classical case, this formalism is equivalent to the point canonical transformation approach. In both frameworks we have illustrated, with some examples, how this formalism works for different types of the mass function.

Acknowledgments

This work is partially supported by the Spanish MEC (MTM2005-09183 and FIS2005-03989) and Junta de Castilla y León (Excellence project VA013C05). SCyC thanks the members of the Theoretical Physics Department, Universidad Valladolid (Spain) for kind hospitality. The support of Conacyt project 24233-50766-F, COFAA-IPN and SIP-IPN, Mexico, is acknowledged.

References

[1] Wannier G H 1937 *Phys. Rev.* **52** 191
[2] Slater J C 1949 *Phys. Rev.* **76** 1592
[3] Bastard G 1998 *Wave Mechanics Applied to Semiconductor Heterostructures*(Les Ulis: Editions de Physique)
[4] Bastard G 1981 *Phys. Rev.* B **24** 5693
[5] von Roos M 1983 *Phys. Rev.* B **27** 7547
[6] Morrow R A and Brownstein K R 1984 *Phys. Rev.* B **30** 678
[7] Morrow R A 1987 *Phys. Rev.* B **35** 8074
[8] Einevoll G T and Hemmer P C 1988 *J. Phys. C: Solid State Phys.* **21** L1193
[9] Serra L and Lipparini E 1997 *Europhys. Lett.* **40** 667
[10] Arias de Saavedra F, Boronat J, Polls A and Fabrocini A 1994 *Phys. Rev.* B **50** 4248
[11] Lakshmanan M 1979 *J. Sound Vib.* **64** 458
[12] Cariñena J F, Rañada M F and Santander M 2004 *Rep. Math. Phys.* **54** 285
[13] Lévy-Leblond J M 1995 *Phys. Rev.* A **52** 1845
[14] Ganguly A, Kuru Ş, Negro J and Nieto L M 2006 *Phys. Lett.* A **360** 228
[15] Kuru Ş and Negro J 2008 *Ann. Phys.* **323** 413
[16] Cruz y Cruz S, Negro J and Nieto L M 2007 *Phys. Lett.* A **369** 400
[17] Plastino A R, Casas M, Gracias F and Plastino A (1999) *Phys. Rev.* A **60** 4318
[18] Roy B and Roy P 2005 *Phys. Lett.* A **340** 70
[19] Gönül B, Özer O, Gönül B, and Üzgün F 2002 *Mod. Phys. Lett.* A **17** 2453
[20] Bagchi B, Gorain P, Quesne C, Roychoudhury R 2004 *Mod. Phys. Lett.* A **19** 2765
[21] Alhaidari A D 2003 *Int. J. Theor. Phys.* **42** 2999
[22] Balian R, Bessis D and Mezincescu 1995 *Phys. Rev.* B **51** 17624
[23] Alhaidari A D 2002 *Phys. Rev.* A **66** 042116

Journal of Physics: Conference Series **128** (2008) 012054 doi:10.1088/1742-6596/128/1/012054

Contractions of affine Kac-Moody algebras[1]

J Daboul[2], C Daboul, M de Montigny[3]

[2] Physics Department, Ben Gurion University of the Negev, 84105 Beer Sheva, Israel
[3] Campus Saint-Jean and Theoretical Physics Institute, University of Alberta, 8406 - 91 Street, Edmonton, Alberta, Canada T6C 4G9

E-mail: jdaboul@gmail.com, claudia@ibp.de, montigny@phys.ualberta.ca

Abstract. I review our recent work on contractions of affine Kac-Moody algebras (KMA) and present new results. We study generalized contractions of KMA with respect to their twisted and untwisted KM subalgebras. As a concrete example, we discuss contraction of $D_4^{(1)}$ and $D_4^{(3)}$, based on $\mathbb{Z}_3$-grading. We also describe examples of 'level-dependent' contractions, which are based on $\mathbb{Z}$-gradings of KMA. Our work generalizes the Inönü-Wigner contraction of P. Majumdar in several directions. We also give an algorithm for constructing Kac-Moody-like algebras $\widehat{\mathfrak{g}}$ for *any* Lie algebra $\mathfrak{g}$.

1. A historical introduction

In 1991 I was spending part of my sabbatical at the Instituto de Física Teórica (IFT) in San Paulo - Brazil. There, Prof. Abraham H. Zimerman and his collaborators were working on Kac-Moody construction of Toda field models. When I wanted to read about Kac-Moody algebra (KMA), I was referred to the review article of Goddard and Olive [1]. At that time this review was the standard initiation article to the subject for physicists. I found the section on the twisted KMA quite confusing, and I did not understand why the half integers and the third integers indices are necessary for describing the *twisted* Kac Moody algebras of the type $\mathfrak{g}^{(2)}$ and $\mathfrak{g}^{(3)}$, respectively. Later, I realized that mathematicians use only integer indices for the twisted algebras [2, 3]. And for a very good reason: the twisted KMA are nothing but *subalgebras* of the affine KMA $\widehat{\mathfrak{g}} \equiv \mathfrak{g}^{(1)}$.

I gained this insight when I worked with my daughter and her thesis adviser, the late Prof. Peter Slodowy, on identifying the symmetry algebra of the 3-dimensional hydrogen atom. We denoted this algebra by $\mathbb{H}_3$, and identified it as a twisted KMA $\mathfrak{so}(4)^{(2)+}$ or $D_2^{(2)+}$ [4], where $\widehat{\mathfrak{g}}^+$ denotes the *positive subalgebra* of $\widehat{\mathfrak{g}}$, as defined in (44) below. However, later, as we generalized our identification to N-dimensional hydrogen atoms [5], we realized that $D_2^{(2)+}$ can be 'untwisted' [1], so that $\mathbb{H}_3$ is isomorphic to $\widehat{A_1}^+ \equiv A_1^{(1)+}$.

Actually, I was not sure whether I should write up our identification of $\mathbb{H}_3$, especially since the *positive* part of KMA, cannot have the central term κ. However, Dr. Svetlana Nussinov, who had an office next door at Ben Gurion University, encouraged me to write up the article, saying that most physicists are not familiar with twisted KMA, even those who have been working with untwisted KMA for many years. It is possible that physicists were scared off by the unnatural

[1] Based on a talk by J Daboul at the 5th International Symposium on Quantum Theory and Symmetries (QTS5), July 22-28, 2007, Vallodolid (Spain)

Journal of Physics: Conference Series **128** (2008) 012054 doi:10.1088/1742-6596/128/1/012054

notation of Goddard and Olive. Moreover, this notation camouflaged the fact that the twisted KMA are subalgebras of the untwisted ones.

When I met in 2005 Marc de Montigny at the symmetry conference in Kiev, he had been working on contractions. So I suggested that we contract KMA, relative to their twisted subalgebras, mainly to emphasize that the twisted KMA are the subalgebras of the untwisted. He knew of a paper by P. Majumdar [6] who studied Inönü-Wigner contraction of KMA. This paper will be reviewed in Sec. 4.2 below.

Our investigation led to several types of general contractions of KMA, which are based on appropriate choice of gradings [8]. In particular, we studied contractions based on $\mathbb{Z}_3$ grading and on infinite $\mathbb{Z}$ gradings, which we call 'level gradings'.

These contractions yield non-standard 'Kac-Moody-like' algebras, which have abelian ideals. To describe these algebras it is necessary to use the general commutation relations of KMA, given (2) below, instead of the more familiar commutation relations in (1). These general commutation relations are defined in terms of the Killing form. Here the expertise of my daughter, who did her Ph.D. on Kac-Moody algebras and contractions [7], became very helpful.

The present paper is organized as follows: In Sec. 2 we give the two definitions of untwisted KMA algebras, and illustrate, by using the example of Z_3 grading, why it is necessary to use the general commutation relations (2). We then describe an algorithm for constructing Kac-Moody-like algebras for any Lie algebra $\mathfrak{g}$. In Sec. 3 we define gradings and the corresponding automorphisms and emphasize the distinction between normal and twist gradings of KMA, and define twisted KMA. In Sec. 4, we give several generalized contractions of KMA, and review the IW contraction of Majumdar. In Sec. 5 we study *'level contractions'*, which by definition depend on the *'level'*, n, of the KM basis elements $T_a^n \in \widehat{\mathfrak{g}}$. Finally, in Sec. 6 we give concluding remarks.

2. Affine Kac-Moody algebras and Killing form

Consider a finite-dimensional Lie algebra $\mathfrak{g}$, defined by the commutation relations:

$$[T_a, T_b] = c_{ab}^c T_c, \quad a, b, c = 1, 2, \ldots, \dim \mathfrak{g},$$

where summation over repeated indices is implied.

The corresponding *affine Kac-Moody algebras (KMA)* are often defined by the following commutation relations: [1, 6]

$$\begin{aligned} [T_a^m, T_b^n] &= c_{ab}^c \, T_c^{m+n} + m\,\kappa\,\delta_{m+n,0}\,\delta_{a,b}\,, \\ [\kappa, T_a^m] &= 0, \qquad\qquad \text{for} \quad -\infty < m, n < \infty\,, \\ [d, T_a^m] &= m\, T_a^m. \end{aligned} \tag{1}$$

The κ in (1) is called the *central term or central charge*. Henceforth, we shall omit the *derivation operator* d, since it is irrelevant for our purposes.

The commutation relations in (1) hold for totally antisymmetric structure constants c_{ab}^c. But for many gradings (see Sec. 3) these commutation relations are inconsistent with the choice of totally antisymmetric c_{ab}^c. For example, let $\mathfrak{g}$ have $\mathbb{Z}_3$-grading, such that

$$T_a, T_b \in \mathfrak{g}_1, \qquad \text{with} \quad c_{a,b}^c \neq 0 \text{ for some } T_c \in \mathfrak{g}_2.$$

Since by assumption $c_{c,a}^b = c_{a,b}^c \neq 0$, it follows that $[T_c, T_a]$ must have at least one nonzero element $c_{c,a}^b T_b$ in $\mathfrak{g}_1$. But this contradicts the grading assumption, namely $[T_c, T_a] \in \mathfrak{g}_0$.

The commutation relations for using general basis are obtained by replacing the δ_{ab} in (1) by the corresponding *Killing form* $K(T_a, T_b)$. We obtain

$$\begin{aligned} [T_a^m, T_b^n] &= c_{ab}^c\, T_c^{m+n} \;+\; m\,\kappa\,\delta_{m+n,0}\, K(T_a, T_b)\ , \\ [\kappa, T_a^m] &= 0, \qquad\qquad \text{for} \quad -\infty < m, n < \infty\ , \end{aligned} \tag{2}$$

where

$$K(x, y) := \frac{1}{2} Tr(\mathrm{ad}(x)\mathrm{ad}(y)), \quad \forall\, x, y \in \mathfrak{g}\ , \tag{3}$$

and $\mathrm{ad}(x)$ is the *adjoint representation* of $x \in \mathfrak{g}$.

2.1. Algorithm for constructing Kac-Moody-like algebras from any finite Lie algebras
The *'standard affine KMA'* $\mathfrak{g}^{(1)}$ are usually constructed for *semisimple finite-dimensional Lie algebras* $\mathfrak{g}$, by starting with the Cartan matrices [2, 3].

However, we can reproduce the standard KMA and in addition construct similar 'Kac-Moody-like' algebras from *any* Lie algebra $\mathfrak{g}$ by using the following *algorithm*:

For *every* basis element $T_a \in \mathfrak{g}$ define formally an infinite set of elements, $\{T_a^n \mid -\infty < n < \infty\}$, by adding an index n, which we call *level*. Among these $\{T_a^n\}$ we formally define commutation relations, exactly as in (2). We now give a simple proof that the above algorithm yields Lie algebras, which we shall also call *'Kac-Moody algebras'* and denote them by $\widehat{\mathfrak{g}} \equiv \mathfrak{g}^{(1)}$:

Proof: The commutators in (2) are clearly antisymmetric. We now prove the Jacobi identity: Writing (2) in a basis-independent form, we obtain

$$[x^m, y^n] = [x, y]^{m+n} \;+\; m\,\kappa\,\delta_{m+n,0}\,(x|y)\ , \tag{4}$$

where $(x|y) \equiv K(x, y)$. It follows that

$$\begin{aligned} [x^s, [y^m, z^n]] &= [x^s, [y, z]^{m+n}] \\ &= [x, [y, z]]^{s+m+n} \;+\; s\,\kappa\,\delta_{m+n+s,\,0}\,(x|[y, z])\ . \end{aligned} \tag{5}$$

Hence, the Jacobi identity is equivalent to

$$\begin{aligned} [x^s, [y^m, z^n]] + \text{cyclic} &= \kappa\,\delta_{m+n+s,0}\,\{s(x|[y,z]) + n(z|[x,y]) + m(y|[z,x])\} \\ &= \kappa\,\delta_{m+n+s,0}(s + n + m)(x|[y, z]) = 0 \end{aligned} \tag{6}$$

where we used the symmetry of the Killing form: $(x|[y, z]) = (z|[x, y]) = (y|[z, x])$. □

3. Gradings of affine and twisted Kac-Moody algebras

3.1. Gradings of finite-dimensional and KM Lie algebras
We begin this section by describing some gradings of KMA, since they are useful for classifying and contracting KMA. A $\mathbb{Z}_M$*-grading* is a decomposition of a Lie algebra $\mathfrak{g}$,

$$\mathfrak{g} = \mathfrak{g}_0 \oplus \mathfrak{g}_1 \oplus \cdots \oplus \mathfrak{g}_{M-1}, \tag{7}$$

which satisfies

$$[\mathfrak{g}_i, \mathfrak{g}_j] \subseteq \mathfrak{g}_{i+j}\ , \qquad i, j, i + j\ (M)\ , \tag{8}$$

where (M) denotes *'modulo M'*. If M in (8) is equal to ∞, then we speak of $\mathbb{Z}^+$-grading. On the other hand, a $\mathbb{Z}$-grading is defined as in (9) below.

Journal of Physics: Conference Series **128** (2008) 012054 doi:10.1088/1742-6596/128/1/012054

Note that $\mathfrak{g}_0$ is a subalgebra of $\mathfrak{g}$. All the other sets $\mathfrak{g}_i$, $i \neq 0$, however, are not Lie subalgebras, but they constitute representation spaces of $\mathfrak{g}_0$.

Affine KMA have an inherent $\mathbb{Z}$-grading:

$$\widehat{\mathfrak{g}} = \bigoplus_{n=-\infty}^{\infty} \widehat{\mathfrak{g}}^{(n)}, \tag{9}$$

where

$$\widehat{\mathfrak{g}}^{(n)} := \{T_a^n \mid T_a \in \mathfrak{g}\} \oplus (\kappa \oplus d)\, \delta_{n,0}\,, \quad n \in \mathbb{Z}\,. \tag{10}$$

Since we refer to the upper index n of T_a^n as its *level*, the above $\mathbb{Z}$-grading is called *level grading*.

3.2. $\mathbb{Z}_M$ normal and twist gradings of affine KMA and their automorphisms

For any $\mathbb{Z}_M$-grading in Eq. (7) one can define the following automorphism:

$$\tau(T_a) = (\lambda_M)^k\, T_a, \qquad \text{for} \quad T_a \in \mathfrak{g}_k, \quad \text{and} \quad k = 0, 1, \ldots, M-1, \tag{11}$$

where $\tau^M = 1$ and

$$\lambda_M := \exp(2\pi \mathrm{i}\,/M).$$

We see from (11) that the subspaces $\mathfrak{g}_k$ become eigenspaces of τ. Conversely, every automorphism τ as in Eq. (11) yields a $\mathbb{Z}_M$-grading.

We can extend the automorphism τ for $\mathfrak{g}$ to the corresponding affine KM algebra $\widehat{\mathfrak{g}}$ in two ways, as follows:

$$\begin{aligned} \tau_0\,(T_a^n) &= \lambda_M^k\, T_a^n, \qquad \text{for} \quad T_a \in \mathfrak{g}_k, \quad k = 0, 1, \ldots, M-1\,, \\ \tau_s\,(T_a^n) &= \lambda_M^{k-n}\, T_a^n, \qquad \text{and} \quad -\infty \leq n \leq \infty\,, \end{aligned} \tag{12}$$

The τ_0 ignores the level index n. In contrast, τ_s gives the level index n the same importance as to the k index. We shall call τ_0 and τ_s **normal and twist automorphisms**, respectively.

By defining the following eigenspaces of τ_0 and τ_s

$$\begin{aligned} \widehat{\mathfrak{g}}_j &:= \bigoplus_{n=-\infty}^{\infty} \{T_a^n \mid T_a \in \mathfrak{g}_j\} \oplus \kappa\delta_{j0}\,, \quad 0 \leq j < M, \\ \widehat{\mathfrak{g}}_{[j]} &:= \bigoplus_{k=0}^{M-1} \left(\bigoplus_{n=-\infty}^{\infty} \{T_a^{(nM+k-j)} \mid T_a \in \mathfrak{g}_k\} \right) \oplus \kappa\delta_{j\,0}\,, \quad 0 \leq j < M, \end{aligned} \tag{13}$$

we obtain two Z_M gradings, which we shall call **'normal and twist gradings'** :

$$\widehat{\mathfrak{g}} = \bigoplus_{j=0}^{M-1} \widehat{\mathfrak{g}}_j\,, \qquad \text{with} \qquad \tau_0\,(\widehat{\mathfrak{g}}_j) = \lambda_M^j\, \widehat{\mathfrak{g}}_j, \tag{14}$$

$$\widehat{\mathfrak{g}} = \bigoplus_{j=0}^{M-1} \widehat{\mathfrak{g}}_{[j]}\,, \qquad \text{with} \quad \tau_s\left(\widehat{\mathfrak{g}}_{[j]}\right) = \lambda_M^j\, \widehat{\mathfrak{g}}_{[j]}. \tag{15}$$

The $\widehat{\mathfrak{g}}_0 = (\widehat{\mathfrak{g}})_0$ in (14) is an *affine* KM subalgebra of $\widehat{\mathfrak{g}}$, which is equal to $\widehat{(\mathfrak{g}_0)} \equiv \mathfrak{g}_0^{(1)}$.

The subspace $\widehat{\mathfrak{g}}_{[0]}$ of the twist grading (15) is a *subalgebra* of $\widehat{\mathfrak{g}}$, which is called **'twisted KMA'** and denoted as $\mathfrak{g}^{(M)}$:

$$\mathfrak{g}^{(M)} \equiv \widehat{\mathfrak{g}}_{[0]} := \bigoplus_{k=0}^{M-1} \left(\bigoplus_{n=-\infty}^{\infty} \{T_a^{(nM+k)} \mid T_a \in \mathfrak{g}_k\} \right) \oplus \kappa\,. \tag{16}$$

3.3. Conditions on $K(T_a, T_b)$ by using automorphisms

The Killing form is *invariant* under all automorphisms τ of $\mathfrak{g}$ [9], i.e.

$$K(\tau(x), \tau(y)) = K(x, y), \quad \forall\, x, y \in \mathfrak{g}. \tag{17}$$

This invariance follows from the symmetry property of the trace, and the relation $\mathrm{ad}(\tau(x)) = \tau \circ \mathrm{ad}x \circ \tau^{-1}$, which in turn follows from

$$\mathrm{ad}(\tau(x))(y) := [\tau(x), y] = \tau(\,[x, \tau^{-1}(y)]\,).$$

By applying this invariance equality to a Lie algebra $\mathfrak{g}$ which has a Z_M-grading, we obtain

$$K(x, y) = 0 \ \text{ if } \ j + k \neq 0\ (M), \qquad \text{for} \quad x \in \mathfrak{g}_j \quad \text{and} \quad y \in \mathfrak{g}_k\ . \tag{18}$$

This condition is important: it tells us which commutators of KMA cannot have a central term κ.

3.4. Example: Normal and twist $\mathbb{Z}_3$-gradings of $D_4^{(1)}$

It is important to distinguish between normal and twist gradings of affine KMA. As an instructive example, let us study the twisted $\mathbb{Z}_3$ grading of $D_4^{(1)}$ in some details, because $D_4^{(3)}$ is the only genuinely twisted KMA $\mathfrak{g}^{(M)}$ with $M \neq 2$.

Let $\mathfrak{g}$ have a Z_3-grading

$$\mathfrak{g} = \mathfrak{g}_0 \oplus \mathfrak{g}_1 \oplus \mathfrak{g}_2 := \{x_a\} \oplus \{y_a\} \oplus \{z_a\}$$

For example, D_4 and G_2 have the following $\mathbb{Z}_3$-gradings:

$$D_4 = G_2 + \mathbf{7} + \bar{\mathbf{7}} \quad \text{and} \quad G_2 = \mathfrak{su}(3) + \mathbf{3} + \bar{\mathbf{3}}. \tag{19}$$

The corresponding affine KMA $\widehat{\mathfrak{g}} = \mathfrak{g}_0^{(1)}$ have two types of Z_3-gradings:

(i) **normal grading:** defined by $\widehat{\mathfrak{g}}_i = \{T_a^n | T_a \in \mathfrak{g}_i\}$, so that

$$\widehat{\mathfrak{g}} = \widehat{\mathfrak{g}}_0 \oplus \widehat{\mathfrak{g}}_1 \oplus \widehat{\mathfrak{g}}_2 = \{x_a^n\} \oplus \{y_a^n\} \oplus \{z_a^n\}. \tag{20}$$

(ii) **twist grading:** defined by $\widehat{\mathfrak{g}} = \widehat{\mathfrak{g}}_{[0]} \oplus \widehat{\mathfrak{g}}_{[1]} \oplus \widehat{\mathfrak{g}}_{[2]}$, where

$$\begin{aligned}
\mathfrak{g}^{(3)} = \widehat{\mathfrak{g}}_{[0]} = \{X_a^n\} &:= \quad \{x_a^{3n}, y_a^{3n+1}, z_a^{3n+2}\},\\
\widehat{\mathfrak{g}}_{[1]} = \{Y_a^n\} &:= \quad \{x_a^{3n+2}, y_a^{3n}, z_a^{3n+1}\},\\
\widehat{\mathfrak{g}}_{[2]} = \{Z_a^n\} &:= \quad \{x_a^{3n+1}, y_a^{3n+2}, z_a^{3n}\}.
\end{aligned}$$

The *normal* $\mathbb{Z}_3$ grading in (20) can be used to define IW or generalized contractions with respect to *affine KMA* $\widehat{\mathfrak{g}}_0 \equiv \mathfrak{g}_0^{(1)}$. For example, $D_4^{(1)}$ can be contracted with respect to $G_2^{(1)}$.

Similarly, the *twist* $\mathbb{Z}_3$ grading as in (21) is necessary to contract $D_4^{(1)}$, with respect to its twisted subalgebra $D_4^{(3)}$.

4. Contractions of Kac-Moody algebras

A (generalized) **contraction** is a limit process through which a Lie algebra $\mathfrak{g}$ is transformed into a non-isomorphic Lie algebra $\mathfrak{g}'$, with the same dimension. The commutation relations of the *'contracted Lie algebra'* $\mathfrak{g}'$ are defined by [10]

$$[x, y]' \equiv \lim_{\varepsilon \to 0} U_\varepsilon^{-1}([U_\varepsilon(x), U_\varepsilon(y)]), \tag{21}$$

where U_ε is an ε-dependent linear transformation of $\mathfrak{g}$.

The simplest contraction [11, 12] is called *Inönü-Wigner (IW) contraction*. It is defined by decomposing $\mathfrak{g}$ into two parts:

$$\mathfrak{g} := \{T_a\} = \mathfrak{g}_R \oplus \mathfrak{g}_C =: \{T_\alpha\} \oplus \{T_j\} \ , \tag{22}$$

where $\mathfrak{g}_R$ is a Lie *subalgebra* of $\mathfrak{g}$. The complement $\mathfrak{g}_C$ is simply a vector subspace. Henceforth, we employ the three types of indices defined in (22) to distinguish the generators encountered in IW contractions, which are defined by the following map

$$U_\varepsilon(T_\alpha) = T_\alpha, \qquad U_\varepsilon(T_j) = \varepsilon T_j. \tag{23}$$

The resulting commutation relations are

$$[T_\alpha, T_\beta]_\varepsilon = c_{\alpha\beta}^{\gamma}\, T_\gamma, \tag{24}$$

$$[T_\alpha, T_j]_\varepsilon = \varepsilon\, c_{\alpha j}^{\beta}\, T_\beta + c_{\alpha j}^{k}\, T_k \Rightarrow c_{\alpha j}^{k}\, T_k, \quad \text{and} \tag{25}$$

$$[T_i, T_j]_\varepsilon = \varepsilon^2\, c_{ij}^{\gamma}\, T_\gamma + \varepsilon\, c_{ij}^{k}\, T_k \Rightarrow 0. \tag{26}$$

The commutation relations in (24) are those of the invariant subalgebra $\mathfrak{g}_R$. Eq. (26) show that the *subspace* $\mathfrak{g}_C$ becomes the following abelian *subalgebra* of $\mathfrak{g}'$:

$$\mathfrak{g}_C' = \mathbf{abel}(\mathfrak{g}_C) := \{\breve{T}_j\}, \quad \text{with} \quad [\breve{T}_j, \breve{T}_k] = 0\ . \tag{27}$$

Moreover, the commutation relations in (25) tell us that $\mathfrak{g}_C'$ is an *ideal* of the contracted algebra $\mathfrak{g}'$. Hence, the above IW contraction yields

$$\mathfrak{g} \searrow \mathfrak{g}' = \mathfrak{g}_R \ltimes\!\!\!+ \mathfrak{g}_C' = \mathfrak{g}_R \ltimes\!\!\!+ \mathbf{abel}(\mathfrak{g}_C). \tag{28}$$

where we use $\mathfrak{g} \searrow \mathfrak{g}'$ to represent the statement *'the Lie algebra $\mathfrak{g}$ is contracted to the Lie algebra $\mathfrak{g}'$'*.

4.1. IW contraction of KMA relative to any subalgebra

We can apply IW contraction to *any* KMA $\widehat{\mathfrak{g}}$, if it has a subalgebra $\widehat{\mathfrak{g}}_R$. Using the familiar decomposition

$$\widehat{\mathfrak{g}} := \widehat{\mathfrak{g}}_R \oplus \widehat{\mathfrak{g}}_C, \tag{29}$$

and the map

$$U_\varepsilon(\widehat{\mathfrak{g}}_R) = \widehat{\mathfrak{g}}_R, \quad U_\varepsilon(\kappa) = \kappa, \quad \text{and} \quad U_\varepsilon(\widehat{\mathfrak{g}}_C) = \varepsilon\widehat{\mathfrak{g}}_C\ ,$$

we obtain the following commutation relations in the limit $\varepsilon \to 0$:

$$[T_\alpha^m, T_\beta^n]_0 = c_{\alpha\beta}^{\gamma}\, T_\gamma^{m+n} + m\kappa\delta_{m+n,0}\, K(T_\alpha, T_\beta), \tag{30}$$

$$[T_\alpha^m, T_j^n]_0 = c_{\alpha j}^{k}\, T_k^{m+n}, \tag{31}$$

$$[T_i^m, T_j^n]_0 = 0. \tag{32}$$

The above IW contraction for $\widehat{\mathfrak{g}}$ yields a semidirect sum,

$$\widehat{\mathfrak{g}} \searrow \widehat{\mathfrak{g}}' = \widehat{\mathfrak{g}}_R \oplus \widehat{\mathfrak{g}}'_C := \widehat{\mathfrak{g}}'_R \oplus \mathbf{abel}(\widehat{\mathfrak{g}}_C) , \tag{33}$$

exactly as the IW contraction in (28) of finite-dimensional Lie algebras $\mathfrak{g}$.

So far we did not specify the subalgebra $\widehat{\mathfrak{g}}_R$. We consider four choices. The first two are:

(i) $\widehat{\mathfrak{g}}_R$ is the 0-level subalgebra $\widehat{\mathfrak{g}}_R = \{T_a^0 | T_a \in \mathfrak{g}\}$of $\widehat{\mathfrak{g}}$,.

(ii) $\widehat{\mathfrak{g}}_R$ is the *positive* subalgebra of $\widehat{\mathfrak{g}}$, i.e. $\widehat{\mathfrak{g}}_R = \widehat{\mathfrak{g}}^+ := \{T_a^n \mid T_a \in \mathfrak{g}, n \geq 0\}$.

The other two subalgebras will be considered in the next two subsections.

4.2. IW Contraction of affine KMA relative to their affine subalgebras: the IW Contraction by Majumdar

Let $\mathfrak{g} = \mathfrak{g}_R \oplus \mathfrak{g}_C$ is a decomposition as in (29), where $\mathfrak{g}_R$ is a subalgebra. The *'affinization'* of $\mathfrak{g}$ yields

$$\widehat{\mathfrak{g}} = \{T_a^n\} \oplus \kappa = (\{T_\alpha^n\} \oplus \kappa) \oplus \{T_j^n\} =: \widehat{\mathfrak{g}}_R \oplus \widehat{\mathfrak{g}}_C, \tag{34}$$

Majumdar [6] applied IW contraction to $\mathfrak{g}$ and $\widehat{\mathfrak{g}}$, and showed by using the Jacobi identity, that

$$\begin{aligned} \widehat{\mathfrak{g}} \searrow \widehat{\mathfrak{g}}' &= \widehat{\mathfrak{g}}_R \oplus \mathbf{abel}(\widehat{\mathfrak{g}}_C) \\ &= \widehat{\mathfrak{g}'} \equiv (\mathfrak{g}')^{(1)} := (\, \mathfrak{g}_R \oplus \mathbf{abel}(\mathfrak{g}_C)\,)^{(1)} . \end{aligned} \tag{35}$$

In particular, he proved that the affinization of $\mathfrak{g}'$ yields $[(\breve{T}_j)^m, (\breve{T}_k)^n] = 0$, in accordance with $[T_i^m, T_j^n]_0 = 0$ in (32); i.e. he showed that the commutators between the $(\breve{T}_j)^n$ are not proportional to the central term κ. This result follows automatically from our algorithm in Sec. 2.1, since $K(\breve{T}_j, \breve{T}_k) = 0$.

4.3. Contractions of affine KMA relative to their twisted KM subalgebras

The subalgebra $\widehat{\mathfrak{g}}_R$ of the KMA $\widehat{\mathfrak{g}}$ in Sec. 4.1 may be a twisted KM subalgebra $\mathfrak{g}^{(M)}$. We now illustrate the simplest case $M = 2$ explicitly:

$$\widehat{\mathfrak{g}} = \widehat{\mathfrak{g}}_R \oplus \widehat{\mathfrak{g}}_C := (\{T_\alpha^{2n}, T_i^{2n+1}\} \oplus \kappa) \oplus \{T_\alpha^{2n+1}, T_i^{2n}\}, \tag{36}$$

where we used the index notation in Eq. (22). The IW contraction yields

$$\widehat{\mathfrak{g}} = \mathfrak{g}^{(1)} \searrow \widehat{\mathfrak{g}}' = \mathfrak{g}^{(1)'} = \mathfrak{g}^{(2)} \oplus \mathbf{abel}(\widehat{\mathfrak{g}}_C) .$$

Note that unlike the IW contractions of Majumdar, the contractions of $\mathfrak{g}^{(1)}$ relative to *twisted* subalgebras, $\mathfrak{g}^{(M)}, M \geq 2$, cannot be reproduced by first contracting $\mathfrak{g}$ to some $\mathfrak{g}'$ and then affinize $\mathfrak{g}'$. This is because there is no finite algebra $\mathfrak{g}'$, such that $\widehat{\mathfrak{g}'} = \mathfrak{g}^{(M)} \oplus \mathbf{abel}(\widehat{\mathfrak{g}}_C)$.

4.4. Contracting twisted KMA relative to their affine subalgebras

Twisted algebras $\widehat{\mathfrak{g}}^{(M)}$ admit a natural Z_M-grading given by (see Eq. (16)):

$$\mathfrak{g}^{(M)} = \bigoplus_{k=0}^{M-1} \mathfrak{g}_k^{(M)} , \qquad \text{where} \quad \mathfrak{g}_k^{(M)} := \bigoplus_{n=-\infty}^{\infty} \{T_a^{(nM+k)} \mid T_a \in \mathfrak{g}_k\} \oplus \kappa\, \delta_{k0} .$$

Journal of Physics: Conference Series **128** (2008) 012054 doi:10.1088/1742-6596/128/1/012054

We can use this grading to contract $\mathfrak{g}^{(M)}$ relative to its KM subalgebra $\mathfrak{g}_0^{(M)}$. For instance, applying an IW contraction to the twisted KMA $D_4^{(3)}$, relative to its untwisted subalgebra $G_2^{(1)}$, we obtain

$$\begin{aligned} D_4^{(3)} &= (D_4)_0^{(3)} \oplus ((D_4)_1^{(3)} \oplus (D_4)_2^{(3)}) =: \mathfrak{g}_R^{(3)} \oplus \mathfrak{g}_C^{(3)} \\ &\searrow \; D_4^{(3)'} = G_2^{(1)} \oplus \; \mathbf{abel}(\mathfrak{g}_C^{(3)}). \end{aligned}$$

where we used

$$(D_4)_0^{(3)} = \bigoplus_{n=-\infty}^{\infty} \{T_a^{(3n)} \mid T_a \in (D_4)_0 \simeq G_2\} \oplus \kappa \simeq G_2^{(1)} \; .$$

4.5. $\mathbb{Z}_M$-graded contraction of affine KMA relative to affine KMA

Let $\mathfrak{g}$ be a finite-dimensional algebra with a Z_M-grading. Then we can define (generalized) *$\mathbb{Z}_M$-graded contractions* of the corresponding affine KMA $\widehat{\mathfrak{g}}$, by using the following contraction map

$$\widehat{\mathfrak{g}}_i \to \varepsilon^{n_i} \; \widehat{\mathfrak{g}}_i \; , \qquad \text{for} \quad i = 0, 1, \ldots, M-1 \; . \tag{37}$$

The commutation relations of the contracted KMA $\widehat{\mathfrak{g}}'$ are obtained by taking the limit of the following commutation relations, as $\varepsilon \to 0$:

$$[T_a^m, T_b^n] = \varepsilon^{n(a)+n(b)-n(c)} c_{ab}^c \; T_c^{m+n} + \varepsilon^{n(a)+n(b)} m \; \kappa \delta_{m+n,0} \; , \tag{38}$$

where

$$n(a) = n_i \geq 0, \quad \text{if} \;\; T_a \in \mathfrak{g}_i \; , \quad \text{etc.}$$

In order for $\widehat{\mathfrak{g}}_0$ to be the invariant subalgebra under the above contraction, we define $n_0 = 0$. Moreover, in order for the contraction factors $\varepsilon^{n(a)+n(b)-n(c)}$ in (38) to be finite for $\varepsilon \to 0$, we must require that

$$n_i + n_j - n_k \geq 0 \; , \quad \forall \quad 0 \leq i, j, k \leq M-1, \quad i + j \equiv k \; (M).$$

Clearly, different choices of the $\{n_i\}$ yield different contractions.

In particular, by choosing $n_0 = 0$ and $n_i = 1$ for $1 \leq i \leq M-1$ we obtain IW contraction.

4.6. The 'modulo' contraction

Another obvious possibility is $n_i = i$ for $0 \leq i \leq M-1$. For the commutator of the contracted algebra we get for $x_i \in \mathfrak{g}_i$ and $y_j \in \mathfrak{g}_j$ with $i, j < M$:

$$[x_i, y_j]' = [x_i, y_j], \quad \text{if } i + j < M \tag{39}$$

$$[x_i, y_j]' = 0, \quad \text{if } i + j \geq M. \tag{40}$$

Thus the subspace

$$\mathfrak{n} := \mathfrak{g}_1 \oplus \cdots \oplus \mathfrak{g}_{M-1}$$

becomes a nilpotent ideal of the contracted Lie algebra $\mathfrak{g}'$, which becomes a semidirect sum of $\mathfrak{g}_0$ and $\mathfrak{n}$:

$$\mathfrak{g}' = \mathfrak{g}_0 \oplus\!\!\!\!+ \mathfrak{n}.$$

Note the difference to Inönü-Wigner contraction, with respect to $\mathfrak{g}_0$: It lies only in the structure of the ideal, which is abelian for the IW contraction, but nilpotent, and in general not abelian, for our contraction.

We call the above contraction the *'modulo contraction'*, because it takes advantage of the '(M) := modulo M', when calculating the powers of ε.

For example, if $M=5$, then the modulo map (37) would yield

$$[U_\varepsilon(x_3), U_\varepsilon(x_4)] = \varepsilon^{3+4-2}[x_3, x_4] \Rightarrow 0, \quad \text{for } \varepsilon \to 0, \quad x_i \in \mathfrak{g}_i.$$

The modulo contraction can be applied to finite Lie algebras $\mathfrak{g}$ and also to KMA $\mathfrak{g}^{(M)}$.

4.7. Example: Modulo contraction of KMA with a Z_3-twist gradings

As an example of the contractions discussed in the previous subsection, we now carry out a twist contraction explicitly, for the simple but important special case of $M=3$. The map $U_\varepsilon(\widehat{\mathfrak{g}}_{[i]}) = \varepsilon^i\ \widehat{\mathfrak{g}}_{[i]}$, $i=0,1,2$ yields the following contraction:

$$\begin{aligned}
[X_a^m, X_b^n] &= c_{ab}^c\ X_c^{(m+n)} + m\ \kappa\delta_{m+n,0}\ K(X_a, X_b),\\
[X_a^m, Y_b^n] &= c_{ab}^c\ Y_c^{(m+n)},\\
[X_a^m, Z_b^n] &= c_{ab}^c\ Z_c^{(m+n)},\\
[Y_a^m, Y_b^n] &= c_{ab}^c\ Z_c^{(m+n)},\\
[Y_a^m, Z_b^n] &= \varepsilon^3 c_{ab}^c\ X_c^{(m+n)} + \varepsilon^3 m\ \kappa\delta_{m+n,0}\ K(Y_a, Z_b) \Rightarrow 0,\\
[Z_a^m, Z_b^n] &= \varepsilon^3 c_{ab}^c\ Y_c^{(m+n)} \Rightarrow 0,
\end{aligned}$$

where we noted the condition (18) when we added the central term.

We see that the vector space

$$\mathfrak{n} := \widehat{\mathfrak{g}}_{[1]}{}' \oplus \widehat{\mathfrak{g}}_{[2]}{}' = \{Y_a^n\} \oplus\ \mathbf{abel}(Z_a^n).$$

is a nilpotent ideal of $\widehat{\mathfrak{g}}'$. Hence,

$$\widehat{\mathfrak{g}} \searrow \widehat{\mathfrak{g}}' = \widehat{\mathfrak{g}}_{[0]} \niplus \mathfrak{n} = \widehat{\mathfrak{g}}^{(3)} \niplus \mathfrak{n}\ .$$

This contraction can be applied to $D_4^{(1)}$ to obtain $D_4^{(3)}$. Of course, $D_4^{(3)}$ can also be obtained by IW contraction.

5. Level-dependent contractions of Kac-Moody algebras

In this section, we turn our attention to a class of contractions which can be defined for infinite-dimensional algebras only. We shall refer to them as *level contractions.* They are defined by the map

$$T_a^n \to \varepsilon^{f(a,n)} T_a^n \quad \text{and} \quad \kappa \to \kappa, \tag{41}$$

in which the power function $f(a,n)$ depends *explicitly* on the level n of the generators $T_a^n \in \mathfrak{g}^{(M)}$. The map (41) yields the following commutation relations :

$$\begin{aligned}
[T_a^m, T_b^n]_\varepsilon &= \varepsilon^{f(a,m)+f(b,n)-f(c,m+n)}\ c_{ab}^c\ T_c^{m+n}\\
&\quad + \varepsilon^{f(a,m)+f(b,n)}\ m\kappa\,\delta_{m+n,0}\,K(T_a, T_b)\ .
\end{aligned} \tag{42}$$

In order to keep the right-hand side of (42) finite in the contraction limit $\varepsilon \to 0$, the power function $f(a,n)$ must satisfy the following two *'power conditions'*:

$$\begin{aligned}
f(a,m) + f(b,n) - f(c,m+n) &\geq 0, \quad \text{for } c_{ab}^c \neq 0\ , \text{ and}\\
f(a,m) + f(b,-m) &\geq 0, \quad \text{for } K(T_a, T_b) \neq 0.
\end{aligned} \tag{43}$$

Journal of Physics: Conference Series **128** (2008) 012054 doi:10.1088/1742-6596/128/1/012054

In the following sections, we shall carry out unusual contractions of various affine KMA and their subalgebras by using different forms of $f(a, n)$. Such *level contractions* have not been studied before, because contractions have dealt mostly with finite Lie algebras.

The most interesting level contractions can be defined only for positive (resp. negative) subalgebras $\mathfrak{g}^{(M)+}$ (resp. $\mathfrak{g}^{(M)-}$) of KMA [?]. These subalgebras contain positive levels only, i.e.

$$\mathfrak{g}^{(M)+} := \{T_a^n \mid T_a^n \in \mathfrak{g}^{(M)}, n \geq 0\}. \tag{44}$$

Note that *positive (or negative) KMA do not have a central element* κ, because of the factor $m\delta_{m+n,0}$ in the definition of KMA (see Eq. (2)). Hence, positive KM subalgebras can be regarded simply as loop algebras [1, 2, 4].

5.1. Level contractions of $\mathfrak{g}^{(M)+}$ *by the power function* $f_s(n)$

In this section, we contract *positive* KMA $\mathfrak{g}^{(M)+}$ by using the following interesting power function for $s \in \mathbb{N}^+$:

$$f_s(n) := \begin{cases} n & \text{for} \quad 0 \leq n \leq s, \quad s \geq 1, \\ \sqrt{n} & \text{for} \quad s+1 \leq n, \end{cases} \tag{45}$$

Since $f_s(0) = 0$, contracting $\mathfrak{g}^{(M)+}$, $M \geq 1$ with $f_s(n)$ yields the following commutation relations in the limit $\varepsilon \to 0$:

$$\begin{aligned} [T_a^0, T_b^n]_0 &= c_{ab}^c T_c^n && \text{for} \quad n \geq 0 \; , \\ [T_a^m, T_b^n]_0 &= c_{ab}^c T_c^{m+n} && \text{for} \quad m+n \leq s \; , \\ [T_a^m, T_b^n]_0 &= 0, && \text{otherwise} \; . \end{aligned}$$

The contracted algebra $\mathfrak{g}^{(M)+'}$ can be written as

$$\begin{aligned} (\mathfrak{g}^{(M)+})' &= \{T_a^0\} \oplus \{T_a^n \mid n \geq 1\}, && (46) \\ &= (\mathfrak{g}^{(M)+})^{[s]} \oplus \mathbf{abel}(\{T_a^n \mid n \geq s+1\}) \; , && (47) \end{aligned}$$

where

$$(\mathfrak{g}^{(M)+})^{[s]} = \{T_a^n \mid 0 \leq n \leq s\} \, . \tag{48}$$

is a finite-dimensional nonabelian subalgebra of the contracted algebra $\mathfrak{g}^{(M)+'}$. For $s \geq 1$ the $\mathfrak{g}^{(M)+'}$ is *not* a subalgebra of the the original algebra $\mathfrak{g}^{(M)+}$. Hence it is not an invariant algebra, except for $s = 1$; the *invariant subalgebras for all* $s \geq 1$ is the 0-level subalgebra of $\mathfrak{g}^{(M)+}$, as we emphasized in (46).

In (47) we write $(\mathfrak{g}^{(M)+})'$ as a semidirect sum of the subalgebra $(\mathfrak{g}^{(M)+})^{[s]}$ and an infinite-dimensional abelian ideal.

For *untwisted* algebras, $\mathfrak{g}^{(1)+}$, i.e. for $M = 1$, we use for simplicity the notation $\mathfrak{g}^{[s]}$ instead of $(\mathfrak{g}^{(1)+})^{[s]}$. Its dimension is given by

$$|\,\mathfrak{g}^{[s]}\,| = (s+1)|\,\mathfrak{g}| \qquad \text{for} \quad s \geq 1 \; , \tag{49}$$

since for these algebras all n-levels $n \leq s$ have the same number of elements, namely $|\mathfrak{g}|$.

Thus, as s increases, the subalgebras $(\mathfrak{g}^{(M)+})^{[s]}$ in Eq. (48) become larger. In particular, for $s = \infty$ we obtain $(\mathfrak{g}^{(M)+})^{[\infty]} = \mathfrak{g}^{(M)+}$, so that $f_\infty(n) := n$ yields no contraction whatsoever.

In contrast, $f_1(n) = \sqrt{n}$ yields the 'maximal contraction', since $(\mathfrak{g}^{(M)+})^{[1]} = \{T_a^0\}$.

Journal of Physics: Conference Series **128** (2008) 012054 doi:10.1088/1742-6596/128/1/012054

5.2. Constructing 'finite-dimensional Kac-Moody-like' algebras via nilpotent operators
The above non-abelian subalgebras $\mathfrak{g}^{[s]}$ can be represented by using an abelian nilpotent operator p, as follows

$$\mathfrak{g}^{[s]} \simeq \{T_a^n \simeq p^n \otimes T_a \mid T_a \in \mathfrak{g}\} \,, \qquad \text{where} \quad p^{s+1} = 0 \,. \tag{50}$$

In fact, we can apply the above definition to *any* Lie algebra $\mathfrak{g}$. In this way, we obtain an infinite number of *'finite-dimensional Kac-Moody-like'* Lie algebras, $\mathfrak{g}^{[s]}$, one for every $s \geq 1$.

One can also represent the subalgebras $(\mathfrak{g}^{(M)+})^{[s]}$ for $M \geq 2$ by nilpotent operators (see [?]).

We may think of these $\mathfrak{g}^{[s]}$ as *'finite Loop algebras'*. They may become useful for constructing some cutoff procedures.

5.3. Level contractions with $f(a, n) \neq f(n)$
In this section we turn to contractions in which the power functions $f(a, n)$ *depend explicitly on both a and n*. First, we consider a positive KMA with a $\mathbb{Z}_2 \times \mathbb{Z}$ grading:

$$\widehat{\mathfrak{g}}^+ := \widehat{\mathfrak{g}}_0 + \widehat{\mathfrak{g}}_1 := \{X_\alpha^m\} + \{Y_i^n\} \tag{51}$$

On $\widehat{\mathfrak{g}}^+$ we define the following change of basis,

$$X^n \to X^n, \qquad Y^n \to \varepsilon^{f(n)} Y^n, \qquad \text{with} \quad 0 < f(n+1) < f(n) \,,$$

This map yields the following commutation relations :

$$[X_a^m, X_b^n]_\varepsilon = c_{ab}^c \, X_c^{m+n} \,, \tag{52}$$

$$[X_a^m, Y_b^n]_\varepsilon = \varepsilon^{f(n)-f(n+m)} c_{ab}^c \, Y_c^{m+n}, \tag{53}$$

$$\Rightarrow \begin{cases} c_{ab}^c \;\; Y_c^n, & \text{for} \quad m = 0, \\ 0, & \text{for} \quad m > 0, \end{cases} \tag{54}$$

$$[Y_a^m, Y_b^n]_\varepsilon = \varepsilon^{f(n)+f(m)} \; c_{ab}^c \; X_c^{m+n} \Rightarrow 0. \tag{55}$$

From (53) we see that the right-hand side vanishes, in the limit $\varepsilon \to 0$, if $m \geq 1$ and if $f(n)$ is a *strictly monotonically decreasing* function of $n \geq 0$. A simple example of such function is $f(n) = 1/(n + c)$, $c > 0$. The above commutation relations show that the contracted (infinite-dimensional) algebras $\widehat{\mathfrak{g}}'$ contain several interesting subalgebras:

(i) The *affine* KMA $\widehat{\mathfrak{g}}_0 = \{X_a^m\}$, defined in Eq. (52), is an invariant subalgebra of $\widehat{\mathfrak{g}}^+$.

(ii) The abelian subalgebra $I_Y := \{Y_a^n | n \geq 0\}$ is an ideal of $(\widehat{\mathfrak{g}}^+)'$. Thus, the contracted algebra $(\widehat{\mathfrak{g}}^+)'$ has the familiar structure of being a semidirect sum of an invariant subalgebra and an abelian ideal.

$$(\widehat{\mathfrak{g}}^+)' = \widehat{\mathfrak{g}}_0 \oplus \mathbf{abel}(Y_a^n) \,. \tag{56}$$

(iii) The level 0 of the contracted algebra in (56) yields a contraction of the underlying algebra $\mathfrak{g} = \mathfrak{g}_0 + \mathfrak{g}_1$:

$$\mathfrak{g}' = \mathfrak{g}_0 \oplus \mathbf{abel}(\mathfrak{g}_1) = \{X_a^0\} \oplus \mathbf{abel}(Y_a^0).$$

6. Concluding remarks

We contracted affine KMA relative to affine and twisted KMA by using IW contractions and generalized contractions based on $\mathbb{Z}_M$ and $\mathbb{Z}_M \times \mathbb{Z}$ gradings. These contractions may become useful in the study of representations of KMA.

There is another aspect which we also find interesting: in the case of finite-dimensional Lie algebras, one can obtain different limits of these algebras and their representations by applying two contractions in different order. This was done, for example, by first 'deforming' a harmonic-oscillator Hamiltonian $H_1(k)$, by adding to it a constant force term $-fx$. The symmetry algebra of the resulting two-parameter Hamiltonian, $H_2(k,f) := H_1(k) - fx$, is $\mathfrak{so}(3)$. This algebra was contracted by taking the limits $(k,f) \longrightarrow 0$ in different orders: (1) taking the limits $f \longrightarrow 0$ followed by $k \longrightarrow 0$ yields the Euclidean algebra $\mathfrak{e}(2)$, whereas (2) taking the limits $k \longrightarrow 0$ followed by $f \longrightarrow 0$ yields the Heisenberg-Weyl algebra $\mathfrak{w}_1$. This phenomenon was called the *Deformation-Contraction (DC) hysteresis* [13, 14, 15]. It should be interesting if one can construct a model which yields a DC hysteresis for affine KM algebras.

Many more contractions of KMA can be constructed by various choices of gradings. In [7] graded contractions were studied which made use of gradings of Kac Moody algebras based on their root lattices.

Acknowledgments
JD and MdM are grateful to the *Mathematisches Forschungsinstitut Oberwolfach* (Germany) where we discussed this project. JD thanks the hospitality of Professor Jean Pierre Gazeau, of the Laboratoire Astroparticules et Cosmologie APC, Université Paris 7, where a major part of this manuscript was written. MdM acknowledges partial financial support from the Natural Sciences and Engineering Research Council (Canada).

[1] Goddard P and Olive D 1988 Kac-Moody and Virasoro algebras in relation to quantum physics *Int. J. Mod. Phys.* **A1**, 303-414
[2] Kac V G 1990 *Infinite dimensional Lie algebras* 3rd Ed. (Cambridge: Cambridge University Press)
[3] Kass S, Moody R V, Patera J and Slansky R 1990 *Affine Lie Algebras, Weight Multiplicities and Branching Rules* (Berkeley: Univ. of California Press)
Cornwell J F 1989 *Group Theory in Physics*, Vol. 3 (New York: Academic Press)
Wan Z X 1991 *Introduction to Kac-Moody algebra* (Singapore: World Scientific)
Fuchs J and Schweigert C 1997 *Symmetries, Lie Algebras and Representations* (New York: Cambridge University Press)
Carter R 2005 *Lie Algebras of Finite and Affine Type*, Series: Cambridge Studies in Advanced Mathematics (No. 96)
[4] Daboul J, Slodowy P and Daboul C 1993 The hydrogen algebra as centerless twisted Kac-Moody algebra *Phys. Lett.* B **317**, 321-8
[5] Daboul C and Daboul J 1998 From hydrogen atom to generalized Dynkin diagrams *Phys. Lett.* B **425**, 135-44
[6] Majumdar P 1993 Inönü-Wigner contraction of Kac-Moody algebras *J. Math. Phys.* **34**, 2059-65
[7] C. Daboul, *Deformationen und Degenerationen von Liealgebren und Liegruppen*, Doctoral dissertation, University of Hamburg, 1999 (in German). URL: `http://www.sub.uni-hamburg.de/opus/volltexte/1999/30/`
[8] Daboul C, Daboul J and de Montigny M 2008 Gradings and Contractions of Twisted and Untwisted Kac-Moody algebras, submitted to J. Math. Phys.
[9] Samelson H 1990 *Notes on Lie Algebras, Springer*, 3. Edition; this book can be downloaded from `http://math.stanford.edu/~ykatznel/samelson/liealgeb.pdf`
[10] Saletan E 1961 Contraction of Lie groups *J. Math. Phys.* **2**, 1-21
Gilmore R 1974 *Lie Groups, Lie Algebras, and Some of Their Applications* (New York: Wiley)
Izmest'ev A A, Pogosyan G S , Sissakian A N and Winternitz P 2001 Contractions of Lie algebras and the separation of variables *J. Phys.* A: Math. Gen. **34**, 521-554
[11] Segal I E 1951 A class of operator algebras determined by groups *Duke Math. J.* **18**, 221-265
[12] Inönü E and Wigner E P 1953 On the contraction of groups and their representations *Proc. Nat. Acad. Sci. US* **39**, 510-24
[13] Daboul J 2006 Contraction of broken symmetries via Kac-Moody formalism *J. Math. Phys.* **47**, 083507-20
[14] Daboul J and Wolf K B 2006 Noncommuting contractions of oscillators with constant force *J. Phys.* A: Math. Gen. **39**, 4173-4180
[15] Daboul J, Pogosyan G S and Wolf K B 2007 Noncommuting Limits Of Oscillator Wave Functions *Physics of Atoms and Nuclei* **70**, 513-519

Journal of Physics: Conference Series **128** (2008) 012055 doi:10.1088/1742-6596/128/1/012055

An approach to quantum anharmonic oscillators via Lie algebra

M Jafarpour and D Afshar

Physics Department, Shahid Chamran University, Ahvaz, Iran

E-mail: mojtaba_jafarpour@hotmail.com and da_afshar@yahoo.com

Abstract. We present a new Lie algebraic method to study the eigenvalues and eigenfunctions of quantum anharmonic oscillators. We consider the Hamiltonians for the simple harmonic and anharmonic oscillator as the two generators of a Lie algebra, whose other generators may be found exactly or up to any order of the parameter involved. Specifically, the closed commutator algebra for the quartic anharmonic oscillator is established in a perturbation sense. An element of this Lie group, turning out to be the four-photon operator, transforms the quartic anharmonic oscillator Hamiltonian to the harmonic one in a perturbation sense; thus, facilitating the calculation of the eigenvalues and eigenfunctions of the former.

1. Introduction

Simple harmonic oscillator is an idealized model to describe many phenomena in physics and chemistry. Anharmonic oscillator is a deviation from this idealized model to a realistic one. Rayleigh-Schrödinger perturbation theory has been widely used, providing the energy of an anharmonic oscillator, as a formal power series of the perturbation parameter involved [1, 2, 3, 4]. However, the power series diverges even for small coupling constants; thus, appropriate techniques must be applied to alleviate this problem [5, 6, 7, 8, 9, 10, 11, 12]. Several alternative approaches have also been developed; among them, a Lie algebraic method using canonical transformations [13], multiple scale method [14, 15, 16], quasilinearization method [17, 18], a variational method based on the squeezed states [19, 20] and Naundorf method [21] are mentioned.

In this work, we use a Lie algebra method to relate the Hamiltonian of an anharmonic oscillator to that of a harmonic one, in a perturbation sense. This is achieved via a canonical transformation, which is unique to the structure of the specific Hamiltonian under study. We introduce our method in section 2, where we apply it to two prototype trivial examples: a harmonic oscillator with a linear perturbation term and one with a quadratic perturbation term that just brings about a frequency shift. In section 3, a unitary transformation is found that relates the Hamiltonian of a quartic anharmonic oscillator to that of a harmonic one. It is observed that the matrix representation of the transformed Hamiltonian is diagonal, up to the first order in the perturbation parameter involved. The eigenvalues of the anharmonic oscillator are evaluated and the results are compared with those obtained from Rayleigh-Schrödinger perturbation theory. Finally, we deal with the discussion and conclusions in section 4.

Journal of Physics: Conference Series **128** (2008) 012055 doi:10.1088/1742-6596/128/1/012055

2. Our Lie algebra method

The anharmonic oscillator, described by the Hamiltonian

$$H_n = H_0 + \lambda x^n = \frac{p^2}{2} + \frac{x^2}{2} + \lambda x^n, \tag{1}$$

is a deviation from the harmonic one, described by the Hamiltonian H_0. If N hermitian generators L_i, including $L_1 = H_0$ and $L_2 = H_n$, satisfying

$$[L_i, L_j] = \sum_k c_{ijk} L_k \ , \tag{2}$$

form a closed commutator algebra, the operators L_i specify a Lie group, whose generators they are. The elements of this group are given by

$$U = exp(-\mathrm{i}\sum_{i=1}^{N} \alpha_i L_i), \tag{3}$$

where α_i's are real parameters. Now, an element of the group may be found that transforms H_0 to H_n. This coming about, the eigenvalues and the eigenfunctions of the anharmonic oscillator can be obtained in terms of those of a harmonic one in a simple manner.

Now, let's assume $n = 1$ in equation (1); we have

$$H_1 = \frac{p^2}{2} + \frac{x^2}{2} + \lambda x = a^\dagger a + \frac{1}{2} + \lambda(\frac{a + a^\dagger}{\sqrt{2}}), \tag{4}$$

which may be considered as a harmonic oscillator in an external electric field. a and $a^\dagger$ are the ordinary annihilation and creation operators. Considering the commutator $[H_0, H_1] = \lambda(a^\dagger - a)/\sqrt{2}$, we realize that the operators H_0, H_1, $\mathrm{i}(a^\dagger - a)$ and the identity operator I form a closed commutator algebra and are the generators of a Lie group. The Glauber operator $U_1 = exp(\alpha a^\dagger - \alpha^* a)$, with α assumed real in this section, is an element of this group. The transformation of H_0 to H_1 is carried out in the following manner

$$H_1 = U_1^\dagger H_0 U_1 - \frac{\lambda^2}{2} \ , \tag{5}$$

where, α has been replaced by $\lambda/\sqrt{2}$ in U_1. Using (5) we find

$$H_1[U_1^\dagger |n\rangle_a] = [(n + \frac{1}{2}) - \frac{\lambda^2}{2}][U_1^\dagger |n\rangle_a] \ ,$$

where $\{|n\rangle_a\}$ are a-mode number states; meaning that the eigenvalues of H_1 are given by

$$E_n = (n + \frac{1}{2}) - \frac{\lambda^2}{2} \ , \tag{6}$$

and the eigenfunctions, called the displaced number states [22] are expressed by

$$U_1^\dagger |n\rangle_a = \mathrm{e}^{\frac{\lambda}{\sqrt{2}}(a - a^\dagger)} |n\rangle_a \ . \tag{7}$$

Now, we consider our second prototype trivial example. The Hamiltonian of a shifted-frequency harmonic oscillator, the case $n = 2$ in equation (1), is given by

$$H_2 = \frac{p^2}{2} + \frac{x^2}{2} + \lambda x^2 = a^\dagger a + \frac{1}{2} + \lambda(\frac{a + a^\dagger}{\sqrt{2}})^2. \tag{8}$$

Journal of Physics: Conference Series **128** (2008) 012055 doi:10.1088/1742-6596/128/1/012055

In this case, we have $[H_0, H_2] = \lambda(a^{\dagger 2} - a^2)$; therefore, the Hermitian operators H_0, H_2, $\mathrm{i}(a^{\dagger 2} - a^2)$ and I, form a closed commutator algebra and are the generators of a Lie group. The squeezing operator $U_2 = exp(\beta a^{\dagger 2} - \beta^* a^2)$, with β assumed real, is an element of this group. The transformation of the fundamental mode operators a and $a^\dagger$ into the new ones, b and $b^\dagger$ under U_2, is carried out by the following Bogoliubov transformations

$$b = U_2^\dagger a U_2 = \mu a + \nu a^\dagger,$$
$$b^\dagger = U_2^\dagger a^\dagger U_2 = \mu a^\dagger + \nu a,$$

where, $\mu = \cosh(2\beta)$ and $\nu = \sinh(2\beta)$. As $\mu^2 - \nu^2 = 1$, we have also $[b, b^\dagger] = 1$. The transformed Hamiltonian under U_2 is given by

$$H_2 = \sqrt{1+2\lambda}\, U_2^\dagger H_0 U_2 = \sqrt{1+2\lambda}(b^\dagger b + \frac{1}{2}), \tag{9}$$

where, $\lambda = 2\mu\nu/(\mu - \nu)^2$ has been assumed. Using equation (9) we find

$$H_2[U_2^\dagger|n\rangle_a] = \sqrt{1+2\lambda}(n + \frac{1}{2})[U_2^\dagger|n\rangle_a] \ ,$$

or

$$H_2|n\rangle_b = \sqrt{1+2\lambda}(n + \frac{1}{2})|n\rangle_b \ ,$$

implying that eigenfunctions of the shifted-frequency harmonic oscillator, the b-mode number states, or squeezed number states [22], are given by $|n\rangle_b = U_2^\dagger|n\rangle_a$, and the corresponding energies by $\sqrt{1+2\lambda}(n + \frac{1}{2})$.

3. Quartic anharmonic oscillator

Now we consider the quartic anharmonic oscillator. Its Hamiltonian in terms of the fundamental mode operators a and $a^\dagger$, may be given by

$$H_4 = a^\dagger a + \frac{1}{2} + \lambda(\frac{a + a^\dagger}{\sqrt{2}})^4. \tag{10}$$

We can show that

$$[H_0, H_4] = -\mathrm{i}L_3 - 3\mathrm{i}L_4 - 2\mathrm{i}L_5, \tag{11}$$

where,

$$L_3 = \mathrm{i}\lambda(a^{\dagger 4} - a^4),$$
$$L_4 = \mathrm{i}\lambda(a^{\dagger 2} - a^2),$$
$$L_5 = \mathrm{i}\lambda(a^{\dagger 3} a - a^\dagger a^3).$$

Inspecting the mutual commutators between the operators H_0, H_4, L_3, L_4 and L_5, we are finally led to prove that H_0, H_4, $\mathrm{i}\lambda(a^{\dagger 4} - a^4)$, $\mathrm{i}\lambda(a^{\dagger 2} - a^2)$, $\mathrm{i}\lambda(a^{\dagger 3}a - a^\dagger a^3)$, $\lambda(a^{\dagger 4} + a^4)$, $\lambda(a^{\dagger 2} + a^2)$, $\lambda(a^{\dagger 3}a + a^\dagger a^3)$ and I form a closed commutator algebra up to the first order, if the perturbation parameter λ is small, and therefore they are the generators of a Lie group, up to that order. We now focus on the following element of this group, known as the four photon operator [23]

$$U_4 = exp[A\lambda(a^{\dagger 4} - a^4) + B\lambda(a^{\dagger 2} - a^2) + C\lambda(a^{\dagger 3}a - a^\dagger a^3)], \tag{12}$$

Journal of Physics: Conference Series **128** (2008) 012055 doi:10.1088/1742-6596/128/1/012055

where A, B and C are real parameters. This operator transforms the fundamental mode operators, a and $a^\dagger$ to the new ones c and $c^\dagger$. Using Baker-Campbell-Hausdorff relation up to the first order, they may be expressed by the following expressions

$$c = U_4^\dagger a U_4 = a + \lambda(4Aa^{\dagger 3} + 2Ba^\dagger + 3Ca^{\dagger 2}a - Ca^3), \tag{13}$$

$$c^\dagger = U_4^\dagger a^\dagger U_4 = a^\dagger + \lambda(4Aa^3 + 2Ba + 3Ca^\dagger a^2 - Ca^{\dagger 3}). \tag{14}$$

It should be emphasized that c and $c^\dagger$ also obey the canonical commutation relation $[c, c^\dagger] = 1$; therefore, we have introduced a new class of Bogoliubov transformations by (13) and (14). Assuming $A = 1/16$, $B = 3/4$ and $C = 1/2$, the four photon operator transforms H_0 to H_4, up to the first order in the parameter λ, as follows

$$U_4^\dagger H_0 U_4 = c^\dagger c + \frac{1}{2} = H_4 - \frac{3\lambda}{2}(N_a^2 + N_a) - \frac{3\lambda}{4} ,$$

$$H_0 = U_4 H_4 U_4^\dagger - \frac{3\lambda}{2} N_a(N_a + 1) - \frac{3\lambda}{4} , \tag{15}$$

where, $N_a = a^\dagger a$ is the normal number operator. Using equation (15), we find

$$H_4[U_4^\dagger |n\rangle_a] = [n + \frac{1}{2} + \frac{3\lambda}{4} + \frac{3\lambda}{2} n(n+1)][U_4^\dagger |n\rangle_a] .$$

Thus, the eigenstates of the quartic anharmonic oscillator, our c-mode number states, are given by

$$|n\rangle_c \equiv U_4^\dagger |n\rangle_a = \mathrm{e}^{-\lambda[\frac{1}{16}(a^{\dagger 4} - a^4) + \frac{3}{4}(a^{\dagger 2} - a^2) + \frac{1}{2}(a^{\dagger 3}a - a^\dagger a^3)]} |n\rangle_a, \tag{16}$$

and the first order perturbed energies are expressed by

$$E_n = n + \frac{1}{2} + \frac{3\lambda}{4} + \frac{3\lambda}{2} n(n+1) . \tag{17}$$

The above results are in complete agreement with those obtained, using first order Rayleigh-Schrödinger perturbation theory [24].

Furthermore, although (16) and (17)are formally derived up to first order in the parameter λ, it is worthwhile to investigate the contribution of higher order terms to the energy using (16). For example, we calculate the expectation value of H_4 in the c-mode number states basis up to the fourth order; we find

$$E_0 = 0.5 + 0.75\lambda - 2.625\lambda^2 + 20.8125\lambda^3 - 104.098\lambda^4 + \ldots , \tag{18}$$

Bender and Wu also give the following perturbation series, for the ground state energy $E_{0,B}$, of the quartic anharmonic oscillator which we write down up to the fourth order also [1]

$$E_{0,B} = 0.5 + 0.75\lambda - 2.625\lambda^2 + 20.8125\lambda^3 - 241.289\lambda^4 + \ldots . \tag{19}$$

We note that the result (18) agrees with (19), up to the third order in the parameter λ. We have also ascertained that our result (18) is in agreement, up to the third order in the parameter λ, with those obtained from multiple scale method [16], and also those from Rayleigh-Schrödinger perturbation theory [24].

The agreement between (18) and (19) should not surprise the reader. We should remember that this situation is similar to the one we encounter in Rayleigh - Schrödinger perturbation theory. Moreover, both series are divergent even for small values of λ, and one has to truncate the series, to get any physical result. Obviously, a judicious choice of the truncation order is essential to the finding of good results [12]. For example, using (18) we calculate the ground state energy up to the fourth order; for $\lambda = 0.1$ we find $E_0 = 0.5591527$. Comparing this result with the exact value $E_0 = 0.559146327183519576$ obtained by Vinette and Čížek [25], the error is about 0.001 percent.

4. Discussion

We have introduced a concise method to study anharmonic oscillators. It is based on the construction of a Lie group, whose generators include H_0 and H_n. The generator of the group H_0 may be transformed to the generator H_n by a unitary transformation, endowed by an element of the group. If a closed commutator algebra is formed exactly, the method can be used to obtain exact eigenvalues and eigenfunctions of H_n. However, the commutator algebra is generally obeyed up to some order of the parameter λ, thus the eigenvalues and eigenfunctions of H_n are found approximately. We applied the method to obtain the energy eigenvalues of a quartic anharmonic oscillator; the results are in agreement with the other methods, up to third order in the parameter λ.

References

[1] Bender C M and Wu T T 1969 *Phys. Rev.* **184** 1231-60
[2] Bender C M and Wu T T 1971 *Phys. Rev. Lett.* **27** 461-5
[3] Bender C M and Wu T T 1973 *Phys. Rev. D* **7** 1620-36
[4] Banks T, Bender C M and Wu T T 1973 *Phys. Rev. D* **8** 3346-66
[5] Seznec R and Zinn-Justin J 1979 *J. Math. Phys.* **20** 1398-408
[6] Fernández F M 1995 *J. Math. Phys.* **36** 3922-30
[7] Janke W and Kleinert H 1995 *Phys. Rev. Lett.* **75** 2787-91
[8] Weniger E J 1996 *Phys. Rev. Lett.* **77** 2859-62
[9] Ivanov I A 1996 *Phys. Rev. A* **54** 81-6
[10] Ivanov I A 1998 *J. Phys. A:Math. Gen.* **31** 6995-7003
[11] Skála L, Čížek J, Weniger E J and Zamastil J 1999 *Phys. Rev. A* **59** 102-6
[12] Weniger E J. 1997 *Phys. Rev. A* **56** 5165-68
[13] Fernández F M 1992 *Phys. Rev. A* **45** 1333-8
[14] Bender C M and Bettencourt L M A 1996 *Phys. Rev. Lett.* **77** 4114-7
[15] Bender C M and Bettencourt L M A 1996 *Phys. Rev. D* **54** 7710-23
[16] Auberson G and CapdequiPeyranere M 2002 *Phys. Rev. A* **65** 032120
[17] Mandelzweig V B 1999 *J. Math. Phys.* **40** 6266-91
[18] Liverts E Z, Mandelzweig V B and Tabakin F 2006 *J. Math. Phys.* **47** 062109
[19] Jafarpour M and Afshar D 2002 *J. Phys. A:Math. Gen.* **35** 87-92
[20] Jafarpour M and Afshar D 2005 *Nuovo Cimento B* **120** 335-344
[21] Gómez F J and Sesma J 2005 *J. Phys. A:Math. Gen.* **38** 3193-202
[22] Nieto M M 1997 *Phys. Lett. A* **229** 135-43
[23] Fisher R A, Nieto M M and Sandberg V D 1984 *Phys. Rev. D* **29** 1107-10
[24] McRae S M and Vrscay E R 1997 *J. Math. Phys.* **38** 2899-921
[25] Vinette F and Čížek J 1991 *J. Math. Phys.* **32** 3392-404

Journal of Physics: Conference Series **128** (2008) 012056 doi:10.1088/1742-6596/128/1/012056

Quantum mechanics and umbral calculus

J E López-Sendino, J Negro, M A del Olmo and E Salgado

Departamento de Física Teórica, Universidad de Valladolid, E-47011, Valladolid, Spain

E-mail: jels@metodos.fam.cie.uva.es, jnegro@fta.uva.es, olmo@fta.uva.es

Abstract. In this paper we present the first steps for obtaining a discrete Quantum Mechanics making use of the Umbral Calculus. The idea is to discretize the continuous Schrödinger equation substituting the continuous derivatives by discrete ones and the space-time continuous variables by well determined operators that verify some Umbral Calculus conditions. In this way we assure that some properties of integrability and symmetries of the continuous equation are preserved and also the solutions of the continuous case can be recovered discretized in a simple way. The case of the Schrödinger equation with a potential depending only in the space variable is discussed.

1. Introduction

In the last years the interest in physics for discrete space-time theories has increased mainly in relation with quantum gravity [1]. In a scenario where the physical space-time may be discrete (i.e. there is a fundamental length in the space-time –let σ be the fundamental space length and τ the fundamental time length – that could be related with Planck's constant), continuous theories would only be approximations to the reality. Hence, it seems interesting the study of discrete physical theories such that in the limit of σ going to zero we recover the well known and established continuous physical theories. One way to do that is to discretize continuous physical theories, in particular quantum theories. It is obvious, that after discretization some properties will be conserved and other ones will disappear. A detailed study about when, how and why this happens would be pertinent and very interesting. This is the aim of this work: to study how to discretize a physical theory in such a way that properties related with symmetries in the continuous case can be preserved in the process, and so, to analyse the behaviour of quantum mechanics after discretization.

The Umbral Calculus is the mathematical tool that we will use to discretize Quantum Mechanics. It has been used recently to provide discrete representations of canonical commutation relations (i.e. $[\partial_x, x] = 1$) in order to discretize linear differential equations [2]–[9], in such a way that the continuous point symmetries are preserved the most cases.

In this paper we will consider a fixed lattice and use the umbral correspondence to obtain discrete solutions maintaining certain commutation relations [9].

The history of the Umbral Calculus goes back to the 17th century: Taylor, Newton, Barrow, Vandermonde, etc. [10] found some theorems and expressions relating powers with polynomials. However, in the second half of the 19th century through Sylvester, Cayley and Blissard [11]–[13] appeared the terms 'umbrae' (shadow in Latin) and 'Umbral Calculus' for the first time in relation with a set of 'magic' rules of lowering and raising indices. Some mathematical developments, like the theory of Sheffer polynomials (called Appell polynomials or more generally

Journal of Physics: Conference Series **128** (2008) 012056 doi:10.1088/1742-6596/128/1/012056

polynomials of type zero) [14]–[17] and the theory of abstract linear operators by Pincherle [15], converged to Umbral Calculus. Finally, in the second half of the 20th century Rota, Roman and collaborators [18]–[23] developed the Umbral Calculus as a linear algebra of operators. Thus, they formulated the theory of the umbral algebra and its dual in the space of formal series. A large bibliography about Umbral Calculus and its applications can be found in Ref. [24].

About recent results, we can mention that new methods in Umbral Calculus have been introduced for computation of series by integration [25]. Also Ref. [26]–[30] study the symmetries of nonlinear difference equations or multidimensional ones related with Quantum Mechanics.

This paper is organised as follows. In the next section we introduce the basic concepts of Umbral Calculus in terms of the theory of linear operators. In section 3 we show the main facts relative to the discretization of the (continuous) Quantum Mechanics. The most usual representations for the discrete derivative operator (right, left and symmetric discrete derivatives) are analysed in section 4. In order to see how the above umbral correspondences work we present in section 5 their action on some non-polynomial functions: the exponential and the trigonometric functions. The next section is devoted to present the basis for constructing a discrete Quantum Mechanics founded on the Umbral Calculus. Thus we present an umbral version of the Schrödinger equation for some potentials. Finally some conclusions close the paper.

2. Introduction to Umbral Calculus

This introductory exposition of the modern theory of the Umbral Calculus, developed mainly by Rota and Roman, follows mainly Ref. [23]. The *umbral algebra*, $\mathcal{L}$, is the algebra of linear operators acting on the algebra, $\mathcal{F}$, of formal power series in a variable or acting on the algebra of real or complex polynomials in a variable, $\mathcal{P}$. The elements of $\mathcal{F}$ are of the form

$$f(x) = \sum_{n=0}^{\infty} a_n x^n, \quad a_n \in \mathbb{F},$$

and those of $\mathcal{P}$ are

$$p_n(x) = \sum_{i=0}^{n} a_i x^i, \quad a_i \in \mathbb{F},\ n \in \mathbb{N},$$

being $\mathbb{F}$ a field of characteristic zero (in our case it will be identified with the real numbers $\mathbb{R}$, but it could be, for example, the complex numbers).

The linear operators of $\mathcal{L}$ called *shift operator*, T, and *coordinate operator*, X, act on the elements of $\mathcal{F}$ or $\mathcal{P}$ as follows

$$\begin{aligned} T_\sigma \cdot p(x) &= p(x+\sigma), \\ X \cdot p(x) &= x\, p(x). \end{aligned} \tag{1}$$

All the results of this work can be extended to separable equations of any number of variables. For non separable equations the method is the same but the results may be quite different.

An operator $O \in \mathcal{L}$ is said *shift invariant* if and only if commutes with the shift operator, i.e.,

$$[O, T_\sigma] = OT_\sigma - T_\sigma O = 0, \qquad \forall\, \sigma \in \mathbb{F}.$$

The so-called *delta operators*, Δ, are those operators of $\mathcal{L}$ such as

$$\begin{aligned} &\Delta \cdot x = c \neq 0, \qquad c \in \mathbb{F}, \\ &[\Delta, T_\sigma] = \Delta T_\sigma - T_\sigma \Delta = 0, \qquad \forall\, \sigma \in \mathbb{F}, \end{aligned} \tag{2}$$

where $c \in \mathbb{F}$ is any fixed constant. It is easy to prove that

$$\Delta \cdot a = 0, \qquad \forall\, a \in \mathbb{F}.$$

For every delta operator Δ there is a *series of basic polynomials* $\{p_0(x), p_1(x), p_2(x), \ldots, p_n(x)\}$ of $\mathcal{P}$ verifying

$$\begin{aligned} &p_0(x) = 1, \\ &p_n(0) = 0, \quad \forall\, n > 0, \\ &\Delta \cdot p_n(x) = n\; p_{n-1}(x). \end{aligned} \tag{3}$$

We see that Δ acts as a lowering (index) operator on its associated polynomial. A very important fact of the theory is that each pair $\{\Delta, p_n(x)\}$ defines a unique realization or representation of the umbral algebra. There is a bijective correspondence $\mathfrak{J}$, between basic series of polynomials and their Δ operators. Hence, $\mathfrak{J}$ may be defined with only one element of the pair.

The commutator of an operator O with the coordinate operator X

$$O' = [O, X] = OX - XO \tag{4}$$

is called the *Pincherle derivative* of O.

Given a delta operator Δ we can find an associated operator ξ such that

$$[\Delta, \xi] = 1. \tag{5}$$

This operator ξ is not shift invariant and together with Δ and 1 close the Heisenberg-Weyl algebra or, in other words, expression (5) is a Heisenberg-like relation between Δ and its conjugate ξ. It is worthy to note that ξ is associated with the corresponding basic series of the operator Δ by

$$\xi^n \cdot 1 = p_n(x).$$

Although only in a determined correspondence $\xi^n \cdot 1$ coincides with the Pockhammer symbol $x^{(n)}$, from now on we will use the following notation

$$\xi^n \cdot 1 = p_n(x) \equiv x^{(n)}, \tag{6}$$

which is easy to prove taking into account that

$$[\Delta, \xi^n] = n\; \xi^{n-1}.$$

Obviously, there is only one way to define ξ if its associated polynomial series obeys all the relations (3). Given an operator Δ its associated operator ξ must be

$$\xi = X\beta,$$

where β^{-1} is the first Pincherle derivative of Δ. It is easy to demonstrate that β^{-1} is invertible and β is shift invariant.

Let Δ_1 and Δ_2 be two delta operators with associated ξ-operators ξ_1 and ξ_2, respectively. A map $\mathcal{R} : \mathcal{L} \to \mathcal{L}$ is called an *umbral correspondence* if

$$\mathcal{R} : \xi_1^n \;\to\; \xi_2^n, \qquad \forall n \in \mathbb{N}.$$

The umbral correspondence also induces a correspondence $\mathcal{R}$ between two delta operators. Summarising, the following diagram is commutative

$$\begin{array}{ccc} \xi_1^n & \overset{\mathcal{R}}{\longleftrightarrow} & \xi_2^n \\ \mathfrak{J}\updownarrow & & \mathfrak{J}\updownarrow \\ \Delta_1 & \overset{\mathcal{R}}{\longleftrightarrow} & \Delta_2 \\ \mathfrak{J}\updownarrow & & \mathfrak{J}\updownarrow \\ \xi_1^n\cdot 1 = x_1^{(n)} & \overset{\mathcal{R}}{\leftrightarrow} & \xi_2^n \cdot 1 = x_2^{(n)} \end{array}$$

From the above diagram one easily sees that there is a triplet $\{\Delta, \xi, p_n(x)\}$ that defines a unique representation of the umbral algebra. Moreover through $\mathfrak{J}$ we can go in a bijective way from one element to another of the triplet in such a way that only one of these elements is necessary and sufficient to define the umbral representation. The bijective application $\mathcal{R}$ bring us from a umbral representation to another one.

3. Umbral correspondence and discretized quantum mechanics

As we mentioned before, the main goal of this work is the application of the umbral correspondence to discretize Quantum Mechanics. In fact, the 1-dimensional Quantum Mechanics is an umbral realization via the identification

$$\Delta = \partial_x, \qquad \xi = X, \qquad p_n(x) = x^n.$$

Since we are interested in the discretization of Quantum Mechanics we will use the umbral correspondence to relate the triplet $(\partial_x,\ X,\ x^n)$ with another one defined on the lattice field $\mathbb{F}_\sigma = \{x \in \mathbb{R}\ |x = m\sigma,\ m \in \mathbb{Z}\}$ with lattice parameter $\sigma \in \mathbb{R}$.

Let us start defining a *general difference operator*, Δ, as a delta operator such that in the limit when σ goes to zero becomes the differential operator ∂_x. Supposing that Δ is a linear difference operator then it will have a polynomial dependence on the shift operator T, i.e.,

$$\Delta = \frac{1}{N\sigma}\sum_{n=-j}^{k} a_n T^n, \tag{7}$$

where $N \in \mathbb{N}$, $a_n \in \mathbb{R}$ and $\sigma \in \mathbb{R}$ is the lattice parameter. Imposing condition (2a), i.e. $\Delta\cdot x = 1$, since condition (2b) is trivially verified we get

$$\begin{aligned} \textstyle\sum_{n=-j}^{k} a_n &= 0, \\ \textstyle\sum_{n=-j}^{k} n a_n &= N. \end{aligned} \tag{8}$$

These conditions (8) guarantee that in the limit when σ goes to zero Δ goes to ∂_x as it is easy to prove.

Recalling that the pair of Quantum Mechanics operators (∂_x, X) verifies the Heisenberg relation

$$[\partial_x, X] = 1,$$

we call *discrete position operator* to a non-shift invariant operator ξ such that with the difference operator Δ verifies a similar Heisenberg relation, i.e.,

$$[\Delta, \xi] = 1.$$

Journal of Physics: Conference Series **128** (2008) 012056 doi:10.1088/1742-6596/128/1/012056

The umbral correspondence allows us to relate continuous linear equations h and their solutions f if they can be developed in power series, with umbral discrete versions $\hat{h}$ and $\hat{f}$, respectively, as follows

$$\begin{aligned} h\left(\partial_x^n f(x), x^s\right) = 0 \quad &\longleftrightarrow \quad \hat{h}\left(\Delta^n \hat{f}(\xi), \xi^s\right) \cdot 1 = 0, \qquad n, s \in \mathbb{N}; \\ f(x) = \sum_{n=0}^{\infty} f_n x^n \quad &\longleftrightarrow \quad \hat{f}(\xi) \cdot 1 = \sum_{n=0}^{\infty} f_n x^{(n)}. \end{aligned} \tag{9}$$

This is the main result of Umbral Calculus that we will exploit here for ours purposes. Umbral Calculus allows us not only to discretize continuous differential equations obtaining discrete equations but also to get solutions of these equations if their continuous counterparts are polynomial solutions or may be developed in power series which converge after the correspondence. Note, however, that with the umbral correspondence we obtain an operator which must be applied to 1 to get a difference equation or its solution.

For the particular case of a *symmetric difference operator*, i.e., a difference operator Δ_σ such that

$$\Delta_\sigma \cdot f(x) = -\Delta_{-\sigma} \cdot f(x),$$

we have that the general expression (7) reduces to

$$\Delta_\sigma = \frac{1}{N\sigma} \sum_{n=1}^{j} a_n (T^n - T^{-n}). \tag{10}$$

4. Differential operators for discretized quantum mechanics

In the following we will use some particular expressions (or representations) for the differential operator Δ, all of them of low (first and second) order on the shift operator T. Let us consider the two possible representations for Δ of first order, the right discrete derivative Δ_+ and the left discrete derivative Δ_-, and the (only) symmetric discrete derivative of second order Δ_s.

The *right-correspondence* is determined by the couple of operators (Δ_+, ξ_+) given by

$$\Delta_+ := \frac{1}{\sigma}(T - 1), \qquad \beta_+ = T^{-1},$$

and the *left-correspondence* (Δ_-, ξ_-) by

$$\Delta_- := \frac{1}{\sigma}(1 - T^{-1}), \qquad \beta_- = T.$$

Note that both are connected by the change $\sigma \leftrightarrow -\sigma$.

Journal of Physics: Conference Series **128** (2008) 012056 doi:10.1088/1742-6596/128/1/012056

The basic polynomial series of the right-correspondence is related with the Pochhammer symbols. Its explicit expression is

$$\begin{aligned}(X\beta_+)^n \cdot 1 = x_+^{(n)} &= x(x-\sigma)(x-2\sigma)\dots(x-(n-1)\sigma) \\ &= \prod_{i=0}^{n-1}(x-i\sigma) = \sigma^n \prod_{i=0}^{n-1}(m-i) \\ &= \begin{cases} (-\sigma)^n \dfrac{(-m+n-1)!}{(-m-1)!} & \text{if } m<0 \\ 0 & \text{if } \begin{cases} m \geq 0 \\ m < n \end{cases} \\ \sigma^n \dfrac{m!}{(m-n)!} & \text{if } \begin{cases} m \geq 0 \\ m \geq n \end{cases} \end{cases}\end{aligned} \tag{11}$$

where $x = m\sigma$ with $m \in \mathbb{Z}$. This series of polynomials has some interesting properties. First of all we can see that the n-th order zero of x^n splits into n first order zeros of $x_+^{(n)}$ at the right of the origin. If the right-correspondence is applied to a Taylor power series then, for the positive points of the domain, it will be truncated and therefore will converge, but not for the negative ones, and therefore could diverge.

The left-correspondence $\{\Delta_-, \xi_-\}$ is nearly similar to the right one. The basic sequence is related with the rising factorial (instead of the falling factorial of the previous case), and it is mirror symmetric respect to the y-axis to the right-correspondence: in fact, the negative points of a Taylor series discretized in this way, always converge.

According to the formula (10) there is only one symmetric representation of second order on T (S-representation) determined by the pair $\{\Delta_s, \xi_s\}$, where

$$\Delta_s = \frac{1}{2\sigma}(T - T^{-1}), \qquad \beta_s = 2(T + T^{-1})^{-1},$$

and its basic sequence is

$$\begin{aligned}(X\beta_s)^n \cdot 1 &= x_s^{(n)} = x[x-(n-2)\sigma][x-(n-4)\sigma]\dots[x+(n-4)\sigma][x+(n-2)\sigma] \\ &= x\prod_{i=0}^{n-2}[x+(2i-(n-2))\sigma] = \sigma^n m \prod_{i=0}^{n-2}[m+2i-(n-2)] \\ &= \begin{cases} (\operatorname{sign}(m)\sigma)^n |m| \dfrac{(|m|+n-2)!!}{(|m|-n)!!} & \text{if } n \leq |m| \\ 0 & \text{if } \begin{cases} n > |m| \\ \mathrm{p}(n-m) = +1 \end{cases} \\ \begin{array}{l}(-1)^{\frac{1}{2}(n-|m|-1)}(\operatorname{sign}(m)\sigma)^n m \\ \quad\times(|m|+n-2)!!(n-|m|-2)!!\end{array} & \text{if } \begin{cases} n > |m| \\ \mathrm{p}(n-m) = -1 \end{cases} \end{cases}\end{aligned} \tag{12}$$

where $\mathrm{p}(n) = (-1)^n$ is the parity function and the double factorial is defined as $n!! = n(n-2)!!$ with $n!! = 1$ when n is 0 or 1.

The behaviour of this basic sequence is much more complicated than those of the right and left-correspondences. The zero at the origin of order n of the continuous power function of degree n splits into n simple zeros after discretization. These zeros appear at the points $x = m\sigma$ with the conditions

$$\mathrm{p}(n-m) = +1, \qquad n > |m|. \tag{13}$$

Journal of Physics: Conference Series **128** (2008) 012056 doi:10.1088/1742-6596/128/1/012056

Thus, applying the S-correspondence to a Taylor series we find that at the points obeying the conditions (13) there is a cut-off in the series and, hence, the sums will always converge in case the function has a well defined parity.

In the following diagram we display the basis elements $(\xi^n, \Delta, x^{(n)})$ of the three usual representations:

Right representation		**Left representation**		**Symmetric representation**
$\xi_+^n = (XT^{-1})^n$	$\overset{\mathcal{R}}{\longleftrightarrow}$	$\xi_-^n = (XT)^n$	$\overset{\mathcal{R}}{\longleftrightarrow}$	$\xi_s^n = \left(2X\left(T+T^{-1}\right)^{-1}\right)^n$
$\mathfrak{J}\updownarrow$		$\mathfrak{J}\updownarrow$		$\mathfrak{J}\updownarrow$
$\Delta_+ = \frac{1}{\sigma}(T-1)$	$\overset{\mathcal{R}}{\longleftrightarrow}$	$\Delta_- = \frac{1}{\sigma}\left(1-T^{-1}\right)$	$\overset{\mathcal{R}}{\longleftrightarrow}$	$\Delta_s = \frac{1}{2\sigma}\left(T-T^{-1}\right)$
$\mathfrak{J}\updownarrow$		$\mathfrak{J}\updownarrow$		$\mathfrak{J}\updownarrow$
$x_+^{(n)} = \prod_{i=0}^{n-1}(x-i\sigma)$	$\overset{\mathcal{R}}{\longleftrightarrow}$	$x_-^{(n)} = \prod_{i=0}^{n-1}(x+i\sigma)$	$\overset{\mathcal{R}}{\longleftrightarrow}$	$x_s^{(n)} = x\prod_{i=0}^{n-2}(x+(2i-n+2)\sigma).$

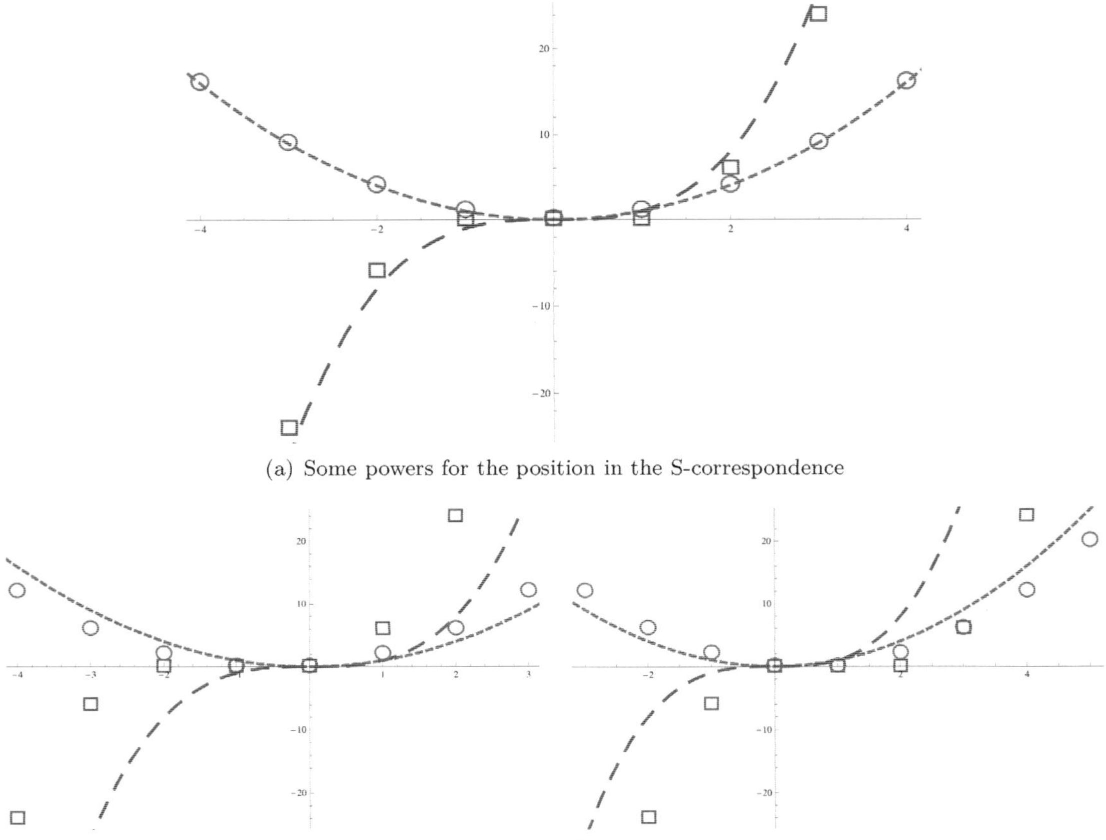

(a) Some powers for the position in the S-correspondence

(b) Some powers for the position in the left-correspondence

(c) Some powers for the position in the right-correspondence

Figure 1. Represented in dashed line the continuous powers and in dots the powers for the discrete position with red circles for x^2 and blue squares for x^3. The number of zeros is equal to the power n for the right and the left cases.

Journal of Physics: Conference Series **128** (2008) 012056 doi:10.1088/1742-6596/128/1/012056

5. The discretized exponential function and the trigonometric functions

We start with the exponential function which in some sense, is the simplest case. The continuous exponential function, as it is well known, is defined through the following differential equation

$$\frac{\mathrm{d}}{\mathrm{d}x}\, y_k = k y_k, \quad y_k(0) = 1, \qquad k \in \mathbb{R},$$

and its power expansion series around the origin is

$$y_k(x)) = e^{kx} = \sum_{n=0}^{\infty} \frac{(kx)^n}{n!}.$$

As we mentioned before, the umbral correspondence gives rise to the same function either by solving the discrete difference equation, obtained by applying the umbral correspondence to the continuous differential equation, or by applying the umbral correspondence to its solutions as long as their power series converge. Hence, we have to calculate

$$\exp_U(k, x) = \sum_{n=0}^{\infty} \frac{k^n}{n!} \xi^n \cdot 1 \,. \tag{14}$$

If we use the right-correspondence (11) the umbral-exponential (14) becomes

$$\exp_+(k, \sigma m) = \left\{ \begin{array}{ll} \displaystyle\sum_{n=0}^{m} (k\sigma)^n \frac{m!}{n!(m-n)!} & \text{if } m > 0 \\ \displaystyle\sum_{n=0}^{\infty} (-k\sigma)^n \frac{(-m+n-1)!}{n!(-m-1)!} & \text{if } m < 0 \end{array} \right\} = (1 + k\sigma)^m.$$

For the left-correspondence we get

$$\exp_-(k, \sigma m) = \left\{ \begin{array}{ll} \displaystyle\sum_{n=0}^{\infty} (k\sigma)^n \frac{(m+n-1)!}{n!(m-1)!} & \text{if } m > 0 \\ \displaystyle\sum_{n=0}^{-m} (-k\sigma)^n \frac{(-m)!}{n!(-m-n)!} & \text{if } m < 0 \end{array} \right\} = (1 - k\sigma)^{-m}.$$

In Fig. 2(a) it is displayed the continuous exponential function together with its discrete version by the right-correspondence. It is worthy to note that if in the continuous case we found that $\exp(k\ x) = \exp((-k)(-x))$, now in the discrete case we have that $\exp_+(k, x) = \exp_-(-k, -x)$.

Another possibility is to apply the S-correspondence (12). So, we obtain

$$\exp_s(k, m\sigma) = \sum_{n=0}^{\infty} \frac{(k\sigma)^n}{n!} m \prod_{i=0}^{n-2} (m + 2i - (n-2)) = \left(k\sigma + \sqrt{(k\sigma)^2 + 1}\right)^m.$$

These discrete exponentials only converge when the continuous variable k obeys, $(k\sigma)^2 < 1$, for all the three correspondences studied in the previous sections.

Also, we can easily study the trigonometric functions (as well as the hyperbolic functions) taking into account their expressions in terms of the exponential functions. In this way, we can obtain the discrete trigonometric functions from the discrete exponential function. We display in Fig. 2(b) the sine function.

We can see that the symmetric discrete trigonometric functions maintain the periodic behaviour of the continuous ones. However the period λ_s is different to the continuous function $\lambda = 2\pi/k$. We find it easier in the case of the sine function through $\sin_s(k, l\sigma) = \sin_s(k, 0\sigma)$. We have fixed $\lambda = l\sigma$ with $l \in \mathbb{Z}$, but it can be generalized without problems to $l \in \mathbb{R}$.

$$\sin_s(k, l\sigma) = \frac{1}{2\mathrm{i}}\left(\left(\sqrt{1-(k\sigma)^2}+\mathrm{i}k\sigma\right)^l - \left(\sqrt{1-(k\sigma)^2}-\mathrm{i}k\sigma\right)^l\right) = 0 = \sin_s(k, 0\sigma)$$
$$\Rightarrow \left(\sqrt{1-(k\sigma)^2}+\mathrm{i}k\sigma\right)^l = \left(\sqrt{1-(k\sigma)^2}-\mathrm{i}k\sigma\right)^l$$

to solve this identity we define

$$\sqrt{1-(k\sigma)^2}+\mathrm{i}k\sigma = re^{\mathrm{i}\theta} \Rightarrow \begin{cases} \theta = & \arctan\left(\frac{k\sigma}{\sqrt{1-(k\sigma)^2}}\right) \\ r = & 1 \end{cases}$$

and this guide us to the solution

$$\sin(\theta l) = 0 \Rightarrow \theta_n = \frac{n\pi}{l} \quad \Rightarrow \quad k_n = \frac{1}{\sigma}\sin\frac{n\pi}{l}.$$

As there are two zeros in each period of the function, we can find the relation between the number of waves and the wave length. We have added an superscript to remark that these relations are different for each representation, then

$$k^s = \frac{1}{\sigma}\sin\left(\frac{2\pi}{l}\right), \quad \lambda^s = \frac{2\pi\sigma}{\arcsin(k^s\sigma)}. \tag{15}$$

It is easy to demonstrate the periodic behaviour of the symmetric trigonometric functions, in fact introducing (15) into de symmetric sinus

$$\sin_s(k^s, m\sigma) = \sin\left(m\frac{2\pi}{l}\right).$$

For the right and left correspondences the trigonometric functions are not periodic. However the extremums of the functions and the intersection points with the abscissa axis appear periodically and then we can define a wave function through this property. With a similar computation than in the symmetric case, we find

$$k^+ = \frac{1}{\sigma}\tan\left(\frac{2\pi}{l}\right), \quad \lambda^+ = \frac{2\pi\sigma}{\arctan(k^+\sigma)}. \tag{16}$$

The wave length is bigger in the right and left discretizations and smaller in the symmetric discretization than in the continuous trigonometric functions. Also in these three correspondences the trigonometric functions have a minimum period, it appears in the convergence limit, when $k\sigma = 1$. For the right and the left discretizations we have $\lambda^+_{min} = 8\sigma$, in fact it is a 8 point wave, and for the symmetric one $\lambda^s_{min} = 4\sigma$, in fact it is a 4 point wave.

We set $\lambda = l\sigma$ and without loss of generality we fix l integer. Then it is easy to calculate the change in the amplitude in the right and left correspondences through (16)

$$\sin_+(k^+, m\sigma) = \frac{1}{2\mathrm{i}}\left((1+\mathrm{i}k^+\sigma)^m - (1-\mathrm{i}k^+\sigma)^m\right) = \cos^{-m}\left(\frac{2\pi}{l}\right)\sin\left(m\frac{2\pi}{l}\right)$$

and

$$\sin_+(k, m\sigma)A_n(l) = \sin_+(k, (m+nl)\sigma)$$

Hence, the non periodic part is then the multiplicative factor

$$A_n(l) = \left(\sec^l\left(\frac{2\pi}{l}\right)\right)^n. \tag{17}$$

This is also true for the cosine function and for any $l \in \mathbb{R}$. Although we have a wave length $\lambda = l\sigma$ for the trigonometric functions in the right and left representations, we can see that they are not periodic.

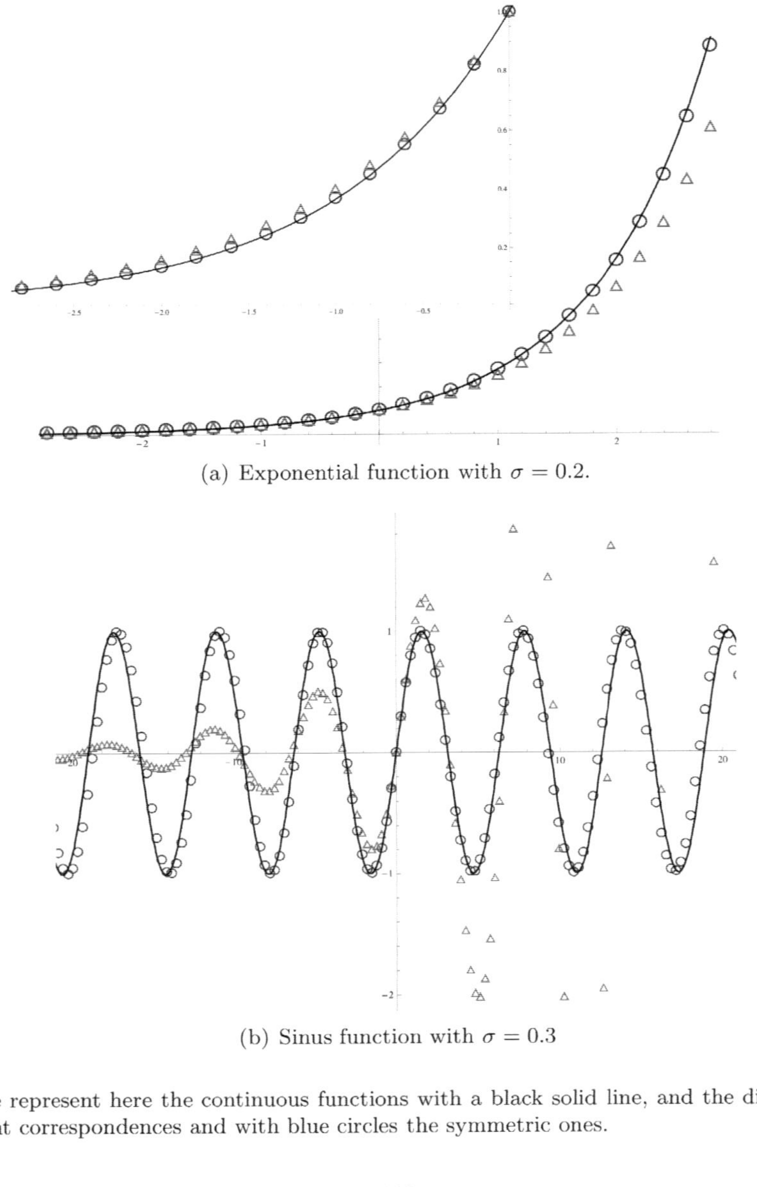

(a) Exponential function with $\sigma = 0.2$.

(b) Sinus function with $\sigma = 0.3$

Figure 2. We represent here the continuous functions with a black solid line, and the discrete ones with red triangles the right correspondences and with blue circles the symmetric ones.

It is important to note that all the properties that obey the exponential and the trigonometric functions are maintained after the discretization with an exception: any property that includes two or more different constants k, because they give similar effects than a different lattice spacing σ, nor a different x like in the continuous case. Thus, for example

$$\exp_U(k, m\sigma)\exp_U(k, n\sigma) = \exp_U(k,(m+n)\sigma),$$
$$\exp_U(k, m\sigma)\exp_U(k', m\sigma) \neq \exp_U(k+k', m\sigma).$$

6. Umbral Schrödinger equation

Let us go to discretize the Schrödinger equation with a potential $V(x)$. We will profit the well known fact that if the potential is independent of the time then the continuous Schrödinger equation is separable. Thus, from the Schrödinger equation

$$\mathrm{i}\hbar\partial_t\phi(x,t) = -\frac{\hbar^2}{2m}\partial_x^2\phi(x,t) + V(x)\phi(x,t),$$

and considering $\phi(x,t) = \psi(x)\varphi(t)$ we obtain two uncoupled ordinary differential equations

$$-\frac{\hbar^2}{2m}\partial_x^2\psi(x) + V(x)\psi(x) = E\psi(x),$$
$$\mathrm{i}\hbar\partial_t\varphi(t) = E\varphi(t),$$

that we can rewrite with natural units as

$$\begin{aligned} \partial_x^2\psi(x) + V(x)\psi(x) &= E\psi(x), \\ -\mathrm{i}\partial_t\varphi(t) &= E\varphi(t). \end{aligned} \tag{18}$$

The solution of the equation (18b) is

$$\varphi(t) = Ce^{\mathrm{i}Et}. \tag{19}$$

6.1. Constant potential

In relation with the first equation of (18), the simplest case appears when the potential is constant ($V(x) = V_0$). The solution is

$$\begin{aligned} \psi(x) &= Ae^{\mathrm{i}kx} + Be^{-\mathrm{i}kx} \qquad \text{if } E > V_0, \\ \psi(x) &= Ae^{kx} + Be^{-kx} \qquad \text{if } E < V_0, \end{aligned} \tag{20}$$

with $k = \sqrt{|E - V_0|}$. Therefore, the solution $\phi(x,t)$ is a linear combination of trigonometric or exponential functions and, hence, their discretizations have been studied in section 5.

The energy, for a fixed momentum k, is the same than in the continuous case for any discretization as we can see:

$$\begin{aligned} H\psi &= \left(-\Delta^2 + V_0\right)\sum_{n=0}^{\infty}\frac{(\mathrm{i}k)^n}{n!}\left(A + (-1)^n B\right)\xi^n = \\ &= \left(k^2 + V_0\right)\sum_{n=0}^{\infty}\frac{(\mathrm{i}k)^n}{n!}\left(A + (-1)^n B\right)\xi^n = \left(k^2 + V_0\right)\psi. \end{aligned}$$

Journal of Physics: Conference Series **128** (2008) 012056 doi:10.1088/1742-6596/128/1/012056

In spite of this result, the relation between the momentum and the wave length $\lambda = l\sigma$ is different in the discrete case:

$$k = \frac{2\pi}{\lambda}, \quad k_{\pm} = \frac{1}{\sigma}\tan\left(\frac{2\pi}{l}\right), \quad k_s = \frac{1}{\sigma}\sin\left(\frac{2\pi}{l}\right).$$

Then the phase velocity changes, although the group velocity is the same as in the continuous case and, for this reason, the relation between the energy and the velocity.

Let be $t = \tilde{n}\tau$ where $\tilde{n} \in \mathbb{Z}$ the discrete time, and τ the fundamental time length. Following the limits of convergence of the discrete exponentials we find two upper limits for the energy, one in relation with the temporal part of the wave function (19) and the other one in relation with the spatial part of the wave function (20).

$$E_{max_t} = \frac{\hbar}{\tau}, \quad E_{max_l} = \frac{\hbar^2}{2m\sigma^2}. \tag{21}$$

In fact, we have to take the minor of them in each situation. Taking $\sigma \approx l_{Planck} = 1.62\ 10^{-35}m$ and $\tau \approx t_{Planck} = 5.39\ 10^{-44}s$ we find values of the maximum energy for the time wave E_{max_t} and for the spatial wave for a proton E_{p^+,max_l} and an electron E_{e^-,max_l}

$$E_{max_t} \approx E_{Planck} = 1.22\ 10^{28}eV \quad , \quad E_{max_l} = 1.38\ 10^{20}\frac{1}{m}eV$$
$$E_{p^+,max_l} = 7.94\ 10^{46}eV \quad , \quad E_{e^-,max_l} = 1.46\ 10^{50}eV.$$

Remark that these are non-relativistic effects. We can study also the change in the amplitude for the right and left correspondences. The factor of expansion or contraction is

$$A_n(l) = \left(\sec^l\left(\frac{2\pi}{l}\right)\right)^n \simeq \left(1 + \frac{2\pi^2}{l}\right)^n$$

where l is the number of points into the wave length and n is the number of wave lengths along the wave. For the temporal part of the wave function of a particle which travel into the LEP, that is during 910^-5s with an energy of $1.9910^{14}eV$, we find that the factor is $A_n(l) \sim 10^{10^5}$. For the spatial part of the same particle, an electron, the factor is of order $A_n(l) \sim 3$. These factors grow a lot with the energy or the mass of the particle. As at this energy the relativistic effects are important and we are using now non relativistic quantum mechanics, it is comprehensible to find these too big numbers.

6.2. Infinite potential well

A little more complicated is the infinite potential well, although it is very similar to a constant potential, it reduces the space to a finite region and due to this fact new differences will appear between the continuous and the discrete case. The potential is

$$V(x) = \begin{cases} 0 & \text{if } x \in [0, L] \\ \infty & \text{if } x \notin [0, L] \end{cases}$$

The wave functions must obey the boundary conditions $\psi(0) = \psi(L) = 0$ which lead to a countable number of solutions:

$$\psi_n(x) = \sin\left(n\pi\frac{x}{L}\right), \quad k = \frac{n\pi}{L}, \quad n \in \mathbb{Z},$$

Journal of Physics: Conference Series **128** (2008) 012056 doi:10.1088/1742-6596/128/1/012056

in the same way we apply the boundary conditions to all our discretizations. We can find the quantum rule for both the right and left correspondences following the computation of (15). So

$$k_n^+ = \frac{1}{\sigma} \tan \pi \frac{n}{M}$$

where $L = M\sigma$. Therefore there are only $M/2$ possible states instead of the infinite states of the continuous case, with a degeneration relative to the parity. The energy is $E_n = k_n^2$ and it is bounded at $n = (M-1)/2$. So the maximum energy is

$$E_{max}^+ = \frac{\hbar^2}{2m\sigma^2} \tan^2 \left(\pi \left[\frac{M-1}{2M} \right] \right)$$

At $n = M/2$ we find a non physical state of infinite energy and infinite wave function amplitude. We can do a small computation finding that an electron in an infinite well of wide the Bohr radius ($M \sim 10^{27}$) will have a maximum state energy of $E_{max}^+ \sim 10^{77} eV$. The maximum energy grows when increasing the number of points M and also when decreasing the mass of the particle. But the wave function, as we saw previously, has a convergence range that limits the number of states to $n \in (-M/4, M/4)$ because $k < 1/\sigma$. Hence, the maximum energy is much more smaller.

We have that the quantum rule for the S-correspondence is

$$k_n^s = \frac{1}{\sigma} \sin \pi \frac{n}{M}.$$

Again we have a finite number of states, $M/2$, rounding down, with a degeneration relative to the parity also. The energy is always bounded by $\hbar/(2m\sigma^2)$, just as we found with the convergence condition

$$E_n^s = (k_n^s)^2 = \frac{1}{\sigma^2} \sin^2 \pi \frac{n}{M}.$$

In Figure 3 are drawn the energy levels and the wave functions of the ground state and the two first excited states for the continuous case and the three discretizations. We can see the great fit of the symmetric one for the first excited states, but remarking that as its number of states is finite, $M/2$, then upper these wave functions there are not more states. The right and the left ones have an asymmetry to the left and right directions of the space and then are much more different to the continuous wave functions.

7. Conclusions

In this paper we have shown how to apply the techniques of the Umbral Calculus for the discretization of equations preserving the point symmetries. We can see the advantage of a tool that may be used not only on the equations, but also over their solutions. Due to this fact we have not to solve difference equations because we alternatively can sum series. Moreover, in the most difficult cases we can use numerical methods to get a solution.

In the umbral discretization we can choose between infinite umbral representations. However, still in the simplest cases they present very different behaviours. We can separate all of them in two classes, one preserving the parity of the equations and solutions (like the S-representation) and the other one that does not (like the right and left representations). From the point of view of the local symmetries in Ref ([2]–[9], [26]) it was proved that independently of the correspondence we get the same discrete point symmetries.

The study of the discrete Schrödinger equation together with other potentials (Coulomb, harmonic oscillator, ...) are in progress [31]–[32] and the results will be published elsewhere.

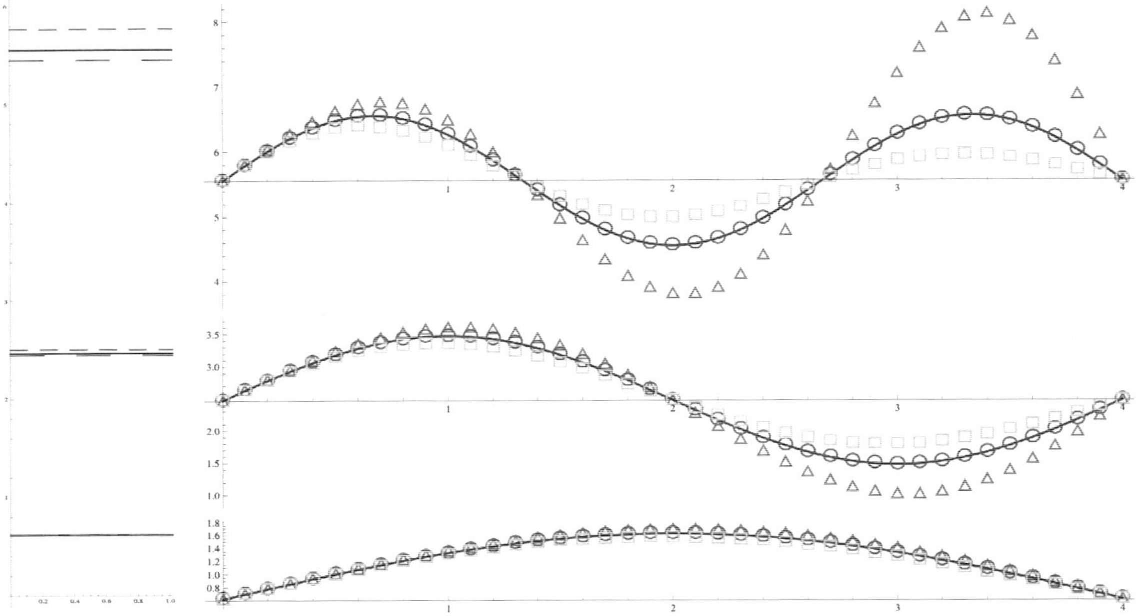

Figure 3. Energy levels and wave functions of the infinite well. The energy levels are in the left, ones in solid line for the continuous case and dashed for the discrete case, the shorter red lines are the right/left case energies and the longer blue lines are the symmetric case energies. Wave functions are in the right, in black solid line are for the continuous case, blue circled points are for the S-correspondence, red triangled ones for the right correspondence and green squared ones the left correspondence.

Acknowledgments
We thank to D. Levi for stimulating interesting discussions and the Physics Dp. of Roma Trè Univ. for their hospitality. This work was partially supported by the Ministerio de Educación y Ciencia of Spain (Projects FIS2005-03989 and by the Junta de Castilla y León (Project VA013C05). JLS thanks the Universidad de Valladolid and the Banco Santander Central Hispano by the PhD grant of the program "Ayudas de la Universidad de Valladolid cofinanciadas por el BSCH".

References
[1] Gibbs P 1995 *The small scale structure of space-time: a bibliographical review* hep-th/9506171
[2] Dimakis A, Müler-Hoissen F and Striker T 1996 Umbral calculus, discretization, and quantum mechanics on a lattice *J. Phys.* A **29** 6861
[3] Levi D, Vinet L and Winternitz P 1997 Lie group formalism for difference equations *J. Phys.* A **30** 633
[4] Levi D, Negro J and del Olmo M A 2001 Discrete derivatives and symmetries of difference equations *J. Phys.* A **34** 2023
[5] Levi D, Negro J and del Olmo M A 2001 Lie symmetries of difference equations *Czech. J. Phys.* **51** 341
[6] Senosiain M J 2002 *Teoría de las Ecuaciones Genéricas: Una Introducción al Cálculo Umbral* Memoria de Grado Universidad de Salamanca
[7] Levi D, Negro J and del Olmo M A 2004 Discrete q-derivatives and symmetries of q-difference equations *J. Phys.* A **37** 3459–3473
[8] Levi D, Tempesta P and Winternitz P 2004 Umbral calculus, difference equations and the discrete Schrödinger equation *J. Math. Phys.* **45** 4077
[9] Salgado E 2004 *Mecánica Cuántica Discreta* Trabajo Fin de Carrera de Físicas Univ. de Valladolid
[10] Mullin R and Rota G C 1970 *On the foundations of combinatorial theory III. Theory of binomial enumeration* (*Graph theory and its applications*) Ed. B. Harris (Academic Press) p 167-213
[11] Blissard J 1861 Theory of generic equations *Quart. J. Pure Appl. Math.* **4** 279
[12] Blissard J 1862 Theory of generic equations *Quart. J. Pure Appl. Math.* **5** 58, 184

[13] Bell E T 1938 The history of Blissard's symbolic calculus, with a sketch of the inventor's life *Amer. Math. Monthly* **45** 414
[14] Appell P 1880 Sur une classe de polynômes *Ann. Sci. Ecole Norm. Sup. (2)* **9** 119
[15] Pincherle S and Amaldi S 1901 *Le Operazioni Distributive e le loro Applicazioni all'Analisi* (Bologna: N. Zanichelli)
[16] Davis H T, 1936, *The Theory of Linear Operators* (Bloomington: Principia Press)
[17] Sheffer I M 1939 Some properties of polynomial sets of type zero *Duke Math. J.* **5** 590
[18] Rota G C, Kahaner D and Odlyzko A 1973 On the foundations of combinatorial theory VII. Finite operator calculus *J. Math. Anal. Appl* **42** 684
[19] Rota G C 1975 *Finite Operator Calculus* (San Diego: Academic)
[20] Roman S M 1982 The theory of umbral calculus I *J. Math. Anal. Appl.* **87** 58
[21] Roman S M 1982 The theory of umbral calculus II *J. Math. Anal. Appl.* **89** 290
[22] Roman S M 1983 The theory of umbral calculus III *J. Math. Anal. Appl.* **95** 528
[23] Roman S M 1975 *The Umbral Calculus* (San Diego: Academic)
[24] Di Bucchianico A and Loeb D 2000 A selected survey of umbral calculus *Electr. J. Combin.* DS3
[25] Zachos C K 2007 Umbral deformations on discrete spacetime *Preprint* hep-th/07102306
[26] Levi D and Winternitz P 2005 Continuous symmetries of difference equations *J. Phys.* A **39** R1–R63
[27] Levi D and Petrera M 2007 Continuous symmetries of the lattice potential KdV equation *J. Phys.* A **40** 4141–4159
[28] Levi D, Tremblay S and Winternitz P 2007 Lie point symmetries of difference equations and lattices *Preprint* math-ph/07093112
[29] Levi D, Tremblay S and Winternitz P 2007 Lie symmetries of multidimensional difference equations *Preprint* math-ph/07093238
[30] Hernandez Heredero R, Levi D, Petrera M and Scimiterna C 2007 Multiscale expansion on the lattice and integrability of partial difference equations *Preprint* math-ph/07105299
[31] López-Sendino J E, Negro J, del Olmo M A and Salgado E 2008 Discretizing quantum mechanics using umbral calculus *Preprint*
[32] López-Sendino J E, Negro J and del Olmo M A 2008 Discrete harmonic oscillator and umbral calculus *Preprint*

Journal of Physics: Conference Series **128** (2008) 012057 doi:10.1088/1742-6596/128/1/012057

On continuous symmetries of second-order homogeneous linear ordinary differential equations

Adelio R Matamala[1], **Cristian A Salas**[2], **Marcelo J Alid**[2,3]
[1]Facultad de Ciencias Químicas, Universidad de Concepción, Casilla, 160-C, Concepción, Chile.
[2]Departamento de Física, Universidad de Concepción, Casilla, 160-C, Concepción, Chile.
[3]Departamento de Matemáticas, Estadística y Física, Universidad de Las Américas, Chile.

E-mail: amatamal@udec.cl

Abstract. In this work, a method to extract continuous symmetries of general second-order linear ordinary differential equation is presented. The formalism is illustrated by two examples.

1. Introduction

Ordinary differential equations (ODEs) appear in many fields of Physics [1] and the analysis of their symmetries plays an important role to extract information about the solutions of those equations [2-4]. In this work, we have applied the Anderson-Kumei-Wulfman method [5-8] to extract continuous symmetries of general second-order linear ordinary differential equation.

In general, we consider general homogeneous linear ODEs represented by the action of the differential operator $\hat{A}(x)$ on a function $f(x)$

$$\hat{A}(x)f(x) = 0, \tag{1}$$

where the differential operator $\hat{A}(x)$ is give by

$$\hat{A}(x) = \alpha_0(x) + \alpha_1(x)\partial_x + \alpha_2(x)\partial_{xx} + ... \tag{2}$$

In this work we are interested in second order ODEs, then $\alpha_i = 0$, for $i \geq 3$.

A symmetry operator $\hat{Q}$ of equation (1) is defined as a differential operator that maps solution of equation (1) into solution of the same equation, i.e.,

$$\hat{A}(x)g(x) = 0, \tag{3}$$

where

$$g(x) = \hat{Q}(x)f(x). \tag{4}$$

After assuming a particular form for the operators $\hat{Q}$, condition (3) under constrain (1) defines the symmetries of the ODE (1).

Journal of Physics: Conference Series **128** (2008) 012057 doi:10.1088/1742-6596/128/1/012057

2. Symmetry extraction

Let us start by considering a general second-order homogeneous linear ODE:

$$\frac{d^2y}{dx^2} + p(x)\frac{dy}{dx} + q(x)y(x) = 0. \tag{5}$$

After substituting

$$y(x) = f(x)\exp\left[-\frac{1}{2}\int^x p(\xi)d\xi\right] \tag{6}$$

in equation (5), it is easy to see that the function $f(x)$ satisfies the following differential equation:

$$\frac{d^2f}{dx^2} + v(x)f(x) = 0, \tag{7}$$

where

$$v(x) = q(x) - \frac{1}{4}p^2(x) - \frac{1}{2}\frac{dp}{dx}. \tag{8}$$

We are interested in continuous symmetry generators of the form

$$\hat{Q}(x) = \alpha(x) + \beta(x)\frac{d}{dx} \tag{9}$$

for the equation (7). In other words, if $f(x)$ is a solution of equation (7) then we need to find the functions $\alpha(x)$ and $\beta(x)$ so that:

$$g(x) = \hat{Q}(x)f(x) \tag{10}$$

is also a solution of equation (7), i.e.

$$\frac{d^2g}{dx^2} + v(x)g(x) = 0. \tag{11}$$

Then differentiating equation (10), replacing in (11) and using that $f(x)$ is lineally independent, we obtain the following equation system:

$$\alpha'' - v'\beta - 2v\beta' = 0, \tag{12}$$
$$2\alpha' + \beta'' = 0. \tag{13}$$

The above system determines the conditions for $\alpha(x)$ and $\beta(x)$ function.
Working on equations (12) and (13), we obtain:

$$\beta''' + 4v\beta' + 2v'\beta = 0. \tag{14}$$

Equation (14) may be solved if $v(x)$ function is known.
Let us assume that we know two particular solutions $u_1(x)$ and $u_2(x)$ for equation (7), i.e.

$$u_1'' + vu_1 = 0 \quad \text{and} \quad u_2'' + vu_2 = 0. \tag{15}$$

Now, if we consider the following definition

$$\phi(x) = C_1u_1^2 + C_2u_2^2 + C_3u_1u_2, \tag{16}$$

It is easy to show that $\phi(x)$ function satisfies the equation:

$$\phi''' + 4v\phi' + 2v'\phi = 0. \tag{17}$$

Journal of Physics: Conference Series **128** (2008) 012057 doi:10.1088/1742-6596/128/1/012057

Therefore, equation (14) defines the $\beta(x)$ function as

$$\beta(x) = C_1 u_1^2 + C_2 u_2^2 + C_3 u_1 u_2, \tag{18}$$

where the functions $u_1(x)$ and $u_2(x)$ are two lineally independent particular solutions of equation (7):

$$u_1'' + v u_1 = 0 \quad \text{and} \quad u_2'' + v u_2 = 0.$$

However, it is only necessary to know one solution since the second linear independent solution of equation (7) is obtained by the relation

$$u_2(x) = u_1(x) \int^x \frac{d\xi}{[u_1(\xi)]^2}. \tag{19}$$

Finally, having $\beta(x)$ it is possible to obtain $\alpha(x)$ expansion coefficient by:

$$\alpha'' = -\frac{1}{2}\beta'''. \tag{20}$$

3. Examples

The symmetry extaction method will be exemplified by two simple equations:

3.1. *Example 1*

Let us consider the simplest second order ODE:

$$f_{xx} = 0. \tag{21}$$

Step 1: Find a particular solution of (21)

$$u_1 = C_1 x. \tag{22}$$

Step 2: Use equation (19), to build a second independent solution

$$u_2 = C_1 x \int^x \frac{d\xi}{(C_1\xi)^2} = C_2. \tag{23}$$

Step 3: Use equation (18) to obtain $\beta(x)$ function

$$\beta(x) = B_1 x^2 + B_2 + B_3 x. \tag{24}$$

Step 4: Use equation (13) to obtain $\alpha(x)$ function

$$\alpha(x) = -\frac{1}{2}\int^x \beta''(\xi)d\xi = -B_1 x. \tag{25}$$

Step 5: Build the symmetry generators

$$\hat{Q}_1 = \frac{d}{dx}, \quad \hat{Q}_2 = x\frac{d}{dx}, \quad \hat{Q}_3 = x - x^2\frac{d}{dx}. \tag{26}$$

Step 6: Check the symmetry property

$$\hat{A}\hat{Q}_1 f = 0, \quad \hat{A}\hat{Q}_2 f = 0, \quad \hat{A}\hat{Q}_{.3} f = 0 \tag{27}$$

Step 7: Find the algebra

$$[\hat{Q}_1, \hat{Q}_2] = \hat{Q}_1, \quad [\hat{Q}_1, \hat{Q}_3] = I - 2\hat{Q}_2, \quad [\hat{Q}_2, \hat{Q}_3] = \hat{Q}_3. \tag{28}$$

And, introducing the new definition:

$$\hat{A}_0 = \hat{Q}_2 - 1/2, \quad \hat{A}_- = \hat{Q}_1, \quad \hat{A}_+ = \hat{Q}_3, \tag{29}$$

the following commutation relations are obtained:

$$[\hat{A}_0, \hat{A}_\pm] = \pm\hat{A}_\pm, \quad [\hat{A}_+, \hat{A}_-] = -2\hat{A}_0. \tag{30}$$

Journal of Physics: Conference Series **128** (2008) 012057 doi:10.1088/1742-6596/128/1/012057

3.1.1. ***Symmetry Visualization:*** With the symmetry generators we can obtain the action of this generators on the solution of the original ODE.

The general solution of ODE (21) is:

$$y = a + bx, \tag{31}$$

where a and b are arbitrary constants. Then the actions of $\hat{A}_+$, $\hat{A}_-$ and $\hat{A}_0$ over (31) are

$$\begin{aligned}
\exp(\theta \hat{A}_0)(a + bx) &= \bar{a} + \bar{b}x, & \bar{a} &= e^{-\theta/2}a, & \bar{b} &= e^{\theta/2}b \\
\exp(\theta \hat{A}_-)(a + bx) &= \bar{a} + bx, & \bar{a} &= a + \theta b \\
\exp(\theta \hat{A}_+)(a + bx) &= a + \bar{b}x, & \bar{b} &= b + \theta a
\end{aligned}$$

Figure 1 shows the plot of (31) for particular values of a and b constants. Figures 2, 3 and 4 show the action of $\hat{A}_+$, $\hat{A}_-$ and $\hat{A}_0$ symmetry operators on solution (31) respectively.

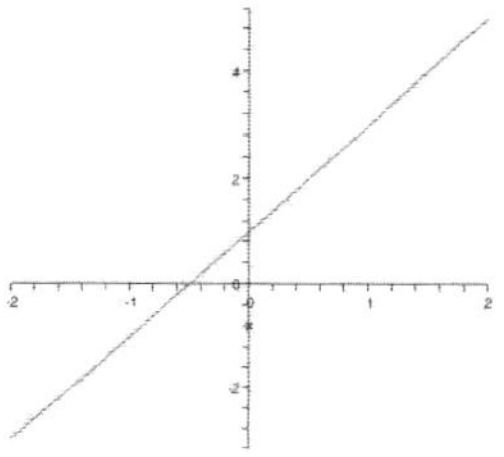

Figure 1

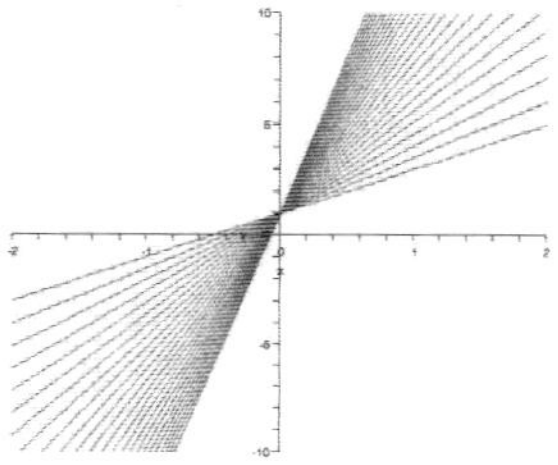

Figure 2

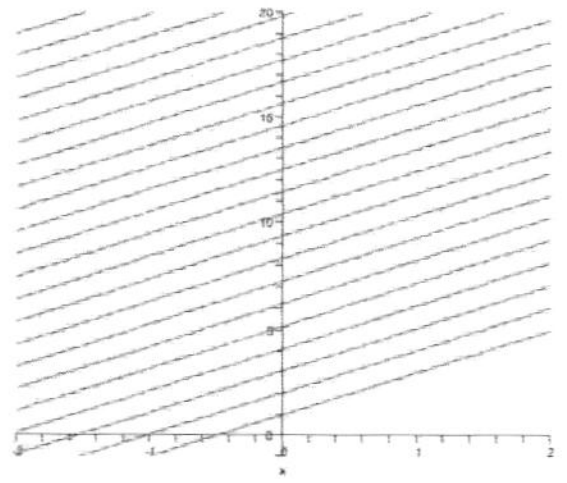

Figure 3

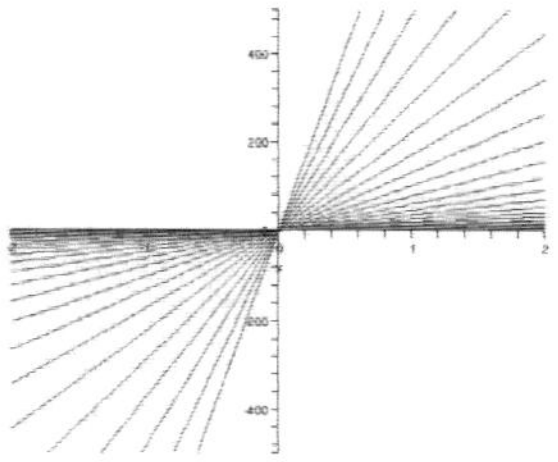

Figure 4

3.2. Example 2

Let us consider the second order ODE:

$$f_{xx} + k^2 f = 0. \tag{32}$$

Step 1: Find a particular solution of (32)

$$u_1 = \cos(kx). \tag{33}$$

Step 2: Use equation (19) to build a second independent solution

$$u_2 = \cos(kx) \int^x \frac{d\xi}{[\cos(kx)]^2} = \frac{\sin(kx)}{k}. \tag{34}$$

Step 3: Use equation (18) to obtain $\beta(x)$ function

$$\beta(x) = C_1 \cos^2(kx) + C_2 \frac{\sin^2(kx)}{k^2} + C_3 \frac{\sin(kx)}{k} \cos(kx). \tag{35}$$

Step 4: Use equation (13) to obtain $\alpha(x)$ function

$$\alpha(x) = C_1 \frac{k \sin(2kx)}{2} - C_2 \frac{\sin(2kx)}{2k} + C_3 \frac{1}{2}(1 - \cos(2kx)) + C_4. \tag{36}$$

Step 5: Build the symmetry generators

$$\hat{Q}_1 = \tfrac{k}{2} \sin(2kx) + \tfrac{1}{2}(1 + \cos(2kx)) \tfrac{d}{dx}, \tag{37}$$
$$\hat{Q}_2 = \tfrac{1}{2} - \tfrac{1}{2} \cos(2kx) + \tfrac{1}{2k} \sin(2kx) \tfrac{d}{dx}, \tag{38}$$
$$\hat{Q}_3 = \tfrac{1}{2k} \sin(2kx) + \tfrac{1}{2k^2}(\cos(2kx) - 1) \tfrac{d}{dx}. \tag{39}$$

Step 6: Check the symmetry property

$$\hat{A}\hat{Q}_1 f = 0 \quad \hat{A}\hat{Q}_2 f = 0, \quad \hat{A}\hat{Q}_3 f = 0. \tag{40}$$

Step 7: Find the algebra

$$[\hat{Q}_1, \hat{Q}_2] = \hat{Q}_1, \quad [\hat{Q}_1, \hat{Q}_3] = I - 2\hat{Q}_2, \quad [\hat{Q}_2, \hat{Q}_3] = \hat{Q}_3. \tag{41}$$

And, introducing the new definitions:

$$\hat{A}_0 = \hat{Q}_2 - 1/2, \quad \hat{A}_- = \hat{Q}_1, \quad \hat{A}_+ = \hat{Q}_3, \tag{42}$$

the following commutation relations are obtained:

$$[\hat{A}_0, \hat{A}_\pm] = \pm \hat{A}_\pm, \quad [\hat{A}_+, \hat{A}_-] = -2\hat{A}_0. \tag{43}$$

3.2.1. ***Matching examples 1 and 2:*** From the equation (33) and (34) we have the general solution of EDO (32)

$$y = C_1 \cos(kx) + \frac{C_2}{k} \sin(kx). \tag{44}$$

Taking the limit $k \to 0$ on equation (32), we obtain equation (21). In same way, this limit reduces solution (44) to the solution of equation (21).

$$y = C_1 + C_2 x. \tag{45}$$

On the other hand, taking the limit $k \to 0$ on symmetry generators(37), (38) and (39), we obtain

$$\begin{aligned}
\lim_{k \to 0} (\hat{Q}_1) &= \lim_{k \to 0} \left(\frac{k}{2} \sin(2kx) + \frac{1}{2}(1 + \cos(2kx)) \frac{d}{dx} \right) = \frac{d}{dx}, \\
\lim_{k \to 0} (\hat{Q}_2) &= \lim_{k \to 0} \left(\frac{1}{2} - \frac{1}{2} \cos(2kx) + \frac{1}{2k} \sin(2kx) \frac{d}{dx} \right) = x \frac{d}{dx}, \\
\lim_{k \to 0} (\hat{Q}_3) &= \lim_{k \to 0} \left(\frac{1}{2k} \sin(2kx) + \frac{1}{2k^2}(\cos(2kx) - 1) \frac{d}{dx} \right) = x - x^2 \frac{d}{dx}.
\end{aligned}$$

Showing the compatibility between the symmetry operators obtained in example 1 and 2 respectively. It is noteworthy that commutations relations (28) remain invariant under the limit $k \to 0$.

Journal of Physics: Conference Series **128** (2008) 012057 doi:10.1088/1742-6596/128/1/012057

4. Conclusion

A general method to find continuous symmetries of second-order linear ODEs has been presented. To obtain the symmetry generators a particular solution of the ODE under study is required. The method has been illustrated by two examples.

Acknowledgments

The authors thank the organizers of 5th International Symposium on Quantum Theory and Symmetries, (Valladolid, Spain), for their warm hospitality. A.R.M. thanks support from FONDECYT (Chile) under Grant No 1050536. Also C.A.S. thanks support from MECESUP (Chile) program under Grant No UCO-0209.

References

[1] Morse P M and Feshbach H 1953 *Methods of Theoretical Physics. Part I and II* (New York: McGraw-Hill)
[2] Ibragimov N H 1999 *Elementary Lie Group Analysis and Ordinary Differential Equations* (New York: John Wiley and Sons)
[3] Hydon P E 2000 *Symmetry Methods for Differential Equations* (New York: Cambridge University Press)
[4] Cantwell B J 2002 *Introduction to Symmetry Analysis* (New York: Cambridge University Press)
[5] Anderson R L, Kumei S, and Wulfman C E 1972 *Rev. Mex. Fis.* **21** 1-33
[6] Anderson R L, Kumei S, and Wulfman C E 1972 *Rev. Mex. Fis.* **21** 35-57
[7] Anderson R L, Kumei S, and Wulfman C E 1972 *Phys. Rev. Lett.* **28** 988-991
[8] Anderson R L, Kumei S, and Wulfman C E 1973 *J. Math. Phys.* **14** 1527-1531

Journal of Physics: Conference Series **128** (2008) 012058 doi:10.1088/1742-6596/128/1/012058

Application of Lie point symmetry and Adomain decomposition techniques to thermal-storage nonlinear diffusion models

Raseelo Joel Moitsheki[1] **and Oluwelo Daniel Makinde** [2]

[1]Department of Mathematics, Vaal University of Technology, Private bag X021, Vanderbijlpark 1900

[2] Department of Applied Mathematics, University of Limpopo Turfloop campus, Private Bag X1106, Sovenga 0727, South Africa

E-mail: joelm@vut.ac.za

E-mail: makindeo@ul.ac.za

Abstract. Classical Lie point symmetry techniques are employed to time dependent nonlinear heat diffusion equations describing thermal energy storage in a medium subjected to a convective heat transfer to the surrounding environment at the boundary through a variable heat transfer coefficient. Exponential temperature-dependent thermal conductivity and heat capacity are assumed. Group classification for the source term is performed and some exciting large symmetry algebras are admitted. It turns out that the principal Lie algebra extends when the source term vanishes and when it is given as the exponential function of temperature. Reduction by one of the independent variables is performed for some realistic choices of the source term. In some case the resulting nonlinear ordinary differential equation with appropriate corresponding conditions are solved using Adomian decomposition method.

1. Introduction

The operation of hot water storage tanks for thermal energy storage is classified as static mode and dynamic mode [7, 12, 8]. The static mode is further classified into fully charged and partially charged mode. Dynamic mode of operation includes charging and discharging cycle. In static fully charged mode, the storage tank is initially completely filled with hot water at a constant temperature and is subjected to convective heat loss from the tank walls to the ambient [18]. In static partially charged mode, the storage tank is initially charged with different levels of hot and cold water separated by thermocline [12]. The heat loss from the stored fluid to the ambient decreases the temperature of the fluid near to the tank wall, thereby increasing its density. Therefore, accurate and efficient design of solar heating or cooling thermal energy storage tanks generally requires adequate theoretical framework from which insight might be gleaned into an inherently complex physical process [18]. The solution of the transient nonlinear heat conduction for thermal energy storage applicable in rectangular, cylindrical and spherical coordinates remains an extremely important problem of practical relevance in the engineering sciences [6, 3]. The mathematical techniques employed in the present investigation include the parameter-group transformation [16] and Adomain decomposition method [2]. The foundation of the group-theoretical method is contained in the general theory of continuous transformation

Journal of Physics: Conference Series **128** (2008) 012058 doi:10.1088/1742-6596/128/1/012058

groups that were introduced and treated extensively by Lie (see e.g. [10]). This paper is arranged as follows, Section 2 provides a mathematical formulation of the problem, In Section 3 a brief theory of symmetry analysis of differential equations is discussed. Solutions to the problem is discussed in section 4, and Section 5 and 6 deals with results and conclusions respectively.

2. Mathematical models

The transient heat conduction equation with heat source term modelling the thermal storage problem in a rectangular, cylindrical or spherical coordinate system is given as [12, 3];

$$\rho C(T)\frac{\partial T}{\partial t} = \frac{1}{r^m}\frac{\partial}{\partial r}\left(K(T)r^m\frac{\partial T}{\partial r}\right) + S(T), \tag{1}$$

with the initial condition

$$T(r,t) = T_0 \quad \text{at} \quad t = 0, \tag{2}$$

and the boundary condition

$$\left.\begin{aligned} &\frac{\partial T}{\partial r} = 0; \quad \text{at} \quad r = 0 \\ &K(T)\frac{\partial T}{\partial r} = -h(t)\left(T - T_\infty\right); \quad \text{at} \quad r = a \end{aligned}\right\} \tag{3}$$

where T is the temperature, ρ is the density, r is the space variable, t is time, T_0 is the initial temperature of the body and $S(T)$ is the temperature-dependent heat source term. ρ is the density and $h(t) = h_0 f(t)$ is the time-dependent heat transfer coefficient. Following [3], the exponential temperature-dependent thermal conductivity and heat capacity are taken as

$$K(T) = K_0 \exp\left\{p\left(\frac{T - T_\infty}{T_0 - T_\infty}\right)\right\} \quad \text{and} \quad C(T) = C_0 \exp\left\{q\left(\frac{T - T_\infty}{T_0 - T_\infty}\right)\right\},$$

where K_0, C_0, p and q are constants, T_0 is the initial temperature of the body, T_∞ is the ambient temperature of the surrounding environment. The geometry of the body is specified by $m = 0, 1, 2$ representing rectangular, cylindrical and spherical coordinates respectively. Equations (1-2) are made dimensionless by introducing the following quantities:

$$\bar{r} = \frac{r}{a}, \quad \bar{t} = \frac{K_0 t}{\rho C_0 a^2}, \quad \bar{T} = \frac{T - T_\infty}{T_0 - T_\infty}, \quad \bar{S} = \frac{a^2 S}{K_0(T_0 - T_\infty)}, \quad Bi = \frac{a h_0}{K_0}$$

Neglecting the bar symbol for clarity, the dimensionless governing equations become

$$e^{pT}\frac{\partial T}{\partial t} = \frac{1}{r^m}\frac{\partial}{\partial r}\left(e^{qT}r^m\frac{\partial T}{\partial r}\right) + S(T). \tag{4}$$

$$\left.\begin{aligned} &\frac{\partial T}{\partial r} = 0; \quad \text{at} \quad r = 0 \\ &e^{qT}\frac{\partial T}{\partial r} = -Bif(t)T; \quad \text{at} \quad r = 1 \end{aligned}\right\} \tag{5}$$

Journal of Physics: Conference Series **128** (2008) 012058 doi:10.1088/1742-6596/128/1/012058

3. Symmetry techniques for differential equations

The theory and applications of continuous symmetry groups were founded by Lie in the 17th century [13]. The modern accounts of this theory may be found in excellent text such as those of [4, 14]. We restrict our discussion to classical point symmetries, since we will only use such symmetries. The reader is referred to [4, 10, 14, 16] for more details on this theory. Given a continuous one parameter symmetry group, it is possible to reduce the number of independent variables by one. Lie's fundamental result is that the whole of one parameter group can be determined from the transformation laws up to the first degree of the parameter ϵ, i.e. determination of symmetry groups involve finding transformation of the form

$$\begin{aligned}\overline{x}_j &= x_j + \epsilon X_j(\mathbf{x}, u) + \mathbf{O}(\epsilon^2)\\ \overline{u} &= u + \epsilon U(\mathbf{x}, u) + \mathbf{O}(\epsilon^2)\end{aligned}$$

that leave the mth order governing partial differential equation,

$$\Delta(\mathbf{x}, u, u^{(1)}, ..., u^{(m)}) = 0,$$

invariant. Here, $\mathbf{x}$ is set of n independent variables $(x_1, x_2, ..., x_n)$ and $u^{(k)}$ denotes a set of coordinates corresponding to all the kth order partial derivatives of u with respect to $x_1, x_2, ..., x_n$. The coefficients X_j and U are the components of the infinitesimal symmetry generator, which is one of the vector fields

$$\Gamma = X_j(\mathbf{x}, u)\frac{\partial}{\partial x_j} + U(\mathbf{x}, u)\frac{\partial}{\partial u},$$

which span the associated Lie algebra. Here we sum over repeated index (see e.g. [4]). Calculations of these generators or symmetry groups are very long and tedious. However, we use the freely available program Dimsym [17], which is written as a subprogram for the computer algebra package Reduce [9] to construct the admitted symmetries.

If a differential equation is invariant under some point symmetry, one can often construct similarity solutions which are invariant under some subgroup of the full group admitted by the equation in question. These solutions result form solving a reduced equation in fewer variables.

More often, differential equations arising in real world problem involve one or more functions depending on either the independent variable or on the dependent variables. It is possible by symmetry techniques to determine the cases which allow the equation in question to admit extra symmetries. The exercise of searching for the forms of arbitrary functions that extend the principal Lie algebra is called *group classification*. The problem of group classification was introduced by Ovsiannikov [15] and recent accounts on this topic may be found for example in [5, 11, 19, 21]. We adopt methods in [5] to perform group classification.

4. Solution to the problem

Firstly, we reduce the boundary value problem (BVP) (4)-(5) to a system of ordinary differential equations using the Lie group methods. This will be followed by the application of Adomian decomposition method [1, 2] in order to obtain a semi-analytical non-perturbative approximate solution to the problem. Instead of determining symmetries for the entire boundary value problem, we consider the governing equation (4). It is well known (e.g. [5]) that symmetry analysis of the entire system such as (4)-(5) may result in reduced admitted group of transformations.

Classical Lie symmetry reductions

In the initial symmetry analysis of equation (4) where all the constants are arbitrary, we obtain nothing beyond translation in time t. It turns out using Dimsym [17] that $p = q$. Adopting techniques in [5] we observe that extra Lie point symmetries are admitted for the cases of the source term listed below;

Case (a) S=0. In this case Equation (4) admits a finite four dimensional Lie algebra

$$\left.\begin{aligned} \Gamma_1 &= -\frac{1}{4q}\left(r^2 - 2(m+1)t\right)\frac{\partial}{\partial T} + rt\frac{\partial}{\partial r} + t^2\frac{\partial}{\partial t}; \\ \Gamma_2 &= r\frac{\partial}{\partial r} + 2t\frac{\partial}{\partial t}; \quad \Gamma_3 = \frac{\partial}{\partial t}; \quad \Gamma_4 = \frac{\partial}{\partial T} \end{aligned}\right\} \tag{6}$$

and the infinite symmetry generator

$$\Gamma_\infty = \frac{H(r,t)}{qe^{qT}}\frac{\partial}{\partial T},$$

where H is a function of r and t satisfies the equation

$$\frac{\partial^2 H}{\partial r^2} + \frac{m}{r}\frac{\partial H}{\partial r} = \frac{\partial H}{\partial t}.$$

Case (b) S=0, $m = 2$. Other than the infinite symmetry generator, Equation (4) admits a six dimensional finite Lie algebra for this case, namely;

$$\left.\begin{aligned} \Gamma_1 &= -\frac{1}{4q}\left(r^2 - 2(m+1)t\right)\frac{\partial}{\partial T} + rt\frac{\partial}{\partial r} + t^2\frac{\partial}{\partial t}; \\ \Gamma_2 &= r\frac{\partial}{\partial r} + 2t\frac{\partial}{\partial t}; \quad \Gamma_3 = \frac{\partial}{\partial t}; \quad \Gamma_4 = \frac{\partial}{\partial T}; \quad \Gamma_5 = -\frac{1}{qr}\frac{\partial}{\partial T} + \frac{\partial}{\partial r}; \\ \Gamma_6 &= -\frac{1}{2qr}\left(r^2 + 2t\right)\frac{\partial}{\partial T} + t\frac{\partial}{\partial r} \end{aligned}\right\} \tag{7}$$

and

$$\Gamma_\infty = \frac{H(r,t)}{qe^{qT}}\frac{\partial}{\partial T};$$

where H satisfies

$$\frac{\partial^2 H}{\partial r^2} + \frac{2}{r}\frac{\partial H}{\partial r} = \frac{\partial H}{\partial t}.$$

Case (c) S=0, $m = 0$. Similar to case (b), the governing equation admits a six dimensional Lie algebra,

$$\left.\begin{aligned} \Gamma_1 &= -\frac{1}{4q}\left(r^2 - 2t\right)\frac{\partial}{\partial T} + rt\frac{\partial}{\partial r} + t^2\frac{\partial}{\partial t}; \\ \Gamma_2 &= r\frac{\partial}{\partial r} + 2t\frac{\partial}{\partial t}; \quad \Gamma_3 = \frac{\partial}{\partial t}; \quad \Gamma_4 = \frac{\partial}{\partial T}; \quad \Gamma_5 = \frac{\partial}{\partial r}; \\ \Gamma_6 &= -\frac{r}{2qr}\frac{\partial}{\partial T} + t\frac{\partial}{\partial r} \end{aligned}\right\} \tag{8}$$

Journal of Physics: Conference Series **128** (2008) 012058 doi:10.1088/1742-6596/128/1/012058

and

$$\Gamma_\infty = \frac{H(r,t)}{qe^{qT}}\frac{\partial}{\partial T};$$

where H satisfies

$$\frac{\partial^2 H}{\partial r^2} = \frac{\partial H}{\partial t}.$$

Case (d) $S = Ae^{kT}$.

$$\left.\begin{aligned} \Gamma_1 &= -\frac{1}{(k-q)}\frac{\partial}{\partial T} + \frac{r}{2}\frac{\partial}{\partial r} + t\frac{\partial}{\partial t}; \\ \Gamma_2 &= \frac{\partial}{\partial t} \end{aligned}\right\} \tag{9}$$

Case (e) $S = Ae^{kT}, \quad k = q$. In this case the admitted finite algebra is four dimensional;

$$\left.\begin{aligned} \Gamma_1 &= \frac{1}{4q}\left(4qAt^2 - r^2 - 2(m+1)t\right)\frac{\partial}{\partial T} + rt\frac{\partial}{\partial r} + t^2\frac{\partial}{\partial t}; \\ \Gamma_2 &= 2At\frac{\partial}{\partial T} + r\frac{\partial}{\partial r} + 2t\frac{\partial}{\partial t}; \quad \Gamma_3 = \frac{\partial}{\partial t}; \quad \Gamma_4 = \frac{\partial}{\partial T}; \end{aligned}\right\} \tag{10}$$

and

$$\Gamma_\infty = \frac{H(r,t)}{qe^{qT}}\frac{\partial}{\partial T};$$

where H satisfies

$$AqH + \frac{\partial^2 H}{\partial r^2} + \frac{m}{r}\frac{\partial H}{\partial r} = \frac{\partial H}{\partial t}.$$

Case (f) $S = Ae^{kT}, \quad k = q, \quad m = 2$.

$$\left.\begin{aligned} \Gamma_1 &= \frac{1}{4q}\left(4qAt^2 - r^2 - 6t\right)\frac{\partial}{\partial T} + rt\frac{\partial}{\partial r} + t^2\frac{\partial}{\partial t}; \\ \Gamma_2 &= 2At\frac{\partial}{\partial T} + r\frac{\partial}{\partial r} + 2t\frac{\partial}{\partial t}; \quad \Gamma_3 = \frac{\partial}{\partial t}; \quad \Gamma_4 = \frac{\partial}{\partial T}; \\ \Gamma_5 &= -\frac{1}{qr}\frac{\partial}{\partial T} + \frac{\partial}{\partial r}; \quad \Gamma_6 = -\frac{1}{2qr}\left(r^2 + 2t\right)\frac{\partial}{\partial T} + t\frac{\partial}{\partial r}; \end{aligned}\right\} \tag{11}$$

and

$$\Gamma_\infty = \frac{H(r,t)}{qe^{qT}}\frac{\partial}{\partial T};$$

where H satisfies

$$AqH + \frac{\partial^2 H}{\partial r^2} + \frac{2}{r}\frac{\partial H}{\partial r} = \frac{\partial H}{\partial t}.$$

Case (f) $S = Ae^{kT}, \quad k = q, \quad m = 0.$

$$\left.\begin{array}{l}\Gamma_1 = \dfrac{1}{4q}\left(4qAt^2 - r^2 - 2t\right)\dfrac{\partial}{\partial T} + rt\dfrac{\partial}{\partial r} + t^2\dfrac{\partial}{\partial t};\\ \\ \Gamma_2 = 2At\dfrac{\partial}{\partial T} + r\dfrac{\partial}{\partial r} + 2t\dfrac{\partial}{\partial t}; \quad \Gamma_3 = \dfrac{\partial}{\partial t}; \quad \Gamma_4 = \dfrac{\partial}{\partial T};\\ \\ \Gamma_5 = \dfrac{\partial}{\partial r}; \quad \Gamma_6 = -\dfrac{r}{2q}\dfrac{\partial}{\partial T} + t\dfrac{\partial}{\partial r};\end{array}\right\} \qquad (12)$$

and

$$\Gamma_\infty = \frac{H(r,t)}{qe^{qT}}\frac{\partial}{\partial T};$$

where H satisfies

$$AqH + \frac{\partial^2 H}{\partial r^2} = \frac{\partial H}{\partial t}.$$

Variable reductions
In most cases, one may reduce the number of independent variables of the partial differential equation by one, using the symmetries admitted by such a PDE. In some cases it is also possible to reduce or transform the imposed boundary conditions. We concentrate on the following examples;

Example 1(a)
In case (d) where S is also given by an exponential, we observe that the Γ_1- invariant solution is of the functional form

$$T = -\frac{1}{k-q}\ln t + G(\gamma)$$

where $\gamma = \frac{r}{\sqrt{t}}$ is the invariant or the Boltzman similarity variable, and G satisfies the O.D.E.

$$G'' + \frac{m}{\gamma}G' + q(G')^2 + \frac{\gamma}{2}G' + \left(A + \frac{1}{k-q}\right) = 0$$

It appears the boundary conditions are impossible to transform except Equation (5(a)) becomes $G' = 0, \quad \gamma = 0.$

Example 1(b)
In case (d), the operator Γ_2 suggests that both T and r are invariants. In fact the functional form of the invariant solution is simply $T = G(r)$ (see e.g. [20]), where G satisfies the O.D.E.

$$\frac{1}{r^m}\frac{d}{dr}\left(e^{qG}r^m\frac{dG}{dr}\right) + Ae^{kG} = 0,$$

and the boundary conditions transforms to

$$\frac{dG}{dr} = 0, \quad r = 0 \qquad \text{and} \qquad G = 0, \quad r = 1 \quad (\text{i.e. as } Bi \to \infty).$$

Example 2
In this example we consider the Γ_2 reduction in case (a) $S = 0$. Γ_2 leads to the solution $T = G(\gamma)$ where $\gamma = \frac{r}{\sqrt{t}}$ and G satisfies the O.D.E.

$$-\frac{1}{2}\gamma G' = q(G')^2 + \frac{m}{\gamma}G' + G''.$$

The initial condition (2) reduces to $G \to G_a$ as $\gamma \to \infty$, with G_a being a constant and the boundary conditions become

$$\frac{dG}{d\gamma} = 0, \quad \gamma = 0.$$

The boundary condition (5b) has proven to be difficult to transform in terms of the obtained invariant γ. One may note that the invariant (similarity) solution to the above reduced O.D.E. subject to the transformed conditions is merely trivial i.e. $G =$ constant.

Numerical methods

In this section, Adomian decomposition technique is employed in order to perform a numerical experiment on the above transformed nonlinear ordinary differential equations. The advantage of this method is that it provides a direct scheme for solving the problem, i.e., without the need for linearization, perturbation or massive computation. For illustration purpose, we consider the transformed nonlinear ordinary differential equation in example 1(b) above depicting the energy storage problem subject to exponentially increasing heat source with very large Biot number;

$$\frac{1}{r^m}\frac{d}{dr}\left(r^m\frac{dG}{dr}\right) + q\left(\frac{dG}{dr}\right)^2 + A\mathrm{e}^{(k-q)G} = 0, \tag{13}$$

with

$$\frac{dG}{dr} = 0, \quad r = 0 \qquad \text{and} \qquad G = 0, \quad r = 1, \quad (\text{i.e. as } \; Bi \to \infty) \tag{14}$$

Following [7], we rewrite Equation (13) with respect to Equation (14) in the form

$$L_r G = -q\,(G_r)^2 - A\mathrm{e}^{(k-q)G}, \tag{15}$$

where the subscript r represents derivatives with respect to r and the differential operator employs the first two derivatives in the form

$$L_r = \frac{1}{r^m}\frac{d}{dr}\left(r^m\frac{d}{dr}\right), \tag{16}$$

in order to overcome the singularity behaviour at the point $r = 0$. In view of Equation (16), the inverse operator L_r^{-1} is considered a twofold integral operator defined by

$$L_r^{-1} = \int_0^r r^{-m}\int_0^r r^m(.)drdr. \tag{17}$$

Applying L_r^{-1} to both sides of Equation (15), using the boundary conditions in Equation (14), we obtain

$$G(r) = G(0) - qL_r^{-1}\left(G_r^2\right) - AL_r^{-1}\left(\mathrm{e}^{(k-q)G}\right). \tag{18}$$

As usual in Adomian decomposition method the solution of Equation (18) is approximated as an infinite series

$$G(r) = \sum_{j=0}^{\infty} G_j, \tag{19}$$

and the nonlinear terms are decomposed as follows:

$$G_r^2 = \sum_{j=0}^{\infty} H_j, \quad \mathrm{e}^{(k-q)G} = \sum_{j=0}^{\infty} M_j, \tag{20}$$

Journal of Physics: Conference Series **128** (2008) 012058 doi:10.1088/1742-6596/128/1/012058

where H_j and M_j are polynomials (called Adomian polynomials) given by

$$H_j = \frac{1}{j!}\frac{d^j}{d\lambda^j}\left[\left(\sum_{i=0}^{\infty} G_{ri}\lambda^i\right)^2\right]_{\lambda=0}, \quad M_j = \frac{1}{j!}\frac{d^j}{d\lambda^j}\left[\left(\mathrm{e}^{(k-q)\left(\sum_{i=0}^{\infty} G_{ri}\lambda^i\right)}\right)\right]_{\lambda=0}. \tag{21}$$

Thus, we can identify

$$\left.\begin{aligned} &G_0 = G(0), \\ &G_{j+1} = -qL_r^{-1}(H_j) - AL_r^{-1}(M_j), \quad \text{for } j \geq 0. \end{aligned}\right\} \tag{22}$$

Substituting Equations (20)-(22) into Equation (19), and using MAPLE we obtained a few terms approximation to the solution as

$$\Psi_N = \sum_{n=0}^{N} G_n, \tag{23}$$

where $G(r) = \lim_{N\to\infty} (\Psi_N)$. Using the above procedure, we obtain

$$\left.\begin{aligned} G(r) := G_0 &- \frac{1}{2}\frac{A\mathrm{e}^{kG_0}r^2}{(m+1)\mathrm{e}^{qG_0}} + \frac{1}{8}\frac{A^2(k-q)r^4\mathrm{e}^{2(k-q)G_0}}{(m+1)(m+3)} \\ &- \frac{1}{4}\frac{qA^2\left(\mathrm{e}^{kG_0}\right)^2 r^4}{\left(\mathrm{e}^{qG_0}\right)^2(m+1)(m^2+4m+3)} \\ &- \frac{1}{24}\frac{A^3(mk^2-2kmq+mq^2+2k^2-4kq+2q^2)\left(\mathrm{e}^{kG_0}\right)^3 r^6}{(m+1)^2(m^2+8m+15)\left(\mathrm{e}^{qG_0}\right)^3} + \ldots \end{aligned}\right\} \tag{24}$$

By applying the boundary conditions in Equation (14) to the expression for in Equation (24), we obtain approximately the values for G_0 as shown graphically in the following section. Usually, the decomposition method yields rapidly convergent series solutions by using a few iterations for the nonlinear deterministic equations [2].

5. Results and discussion

The approximate solution given in Equation (24) is valid for energy storage systems in an interval $0 \leq r \leq 1$ under exponentially increasing heat source. For the numerical validation of our results we have chosen physically meaningful values of the parameters for the problem. Unless otherwise stated we have taken: $k = q = 1; Bi \to \infty$. In Figures 1 and 2, we depict the effect of material parameters on the medium temperature profiles. For all different geometrical regions $(m = 0, 1, 2)$, we observe that the temperature profiles decrease transversely with maximum value along the central region and minimum value at the surface as shown in Figure 2. The minimum value at the surface temperature can be attributed to the convective cooling at the region. However, it is noteworthy that the medium temperature is highest with rectangular configuration, followed by that of cylindrical geometry and the lowest is observed with spherical geometry. Figure 2 illustrates the effects of increasing values of heat source parameter (A) on the temperature profile. It is interesting to note that the medium temperature increases as the heat source parameter increases due to increasing heat generation as expected.

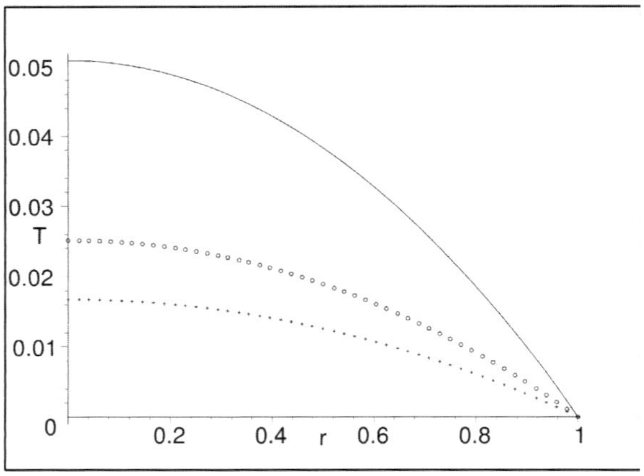

Figure 1. Medium temperature profile as $Bi \to \infty$, $k = q = 1$, $A = 0.1; - - - - - m = 0$; ooooo$m = 1$; $+ + + + + + m = 2$

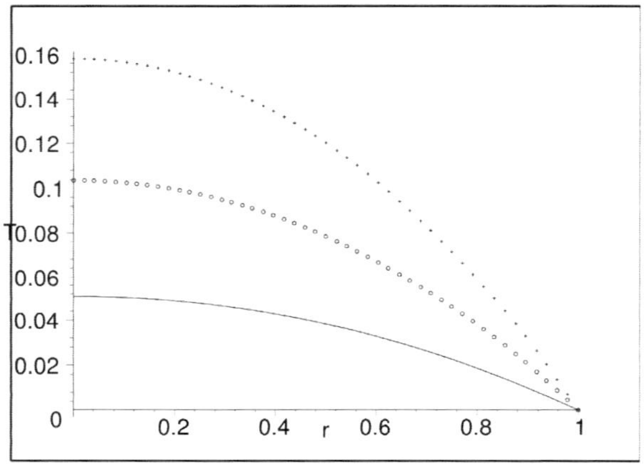

Figure 2. Medium temperature profile as $Bi \to \infty$, $k = q = 1$, $m = 0; - - - - - A = 0.1$; ooooo$A = 0.2$; $+ + + + + + A = 0.3$

6. Conclusion

Classical Lie point symmetry analysis resulted in some large symmetries being admitted for special cases of the source term. It is possible to reduce the governing P.D.E. to O.D.E.s using any linear combination of the admitted symmetries. In fact one may determine the optimal systems of sub algebras (symmetries) obtained in each case, to find reductions that are nonequivalent, and so one may classify group invariant solutions (see e.g. [14]). In Section 4, reductions were performed for some examples. Not surprisingly, the boundary condition seemed difficult to be transformed in terms of the new variables for many examples. However, in Example 1(b) the problem resulted in an exiting transformed system of O.D.E.s which was completely solved using Adomain decomposition techniques.

Acknowledgements

The authors wish to thank the National Research Foundation of South Africa under Thuthuka program for the financial support. RJM is grateful to the University of Valladolid for the grant awarded in support of conference (QTS5) participation and to the Vaal University of Technology for the continued generous financial support.

7. References

[1] Abboui K and Cherruault Y 1995 New ideas for proving convergence of decomposition methods, *Comput. Appl. Math.* **29** (7), 103-105.
[2] Adomian G 1994 Solving frontier problems of physics: The Decomposition method, Kluwer Academic Publishers, Dordecht.
[3] Badran N A and Abd-el-Malek M B 1995 Group analysis of nonlinear heat-conduction problem for a semi-infinite body, *Nonlinear Math. Phys.* **2**(3-4), 319-328.
[4] Bluman G W and Anco S C 2002 Symmetry and integration methods for differential equations, New York: Springer-Verlag.
[5] Bluman G W and Kumei S 1989 Symmetry and differential equations, Springer-Verlag, New York.
[6] Davies T W 1985 Transient conduction in a plate with counteracting convection and thermal radiation at the boundary, *Appl. Math. Model.* **9**, 337-340.
[7] Duffie J A and Becham W A 1980 Solar Engineering of thermal processes, 485, Wiley.
[8] Hawlader M N A and Brinkworth B J 1981 An analysis of the non-convective solar pond, Solar Energy **27** (3), 195.
[9] Hearn A C 1985 Reduce user's manual version 3.4, Santa Monica, California: Rand Publication CP78, The Rand Corporation.
[10] Ibragimov N H 1999 Elementary Lie group analysis and ordinary differential equations, John Wiley & sons, New York.
[11] Ivanova N M and Sophocleous C 2006 On the group classification of variable nonlinear diffusion-convection equations, *J. Comput. Appl. Math.*, **197**(2), 322-344.
[12] Jahiria Y and Gupta S K 1982 Decay of thermal stratification in a water body for solar energy storage, Solar Energy **28** (2) 137.
[13] Lie S 1880 Theorie der Transformationgruppen, *Math. Ann.*, **16**, 441-528.
[14] Olver P J 1972 Applications of Lie groups to differential equations, New York: Springer-Verlag.
[15] Ovsiannikov L V 1959 Group relations of the equation of non-linear heat conductivity, *Dokl. Akad. Nauk SSSR.* **125**, 492-495.
[16] Ovsiannikov L V 1982 Group analysis of differential equations, New York: Academic Press.
[17] Sherring J 1993 Dimsym: symmetry determination and linear differential equation package, Latrobe University Mathematics Dept., Melbourne.
[18] Shin M S, Kim H S, Jang D S, Lee S N and Yoon H G 2004 Numerical and experimental study on the design of a stratified thermal storage system, Applied Thermal Engineering, **24**, 17-27.
[19] Sophocleous C 2003 Symmetries and form-preserving transformation of generalised inhomogeneous nonlinear diffusion equations *Physica A.*, **324**, 509-529.
[20] Stephani H 1989 Differential equations, their solution using symmetries, Cammbridge University Press, New York.
[21] Vaneeva O O, Johnpillai A G, Popovych R O and Sophocleous C 2007 Enhanced group analysis and conservation laws of variable coefficient reaction-diffusion equations with power nonlinearities, *J. Math. Anal. Appl.* **330**(2): 1363-1386.

Journal of Physics: Conference Series **128** (2008) 012059 doi:10.1088/1742-6596/128/1/012059

Quadratic algebras and position-dependent mass Schrödinger equations

Christiane Quesne

Physique Nucléaire Théorique et Physique Mathématique, Université Libre de Bruxelles, Campus de la Plaine CP229, Boulevard du Triomphe, B-1050 Brussels, Belgium

E-mail: cquesne@ulb.ac.be

Abstract. During recent years, exact solutions of position-dependent mass Schrödinger equations have inspired intense research activities, based on the use of point canonical transformations, Lie algebraic methods or supersymmetric quantum mechanical techniques. Here we highlight the interest of another approach to such problems, relying on quadratic algebras. We illustrate this point by constructing spectrum generating algebras for a class of d-dimensional radial harmonic oscillators with $d \geq 2$ (including the one-dimensional oscillator on the line via some minor changes) and a specific mass choice. This provides us with a counterpart of the well-known su(1,1) Lie algebraic approach to the constant-mass oscillators.

1. Introduction

During recent years, quantum mechanical systems with a position-dependent (effective) mass (PDM) have attracted a lot of attention and inspired intense research activites (see [1] for a list of references). They are indeed very useful in the study of many physical systems, such as electronic properties of semiconductors and quantum dots, nuclei, quantum liquids, ^{3}He clusters, metal clusters, etc. Furthermore, the PDM presence in quantum mechanical problems may reflect some other unconventional effects, such as a deformation of the canonical commutation relations or a curvature of the underlying space. Hermitian PDM Hamiltonians may also be equivalent to some $\mathcal{PT}$-symmetric systems with constant mass at lowest order of perturbation theory.

To find exact solutions for some PDM Schrödinger equations, use has been made of point canonical transformations, Lie algebras or supersymmetric quantum mechanical techniques by extending methods known for constant mass. Very recently, nonlinear algebras, especially quadratic ones, have started to be employed. The presence of one of them has been signalled in a one-dimensional PDM problem [2]. Then the quadratic algebra generated by the integrals of motion of a two-dimensional superintegrable PDM system has been studied [3]. Finally, a quadratic algebra approach has allowed the construction of spectrum generating algebras for a class of d-dimensional radial harmonic oscillators with specific PDM. In the present contribution, we briefly summarize the results of the last work (for more details, see [1]).

2. Constant-mass d-dimensional radial harmonic oscillator

In units wherein $\hbar = 1$ and the mass $m_0 = 1/2$, the radial Schrödinger equation for the d-dimensional harmonic oscillator ($d \geq 2$) can be written as

$$\left(-\frac{d^2}{dr^2} + \frac{L(L+1)}{r^2} + \frac{1}{4}\omega^2 r^2\right)\psi(r) = E\psi(r), \tag{1}$$

where r runs on the half-line $0 < r < \infty$, L is defined by $L = l + (d-3)/2$ in terms of the angular momentum quantum number l and the radial wavefunction is actually $r^{-(d-1)/2}\psi(r)$.

Equation (1) has an infinite number of bound-state solutions

$$\psi_{n,L}(r) = \mathcal{N}_{n,L} r^{L+1} L_n^{(L+\frac{1}{2})}(\tfrac{1}{2}\omega r^2) e^{-\frac{1}{4}\omega r^2}, \qquad n = 0, 1, 2, \ldots, \tag{2}$$

corresponding to the energy eigenvalues

$$E_{n,L} = \omega(2n + L + \tfrac{3}{2}). \tag{3}$$

In (2), $L_n^{(\alpha)}(y)$ denotes a Laguerre polynomial and $\mathcal{N}_{n,L}$ is a normalization coefficient.

All the wavefunctions (2), corresponding to a given value of L and $n = 0, 1, 2, \ldots$, belong to a single positive-discrete series unitary irreducible representation D_k^+ of an su(1,1) Lie algebra. The latter is generated by the operators

$$\begin{aligned} K_0 &= \frac{1}{2\omega}\left(-\frac{d^2}{dr^2} + \frac{L(L+1)}{r^2} + \frac{1}{4}\omega^2 r^2\right), \\ K_\pm &= \frac{1}{2\omega}\left[\frac{d^2}{dr^2} - \frac{L(L+1)}{r^2} + \frac{1}{4}\omega^2 r^2 \mp \omega\left(r\frac{d}{dr} + \frac{1}{2}\right)\right], \end{aligned} \tag{4}$$

satisfying the commutation relations $[K_0, K_\pm] = \pm K_\pm$, $[K_+, K_-] = -2K_0$, and the Hermiticity properties $K_0^\dagger = K_0$, $K_\pm^\dagger = K_\mp$, while its Casimir operator reads $C = -K_+K_- + K_0(K_0 - 1)$. The lowest weight characterizing the irreducible representation is here $k = \frac{1}{2}\left(L + \frac{3}{2}\right)$.

The wavefunctions $\psi_{n,L}(r)$, $n = 0, 1, 2, \ldots$, are simultaneous eigenfunctions of C and K_0,

$$\begin{aligned} C\psi_{n,L}(r) &= k(k-1)\psi_{n,L}(r) = \tfrac{1}{4}\left(L + \tfrac{3}{2}\right)\left(L - \tfrac{1}{2}\right)\psi_{n,L}(r), \\ K_0\psi_{n,L}(r) &= \mu\psi_{n,L}(r) = (k+n)\psi_{n,L}(r) = \frac{1}{2\omega}E_{n,L}\psi_{n,L}(r). \end{aligned} \tag{5}$$

Furthermore, K_+ and K_- act on them as

$$\begin{aligned} K_+\psi_{n,L}(r) &= [(\mu - k + 1)(\mu + k)]^{1/2}\psi_{n+1,L}(r) = \left[(n+1)\left(n + L + \tfrac{3}{2}\right)\right]^{1/2}\psi_{n+1,L}(r), \\ K_-\psi_{n,L}(r) &= [(\mu - k)(\mu + k - 1)]^{1/2}\psi_{n-1,L}(r) = \left[n\left(n + L + \tfrac{1}{2}\right)\right]^{1/2}\psi_{n-1,L}(r). \end{aligned} \tag{6}$$

This shows that su(1,1) is a spectrum generating algebra for the d-dimensional radial harmonic oscillator.

3. PDM d-dimensional radial harmonic oscillator

Let us now consider a PDM d-dimensional harmonic oscillator, whose radial Schrödinger equation is obtained by replacing in (1) the radial momentum $p_r = -\mathrm{i}d/dr$ by some deformed one, $\pi_r = \sqrt{f(\alpha; r)}\, p_r \sqrt{f(\alpha; r)}$, where $f(\alpha; r) = 1 + \alpha r^2$ and α is a positive real constant. The result of this substitution reads

$$\left(\pi_r^2 + \frac{L(L+1)}{r^2} + \frac{1}{4}\omega^2 r^2\right)\psi^{(\alpha)}(r) = E^{(\alpha)}\psi^{(\alpha)}(r), \tag{7}$$

which is equivalent to

$$\left(-\frac{d}{dr}\frac{1}{M(\alpha;r)}\frac{d}{dr} + V_{\text{eff}}(\alpha;r)\right)\psi^{(\alpha)}(r) = E^{(\alpha)}\psi^{(\alpha)}(r), \tag{8}$$

where $M(\alpha;r)$ and $V_{\text{eff}}(\alpha;r)$ denote the PDM

$$M(\alpha;r) = \frac{1}{f^2(\alpha;r)} = \frac{1}{(1+\alpha r^2)^2} \tag{9}$$

and the effective potential

$$V_{\text{eff}}(\alpha;r) = \frac{L(L+1)}{r^2} + \frac{1}{4}(\omega^2 - 8\alpha^2)r^2 - \alpha, \tag{10}$$

respectively.

It is known [4] that equation (7) has an infinite number of bound states giving rise to a quadratic energy spectum

$$E^{(\alpha)}_{n,L} = \alpha\left(4n^2 + 4n(L+1) + L + 1 + (4n+2L+3)\frac{\lambda}{\alpha}\right), \qquad n = 0,1,2,\ldots, \tag{11}$$

where $\lambda = \frac{1}{2}(\alpha+\Delta)$ and $\Delta = \sqrt{\omega^2+\alpha^2}$. Furthermore, the corresponding eigenfunctions can be written as

$$\psi^{(\alpha)}_{n,L}(r) = \mathcal{N}^{(\alpha)}_{n,L} r^{L+1} P_n^{\left(\frac{\lambda}{\alpha}-\frac{1}{2},L+\frac{1}{2}\right)}(t) f^{-[\lambda+(L+2)\alpha]/(2\alpha)}, \tag{12}$$

where $P_n^{\left(\frac{\lambda}{\alpha}-\frac{1}{2},L+\frac{1}{2}\right)}(t)$ is a Jacobi polynomial in the variable $t = 1 - 2/f = (-1+\alpha r^2)/(1+\alpha r^2)$ and $\mathcal{N}^{(\alpha)}_{n,L}$ is some normalization coefficient.

The Hamiltonian in the PDM Schrödinger equation (7), the variable t appearing in the wavefunctions (12) and their commutator can be taken as the generators

$$\tilde{K}^{(\alpha)}_1 = \pi_r^2 + \frac{L(L+1)}{r^2} + \frac{1}{4}\omega^2 r^2, \qquad \tilde{K}^{(\alpha)}_2 = t, \qquad \tilde{K}^{(\alpha)}_3 = -4\mathrm{i}\alpha\left(2\frac{r}{f}\pi_r + \mathrm{i}t\right) \tag{13}$$

of a quadratic algebra, which can be identified with a quantum Jacobi algebra QJ(3) [5]. Their commutation relations are indeed given by

$$\begin{aligned}
[\tilde{K}^{(\alpha)}_1, \tilde{K}^{(\alpha)}_2] &= \tilde{K}^{(\alpha)}_3, \\
[\tilde{K}^{(\alpha)}_2, \tilde{K}^{(\alpha)}_3] &= 8\alpha(1 - \tilde{K}^{(\alpha)2}_2), \\
[\tilde{K}^{(\alpha)}_3, \tilde{K}^{(\alpha)}_1] &= -8\alpha\{\tilde{K}^{(\alpha)}_1, \tilde{K}^{(\alpha)}_2\} - 16\alpha^2\left[\frac{\lambda}{\alpha}\left(\frac{\lambda}{\alpha}-1\right) + L(L+1) - 1\right]\tilde{K}^{(\alpha)}_2 \\
&\quad - 16\alpha^2\left[\frac{\lambda}{\alpha}\left(\frac{\lambda}{\alpha}-1\right) - L(L+1)\right].
\end{aligned} \tag{14}$$

It can be shown that this algebra has a single positive-discrete series unitary irreducible representation $D^+_{p_0}$ with $p_0 = \frac{1}{2}\left(\frac{\lambda}{\alpha} + L\right)$ related to the lowest energy eigenvalue $E^{(\alpha)}_{0,L}$.

It is possible to construct another basis $\left(K^{(\alpha)}_0, K^{(\alpha)}_+, K^{(\alpha)}_-\right)$ of this quadratic algebra, satisfying the following three properties:

Journal of Physics: Conference Series **128** (2008) 012059 doi:10.1088/1742-6596/128/1/012059

(i) $K_0^{(\alpha)}$ is proportional to the Hamiltonian of the problem, while $K_+^{(\alpha)}$ (resp. $K_-^{(\alpha)}$) is a raising (resp. lowering) ladder operator, which means that, up to some multiplicative factor, it transforms $\psi_{n,L}^{(\alpha)}$ into $\psi_{n+1,L}^{(\alpha)}$ (resp. $\psi_{n-1,L}^{(\alpha)}$) for any $n \in \mathbb{N}$ (resp. $n \in \mathbb{N}^+$) with the additional condition that $K_-^{(\alpha)}$ annihilates $\psi_{0,L}^{(\alpha)}$.

(ii) The operators $K_0^{(\alpha)}$, $K_+^{(\alpha)}$, $K_-^{(\alpha)}$ satisfy the same Hermiticity properties as K_0, K_+, K_-, i.e., $K_0^{(\alpha)\dagger} = K_0^{(\alpha)}$ and $K_\pm^{(\alpha)\dagger} = K_\mp^{(\alpha)}$.

(iii) In the $\alpha \to 0$ limit, they go over to the su(1,1) generators K_0, K_+, K_-.

Such properties explicitly show (i) that the PDM d-dimensional radial harmonic oscillator Schrödinger equation (7) admits a spectrum generating algebra, and (ii) that the latter is a deformed su(1,1) algebra.

The operators $K_0^{(\alpha)}$, $K_+^{(\alpha)}$, $K_-^{(\alpha)}$ can indeed be expressed in terms of $\tilde{K}_1^{(\alpha)}$, $\tilde{K}_2^{(\alpha)}$, $\tilde{K}_3^{(\alpha)}$ as

$$
\begin{aligned}
K_0^{(\alpha)} &= \frac{1}{4\lambda}\tilde{K}_1^{(\alpha)}, \\
K_\pm^{(\alpha)} &= \pm\frac{1}{16\lambda}A_\pm^{(\alpha)}(\delta \pm 1)\sqrt{\frac{\delta \pm 2}{\delta}} = \pm\frac{1}{16\lambda}(\delta \mp 1)\sqrt{\frac{\delta}{\delta \mp 2}}A_\pm^{(\alpha)},
\end{aligned}
\tag{15}
$$

where

$$
\begin{aligned}
A_\pm^{(\alpha)} &= \tilde{K}_3^{(\alpha)} - 4\alpha\tilde{K}_2^{(\alpha)}(1 \mp \delta) + 4\alpha\frac{\left(\frac{\lambda}{\alpha} - L - 1\right)\left(\frac{\lambda}{\alpha} + L\right)}{1 \pm \delta}, \\
\delta &= \sqrt{\frac{\tilde{K}_1^{(\alpha)}}{\alpha} + \frac{\lambda}{\alpha}\left(\frac{\lambda}{\alpha} - 1\right) + L(L+1)},
\end{aligned}
\tag{16}
$$

and their action on $\psi_{n,L}^{(\alpha)}(r)$ is given by

$$
\begin{aligned}
K_0^{(\alpha)}\psi_{n,L}^{(\alpha)} &= \frac{1}{4\lambda}E_{n,L}^{(\alpha)}\psi_{n,L}^{(\alpha)}, \\
K_+^{(\alpha)}\psi_{n,L}^{(\alpha)} &= \frac{\alpha}{\lambda}\left[(n+1)\left(n+L+\frac{3}{2}\right)\left(n+\frac{\lambda}{\alpha}+L+1\right)\left(n+\frac{\lambda}{\alpha}+\frac{1}{2}\right)\right]^{1/2}\psi_{n+1,L}^{(\alpha)}, \\
K_-^{(\alpha)}\psi_{n,L}^{(\alpha)} &= \frac{\alpha}{\lambda}\left[n\left(n+L+\frac{1}{2}\right)\left(n+\frac{\lambda}{\alpha}+L\right)\left(n+\frac{\lambda}{\alpha}-\frac{1}{2}\right)\right]^{1/2}\psi_{n-1,L}^{(\alpha)}.
\end{aligned}
\tag{17}
$$

The three deformed su(1,1) generators $K_0^{(\alpha)}$, $K_+^{(\alpha)}$ and $K_-^{(\alpha)}$ satisfy the commutation relations

$$
\begin{aligned}
\left[K_0^{(\alpha)}, K_\pm^{(\alpha)}\right] &= \pm\frac{\alpha}{\lambda}K_\pm^{(\alpha)}(\delta \pm 1) = \pm\frac{\alpha}{\lambda}(\delta \mp 1)K_\pm^{(\alpha)}, \\
\left[K_+^{(\alpha)}, K_-^{(\alpha)}\right] &= -\frac{\alpha\delta}{\lambda}\left(2K_0^{(\alpha)} + \frac{\alpha}{4\lambda}\right),
\end{aligned}
\tag{18}
$$

and the algebra admits a Casimir operator

$$
C^{(\alpha)} = -K_+^{(\alpha)}K_-^{(\alpha)} + K_0^{(\alpha)2} - \frac{\alpha}{\lambda}\left(\delta - \frac{5}{4}\right)K_0^{(\alpha)} - \frac{\alpha^2}{8\lambda^2}\delta, \tag{19}
$$

with eigenvalues given in

$$
C^{(\alpha)}\psi_{n,L}^{(\alpha)} = \left[\frac{1}{4}\left(1 - \frac{\alpha}{\lambda}\right)\left(L + \frac{3}{2}\right)\left(L - \frac{1}{2}\right) - \frac{3\alpha^2}{16\lambda^2}L(L+1)\right]\psi_{n,L}^{(\alpha)}. \tag{20}
$$

Journal of Physics: Conference Series **128** (2008) 012059 doi:10.1088/1742-6596/128/1/012059

4. Final remarks

We have shown that a quadratic algebra approach provides us with a useful tool for constructing spectrum generating algebras for a class of d-dimensional radial harmonic oscillators with specific PDM depending on some parameter α. The well-known su(1,1) algebra corresponding to the constant-mass limit is retrieved for $\alpha \to 0$. The case of the one-dimensional harmonic oscillator on the full line (either with a constant mass or with a similar PDM) can be derived from that of the radial oscillator by formally replacing L by -1 or 0. The two resulting unitary irreducible representations of the spectrum generating algebra are spanned by even- and odd-parity wavefunctions, respectively.

References

[1] Quesne C 2007 *J. Phys. A: Math. Theor.* **40** 13107
[2] Roy B and Roy P 2005 *Phys. Lett.* A **340** 70
[3] Quesne C 2007 *SIGMA* **3** 067
[4] Bagchi B, Banerjee A, Quesne C and Tkachuk V M 2005 *J. Phys. A: Math. Gen.* **38** 2929
[5] Granovskii Ya I, Lutzenko I M and Zhedanov A S 1992 *Ann. Phys., NY* **217** 1

Journal of Physics: Conference Series **128** (2008) 012060 doi:10.1088/1742-6596/128/1/012060

Ternary algebras and groups

Michel Rausch de Traubenberg

Laboratoire de Physique Théorique, CNRS UMR 7085, Université Louis Pasteur, 3 rue de l'Université, 67084 Strasbourg Cedex, France

E-mail: rausch@lpt1.u-strasbg.fr

Abstract. We construct explicitly groups associated to specific ternary algebras which extend the Lie (super)algebras (called Lie algebras of order three). It turns out that the natural variables which appear in this construction are variables which generate the three-exterior algebra. An explicit matrix representation of a group associated to a peculiar Lie algebra of order three is constructed considering matrices with entry which belong to the three exterior algebra.

1. Introduction

Most of the laws of physics are based on algebras *i.e.* on mathematical structures equipped with a binary product. That is, if one considers an algebra $\mathcal{A}$ with a product m_2 we have for any $a \otimes b \in \mathcal{A} \otimes \mathcal{A}, m_2(a \otimes b) \in \mathcal{A}$. Among algebras the Lie algebras and the Lie superalgebras are of special importance. The former allow a description of the space-time and internal symmetries in particle physics although the latter lead to a supersymmetric extension of symmetries. In both cases the Lie (super)algebras correspond to the symmetries of a physical system at the infinitesimal level. However, in these two cases one is able to define symmetries at a finite level. The corresponding symmetries are associated to Lie groups or Lie supergroups.

The ternary algebras have been considered in physics only occasionally (see for instance [1, 2, 3, 4, 5, 6] and references therein). For some mathematical references one can see [7, 8, 9]. A ternary algebra $\mathcal{A}$ is an algebraic structure equipped with a ternary product $m_3 : \mathcal{A} \otimes \mathcal{A} \otimes \mathcal{A} \to \mathcal{A}$.

In [10, 11] an $F-$arry algebra which can be seen as a possible generalisation of Lie (super)algebras have been considered and named Lie algebra of order F. A Lie algebra of order F admits a $\mathbb{Z}_F-$grading ($F = 3$ in this paper), the zero-graded part being a Lie algebra. An $F-$fold symmetric product (playing the role of the anticommutator in the case $F = 2$) expresses the zero graded part in terms of the non-zero graded part. This means that part of the algebra is a (binary) algebra and part of the algebra is a ternary algebra (when $F = 3$). Subsequently, a specific Lie algebra or order 3, leading to a non-trivial extension of the Poincaré algebra, has been studied together with its implementation in Quantum Field Theory [12, 13, 14]. A general study of the possible non-trivial extensions of the Poincaré algebra in $(1 + 3)-$dimensions has been undertaken in [15, 16]. However, all these mathematical structures have been considered at the level of algebras *i.e.* at the level of infinitesimal transformations and no groups associated to Lie algebras of order 3 were considered. At a first glance these two structures seem to be incompatible since for a Lie algebra of order 3 for some elements only the product of three elements is defined although for a group the product of two elements is always defined.

In this paper we show that in fact a specific ternary group that we might call a Lie group of order 3 can be associated to a Lie algebra of order 3. We show that one possible way to

Journal of Physics: Conference Series **128** (2008) 012060 doi:10.1088/1742-6596/128/1/012060

associate a group to a given Lie algebra of order 3 is to endow the universal enveloping algebra of a Lie algebra of order 3 with a Hopf algebra structure. Then we identify the parameters of the transformation. It turns out that these parameters are strongly related to the 3−exterior algebra. The 3−exterior algebra being the algebra generated by canonical generators θ^i satisfying [17]

$$\theta^i\theta^j\theta^k + \theta^j\theta^k\theta^i + \theta^k\theta^i\theta^j + \theta^i\theta^k\theta^j + \theta^j\theta^i\theta^k + \theta^k\theta^j\theta^i = 0.$$

The content of this paper is the following. In section 2 we recall the definition of Lie algebras of order 3 and give some examples. We then define its universal enveloping algebra and endow it with a Hopf algebra structure. Then the consideration of the Hopf dual enable us to define the parameters of the transformations which turn out to be related to the 3−exterior algebra [17]. In section 3 we construct explicitly a group associated to a specific Lie algebra of order 3 by considering matrices with entries which belong to the three exterior algebra.

2. Lie algebras of order 3

2.1. Definition and examples

The general definition of Lie algebras of order F, was given in [10, 11] together with an inductive way to construct Lie algebras of order F associated with *any* Lie algebra or Lie superalgebra. We recall here the main results useful for the sequel. Let F be a positive integer and define $q = e^{\frac{2\pi i}{F}}$. We consider $\mathfrak{g}$ a complex vector space and ε an automorphism of $\mathfrak{g}$ satisfying $\varepsilon^F = 1$. Set $\mathfrak{g}_i \subseteq \mathfrak{g}$ the eigenspace corresponding to the eigenvalue q^i of ε. Then, we have $\mathfrak{g} = \mathfrak{g}_0 \oplus \cdots \oplus \mathfrak{g}_{F-1}$.

Definition 2.1 *Let $F \in \mathbb{N}^*$. A $\mathbb{Z}_F$-graded $\mathbb{C}-$vector space $\mathfrak{g} = \mathfrak{g}_0 \oplus \mathfrak{g}_1 \oplus \mathfrak{g}_2 \cdots \oplus \mathfrak{g}_{F-1}$ is called a complex Lie algebra of order F if*

(i) $\mathfrak{g}_0$ *is a complex Lie algebra.*

(ii) For all $i = 1, \ldots, F-1$, $\mathfrak{g}_i$ is a representation of $\mathfrak{g}_0$. If $X \in \mathfrak{g}_0, Y \in \mathfrak{g}_i$ then $[X,Y]$ denotes the action of $X \in \mathfrak{g}_0$ on $Y \in \mathfrak{g}_i$ for all $i = 1, \cdots F-1$.

(iii) For all $i = 1, \ldots, F-1$ there exists an $F-$linear, $\mathfrak{g}_0-$equivariant map

$$\{\cdots\} : \mathcal{S}^F(\mathfrak{g}_i) \to \mathfrak{g}_0,$$

where $\mathcal{S}^F(\mathfrak{g}_i)$ denotes the $F-$fold symmetric product of $\mathfrak{g}_i$, satisfying the following (Jacobi) identity

$$\sum_{j=1}^{F+1} [Y_j, \{Y_1, \ldots, Y_{j-1}, Y_{j+1}, \ldots, Y_{F+1}\}] = 0,$$

for all $Y_j \in \mathfrak{g}_i$, $j = 1, .., F+1$.

By definition when $F = 1$, $\mathfrak{g} = \mathfrak{g}_0$ is a Lie algebra and when $F = 2$, $\mathfrak{g} = \mathfrak{g}_0 \oplus \mathfrak{g}_1$ is a Lie superalgebra. Thus Lie algebra of order three can be seen as some possible generalisation of Lie (super)algebras. It can be noticed that for any $k = 1, \ldots, F-1$, the $\mathbb{Z}_F-$graded vector spaces $\mathfrak{g}_0 \oplus \mathfrak{g}_k$ satisfy all the properties in Definition 2.1 and thus are Lie algebras of order F. We call these types of algebras *elementary Lie algebras of order F*.

Definition 2.2 *A representation of an elementary Lie algebra of order F is a linear map $\rho : \mathfrak{g} = \mathfrak{g}_0 \oplus \mathfrak{g}_1 \to \mathrm{End}(V)$, such that (for all $X_i \in \mathfrak{g}_0, Y_j \in \mathfrak{g}_1$)*

$$\begin{array}{l}\rho\left([X_1,X_2]\right)=\rho(X_1)\rho(X_2)-\rho(X_2)\rho(X_1)\\ \rho\left([X_1,Y_2]\right)=\rho(X_1)\rho(Y_2)-\rho(Y_2)\rho(X_1)\\ \rho\left(\{Y_1.\cdots,Y_F\}\right)=\sum\limits_{\sigma\in S_F}\rho\left(Y_{\sigma(1)}\right)\cdots\rho\left(Y_{\sigma(F)}\right)\end{array}\tag{1}$$

(S_F being the group of permutations of F elements).

By construction the vector space V is $\mathbb{Z}_F-$graded $V = V_0 \oplus \cdots \oplus V_{F-1}$ and for all $a \in \{0,\cdots,F-1\}$, V_a is a $\mathfrak{g}_0-$module and we have $\rho(\mathfrak{g}_1)(V_a) \subseteq V_{a+1}$.

From now on we only consider elementary Lie algebras of order 3. It can be observed that part of the algebra is binary and part of the algebra is ternary. Indeed we have by Definition 2.1 the following products: $m_1 : \mathfrak{g}_0 \otimes \mathfrak{g}_0 \to \mathfrak{g}_0$, $m_2 : \mathfrak{g}_0 \otimes \mathfrak{g}_1 \to \mathfrak{g}_1$, $m_3 : \mathfrak{g}_1 \otimes \mathfrak{g}_1 \otimes \mathfrak{g}_1 \to \mathfrak{g}_0$. Furthermore, there is essentially two types of Lie algebras of order 3. The first types are the Lie algebras of order 3 related to Lie algebras of order 3 obtained from a Lie (super)algebra through the inductive process [11]. One example of this types of algebras is the cubic extension of the Poincaré algebra

Example 2.3 *Let $\mathfrak{g}_0 = \langle L_{\mu\nu}, P_\mu\rangle$ be the Poincaré algebra in $D-$dimensions and $\mathfrak{g}_1 = \langle V_\mu\rangle$ be the $D-$dimensional vector representation of $\mathfrak{g}_0$. The brackets*

$$\begin{array}{rcl}[L_{\mu\nu},L_{\rho\sigma}] &=& \eta_{\nu\sigma}L_{\rho\mu}-\eta_{\mu\sigma}L_{\rho\nu}+\eta_{\nu\rho}L_{\mu\sigma}-\eta_{\mu\rho}L_{\nu\sigma},\\ [L_{\mu\nu},P_\rho] &=& \eta_{\nu\rho}P_\mu-\eta_{\mu\rho}P_\nu,\ [L_{\mu\nu},V_\rho]=\eta_{\nu\rho}V_\mu-\eta_{\mu\rho}V_\nu,\ [P_\mu,V_\nu]=0,\\ \{V_\mu,V_\nu,V_\rho\} &=& \eta_{\mu\nu}P_\rho+\eta_{\mu\rho}P_\nu+\eta_{\rho\nu}P_\mu,\end{array}$$

with the metric $\eta_{\mu\nu} = \mathrm{diag}(1,-1,\cdots,-1)$, endow $\mathfrak{g} = \mathfrak{g}_0 \oplus \mathfrak{g}_1$ with an elementary Lie algebra of order 3 structure.

The second types of Lie algebras of order 3 are obtained as a sub-algebra of Lie algebras of order 3 defined by some matrix representation.

Example 2.4 *Let $\mathfrak{gl}(m_1,m_2,m_3)$ and $\mathfrak{gl}_{\rm el}(m_1,m_2,m_3)$ be the set of $(m_1+m_2+m_3)\times(m_1+m_2+m_3)$ matrices of the form*

$$\mathfrak{gl}_{\rm el}(m_1,m_2,m_3)=\left\{\begin{pmatrix}a_0&b_1&0\\0&a_1&b_2\\b_0&0&a_2\end{pmatrix}\right\},\quad \mathfrak{gl}(m_1,m_2,m_3)=\left\{\begin{pmatrix}a_0&b_1&c_2\\c_0&a_1&b_2\\b_0&c_1&a_2\end{pmatrix}\right\},\tag{2}$$

with $a_0\in\mathfrak{gl}(m_1), a_1\in\mathfrak{gl}(m_2), a_2\in\mathfrak{gl}(m_3)$, $b_1\in\mathcal{M}_{m_1,m_2}(\mathbb{C}), b_2\in\mathcal{M}_{m_2,m_3}(\mathbb{C}), b_0\in\mathcal{M}_{m_3,m_1}(\mathbb{C})$, and $c_0\in\mathcal{M}_{m_2,m_1}(\mathbb{C}), c_1\in\mathcal{M}_{m_3,m_2}(\mathbb{C}), c_2\in\mathcal{M}_{m_1,m_3}(\mathbb{C})$. A basis of these sets of matrices can be constructed as follow. Consider the $(m_1+m_2+m_3)^2$ canonical matrices ${e_I}^J$, $1\le I,J\le m_1+m_2+m_3$. With the following convention for the indices $1\le i,j\le m_1$, $m_1+1\le i',j'\le m_1+m_2$, $m_1+m_2+1\le i'',j''\le m_1+m_2+m_3$, the generators are given by

$$\begin{array}{lll}{e_i}^j \text{ for } \mathfrak{gl}(m_1), & {e_{i'}}^{j'} \text{ for } \mathfrak{gl}(m_2), & {e_{i''}}^{j''} \text{ for } \mathfrak{gl}(m_3),\\ {e_i}^{j'} \text{ for } \mathcal{M}_{m_1,m_2}(\mathbb{C}), & {e_{i'}}^{j''} \text{ for } \mathcal{M}_{m_2,m_3}(\mathbb{C}), & {e_{i''}}^{j} \text{ for } \mathcal{M}_{m_3,m_1}(\mathbb{C}),\\ {e_{i'}}^{j} \text{ for } \mathcal{M}_{m_2,m_1}(\mathbb{C}), & {e_{i''}}^{j'} \text{ for } \mathcal{M}_{m_3,m_2}(\mathbb{C}), & {e_i}^{j''} \text{ for } \mathcal{M}_{m_1,m_3}(\mathbb{C}).\end{array}$$

Writing $\mathfrak{gl}(m_1,m_2,m_3) = \mathfrak{gl}(m_1,m_2,m_3)_0 \oplus \mathfrak{gl}(m_1,m_2,m_3)_1 \oplus \mathfrak{gl}(m_1,m_2,m_3)_2$ and $\mathfrak{gl}_{\rm el}(m_1,m_2,m_3) = \mathfrak{gl}_{\rm el}(m_1,m_2,m_3)_0 \oplus \mathfrak{gl}_{\rm el}(m_1,m_2,m_3)_1$ we denote generically by ${X_I}^J$ the generators of grade zero ${Y_I}^J$ the generators of grade one and ${Z_I}^J$ those of grade two, and the brackets read

$$
\begin{array}{rcl}
[X_I{}^J, X_K{}^L] &=& \delta^J{}_K X_I{}^L - \delta^L{}_I X_K{}^J, \\
\left[X_I{}^J, Y_K{}^L\right] &=& \delta^J{}_K Y_I{}^L - \delta^L{}_I Y_K{}^J, \\
\left[X_I{}^J, Z_K{}^L\right] &=& \delta^J{}_K Z_I{}^L - \delta^L{}_I Z_K{}^J, \\
\left\{Y_I{}^J, Y_K{}^L, Y_M{}^N\right\} &=& \delta^J{}_K \delta^L{}_M X_I{}^N + \delta^N{}_I \delta^J{}_K X_M{}^L + \delta^L{}_M \delta^N{}_I X_K{}^J \\
&+& \delta^J{}_M \delta^N{}_K X_I{}^L + \delta^N{}_K \delta^L{}_I X_M{}^J + \delta^L{}_I \delta^J{}_M X_K{}^N, \\
\left\{Z_I{}^J, Z_K{}^L, Z_M{}^N\right\} &=& \delta^J{}_K \delta^L{}_M X_I{}^N + \delta^N{}_I \delta^J{}_K X_M{}^L + \delta^L{}_M \delta^N{}_I X_K{}^J \\
&+& \delta^J{}_M \delta^N{}_K X_I{}^L + \delta^N{}_K \delta^L{}_I X_M{}^J + \delta^L{}_I \delta^J{}_M X_K{}^N,
\end{array} \tag{3}
$$

this shows that $\mathfrak{gl}(m_1, m_2, m_3)$ (resp. $\mathfrak{gl}_{\rm el}(m_1, m_2, m_3)$) is endowed with the structure of Lie algebra of order three (resp. the structure of elementary Lie algebra of order three).

The algebra of Example 2.3 was firstly introduced in [10] and its implementation in Quantum Field Theories was realised in 4−dimensions in [12, 13] and in D−dimensions in relations to generalised gauge fields or p−forms in [14]. A general classification of Lie algebra of order 3 extending non-trivially the Poincaré algebra in 4−dimensions was undertaken in [15, 16]. However, even if invariant Lagrangian were constructed all was done at the level of algebras *i.e.* of infinitesimal transformations and nothing were said at the level of groups. Indeed, by definition, a Lie algebra of order 3 is partially a ternary algebra since the product of three elements of $\mathfrak{g}_1$ is defined and not the product of two. This seem to be *a priori* in contradiction with groups since for a group a product of two elements is always defined. We will see in the next subsection how one can evade this contradiction and construct a group associated to the ternary algebras we are considering.

2.2. Hopf algebras associated to elementaries Lie algebras of order 3

In this subsection, we show that the introduction of Hopf algebras is a possible way to construct a group associated to a given Lie algebra of order 3. We also show that the parameters of the transformation associated to an element of $\mathfrak{g}_1$ belong to the 3−exterior algebra (see the appendix for definition).

Let $\mathfrak{g} = \mathfrak{g}_0 \oplus \mathfrak{g}_1$ be a given Lie algebra of order 3. We denote generically by X the elements of $\mathfrak{g}_0$ and by Y the elements of $\mathfrak{g}_1$. To endow $\mathfrak{g}$ with a Hopf algebra structure we need first to define $\mathcal{U}(\mathfrak{g})$ the enveloping algebra associated to $\mathfrak{g}$. Consider the tensor algebra

$$T(\mathfrak{g}) = \mathbb{C} \oplus \mathfrak{g} \oplus (\mathfrak{g} \otimes \mathfrak{g}) \oplus \cdots \ .$$

This is an associative algebra with multiplication given for any $t_1, t_2 \in T(\mathfrak{g})$ by $t_1 \otimes t_2 \in T(\mathfrak{g})$. Now, in order to assume the multiplication given in Definition 2.1 we define (for any $X_1, X_2 \in \mathfrak{g}_0, Y_1, Y_2, Y_3 \in \mathfrak{g}_1$) the two-sided ideal $\mathcal{I}$ generated by

$$
\begin{array}{l}
X_1 \otimes X_2 - X_2 \otimes X_1 - [X_1, X_2], \\
X_1 \otimes Y_2 - Y_2 \otimes X_1 - [X_1, Y_2], \\
Y_1 \otimes Y_2 \otimes Y_3 + Y_2 \otimes Y_3 \otimes Y_1 + Y_3 \otimes Y_1 \otimes Y_2 + \\
Y_1 \otimes Y_3 \otimes Y_2 + Y_2 \otimes Y_1 \otimes Y_3 + Y_3 \otimes Y_2 \otimes Y_1 - \{Y_1, Y_2, Y_2\}.
\end{array}
$$

The enveloping algebra of $\mathfrak{g}$ (in fact it is universal but this will be shown elsewhere [18]) is then defined by $\mathcal{U}(\mathfrak{g}) = T(\mathfrak{g})/\mathcal{I}$. This means in particular that in the associative algebra $\mathcal{U}(\mathfrak{g})$ the relation given in Definition 2.1 are satisfied. Furthermore, as in the Lie (super)algebra cases, a basis of $\mathcal{U}(\mathfrak{g})$ can be given. This is an analogous of the the Poincaré-Birkhoff-Witt theorem in this case. Indeed, it can be proven [18] that we have the following isomorphism of vector space

$$\mathcal{U}(\mathfrak{g}) \cong S(\mathfrak{g}_0) \otimes \Lambda_3(\mathfrak{g}_1) \tag{4}$$

where $S(\mathfrak{g}_0)$ is the symmetric algebra over $\mathfrak{g}_0$ and $\Lambda_3(\mathfrak{g}_1)$ is the three-exterior algebra on $\mathfrak{g}_1$ (see the appendix for definition). Since the composition of the natural map $\mathfrak{g} \to T(\mathfrak{g})$ with the canonical projection $T(\mathfrak{g}) \to \mathcal{U}(\mathfrak{g})$ gives $\mathfrak{g} \subset \mathcal{U}(\mathfrak{g})$ we identify $\mathfrak{g}$ with its image under this map. Thus, if we denote $\{X_a, 1 \le a \le \dim \mathfrak{g}_0 = n_0\}$ (resp. $\{Y_i, 1 \le i \le \dim \mathfrak{g}_i = n_1\}$ a basis of $\mathfrak{g}_0$ (resp. $\mathfrak{g}_1$) a basis of $\mathcal{U}(\mathfrak{g})$ is given by the elements of the form [18]

$$g_{\vec{a},I_\ell} = \frac{X_1^{a_1}}{a_1!} \cdots \frac{X_{n_0}^{a_{n_0}}}{a_{n_0}!} Y_{I_\ell}, \quad (a_1, \cdots, a_{n_0}) \in \mathbb{N}^{n_0}, \quad \ell \in \mathbb{N}, \quad I_\ell \in \{1, \cdots, n_1\}^\ell \setminus I_{\ell,n_1}. \tag{5}$$

where $Y_{I_\ell} = Y_{(i_1,\cdots,i_\ell)} = Y_{i_1} \cdots Y_{i_\ell}$. (See the appendix for the definition of I_{ℓ,n_1}.) An element Y_{I_ℓ} with $I_\ell \in \{1, \cdots, n_1\}^\ell \setminus I_{\ell,n_1}$ is called a Roby element.

As in the usual Lie (super)algebra cases, $\mathcal{U}(\mathfrak{g})$ can be endowed with a Hopf algebra structure. Although the algebra structure depends on the brackets defining the Lie algebra of order 3 we are considering, the co-product not. Since $\mathfrak{g}$ is a graded vector space it is natural to define the co-product Δ to be an homorphism from $\mathcal{U}(\mathfrak{g})$ to $\mathcal{U}(\mathfrak{g}) \hat{\otimes} \mathcal{U}(\mathfrak{g})$, where $\hat{\otimes}$ is the $\mathbb{Z}_3-$graded tensor product. This means in particular that for homogeneous elements $a_1, a_2, b_1, b_2 \in \mathcal{U}(\mathfrak{g})$ we have $(a_1 \hat{\otimes} a_2)(b_1 \hat{\otimes} b_2) = q^{\mathrm{gr}(a_2)\mathrm{gr}(b_1)} a_1 b_1 \hat{\otimes} a_2 b_2$, with $\mathrm{gr}(a)$ the grade of a. We define

$$\Delta 1 = 1 \hat{\otimes} 1, \ \Delta g = g \hat{\otimes} 1 + 1 \hat{\otimes} g, \text{ for any } g \in \mathfrak{g}. \tag{6}$$

The co-product extends to any elements of $\mathcal{U}(\mathfrak{g})$ using (5) and $\Delta(g_1 g_2) = \Delta(g_1)\Delta(g_1)$. In particular this means for any $Y_1, Y_2, Y_3 \in \mathfrak{g}_1$, from $\Delta(Y_1Y_2Y_3) = \Delta(Y_1)\Delta(Y_2)\Delta(Y_3) = Y_1Y_2Y_3 \hat{\otimes} 1 + Y_1Y_2 \hat{\otimes} Y_3 + qY_1Y_3 \hat{\otimes} Y_2 + q^2 Y_2Y_3 \hat{\otimes} Y_1 + Y_1 \hat{\otimes} Y_2Y_3 + qY_2 \hat{\otimes} Y_1Y_3 + q^2 Y_3 \hat{\otimes} Y_1Y_2 + 1 \hat{\otimes} Y_1Y_2Y_3$, that the co-product (6) is compatible with the Definition 2.1 *i. e.* we have

$$\{\Delta Y_1, \Delta Y_2, \Delta Y_3\} = \Delta\{Y_1, Y_2, Y_3\}. \tag{7}$$

The antipode is defined by

$$S(1) = 1, \ S(g) = -g, \text{ for any g } \in \mathfrak{g}, \tag{8}$$

and the co-unit by

$$\epsilon(1) = 1, \ \epsilon(g) = 0, \text{ for any g } \in \mathfrak{g}. \tag{9}$$

It should be mentioned that a different Hopf algebra associated to Lie algebras of order three has been introduced in [19], where the co-product was defined by the usual tensorial product and the twist (necessary to ensure (7)) was generated by an additional element.

Now, if for $g \in \mathfrak{g}$ we consider the formal element e^g of $\mathcal{U}(\mathfrak{g})$ (for the moment, we do not take care of any convergence problem and e^g is defined as a formal series), then, one can show that e^g is a group-like element and we have

Journal of Physics: Conference Series **128** (2008) 012060 doi:10.1088/1742-6596/128/1/012060

$$\Delta e^g = e^g \otimes e^g, \ S(e^g) = e^{-g}, \ \epsilon(e^g) = 1. \tag{10}$$

This means that e^g is a good candidate to define the group associated to the Lie algebra of order three $\mathfrak{g}$. However, before identifying e^g to an element of the group associated to $\mathfrak{g}$, the parameters of the transformation should be firstly identified. It is well known that the parameters associated to a Lie superalgebra are commuting numbers for the even part and Grassmann numbers for the odd part. Similarly, we now show that here the parameters associated to $\mathfrak{g}_0$ are commuting numbers although the parameters associated to $\mathfrak{g}_1$ belongs to the 3−exterior algebra.

To identify the parameters of the transformation one needs first to define $\mathcal{U}(\mathfrak{g})^*$ the dual of the Hopf algebra $\mathcal{U}(\mathfrak{g})$. We define the dual basis to be $g^{\vec{b},I_\ell}, \vec{b} \in \mathbb{N}^{n_0}, \ell \in \mathbb{N}$ and $I_\ell \in \{1, \cdots, n_1\}^\ell \backslash I_{\ell,n_1}$ with the natural pairing:

$$\left\langle g^{\vec{b},I_\ell}, g_{\vec{a},I_k} \right\rangle = \delta^{\vec{b}}{}_{\vec{a}} \delta^\ell{}_k \delta^{I_\ell}{}_{I_k}. \tag{11}$$

Denote α^a the dual of X_a and θ^i the dual of Y_i. The co-product (6) on $\mathcal{U}(\mathfrak{g})$ induces the product on $\mathcal{U}(\mathfrak{g})^*$. Recall that if e_i constitute a basis of a Hopf algebra H with co-product $\Delta e_i = \delta_i{}^{jk} e_j \otimes e_k$, for the dual space $H^\star$ with dual basis e^i we have $e^j e^k = \delta_i{}^{jk} e^i$. Using (6) it is easy to see that the variables α^a commute between themselves and with the variables θ^i. Furthermore, a little algebra gives that the variables θ generate the 3−exterior algebra and thus, as we now show, (A.1) is satisfied for any $\theta^i, \theta^j, \theta^k$.

If we denote $\theta^1, \theta^{(1,1)}$ the dual variables of Y_1 and $Y_1 Y_1$ the co-product (6) gives

$$\theta^1\theta^1 = -q^2\theta^{(1,1)}, \quad \theta^1\theta^{(1,1)} = \theta^{(1,1)}\theta^1 = 0, \quad \text{and thus} \quad (\theta^1)^3 = 0.$$

The case with two different variables, say θ^1, θ^2 is more involved. We introduce $\theta^1, \theta^2, \theta^{(1,2)}, \theta^{(2,1)}, \theta^{(1,2,1)}, \theta^{(2,1,1)}$ the dual variables of $Y_1, Y_2, Y_1Y_2, Y_2Y_1, Y_1Y_2Y_1, Y_2Y_1Y_1$ (since $Y_1Y_1Y_2$ is not a Roby element $-\ 1 \le 1 \le 2\ -$ there is no need to introduce the element $\theta^{(1,1,2)}$). Using

$$\Delta(Y_iY_j) = Y_iY_j \hat{\otimes} 1 + Y_i \hat{\otimes} Y_j + qY_j \hat{\otimes} Y_i + 1 \hat{\otimes} Y_iY_j.$$

we obtain

$$\theta^1\theta^2 = \theta^{(1,2)} + q\theta^{(2,1)}, \quad \theta^2\theta^1 = \theta^{(2,1)} + q\theta^{(1,2)}.$$

Similarly,

$$\begin{aligned}
\Delta(Y_1Y_2Y_1) &= Y_1Y_2Y_1 \hat{\otimes} 1 + Y_1Y_2 \hat{\otimes} Y_1 + qY_1Y_1 \hat{\otimes} Y_2 + q^2Y_2Y_1 \hat{\otimes} Y_1 \\
&+ Y_1 \hat{\otimes} Y_2Y_1 + qY_2 \hat{\otimes} Y_1Y_1 + q^2Y_1 \hat{\otimes} Y_1Y_2 + 1 \hat{\otimes} Y_1Y_2Y_1, \\
\Delta(Y_2Y_1Y_1) &= Y_2Y_1Y_1 \hat{\otimes} 1 - q^2Y_2Y_1 \hat{\otimes} Y_1 + q^2Y_1Y_1 \hat{\otimes} Y_2 \\
&+ Y_2 \hat{\otimes} Y_1Y_1 - Y_1 \hat{\otimes} Y_1Y_2 + 1 \hat{\otimes} Y_2Y_1Y_1,
\end{aligned}$$

give

$$\begin{aligned}
\theta^1\theta^1\theta^2 &= -\theta^{(1,2,1)} - q\theta^{(2,1,1)} \\
\theta^1\theta^2\theta^1 &= 2\theta^{(1,2,1)} - \theta^{(2,1,1)}) \\
\theta^2\theta^1\theta^1 &= -\theta^{(1,2,1)} - q^2\theta^{(2,1,1)}
\end{aligned}$$

leading to $\theta^1\theta^1\theta^2 + \theta^1\theta^2\theta^1 + \theta^2\theta^1\theta^1 = 0$.

A similar, but more tedious calculus, gives along the same lines that for $i \neq j \neq k$ we have the relation (A.1). Thus, as an algebra we have the following isomorphism:

$$\mathcal{U}(\mathfrak{g})^* \cong \mathbb{C}[\alpha^1, \cdots, \alpha^{n_0}] \otimes \Lambda_3(\mathbb{C}^{n_1}).$$

This means in particular that if one wants to associate a group to the Lie algebra of order 3 $\mathfrak{g} = \mathfrak{g}_0 \oplus \mathfrak{g}_1$ the parameters of the transformation should be commuting numbers for the elements of $\mathfrak{g}_0$ and should belong to the 3−exterior algebra for the elements of $\mathfrak{g}_1$. This means, in relation with (10) that one might take something like

$$e^{\alpha^a X_a + \theta^i Y_i}, \tag{12}$$

(with α^a commuting numbers and θ^i the canonical generators of the 3−exterior algebra) to obtain a group. However, here some care as to be taken. Firstly developing $e^{\theta^i Y_1}$ above since $\Lambda_3(\mathbb{C}^{n_1})$ is infinitely generated (see the appendix) there is an infinite number of terms in the series, and we do not know whether or not the series converge since no convergence properties have been established on $\Lambda_3(\mathbb{C}^{n_1})$, even if Y_a are represented by finite-dimensional matrices. Secondly since there is no Baker-Campbell-Hausdorff formulæ for elements of $\Lambda_3(\mathbb{C}^{n_1})$. This means that the product of two elements of the type (12) is not of the same type. These problems have been solved in [18], but here we propose another way to construct a group associate to Lie algebras of order 3.

3. Group associated to Lie algebra of order 3 – the $\mathfrak{g}_{\text{el}}(m_1, m_2, m_3)$ case

In this section we are constructing explicitly a group associated to the Lie algebra of order 3 $\mathfrak{g}_{\text{el}}(m_1, m_2, m_3)$ in a straight analogy with the construction of the Lie supergroup associated to the Lie superalgebra $\mathfrak{gl}(m, n)$. Using the results of subsection 2.2, since the parameters of the transformation are strongly related with the 3−exterior algebra, we are considering matricial groups whose entries belong to the 3−exterior algebra.

First of all we introduce some set of matrices with coefficients in $\Lambda_3(\mathbb{C}^p)$:

$$\begin{aligned} \mathcal{M}_k(\Lambda_0) &= \mathcal{M}_k(\mathbb{C}) \otimes \Lambda(\mathbb{C}^p)_0, \\ \mathcal{M}_{k,\ell}(\Lambda_1) &= \mathcal{M}_{k,\ell}(\mathbb{C}) \otimes \Lambda(\mathbb{C}^p)_1, \\ \mathcal{M}_{k,\ell}(\Lambda_2) &= \mathcal{M}_{k,\ell}(\mathbb{C}) \otimes \Lambda(\mathbb{C}^p)_2. \end{aligned} \tag{13}$$

A matrix, *e.g.* in $\mathcal{M}_k(\Lambda_0)$ is given by

$$A(\Lambda_0) = A_{(0)} + A_{(3)ijk}\theta^i\theta^j\theta^k + A_{(6)ijk\ell mn}\theta^i\theta^j\theta^k\theta^\ell\theta^m\theta^n + \cdots \tag{14}$$

where the sum above is finite and taken only over Roby elements.

Proposition 3.1 *The matrix given in* (14) *is invertible if and only if* $A_{(0)}$ *is invertible.*

Proof: We introduce $B(\Lambda_0) = B_{(0)} + B_{(3)ijk}\theta^i\theta^j\theta^k + B_{(6)ijk\ell mn}\theta^i\theta^j\theta^k\theta^\ell\theta^m\theta^n + \cdots$, with the sum taken over Roby elements. Solving the equation $A(\Lambda_0)B(\Lambda_0) = 1$, gives the coefficients of $B(\Lambda_0)$ terms by terms. We thus get for the two first terms $B_{(0)} = A_{(0)}^{-1}$, $B_{(3)ijk} = -A_{(0)}^{-1}A_{(3)ijk}A_{(0)}^{-1}$.

Journal of Physics: Conference Series **128** (2008) 012060 doi:10.1088/1742-6596/128/1/012060

However, for the higher order terms some care as to be taken. Indeed, the terms of degree six give:

$$(A_{(0)}B_{(6)ijk\ell mn} + A_{(6)ijk\ell mn}B_0)\theta^i\theta^j\theta^k\theta^\ell\theta^m\theta^n + A_{(3)i'j'k'}B_{(3)i''j''k''}\theta^{i'}\theta^{j'}\theta^{k'}\theta^{i''}\theta^{j''}\theta^{k''} = 0.$$

But even though $\{\theta^i\theta^j\theta^k\theta^\ell\theta^m\theta^n\}$ are by definition of $A(\Lambda_0)$ and $B(\Lambda_0)$ independent elements of $\Lambda_3(\mathbb{C}^p)$, in general it is not true for $\left\{\theta^{i'}\theta^{j'}\theta^{k'}\theta^{i''}\theta^{j''}\theta^{k''}\right\}$. Indeed, if the sequences $(i',j',k'),(i'',j'',k'') \in \{1,\cdots,p\}^3 \setminus I_{3,p}$ this does not mean that the sequence $(i',j',k',i'',j'',k'') \in \{1,\cdots,p\}^6 \setminus I_{6,p}$, as can be seen with $(1,2,1)$ and $(1,2,1)$. Decomposing this last terms in the basis of $\Lambda_3(\mathbb{C}^p)$ we have

$$\theta^{i'}\theta^{j'}\theta^{k'}\theta^{i''}\theta^{j''}\theta^{k''} = C^{i'j'k'i''j''k''}{}_{ijk\ell mn}\theta^i\theta^j\theta^k\theta^\ell\theta^m\theta^n$$

(where the sum is taken over $(ijk\ell mn) \in \{1,\cdots,p\}^6 \setminus I_{6,3}$) and we get

$$B_{(6)ijk\ell mn} = -A_0^{-1}A_{(6)ijk\ell mn}A_{(0)}^{-1} - C^{i'j'k'i''j''k''}{}_{ijk\ell mn}A_{(0)}^{-1}A_{(3)i'j'k'}A_{(0)}^{-1}A_{(3)i''j''k''}A_0^{-1}.$$

This procedure extends simply order by order. QED.

We denote by $GL(m_1,\Lambda_0)$ the set of invertible matrices of $\mathcal{M}_{m_1}(\mathbb{C},\Lambda_0)$.

Proposition 3.2 *Let* $A_0 \in GL(m_1,\Lambda_0), A_1 \in GL(m_2,\Lambda_0), A_2 \in GL(m_3,\Lambda_0)$, $B_0 \in \mathcal{M}_{m_3,m_1}(\Lambda_1), B_1 \in \mathcal{M}_{m_1,m_2}(\Lambda_1), B_2 \in \mathcal{M}_{m_2,m_3}(\Lambda_1)$ *and* $C_0 \in \mathcal{M}_{m_2,m_1}(\Lambda_2), C_1 \in \mathcal{M}_{m_3,m_2}(\Lambda_2), C_2 \in \mathcal{M}_{m_1,m_3}(\Lambda_2)$, *then the* $m_1\times m_2\times m_3$ *matrix with coefficients in* $\Lambda_3(\mathbb{C}^p)$

$$M = \begin{pmatrix} A_0 & B_1 & C_2 \\ C_0 & A_1 & B_2 \\ B_0 & C_1 & A_2 \end{pmatrix}, \tag{15}$$

is invertible.

Proof: By definition, it is obvious to see that if M,N are two matrices of the type above then the product MN is also as above. We now show that the matrix M is invertible. Define

$$N = \begin{pmatrix} A'_0 & B'_1 & C'_2 \\ C'_0 & A'_1 & B'_2 \\ B'_0 & C'_1 & A'_2 \end{pmatrix},$$

and solving MN$=1$, we get

$$
\begin{aligned}
A'_0 &= \Big\{A_0 - C_2(A_2 - C_1A_1^{-1}B_2)^{-1}(B_0 - C_1A_1^{-1}C_0) \\
&\qquad - B_1A_1^{-1}\left[C_0 - B_2(A_2 - C_1A_1^{-1}B_2)^{-1}(B_0 - C_1A_1^{-1}C_0)\right]\Big\}^{-1}, \\
B'_0 &= -(A_2 - C_1A_1^{-1}B_2)^{-1}(B_0 - C_1A_1^{-1}C_0)A'_0, \\
C'_0 &= -A_1^{-1}\Big[C_0 - B_2(A_2 - C_1A_1^{-1}B_2)^{-1}(B_0 - C_1A_1^{-1}C_0)\Big]A'_0, \\
A'_1 &= \Big\{A_1 - C_0(A_0 - C_2A_2^{-1}B_0)^{-1}(B_1 - C_2A_2^{-1}C_1) \\
&\qquad - B_2A_2^{-1}\left[C_1 - B_0(A_0 - C_2A_2^{-1}B_0)^{-1}(B_1 - C_2A_2^{-1}C_1)\right]\Big\}^{-1}, \\
B'_1 &= -(A_0 - C_2A_2^{-1}B_0)^{-1}(B_1 - C_2A_2^{-1}C_1)A'_1, \\
C'_1 &= -A_2^{-1}\Big[C_1 - B_0(A_0 - C_2A_2^{-1}B_0)^{-1}(B_1 - C_2A_2^{-1}C_1)\Big]A'_1, \\
A'_2 &= \Big\{A_2 - C_1(A_1 - C_0A_0^{-1}B_1)^{-1}(B_2 - C_0A_0^{-1}C_2) \\
&\qquad - B_0A_0^{-1}\left[C_2 - B_1(A_1 - C_0A_0^{-1}B_1)^{-1}(B_2 - C_0A_0^{-1}C_2)\right]\Big\}^{-1}, \\
B'_2 &= -(A_1 - C_0A_0^{-1}B_1)^{-1}(B_2 - C_0A_0^{-1}C_2)A'_2, \\
C'_2 &= -A_0^{-1}\Big[C_2 - B_1(A_1 - C_0A_0^{-1}B_1)^{-1}(B_2 - C_0A_0^{-1}C_2)\Big]A'_2.
\end{aligned}
\tag{16}
$$

The formulæ above make sense because of Proposition 3.1. QED

However, it can happen that the sum defining the inverse matrix of a given matrix $A(\Lambda_0)$ is infinite even if $A(\Lambda_0)$ is defined by a finite sum. Indeed, we have $(A_0 + A_1\theta^3\theta^2\theta^1)^{-1} = \sum_{k\geq 0}(-A_0^{-1}A_1)^k A_0^{-1}(\theta^3\theta^2\theta^1)^k$, we thus restrict ourselves to the set of matrices such that $A(\Lambda_0)$ and its inverse $A^{-1}(\Lambda_0)$ are given by finite sums. Looking to the sum above, $A(\Lambda_0)^{-1}$ becomes a finite sum if there is some nilpotent elements. For instance this happens if $A_0^{-1}A_1$ is nilpotent. It can also happen that some power of elements in $\Lambda_3(\mathbb{C}^p)$ vanishes. For instance, since $(\theta^2\theta^2\theta^1)^2 = -\theta^2\theta^2(\theta^2\theta^1\theta^2 + \theta^2\theta^2\theta^1)\theta^1 = 0$, we have $(A_0 + A_1\theta^2\theta^2\theta^1)^{-1} = A_0^{-1} - A_0^{-1}A_1A_0^{-1}\theta^2\theta^2\theta^1$. Thus in order that the formulæ above make sense we consider the set of matrix (15) such that the various matrices together with their inverse (see (16)) are given by finite sum.

Definition 3.3 *The set of matrices of the forme*

$$
M = \begin{pmatrix} A_0 & B_1 & C_2 \\ C_0 & A_1 & B_2 \\ B_0 & C_1 & A_2 \end{pmatrix}, \tag{17}
$$

where the matrices $A_0 \in GL(m_1, \Lambda_0), A_1 \in GL(m_2, \Lambda_0), A_2 \in GL(m_3, \Lambda_0)$, $B_0 \in \mathcal{M}_{m_3,m_1}(\Lambda_1), B_1 \in \mathcal{M}_{m_1,m_2}(\Lambda_1), B_2 \in \mathcal{M}_{m_2,m_3}(\Lambda_1)$, $C_0 \in \mathcal{M}_{m_2,m_1}(\Lambda_2), C_1 \in \mathcal{M}_{m_3,m_2}(\Lambda_2), C_2 \in \mathcal{M}_{m_1,m_3}(\Lambda_2)$, *together with the matrices* $A'_0 \in GL(m_1, \Lambda_0), A'_1 \in GL(m_2, \Lambda_0), A'_2 \in GL(m_3, \Lambda_0)$, $B'_0 \in \mathcal{M}_{m_3,m_1}(\Lambda_1), B'_1 \in \mathcal{M}_{m_1,m_2}(\Lambda_1), B'_2 \in \mathcal{M}_{m_2,m_3}(\Lambda_1)$, $C'_0 \in \mathcal{M}_{m_2,m_1}(\Lambda_2), C'_1 \in \mathcal{M}_{m_3,m_2}(\Lambda_2), C'_2 \in \mathcal{M}_{m_1,m_3}(\Lambda_2)$, *(see* (16)*) are given by finite sums is a group we denote* $GL_f(m_1, m_2, m_3, \Lambda)$.

We now show, in relation with the previous section, that this group in non-empty. Indeed, if we take

$$g = e^A \prod_k e^{B_k},$$

with A an arbitrary matrix of $\mathfrak{gl}(m_1) \times \mathfrak{gl}(m_2) \times \mathfrak{gl}(m_3)$ and B_k an arbitrary nilpotent matrix of $\mathfrak{gl}(m_1, m_2, m_3)_1 \otimes \Lambda_3(\mathbb{C}^p)_1$ then $g \in GL_f(m_1, m_2, m_3, \Lambda)$. There is several type of matrix B_k. For instance one can take $B_k = YP(\theta)$ with (i) Y a nilpotent matrix of $\mathfrak{gl}(m_1, m_2, m_3)_1$ and P an arbitrary polynomial of degree 1 mod 3 in θ or (ii) Y an arbitrary element of $\mathfrak{gl}(m_1, m_2, m_3)_1$ and P a nilpotent polynomial of degree 1 mod 3 in θ (some power of P vanishes). More details will be given in [18].

To close this section, we now show that the algebra corresponding to the group $GL_f(m_1, m_2, m_3, \Lambda)$ is $\mathfrak{gl}_{\rm el}(m_1, m_2, m_2)$. Indeed, if we take the infinitesimal version of $GL_f(m_1, m_2, m_3, \Lambda)$ *i.e.* we assume that the parameters of the transformation characterised by θ goes to zero we have

$$\begin{array}{llll} A_0 & = A_{0(0)} + A_{0(3)ijk}\theta^i\theta^j\theta^k + \cdots & \to & A_{0(0)} + \mathcal{O}((\theta)^2) \\ B_0 & = B_{0(1)i}\theta^i + A_{0(4)ijk\ell}\theta^i\theta^j\theta^k\theta^\ell + \cdots & \to & A_{0(1)i}\theta^i + \mathcal{O}((\theta)^2) \\ C_0 & = C_{0(2)ij}\theta^i\theta^j + C_{0(5)ijk\ell m}\theta^i\theta^j\theta^k\theta^\ell\theta^m + \cdots & \to & 0 + \mathcal{O}((\theta)^2) \end{array} \tag{18}$$

with $A_{0(0)} \in GL(m_1)$ (and similar relations for $A_1, A_2, B_1, B_2, C_1, C_2$). Since $\mathfrak{gl}(m_1)$ is the Lie algebra of $GL(m_1)$, the relations (18) ensure that the algebra associated to $GL_f(m_1, m_2, m_2, Lambda)$ is the Lie algebra of order 3 $\mathfrak{gl}_{\rm el}(m_1, m_2, m_3)$. In fact the Hopf algebra technics of the previous sub-section gives the converse correspondence *i.e.* enables us to establish that the group associated to $\mathfrak{gl}_{\rm el}(m_1, m_2, m_3)$ is $GL_f(m_1, m_2, m_3, \Lambda)$ [18].

Acknowledgments
J. Lukierski is gratefully acknowledge for discussions.

Appendix A. The three-exterior algebra

The three-exterior algebra has been introduced by Roby [17] in 1970 as a possible (cubic) generalisation of the Grasmann algebra. The three-exterior algebra $\Lambda_3(\mathbb{K}^n)$ is the unitary (with unit denoted by 1) $\mathbb{K}-$algebra ($\mathbb{K} = \mathbb{R}$ or $\mathbb{K} = \mathbb{C}$) generated by n canonical generators $\theta^1, \cdots, \theta^n$ submitted to the relation

$$\theta^i\theta^j\theta^k + \theta^j\theta^k\theta^i + \theta^k\theta^i\theta^j + \theta^i\theta^k\theta^j + \theta^j\theta^i\theta^k + \theta^k\theta^j\theta^i = 0. \tag{A.1}$$

This algebra which can be seen as a possible generalisation of the Grassmann algebra is indeed very different. Because the algebra $\Lambda_3(\mathbb{K}^n)$ is defined through cubic relations the number of independent monomials increases with polynomial's degree (for instance, $(\theta^1)^2\theta^2$ and $\theta^1\theta^2\theta^1$ are independent). This means that we do not have enough constraints among the generators to order them in some fixed way and, as a consequence, $\Lambda(\mathbb{K}^n)$ turns out to be an infinite dimensional algebra. To characterise precisely $\Lambda(\mathbb{K}^n)$ it should be interesting to obtain a basis. This is complicated by the fact that we have some cubic relations among the generators and for instance the three element $(\theta^1)^2\theta^2, \theta^1\theta^2\theta^1$ and $\theta^2(\theta^1)^2$ are not independent since there sum is zero. To characterise the set of independent elements one needs a definition.

Definition A1 *The sequence* $(i_1, i_2, \cdots i_k) \in \{1, \cdots, n\}^k$, $k \geq 3$ *has a rise of length* 3 *if there exists* $0 \leq \ell \leq k-3$ *such that* $i_{\ell+1} \leq i_{\ell+2} \leq i_{\ell+3}$. *We denote by* $I_{k,n}$ *the set of* k *indices which has a rise of length* 3.

Journal of Physics: Conference Series **128** (2008) 012060 doi:10.1088/1742-6596/128/1/012060

Theorem A2 *(N. Roby [17]) The elements*

$$\left\{\theta^{i_1}\theta^{i_2}\cdots\theta^{i_k}, k\in\mathbb{N}, (i_1,\cdots,i_k)\in\{1,\cdots n\}^k\setminus I_{k,n}\right\},$$

constitute a basis of $\Lambda_3(\mathbb{K}^n)$.

For instance if $n=2$ a basis of $\Lambda_3(\mathbb{C}^2)$ is given by

$$\begin{array}{l}
1,\\
\theta^i,\quad i=1,2,\\
\theta^i\theta^j,\quad i,j=1,2,\\
\theta^i(\theta^2\theta^1)^n,\quad (\theta^2\theta^1)^n\theta^i,\quad i=1,2,\quad n\geq 1,\\
\theta^i(\theta^2\theta^1)^{n-1}\theta^j,\quad i,j=1,2,\quad (\theta^2\theta^1)^n,\quad n\geq 2.
\end{array}$$

From now on we call Roby elements the elements $\theta^{I_k}=\theta^{i_1}\cdots\theta^{i_k}$ such that $I_k\in\{1,\cdots,n\}^k\backslash I_{k,n}$.

The $3-$exterior algebra is a $\mathbb{Z}_3-$graded algebra. Indeed, if we define $\Lambda(\mathbb{K}^n)_i, i=0,1,2$ the set of elements of degree i mod 3 we have $\Lambda_3(\mathbb{K}^n)_i\Lambda_3(\mathbb{K}^n)_j\subseteq\Lambda_3(\mathbb{K}^n)_{i+j}$ and $\Lambda_3(\mathbb{K}^n)=\Lambda_3(\mathbb{K}^n)_0\oplus\Lambda_3(\mathbb{K}^n)_1\oplus\Lambda_3(\mathbb{K}^n)_2$.

Finally let us mention that the results given in this appendix extend to any $F>2$ [17].

[1] Bars I and Gunaydin M 1979 *J. Math. Phys.* **20** 1977-1993.
[2] Bars I and Gunaydin M 1980 *Phys. Rev. D* **22** 1403-1413.
[3] Vainerman L and Kerner R 1995 *J. Math. Phys* **37** 2553-2565.
[4] Kerner R 1997 *Class. and Quantum Grav.* **14 (1A)** A203-A225.
[5] Borowiec A, Bazunova N and Kerner R 2004 *Lett. Math. Phys.* **67** 195-206.
[6] Rausch de Traubenberg M, *Clifford algebras, supersymmetry and $\mathbb{Z}_n-$symmetries: Applications in field theory,* arXiv:hep-th/9802141 (Habilitation Thesis).
[7] Filipov V. T. 1985 *Sibirsk. Math. Zh* **26** 126-140.
[8] Gnedbaye A. V. 1995 *C. R. Acad. Sci. Paris Sér. I Math.* **321** 147-152.
[9] Michor P. W. and Vinogradov A. M. 1996 *Rend. Sem. Mat. Univ. Politec. Torino* **54** 373-392.
[10] Rausch de Traubenberg M and Slupinski M J 2000 *J. Math. Phys.* **41** 4556-4579, [arXiv:hep-th/9904126].
[11] Rausch de Traubenberg M and Slupinski M J 2002 *J. Math. Phys.* **43** 5145-5160, [arXiv:hep-th/0205113].
[12] Mohammedi N, Moultaka G and Rausch de Traubenberg M 2004 *Int. J. Mod. Phys.* **A19** 5585-5608, [arXiv:hep-th/0305172].
[13] Moultaka G, Rausch de Traubenberg M and Tanasa A (2005) *Int. J. Mod. Phys.* **A20** 5779-5806, [arXiv:hep-th/0411198].
[14] Moultaka G, Rausch de Traubenberg M and Tanasa A 2004 Proceedings of the XIth International Conference Symmetry Methods in Physics, Prague 21-24 June 2004, [arXiv:hep-th/0407168].
[15] Rausch de Traubenberg M 2006 *to appear in the Proocedings of the International Conference on Symmetry Methods in Physics (SYMPHYS-12), Yerevan, Armenia, 3-8 Jul 2006*, [arXiv:hep-th/0612204].
[16] Goze M, Rausch de Traubenberg M and Tanasa A 2007 *J. Math. Phys.* **48** 093507, [arXiv:math-ph/0603008].
[17] Roby N 1970 *Bull. Sc. Math.* **94** 49-57.
[18] Goze M and Rausch de Traubenberg M, *in preparation.*
[19] Ahmedov H, Yildiz A and Ucan Y 2001 *J. Phys.* **A 34** 6413-6423, [arXiv:math.RT/0012058].

Journal of Physics: Conference Series **128** (2008) 012061 doi:10.1088/1742-6596/128/1/012061

Algebraic generalization of quantum statistics

N.I. Stoilova [1] **and J. Van der Jeugt**

Department of Applied Mathematics and Computer Science,
Ghent University, Krijgslaan 281-S9, B-9000 Gent, Belgium.

E-mail: `Neli.Stoilova@UGent.be, Joris.VanderJeugt@UGent.be.`

Abstract. Generalized quantum statistics such as para-Bose and para-Fermi statistics are related to the basic classical Lie superalgebras $B(0|n)$ and B_n. We give a quite general definition of "a generalized quantum statistics associated to a Lie superalgebra G". This definition is closely related to a certain $\mathbb{Z}$-grading of G. The generalized quantum statistics is determined by a set of root vectors (the creation and annihilation operators of the statistics) and the set of algebraic relations for these operators. Then we give a complete classification of all generalized quantum statistics associated to the Lie superalgebras A_n, B_n, C_n, D_n, G_2, F_4, E_6, E_7, E_8, $A(m|n)$, $B(m|n)$, $C(n)$, $D(m|n)$, $G(3)$, $F(4)$ and $D(2,1;\alpha)$.

1. Introduction

Green [1] extended the ordinary Bose and Fermi statistics to parastatistics. Both for para-Bose and para-Fermi operators, the bilinear commutators or anticommutators for bosons and fermions are replaced by trilinear relations. These trilinear relations are closely related to a Lie algebra or Lie superalgebra. For example, a set of $2n$ para-Fermi operators f_i^ξ $(\xi = \pm,\ i = 1, \ldots, n)$

$$[[f_j^\xi, f_k^\eta], f_l^\epsilon] = \frac{1}{2}(\epsilon - \eta)^2 \delta_{kl} f_j^\xi - \frac{1}{2}(\epsilon - \xi)^2 \delta_{jl} f_k^\eta,\ \ \xi, \eta, \epsilon = \pm \text{ or } \pm 1;\ j, k, l = 1, \ldots, n \quad (1)$$

generates the Lie algebra $B_n = so(2n+1)$ [2], whereas a set of $2n$ para-Bose operators b_i^ξ $(\xi = \pm,$ $i = 1, \ldots, n)$

$$[\{b_j^\xi, b_k^\eta\}, b_l^\epsilon] = (\epsilon - \xi)\delta_{jl} b_k^\eta + (\epsilon - \eta)\delta_{kl} b_j^\xi,\ \ \xi, \eta, \epsilon = \pm \text{ or } \pm 1;\ j, k, l = 1, \ldots, n \quad (2)$$

considered as odd elements, generates the orthosymplectic Lie superalgebra $B(0|n) = osp(1|2n)$ [3]. Therefore parastatistics can be associated with representations of the Lie superalgebras of class B [4]. Then it is natural to expect that new types of generalized quantum statistics (GQS) can be associated with other Lie superalgebras. Note that, for conveniece, we often consider Lie algebras as a subset of Lie superalgebras (namely as Lie superalgebras with even elements only).

Examples of alternative types of GQS have been considered by Palev [5]-[13]. Inspired by the examples of parastatistics and Palev's examples we give a classification of all GQS associated

[1] Permanent address: Institute for Nuclear Research and Nuclear Energy, Boul. Tsarigradsko Chaussee 72, 1784 Sofia, Bulgaria

with the Lie superalgebras A_n, B_n, C_n, D_n, G_2, F_4, E_6, E_7, E_8, $A(m|n)$, $B(m|n)$, $C(n)$, $D(m|n)$, $G(3)$, $F(4)$ and $D(2,1;\alpha)$. More details of the classification presented here can be found in the series of papers [14]-[16].

GQS associated to a Lie superalgebra G is related to a $\mathbb{Z}$-grading of G of the type $G = G_{-2} \oplus G_{-1} \oplus G_0 \oplus G_{+1} \oplus G_{+2}$. Such a $\mathbb{Z}$-grading is said to be of length 5 if $G_{\pm 2} \neq 0$; if $G_{\pm 2} = 0$, but $G_{\pm 1} \neq 0$, then the $\mathbb{Z}$-grading is of length 3. These $\mathbb{Z}$-gradings imply that one is dealing with Lie (super)triple systems (in the case of Lie algebras, see [17] and in the case of Lie superalgebras, see [18]). Such systems have also been referred to as "(super)ternary algebras" in [19]. Another mathematical structure, F-Lie algebra, also called ternary algebra for $F = 3$ was introduced in [20]. However the systems considered here are in the sense of [19]. In the last paper, examples and explicit constructions of (super)ternary algebras are given for many simple Lie algebras and Lie superalgebras. In section VII of [19] one notes: "It would be interesting to embark on a complete classification of ternary algebras and superternary algebras and provide a list of all possible constructions of a given Lie (super)algebra from (super)ternary algebras." The results of the present paper provide such a complete classification.

In the next section we give a definition of generalized quantum statistics associated with a Lie superalgebra G and the corresponding creation and annihilation operators. This notion is closely related to gradings of G, and to regular subalgebras of G. Following the definition, we describe the classification method. In the process of investigations Dynkin diagram techniques [21, 22] play a crucial role. In the remaining sections of the paper, the classification results are summarised.

2. Definition and classification method

Let G be a Lie superalgebra [4], with bracket $[\![x, y]\!]$, where (in $U(G)$)

$$[\![x, y]\!] = xy - (-1)^{\deg(x)\deg(y)} yx,$$

if x and y are homogeneous (for a Lie algebra, all elements are even and have degree 0; for a Lie superalgebra, homogeneous elements are even or odd, having degree 0 or 1).

Definition 1 *Let G be a Lie superalgebra, with antilinear anti-involutive mapping ω. A set of $2N$ root vectors $x_i^{\pm}$ ($i = 1, \ldots, N$) is called a set of creation and annihilation operators (CAOs) for G if:*

- $\omega(x_i^{\pm}) = x_i^{\mp}$,
- $G = G_{-2} \oplus G_{-1} \oplus G_0 \oplus G_{+1} \oplus G_{+2}$ *is a $\mathbb{Z}$-grading of G, with $G_{\pm 1} = span\{x_i^{\pm},\ i = 1 \ldots, N\}$ and $G_{j+k} = [\![G_j, G_k]\!]$.*

The algebraic relations $\mathcal{R}$ satisfied by the operators $x_i^{\pm}$ are the relations of a generalized quantum statistics (GQS) associated with G.

A consequence of this definition is that G is generated by G_{-1} and G_{+1}, i.e. by the set of CAOs, and since $G_{j+k} = [\![G_j, G_k]\!]$, it follows that

$$G = \mathrm{span}\{x_i^{\xi},\ [\![x_i^{\xi}, x_j^{\eta}]\!];\quad i, j = 1, \ldots, N,\ \xi, \eta = \pm\}. \tag{3}$$

The second condition in Definition 1 implies that G_0 itself is a subalgebra of G spanned by root vectors of G. So G_0 is a regular subalgebra containing the Cartan subalgebra H of G. By the adjoint action, the remaining G_i's are G_0-modules. This implies the following technique in order to obtain a complete classification of all GQS associated with G:

(i) Determine all regular subalgebras G_0 of G. If not yet contained in G_0, replace G_0 by $G_0 + H$.

Journal of Physics: Conference Series **128** (2008) 012061 doi:10.1088/1742-6596/128/1/012061

(ii) For each regular subalgebra G_0, determine the decomposition of G into simple G_0-modules g_k $(k = 1, 2, \ldots)$.

(iii) Investigate whether there exists a $\mathbb{Z}$-grading of G of the form

$$G = G_{-2} \oplus G_{-1} \oplus G_0 \oplus G_{+1} \oplus G_{+2}, \tag{4}$$

where each G_i is either directly a module g_k or else a sum of such modules $g_1 \oplus g_2 \oplus \cdots$, such that $\omega(G_{+i}) = G_{-i}$.

To find regular subalgebras one can use the method of deleting nodes from (extended) Dynkin diagrams [21, 22]. The second stage is straightforward by means of representation theoretical techniques. The third stage requires most of the work: one must try out all possible combinations of the G_0-modules g_k, and see whether it is possible to obtain a grading of the type (4). In this process, if one of the simple G_0-modules g_k is such that $\omega(g_k) = g_k$, then it follows that this module should be part of G_0. In other words, such a case reduces essentially to another case with a larger regular subalgebra.

In the following sections we shall give a summary of the classification process for the simple Lie algebras and basic classical Lie superalgebras.

3. Simple Lie algebras

3.1. The Lie algebra A_n

Let G be the special linear Lie algebra $sl(n+1) \equiv A_n$, consisting of traceless $(n+1) \times (n+1)$ matrices. The Cartan subalgebra H of G is the subspace of diagonal matrices. The root vectors of G are the elements e_{jk} $(j \neq k = 1, \ldots, n+1)$, where e_{jk} is a matrix with zeros everywhere except a 1 on the intersection of row j and column k. The corresponding root is $\epsilon_j - \epsilon_k$, in the usual basis. The anti-involution is such that $\omega(e_{jk}) = e_{kj}$. The simple roots, the Dynkin diagram and the extended Dynkin diagram of A_n are given in Table 1.

In order to find regular subalgebras of $G = A_n$, one should delete nodes from the Dynkin diagram of G or from its extended Dynkin diagram.

Step 1. Delete node i from the Dynkin diagram. The corresponding diagram is the Dynkin diagram of $sl(i) \oplus sl(n-i+1)$, so $G_0 = H + sl(i) \oplus sl(n-i+1)$. In this case, there are only two G_0 modules and we can put

$$G_{-1} = \text{span}\{e_{kl};\ k = 1, \ldots, i,\ l = i+1, \ldots, n+1\}, \quad G_{+1} = \omega(G_{-1}). \tag{5}$$

Therefore $sl(n+1)$ has the following grading:

$$sl(n+1) = G_{-1} \oplus G_0 \oplus G_{+1}, \tag{6}$$

the number of creation and annihilation operators is $N = i(n-i+1)$. Note that the cases i and $n+1-i$ are isomorphic.

The most interesting cases are those with $i = 1$ and $i = 2$, for which we shall explicitly give the relations $\mathcal{R}$ between the CAOs.

For $i = 1$, $N = n$, the rank of A_n. Putting

$$a_j^- = e_{1,j+1}, \qquad a_j^+ = e_{j+1,1},\ j = 1, \ldots, n, \tag{7}$$

the corresponding relations $\mathcal{R}$ read $(j, k, l = 1, \ldots, n)$:

$$\begin{aligned} &[a_j^+, a_k^+] = [a_j^-, a_k^-] = 0, \\ &[[a_j^+, a_k^-], a_l^+] = \delta_{jk} a_l^+ + \delta_{kl} a_j^+, \\ &[[a_j^+, a_k^-], a_l^-] = -\delta_{jk} a_l^- - \delta_{jl} a_k^-. \end{aligned} \tag{8}$$

Table 1. Simple Lie algebras, their (extended) Dynkin diagrams with a labelling of the nodes and the corresponding simple roots.

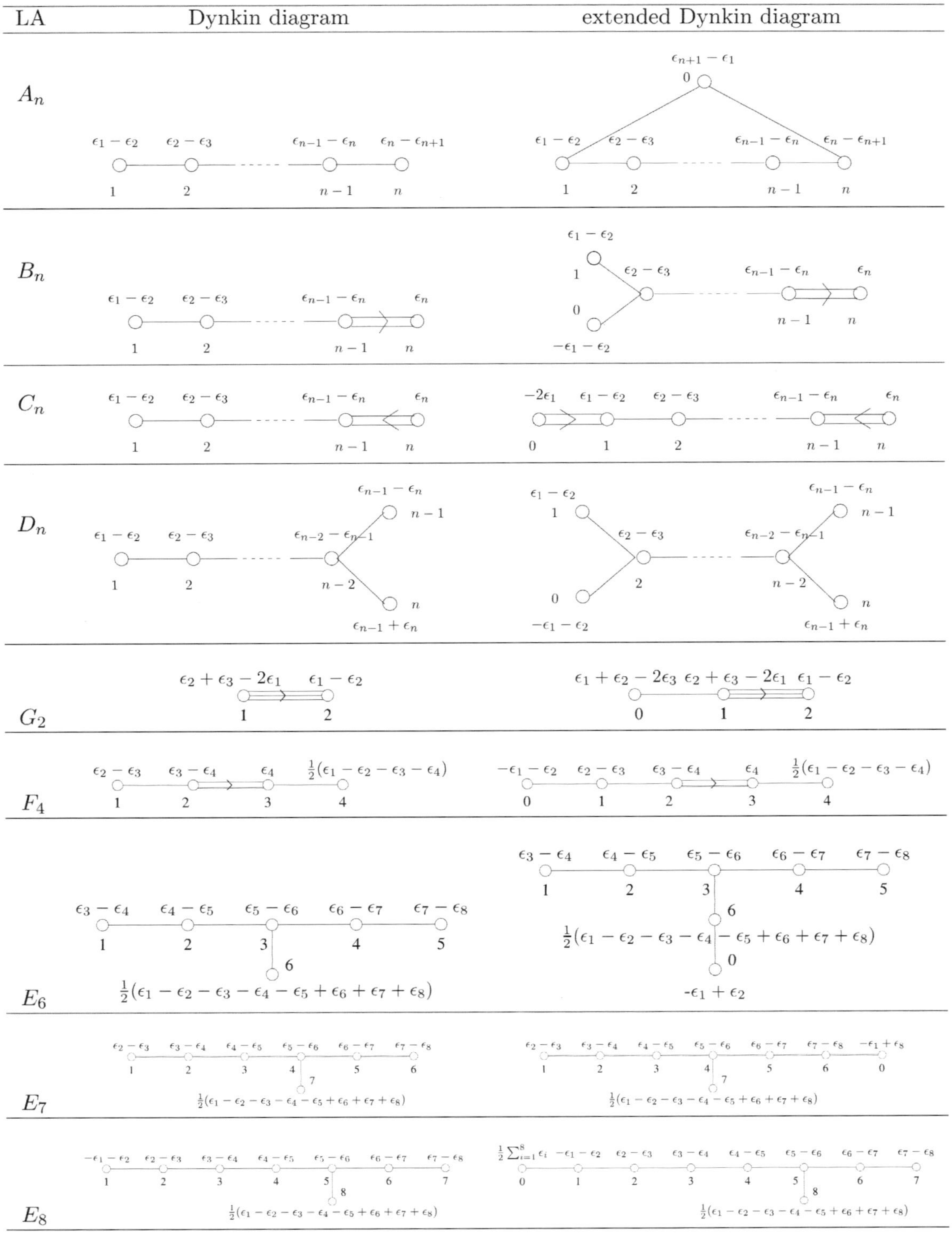

Journal of Physics: Conference Series **128** (2008) 012061 doi:10.1088/1742-6596/128/1/012061

These are the relations of the so called A-statistics [5]-[6], [9]-[11].

For $i = 2$, $N = 2(n-1)$, let

$$\begin{aligned} a^-_{-j} &= e_{1,j+2}, \quad a^-_{+j} = e_{2,j+2}, \qquad j = 1,\dots,n-1, \\ a^+_{-j} &= e_{j+2,1}, \quad a^+_{+j} = e_{j+2,2}, \qquad j = 1,\dots,n-1. \end{aligned} \tag{9}$$

Now the corresponding relations are ($\xi, \eta, \epsilon = \pm;\ j,k,l = 1,\dots,n-1$):

$$\begin{aligned} &[a^+_{\xi j}, a^+_{\eta k}] = [a^-_{\xi j}, a^-_{\eta k}] = 0, \\ &[a^+_{\xi j}, a^-_{-\xi k}] = 0, \qquad j \neq k, \\ &[a^+_{-j}, a^-_{-k}] = [a^+_{+j}, a^-_{+k}], \qquad j \neq k, \\ &[a^+_{+j}, a^-_{-j}] = [a^+_{+k}, a^-_{-k}], \\ &[a^+_{-j}, a^-_{+j}] = [a^+_{-k}, a^-_{+k}], \\ &[[a^+_{\xi j}, a^-_{\eta k}], a^+_{\epsilon l}] = \delta_{\eta\epsilon}\delta_{jk} a^+_{\xi l} + \delta_{\xi\eta}\delta_{kl} a^+_{\epsilon j}, \\ &[[a^+_{\xi j}, a^-_{\eta k}], a^-_{\epsilon l}] = -\delta_{\xi\epsilon}\delta_{jk} a^-_{\eta l} - \delta_{\xi\eta}\delta_{jl} a^-_{\epsilon k}. \end{aligned} \tag{10}$$

These relations are already more complicated than (8). But they are still defining relations for the Lie algebra A_n.

Step 2. Delete node i and j from the Dynkin diagram. By the symmetry of the Dynkin diagram, it is sufficient to consider $1 \le i \le \lfloor \frac{n}{2} \rfloor$ and $i < j < n+1-i$. We have $G_0 = H + sl(i) \oplus sl(j-i) \oplus sl(n+1-j)$. In this case, there are six simple G_0-modules. All the possible combinations of these modules give rise to gradings of the form

$$sl(n+1) = G_{-2} \oplus G_{-1} \oplus G_0 \oplus G_{+1} \oplus G_{+2}.$$

There are essentially three different ways in which these G_0-modules can be combined. To characterize these three cases, it is sufficient to give only G_{-1}:

$$\begin{aligned} G_{-1} &= \text{span}\{e_{kl}, e_{lp};\ k = 1,\dots,i,\ l = i+1,\dots,j,\ p = j+1,\dots,n+1\}, && (11)\\ &\quad \text{with } N = (j-i)(n+1-j+i); \\ G_{-1} &= \text{span}\{e_{kl}, e_{pk};\ k = 1,\dots,i,\ l = i+1,\dots,j,\ p = j+1,\dots,n+1\}, && (12)\\ &\quad \text{with } N = i(n+1-i); \\ G_{-1} &= \text{span}\{e_{kl}, e_{lp};\ k = 1,\dots,i,\ p = i+1,\dots,j,\ l = j+1,\dots,n+1\}, && (13)\\ &\quad \text{with } N = j(n+1-j). \end{aligned}$$

It turns out that the sets of CAOs corresponding to (12) and (13) are isomorphic to (11), so it is sufficient to consider only (11). Each case of (11) with $1 \le i \le \lfloor \frac{n}{2} \rfloor$ and $i < j < n+1-i$ gives rise to a distinct GQS. For reasons explained earlier, we shall give the corresponding set of relations explicitly only for small N. In this case, it is interesting to give $\mathcal{R}$ for $j - i = 1$, because then the number of creation or annihilation operators is $N = n$. One can label the CAOs as follows:

$$\begin{aligned} a^-_k &= e_{k,i+1}, \quad a^+_k = e_{i+1,k}, \qquad k = 1,\dots,i; \\ a^-_k &= e_{i+1,k+1}, \quad a^+_k = e_{k+1,i+1}, \qquad k = i+1,\dots,n. \end{aligned} \tag{14}$$

Using

$$\langle k \rangle = \begin{cases} 0 & \text{if } k = 1,\dots,i \\ 1 & \text{if } k = i+1,\dots,n \end{cases} \tag{15}$$

the quadratic and triple relations read:

$$\begin{aligned}
&[a_k^+, a_l^+] = [a_k^-, a_l^-] = 0, \qquad k,l = 1,\ldots,i \text{ or } k,l = i+1,\ldots,n,\\
&[a_k^-, a_l^+] = [a_k^+, a_l^-] = 0, \qquad k = 1,\ldots,i,\ l = i+1,\ldots,n,\\
&[[a_k^+, a_l^-], a_m^+] = (-1)^{\langle l\rangle+\langle m\rangle}\delta_{kl}a_m^+ + (-1)^{\langle l\rangle+\langle m\rangle}\delta_{lm}a_k^+,\ k,l = 1,\ldots,i \text{ or } k,l = i+1,\ldots,n,\\
&[[a_k^+, a_l^-], a_m^-] = -(-1)^{\langle l\rangle+\langle m\rangle}\delta_{kl}a_m^- - (-1)^{\langle l\rangle+\langle m\rangle}\delta_{km}a_l^-,\ k,l = 1,\ldots,i \text{ or } k,l = i+1,\ldots,n,\\
&[[a_k^\xi, a_l^\xi], a_m^{-\xi}] = -\delta_{km}a_l^\xi + \delta_{lm}a_k^\xi,\ k = 1,\ldots,i,\ l = i+1,\ldots,n,\\
&[[a_k^\xi, a_l^\xi], a_m^\xi] = 0, \qquad (\xi = \pm;\ k,l,m = 1,\ldots,n).
\end{aligned} \tag{16}$$

The existence of the set of CAOs (14) is pointed out in [5] as a possible example. The relations (16) with $n = 2m$ and $i = m$ are the commutation relations of the so called causal A-statistics investigated in [8].

Step 3. If we delete 3 or more nodes from the Dynkin diagram, the resulting $\mathbb{Z}$-gradings of $sl(n+1)$ are no longer of the form $sl(n+1) = G_{-2} \oplus G_{-1} \oplus G_0 \oplus G_{+1} \oplus G_{+2}$, but there would be non-zero G_i with $|i| > 2$, so these cases are not relevant for our classification.

Step 4. Next, we move on to the extended Dynkin diagram of G. If we delete node i from the extended Dynkin diagram, then remaining diagram is again of type A_n, so $G_0 = G$, and there are no CAOs.

Step 5. If we delete node i and j from the extended Dynkin diagram ($0 \le i < j \le n+1$), then $sl(n+1) = G_{-1} \oplus G_0 \oplus G_{+1}$ with $G_0 = H + sl(j-i) \oplus sl(n-j+i+1)$, and

$$G_{-1} = \text{span}\{e_{kl};\ k = i+1\ldots,j,\ l \neq i+1,\ldots,j\}.$$

The number of annihilation operators is $N = (j-i)(n+1-j+i)$. It is not difficult to see that all these cases are isomorphic to those of Step 1. This can also be deduced from the symmetry of the Dynkin diagram.

Step 6. If we delete nodes i, j and k from the extended Dynkin diagram ($i < j < k$), then the corresponding $\mathbb{Z}$-gradings are of the form

$$sl(n+1) = G_{-2} \oplus G_{-1} \oplus G_0 \oplus G_{+1} \oplus G_{+2}.$$

All the corresponding CAOs, however, are isomorphic to those of Step 2 (which can again be seen from the remaining Dynkin diagram).

Step 7. If we delete 4 or more nodes from the extended Dynkin diagram, the corresponding $\mathbb{Z}$-grading of $sl(n+1)$ has no longer the required properties (i.e. there are non-zero subspaces G_i with $|i| > 2$).

3.2. The Lie algebras B_n, C_n and D_n

$G = so(2n+1) \equiv B_n$ is the subalgebra of $sl(2n+1)$ consisting of matrices of the form:

$$\begin{pmatrix} a & b & c \\ d & -a^t & e \\ -e^t & -c^t & 0 \end{pmatrix}, \tag{17}$$

where a is any $(n \times n)$-matrix, b and d are antisymmetric $(n \times n)$-matrices, and c and e are $(n \times 1)$-matrices. The Cartan subalgebra H of G is again the subspace of diagonal matrices.

Journal of Physics: Conference Series **128** (2008) 012061 doi:10.1088/1742-6596/128/1/012061

The root vectors and corresponding roots of G are given by:

$$
\begin{aligned}
e_{jk} - e_{k+n,j+n} &\leftrightarrow \epsilon_j - \epsilon_k, && j \neq k = 1,\ldots,n,\\
e_{j,k+n} - e_{k,j+n} &\leftrightarrow \epsilon_j + \epsilon_k, && j < k = 1,\ldots,n,\\
e_{j+n,k} - e_{k+n,j} &\leftrightarrow -\epsilon_j - \epsilon_k, && j < k = 1,\ldots,n,\\
e_{j,2n+1} - e_{2n+1,j+n} &\leftrightarrow \epsilon_j, && j = 1,\ldots,n,\\
e_{n+j,2n+1} - e_{2n+1,j} &\leftrightarrow -\epsilon_j, && j = 1,\ldots,n.
\end{aligned}
$$

$G = sp(2n) \equiv C_n$ is the subalgebra of $sl(2n)$ consisting of matrices of the form:

$$\begin{pmatrix} a & b \\ c & -a^t \end{pmatrix}, \tag{18}$$

where a is any $(n\times n)$-matrix, and b and c are symmetric $(n\times n)$-matrices. The Cartan subalgebra H consist of the diagonal matrices, and the root vectors and corresponding roots of G are:

$$
\begin{aligned}
e_{jk} - e_{k+n,j+n} &\leftrightarrow \epsilon_j - \epsilon_k, && j \neq k = 1,\ldots,n,\\
e_{j,k+n} + e_{k,j+n} &\leftrightarrow \epsilon_j + \epsilon_k, && j \leq k = 1,\ldots,n,\\
e_{j+n,k} + e_{k+n,j} &\leftrightarrow -\epsilon_j - \epsilon_k, && j \leq k = 1,\ldots,n.
\end{aligned}
$$

$G = so(2n) \equiv D_n$ is the subalgebra of $sl(2n)$ consisting of matrices of the form:

$$\begin{pmatrix} a & b \\ c & -a^t \end{pmatrix}, \tag{19}$$

where a is any $(n \times n)$-matrix, and b and c are antisymmetric $(n \times n)$-matrices. The Cartan subalgebra H consist of the diagonal matrices, and the root vectors and corresponding roots of G are:

$$
\begin{aligned}
e_{jk} - e_{k+n,j+n} &\leftrightarrow \epsilon_j - \epsilon_k, && j \neq k = 1,\ldots,n,\\
e_{j,k+n} - e_{k,j+n} &\leftrightarrow \epsilon_j + \epsilon_k, && j < k = 1,\ldots,n,\\
e_{j+n,k} - e_{k+n,j} &\leftrightarrow -\epsilon_j - \epsilon_k, && j < k = 1,\ldots,n.
\end{aligned}
$$

The simple roots, Dynkin diagrams and extended Dynkin diagrams of B_n, C_n and D_n are given in Table 1. The anti-involution is such that $\omega(e_{jk}) = e_{kj}$. Just as for A_n, we have performed the process of deleting nodes from the Dynkin diagrams and from the extended Dynkin diagrams of B_n, C_n and D_n. The results are summarized in Table 2.

3.3. The exceptional Lie algebras

The Lie algebra G_2 of rank 2 has dimension 14. In terms of the orthonormal vectors $\epsilon_1, \epsilon_2, \epsilon_3$ such that $\epsilon_1 + \epsilon_2 + \epsilon_3 = 0$, the root system is given by

$$\Delta = \{\epsilon_i - \epsilon_j,\ \epsilon_i + \epsilon_j - 2\epsilon_k\ \ (1 \leq i \neq j \neq k \leq 3)\}. \tag{20}$$

The simple root system is

$$\Pi = \{\alpha_1 = \epsilon_2 + \epsilon_3 - 2\epsilon_1,\ \alpha_2 = \epsilon_1 - \epsilon_2\} \tag{21}$$

and the corresponding Dynkin diagram and extended Dynkin diagram are given in Table 1. The process described in the previous section, deleting nodes from the (extended) Dynkin diagram, leads to the results in Table 2.

Table 2. Summary of the classification for simple Lie algebras: all non-isomorphic GQS are given. For each GQS, we list: the subalgebra G_0 (each G_0 contains the complete Cartan subalgebra H, so we only list the remaining part of $G_0 = H + \cdots$); the length l of the $\mathbb{Z}$-grading (3 or 5), $\dim G_1 = \dim G_{-1}$, which is also the number N of creation and annihilation operators.

Lie algebra	$G_0 = H+$	ℓ	dim $G_1 = N$
A_n	$sl(i) \oplus sl(n-i+1), i \leq \lfloor \frac{n+1}{2} \rfloor$	3	$i(n+1-i)$
	$sl(i) \oplus sl(j-i) \oplus sl(n+1-j), (i \leq \lfloor \frac{n}{2} \rfloor, i < j < n+1-i)$	5	$(j-i)(n+1-j+i)$
B_n	$so(2n-1)$	3	$2n-1$
	$sl(i) \oplus so(2(n-i)+1), (2 \leq i \leq n)$	5	$2i(n-i)+i$
C_n	$sl(i) \oplus sp(2(n-i)), (1 \leq i \leq n-1)$	3	$2i(n-i)$
	$sl(n)$	3	$\frac{n(n+1)}{2}$
D_n	$so(2n-2)$	3	$2(n-1)$
	$sl(i) \oplus so(2(n-i)), (2 \leq i \leq n-3)$	5	$2i(n-i)$
	$sl(n-2) \oplus sl(2) \oplus sl(2)$	5	$4(n-2)$
	$sl(n-1)$	5	$2(n-1)$ and $\frac{n(n-1)}{2}$
	$sl(n)$	3	$\frac{n(n-1)}{2}$
G_2	$sl(2)$	5	4
F_4	$sp(6)$	5	14
	$so(7)$	5	8
E_6	$so(10)$	3	16
	$sl(2) \oplus sl(5)$	5	20
	$sl(6)$	5	20
	$so(8)$	5	16
E_7	E_6	3	27
	$sl(2) \oplus so(10)$	5	32
	$so(12)$	5	32
	$sl(7)$	5	35
E_8	E_7	5	56
	$so(14)$	5	64

Let us next consider the Lie algebra F_4, of rank 4 and dimension 52. In terms of the orthonormal vectors $\epsilon_1, \epsilon_2, \epsilon_3, \epsilon_4$ the root system is given by

$$\Delta = \{\pm\epsilon_i \pm \epsilon_j \ (1 \leq i \neq j \leq 4);\ \pm\epsilon_j \ (1 \leq j \leq 4);\ \frac{1}{2}(\pm\epsilon_1 \pm \epsilon_2 \pm \epsilon_3 \pm \epsilon_4)\}. \tag{22}$$

The simple root system is

$$\Pi = \{\alpha_1 = \epsilon_2 - \epsilon_3,\ \alpha_2 = \epsilon_3 - \epsilon_4,\ \alpha_3 = \epsilon_4,\ \alpha_4 = \frac{1}{2}(\epsilon_1 - \epsilon_2 - \epsilon_3 - \epsilon_4)\} \tag{23}$$

and the corresponding Dynkin diagram and the extended Dynkin diagram are given in Table 1. We have considered now the various ways of deleting nodes from these diagrams, and investigate whether they give rise to $\mathbb{Z}$-gradings of the type (4) giving the result in Table 2.

For the remaining exceptional Lie algebras E_6, E_7 and E_8, our labeling of simple roots is again the usual one. But our choice of (simple) roots in terms of vectors ϵ_i is slightly different. For E_8, our choice is essentially the same as in [23, Table 1], except that we work with an independent basis ϵ_i in R^8 (and not a redundant basis in R^9). The roots for E_8 are the same as in [24] (but in this last textbook the choice of simple roots is different). For E_7 and E_6 it is

Journal of Physics: Conference Series **128** (2008) 012061 doi:10.1088/1742-6596/128/1/012061

convenient for us to take the same root space as for E_8. The simple roots of E_7 are then those of E_8 with the first one deleted, and the simple roots of E_6 are then those of E_7 with the first one deleted.

The Lie algebra E_6 of rank 6 has dimension 78. We will use the following root system of E_6. Consider the 8-dimensional real vector space R^8 with orthonormal basis vectors ϵ_i $(i = 1, \ldots, 8)$. The roots of E_6 are elements of the 6-dimensional subspace V of R^8 consisting of those elements $\sum_{i=1}^{8} c_i\epsilon_i$ with $c_1 + c_2 = 0$ and $\sum_{i=3}^{8} c_i = 0$. A set of simple roots of E_6 is then given by the elements

$$\alpha_i = \epsilon_{i+2} - \epsilon_{i+3} \;\; (i = 1, \ldots, 5), \;\; \alpha_6 = \frac{1}{2}(\epsilon_1 - \epsilon_2 - \epsilon_3 - \epsilon_4 - \epsilon_5 + \epsilon_6 + \epsilon_7 + \epsilon_8). \tag{24}$$

All 72 nonzero roots are given by

$$\begin{aligned} &\pm(\epsilon_i - \epsilon_j), \;\; (1 \leq i \neq j \leq 2 \text{ or } 3 \leq i \neq j \leq 8) \\ &\frac{1}{2}(\sum_{i=1}^{8}(-1)^{a_i}\epsilon_i), \;\; (a_i \in \{0, 1\}; \sum_{i=1}^{2} a_i = 1, \; \sum_{i=3}^{8} a_i = 3). \end{aligned} \tag{25}$$

The corresponding Dynkin diagram and the extended Dynkin diagram are given in Table 1.

The roots of E_7 are elements of the 7-dimensional subspace V' consisting of elements $\sum_{i=1}^{8} c_i\epsilon_i$ with $\sum_{i=1}^{8} c_i = 0$. A set of simple roots of E_7 consists of the six simple roots α_i $(i = 1, \ldots, 6)$ of E_6 plus the extra root

$$\epsilon_2 - \epsilon_3. \tag{26}$$

The corresponding Dynkin diagram and the extended Dynkin diagram are given in Table 1.

By construction, the E_6 subsystem of E_7 is evident. The nonzero roots of E_7 consist of

$$\begin{aligned} &\pm(\epsilon_i - \epsilon_j), \;\; (1 \leq i < j \leq 8); \\ &\frac{1}{2}(\sum_{i=1}^{8}(-1)^{a_i}\epsilon_i), \;\; (a_i \in \{0, 1\}; \sum_{i=1}^{8} a_i = 4). \end{aligned} \tag{27}$$

Note that the 72 nonzero roots of E_6 are indeed part of the 126 nonzero roots (27).

In terms of the orthonormal vectors ϵ_i, $i = 1, \ldots, 8$ the root system of E_8 is given by

$$\begin{aligned} &\pm\epsilon_i \pm \epsilon_j, \;\; (1 \leq i < j \leq 8); \\ &\frac{1}{2}\sum_{i=1}^{8}\xi_i\epsilon_i, \;\; \xi_i = \pm 1 \text{ and the number of } \xi_i = +1 \text{ is even.} \end{aligned} \tag{28}$$

A set of simple roots of E_8 consists of the seven simple roots α_i $(i = 1, \ldots, 7)$ of E_7 plus the extra root

$$-\epsilon_1 - \epsilon_2. \tag{29}$$

The corresponding Dynkin diagram and the extended Dynkin diagram are given in Table 1, whereas summary of the GQSs associated with the Lie algebras E_6, E_7 and E_8 are presented in Table 2.

4. The basic classical Lie superalgebras

Let us note first that for the basic classical Lie superalgebras, the description by means of a Dynkin diagram is not unique: besides the so-called distinguished Dynkin diagram, other non-equivalent Dynkin diagrams exist [4], [25]. This feature makes it harder to obtain a complete classification of all generalized quantum systems since one should repeat the process of investigation for all non distinguished Dynkin diagrams and their extensions. In Table 3 we give all (extended) Dynkin diagrams of basic classical Lie superalgebras, relevant for the classification process. All GQS associated with the basic classical Lie superalgebras are listed in Table 4.

Table 3. Basic classical Lie superalgebras, the (extended) Dynkin diagrams, relevant for the classification process, with a labeling of the nodes and the corresponding simple roots.

LSA	Dynkin diagram and extended Dynkin diagram
$A(m\|n)$	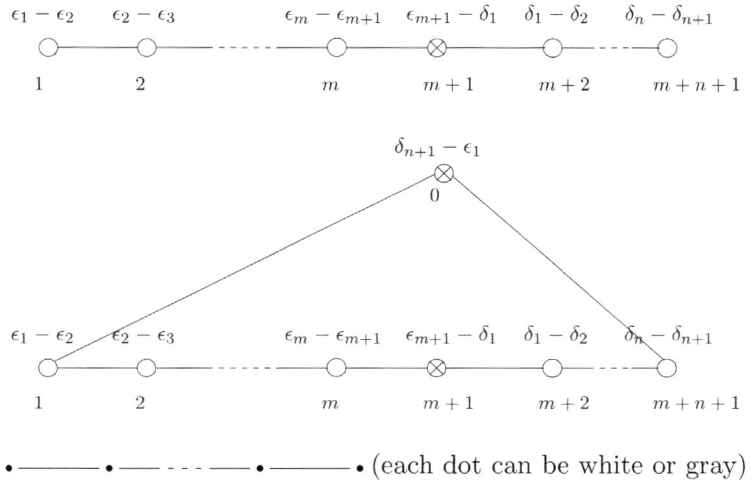
$B(m\|n)$	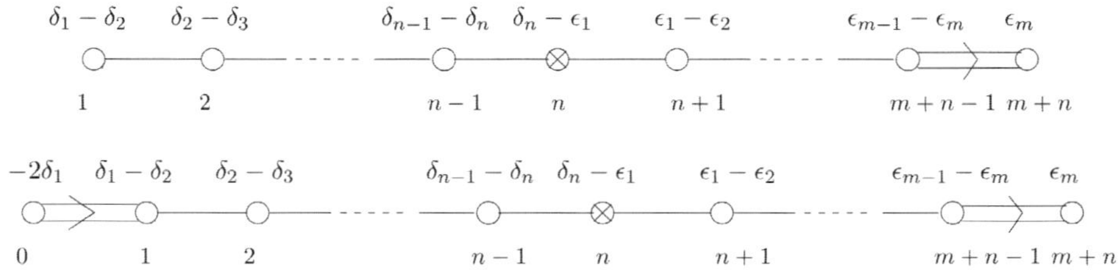
$B(0\|n)$	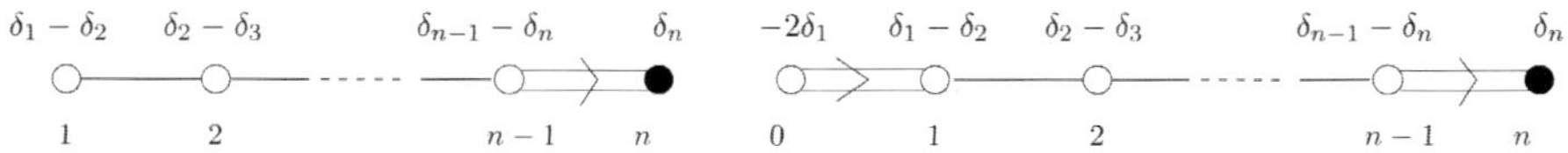
$D(m\|n)$	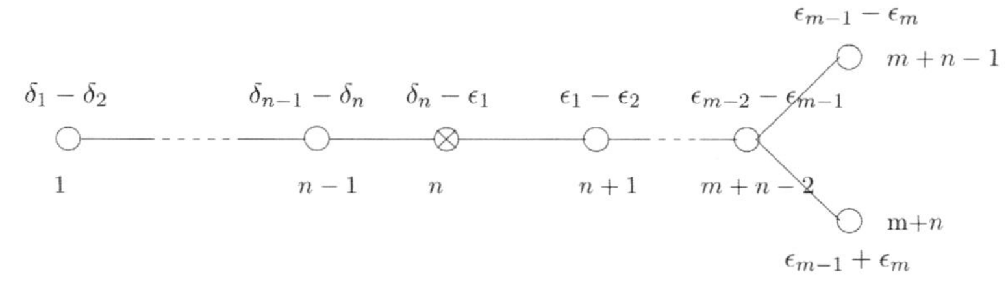

Journal of Physics: Conference Series **128** (2008) 012061 doi:10.1088/1742-6596/128/1/012061

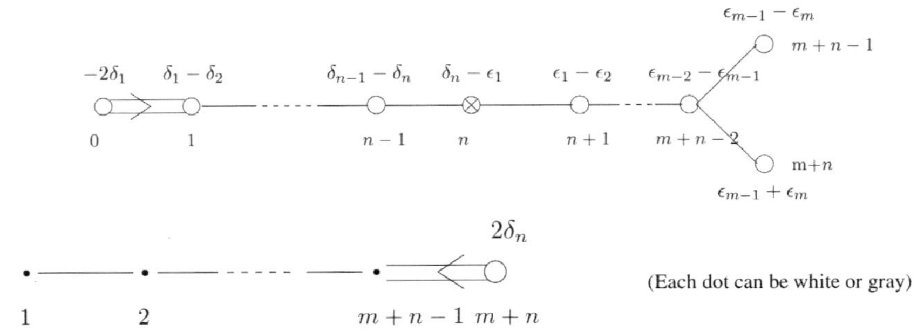

$C(n)$

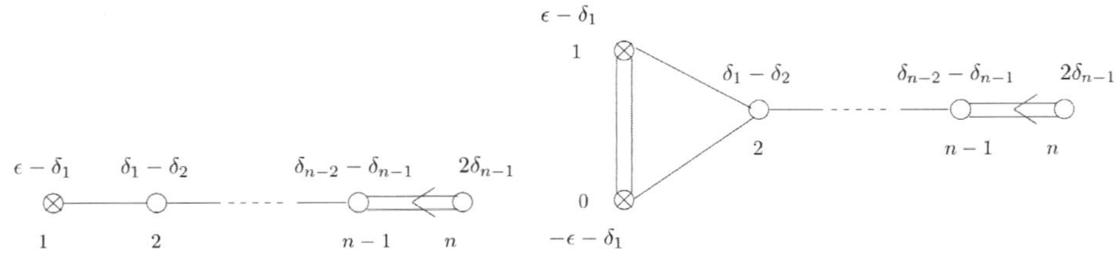

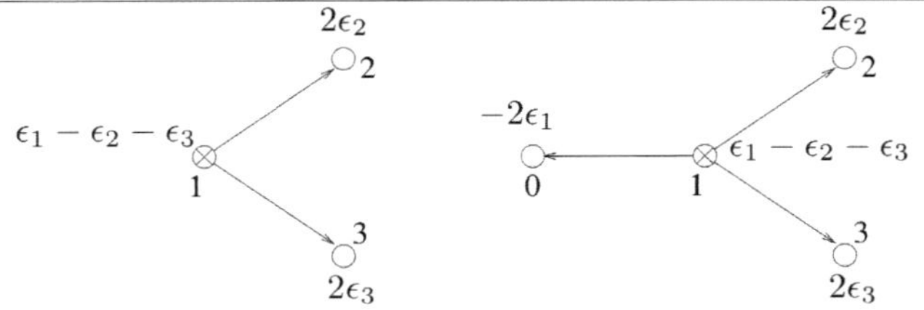

$D(2,1;\alpha)$

$G(3)$

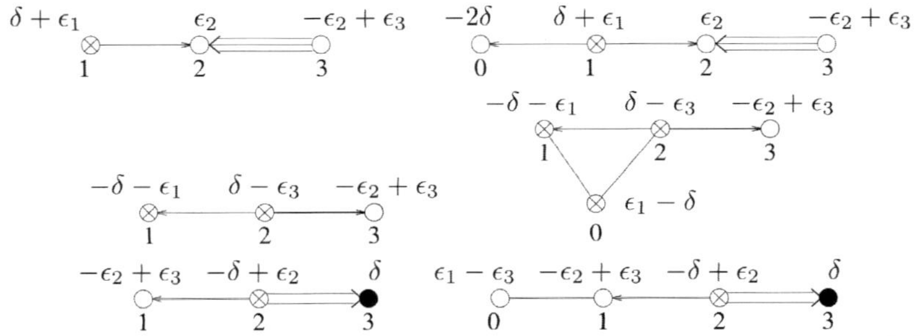

$F(4)$

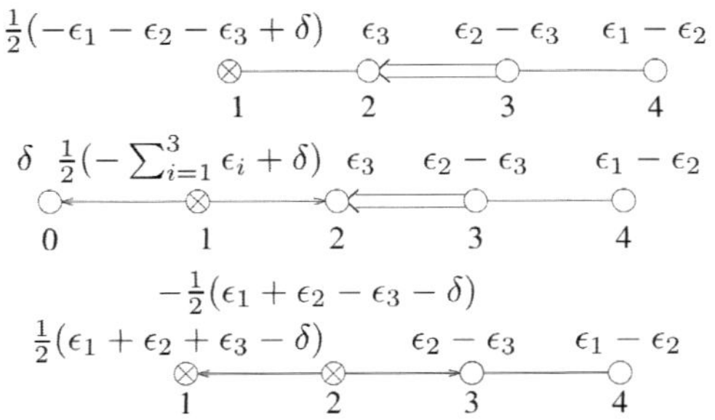

Journal of Physics: Conference Series **128** (2008) 012061 doi:10.1088/1742-6596/128/1/012061

Table 4. Summary of the classification for basic classical Lie superalgebras: all non-isomorphic GQS are given. For each GQS, we list: the subalgebra G_0 (each G_0 contains the complete Cartan subalgebra H, so we only list the remaining part of $G_0 = H + \cdots$); the length ℓ of the $\mathbb{Z}$-grading; $\dim G_{-1} = \dim G_{+1}$, which is also the number N of creation or annihilation operators

LSA	$G_0 = H + \cdots$	ℓ	$\dim G_1 = N$
$A(m\|n)$	$sl(k\|l) \oplus sl(p\|q)$	3	$(k+l)(p+q)$
	$(k+p=m+1,\ l+q=n+1,$		
	$k+l \neq 0,\ p+q \neq 0)$		
	$sl(k\|l) \oplus sl(p\|q) \oplus sl(r\|s)$	5	$(k+l)(p+q+r+s)$
	$(k+p+r=m+1,$	5	$(p+q)(k+l+r+s)$
	$l+q+s=n+1,$	5	$(r+s)(k+l+p+q)$
	$k+l \neq 0,\ p+q \neq 0,\ r+s \neq 0)$		
$B(m\|n)$	$sl(k\|l) \oplus B(m-k\|n-l)$	5	$(k+l)(2m-2k+2n-2l+1)$,
	$(k=0,\ldots,m;\ l=0,\ldots,n;$		
	$(k,l) \notin \{(0,0),(1,0)\})$		
	$B(m-1\|n)$ $[(k,l)=(1,0)]$	3	$2m+2n-1$
$B(0\|n)$	$sl(i) \oplus B(0\|n-i)$	5	$i(2n-2i+1)$
	$(i=1,\ldots,n)$		
$D(m\|n)$	$sl(k\|l) \oplus D(m-k\|n-l)$	5	$2(k+l)(m+n-k-l)$,
	$(k=0,1,\ldots,m;$		
	$l=0,1,\ldots,n;$		
	$(k,l) \notin \{(0,0),(1,0),(m-1,n),(m,n)\})$		
	$D(m-1\|n)$ $[(k,l)=(1,0)]$	3	$2(m+n-1)$
	$sl(m\|n)$ $[(k,l)=[m,n)]$	3	$(m+n)(m+n+1)/2-m$
	$sl(m-1\|n)$ $[(k,l)=(m-1,n)]$	5	$(m+n)(m+n+1)/2-m$
	$sl(m-1\|n)$ $[(k,l)=(m-1,n)]$	5	$2(m+n-1)$
$C(n)$	$sl(k\|l) \oplus D(1-k\|n-1-l)$	5	$2(k+l)(n-k-l)$
	$(k=0,1;\ l=1,\ldots,n-2)$		
	C_{n-1} $[(k,l)=(1,0)]$	3	$2(n-1)$
	$sl(1\|n-1)$ $[(k,l)=(1,n-1)]$	3	$n(n+1)/2-1$
	$sl(n-1)$ $[(k,l)=(0,n-1)]$	5	$n(n+1)/2-1$
	$sl(n-1)$ $[(k,l)=(0,n-1)]$	5	$2(n-1)$
$D(2,1;\alpha)$	$sl(2) \oplus sl(2)$	5	4
	$sl(1\|2)$	3	4
	$sl(1\|1)$	5	4
$G(3)$	G_2	5	7
	$sl(1\|2)$	5	7
	$sl(3\|1)$	3	8
	$osp(3\|2)$	5	8
	$sl(3) \oplus osp(1\|2)$	3	9
$F(4)$	$so(7)$	5	8
	$sl(1\|2) \oplus sl(2)$	5	10
	$osp(2\|4)$	3	10
	$sl(2) \oplus so(5)$	5	8
	$D(2,1;-1/3)$	5	10
	$sl(3\|1)$	5	8

Acknowledgments

NIS was supported by a project from the Fund for Scientific Research – Flanders (Belgium) and by project P6/02 of the Interuniversity Attraction Poles Programme (Belgian State – Belgian Science Policy).

References

[1] Green H S 1953 *Phys. Rev.* **90** 270
[2] Kamefuchi S and Takahashi Y 1962 *Nucl. Phys.* **36** 177; Ryan C and Sudarshan E C G 1963 *Nucl. Phys.* **47** 207
[3] Ganchev A Ch and Palev T D 1980 *J. Math. Phys.* **21** 797
[4] Kac V G 1977 *Adv. Math.* **26** 8
[5] Palev T D 1976 *Lie algebraical aspects of the quantum statistics,* (Habilitation thesis, Inst. Nucl. Research and Nucl. Energy, Sofia, in Bulgarian)
[6] Palev T D 1977 Lie algebraic aspects of quantum statistics. Unitary quantization (A-quantization) it Preprint JINR E17-10550 and hep-th/9705032
[7] Palev T D 1979 *Czech. J. Phys.* **B 29** 91
[8] Palev T D 1980 *Rep. Math. Phys.* **18** 117; **18** 129
[9] Palev T D 1980 *J. Math. Phys.* **21** 1293
[10] Palev T D, Van der Jeugt J 2002 *J. Math. Phys.* **43** 3850
[11] Jellal A, Palev T D and Van der Jeugt J 2001 *J. Phys. A: Math. Gen.* **34** 10179 (*Preprint* hep-th/0110276)
[12] Palev T D, Stoilova N I and Van der Jeugt J 2000 *J. Phys. A: Math. Gen.* **33**, 2545
[13] Palev T D, Stoilova N I and Van der Jeugt J 2003 *J. Phys. A: Math. Gen.* **36**, 7093
[14] Stoilova N I and Van der Jeugt J 2005 *J. Math. Phys.* **46**, 033501
[15] Stoilova N I and Van der Jeugt J 2005 *J. Math. Phys.* **46** 113504
[16] Stoilova N I and Van der Jeugt J 2007 *J. Math. Phys.* **48**, 043504
[17] Jacobson N 1949 *Am. J. Math.* **71** 149
[18] Okubo S 1994 *J. Math. Phys.* **35** 2785
[19] Bars I and Günaydin M 1979 *J. Math. Phys.* **20** 1977
[20] Rausch de Traubenberg M and Sluoinski M J 2000 *J. Math. Phys.* **41**, 4556
[21] Dynkin E B 1950 *Doklady Akad. Nauk SSSR (N.S.)* **73** 877
[22] Van der Jeugt J 1987 *J. Math. Phys.* **28** 292
[23] Onishchik A L and Vinberg E B (Eds.) 1994 *Lie Groups and Lie Algebras III* (Springer Verlag, Berlin)
[24] Bremner M R, Moody R V and Patera J 1985 *Tables of dominant weight multiplicities for representations of simple Lie algebras* (Marcel Dekker, New York)
[25] Leites D A, Saveliev M V and Serganova V V, 1985 *Serpuhov preprint* **85-81**; Van de Leur J W 1985 *Contragredient Lie superalgebras of finite growth* Utrecht thesis; Frappat L, Sciarrino A and Sorba P 1989 *Commun. Math. Phys.* **121** 457

Energy levels and expectation values via accelerated path integral Monte Carlo

D Stojiljković, A Bogojević and A Balaž

Scientific Computing Laboratory, Institute of Physics, Belgrade, Serbia

E-mail: danica@phy.bg.ac.yu

Abstract. A recently developed method systematically improved the convergence of generic path integrals for transition amplitudes, partition functions, expectation values and energy spectra. This was achieved by analytically constructing a hierarchy of discretized effective actions indexed by a level number p and converging to the continuum limit as $1/N^p$. Here we apply the above general method to numerical calculations using Metropolis Monte Carlo simulations of energy expectation values and energy spectra. We analyze and compare the ensuing increase in efficiency of several orders of magnitude.

Keywords: Path integral, Effective action, Monte Carlo

1. Introduction

In the middle of the last century Feynman introduced an alternate approach to quantum mechanics known as the path integral formalism [1, 2, 3]. This approach provided us with a new intuitive picture for understanding quantum mechanics, it enabled us to make connections and analogies between different areas of physics, and it also provided a new mathematical framework for calculating properties of physical systems. In particular, the formalism made it easy to generalize quantum theories from one particle to many particles and finally to fields in a relatively straightforward way. Unfortunately, the new formalism did not increase the number of analytically solvable systems [4, 5]. The fact that it enabled us to treat models that were previously inaccessible is due to its formulation in terms of discretized quantities, making it directly amenable to numerical treatment.

In numerical simulations, our poor understanding of the inherent mathematical structure of path integrals translates into the slow convergence of the sought-after physical quantities. Substantial increase in efficiency of numerical algorithms for calculating path integrals, can, therefore, come only through the input of new analytical information about path integrals into the calculation schemes. A recent series of papers [6, 7, 8] has investigated the relationship between discretizations of different coarseness in the case of a general quantum theory. The new found analytical results were then used to construct a more efficient Path Integral Monte Carlo (PIMC) SPEEDUP code [9] which increased convergence of generic path integrals from $1/N$ to $1/N^p$. For computational reasons, the level p is currently limited to $p = 13$, however, there are no fundamental barriers to going to even higher levels. This substantial increase in efficiency results in speedup of path integral calculations and has been applied to calculations of amplitudes, partition functions, expectation values, as well as low lying energy spectra [10, 11]. Direct numerical calculations of a variety of different models have confirmed the analytically derived results.

All of the key properties of path integrals can be seen already on the example of one particle systems in one dimension. For this reason the current paper limits itself to systems of this kind. Let us note however that the generalization to more complex systems has also been investigated and confirms that the previously derived increase in efficiency holds for general many particle systems as well [12].

PIMC algorithms can be made more efficient either by getting better convergence to the continuum limit through the use of above mentioned hierarchy of discretized effective actions, or through better generation of relevant trajectories in the MC method. So far all the numerical verifications of improved $1/N^p$ convergence have been implemented using a PIMC code in which paths were generated through a Levy construction. In this paper we numerically investigate the derived speedup using Metropolis [13, 14, 15] method for generating relevant paths. In general the Metropolis technique is optimally suited for calculating expectation values. We have numerically verified the $1/N^p$ convergence, showing explicitly coexistence of improvements obtained through the use of effective actions and the Metropolis path generation method.

2. Path integrals

As previously stated, we focus on the motion of one particle in one dimension. The central object is the (Euclidean) amplitude for the quantum system to go from initial position a to final position b in imaginary time T. Feynman gave us three basic rules for calculating this transition amplitude:

(i) the contribution of each path is determined by the action functional $S[q]$, and is proportional to $e^{-S[q]}$,

(ii) one needs to take into account the contributions of all paths consistent with the boundary conditions,

(iii) contributions of different paths add up linearly, and the ensuing sum is called the part integral.

For a majority of physically interesting cases the action is of the form

$$S = \int_0^T dt \left(\frac{1}{2}\dot{q}^2 + V(q) \right) . \tag{1}$$

Note that for simplicity we are working in units where $\hbar$ and particle mass have been set to unity. The only problem in the outlined procedure is the enumeration of all possible paths. This is done by discretizing the time of propagation T into N equal time steps $\varepsilon = T/N$. The contribution of each piecewise linear trajectory is then determined by a discretized action of the form

$$S_N = \sum_{n=0}^{N-1} \varepsilon \left(\frac{(q_{n+1} - q_n)^2}{\varepsilon^2} + V(\bar{q}_n) \right) , \tag{2}$$

wherethe potential is evaluated at $\bar{q}_n = (q_n + q_{n+1})/2$, corresponding to the mid-point or Weyl ordering prescription of the usual operator formalism. The final transition amplitude is given $A(a, b; T)$ is given as $N \to \infty$ limit of the discretized amplitude $A_N(a, b; T)$

$$A(a, b; T) = (2\pi\varepsilon)^{-N/2} \int dq_1 \cdots dq_{N-1} e^{-S_N[q]} , \tag{3}$$

where q_n are the positions at discrete times $n\varepsilon$, $q_0 = a$, $q_N = b$, and $(2\pi\varepsilon)^{-N/2}$ is the appropriate normalization factor.

Journal of Physics: Conference Series **128** (2008) 012062 doi:10.1088/1742-6596/128/1/012062

The above discretized expression represents an $N-1$-fold integral and is directly amenable to numerical treatment. In general, for large N, such expressions are nest handled using Monte Carlo techniques [13, 14, 15]. Before we proceed with this, however, let us note that the transition from continuum to discrete theory is far from unique. Said another way, there exists an infinity of discretized actions that, in the continuum limit, give the same transition amplitude. The naively discretized action given in equation (2) is just the simplest representative. While the choice of different discretized actions does not affect the final continuum amplitude, it may substantially affect the speed of convergence to that continuum limit. The naive action typically leads to $1/N$ convergence. In a previous series of papers [6, 7, 8] we have constructed an explicit procedure for determining a hierarchy of equivalent discretized actions $S^{(p)}$ which lead to improved convergence of generic amplitudes as $1/N^p$. Explicit expression for elements of the hierarchy has so far been obtained for $p \leq 13$ and are available on our web site [9]. There are no practical impediments to going to higher values of p, the problem of determining the appropriate effective actions just gets algebraically more complex and requires the use of some package for symbolic calculus (e.g. MATHEMATICA).

Most often, however, one is interested in calculating not amplitudes but partition functions. The relation between the two is made apparent in the coordinate basis. As a result, the partition function can directly be written as a path integral. It is now an $N \to \infty$ limit of discretized partition function

$$Z_N(T) = \int dq_0 A_N(q_0, q_0; T) . \tag{4}$$

Note therefore that $Z(T)$ is given as a limit of an N-fold integral over periodic piecewise linear trajectories. The partition function contains all thee information about the statistical properties of the system. In particular, we can use it to determine thermodynamic potentials, such as the free energy $F = -\frac{1}{T}\ln Z$. The free energy is also the ideal starting point for evaluating the energy spectrum of a given theory. From evaluating the partition function in the energy eigen-basis it follows that, in the large T limit, the free energy tends to the ground state energy E_0. Similarly, one can introduce auxiliary functions $F^{(n)}$

$$F^{(n)} = -\frac{1}{T}\ln\left(Z - \sum_{i=0}^{n-1} d_i\, e^{-TE_i}\right) . \tag{5}$$

Note that $F^{(n)}$ tends to E_n in the large T limit. In this way, it is possible to use the free energy and the above auxiliary functions to numerically evaluate the low lying energy levels. The fact that $F^{(n)}$ depends on all the lower energy levels and degeneracies results in an accumulation of numerical error as one looks at higher and higher energy levels. This is illustrated in figure 1 on the case of a particle moving in the quartic potential $V(q) = q^2/2 + gq^4/24$.

In addition to amplitudes and partition functions, path integrals are also used to evaluate expectation values of physical variables. The thermal expectation values of an observable $\mathcal{O}$ is given by

$$\langle \mathcal{O} \rangle = \mathrm{Tr}\left(e^{-T\hat{H}}\hat{\mathcal{O}}\right) / \,\mathrm{Tr}\, e^{-T\hat{H}} . \tag{6}$$

This can be directly written in the form of path integrals as the $N \to \infty$ limit of the discretized expectation values

$$\langle \mathcal{O} \rangle_N = (2\pi\varepsilon Z_N(T))^{-N/2} \int dq_0 dq_1 \ldots dq_{N-1} \mathcal{O}_N e^{-S_N[q]} . \tag{7}$$

In the above expression $\mathcal{O}_N$ stands for the discretized estimator of the corresponding physical variable. One needs to be careful in how one chooses estimators. They are not just simple discretizations of the continuum expressions. In addition, the estimator must be consistently

Journal of Physics: Conference Series **128** (2008) 012062 doi:10.1088/1742-6596/128/1/012062

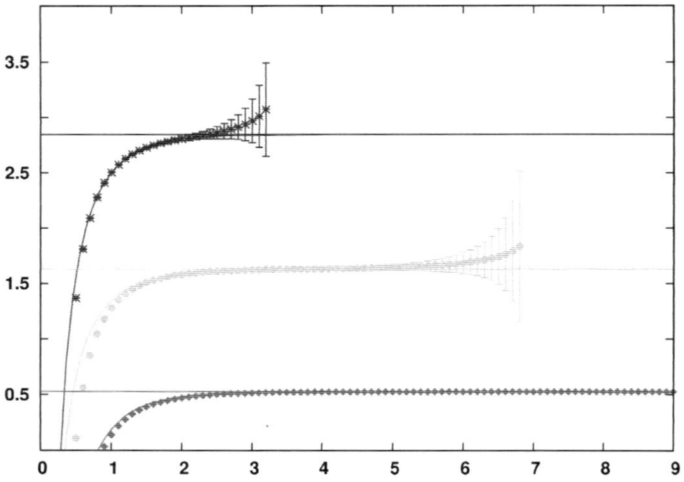

Figure 1. Metropolis implementation of PIMC algorithm for the calculation of low lying energy levels of a quartic potential. The plot shows the dependence of the free energy and the auxiliary function $F^{(1)}$ and $F^{(2)}$ as functions of T. The asymptotes are the corresponding energy levels E_0, E_1 and E_2. The parameters of the theory are $g = 1$, $N_{MC} = 10^7$, $N = 256$. The simulations were performed with $p = 9$ level improved effective actions.

paired with the discretized effective action used in order for expectation values to profit from the same increase in convergence as transition amplitudes. As a particular example, let us look at energy estimators. The continuum expression $\mathcal{E} = \dot{q}^2/2 + V(q)$ would naively be discretized as $\mathcal{E}_N = (q_{n+1} - q_n)^2/2\varepsilon^2 + V(\bar{q}_n$ for any n. Al alternate, more symmetrical estimator would be $\mathcal{E}_N = N^{-1}\sum\left[(q_{n+1} - q_n)^2/2\varepsilon^2 + V(\bar{q}_n)\right]$, where we have made use of the fact that the energy is a conserved quantity. Both of these estimators lead to problems in the continuum limit and give divergent results. This is easily understood if we recall that for short times of propagation each theory is well approximated by a free particle (random walker), satisfying the diffusion relation $\langle(q_{n+1} - q_n)^2\rangle \sim \varepsilon$. As a result, the above naive estimators have a dominant term diverging as $1/\varepsilon$. The source of the problem is in the T-dependence of the normalization of the above expectation value. A better way to derive the energy estimator is to use the relation

$$\langle\mathcal{E}\rangle = -\frac{\partial}{\partial T}\ln Z(T)\,. \tag{8}$$

If we define the discretized energy expectation value to satisfy the same kind of relation with the discretized partition function, it follows that the consistent energy estimator is given by

$$\mathcal{E}_N = \frac{N}{2T} - \frac{1}{N}\sum_{n=0}^{N-1}\frac{(q_{n+1} - q_n)^2}{2\varepsilon^2} + \frac{1}{N}\sum_{n=0}^{N-1} V(\bar{q}_n)\,. \tag{9}$$

The first two terms in the above estimator (9) both diverge in the continuum limit. However, taken together divergences cancel out and one obtains a finite result. Irrespective of this, the above (so called kinetic) estimator does not represent a good choice, as it contains within it the difference of two large numbers, making its standard deviation divergent and such numerical calculations nontractable. The standard way around this problem is to use the virial theorem

$$\left\langle\frac{\hat{p}^2}{2}\right\rangle = \left\langle\frac{1}{2}\hat{x}V'(\hat{x})\right\rangle\,. \tag{10}$$

Using it we obtain what is called the virial energy estimator [16]

$$\mathcal{E}_N^{vir} = \frac{1}{2N}\sum_{n=0}^{N-1} \bar{q}_n V'(\bar{q}_n) + \frac{1}{N}\sum_{n=0}^{N-1} V(\bar{q}_n)\,. \tag{11}$$

Each term of this estimator is well behaved and has finite continuum limit and finite standard deviation. It is important to note that the final form of the energy estimator used follows directly from the form of the discretized action. As a result, by changing the discretized action to $S_N^{(p)}$ we immediately obtain the appropriate p-level generalization of the energy estimator. These generalized estimators have been derived and studied in [11], where it was shown that they have the correct improved $1/N^p$ convergence.

3. Metropolis implementation

PIMC algorithms can be made more efficient either by getting better convergence to the continuum limit through the use of above mentioned hierarchy of discretized effective actions, or through better generation of relevant trajectories in the MC method. So far all the numerical verifications of improved $1/N^p$ convergence have been implemented using a PIMC code in which paths were generated through a Levy construction. The Levy construction [15] samples paths with 2^s time steps through a recursive halving, starting with some boundary conditions (representing the trajectory with 2^0 time steps). In the first step we generate one new node at the moment $T/2$ and get a new trajectory, with 2^1 time steps. The procedure is then repeated recursively for each segment of the trajectory. The new nodes are generated using the free particle approximation: if the coordinates of the boundaries of the segment are R_1 and R_2, and if the current time step is ε, the new node is selected from a Gaussian distribution centered at $(R_1 + R_2)/2$ and with the standard deviation $\sigma^2 = \varepsilon/2$. This method, although powerful and simple, can sometimes require very long runs in order to give results with the desired precision. The generated paths are sampled using a free particle approximation, and for models with strong interactions this way of sampling is far from optimal.

In this section we investigate the analytically derived speedup of path integral calculations within the framework of a PIMC code based on the Metropolis algorithm [13, 14, 15], a path generating technique optimally suited for calculating expectation values. Metropolis rejection algorithm is a special type of Markov process, enabling sampling of arbitrary probability distributions. The desired probability distribution $\pi(q)$ is obtained asymptotically, using a series of transformations (Metropolis moves) of the state of the system. The transformations are characterized by a transition matrix $T(q \to q')$, and the trial configuration q' is accepted according to the following probability

$$A(q \to q') = \min\left\{1, \frac{T(q' \to q)\pi(q')}{T(q \to q')\pi(q)}\right\}\,. \tag{12}$$

Metropolis moves in PIMC implementations are usually chosen to represent random local displacements (of given size) of individual nodes. The trajectories generated by the Metropolis technique are not independent, and we can have large correlations between consecutive paths. The correlations, however, depend on the physical quantity which is being calculated. The measure of such correlations is described through the correlation coefficient

$$c_k = \frac{\langle (O_0 - \langle O\rangle)(O_k - \langle O\rangle)\rangle}{\langle (O_0 - \langle O\rangle)^2\rangle}\,, \tag{13}$$

where O_k is the expectation values of the physical quantity calculated using each k-th configuration. The correlation length is defined as a minimal value of k for which the correlation

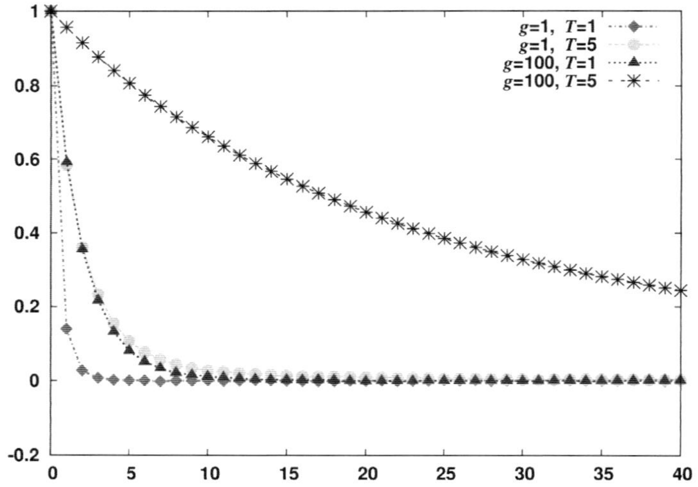

Figure 2. Correlation coefficient as a function of the correlation length k for the Metropolis implementation of the quartic anharmonic oscillator with $g = 100$, $T = 5$, $N_{MC} = 10^6$, $N = 1024$. The simulations were performed with $p = 9$ level improved effective actions.

coefficient c_k is sufficiently small (typically $c_k < 0.1$). Figure 2 illustrates the usual behavior of correlation coefficients for an anharmonic oscillator with quartic coupling.

The described method for correlation reduction is sufficiently good for accurate estimation of expectation values of physical quantities [14]. However, the estimates of MC errors of numerical results (standard deviations) can still be affected by the remaining correlations. This is caused by the deviation of the distribution of MC samples from the Gaussian: the numerically obtained

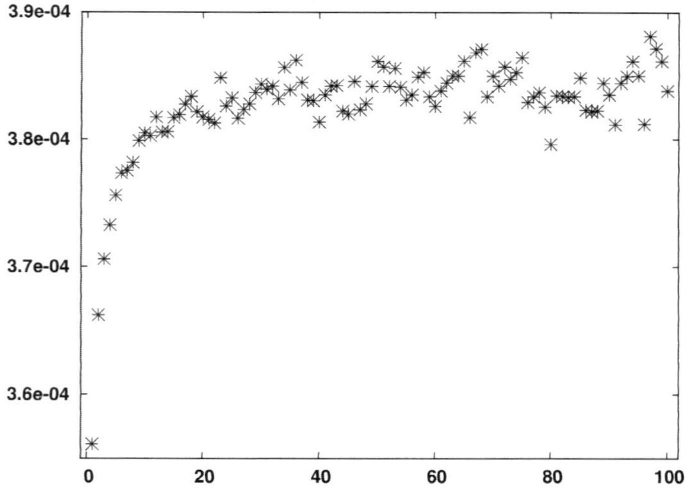

Figure 3. Standard deviation for Metropolis algorithm as a function of block size for a quartic anharmonic oscillator with $g = 1$, $T = 5$, $N_{MC} = 10^6$, $N = 1024$. The simulations were performed with $p = 5$ level improved effective actions.

distribution can be skewed, or have non-zero kurtosis, or both. This is dealt with by dividing the generated set of MC samples into blocks of a chosen size, and then by using just the averages over blocks as a new MC sample. By increasing the block size so that the skew and kurtosis of the obtained new distribution can be neglected, we may correctly estimate the value of MC errors. Note that this procedure does not affect the estimate for the expectation value. Figure 3 gives the dependence of the estimated standard deviation on the block size for a Metropolis implementation of a PIMC calculation of the energy expectation value for a quartic anharmonic oscillator. From this figure wee see that by using small enough block sizes the MC error can be substantially underestimated.

In usual implementation of the Metropolis algorithm, the probability distribution that is to be sampled is given by the exponential of the naively discretized action

$$\pi_N[q(t)] \sim e^{-S_N[q(t)]}\,. \tag{14}$$

In our implementation, the naively discretized action is replaced by one of the improved effective actions in the hierarchy $S_N^{(p)}$, trial paths were sampled using Levy construction, and acceptance of new trajectories was done according to the Metropolis rule (12). Figure 4 illustrates the typical behavior that one uncovers. As expected, the implementation of the Metropolis algorithm does not interfere with the increased convergence obtained through the use of higher level effective actions. Indeed, by using p level effective actions we again find that the numerical results of a new PIMC code display improved convergence of the form $1/N^p$. In fact, far from interfering negatively, the use of higher level effective actions brings about an improved efficiency of the Metropolis algorithm per se through the generation of more relevant trajectories. This is seen through the reduction of the variance of numerical results. A mode detailed investigations of this would be reported elsewhere. In particular, there we will focus on effective actions with $p \geq 10$ and take into consideration the effects of the increased algebraic complexity of these expressions on the computation time. For levels $p \leq 10$ numerical investigations have shown that this accumulated complexity does not have significant effects on the computation time.

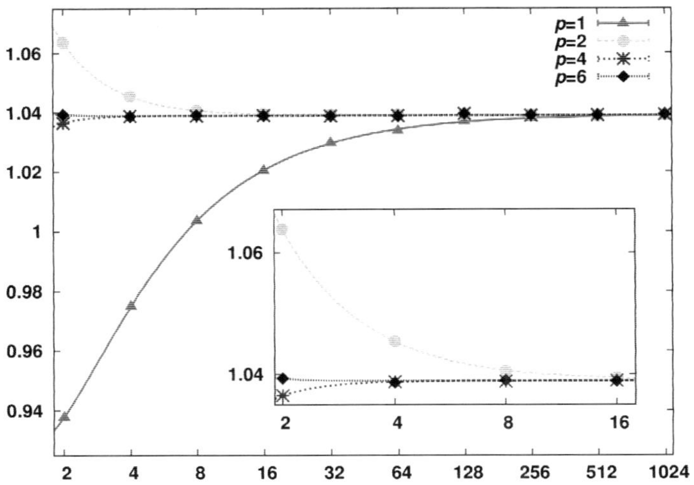

Figure 4. Metropolis implementation of the energy expectation value as a function of the discretization coarseness N for the quartic anharmonic oscillator with $g = 1$, $T = 1$, $N_{MC} = 10^7$. The simulations were performed with level $p = 1, 2, 3, 4$ improved effective actions.

Journal of Physics: Conference Series **128** (2008) 012062 doi:10.1088/1742-6596/128/1/012062

4. Conclusions

We have given a brief overview of the current state of the research effort behind the construction of more efficient PIMC algorithms leading to improved convergence of path integral calculations. In particular, we have outlined how the developed hierarchy of effective actions may be used to calculate transition amplitudes, partition functions and energy spectra. Particular emphasis was given to the calculation of expectation values. We outlined a scheme for the derivation of estimators consistent with the hierarchy of effective actions, i.e. leading to the same increase in convergence. The second part of the paper centers around a new PIMC code encompassing the effective actions and derived estimators, implemented using the Metropolis algorithm. The new PIMC code displays the same $1/N^p$ increase in convergence. Moreover, the use of higher level effective actions brings about an improved efficiency of the Metropolis algorithm through the generation of more relevant trajectories. In the future work we will focus on the detailed investigation of the increase in efficiency of the Metropolis algorithm brought about by the use of higher level effective actions.

Acknowledgments

This work was supported in part by the Ministry of Science of the Republic of Serbia, under project No. OI141035. D. S. also acknowledges support from EU Centre of Excellence grant CX-CMCS. The presented numerical results were obtained on the AEGIS e-Infrastructure whose operation is supported in part by FP6 projects EGEE-II and SEE-GRID-2.

References

[1] Feynman R P 1948 *Rev. Mod. Phys.* **20** 367
[2] Feynman R P and Hibbs A R 1965 *Quantum Mechanics and Path Integrals* (New York: McGraw-Hill)
[3] Feynman R P 1972 *Statistical Mechanics* (New York: W. A. Benjamin)
[4] Steiner F and Grosche C 1997 *A Table of Feynman Path Integrals* (New York: Springer)
[5] Kleinert H 2004 *Path Integrals in Quantum Mechanics, Statistics, Polymer Physics, and Financial Markets* 3rd edition (Singapore: World Scientific)
[6] Bogojević A, Balaž A and Belić A 2005 *Phys. Rev. Lett.* **94** 180403
[7] Bogojević A, Balaž A and Belić A 2005 *Phys. Rev.* B **72** 064302
[8] Bogojević A, Balaž A and Belić A 2005 *Phys. Lett.* A **344** 84
[9] `http://scl.phy.bg.ac.yu/speedup/`
[10] Stojiljković D, Bogojević A and Balaž A 2006 *Phys. Lett* A **360** 217
[11] Grujić J, Bogojević A and Balaž A 2006 *Phys. Lett* A **360** 205
[12] Balaž A, Vidanović I and Bogojević A 2007 Accelerated path integral calculations for many-body systems *QTS-5 Conference* (Valladolid, Spain), to appear in *J. Phys.* A
[13] Metropolis N, Rosenbluth A W, Rosenbluth M N, Teller A H and Teller E 1953 *J. Chem. Phys.* **21** 1087
[14] Kalos M H and Whitlock P A 1986 *Monte Carlo Methods* (New York: John Wiley and Sons)
[15] Ceperley D M 1995 *Rev. Mod. Phys.* **67** 279
[16] Janke W and Sauer T 1997 *J. Chem. Phys.* **107** 5821

AUTHOR INDEX

AUTHOR INDEX